Animal Sonar
Processes and
Performance

NATO ASI Series

Advanced Science Institutes Series

A series presenting the results of activities sponsored by the NATO Science Committee, which aims at the dissemination of advanced scientific and technological knowledge, with a view to strengthening links between scientific communities.

The series is published by an international board of publishers in conjunction with the NATO Scientific Affairs Division

A	**Life Sciences**	Plenum Publishing Corporation
B	**Physics**	New York and London
C	**Mathematical**	Kluwer Academic Publishers
	and Physical Sciences	Dordrecht, Boston, and London
D	**Behavioral and Social Sciences**	
E	**Applied Sciences**	
F	**Computer and Systems Sciences**	Springer-Verlag
G	**Ecological Sciences**	Berlin, Heidelberg, New York, London,
H	**Cell Biology**	Paris, and Tokyo

Recent Volumes in this Series

Series A: Life Sciences

Animal Sonar

Processes and
Performance

Edited by
Paul E. Nachtigall and
Patrick W. B. Moore

Naval Ocean Systems Center
Kailua, Hawaii

Plenum Press
New York and London
Published in cooperation with NATO Scientific Affairs Division

Proceedings of a NATO Advanced Study Institute on
Animal Sonar Systems,
held September 10-19, 1986,
in Helsingor, Denmark

QP
469
.N37
1986

Library of Congress Cataloging in Publication Data

NATO Advanved Study Institute on Animal Sonar Systems (1986: Helsingor, Denmark)
  Animal sonar.

  (NATO ASI series. Series A, Life sciences; vol. 156)
  "Proceedings of a NATO Advanced Study Institute on Animal Sonar Systems, held September 10–19, 1986, in Helsingor, Denmark"—T.p. verso.
  "Published in cooperation with NATO Scientific Affairs Division."
  Bibliography: p.
  Includes indexes.
  1. Echolocation (Physiology)—Congresses. I. Nachtigall, Paul E. II. Moore, Patrick W. B. III. North Atlantic Treaty Organization. Scientific Affairs Division. IV. Title. V. Series: NATO ASI series. Series A, Life sciences; v. 156.
  QP469.N37   1988   591.1'825   88-28886
  ISBN 0-306-43031-2

ORGANIZING COMMITTEE

Chairman

Paul E. Nachtigall
Head, Research Branch
Naval Ocean Systems Center
P.O. Box 997
Kailua, Hawaii 96734 USA

Whitlow Au

Naval Ocean Systems Center
P.O. Box 997
Kailua, Hawaii 96734 USA

William Friedl

Naval Ocean Systems Center
P.O. Box 997
Kailua, Hawaii 96734 USA

Lee Miller

Biologisk Institut
Odense Universitet
Campusvej 55
5230 Odense M, Denmark

Patrick W. B. Moore

Naval Ocean Systems Center
P.O. Box 997
Kailua, Hawaii 96734 USA

Kenneth Norris

Coastal Marine Laboratory
Environmental Studies
University of California
Santa Cruz, California 95064 USA

Homer O. Porter

Naval Ocean Systems Center
San Diego, California 92152 USA

James A. Simmons

Institute of Neuroscience
University of Oregon
Eugene, Oregon 97408 USA

M. Brock Fenton

Faculty of Science
York University
Ontario, Canada M3J 1P3

Jeffrey Haun

Naval Ocean Systems Center
P.O. Box 997
Kailua, Hawaii 96734 USA

Cees Kamminga

Technische Hogeschool Delft
Afdeling Der Elektrotechniek
Mekelweg 4, Postbus 5031
2600 GA Delft, The Netherlands

Bertel Møhl

Department of Zoophysiology
University of Aarhus
DK-8000 Aarhus C, Denmark

Gerhard Neuweiler

Zoologisches Institut Der Universität
Luisenstrasse 14
D-8000 München, Federal Republic of Germany

Bill A. Powell

Naval Ocean Systems Center
San Diego, California 92152 USA

Uli Schnitzler

Institut für Biologie III
Auf der Morgenstelle 28
D-7400 Tübingen, Federal Republic of Germany

ACKNOWLEDGEMENTS

This symposium was sponsored by various organizations. The organizing
committee would like to thank them and their representatives:

Defense Advanced Research Projects Agency, USA

North Atlantic Treaty Organization,
Advanced Study Institutes

Office of Naval Research, USA

National Science Foundation, USA

Danish Ministry of Education
Department of Higher Education, Denmark

PREFACE

The first meeting on biosonar that I had the opportunity to attend was held in 1978 on the Island of Jersey in the English Channel. That meeting, organized by Professor R.G. Busnel and Dr. Jim Fish, was my introduction to an exciting and varied group of hard-working and dedicated scientists studying animal echolocation. They are, by nature, a very diverse group. They tend to publish in different journals and rarely interact despite the fact that they all work on echolocation. When they do interact as a group, as they did in Frascati Italy in 1966, in Jersey in 1978, and during the meeting reported in this volume, the meetings are intense, interesting, and exciting.

This volume is a composition of a series of contributed papers written to foster an interdisciplinary understanding of the echolocation systems of animals. The echolocation pulse production studies in bats and dolphins have recently been concentrated on the ontogeny of infant pulses, other studies, with three-dimensional computer graphics and x-ray computed tomography, have concentrated on finally resolving the old controversy concerning the site of dolphin echolocation click production.

Much has been accomplished on the analysis of bat neural structure and function. The intense effort directed toward understanding the structure, connections, and functional properties of parallel auditory pathways and the parallel and hierarchical processing of information by the mustached bat, has lead to dramatic breakthroughs in understanding brain function. Analogous neurophysiological studies of the dolphin have not been conducted but behavioral studies have demonstrated that the integration window for acoustic reception in the dolphin is less than 300 millionths of a second. Similar anatomical structures that appear to have evolved primarily for very fast acoustic processing have been found in both bat and dolphin brains.

New work on the natural history of echolocating animals has demonstrated closely positioned, yet remarkably different, ecological niches for various species. Much has recently been learned about echolocation by examining the environments and prey species of echolocating bats. Much more must be done to understand the ecology and environment of echolocating dolphins.

Cognitive processes beyond what are usually considered when explaining the performance of echolocating animals have recently come to light. Bats were shown to predict and estimate the trajectory of targets disappearing behind screens and dolphins were shown to have attention processes similar to other species including humans. The bottlenosed dolphin's demonstrated range-gating process allows the experimenter an exciting look at the focus of an animal's attention.

Signal processing strategies for detection, discrimination, time delay estimation, time-frequency processing, filter banks, and a number of other processes were presented. It seems safe to say that there is not, as yet, a totally satisfactory model for predicting the echolocation accuracy of either bats or dolphins.

The organization of any good international scientific meeting is a complex task. The organization of a meeting in Denmark while living in Hawaii required the dedication and hard work of many people. The organizing committee was required to choose among the many outstanding applicants. They chose to emphasize the work of young upcoming scientists. Their work, as demonstrated in the contents of this volume, indicates a strong and vigorous future for the development of a firm understanding of the processes underlying animal echolocation.

While organizations sponsor meetings, individuals within those organizations support the ideas and procedures that allow meetings to occur. Dr. Tony Tether of the Defense Research Projects Agency of the U.S. visited our laboratory at the Naval Ocean Systems Center and became very interested in the echolocation capabilities of dolphins. He was particularly interested in their abilities to discriminate targets and asked what he could do to help our research. I replied that it had been some time since an international meeting on animal sonar had been held and support for such a meeting would be most welcome. After he received a proposal he suggested that an international meeting would be beneficial. At that point, I contacted Dr. Craig Sinclair of the NATO Advanced Studies Institutes. He and the appropriate committees of NATO agreed that a meeting on Animal Sonar was reasonable. Further work with Dr. Donald Woodward of the Office of Naval Research assured the success of the meeting. Assistance from the U.S. National Science Foundation allowed the attendance of a number of active young scientists from the U.S.

Professor Lee Miller of the University at Odense, Denmark, did an outstanding job as the point of contact for the meeting in Denmark. He was also successful at obtaining monies from the Danish Ministry of Education. Professor Gerhard Neuweiler and his fine staff from the University in Munchen administered the funding in a truly professional manner. While all of the section chairpersons contributed to the editing of the sections prior to the meeting.

My colleagues at the Naval Ocean Systems Center deserve special recognition. Original details of the meeting were developed in meetings between Patrick Moore, Bill Friedl, and myself. Whitlow Au and Jeffrey Haun organized the section on theory and applications while Bill Powell and Hop Porter steered me through the necessary steps for approvals within our Laboratory. Winifred Chrismer and Patrick Moore worked tirelessly to keep all of the participant's correspondence files organized and manuscripts in order.

I believe that the success of a meeting is proportional to the number of people working very hard and believing in the value of their contributions. By that criterion, and by its product here enclosed, the third animal sonar systems symposium was an overwhelming success.

Paul E. Nachtigall

CONTENTS

SECTION 3: PERFORMANCE OF ANIMAL SONAR SYSTEMS

SECTION 4: NATURAL HISTORY OF ECHOLOCATION

SECTION 5:   ECHOLOCATION AND COGNITION

SECTION 6:   ECHOLOCATION THEORY AND APPLICATIONS

MEMORIES AND REFLECTIONS ON BIOSONARS

René Guy Busnel

Chemin de la Butte au diable
91570 Bievres, France

Dear Colleagues:

I feel the compulsion to begin this speech with a quote from Hamlet;
"Gentlemen, you are welcome to Elsinor." It is a great pleasure for me to
be here today and I would like to thank the coordinating committee for
inviting me for my contributions in the not so distant past, including the
organization of the first two biosonar symposia in Frascati exactly 20
years ago, and in Jersey, in 1979.

I have been asked to present a few introductory remarks. Instead of
boring you with a scientific lecture, I would like to comment briefly on
the goals, not only of this meeting, but of our research activities as
well. First of all, I would like to honor those colleagues no longer with
us, like Dwight Batteau and Winthrop Kellog, to whom we are all
scientifically indebted. I would also like to take this opportunity to
recall someone whose intellectual caliber and research, under the auspices
of the ONR, both in the USA and abroad, made numerous advances in the field
possible; I'm sure you have recognized Sidney Galler, who along with
another highly esteemed scientist, William MacLean, from San Diego, were
highly influential in the organization of our meetings.

I believe, and I am speaking here as a European, that all of us are
touched by the fact that the Americans have once again chosen to hold this
meeting in Europe rather than in the U.S., although they are financially
responsible for the major part of it. If their intent was to honor the
pioneers in the field, the Italian Spallanzani, the Swiss Jurine, and the
Frenchman Langevin, we can only be flattered, since the true spiritual
leader of echolocation, or Animal Sonar System, was an American, our
friend, Professor Donald Griffin. I take this opportunity to pay a tribute
to him, publicly, and to cite those two other Americans responsible for
extending his discoveries to porpoises: Professors Winthrop Kellog and
William Schevill.

I would first like to take a few moments to review the major events of
the last twenty years. We have published 2348 pages in two volumes,
between 1965 and 1979, covering 975 articles of which 590 are devoted to
porpoises, 342 to bats and 43 to non-bat-non-cetacean species, to use the
expression coined by E. Gould. I have not counted the number of
publications for each member country except for the cetaceans, which due to

1

V. Gurevitch's work, cited in P. Nachtigall's bibliography, raises the number of Soviet publications to 152. Although E. Sh. Airapet'yants and A.I. Konstantinov mention 37 publications in 1973, between 1973 and 1979 the number of Russian publications for cetaceans alone reached approximately 115.

With due respect to the interest inherent to the Soviet literature, I am however afraid that T.H. Bullock's plea, expressed in Jersey (p.512), although certainly noted by our Soviet colleagues, still remains unheeded. I doubt seriously whether they can comply with the standards Bullock put forward, hoping to see them publish: "their full papers, with methods and data, in international journals, more often than they do." Apparently our colleague has no conception of what freedom in the USSR really is, even as regards publishing. The systematic absence of Soviet colleagues from international gatherings, despite having received personal invitations, should be sufficient indication of the value put on freedom in this country. Doubtless Bullock is aware of this.

Over the last twenty years the scientific world has gone through a number of fashions such as molecular biology, biotechnology, sociobiology, cognitive ethology, to name only a few. Despite the introduction of these new fields, which generally engender a bandwagon effect since they go hand in hand with financial support, the Animal Sonar System field has not fared too badly, and has continued to grow, at least up to 1982.

Naturally it was in the U.S. that bat and porpoise experimentation first took place. The U.S. often has two to five times the number of participants as other countries in congresses; this ratio is fairly indicative of their pioneer role in the field. However, we should not overlook the large number of excellent studies on bats conducted in Germany by Prs. G. Neuweiller and H.U. Schnitzler, who in 1966 were the young students of Pr. Mohres, and went on to found and head excellent large research teams. In 1979, a young Canadian team, directed by our dynamic colleague B. Fenton, first attracted international attention.

Although the present situation regarding laboratory research is satisfactory as concerns bats, this is less so for porpoise studies, both in the U.S. as elsewhere. The main cause is lack of funding for these large animals, which are costly to capture and to maintain in captivity, aside from the fact that they may be difficult to locate at sea. For these reasons it is not surprising that the budget cuts affecting most Western nations has also made inroads in funding in this area. As far as I know, in the US, apart from our colleagues in San Diego, Hawaii and Woods Hole, other centers have reduced their activities considerably. In Europe, the number of studies has declined to practically null, or only take place during trips to remote areas, and are mostly for observation rather than for real experimentation. The situation is not very promising, and even our discussing it will probably not change anything, since the administrations who support this type of research have other preoccupations for the moment.

Let us take a look at the goals that were set down at our two last meetings. I am afraid we have not accomplished all the steering committee's resolutions. Aside from benefitting from exchanges we can have amongst ourselves, what has become of the resolution for more joint projects, replications, student exchanges, standardization of techniques and equipment? Each of us can judge for himself. Perhaps the younger generation who is with us today will be more successful in this area. Our policy has always been to encourage collaboration among theoreticians, engineers and biologists, and several remarkable studies, both in the U.S. and in Germany, in the two main areas, have clearly demonstrated the value of these efforts. Certain newcomers in the field have proposed joint projects without realizing that the idea was expressed and even appeared in print as long as ten or twenty years ago. However, my role is not to moralize, and I am glad to see that many areas remain open to further exploration, as you can see from the wealth of topics to be presented at this meeting. A striking number of excellent studies have been conducted in fields as varied as the neurophysiology of the auditory, phonatory and central nervous system, signal processing, target discrimination behavior, comparative interspecies signal studies, etc. Advances have also been made in methods of analysis and types of apparatus, whose transformation over the last twenty years has been amazing. It suffices for the older members of this audience to recall the Kay sonograph! Now we have equipment that in a matter of minutes can calculate auto-correlations, the Fourier transform, spectral density and so on, and are so easy to use that a technician can operate them. The apparatus can be computer-driven and the curves plotted automatically in no time at all. In this context it is not surprizing that R. Altes, W. Au, B. Escudie, R. Floyd, S. Johnson, J.C. Levy and J.D. Pye, amoungst others, have devised new models and theories which have revolutionized experimentation. As regards more recent studies, I must express an admiration for the beauty of the techniques used in experiments in my own field, conducted for the most part in Hawaii and in San Diego by our talented colleagues: W. Au, W. Evans, R.H. Penner, S. Ridgway, R. Schusterman, F.G. Wood and their co-workers.

My enthusiasm for porpoise studies in no way outweighs my admiration for work done by the bat people. They have provided eloquent proof of their capacities as their studies have shown: I am referring here to the German team directed by G. Neuweiler and H.U. Schnitzler, and their students, and the teams in the U.S. initiated by D. Griffin, such as the ones headed by A.D. Grinnel, O.W. Henson, G.D. Pollack, J.A. Simmons, N. Suga, R.A. Suthers, and F. Webster, as well as the Canadian team headed by B. Fenton.

Those who in Jersey were able to test Leslie Kay's "sonar for the blind," were enriched with a new auditory experience, which must have generated new ideas in many of us and may well be mentioned during this meeting.

Let me say how happy I am to be in Denmark with you, since it was here that I set up a tiny laboratory with A. Diedzic, in Middlefart. We encountered a number of obstacles, since at that time the Danish government was not inclined to host a team funded by ONR and NATO. We spent three years here and the most rewarding for me was to have initiated a young researcher, Soren Andersen to the field, and to have encouraged B. Mohl, who has spent some time working with us on sea lions and who is now switching to bats.

It was at this time that I began to prefer working with animals in captivity since my marine expeditions often left me empty-handed. I'll give you several anecdotes. If you, my dear American colleagues, have the good fortune to find porpoises and whales near your shores, as often as they are depicted in Walt Disney movies, in European waters I have spent more than a full month without spotting a single cetacean. On my first trip around Corsica and Sardinia where, according to the fishermen, there were "millions" of porpoises, we spent forty days on a barren sea. It was hard, and expensive. Then suddenly, one night, we were finally able to record signals for a period of eight hours, but we never saw or were able to identify the species of porpoise. The outcome of the mission was thus scientifically null. On another occasion in the Mediteranean, near the Spanish coast, on a calm and transparent sea, we frequently observed the passage of porpoises, swimming at great speed and in a truly depressing silence for a bio-acoustician! But the worst was to finally identify animals and then to have then scared off by noise. This happened to me once when I was diving at night at 2800 Meters in the bathyscaph. The only thing I was able to record was the sound of an American fleet cruising above my head. Another time with Diedzic, in Madeira, after having waited a month for the sperm whales to arrive, the noise of the harpoon boats covered all but a few minutes of our recording. Since I was fed up with the European seas, I tried the New World, and went to South America to study the Rio de la Plata porpoise the <u>Pontoporia blainvillei</u>. I was in Uruguay with B. Brownell, who had located a sector where this porpoise was abundant. He had organized a joint Americano-French-Japanese mission. I spent a month with G. Alcuri in daily expeditions near the Brazilian coast at Punto el Diablo, where there was a village of shark fishers. They used fixed nets and normally captured three to ten of these small porpoises every night. But during the entire month we lived in this tiny fishing village we only saw a Pontoporia once, for thirty seconds, since these porpoises have very small dorsal fins which only extend several centimeters above water; our recording of clicks lasted about 10 seconds. Dissection showed that these animals, extraordinarily silent at this time of the year (December - January) had their bellies full of local fish[1] which at nightfall were extremely noisy themselves - saturating our hydrophones with mating and territorial calls. The porpoises thus did not need echolocation to find their prey; listening was sufficient.

While on this topic, I would like to recount a well-known experiment, that left us completely in the dark at the time and is still one of Nature's best kept secrets. The findings were originally published by F.G. Wood and W. Evans (Jersey volume p.393). A blindfolded porpoise in a large pool, perfectly equipped with a full battery of hydrophones, was able to catch a fish and return to her trainer without emitting any detectable sounds!

In the Amazon on the other hand, aboard the Calypso, as soon as I put my hydrophones under water, and without having sighted any <u>Inia</u> or <u>Sotalia</u> , I heard almost constant clicks for hours on end. This was near Manaos, both in the branches of the Amazon and the Rio Negro, similar to what K. Norris described on his trip on the Alpha Helix. If you have read accounts of observations of <u>Inia</u> in captivity, such as those published by the Caldwel's, R. Penner or W. Evans, these animals are described as being fairly silent.

In fact, as you well know, nothing is simple or straightforward, and laboratory experiments, like marine expeditions, each have their own problems. Let me assure you, however, that not all my marine expeditions were negative! Nor were the excellent studies conducted by the Woods Hole group where W.A. Watkins, following in the footsteps of W. Schevill, generated a series of interesting works. His recent discovery of a coda in sperm whale click series is a new and important advance showing that clicks, as many of us have long suspected, are not only used for echolocation.

I would like to conclude these reminiscences by experiences that have deeply impressed me; I believe I was the first European to have the privilege of a private demonstration of echolocation by Don Griffin. He, as well as others who have become friends, have always offered me the warmest of welcomes, whether it has been in Point Mugu, San Diego, or Hawaii. Nor have I forgotten, my friend K. Norris, that you showed me how a blindfolded porpoise used echolocation, and explained the system of the acoustic window of the inferior maxillary to me. This discovery led to the

---

[1] This sonoriferous fish is <u>Porychtius poroissimus</u>

excellent experiments conducted by V. Varanasi on the phonochemistry of cetacean adipose tissue.

I shall stop here in my recollections of familiar ocean themes for fear of offending my other friends the bat people, who may only have vague interest in these types of problems. Now it's time to get down to work.

To be at Elsinor, that is the question for many of us. It is a privilege to attend a meeting so well organized by Paul Nachtigall fully convinced of his abilities and the quality of your contributions, my dear colleagues, I am fully confident of the success of this gathering.

Thank you.

SECTION I

ECHOLOCATION SIGNALS AND THEIR PRODUCTION

Section Organizer: Bill Friedl

ECHOLOCATION SIGNAL TYPES OF ODONTOCETES
  Cees Kamminga

THE PRODUCTION OF ECHOLOCATION SIGNALS BY BATS AND BIRDS
  Roderick A. Suthers

PROPAGATION OF BELUGA ECHOLOCATION SIGNALS
  Whitlow W. L. Au, Ralph H. Penner and Charles W. Turl

NASAL PRESSURE AND SOUND PRODUCTION IN AN ECHOLOCATING WHITE WHALE,
Delphinapterus leucas
  Sam H. Ridgway and Donald A. Carder

THE STUDY OF THE SOUND PRODUCTION APPARATUS IN THE HARBOUR PORPOISE,
Phocoena phocoena, AND THE JACOBITA, Cephalorhynchus commersoni BY MEANS OF
SERIAL CRYO-MICROTOME SECTIONING AND 3-D COMPUTER GRAPHICS
  Mats Amundin, Erik Kallin, and Sten Kallin

THE ANATOMY OF ACOUSTIC STRUCTURES IN THE SPINNER DOLPHIN FOREHEAD AS SHOWN
BY X-RAY COMPUTED TOMOGRAPHY AND COMPUTER GRAPHICS
  Ted W. Cranford

WHALE HEADS, MAGNETIC RESONANCE IMAGES, RAY DIAGRAMS AND TINY BUBBLES
  R. Stuart Mackay

INDIVIDUAL VARIATION IN VOCAL TRACT RESONANCE MAY ASSIST OILBIRDS IN
RECOGNIZING ECHOES OF THEIR OWN SONAR CLICKS
  Roderick A. Suthers and Dwight H. Hector

MIDBRAIN AREAS AS CANDIDATES FOR AUDIO-VOCAL INTERFACE IN ECHOLOCATING BATS
  Gerd Schuller and Susanne Radtke-Schuller

THE SOUNDS OF SPERM WHALE CALVES
  William A. Watkins, Karen E. Moore, Christopher W. Clark and
  Marilyn E. Dahlheim

APPARENT SONAR CLICKS FROM A CAPTIVE BOTTLENOSED DOLPHIN, Tursiops
truncatus, WHEN 2, 7 AND 38 WEEKS OLD
  Morten Lindhard

ONTOGENY OF VOCAL SIGNALS IN THE BIG BROWN BAT, Eptesicus fuscus
  Cynthia F. Moss

OBSERVATIONS ON THE DEVELOPMENT OF ECHOLOCATION IN YOUNG BOTTLENOSE
DOLPHINS
  Diana Reiss

THE SHORT-TIME-DURATION NARROW-BANDWIDTH CHARACTER OF ODONTOCETE
ECHOLOCATION SIGNALS

Henk Wiersma

ECHOLOCATION SIGNAL TYPES OF ODONTOCETES

Cees Kamminga

Information Theory Group
Delft University of Technology
Delft, the Netherlands

INTRODUCTION

The performance and properties of dolphin sonar systems have been the subject of several investigations during the past 15 years. Facing the manifold questions that have arisen since echolocation was proved in a scientific way, we note substantial progress in the knowledge about the intrinsic properties of the echolocation signals involved in the sonar system and their perception. A closer look at the research carried out following the memorable Jersey meeting in 1979 leaves us, however, with a considerable number of questions as to the why and how of dolphin sonar.

It is interesting to note that most of the studies on echolocation are still being carried out on captive animals in a more or less restricted environment, usually perturbed by a constant higher ambient noise level than in natural circumstances. Furthermore, the favorite animal in captivity still remains the Tursiops truncatus.

Although doing in-depth research with one species is certainly commendable, from the viewpoint of generalisation of sonar behaviour, it would be preferable to extend our knowledge to more species of different families of odontocetes. Only analysis and comparison of many taxonomically varied species of dolphins will provide us with the means to attempt a complete description of dolphin sonar behaviour.

With the information at hand we might postulate that, ecologically speaking, a different behaviour is to be found for pelagic dolphins than for littoral animals, as a response to the different demands imposed on them. Perhaps evolution has influenced the sonar system to resolve different navigational problems. Ranging in open ocean requires only a simple navigational signal in the absence of echoes and reverberations of nearby objects, rocks, sandbanks and underwater flora. On the other hand, we expect to find a more complicated signal if there is an evolutionary component in coastal sonar behaviour. Too few species have been studied in extenso at the present time to supply us with data with which to generally validate this theory.

Besides the tempting reasoning of evolution, one could ponder whether there is also an active adaptation to be found in the echolocation behaviour. The dolphin is perhaps capable of modifying his echolocation signal parameters to create his own meaning. Very little data is available on this point; experimental work in this direction is very time-consuming and delicate and requires a refined experimental set-up. Still, the initial outcomes of such experiments indicate that this question could be solved in the near future and a more generally applicable explanation obtained.

Concentrating again on a specific point of dolphin sonar systems, we present below a recent investigation into the outcomes of reliable echolocation signal types. An attempt to incorporate ecological factors influencing sonar behaviour as well is perhaps speculative, but provides some guideposts for discussion.

A short resume of the open literature available on different species will be offered in the discussion on the unfortunately limited number of species which have thus far been adequately studied. Corrections to and extensions on the latest available (yet incomplete) data list as was given at the conference at Boston in 1983 will also be included.

We are left, however, with a very limited amount of species. Speculations on the sound production system will only be mentioned as far as they can be explained on the basis of the physical appearance of the signal types, obtained from animals in captivity. Statements about different signal types obtained from one animal need to be treated very carefully.

One of the puzzling problems of adaptation from a natural environment to a restrictive, captive environment also remains to be solved in a general way. Although more data will become available in the long run, as far as continuous monitoring and analysis over long periods is concerned, a very tedious continuous effort is required.

Lastly, we also plan to propose an ecological classification to counter the established taxonomical classification made by Van Heel (1981), who states that evolution has given inshore/estuarine dolphins an extra tool for survival acoustically.

LITERATURE SURVEY

In scanning the literature on echolocation phenomena as it is made available in open references with the help of databank abstracts like Biological Abstracts and Animal Behaviour Abstracts, we are supplied by Inspec with a rather limited amount of information published after the Jersey meeting.

Actually, as already mentioned in the introduction, the Atlantic bottlenose dolphin still remains the animal most described in different types of experiments. The characteristics of typical echolocation signals used by T. truncatus in captivity have been examined by Au et al. (1974), Hol & Kamminga (1979), Au (1980), Kamminga & Wiersma (1981) and Poché et al. (1982). The clicks that have been observed under well-controlled conditions have spectral peaks in the 40-60 kHz range, as well as in the 100-130 kHz range.

Au et al. postulated as early as 1974 that Atlantic bottlenosed dolphins in Hawaii typically emit clicks in the range above 100 kHz to compete with the high ambient noise level due to snapping shrimp. However, a lower frequency peak is still present. No discussion is devoted to the possibility that the animal might be capable of switching between the two components, or gradually moving from one to another.

Hol & Kamminga (1979) found spectral peaks in the range of 40-50 kHz in a difficult target detection experiment with an experienced female T. truncatus with a, shall we say, stocky build.

Poché et al. (1982) investigated the echolocation behaviour of Tursiops in a tank lined with anechoic wood panels, to get insight into the adaptation with regard to the noise level in the tank. The noise spectrum level was found to be about 50 dB re: 1 μPa in the range 30-60 kHz, going down to 40 dB in the range above 100 kHz. For comparison, we add that noise measurements in Kaneohe Bay in Hawaii revealed about 55 dB in the spectrum above 100 kHz, while signals in the NRL tank at Orlando showed two spectral peaks of fairly equal amplitude: one around 25 kHz and a broadened one in the range 100-130 kHz (figure 1).

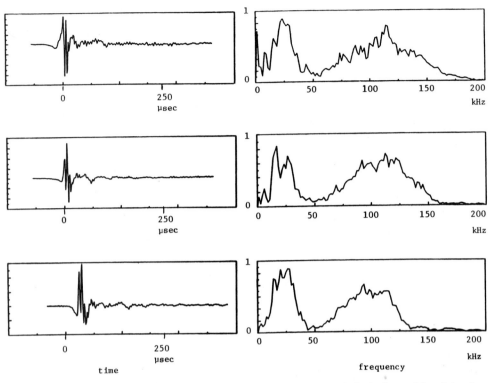

Fig. 1. Echolocation waveform (left) and spectrum (right) in a lined tank. (From Poché et al., 1982).

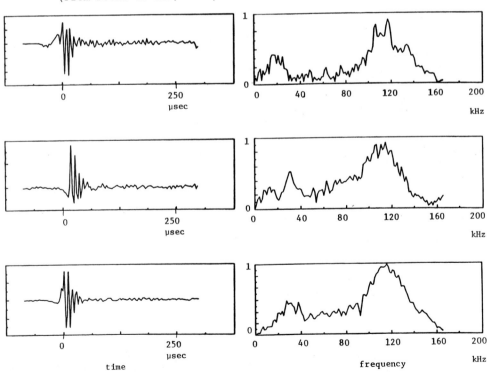

Fig. 2. Echolocation waveform (left) and spectrum (right) in an unlined tank. (From Poché et al., 1982).

11

After the lining was removed in Poché's experiment, the recorded echolocation spectra appeared to be quite similar to those obtained in the anechoic situation, although the low frequency components were of a lower intensity (figure 2).

The outcomes of this two-part experiment leave us still debating about whether they are due to adaptation or shifting or to perhaps two sources simultaneously in use. The highly obvious conclusion would be to consider the low frequency component as a vestigial one. Recently, Au et al. (1985) demonstrated adaptive capabilities of a Beluga whale, explained by the same reasoning of higher ambient noise.

Echolocations tasks with the same animal, first in San Diego and later in Kaneohe Bay, resulted in the dominant frequency's shifting an octave towards the higher part of the spectrum, with a noticeable component at the original frequency around 50 kHz. Bearing in mind that Beluga, living in shallow waters with temperatures lower than about 15° C, is far more an inshore animal than T. truncatus, we are still left with some unceraINTIES about the T. truncatus' capability of perform frequency shifting. In addition, we note that the receiving beam pattern for T. truncatus as measured by Au & Moore (1984) facilitates reception around 120 kHz by having a slightly broader beam pattern than the transmitting beam patterns for echolocation clicks at the same frequency.

It is rather a pity that both in the descriptions of the experiment by Poché et al. as well as the Kaneohe Bay experiments by Au et al. no detailed information about the subjects is given. Taxonomically speaking, it is known that T. truncatus has two populations, an offshore and an inshore/ estuarine one. The subject in the Hol & Kamminga experiments was a typically inshore specimen, while the offshore T. truncatus is slimmer in appearance, tending towards a shape commonly found for pelagic dolphins. It has not been proved much less understood whether the ecological difference plays a role in the sonar behaviour. The slender type of dolphin Skinny, who illustrated echolocation behaviour in the Australian film "stranded" (1981), presents not only the 40 kHz signal, but also a component around 80 kHz, sometimes with an intensity equal to the lower one.

Although we now have at hand a large variety in frequency patterns of echolocation clicks from T. truncatus, the consistency in wave shape is maintained in general during emission. Due to insufficient knowledge of the intrinsic properties of the source of the clicks, a model for the signal type and structure of the click has to be based on an observation of the wave shape alone. Very few attempts to treat the problem from the point of view of mathematical physics have been observed over the past few years. A simple model, based on Gabor's theory of elementary signals, is given by Wiersma (1982) and Wiersma (this volume). It is shown that by using this model, an individual click can be conceived of as the representation of one quantum of information. This quantum approaches the lower bound obtainable for the product of the time duration and frequency bandwidth. This description follows the fundamental uncertainty principle of communication theory. However, the model would benefit from generalisation for different types of echolocation signals.

A modelling of the echolocation click based on the four-parameter Gabor model (dominant frequency, mean epoch, time duration Δt, frequency bandwidth Δf) for two different dominant frequencies is presented in figure 3.

Another animal that has attracted attention for a considerable time is Phocoena phocoena, the harbour porpoise. Although rather difficult to keep in captivity, there are enough recordings and descriptions to discuss the prominent wave shape of this small, coastal animal. The first thorough discussion of Phocoena sounds extending over a far larger spectral range than recording allowed in the sixties used by Busnel et al. in Europe, was by Dubrovskii et al. (1971). Oscillograms of close range echolocation demonstrated spectral amplitudes in the 100 kHz range. It is worthwhile to note the suggestion made by the authors that the animal might actively tune its frequency to adapt to a certain detection problem. This proposition, however, has not been proved. Quantitative data on <u>Phocoena</u> high-frequency clicks came from Møhl & Andersen (1973). In their data the main energy was concentrated between 100 and 150 kHz. Interestingly enough this description did not include the previously known low-frequency component of 2 kHz.

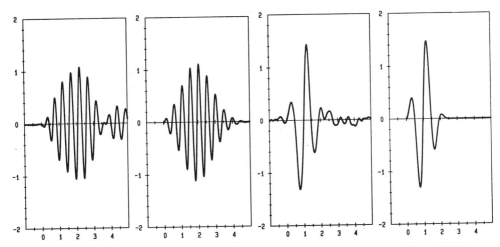

Fig. 3. Modelling of two echolocation clicks for minimum time/bandwidth product with different dominant frequencies based on the Gabor model. Cephalorhynchus (left), Tursiops (right). From Wiersma, this volume.

Later on, Kamminga & Wiersma (1981) confirmed the existence and same wave shape of the high-frequency component in the case of a young <u>Phocoena</u>, brought up in captivity by Andersen (Odense, Denmark). Figure 4 shows a typical example of one of these clicks of young Karl, revealing a dominant frequency of 140 kHz. However, one might wonder why the extremely low-frequency component of about 2 kHz did not show up, in spite of the excellent signal-to-noise ratio. No explanation for this absence is given.

Just when it seemed that several different approaches to signal description were resulting in the same wave shape with a dominant frequency over 100 kHz, another signal type for <u>Phocoena</u> was added by Kamminga & Wiersma (1981). This time the subject was a stranded female, whose sonar activity was recorded shortly after arrival in the dolphinarium. A typical sample of the sonar behaviour, shown in figure 5, features an unusual sonar signal, substantially different from our idea of what a sonar signal from <u>Ph. phocoena</u> should look like.

Not only was the low-frequency component of 2 kHz absent, there was now an observable component wave shape of two frequencies. This two-component sonar character is clearly visible in its time representation, featuring spectral peaks at 20 kHz and 120 kHz. This two-component frequency behaviour was confirmed in 1983 by Kamminga, Wiersma and Andersen (unpublished results) at Odense, Denmark. The underlying mechanism is explained as a linear process of addition, shown in figures 6 and 7, where filtered components are presented individually, with the 20 kHz persisting for 200 μsec. Both frequency components are tightly locked in the time domain over the complete click train.

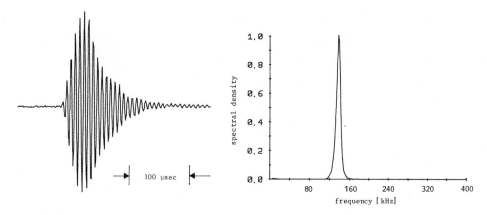

Fig. 4. Example of an echolocation click of a young <u>Phocoena</u>, f_{dom}=140 kHz.

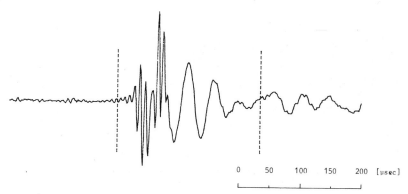

Fig. 5. Sonar click of <u>Ph.phocoena</u>, showing the two-component characteristic.

Fig. 6. Filtered low-frequency component of 20 kHz of the sonar click in fig. 5.

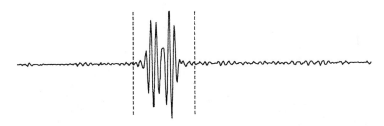

Fig. 7. Filtered high-frequency component of 120 kHz of the sonar click
in fig. 5.

The different propagation and resolution properties of the two dominant frequencies are assumed to be responsible for short range and long
range echolocation. The two-component sonar signal undoubtedly shows that
two different tuned systems, activated in phase by one stimulus, are used
by the dolphin.

Further experimentation on Phocoena is needed to shed light on the
intriguing features of the click structure. Unfortunately, no new information has come flowing forth over the last few years. Moreover, it is not
known if the Russian interest in Phocoena has continued after their noteworthy observations and statements in the seventies.

Another coastal animal whose echolocation behaviour has been investigated in the recent past is Delphinapterus leucas, the Beluga. A well-de-
signed discrimination experiment by Gurevich (1976) supplies us with data
indicating that echolocation performance exceeded that of T. truncatus.
Dominant frequencies up to 80 or 120 kHz were observed. Unfortunately, no
time or frequency plot can be presented in the overview here.

The very vocal Beluga seems to have a very extensive repertoire at
his command. Observations by Kamminga & Wiersma (1981) on the sonar signals of an echolocating captive Beluga at the Duisburg Zoo reveal a two-
component sonar, which can be broken down into a high-frequency component
of 60 kHz, followed after a delay of 0.75 msec by a long duration, low-
frequency component of 1.6 kHz of nearly equal intensity.

The high-frequency part of the sonar click is shown in detail below
in figure 8. Note in particular that the same reverberation phenomenon in
the high-frequency component appears as with the high-freqeuncy part in
Phocoena. No explanation is given for the almost five octave separation
between the two frequency components.

Recently, Au et al. (1985) presented an interesting study on the adaptive echolocation behaviour of a captive Beluga. The animal, a 9-year-old
male weighing 450 kg, was performing in a target-locating experiment. The
clicks of this animal measured in San diego had peak frequencies around
50 kHz with a relative bandwidth of 40%. Average time waveforms and their
accompanying spectra are presented in figure 9.

Two months after being shipped in 1981 to the N.O.S.C., Kaneohe Bay,
the Beluga whale took part in target detection experiments, involving detection of a 7.62 cm. spherical target. The peak frequencies obtained
from this experiment now clearly show the use of a sonar signal about an
octave higher, mostly in the range 100-110 kHz. The mean relative bandwidth remained at the same value as for the earlier-mentioned lower dominant frequency, but this time showed a much larger variance. Figure 9b
gives the waveforms along with their related spectra.

It is noted that the SL in Kaneohe Bay is about 10 dB higher than in
San Diego. The explanation given by the authors for the use of a higher

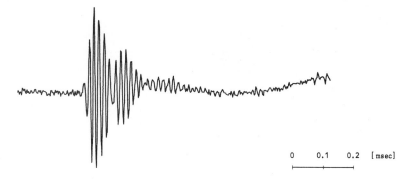

0   0.1   0.2   [msec]

Fig. 8. High-frequency part of an echolocation click of <u>Delphinapterus</u> <u>leucas</u>, as recorded in the Duisburg Zoo, f_{dom} = 60 kHz.

dominant frequency suggests an adaptation to the higher ambient noise environment. The 10 dB higher intensity could be a by-product of the higher frequency, although the waveform plots and the spectra do not clearly indicate this effect.

This interesting experiment with a coastal animal dramatically illustrates the use of different dominant frequencies in different enforced circumstances. The question could be raised of just how the dolphin actively manipulates the sonar sound system at his disposal, whether he is capable of activating more than one signal at will, or is shifting at an arbitrary rate from a single source.

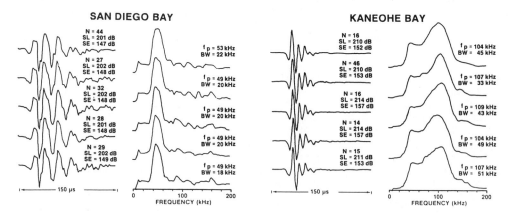

Fig. 9. Averaged time waveforms and spectra for trials in SAN DIEGO Bay and KANEOHE Bay. N = number of clicks, SL = source level, SE = source energy flux density, f_p = peak frequency, BW = 3 dB bandwidth (From Au et al. 1985).

Investigations on this phenomenon should be followed by one on a "not so coastal" dolphin, for example a species that has hitherto revealed only a preference for a single dominant frequency.

With respect to the lesser known and smaller odontocetes, we refer to well-documented recent reports, notably those on Cephalorhynchus commersonii of the Delphinidae and several other members belonging to the family Phocoenidea, including Phocoenoides dalli and Neophocaena phocaenoides.

The analysis and comparison of the sonar sounds of Delphinids and Phocoenids leads to the certain conclusion that the sonar signals of C. commersonii are not significantly different from those of members of Delphinidae. Although earlier bandwidth recordings were limited we now have interesting descriptions and analyses of full-bandwidth recordings from Dall's porpoise (Phocoenoides dalli) by Awbrey et al. (1986).

For Phocoenoides dalli the material consists of field recordings of free-ranging animals. Awbrey et al. recorded click trains that contained both single and double pulses as well as individual single and double pulses, produced at will by the porpoise. Spacing of the pulses of a pair ranged from 0.1 msec to 1 msec. The spectral distribution of the energy showed a rather constant dominant peak around 130 kHz, revealing the typical multiple-cycle waveform of Phocoena.

An example of a double click in both a time and a frequency representation is given in figure 10. Although the bandwidth accompanying this high dominant frequency is not specified, the frequency diagram clearly indicates that the earlier generally accepted description of dolphin clicks as broadband signals obviously has to be revised.

A great deal of individual time duration variability was observed, ranging from 50 μsec to over 1 msec for the single pulse type of sonar click. The complete time duration of this pulse type seems to be split up into two parts. Starting with a bell-shaped form for about 70 μsec, it is followed by an exponentially damped waveform lasting up to more than 1 msec in duration. The pulse shape in the double pulse type sonar click agrees remarkably well in wave shape with Commerson's sonar as described by Kamminga & Wiersma (1981, 1982), Hackbarth et al. (1985) and Evans et al. (1986).

In the particular recordings we are about to discuss we note a possible explanation to the questions raised by Watkins (at the Jersey meeting) concerning the fact that echolocation signals at sea exhibit differences in acoustic parameters than those to be expected during quite other behaviours, such as carefully trained, difficult echolocation tasks. Hackbarth et al., in comparison Commerson's sonar in captivity to that of the free-swimming species, reports that the latter showed a narrower peak in their spectra than the former, due to a minor pronounced FM sweep and phase shifts. According to Hackbarth et al. the detailed intrinsic click structure indicates a downward FM sweep at the beginning which is responsible for a broadening of the spectrum above the dominant frequency of 130 kHz. This FM sweep is not reported by Kamminga & Wiersma, who found a time-bandwidth product too low to indicate a substantial FM content in their recorded clicks.

In reviewing the now abundantly available collection of sounds from the families Phocoenidae and Delphinidae, we can state that a remarkable similarity is present for the wave shapes of C. commersonii, Ph. phocoena and Phocoenoides dalli.

TAXONOMICAL CONSIDERATIONS

A look at the taxonomy reveals that there are up to 6 members of the Phocoenidae family, with not all of them yet acoustically described. The genus Phocoenoides, radically different externally from the other species of the genus Phocoena, is, however, anatomically very similar. Two popula-

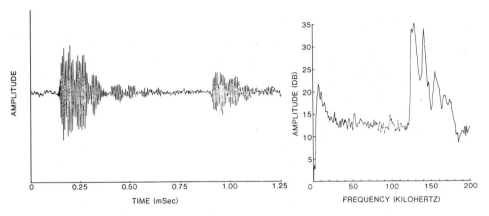

Fig. 10. Time representation of a double click from Phocoenoides dalli.

tions are distinguished, namely an inshore/estuarine and a river population. An acoustical description of the behaviour of the coastal population of Japan is given by Kamminga et al. (1986). The absence of a low-
frequency component is noted. The overall wave shapes of this population,
recorded in captivity at Toba, Japan, show a remarkable similarity to signal types from Ph. phocoena and C. commersonii. Figure 11 depicts a typical example of a sonar click, together with its spectral representation.
The dominant frequency lies between 115 and 130 kHz, with no sign of a low-
frequency component. This description brings the total up to 3 out of 6
members of the family about whom information on their sonar signals is
currently known. Speculations or predictions can now be made as to the behaviour of the missing three species.
　　A very intriguing question could then be raised with regard to the
river population of Neophocaena, who occupies, together with the true river dolphin Lipotes vexillifer, the same habitat in the middle reaches of
the Yangtse River. In fact, this is the only area in which two completely
different dolphins share the same surroundings. Unfortunately, we are
still in the dark as to the acoustic behaviour of the river population of
Neophocaena. Some light has been cast on the sonar signals of Lipotes,
measured by Jing et al. (1981), although these measurements are unfortunately of limited value, because the recording equipment was restricted
to 30 kHz, whereas it is very likely that the dominant frequency of the
Lipotes echolocation signals lies well above that, if compared for example
Inia geoffrensis and Platanista indi. Just as with Neophocaena phocaenoides, there are other examples of dolphins represented by inshore/estuarine
and river populations. Orcaella brevirostris as well as Sotalia fluviatilis
both have these two forms of representation, with O. brevirostris being
found on the rivers Irrawadi, Mekong and Kalimantan, and the inshore representatives being spread along the coasts of India, Burma, Malaya, Thailand, Vietnam and Indonesia up to Northern Australia.
　　We note a description of the echolocation signals of several captive
O. brevirostris from the Mahakam river by Kamminga et al. (1983). Figure
12 points out a very elementary sonar signal, consisting of about 2 sinusoidal cycles, followed optionally by some reverberations which sometimes
interfere with the main sonar pulse. A dominant frequency is observed in
the range of 60-70 kHz, without large excursions on either side. A simila-
rith with sonar from T. truncatus is unmistakable. It is necessary to compare this sonar behaviour with the inshore/estuarine population, however,
there are as yet no recordings available. Succumbing to the temptation to
compare this with Ph. phocoena wer are forced to predict a similar behaviour.

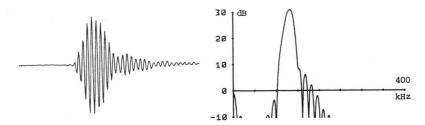

Fig. 11. Sonar pulse from a captive <u>Neophocaena phocaenoides</u> of the coastal
population in Japan, dominant frequency 128 kHz.

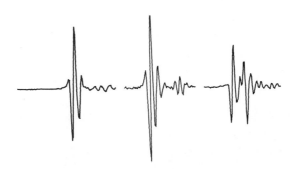

Fig. 12. Three representative waveforms of an echolocation click train
from <u>O. brevirostris</u> of 99 pulses, during 1.15 sec. Dominant
frequency 600 kHz.

The other example having two populations, <u>Sotalia fluviatilis</u>, appears
for the coastal population <u>S.f. guyanensis</u> to possess a sonar signal with
a high-frequency component at 95 kHz and a low-frequency component of quite
the same intensity at 30 khz (Wiersma, 1982). The two individual components
that make up the sonar pulse are pictured in figure 13 together with the
composite pulse and its related frequency spectrum. The separation between
the low- and the high-frequency component here is 2½ octave, much less
than in the caes of <u>Ph. phocoena</u>. Explanations of the nature of the <u>Sotalia</u>
sonar pulse are not discussed any further. Further research on this sound
production system is highly recommended. At first glance, an adition of the
signals of two differently tuned, identical sources would appear to be an
acceptable hypothesis.

In short, what we are faced with is different acoustic behaviour for
populations which are regarded taxonomically as belonging to the same species. Therefore, next to the classical taxonomy, we probably need to set
up another taxonomy which takes into account ecological factors influencing the sonar behaviour (Dudok van Heel, 1981).

CONCLUSIONS

In the foregoing, we have surveyed the literature on sonar behaviour
which has appeared since the Jersey meeting in 1979. Some of the species
which came up for discussion were: <u>Tursiops truncatus, Delphinapterus
leucas, Phocoena phocoena, Neophocoena phocaenoides, Phocoenoides dalli,
Cephalorhynchus commersonii, Sotalia fluviatilis guyanensis, Orcaella bre-</u>

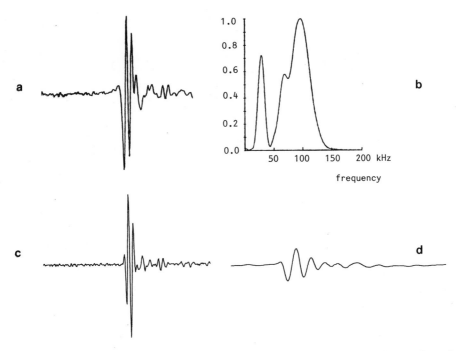

Fig. 13. Sonar pulse of <u>Sotalia fluviatilis guyanensis</u>(a), its related
frequency spectrum (b), the low-frequency component at 30 kHz
(c) and the high-frequency component at 95 kHz (d).

<u>virostris</u> and <u>Lipotes vexillifer</u>. Although a great deal of interesting work
has been performed in this field, there is very clearly a crying need for
research into the echolocation behaviour of species not investigated thus
far as well as that into species for which the data available at the moment
is insufficient to enable the signal types to be compared.

In looking over the collection of adequate descriptions of sonar behaviour of odontocetes described here, it is apparent that many gaps have
to be filled in, especially for the pelagic dolphins. However, among these
species <u>Pseudorca crassidens</u> is making its way into the sonar sounds collection. (Thomas, this volume, Kamminga et al. in press). First impressions
indicate a relatively low dominant frequency around 25-30 kHz with the same
wave sahpe as <u>T. truncatus</u>. Even for the more littoral and river types of
odontocetes there is far too little reliable material at hand to be able
to arrive at thismoment at a generalisation in the description of sonar
signals. Numerous difficulties have to be overcome, especially in recording
echolocation signals at sea (Watkins, 1980).

The difference in acoustic parameters that might be expected during
quite other behaviours than carefully trained, difficult echolocation tasks
needs to be clarified, although numerous difficulties will be encountered
in that type of field research. Today, we have too little and too sparse
descriptions of clicks at sea to answer fundamental questions.

In addition, is proposed a new taxonomy, based on ecological considerations. To this end, the acoustic production of the inshore/estuarine populations of several species has been compared to that of their river counterparts.

ADDENDUM

The author would like to take this opportunity to make a few additional comments and to ride a few of his favourite hobbyhorses.

Clearly, some agreement with respect to recording equipment and analysis techniques among dolphin researchers is needed. It might be wise to follow the example of researchers in pattern recognition, testing analysis techniques on available standardized, well-described data.

It is lamentable to have to state that even today in some exploratory observations of wild odontocetes the acoustic behaviour is neglected due to a lack of adequate recording equipment. Recently, Lockyer and Morris (1986) had the opportunity to study a wild, but sociable T. truncatus along the coast of Cornwall, U.K. Although the sonar sounds are described as numerous click trains of varying frequency (i.e. pulse repetition frequency), no recordings are available.

The benign research in the Indian Ocean whale sanctuary, a project funded to the tune of $ 150,000 (1982, 1983), could have added significantly to our store of knowledge of the sonar behaviour of the sperm whale and other odontocetes in the wild if they had had the foresight, as they did for the outstanding underwater filming, to use the right equipment.

REFERENCES

Au, W.W.L., R.W. Floyd, R.H. Penner, and A.E. Murchison, 1974, Measurement of echolocation signals of the Atlantic bottlenose dolphin, Tursiops truncatus, in open waters, J. Acoust. Soc.Am., 56: 1280.

Au, W.W.L., 1980. Echolocation signals of the Atlantic bottlenose dolphin, Tursiops truncatus in open waters, in: 'Animal Sonar Systems', R.G. Busnel and J.F. Fish, eds., Plenum Press, New York.

Au, W.W.L. and P.W.B. Moore, 1984, Receiving beam patterns and directivity indices of the Atlantic bottlenose dolphin Tursiops truncatus, J. Acoust. Soc. Am., 75, 255: 262.

Au, W.W.L., D.A. Carder, R.H. Penner, and B. Scrouces, 1985. Demonstration of adaptation in Beluga whale echolocation signals, J. Acoust.Soc.Am. 77(2), 726: 730.

Awbrey, F.T., J.C. Norris, A.B. Hubbard and W.E. Evans, 1979. The bioacoustics of the Dall porpoise - salmon driftnet interaction, H/SRWI Technical Report 79-120, San Diego.

Dubrovskii, N.A., P.S. Krasnov and A.A. Titov, 1971, On the emission of echolocation signals by the Azov sea Harbour Porpoise, Soviet Physics Acoustics, 16(4), 444:448.

Dudok van Heel, W.H., 1981. Investigations on Cetacean sonar III, A proposal for an ecological classification of odontocetes in relation with sonar, Aq. Mammals, 8(2), 65:69.

Evans, W.E., F.T. Awbrey, H. Hackbarth, 1985, High frequency pulses of Commerson's dolphin compared to those of Phocoenids, H/SRWI report SC/37/SM, San Diego.

Hackbarth, H., F.T. Awbrey and W.E. Evans, 1985, High frequency sounds in Commerson's dolphins, J. Acoust. Soc. Am., (in press).

Hol, W.A. and C. Kamminga, 1979, Some results on the threshold detection of hollow and solid spheres performed by the Atlantic bottlenose dolphin, Aq. Mammals, 7(3), 41:64.

Jing Xianying, Xiao Youfu and Jing Rongcai, 1981. Acoustic signals and acoustic behaviour of the Chinese river dolphin (Lipotes vexillifer), Scientia Sinica 24, 407:415.

Kamminga, C., and H. Wiersma, 1981, Investigations on Cetacean sonar II, Acoustical similarities and differences in odontocete sonar signals, Aq. Mammals 8(2), 41:62.

Kamminga, C. H. Wiersma, W.H. Dudok van Heel and Tas'an, 1983, Investigations on Cetacean sonar VI, sonar sounds in Orcaella brevirostris of the Mahakam river, Aq. Mammals, 10(3), 83:95.

Kamminga, C., T. Kataoka, and F.J. Engelsma, 1986, Investigations on cetacean sonar VII, Underwater sounds of Neophocaena phocaenoides of the Japanese coastal populations, Aq. Mammals, 12(2), 52:60.

Kamminga, C., J.G. van Velden and W.E. Evans, 1986, Sonar behaviour of Pseudorca crassidens compared with Tursiops truncatus, Aq. Mammals (in press).

Lockyer, Chr. and R.J. Morris, 1986, The history and behaviour of a wild, sociable bottlenose dolphin off the worth coast of Cornwall, Aq. Mammals, 12(1), 3:17.

Møhl, B. and S.A. Andersen, 1973, Echolocation; high-frequency component in the click of the Harbour Porpoise, J. Acoust.Soc. Am., 54, 1368:1373.

Poché, L.B., L.D. Luker and P.H. Rogers, 1982, Some observations of echolocation clicks from free-swimming dolphins in a tank, J. Acoust.Soc. Am., 71, 1036:1039.

"Stranded" (1982), a film on the subject 'stranding of Cetacea', made by Golden Dolphin Productions Pty, Ltd, Australia.

Watkins, W.A. 1980, Click sounds from animals at sea, in: Animal sonar systems, R.G. Busnel and J.F. Fish, ed., Plenum Press, New York.

Wiersma, H., 1982, Investigations on Cetacean sonar IV, A comparison of wave shapes of odontocete sonar signals, Aq. Mammals, 9(2), 57:67.

THE PRODUCTION OF ECHOLOCATION SIGNALS BY BATS AND BIRDS

Roderick A. Suthers

School of Medicine and Department of Biology
Indiana University
Bloomington, IN 47405 U.S.A.

The mechanisms by which echolocating animals produce their sonar signals have generally received less attention than has the detection and processing of these signals by the auditory system.   Since the transmitter and receiver are equal partners in the successful operation of a sonar system, we need to know more about how an animal controls the important information-bearing properties of its echolocative signal and about interactions between its vocal and auditory systems.   This paper reviews recent developments in the laryngeal or syringeal physiology and vocal tract acoustics of echolocating bats and birds.

PRODUCTION OF SONAR PULSES BY BATS

Although the physiology of sound production has been studied in only a few species of echolocating bats, it is already clear that no single model of the microchiropteran larynx will apply to all echolocating bats. Neither can one always safely infer the function of particular muscles in the bat's larynx from their action in the human larynx. The following discussion will concentrate on representatives of three families, the Rhinolophidae, Mormoopidae and Vespertilionidae, for which the most information is available.

The Laryngeal Generator

The microchiropteran larynx, like that of other mammals, is a complex structure consisting of a cartilaginous framework enveloped in several pairs of muscles. All of these but the cricothyroid muscles are innervated by the inferior (or recurrent) laryngeal nerve.   The cricothyroid muscles are innervated by the superior laryngeal nerve. The reader is referred to Elias (1907) Fischer and Vömel (1961) Fischer and Gerken (1961), and Griffiths (1983) for anatomical descriptions of the chiropteran larynx.

One of the most important modifications of the microchiropteran larynx for echolocation is the presence of very thin membranes--the vocal membranes--which lie along the edge of the vocal folds and are unique to echolocating bats (Fig. 1).   These vocal membranes are composed of connective tissue only several microns thick.   The ventricular folds also have a very thin edge, but in Eptesicus fuscus at least, they are not differentiated into as prominent a membrane

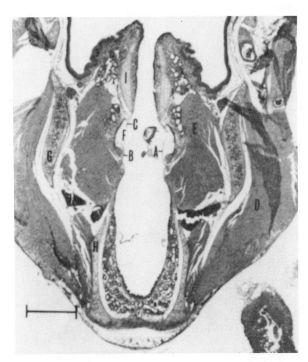

Fig. 1.   Cross section through larynx of Eptesicus fuscus. During
          phonation expiratory air enters ventral chamber from trachea
          and flows upward past vocal membranes toward pharynx. A,
          vocal membranes; B, vocal fold; C, ventricular membrane; D,
          cricothyroid muscle; E, thyroarytenoid muscle; F, ventricle
          of Morgagni; G, thyroid cartilage; H, cricoid cartilage; I,
          arytenoid cartilage. Bar equals 500 microns   (from Suthers
          and Fattu, 1973).

(Griffin, 1958; Suthers and Fattu, 1973). Transverse cuts in both
ventricular folds of Eptesicus fuscus have no effect on its ability to
produce sonar pulses, but similar cuts in the two vocal membranes render
this bat aphonic (Durrant and Suthers, personal observation). After
this treatment the only sound produced is a faint click associated with
the opening or closing of the glottis, which is formed by the vocal
folds (Suthers and Fattu, 1973), and sometimes a fricative hissing
sound.

     It is assumed that ultrasonic vocalizations are produced by
vibration of the vocal membranes which is caused by the action of
Bernoulli forces generated when the membranes are adducted into the
stream of expiratory air. The acoustic function of the ventricular
membranes is not known.

Respiratory Correlates of Phonation

     The need to continually produce rapid, high intensity sonar pulses
places special restrictions on respiratory ventilation; restrictions
which are often least flexible during difficult flight maneuvers
associated with high pulse repetition rates when tissue demands for
oxygen are also highest due to the energetic cost of flight.
Microchiroptera have thus evolved the ability to insert sonar pulses

24

into the expiratory phase of their respiratory cycle with minimal interference to pulmonary ventilation.

Bats which produce long duration sonar pulses at low repetition rates and long duty cycles have solved this problem differently than species emitting short pulses at higher maximum repetition rates. Rhinolophus hildebrandti, a member of the former group, increases its respiration rate in parallel with the rate of sonar pulse emission, so that even at pulse repetition rates as high as 25/s there is rarely more than 1 pulse per respiratory cycle (Fig. 2a). At this repetition rate pulse duration is 25 to 35 ms. Each pulse occupies 80 to 90% of the expiratory phase and is followed by a rapid inhalation resulting in a duty cycle for phonation of about 50%. Only when the interpulse interval (IPI) drops below 10 or 15 ms are two pulses occasionally emitted during one exhalation (Fig. 2b). Somewhat shorter pulses are emitted by Pteronotus parnellii. At high repetition rates of 40 pulses/s this bat produces several 15 to 20 ms long pulses during a single exhalation (Fig. 2c).

Eptesicus fuscus contrasts with both of the preceding long CF-FM bats in the way its short FM orientation sounds are inserted into the respiratory cycle. At low repetition rates of about 10 pulses/s Eptesicus emits one pulse/respiratory cycle as do the preceeding CF bats. Eptesicus increases its pulse repetition rate, however, by inserting more pulses/respiratory cycle so that during a terminal buzz when the repetition rate may exceed 100/s, ten or more pulses are emitted during a single expiration (Fig. 2 in Fattu and Suthers, 1981).

The longest pulses of bats such as Rhinolophus often utilize nearly the full volume of expiratory air available in the lungs. In contrast, individual pulses of Pteronotus and Eptesicus occupy a relatively small portion of the total expiratory air supply. The expiratory volume of these latter species places a limit on the maximum duration of pulse trains that are uninterrupted by a pause for inspiration.

Respiratory Dynamics and Control of Pulse Intensity

The intensity of a sonar pulse is one of the important factors which determines the maximum range at which a target can be detected. Increasing pulse intensity increases the potential range of a sonar system. It is also important that the bat be able to control pulse intensity and thus maintain echo amplitude within optimal limits for both large and small, or near and distant, targets (Kick and Simmons, 1984; Kobler et al., 1985).

The intensity of the emitted pulse depends on a variety of factors including the amplitude of the signal generated by the laryngeal membranes, the acoustic properties of the vocal tract and the degree to which sound radiation is focussed in the forward direction.

The intensity of the laryngeal signal generated by the vocal membranes--the glottal waveform--is related to the rate of airflow through the glottis which in turn depends on the subglottic pressure and glottal resistance. A positive subglottic pressure, developed through activity of the expiratory muscles, provides the driving force for phonation. Pteronotus, Eptesicus and Rhinolophus all emit high intensity pulses and typically develop peak subglottic pressures between about 20 and 50 cm H_2O during phonation. In Eptesicus the maximum intensity of a sonar pulse has an approximately linear relationship to the subglottic pressure at the beginning of the vocalization (Fattu and

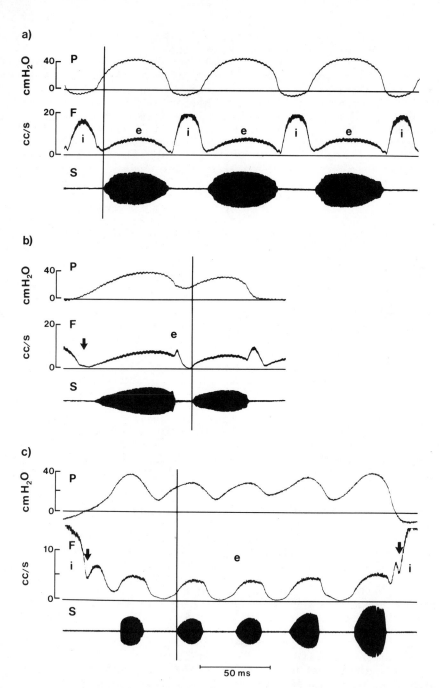

Fig. 2.   Dynamics of glottal airflow and subglottic pressure during
echolocation.   a) Rhinolophus hildebrandti emitting one sonar
pulse/breath.   b) R. hildebrandti emitting two sonar pulses
during one exhalation.   c) Pteronotus parnellii emitting
pulse train during one exhalation.   P, subglottic pressure;
F, rate of tracheal airflow; S, sonar pulses; i, inspiration;
e, expiration. In b and c, arrows mark points of flow
reversal between inspiration and expiration.   In c response
time of microbead thermistor was not fast enough to return to
zero flow at reversal point.   Vertical lines at start of
phonation are to assist in temporal comparisons.

Suthers, 1981). Glottal resistance is highest at sound onset and decreases during the pulse (Durrant, 1986). The peak subglottic pressure attained during a sonar pulse by Pteronotus and the rate of airflow through the glottis are also highly correlated with its maximum sound pressure level (SPL) (Fig. 3a and b). Glottal resistance, on the other hand, remains nearly constant (Fig. 3c), indicating that Pteronotus adducts its vocal folds and membranes to a set phonic position in which the glottal aperture is maintained constant, regardless of fluctuations in subglottic pressure. The observed relationship between sound intensity, subglottic pressure and the rate of glottal air flow is expected if high rates of air flow through the adducted glottis produce larger Bernoulli forces which in turn cause the vocal membranes to vibrate with a greater amplitude.

The intensity of the emitted sonar pulse can be controlled by varying subglottic pressure through adjustments in expiratory effort. The advantages to most echolocating bats of emitting high intensity sonar signals, explains why the subglottic pressure during phonation is quite high compared to that encountered during human speech, where subglottic pressure during normal conversation is about 8 cm H_2O (Ladefoged, 1968) and only reaches 20 to 30 cm H_2O when shouting at maximum intensity (Isshiki, 1964).

The maximum sound intensity that can be generated by increasing subglottic pressure is limited by the necessity of maintaining blood flow through the lung capillaries. If subglottic (i.e. thoracic) pressure exceeds the pressure of venous return to the right atrium, pulmonary circulation will be interrupted. Pressures in the bat's vena cava are not known, but the capillary pressure in the patagium of Tadarida brasiliensis is about 43.5 and 21.8 cm H_2O at the arterial and venous ends, respectively, (Wiederhielm and Weston, 1973). Pressures in the pulmonary circulation should be lower. Even allowing for somewhat higher blood pressure during flight, which is accompanied by a dramatic increase in heart rate (Thomas and Suthers, 1972), it is probable that Microchiroptera have increased their subglottic pressure during phonation as much as their respiratory and circulatory physiology permits. Echolocating bats have a relatively non-compliant lung (Fattu and Suthers, 1981) and it would be interesting to know if pressures in their pulmonary circulation are higher than in non-echolocating mammals. The need to maintain pulmonary circulation is arguably the most important factor ultimately limiting pulse intensity.

The high glottal resistance during phonation, about 5000 dyne sec/cm^5 in P. parnellii, reflects the smallness of the glottal opening which is important for efficient conversion of the fluid subglottic energy into acoustic energy. If the glottal aperture becomes too large in humans, the efficiency of this conversion from fluid to acoustic energy is decreased because the central portion of the expiratory air stream through the glottis is not pulsed by membrane vibration (Cavagna and Margaria, 1965). Echolocating bats must sparingly meter the supply of air in their lungs that is available for phonation. A leaky glottis would reduce the maximum duration of sonar pulses in long CF bats and reduce the length of pulse trains in short FM species. P. parnellii exhales only about 0.16 ml of air during a 15 to 20 ms long pulse. The independence of glottal resistance from pulse intensity suggests Pteronotus maintains its glottis in a fixed configuration for efficient sound production.

Data from Pteronotus suggest echolocating bats have maximized the efficiency with which subglottic fluid energy is converted into radiated acoustic energy. The ratio of the fluid subglottic power to the

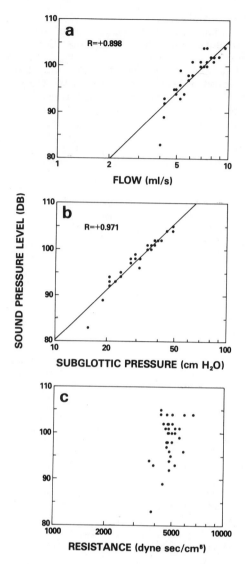

Fig. 3.   Relationship of rate of glottal airflow, subglottic pressure
and glottal resistance to sonar pulse intensity in Pteronotus
parnellii.

radiated acoustic power has been called "vocal efficiency" (Berg,
1956).   Pteronotus develops a subglottic power of 10 to 24 mW when
producing a sonar pulse at 100 dB SPL.   Measurements of the sound
radiation pattern indicate the total radiated acoustic power may be on
the order 1.6 mW, giving a vocal efficiency between about 7 and 16%
(Suthers, 1986a).   These calculations suggest that the vocal efficiency
of Pteronotus is about 2 orders of magnitude more efficient than man
(Berg, 1956; Isshiki, 1964).   Such estimates do not include the energy
used by expiratory or laryngeal muscles during phonation.   A high vocal
efficiency is valuable not only in maximizing pulse intensity, but also

in minimizing the energetic cost of echolocation which may be a significant item in the energy budget of echolocating bats, due to the high subglottic pressure that they must generate.

The vocal efficiency of short FM bats, such as Eptesicus, has not been estimated. The subglottic fluid power of Eptesicus, unlike that of Pteronotus and Rhinolophus, remains nearly constant during an amplitude modulated FM pulse, rather than paralleling changes in SPL, suggesting vocal efficiency may change significantly during the sonar pulse.

Frequency of the Laryngeal Generator

Echolocating bats maintain careful control of the frequency structure of their sonar signals. This is most clearly demonstrated in the species which precisely adjust the emitted frequency to compensate for Doppler shifts of the echo (Schuller et al., 1974), but it is also evident in the ability of other species to control and vary the slope of an FM sweep (e.g. Simmons et al., 1978) or to maintain individually unique frequencies when flying in small groups (Habersetzer, 1981).

The frequency of the sound generated in the Microchiropteran larynx depends on the tension of the vocal membranes which extend between the thyroid and arytenoid cartilages. Unlike the human vocal fold, these membranes contain no muscle fibers. Their tension is controlled by activity of the cricothyroid muscle (CTM) which tenses them by flexing the cricothyroid joint (Novick and Griffin, 1961). The CTM is hypertrophied in Microchiroptera and is divisible into at least two parts--the anterior cricothyroid (aCTM) and posterior cricothyroid (pCTM)--based on anatomical and physiological evidence (Suthers and Fattu, 1973; Griffiths, 1978, 1983, Suthers and Durrant, 1980, Durrant, 1986).

The temporal pattern of CTM activity depends on the type of signal being produced. In Eptesicus fuscus, both portions of the CTM become electrically active some 30-40 ms before the start of the downward sweeping FM pulse at low pulse repetition rates or if it is the first pulse in a train. Within a pulse train, each vocalization is preceeded by short emg starting about 12 ms before the sound. In both caes the emg ceases about 10 ms before phonation begins (Suthers and Fattu, 1973, Durrant, 1986) (Fig. 4a and 7). The mechanical action of the muscle lags several milliseconds behind the electrical events so that muscle force and vocal membrane tension peak just before phonation begins (Suthers and Fattu, 1973). The downward FM sweep is the result of decreasing membrane tension as the CTM relaxes. Bilateral section of the superior laryngeal nerves in E. fuscus prevents the development of tension in the vocal membranes and results in abnormal, roughly CF pulses with a fundamental at about 7.9 kHz, accompanied by a series of harmonics (Suthers and Fattu, 1982).

The timing of activity in the superior laryngeal nerve and CTM of Rhinolophus ferrumequinum differs from that in Eptesicus in a manner predicted by the contrasting pulse types of these species. Schuller and Suga (1976) found that fibers in the caudal portion of the rhinolophid CTM become active before sound emission and continue firing until the end of the long CF pulse. Furthermore, their firing rate is proportional to the emitted CF. Craniolateral portions of the CTM fire during the FM portions of the sonar pulse. Electrical stimulation of the CTM increases the CF frequency. Unilateral denervation of the CTM

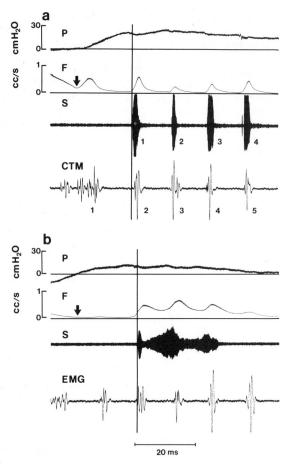

Fig. 4.   Respiratory and muscular events during a train of FM sonar
pulses by <u>Eptesicus</u> <u>fuscus</u>.   a) Intact bat; b) same bat after
bilateral section of recurrent laryngeal nerve.   Despite
normal firing pattern of posterior cricothyroid muscles,
laryngeal gating fails due to paralysis of the ventral
thyroarytenoid muscles and a single long duration
vocalization is produced, the frequency of which varies
sinusoidally with the flow rate. Symbols as in Fig. 2. CTM
and EMG are electrical activity of posterior cricothyroid
muscle. EMG 1 controls starting frequency of pulse 1, etc.

in R. ferrumequinum causes only a slight (1.5 to 4 kHz) drop in the
resting frequency which is normally about 83 kHz. After bilateral
denervation of the CTM the fundamental varies between 12 and 42 kHz from
pulse to pulse and is associated with a series of harmonics. Doppler
shift compensation is unstable after unilateral denervation and absent
after bilateral denervation (Schuller and Suga, 1976).

30

Schuller and Rübsamen (1981) recorded from the superior laryngeal nerve of R. ferrumequinum as the bat adjusted the CF of its emitted pulse to compensate for artificial Doppler shifts which were electronically imposed on the returning echoes. They found that at the bat's resting frequency of 82.5 kHz, 30 to 40 spikes are produced during each sonar pulse, corresponding to a spike frequency of 600 to 800/s. As the bat lowered its CF to compensate for Doppler shifts, the spike count function of these neurons decreased linearly with the CF to 8 or 10 spikes/pulse (160-200 spikes/s) when the resting CF was reduced by 3 kHz (Fig. 5). These authors note that in order to maintain its CF with an observed accuracy of $\pm$ 50 Hz (Schuller et al., 1974), Rhinolophus must regulate the discharge rate of the SLN with an accuracy of $\pm$ 10 spikes per second.

Pteronotus parnellii emits CF/FM sonar pulses of intermediate duration. The pattern of electrical activity in the CTM of Pteronotus parnellii suggests a complex interaction between its anterior and posterior portions. The aCTM typically is active prior to pulse emission, electrically silent during the first half of the pulse and then becomes active again before the end of the pulse. Electrical activity in the pCTM begins 20 to 30 ms before pulse emission and continues until about 10 ms before the end of the pulse (Fig. 6).

The aCTM of Pteronotus is particularly important in controlling frequency at the onset of phonation. After bilateral aCTM denervation, the pulse frequency gradually rises initially before reaching a stable CF about 2 kHz below that normally emitted. There is a similar 2 to 3 kHz drop in CF after bilateral denervation of the pCTM alone. The frequency again rises slightly at the start of the pulse before stabilizing and the terminal downward FM begins much earlier than in the case of an intact larynx. Bilateral denervation of both the aCTM and pCTM causes the fundamental to drop from about 30 kHz down to 10 or 12 kHz with 9 or 10 harmonics. Apparently the aCTM and pCTM can each partially compensate for paralysis of the other portion of this muscle. A similar situation exists in Eptesicus where denervating the aCTM causes only a small drop in emitted frequency (Durrant, 1986).

The Glottal Gate: Control of Pulse Type and Duration

Pye (1967) pointed out that, to a first approximation, the type of pulse (i.e., rising or descending FM, CF, or some combination of these) a bat emits can in theory be achieved simply by turning phonation on and off at appropriate points during the cycle of contraction and relaxation of the CTM. Neuromuscular mechanisms must exist to regulate this glottal gate so that phonation occurs at appropriate times in the cycle of changing vocal membrane tension. In order for vocalization to occur the vocal membranes must be set in a phonic position. Phonation can be terminated either by further adduction, which causes the vocal fold to maintain glottal closure against rising subglottic pressure, or by abducting the vocal membranes out of the phonic position, allowing airflow through the glottis to increase and subglottic pressure to drop. The former method--glottal closure--is used by Eptesicus to terminate pulses within trains at high repetition rates, where subglottic pressure rises and tracheal airflow ceases immediately upon termination of each pulse except for the last pulse of the train (Fig. 4a) (Suthers and Fattu, 1973; Fattu and Suthers, 1981). The latter method--membrane abdution--is employed by R. hildebrandti (Fig. 2b) and on the last member of pulse trains emitted by Eptesicus or Pteronotus (Fig. 2c).

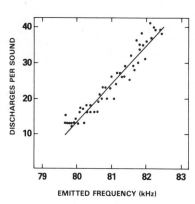

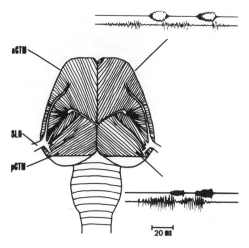

Fig. 5. Discharge rate of
superior laryngeal
nerve of Rhinolophus
ferrumequinum of the
emitted sonar pulse
(from Schuller and
Rubsamen, 1981).

Fig. 6. Pattern of electrical activity
in anterior and posterior
cricothyroid muscles of
Pteronotus parnellii during
production of sonar pulses.
aCTM, anterior cricothyroid
muscle; pCTM, posterior
cricothyroid muscle; SLN,
superior laryngeal nerve.

Pteronotus appears to sometimes use a third technique to gate phonation, in which the glottal resistance is held constant while expiratory muscles are used to increase or decrease laryngeal airflow. This is illustrated in the first four pulses of the pulse train in Fig. 2c where pressure and flow parallel each other. Phonation presumably occurs whenever the rate of airflow through the glottis exceeds a critical value at which Bernoulli forces become large enough to induce vocal membrane vibration. Measurement of expiratory muscle activity is needed to test this hypothetical mechanism which, if true, implies a finer degree of central control over respiratory muscle activity and of their coordination with laryngeal muscles than is necessary when gating is controlled entirely at the laryngeal level. The gating method employed by a particular species probably varies depending on the conditions or physiological context in which a vocalization is produced.

The problem of achieving precise coordination between the muscle activity controlling membrane tension and that controlling the glottal gate is not trivial when one considers that--especially at high pulse repetition rates--millisecond precision is often required. Eptesicus fuscus has solved the problem of gating phonation at the appropriate time during the cycle of vocal membrane tension in part by using the CTM to control glottal resistance as well as vocal membrane tension. Durrant (1986) has shown that the force developed in the CTM of this bat has a vertical and a horizontal vector. The vertical vector flexes the cricothyroid joint and tenses the vocal membranes as described above. If flexing is great enough (about $30°$ of cricothyroid joint rotation) this action also adducts the vocal folds, but observations of the larynx and glottis indicate that during phonation the cricothyroid joint does not rotate enough to adduct the glottis. Instead adduction of the glottis for phonation depends on a second horizontal force vector of the CTM which causes the thyroid lamina to flex inward slightly, toward the midline. This flexion also tenses the vocal membranes, ensuring that membrane tension and glottal resistance rise simultaneously in preparation for a downward sweeping FM pulse as the CTM subsequently relaxes, allowing the glottis to open and membrane to tension drop.

Bilateral denervation of the CTM in E. fuscus causes pulse duration to become highly variable and decreases pulse intensity. Pulse intensity is no longer correlated with subglottic pressure, suggesting that the bat is no longer able to accurately regulate glottal resistance without a functional CTM (Suthers and Fattu, 1982).

The ventral thyroarytenoid muscle (vTAM), innervated by the inferior laryngeal nerve, is also an important glottal adductor in Eptesicus. Griffiths (1983) believed this muscle reduced tension on the vocal folds. Durrant (1986) has shown that it is electrically active concurrently with the CTM before phonation (Fig. 7) and observed that the vocal folds bulge into the laryngeal lumen when it contracts in response to recurrent nerve stimulation.

The dorsal and lateral cricoarytenoid muscles (dCAM and lCAM) of Eptesicus play a role in terminating phonation, especially at low pulse repetition rates when this is accomplished by abducting the glottis. The start of electrical activity in these muscles is closely correlated with the end of phonation, but not with its onset (Fig. 7). The dCAM abducts the arytenoids for inspiration and rotates them dorsally while the lCAM opposes the dorsal rotation of the arytenoids as the dCAM contracts. In both muscles, the emg begins 3 or 5 ms before the end of phonation and continues for 20 to 30 msec. At high pulse repetition rates phonation must be terminated by the adductive action of the CTM and vTAM.

Bilateral section of the inferior laryngeal nerves paralyzes the muscles of adduction and abduction, making its effect on vocalization difficult to interpret. It is surprising that after this radical surgery Eptesicus can still continue to make normal FM pulses, although the maximum subglottic pressures and SPL are reduced. Sometimes, however, the laryngeal gate opens at the wrong time, during contraction of the CTM, with the result that an abnormal rising FM pulse is produced (Suthers and Fattu, 1982). Attempts to produce pulse trains sometimes result in long duration vocalizations with sinusoidally varying frequency that represents the ungated glottal waveform (Fig. 4b).

Electrical activity of muscles innervated by the inferior laryngeal nerve has not been recorded in CF species. Increased activity in the inferior laryngeal nerve of Rhinolophus during inspiration, presumably reflects impulses to abductor muscles. During silent expiration there is little impulse traffic. If a sonar pulse is produced there is a prominent peak of activity 20 ms before phonation, a fairly high sustained activity level during vocalization and a second peak of activity just prior to the terminal FM sweep. These peaks must represent impulses to adductive and abductive muscles controlling the glottal gate (Rübsamen and Schuller, 1981). If the recurrent nerves are cut, Rhinolophus--but not Eptesicus--suffocates due to increased airway resistance (Schuller and Suga, 1976).

Lingual Production of Sonar Clicks

The megachiropteran fruit bat Rousettus agyptiacus often roosts in dark caves where it navigates with echolocative clicks. (Möhres and Kulzer, 1956; Novick, 1958 and Kulzer, 1960). The clicks are produced by moving the side of the tongue against the gum and teeth of the lower jaw lateral and anterior to the Frenulum linguae. Clicks are produced in pairs, one on the right side, the other on the left side of the lower jaw (Kulzer, 1960). Summers (1983) found that the intrapair interval (between members of a pair of clicks) ranged from about 16 to 29 ms and the interpair interval ranged from about 40 to 130 ms in Rousettus trained to fly through wire obstacles. Click repetition rates varied from about 10 to 30/sec with the highest rate occurring at the barrier of vertical wires. Both inter and intrapair intervals were shortened as the bat approached the barrier.

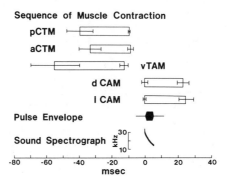

Fig. 7. Sequence of electrical activity in laryngeal muscles of Eptesicus fuscus during production of an FM sonar pulse. Rectangles show mean and horizontal bars indicate standard deviation of start and end of EMG relative to onset of pulse (pCTM, aCTM and vTAM) or end of pulse (dCAM and lCAM). pCTM and aCTM, posterior and anterior cricothyroid muscles, respectively; vTAM, ventral thyroarytenoid muscles; dCAM and lCAM, dorsal and lateral cricoarytenoid muscles, respectively (from Durrant, 1986).

Summers (1983) also used brain stimulation to elicit clicks and found that it is equally likely for the first click of a pair to be generated on the left or right side of the jaw. In one of her bats the amplitude spectra of the left and right clicks were quite distinct, making it possible to determine the side of origin by visual inspection of the spectrum, but in a second bat no consistent differences were evident in the spectra of left compared to right clicks. Unilateral section of the lingual branch of the hypoglossal nerve paralyzes one side of the tongue so that only single clicks are produced from the intact side. Summers (1983) found that postoperative click repetition rates of Rousettus flying in darkness were about one-half that of preoperative flights and their ability to avoid vertical wires evenly spaced across their flight path was severely degraded. There were, however, only minor differences in pre and postoperative amplitude spectra of the clicks.

Further experiments are needed to determine the relationship between click production and respiration. A possible advantage of lingual, in contrast to laryngeal, sound production is that the former may be independent of respiratory air flow. High speed films of flying Rousettus show that at low repetition rates a double click is produced during the downstroke of the wingbeat cycle, but at higher click repetition rates a double click is produced during both the up and downstroke of the wings (Herbert, 1985).

Vocal Tract Filtering: Control of Emitted Frequency

The laryngeal sound produced by the vocal membranes may be significantly modified by filter properties of the vocal tract before it is emitted as a sonar pulse. Vocal tract resonances and filters can provide a means of regulating the harmonic emphasis and spectral content of the sonar signal.

Roberts (1972) found frequency dependent amplitude peaks in the emphasized second harmonic of rhinolophids and hipposiderids. Since the amplitude of the second harmonic can be modulated independently of other harmonic elements, he concluded that these amplitude peaks are not source generated, but depend on the resonant properties of the vocal tract. He further concluded that the bat must be able to vary this resonant tuning.

Additional evidence of vocal tract resonance was obtained from bats breathing light gas mixtures (Roberts, 1973). Since the velocity of sound increases as gas density decreases, this method provides a way of changing the fundamental resonant frequency of the vocal tract without affecting the vibration of the laryngeal membranes. If cavity resonances are important, light gas mixtures will shift the relative amplitude of various harmonics or alter formant frequencies in the emitted pulse. Roberts (1973) demonstrated that when breathing a mixture of 20% oxygen and 80% helium the harmonic emphasis of rhinolophids and hipposiderids changes from the second harmonic to a normally suppressed fundamental and higher harmonics. He concluded that in air these vocal tracts are tuned to transmit the second harmonic and suppress the fundamental. In contrast, Roberts found that light gas has little effect on the amplitude spectrum of the broadband FM pulses emitted by vespertilionids and Desmodus, indicating that these species have broadly tuned vocal tracts.

Light gas experiments thus demonstrate that, at least in a number of Microchiroptera that emit sonar pulses containing long CF portions, the vocal tract is responsible for significant filtering. Very little is yet known about the acoustic mechanism by which such filtering takes place. Do discrete parts of the vocal tract perform specific filter functions? What sort of acoustic coupling exists between them? To what extent can the filter function of the vocal tract be varied or tuned by the bat?

In man, the glottis forms a high impedance termination of the vocal tract. In the source-filter theory of speech production it acts as the source, and its spectrum is modified by the filter or transfer function of the supraglottal vocal tract (Fant, 1970; Lieberman, 1977). In the case of bats, it is not clear what, if any, role subglottal structures have on the spectrum of the emitted sonar pulse. In the following discussion it is nevertheless useful to consider the supra and subglottal portion of the vocal tract separately.

Supraglottal filters. Perhaps the most striking example of vocal tract filtering is the ability of a number of species--including rhinolophids, hipposiderids and mormoopids--to suppress the fundamental in the laryngeal waveform while emitting the second harmonic at a high amplitude.

Rhinolophids and hipposiderids emit sonar pulses through their nostrils. The epiglottis of an adult rhinolophid fits into the nasopharyngeal hole of the soft palate to form a tight laryngonasal junction that isolates the mouth and buccal cavity from the vocal tract (Matsumura, 1979). All of the emitted sound must thus pass through the nasal passages which contain enlarged bony chambers just inside the nares. Suthers and Wenstrup (1986) found that partly filling the enlarged nasal chambers of R. hildebrandti with dental impression medium sometimes altered the harmonic emphasis of the nasally emitted pulse but in other cases had no effect. When both nostrils of R. hildebrandti are sealed the bat is forced to disconnect its laryngonasal junction so it can breath and vocalize through its mouth. Pulses emitted orally under these conditions also show a pronounced increase in the amplitude of the fundamental relative to the second harmonic. The nasal cavities or other portions of the supraglottal vocal tract must thus function as reject filters tuned to suppress the fundamental and 4th harmonic.

The acoustic mechanism by which this nasal filtering occurs is not known. It is possible that the nasal cavities may act as Helmholtz resonators tuned to filter out the fundamental. Such a resonator consists of a rigid chamber with a small opening on one side (Kinsler and Frey, 1962). The tuning of a Helmholtz resonator is determined by the volume of the chamber, the effective length of the neck of the opening and the cross-sectional area of the opening. If the bat is able to vary one of these parameters it may be able to tune its resonator. Roberts (1972) hypothesized that rhinolophids might tune their nasal chambers by varying the dimensions of their external nares. There is at present no evidence to support or refute this hypothesis.

Filtering can also be the result of abrupt changes in the cross-sectional area of the vocal tract which can be considered as being made up of two or more tubes that because of their different diameters are mismatched in their acoustic impedance. The spectrum of the emitted vocalization is then determined by the resonances and anti-resonances of the appropriate compound tube model (Flanagan, 1972).

Little is known about the acoustics of nasal emission in other echolocating bats such as the Phyllostomidae. These bats produce FM signals and do not have a pronounced enlargement of their nasal cavities. At least one of them, Carollia perspicillata, can suppress the fundamental of its FM pulse (D. Hartley, personal communication). It will be interesting to learn if the nasal passages are also a part of this filter.

In addition to its potential as an acoustic filter, nasal emission can provide a means of increasing the directionality of the emitted sound by acoustic interference patterns that depend on the relationship between nostril spacing and wavelength (Mohres, 1953).

Some orally emitting species also suppress their fundamental. The mormoopids, Pteronotus parnellii and P. davyi, for example, emphasize the 2nd harmonic of their long CF/FM pulses. The mechanism by which the

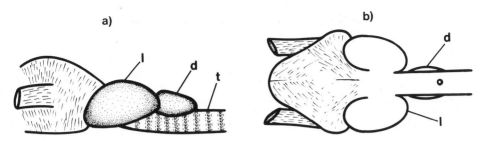

Fig. 8. Schematic lateral (a) and cut-away ventral (b) views of larynx and rigid tracheal pouches in Rhinolophus hildebrandti. l, lateral pouch; d, dorsal pouch; t, trachea.

fundamental is suppressed is unknown. The cranial end of the trachea in these bats is expanded (Fig. 6) and Griffiths (1978, 1983) suggested it may act as the chamber of a Helmholtz resonator with the laryngeal ventricle acting as the neck. Griffiths further postulated that the posterior cricothyroid muscles serve to tune this putative resonator to the frequency of the second harmonic by pulling on the flexible ventricular walls and so varying the cross-sectional area of the neck.

Suthers and Durrant (1980) recorded sonar pulses of P. parnellii in air and in flight with a gas mixture before and after denervation of the posterior CTM. The results do indicate that this muscle affects the vocal tract filtering. There are potential problems, however, with the Helmholtz resonator model as proposed by Griffiths. In Griffiths' model the glottis separates the neck of the resonator from its chamber. Since glottal resistance during phonation is very high (see above), it is questionable if there is adequate acoustic coupling between the supra and subglottal vocal tract. Furthermore, the expanded cranial end of the mormoopid trachea is elastic so that its volume during phonation increases and decreases considerably with fluctuating subglottic pressure. A compound tube model of the supralaryngeal vocal tract may be more appropriate in this case. The laryngeal ventricle may constitute the first element of such a model and the pCTM may affect the filter properties by varying its cross-sectional area.

Subglottal filters. Acoustic filters also exist in the subglottal vocal tract of many echolocating bats. Little is yet known about these filters, but some preliminary data (described below) suggest that the high glottal impedance during phonation acoustically isolates them from the supraglottal vocal tract so that they have quite different functions than do their supraglottal counterparts. Glottal impedance and hence the acoustic isolation between sub and supraglottal vocal tract, is frequency dependent, however. It may also be possible for a bat to control the influence of subglottal filters on its emitted vocalization by altering its glottal impedance.

Rhinolophids and hipposiderids have, in addition to their inflated nasal cavities, rigid cartilaginous out-pouchings of the trachea which form air-filled chambers that are continuous with the tracheal lumen (Elias, 1907, Möhres, 1953). Rhinolophus hildebrandti is typical of other members of its family in having three such chambers consisting of a pair of lateral pouches immediately behind the larynx and a single mid-dorsal pouch connected to the trachea several rings posterior to the lateral pouches by a short tube of small diameter (Fig. 8).

Filling these three tracheal pouches with dental impression medium has no effect on the harmonic emphasis of the emitted sonar signal. It does, however, cause a 15 to 16 dB increase in the amplitude of the fundamental traveling down the trachea toward the lungs. The amplitude of higher harmonics in the trachea is not affected by filling the tracheal pouches (Suthers and Wenstrup, 1986).

Why have rhinolophids evolved prominent tracheal chambers if they do not affect the emitted sound but only attenuate the fundamental component being propagated backwards toward the lungs? The answer is not known but perhaps it is important to prevent the fundamental, which is present at a high intensity in the trachea, from being reflected from the lungs back up the neck and stimulating the cochlea by tissue conduction. It may be important that direct cochlear stimulation by the fundamental arrive by only one pathway with a short delay time. Since these bats suppress the fundamental in their emitted signal, its arrival at the cochlea by tissue conduction could be used to distinguish the emitted pulse from the returning, often overlapping, echo. The presence of the fundamental could, for example, be an important cue in timing pulse-echo delay for range discrimination. Cochlear stimulation via delayed reflections from the thorax might seriously degrade this ability (Suthers and Wenstrup, 1986).

Combination sensitive neurons present in the auditory cortex of Pteronotus, all respond best to selected harmonic components, when they are accompanied by the fundamental (Suga, O'Neill, Kujirai and Manabe, 1983), yet Pteronotus suppresses the fundamental of its emitted signal! Similar combination sensitive neurons have recently been found in Rhinolophus rouxi (Schuller, Radtke-Schuller and O'Neill, pers. comm.) Such neurons should respond differently to a returning echo depending on whether or not it overlaps direct cochlear stimulation from the larynx during phonation. Although Pteronotus lacks rigid tracheal pouches, the enlarged elastic cranial end of its trachea inflates during phonation and may function in a similar manner to attenuate the fundamental of the acoustic wave propagating down the trachea, thus reducing delayed reflections to the cochlea.

PRODUCTION OF SONAR SIGNALS BY ECHOLOCATING BIRDS

Two groups of cave nesting birds, diurnal swiftlets in the genus Collocalia (Medway, 1967) and the nocturnal oilbirds, Steatornis caripensis, (Griffin, 1953) use echolocation to navigate within their often dark breeding colonies. Some authors (Brooke, 1972) place echolocating swiftlets in a separate genus, Aerodramus. Much of the following summary is based on recent studies of the physiology of vocalization in the grey swiftlet, Collocalia spodiopygia (Suthers and Hector, 1982) and in the oilbird (Suthers and Hector, 1985).

Syringeal Mechanisms

Swiftlets and oilbirds both produce broadband click-like sonar signals with most of the acoustic energy between about 1 kHz and 16 kHz. Click repetition rates are seldom higher than about 12/sec. At low click repetition rates, oilbirds produce no more than 1 click during a normal expiration. At higher click rates the bird may produce a group of several clicks during each expiration with successive groups separated by a silent period for inspiration. When a sustained train of clicks is needed, oilbirds change their respiratory pattern to a series of rapid shallow mini-breaths with one click produced on each mini-exhalation.

Grey swiftlets normally emit double sonar clicks which consist of two 3 to 5 msec long clicks separated from each other by a silent period lasting about 18 to 25 ms. Oilbirds also often produce double clicks, but each member of the click pair has a duration about twice that of swiftlets and the silent interval separating members of the double click is similarly about twice as long as in the case of swiftlets. Many clicks emitted by oilbirds are not interrupted by a silent interval to form a double click. These uninterrupted or continuous clicks have a duration of 40 to 80 ms. One swiftlet, C. maxima, also produces continuous clicks.

Despite the obvious similarities between their sonar signals the syrinx of swiftlets and oilbirds have some striking differences. Swiftlets have a tracheobronchial syrinx which lacks intrinsic muscles (Fig. 9a). Syringeal control of vocalization is accomplished by the tracheolateralis and the sternotrachealis muscles, which are innervated by a branch of the XIIth cranial nerve. The oilbirds have a bronchial syrinx which is divided into two semi-syrinxes located along each primary bronchus (Fig. 9b). The sternotrachealis muscles play an important role in the oilbird's vocalization, but the role of the tracheolateralis muscle is taken over by a true syringeal muscle, the broncholateralis, which is not present in swiftlets. The broncholateralis muscles, which are also innervated by the XIIth cranial nerve, always contract just prior to the end of each sonar click, but never in association with social vocalizations such as agonistic squawks. They may thus represent a syringeal specialization for echolocation.

In both swiftlets and oilbirds clicks are preceded by contraction, during expiration, of the sternotrachealis muscles (Fig. 10). These muscles act to pull the posterior end of the trachea caudally and so reduce the tension across the internal and external tympaniform membranes (ITM and ETM). This in turn causes the cartilaginous bronchial semirings bordering the ETM to rotate into the syringeal lumen. In this way a restricted aperture is created between the ITM and ETM. Sound is most likely generated by vibration of the ITM (and/or possibly the ETM) induced by Bernoullii forces when the ETM is adducted to a phonic position, although a whistle mechanism as proposed by Gaunt, Gaunt, and Casey (1982) has not been ruled out. In swiftlets, and sometimes in oilbirds, adduction continues until the ETM touches the ITM, occluding the airway and terminating phonation to produce the first member of the double click. This period of total adduction corresponds to the silent intraclick interval, during which no air flows through the trachea, even though continued expiratory effort causes subsyringeal airsac pressures to rise.

The silent intraclick interval is terminated by contraction of the ETM abductor muscles--i.e., the tracheolateralis in swiftlets; the broncholateralis in oilbirds. The second member of the double click is produced during the initial phase of membrane abduction while the ETM and ITM are still close together. Further abduction terminates phonation. Tracheal pressure varies little during click production showing that the glottis remains open. The oilbirds' continuous click is produced when the ETM is not adducted far enough to block the airway (Fig. 11).

It is interesting to speculate whether or not continuous and double clicks have different functions in echolocation. Double clicks may have a wider bandwidth, whereas continuous clicks may contain more energy. It is not known if oilbirds vary their click type to optimize their sonar pulses for differing conditions or if these click types simply reflect random variations due to imperfect syringeal control mechanisms.

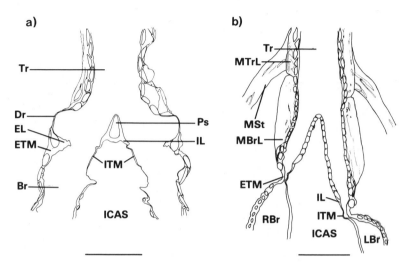

Fig. 9.   Ventral view of a) grey swiftlet and b) oilbird syrinx. Br, primary bronchus; Dr, cartilaginous drum of syrinx; El, external labium; ETM, external tympaniform membrane; ICAS, interclavicular air sac; IL, internal labium; ITM, internal tympaniform membrane; LBr, caudal portion of left primary bronchus; MBrL, broncholateralis muscle; MSt, sternotrachealis muscle; MTrL, tracheolateralis muscle; RBr, caudal portion of right primary bronchus; Ps, pessulus; Tr, trachea. Bar equals 1 mm (swiftlet); 5 mm (oilbird) (from Suthers and Hector, 1982; 1985).

     Simultaneous measurements of airflow through each semi-syrinx show that both contribute to click production. Both sides of the syrinx produce both members of the double click. Grey swiftlets continue to produce double clicks after one primary bronchus is plugged. Although the rate of airflow through the oilbird's two semi-syrinxes is often unequal due to small differences in their dimensions, these flow measurements show that both sides contribute to the beginning and end of each sonar click. The mid-portion of a continuous sonar click, however, is sometimes generated in only one semi-syrinx. This happens when the membranes of only one semi-syrinx are completely adducted to produce a double click while the contralateral semi-syrinx produces a continuous click.

Although vocal tract resonance has been considered to be unimportant in birds (Greenwalt, 1968), resonant properties of the oilbird's vocal tract can have an important influence on the spectrum of sonar clicks and social vocalizations. The vocal tract of this species is prominently asymmetrical, the right semi-syrinx being closer to the trachea than the left. To a first approximation, the vocal tract can be

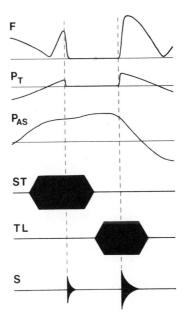

Fig. 10.   Schematic summary of temporal relationships between events during a swiftlet's double click. F, rate of tracheal airflow; P_T, tracheal pressure; P_{AS}, sternal air sac pressure; ST, EMG of sternotrachealis muscle; TL, EMG of tracheolateralis muscle; S, double clicks (from Suthers and Hector, 1982).

modeled as three tubes each of different length and diameter (Suthers and Hector, 1986; Suthers 1986b). Two of these are the cranial portions of the primary bronchi between the semi-syrinx and the trachea, the third is the trachea. Three formants corresponding to the fundamental resonant frequencies of these tubes are evident in many of the sonar clicks. Since the length of the cranial bronchi varies in a seemingly random way between individual oilbirds, the spacing of the upper two formants is potentially unique for each individual and may assist oilbirds in recognizing echoes of their own sonar clicks from those of other individuals (Suthers and Hector, 1986).

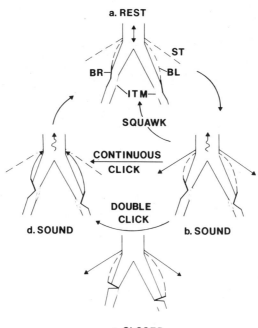

a. REST

ST

BR— —BL

ITM—

SQUAWK

CONTINUOUS
CLICK

DOUBLE
CLICK

d. SOUND   b. SOUND

c. CLOSED

Fig. 11. Schematic representation of the sequence of principle events
controlling the oilbird's bronchial syrinx during
vocalization. a) Condition during silent respiration.
Broncholateralis and sternotrachealis muscles are relaxed
(broken lines). b) Contraction of sternotrachealis muscles
(lateral arrows) causes infolding of external tympaniform
membranes (ETM) constricting syringeal lumen and initiating
phonation. A sustained contraction which maintains the ETM
in a partially adducted position generates a social squawk
which is terminated by relaxation of the sternotrachealis
muscles, returning syrinx to condition a. Sonar clicks are
actively terminated by contraction of the broncholateralis
muscles which rotate outward the bronchial cartilage
supporting the ETM and thus rapidly adduct it (d). If the
initial contraction of the sternotrachealis muscles is strong
enough to fully adduct the ETM, momentarily closing the
syringeal lumen (c), a double click is produced. BL
broncholateralis muscle; BR bronchial cartilage supporting
anterior edge of ETM and on which tendon of broncholateralis
muscle inserts. ITM internal tympaniform membrane; ST
sternotrachealis muscle (from Suthers and Hector, 1985).

ACKNOWLEDGMENTS

    The author thanks G. Durrant and D. Hartley for helpful comments on
the manuscript. Portions of the research reported here were supported
by research grants to the author from the U. S. National Science
Foundation.

REFERENCES

Berg, Jw van den, 1956, Direct and indirect determination of the mean
    subglottic pressure, Folia Phoniatr. (Basel), 8:1-24.

Brooke, R. K., 1972, Generic limits in old world Apodidae and
    Hirundinidae, Bull. Br. Ornithol. Club, 92:52-57.

Cavagna, G. A., and Margaria, R., 1965, An analysis of the mechanics of phonation, J. Appl. Physiol., 20:301-307.

Durrant, G. D., 1986, Laryngeal control of vocalization in the echolocating bat Eptesicus fuscus, Doctoral dissertation, Indiana University, Bloomington, IN, (in preparation).

Elias, H., 1907, Zur Anatomie des Kehlkopfes der Microchiropteran, Morphologisches Jahrbuch, 37:70-118.

Fant, G., 1970, Acoustic Theory of Speech Producion, Mouton and Co., the Hague, Netherlands.

Fattu, J. M., and Suthers, R. A., 1981, Subglottic pressure and the control of phonation by the echolocating bat, Eptesicus fuscus, J. Comp. Physiol., 143:465-475.

Fischer, H., and Gerken, H., 1961, Le larynx de la chauve-souris (Myotis myotis) et le larynx human, Ann. Oto-Laryngol., 78:577-585.

Fischer, H., and Vömel, H. J., 1961, Der Ultraschallapparat des Larynx von Myotis myotis, Gegenbaurs Jahrb. Morphol. Mikr. Anat., Abt 1., 102:200-226.

Flanagan, J. L., 1972, Speech analysis, synthesis and perception, 2nd Ed., Springer Verlag, N. Y.

Gaunt, A. S., Gaunt, S. L. L., Casey, R. M., 1982, Syringeal mechanics reassessed: Evidence from Streptopelia, Auk., 99:474-494.

Greenwalt, C. H., 1968, Bird Song: Acoustics and Physiology, Smithsonian Insitution Press.

Griffin, D. R., 1953, Acoustic orientation in the oilbird, Steatornis, Proc. Nat. Acad. Sci., 39:884-893.

Griffin, D. R., 1958, Listening in the Dark. The Acoustic Orientation of Bats and Men, Yale University Press, New Haven, 413 pp.

Griffiths, T. A., 1978, Modification of M. cricothyroideus and the larynx in the Mormoopidae, with reference to amplification of high-frequency pulses, J. Mammal, 59:724-730.

Griffiths, T. A., 1983, Comparative laryngeal anatomy of the big brown bat, Eptesicus fuscus, and the mustached bat, Pteronotus parnellii, Mammalia, 47:377-394.

Habersetzer, J., 1981, Adaptive echolocation sounds in the bat Rhinopoma hardwickei, a field study, J. Comp. Physiol., 144:559-566.

Herbert, H., 1985, Echolocation in the megachiropteran bat, Rousettus aegyptiacus, Abstracts of 7th Intern. Bat Research Conf., Aberdeen.

Isshiki, N., 1964, Regulatory mechanism of voice intensity variation, J. Speech Hear. Res., 7:17-29.

Kick, S. A., and Simmons, J. A., 1984, Automatic gain control in the bats' sonar receiver and the neuroethology of echolocation, J. Neuroscience, 4:2725-2737.

Kinsler, L. E., and Frey, A. R., 1962, Fundamentals of Acoustics, 2nd Ed., John Wiley and Sons, N. Y.

Kobler, J. B., Wilson, B. S., Henson, O. W., Jr., Bishop, A. L., 1985, Echo intensity compensation by echolocating bats, Hearing Research, 20:99-108.

Kulzer, E., 1960, Physiologishe und Morphologische Untersuchunge über die Erzeugung der Orientierungslaute von Flughunden der Gattung Rousettus, Z. vergl. Physiol., 43:231-268.

Ladefoged, P., 1968, Linguistic aspects of respiratory phenonema, Ann. N. Y. Acad. Sci., 155:(1):141-150.

Lieberman, P., 1977, Speech Physiology and Acoustics Phonetics: An Introduction. MacMillan Publishing Co., Inc., N. Y.

Matsumura, S., 1979, Mother-infant communication in a horseshoe bat (Rhinolophus ferrumequinum nippon: Development of vocalization, J. Mammal., 60:76-84.

Medway, L., 1967, The function of echonavigation among swiftlets, Anim. Behav., 15:416-420.

Möhres, F. P., 1953, Über die Ultraschallorientierung der Hufeisen-nasen (Chiroptera-Rhinolophinae), Z. Vergl. Physiol., 34:547-588.

Möhres, F. P., and Kulzer, E., 1956, Über die Orientierung der Flughunde (Chiroptera, Pteropodidae), Z. Vergl. Physiol., 38:1-29.

Novick, A., 1958, Orientation in paleotropical bats. II. Megachiroptera, J. Exper. Zool., 137:443-462.

Novick, A., Griffin, D. R., 1961, Laryngeal mechanisms in bats for the production of orientation sounds, J. Exper. Zool., 148:125-145.

Pye, J. D., 1967, Synthesizing the waveforms of bats' pulses, in: "Animal Sonar Systems, Biology and Bionics," R.-G. Busnel, ed., pp. 43-64, Laboratoire de Physiologie Acoustique, INRA-CNRZ, Jouy-en-Josas.

Roberts, L. H., 1972, Variable resonance in constant frequency bats, J. Zool. (Lond.), 166:337-348.

Roberts, L. H., 1973, Cavity resonances in the production of orientation cries, Periodicum Biologorum, 75:27-32.

Rübsamen, R., and Schuller, G., 1981, Laryngeal nerve activity during pulse emission in the CF-FM bat, Rhinolophus ferrumequinum. II. The recurrant laryngeal nerve, J. Comp. Physiol., 143:323-327.

Schuller, G., Beuter, K., and Schnitzler, H.-U., 1974, Response to frequency shifted artificial echoes in the bat Rhinolophus ferrumequinum, J. Comp. Physiol., 89:275-286.

Schuller, G., and Rübsamen R., 1981, Laryngeal nerve activity during emission in the CF-FM bat, Rhinolophus ferrumequinum I. Superior laryngeal nerve (External motor branch), J. Comp. Physiol., 143:317-321.

Schuller, G., and Suga, N., 1976, Laryngeal mechanisms for the emission of CF-FM sounds in the Doppler-shift compensating bat, Rhinolophus ferrumequinum, J. Comp. Physiol., 107:253-262.

Simmons, J.A., Lavender, W. A., Lavender, B. A., Childs, J. E., Hulebak, K., Rigden, M. R., Sherman J., and Woolman, B., 1978, Echolocation by free-tailed bats (Tadarida), J. Comp. Physiol., 125:291-299.

Suga, N., O'Neill, W. E., Kujirai, K., and Manabe, T., 1983, Specificity of combination-sensitive neurons for processing of complex biosonar signals in auditory cortex of the mustached bat, J. Neurophysiol., 49:1573-1626.

Summers, C. A., 1983, Acoustic orientation in the megachiropteran bat, Rousettus, Docotral dissertation, Indiana University, Bloomington, IN.

Suthers, R. A., 1986a, Echolocating bats have a high vocal efficiency, in preparation.

Suthers, R. A., 1986b, Avian vocal tract resonance: Structural variation produces individually unique vocalizations in oilbirds. Submitted for publication.

Suthers, R. A., Durrant, G. E., 1980, The role of the anterior and posterior cricothyroid muscles in the production of echolocative pulses by Mormoopidae, in: "Animal Sonar Systems," R.-G. Busnel and J. F. Fish, ed., pp. 995-997, New York: Plenum Press.

Suthers, R. A., Fattu, J. M., 1973, Mechanisms of sound production by echolocating bats, Amer. Zool., 13:1215-1226.

Suthers, R. A., Fattu, J. M., 1982, Selective laryngeal neurotomy and the control of phonation by the echolocating bat, Eptesicus, J. Comp. Physiol., 145:529-537.

Suthers, R. A., and Hector, D. H., 1982, Mechanism for the production of echolocating clicks by the grey swiftlet, Collocalia spodiopygia, J. Comp. Physiol., 148:457-470.

Suthers, R. A., and Hector, D. H., 1985, The physiology of vocalization by the echolocating oilbird, Steatornis caripensis, J. Comp. Physiol. A., 156:243-266.

Suthers, R. A., and Hector, D. H., 1986, Individual variation in vocal tract resonace may assist oilbirds in recognizing echoes of their own sonar clicks, this volume.

Suthers, R. A., Wenstrup, J. J., 1986, Acoustic significance of rigid tracheal pouches and inflated nasal chambers in the echolocating horseshoe bat, Rhinolophus hildebrandti, in preparation.

Thomas, S. P., and Suthers, R. A., 1972, The physiology and energetics of bat flight, J. Exper. Biol., 57:317-335.

Wiederhielm, C. A., Weston, B. V., 1973, Microvascular lymphatic and tissue pressures in the unanesthetized mammal, Am. J. Physiol., 225:992-996.

PROPAGATION OF BELUGA ECHOLOCATION SIGNALS

Whitlow W.L. Au, Ralph H. Penner and Charles W. Turl

Naval Ocean Systems Center
Kailua, Hawaii 96734

The beluga or white whale (Delphinapterus leucas) is very vociferous and one of the first Cetacea to have its underwater sound emissions recorded (Schevill and Lawrence, 1949). Gurevich and Evans (1976) trained a blindfolded beluga to discriminate between complex targets presented in pairs, and found that the animal typically emitted echolocation signals with peak frequencies close to 40 kHz, and secondary peaks at 80 and 120 kHz. Kamminga and Wiersma (1981) reported that belugas in a tank emitted clicks in pairs. The first click typically had peak frequencies close to 60 kHz and the second click had peak frequencies around 1.6 kHz. Au et al. (1985) measured the echolocation signal of a beluga in San Diego Bay and later in Kaneohe Bay, Hawaii. Typical peak frequencies measured in Kaneohe Bay were between 100 and 120 kHz, approximately an octave higher than the 40 to 60 kHz measured in San Diego Bay. Signal intensities measured in Kaneohe Bay were at least 18 dB higher than in San Diego Bay. Au et al. (1985) attributed the beluga's use of high intensity and high frequency signals to the high ambient noise environment of Kaneohe Bay.

In this study, a series of acoustic measurements was made to determine the propagational characteristics of high frequency echolocation signals used by a beluga while performing a target detection task in Kaneohe Bay. Measurements were made to determine the following properties of the whale's sonar system:

(a) The pattern of the transmitted signal in both the vertical and horizontal planes.

(b) The transition region between the acoustic near- and far-fields along the major axis of the transmission beam.

(c) The characteristics of the signal measured at different angles away from the major acoustic axis.

(d) An equivalent planar transducer aperture that represents the whale's echolocation signal projection system.

EXPERIMENTAL CONFIGURATION AND PROCEDURE

The experiment was conducted with a 11-year-old male beluga, designated as DL-575M, that weighed approximately 308 kg and measured 3.0 m in length. A 7.6 x 7.6 x 2.4-m floating pen was used to house the whale and a 6.1 x 6.1-m test pen was located adjacent to the housing pen. The whale was required to extend its head into a 41-cm-diam stationing hoop and bite on a bar attached to the hoop assembly, as shown in Fig. 1. An underwater television camera was used to monitor the animal's position on the stationing assembly. An acoustically opaque screen was situated directly in front of the hoop to prevent the animal from echolocating the target while entering the stationing device. When the screen was lowered, the whale ensonified the target through a clear aperture. A 7.62-cm water-filled stainless steel sphere was used as the target and was located at a fixed range of 80 m from the hoop. The depth of the target, when present, was kept at the same 1 m depth as the bite bar.

Seven Bruel and Kjaer 8103 hydrophones were used in various array configurations to measure the animal's echolocation signals. Each hydrophone had a flat response (+3 dB) to 160 kHz. The transmission beam in the vertical plane was measured with a linear vertical array located approximately 1.9 m from the beluga's nasal plug. The cylindrical hydrophones were oriented with their axes parallel to the water surface and perpendicular to the animal. The sound source was assumed to be in the vicinity of the nasal plugs (Ridgway, this volume) which were estimated to be 0.4 m from the tip of the snout and 0.13 m above the animal's mouth. The transmission beam in the horizontal plane was measured with the hydrophones positioned in an arc approximately 1.9 m from the nasal plugs. Each hydrophone was positioned at a depth corresponding to the major axis of the animal's beam in the vertical plane. The location of the transition region between the acoustic near- and far-fields was determined with an array that extended away from the animal. The output of each hydrophone was connected to specially built amplifiers (400 kHz minimum bandwidth) and recorded on a Ampex PR-2200 portable instrument tape recorder operated at 152.4 cm/s (60 ips).

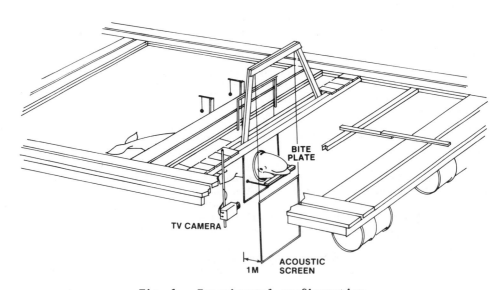

Fig. 1. Experimental configuration.

A multi-channel peak-hold amplifier, multiplexer and digitizer circuit controlled by a Franklin Ace-1000 personal computer was used to analyze the analog tapes. The peak amplitudes of all hydrophones were simultaneously measured, digitized and the results stored on floppy disks.

RESULTS AND DISCUSSION

The signals could be separated in three distinct catagories based on interclick intervals. Mode 1 signals had interclick intervals that were greater than the signals' two-way transit time to the target and back. The amplitudes of Mode 1 signals were always high, with an average peak- to-peak sound pressure level (SPL) of 219 dB re 1 μPa. Mode 2 signals had interclick intervals that were less than the two-way transit time but greater than 5 ms. These signals had mixed amplitudes; some were almost as high, while others were about 12 dB below the Mode 1 signals. The average amplitude of the mode 2 signals was 210 dB. Mode 3 signals had interclick intervals less than 5 ms, typically between 1 and 2 ms. The amplitudes of these signals were always low, with an average SPL of 206 dB. These signals occurred typically towards the beginning or ending portions of click trains. They were probably not used for echolocation since they occurred only in half of the trials and their amplitudes were too low.

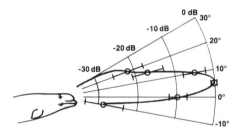

Fig. 2.   Composite vertical beam pattern of the Mode 1 signals from 40 trials.

The composite transmission beam pattern in the vertical plane for Mode 1 signals is shown in Fig. 2. Data from 40 trials were used to determine the composite vertical beam pattern. Only Mode 1 signals were used since the shape of the beam on a click to click basis did not vary much. The major axis of the beam was elevated at an angle of 5°, which is similar to that of Tursiops truncatus (Au et al., 1986). The 3-db beamwidth was approximately 6.5°. The beams for the other signals showed considerable fluctuations from click to click. The beluga's vertical beam is slightly narrower than that of Tursiops by about 3.5°.

The composite transmission beam pattern in the horizontal plane for Mode 1 signals is shown in Fig. 3. Data from 50 trials were used to determine the composite horizontal beam pattern. As in the vertical beam,

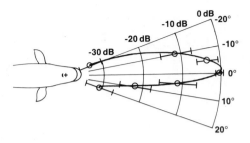

Fig. 3.   Composite horizontal beam pattern of Mode 1 signals from 50
          trials.

only the Mode 1 signals were used. The horizontal beam is pointed directly
forward of the animal and has a 3-dB beamwidth of 6.5°. The beluga
horizontal beam is slightly narrower than the bottlenose dolphin's by
approximately 3° to 4°.

     The results of measurements to determine the location of the transi-
tion region between the acoustic near and far fields are shown in Fig. 4,
for Mode 1 signals. Data from 40 trials were used in this figure. The

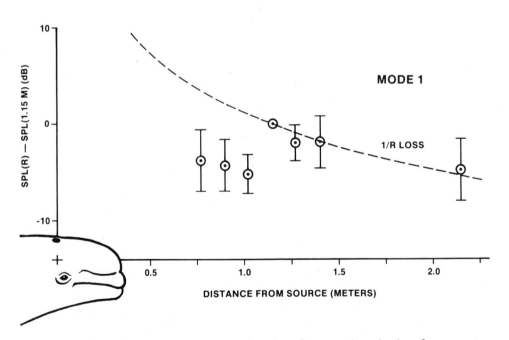

Fig. 4.   Relative sound pressure levels of transmitted signals across
          the acoustic near- and far-fields for Mode 1 signals.

dashed curve represents the spherical spreading loss with the reference at
1.15 m. The data indicate that the transition region between the acoustic
near- and far-fields exists between 1.02 and 1.15 m from the source, or
between 0.62 and 0.75 m from the tip of the beluga's snout.

The acoustic projection system of the beluga whale can be modeled by
an equivalent planar transducer that has the same vertical and horizontal
beamwidths and the same near-field to far-field transition distance. Au et
al. (1978) presented expressions for an equivalent circular aperture based
on beamwidth (eq. 2) and on the location of the transition region (eq. 4).
Using a beamwidth value of 6.5° and a peak frequency of 110 kHz, the
equivalent planar circular aperture has a diameter of 12.8 cm. Using a
transition distance of 1.09 m, the equivalent aperture has a diameter of
13.6 cm. The results of both calculations based on two different sets of
measurements agreed within 8.2%.

REFERENCES

Au, W.W.L., Floyd, R.W., and Haun, J.E., 1978, Propagation of Atlantic
    bottlenose dolphin echolocation signals, J. Acoust. Soc. Am., 64:411-
    422.

Au, W.W.L., Carder, D.A., Penner, R.H., and Scronce, B.L., 1985, Demon-
    stration of Adaptation in Beluga Whale Echolocation Signals,
    J. Acoust. Soc. Am., 77:726-730.

Au, W.W.L., Moore, P.W.B., and Pawloski, D., 1986, Echolocation
    Transmitting Beam of the Atlantic Bottlenose Dolphin, J. Acoust. Soc.
    Am., 79:688-691.

Gurevich, V.S., and Evans, W.E., 1976, Echolocation Discrimination of
    Complex Planar Targets by the Beluga Whale (Delphinapterus leucas),
    J. Acoust. Soc. Am., Suppl. 1, 60:S5 (A).

Kamminga, C., and Wiersma, H., 1981, Investigations of Cetacean Sonar II.
    Acoustical Similarities and Differences in Odontocete Sonar Signals,
    Aquat. Mam., 8:41-62.

Schevill, W.E., and Lawrence, B., 1949, Underwater Listening to the White
    Porpoise (Delphinapterus leucas), Science, 109:143-144.

NASAL PRESSURE AND SOUND PRODUCTION IN AN ECHOLOCATING

WHITE WHALE, Delphinapterus leucas

Sam H. Ridgway and Donald A. Carder

Naval Ocean Systems Center
San Diego, California 92152 USA

INTRODUCTION

At the Jersey conference on Animal Sonar Systems in 1979 we gave strong evidence that dolphins produce sound in the nasal system rather than in the larynx as most mammals do (Ridgway et al. 1980). With electromyography (EMG), we studied the activity of laryngeal muscles and nasal muscles, making comparisons between the two groups of muscles during sound production. Certain muscles of the nasal system were active during all dolphin whistles and click trains while muscles of the larynx were active during respiration but not during sound production. Perhaps more importantly, we measured pressure in the nasal cavities and in the trachea adjacent to the larynx. During sound production, intranasal pressure increased markedly but intratracheal pressure remained unchanged. Subsequently, Amundin and Andersen (1983) replicated the EMG and pressure monitoring aspects of our study in Tursiops and Phocoena.

Finally, we put a catheter or tube in the nasal cavity and sealed the opening of the tube with a thumb, the dolphin could produce sound with this sealed tube in place, but if the thumb was lifted, releasing pressure in the nasal cavity, the animal could not phonate.

In none of the previous studies was the dolphin making an echolocation discrimination during the tests, so we could not be certain that we were studying the actual echolocation generating system. Therefore, we decided to study the Arctic white whale Delphinapterus leucas in a more sophisticated experiment.

MATERIALS AND METHODS

The experimental animal was a male Delphinapterus about 10 years of age that was collected near Churchill, Manitoba, Canada, in the summer of 1977 and had been with our program since that time. At the time of these experiments the whale was 340 cm in length and weighed 450 kg. During training the whale was fed 18 to 20 kg per day of Columbia River smelt, Pacific mackerel, and herring.

The whale was conditioned to allow its trainer to place opaque suction cups over its eyes, to station on a plastic bite bar one meter underwater and behind an acoustically opaque screen made from a closed cell air-foam material, and when the screen was raised to make a discrimination of target present or target absent. The target was a cylinder of 10 X 3.81 cm aluminum pipe with a wall thickness of 0.32 cm suspended on monofilament line one meter in front of the bite bar at one meter depth. The animal was trained to whistle when the target was present and to remain silent for at least 15 seconds when no target was present.

Since this is the first instance we know of in which a cetacean has been trained to vocally report target-present in an echolocation task, perhaps the method used to train this behavior

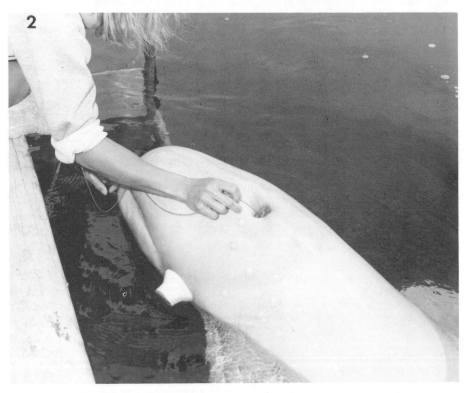

Figure 1. (1) The whale stations at the surface in front of the trainer who is shown inserting the catheter before sending the whale down to the bite bar to wait for the screen to raise for a detection trial. The target is suspended 1 m forward of the screen. (2) A closer view of the trainer inserting the catheter into the whales blowhole. The suction cup blindfolds are made of soft silicone rubber.

deserves special mention. We noticed that when the whale whistled, there was a characteristic movement along the left posterior margin of the blowhole often with the escape of some air. Our trainers quickly induced the whale to repeat vocalizations by tapping with a finger or manipulating the area of the blowhole where movement and air escape had been detected. After whistles were reliably elicited in this manner, the signal was transferred slowly to a simple stroke of the whale's melon. Now, with the blindfolded whale on the bite bar and the target in the water in front of the screen, the trainer raised the screen and then reached down to stroke the whale, eliciting the whistle for which the trainer sounded a bridge signal and gave the whale a fish reward.

Raising the screen invariably resulted in a train of echolocation pulses from the whale. If the target was absent, the trainer did not stroke the whale to elicit the whistle. Gradually the trainer's stroke was omitted as the whale began to whistle when the target was present. When the target was absent, only echolocation clicks were emitted by the whale. The echolocation task was apparently simple for the whale and over 98% correct responses were achieved during several sessions.

Training the catheter insertion was the most difficult step. The trainer touched the whale just behind its blowhole until the animal opened the blowhole. First the whale was rewarded for leaving his blowhole partly open and allowing the trainer to insert the catheter only about 1 cm. Gradually the insertion depth was increased until the catheter could be placed 20 cm into either right or left nasal cavity. Once the whale would allow the catheter to be placed in the nasal system, he was rewarded for sealing his blowhole around it.

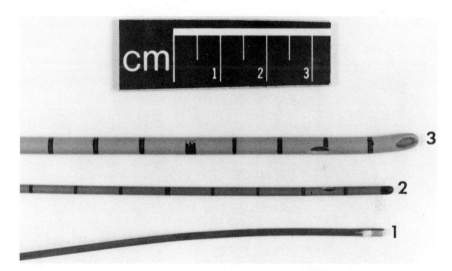

Figure 2. Three catheters were used in the experiment. The closed Millar catheter (1) contains a pressure transducer that rapidly measures pressure (frequency response to 35 kHz) with an electronic sensor under the white area near the tip. A voltage proportional to pressure was amplified from wire leads and recorded. The other two catheters (2) and (3) are intravascular catheters. The larger has an outside diameter of about 3 mm and an inside diameter of 2 mm. The smaller one has an outside diameter of 1 mm and an inside diameter of about 0.5 mm.

Finally, the sequence of trained behaviors was as follows: The whale stationed on the surface near the trainer, the trainer placed the eye cups on the whale, the trainer touched the whale behind the blowhole, the whale opened its blowhole, the trainer inserted the catheter (Figure 1), the whale closed the blowhole, the trainer touched the whale's melon signalling it to go on the bite bar one meter under water, the trainer raised the screen, and the whale pulsed and responded with a whistle when the target was present or no whistle when the target was absent. The screen was then lowered. With a correct response, the trainer sounded a bridge and the whale surfaced for a fish reward. Gradually, the series was extended so that the whale would make three to five responses during one submersion on the bite bar. If the whale's response was

was incorrect, the trainer splashed the water and required the whale to surface and station without reward before she sent the animal back to the bite bar for another series. The target presentation was randomized by referring to a Gellerman series.

Three types of catheters (Figure 2) were used: a 1-mm tube catheter with an opening of about 0.5 mm, a 3-mm tube catheter with an opening of about 2 mm, and a closed Millar pressure transducer catheter with a frequency response of about 35 kHz. All of these catheters were used alone or in combination.

Two hydrophones (B & K 8103) were used to record sounds from the whale. The one for recording echolocation was placed 2 mforward of the bite bar and 10 cm to the left of a line drawn from the blowhole through the target. The hydrophone for recording whistles was placed 50 cm to the left of the whale's melon. The hydrophone outputs, along with FM and analog pressure signals from the Millar catheter and a voice log, were recorded on a Racal instrumentation tape recorder at a speed of 60 inches (152.4 cm) per second. The recorded sound was analyzed on an SD-350 real-time spectrum analyzer with a filter bandwidth of 62.5 Hz or 125 Hz depending on the frequency range setting used.

RESULTS

The whale maintained a high level of correct responses even when all three sealed catheters were placed into the blowhole. With one or more of the catheters in place, the whale responded correctly on more than 90% of the trials after an initial training period of approximately 500 trials was complete. We then recorded 238 trials (106 without a target) with

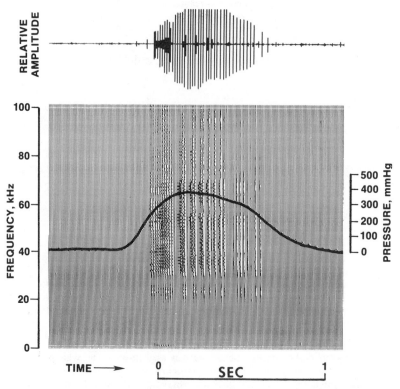

Figure 3. Spectrogram of a typical response by the white whale to a target absent trial. Only echolocation pulses are emitted by the whale. The upper tracing labeled relative amplitude records the peak acoustic amplitude in each 6.5 msec line of the spectrogram. The lower section is a 0 to 100 kHz spectrogram with a bandwidth of 62.5 Hz. The heavy black line indicates pressure. The output of the recorded pressure channel was fed into a voltage-controlled oscillator and calibrated to frequency-modulate a 40 kHz signal in proportion to pressure so that the pressure and acoustic events could appear on the same graph.

catheters in place in either the left or right nasal cavity. The whale responded correctly on 213 trials (89.5%). On 123 trials the whale correctly identified the target as present by whistling after the screen was raised. On 90 trials, the whale correctly responded to the absence of the target by refraining from whistling for 15 seconds after the screen was raised.

High-speed tape recordings of these trials allowed us to directly compare air pressure within the nasal cavities with all sound produced by the whale during each trial. A spectrogram showing a typical response to a target-absent trial is shown in Figure 3. On the target-absent trials only echolocation pulses were emitted by the whale. In contrast, a target-present trial is shown in Figure 4. The intranasal pressure increased just as the screen was raised and just before the whale began to pulse. The pressure increased slightly as the pulse amplitude rose. A typical pulse from this train is shown in Figure 5. The first two high-amplitude peaks were very consistent, lasting 40 or 50 μsec. The second component was smaller and much more variable, lasting 50 to 100 μsec. Peak frequencies measured on our spectrum analyzer ranged from 50 to 80 kHz. With the end of the echolocation pulse train, intranasal pressure increased still more before the whistle began, and remained high during the whistle. During the whistle, pressure fell slightly as the tone swept down in frequency and rose as the tone swept up. Thus increasing frequency occured with rising pressure and decreasing frequency with falling pressure.

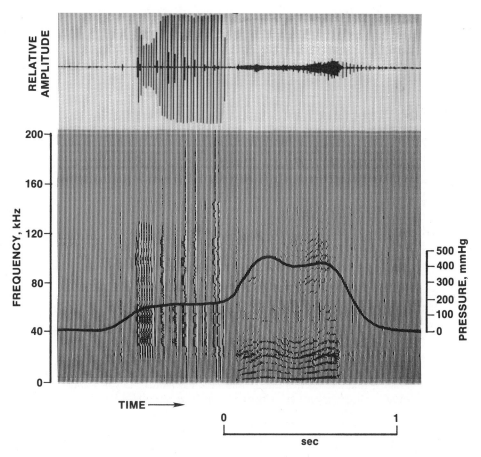

Figure 4. Spectrogram of a typical response by the white whale to a target present trial. Since the measured frequencies were 0 to 200 kHz the frequency bandwidth for this record was 125 Hz.

With the sealed pressure catheter in either the right or left nasal cavity, the whale was able to perform the task correctly. Whenever the whale pulsed or whistled, pressure was always increased. At times during whistles, intranasal pressure reached 800-mm Hg or more than an atmosphere of positive pressure.

57

On some trials we inserted an open catheter along with the pressure catheter. If we held a thumb over the open catheter, effectively sealing it, there was no change in the whale's behavior or sound production. However, if we lifted the thumb as the screen was raised, effectively producing an opening in the nasal cavity, the result was different. With our 1-mm catheter, which had an opening of about 0.5 mm, the whale was able to pressurize and produce a least one train of pulses. However there was often noise or distortion within the 15 to 40 kHz range and the whistle was always truncated and somewhat distorted (Figure 6). Even with the larger catheter in place the whale could pressurize its nasal system, echolocate, whistle, and perform the entire task correctly as long as the catheter was plugged. However, if we unplugged the larger 3-mm tube which had an opening of about 2 mm, we could feel the air rush out as the whale attempted to pressurize for echolocation when the screen was raised. With this larger catheter open in either nasal cavity the whale could not produce pulses or whistles.

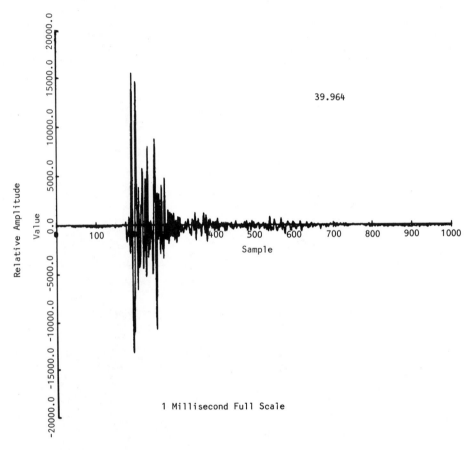

Figure 5. Oscillogram of an echolocation pulse from the target- present trial shown in Figure 4. The pulse was digitized at an effective rate of 480 kHz and plotted. Typical pulse peak frequencies measured on our spectrum analyzer ranged from 50 to 80 kHz.

DISCUSSION

Since toothed whales routinely dive in search of food into waters where very little light penetrates, most species apparently rely on echolocation for finding food. Therefore, the animal must maintain a working sound production system as it dives in the sea and is continually subjected to pressure changes. Air in the thorax is compressed as the whale dives. If sound were produced in the larynx, as it is in most mammals, the resonant characteristics of the dolphin sounds would continuously change with the pressure change. However, the best evidence from the toothed whales that have been studied suggests that Tursiops, Delphinapterus,

Delphinus, Phocoena, and Stenella, at least, produce sound not in the larynx but in the nasal system (Dormer, 1979; Ridgway et al.; 1980, Mackay and Liaw, 1981; Amundin and Andersen, 1983). Since the volume of the nasal system is relatively small, air moved from the lungs can maintain a sufficient volume for sound production as the lung is compressed. Because air pressure and water pressure are equilibrated at any depth, muscles will be able to operate the sacs and valves of the nasal system to produce pressure differentials.

White whales are capable of diving to depths of more than 600 m (Ridgway et al. 1984) where absolute water pressure is greater than 60 atmospheres. As the whale dives, air cavities within the whale are compressed; the whale's thorax is especially flexible to accommodate collapse as air in the lungs is compressed (Ridgway et al., 1969). Since the blood and tissues of the whale, like water, are nearly incompressible, only the air-containing portions of the animal are markedly affected by water pressure. The whale's heart does not have to pump blood against a large differential pressure as the animal dives. Water pressure is applied almost equally over the whale's body and is quite different from differential pressure within the whale's body like that we observed during sound production.

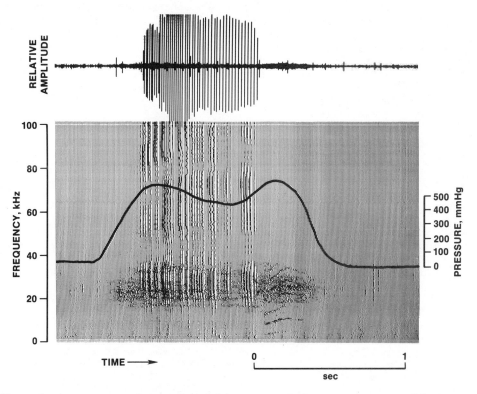

Figure 6. A spectrogram showing the whale's response during a target-present trial when a small (0.5 mm) open catheter was in the left nasal cavity. The whistle given to indicate target-detection is truncated and there is some distortion of the typical sound, especially in the 15 to 40 kHz range of both pulses and the whistle.

Pressure in the nasal cavity is essential for sound production by the white whale. Before sound can be produced, muscles connected to various nasal structures must contract producing pressure inside the nasal cavity. Our studies have shown that this pressure is typically much higher, often four or more times higher, than blood pressure. With such high pressures it is important to isolate the pressure chamber driving sound production from the thoracic cavity. Such pressures must compress adjacent blood vessels and drive blood from adjacent tissue. This would be detrimental to critical circulation in the thorax. The bony nasal cavities, nasal sacs and the fibromuscular nasal plugs have a tough structure that can tolerate such differential pressures during sound production.

When we produced a tiny leak in the whale's nasal cavity by opening the small catheter the whale was still able to produce sound but with some distortion. When we produced a sizable leak by opening the larger catheter the whale could not produce sound. If the sounds were produced below the nasal cavity (i.e. in the larynx) this should not happen. Purves and Pilleri (1983) considered that "all pressure changes during phonation are initiated in the larynx". We think that it is difficult to reconcile this assertion with our results. If the echolocation pulses were "relaxation oscillations" produced in the nasal cavity by jets of air from the larynx as suggested by Purves and Pilleri (1983), we suspect that our experimental manipulations would have had quite different effects. Possibly our large pressure leak would have altered the pulses, but it is difficult to see how this procedure could have eliminated them completely.

Whistles that the whale produced to indicate that the target had been detected were always attended by higher pressures than the echolocation pulse train that preceded. This at first seemed paradoxical since the peak acoustic energy in each echolocation pulse is considerably higher than that of the whistle. However, the pulses are periodic signals lasting at most a few hundred μsec. In Figure 4 the pulse train duration is just under 500 $m$sec and the whistle is just over 500 $m$sec. If we take the average acoustic amplitude of each of the whale's phonations over this period, the whistle is higher in average amplitude because it is a continuous signal whereas the pulses are intermittent. Our finding that intranasal pressure is always greater during whistling suggests to us that continuous signals require more pressure, and possibly that intranasal pressure is correlated more with the total acoustic energy produced than with peak energies. Further, the findings suggests that the echolocation pulse production system is quite efficient, since high-amplitude pulses can be produced with relatively low pressures compared to those measured during whistling.

ACKNOWLEDGEMENTS

We thank Ms. Michelle Jeffries who originally trained our whale and Ms. Joy Ross who continued that training through the experimental period. Mr. F. G. Wood, Dr. R. S. Mackay and Dr. A. S. Gaunt made helpful suggestions on the manuscript.

REFERENCES

Amundin, M. and Andersen, S. H., 1983, Bony nares air pressure and nasal plug muscle activity during click production in the harbor porpoise, Phocoena phocoena, and the bottlenosed dolphin, Tursiops truncatus, J. Exp. Biol. 105:275-282.

Dormer, K., 1979, Mechanism of sound production and air recycling in delphinids: cineradiographic evidence, J. Acoust. Soc. Am. 65:229-239.

Green R. F., Ridgway, S. H. and Evans, W. E., 1980, Functional and descriptive anatomy of the bottlenosed dolphin nasolaryngeal system with special reference to the musculature associated with sound production. In: Animal Sonar Systems, R. G. Busnel and J. F. Fish, Eds. Plenum, New York, pp199-238.

Mackay, R. S. and Liaw, H. M., 1981, Dolphin Vocalization Mechanisms, Science 212:676-678.

Purves, P. E. and Pilleri, G., 1983, Echolocation in Whales and Dolphins, London, Academic Press.

Ridgway, S. H., Carder, D. A., Green, R. F., Gaunt, A. S., Gaunt, S. L. L. and Evans, W. E. 1980, Electromyographic and pressure events in the nasolaryngeal system of dolphins during sound production. In: Animal Sonar Systems, R. G. Busnel and J. F. Fish Eds. Plenum, New York, pp239-249.

Ridgway, S. H., Bowers, C. A., Miller, D., Schultz, M. L., Jacobs, C. A. and Dooley, C. A., 1984, Diving and blood oxygen in the white whale, Can. J. Zool. 62:2349-2351.

Ridgway, S. H., Scronce, B. L. and Kanwisher, J., 1969, Respiration and deep diving in a bottlenose porpoise, Science, 166:1651-1654.

THE STUDY OF THE SOUND PRODUCTION APPARATUS IN THE HARBOUR PORPOISE, PHOCOENA PHOCOENA, AND THE JACOBITA, CEPHALORHYNCHUS COMMERSONI BY MEANS OF SERIAL CRYO-MICROTOME SECTIONING AND 3-D COMPUTER GRAPHICS

Mats Amundin, Erik Kallin, and Sten Kallin*

Dept. of Functional Morphology, Zoological Institute
University of Stockholm, Sweden and Kolmardens Djurpark

* IBM Sweden, Stockholm, Sweden

INTRODUCTION

This preliminary report gives the first results from a study using high resolution 3-D computer reconstructions to describe the sound production apparatus of the harbour porpoise, Phocoena phocoena and the jacobita or the Commerson's dolphin, Cephalorhynchus commersoni. As the majority of investigators seem to agree that the sound source lies in the upper nasal passage (Norris et al., 1971; Diercks et al., 1971; Hollien et al., 1976; Dormer, 1979; Ridgway et al., 1980; Amundin and Andersen, 1983), the present study was restricted to the area from the bony nares to the blowhole.

A comparison of the sound production apparatus of Phocoena and the jacobita is interesting, as their high frequency sonar clicks show remarkable resemblance and stereotype in wave pattern and spectrum (Kamminga and Wiersma, 1981 and 1982). The clicks of the Phocoena and the jacobita consist of several, almost monochromatic cycles within a two part envelope. The click bandwidth is only 11-19 kHz, with peak frequency around 120 to 140 kHz (Møhl and Andersen, 1973; Kamminga and Wiersma, 1981 and 1982). The Phocoena does not produce any of the whistling sounds which are so prominent in many dolphins (Schevill et al., 1969; Møhl and Andersen, 1973; Kamminga and Wiersma, 1981 and 1882). According to Kamminga and Wiersma (1981) the dominant frequency of the click spectrum in Phocoena seems to drop with increasing body size - 140 kHz in a subadult male, to 120 kHz in an adult female. This, together with the observation (Amundin and Andersen in prep) that the high frequency click spectrum in Phocoena is unaffected by the animal breathing helox (a mixture of helium and oxygen, which causes an increased dominant frequency in sounds being filtered in a gas filled cavity) suggest that some acoustically demarcated and fixed tissue structure(s) might be responsible for the shaping of this click component.

There are many reports on the anatomy of the dolphin upper nasal passage (Lawrence and Schevill, 1956; Schenkkan, 1973; Hollien et al., 1976; Dormer, 1979; Green et al., 1980; Purves and Pilleri, 1983). However, the complexity of the structures involved in sound production is such, that their 3-dimensional configuration, exact size and detailed shape remain obscure. The forehead anatomy of Phocoena shows several features which differ from other dolphin species (Gruhl, 1911; cited in Schenkkan, 1973;

61

Purves, 1967; Schenkkan, 1973). It has not been possible from these
reports to deduce the functional consequences of these deviations for the
sound production. No reports on the forehead anatomy of the jacobita have
been found, only of it's close relative <u>Cephalorhynchus</u> <u>hectori</u> (Schenkkan,
1973). Schenkkan's schematic figures indicate that <u>C</u>. <u>hectori</u> have some
similar features to <u>Phocoena</u>, but also show major differences. A
comparison of the <u>Phocoena</u> and the jacobita should cast some light on
possible common anatomical structures responsible for shaping the click.

MATERIALS AND METHODS

   One <u>Phocoena</u> and one jacobita head have been subjected to serial
sectioning. An attempt to inflate the air sacs in the <u>Phocoena</u> head was
only partially successful. Also the nasal plugs were slightly displaced
dorsally. In the jacobita the melon and upper nasal tract were carefully
removed prior to sectioning, as the skull had to be preserved intact. The
tissues were arranged, in as close as possible the correct shape, and were
then refrozen.

   Sample preparation, cryo-sectioning and photographing followed the
method described by Ullberg (1977) and Rauschning (1979). A Pentax NE
Super camera with a wide angle lens (28 mm) was used. The cut surfaces
were photographed at 1 mm intervals.

   A 3-D graphics program package, containing one set for digitizing and
one for 3-D reconstruction, was developed for an IBM PC/XT. A special
database organization was designed, to allow rapid selection of structures
to be displayed. An IBM Professional Graphics Display (PGD), with a
resolution of 640 pixel by 480 lines, was used as feedback during the

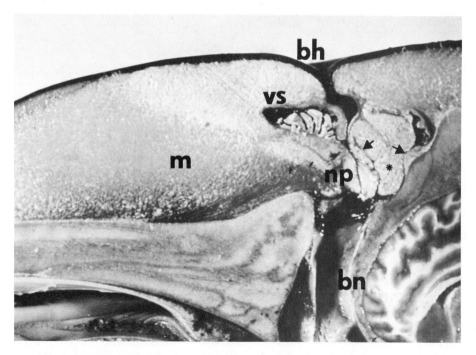

Fig 1. <u>Phocoena</u>: The "hintere Klappe" (*) are outgrowths from the roof of
the posterior part of the nasofrontal sacs (arrows). np=nasal plug,
m=melon, bn=bony nares. bh=blowhole, vs=vestibular sac.

digitization and to display the high resolution 3-D reconstructions. The
serial section diapositives were projected on a Hipad EDT-11H, 300 by 300
mm digitizing tablet (Houston Instruments) attached to the PC/XT. Selected
structures (air sacs, skull contours etc.) were then digitized manually.
In Phocoena two plastic sticks, perpendicular to the sagittal plane, and in
jacobita one side of the Carboxy Methyl Cellulose block (in which the
sample was embedded) were used for registration and scaling purposes. At
the reconstruction, the sections, structure(s), angle of projection,
colour, scale etc., were chosen from a menu on the Hipad tablet. The
sections were then stacked upon each other on the, PGD screen, forming a 3-D
reconstruction, with the picture depth indicated by means of colour shades.

RESULTS

The Phocoena

    The posterior parts of both nasofrontal sacs in Phocoena are expanded
laterally and divided into a rostral and caudal part by means of a
transverse outgrowth from the roof (fig 1 and 2). It was termed "hintere
Klappe" by Gruhl (1911), (cited in Schenkkan, 1973). There is one "hintere
Klappe" (hk) in the left and one in the right nasofrontal sac. Except in
their dorsal attachment, the hk's are completely surrounded by the
flattened posterior nasofrontal sacs. Their widths are 23-25 mm. Their
caudal surfaces follow smooth concavities of the skull, caudo-laterally of
the nares (fig 2) and are lined with the caudal part of the nasofrontal
sacs. Each of these sacs end blindly with a small cavity, full of small
diverticule, just below and on either sides of the supra-occipital notch.
The rostral parts of the nasofrontal sacs constitute the slit-like,
caudally curved entrances to the anterior horns of the nasofrontal sacs,

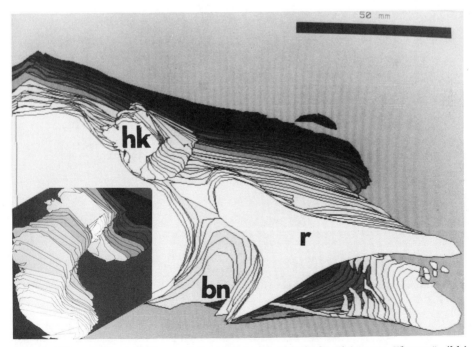

Fig 2. Phocoena: A 3-D reconstruction of the left "hintere Klappe" (hk) in
correct position on the left skull half. bn=bony nares, r=rostrum.
Inserted picture: both "hintere Klappe" in the same angle of projection.

which surround the main nasal passage. The lateral lips of the nasal plugs
are comparably small. They seem to rest dorsally of the entrances from the
inferior vestibule to the nasofrontal sacs, which traverse the full width
of the nares and obviously cannot block them. This may be an artifact as
the nasal plugs were displaced dorsally during the preparation of the head.

The Jacobita

No "hintere Klappe" were found in this species. The lateral lips of the
nasal plugs were well developed on both sides and fitted into grooves in
the caudal wall of the main passage. In the bottom of these grooves were
the tube-like entrances to the nasofrontal sacs (see fig 3). The lateral
widths of the entrances were 6 (left side) and 9 (right side) mm
respectively.

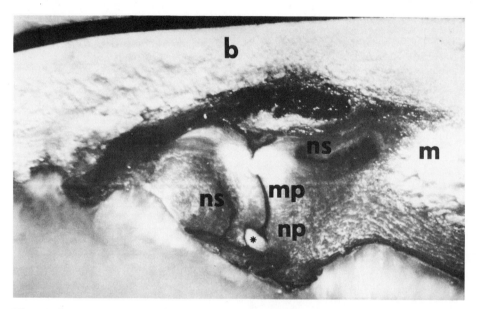

Figure 3. Jacobita: The lateral lips of the nasal plug (*) fit into
grooves, in the bottom of which are the entrances to the nasofrontal sacs
(ns). m=melon, mp=main nasal passage, np=nasal plug, b=blubber.

DISCUSSION

The rough comparison of Phocoena and jacobita anatomy presented here,
revealed no obvious common sound generation design. On the contrary, a
major difference was found: the "hintere Klappe" in Phocoena was not found
in the jacobita. At present the possible role of these conspicuous
structures in sound generation or shaping is not clear. A more
comprehensive description of the Phocoena and jacobita forehead anatomy is
necessary.

Serial Cryo-Microtome sectioning, combined with 3-D computer graphics,
proved to be a useful way of studying the complex anatomy of the delphinid
sound production mechanism. The high resolution 3-D reconstructions created
by the computer program package used in this study revealed the internal
geometry and true shape of the anatomical structures. The classification
at the digitization allowed structures of interest to be selected for

display. The resulting 3-D image could then be rotated until the most instructive angle of projection was reached.

On the CT-images reported by Cranford (this volume) tissues of different densities could easily be separated. The "dorsal bursae" and their connection with the melon were especially evident. The air sacs were more difficult to interpret, as they mostly were collapsed, and the densities of the surrounding tissues were very similar. On the Cryo-microtome sections, however, the sacs were easily discernible, even if completely collapsed, as most of them were lined with black pigmented epithelium. A combination of the two methods should reveal the crucial correlation between e.g. the "dorsal bursae" found by Cranford and the air sac complex described by us. Such a study is under way.

ACKNOWLEDGEMENTS

This study was sponsored by IBM Sweden and Kolmardens Djurpark. Thanks Uppsala University and to Anders lngvast and Ulf Andersson, Ullerakers Sjukhus, Uppsala, for letting us use the Cryo-Mictrotome of their respective departments. Thanks also to Dr Wolfgang Gewalt, Duisburg Zoo, West Germany, for providing jacobita tissue material.

REFERENCES

Amundin, M., and Andersen, S.H., 1983, bony nares air pressure and nasal plug muscle activity during sound production in the harbour porpoise, Phocoena phocoena and the bottlenosed dolphin, Tursiops truncatus, J. Exp. Biol., 105:275.

Amundin, M., and Andersen. S.H. (in prep), The effect of substituting air with a Helium/Oxygen mixture on the click spectrum of the harbour porpoise, Phocoena phocoena.

Cranford, T.W., (this volume), Anatomical basis for high frequency sound transduction in the dolphin head using X-ray computed tomography and computer graphics.

Diercks, K.J., Trochta, R.T., and Greenlaw, C.F., 1971, recording and analysis of dolphin echolocation signals, J. Acoust. Soc. Am., 49(6):1730.

Dormer, K.J., 1979, Mechanisms of sound production end air recycling in delphinids: Cineradiographic evidence, J. Acoust. Soc. Am., 65(1):229.

Green, R.F., Ridgway, S.H, and Evans, W.E., 1980, Functional and descriptive anatomy of the bottlenosed dolphin nasolaryngeal system with special reference to the musculature associated with sound production, in: "Animal Sonar Systems", R.- G. Busnel and J.F. Fish, ed., Plenum Publ. Corp., New York, N.Y.

Hollien, H., Hollien, P., Caldwell, D.K., and Caldwell, M.C., 1976, Sound production by the Atlantic bottlenosed dolphin, Tursiops truncatus, Cetology, 26:1.

Kamminga, C., and Wiersma, H., 1981, Investigations on Cetacean sonar. II. Acoustical similarities and differences in Odontocete sonar signals, Aquatic Mammals, 8(2):41

Kamminga, C., and Wiersma, H., 1982, Investigations on Cetacean sonar V. The true nature of the sonar sounds of Cephalorhyncus commersoni, Aquatic Mammals, 9(3):95.

Lawrence, B., and Schevill, W.E., 1956, The functional anatomy of the delphinid nose, Mus. Como. Zool. Bull., 114(4):103.

Møhl, B., and Andersen, S.H., 1973, Echolocation: high frequency component in the click of the harbour porpoise (Phocoena phocoena L.), J. Acoust. Soc. Am., 54:1368.

Norris, K.S., Dormer, K.J., Pegg. J., and Liese, G.J., 1971, The mechanism of sound production and air recycling in porpoises: a preliminary report, in: "Proc. 8th Annual Conf. Biol. Sonar and Diving Mammals", p:113.

Purves, P.E., 1967, Anatomical and experimental observations on the
    Cetecean sonar system, in: "Animal Sonar Systems, Biology and Bionics",
    R.- G. Busnel, ed., Imprimerie Louis Jean, GAP (Hautes Alpes), France.
Purves, P.E., and Pilleri, G.E., 1983, "Echolocation in whales and
    dolphins", Academic Press, London.
Reuschning, W., 1979, Serial cryosectioning of human knee-joint specimen
    for a study of functional anatomy, Science Tools, 26(3):47.
Ridgway, S.H., Carder, D.A., Green, R.F., Gaunt, A.S., Gaunt, S.L.L., and
    Evans, W.E., 1980, Electromyography and pressure events in the
    nasolaryngeal system of dolphins during sound production, in: "Animal
    Sonar Systems", R-G Busnel and J.F. Fish, ed., Plenum Publ. Corp., New
    York, N.Y.
Schenkkan, E.J., 1973, On the comparative anatomy and function of the nasal
    tract in Odontocetes (Mammalia, Cetacea), Bijdr. Dierk., 43(2):127.
Ullberg, S., 1977, The technique of whole body autoradiography
    cryosectioning of large specimen, Science Tools, The LKB Instrument
    Journal, Special Issue.

THE ANATOMY OF ACOUSTIC STRUCTURES IN THE SPINNER DOLPHIN FOREHEAD AS SHOWN

BY X-RAY COMPUTED TOMOGRAPHY AND COMPUTER GRAPHICS

Ted W. Cranford

Long Marine Laboratory
Institute of Marine Sciences
University of California, Santa Cruz, CA 95064, USA

INTRODUCTION

The dolphin's unusual skull is characterized by profound
displacement and modification of the bones composing the snout and nares,
resulting from an evolutionary process referred to as "telescoping"
(Miller, 1923). Deep within the forehead and acting as valves in the
superior bony nares are two muscled nasal plugs. These both have
connective tissue flaps (lips) on their lateral margins that may be
involved in sound generation (Evans and Prescott, 1962). Above the nasal
plugs the nasal passages empty into a single airway (spiracular cavity)
which exits at the blowhole on top of the head. Along this airway are at
least six blind-ended air sacs which are presumed to act as nearly perfect
reflectors of sound propagated in the tissues of the forehead (Norris,
1964). The fatty melon is an ellipsoid structure anterior to the nasal
passages that grades from a tough exterior shell, rich in connective
tissue, to a pellucid fatty core.

Although there have been detailed treatises of delphinid cephalic
anatomy (Lawrence and Schevill, 1956; Fraser and Purves, 1960; Lawrence and
Schevill, 1965; Hosokawa and Kamiya, 1965; Purves, 1966; Schenkkan, 1973;
Mead, 1975; Green et al., 1980), none have attempted to outline the
anatomic geometry. This is the first study which seeks to describe the
geometric relationships that will form the basis of an accurate model for
sound generation and propagation mechanism(s); this model will be reported
in detail elsewhere.

The geometric configurations and sound propagation characteristics of
tissue structures and interfaces in the dolphin forehead are likely to be
key functional components in the formation of a narrow forwardly directed
sonar beam. Previous workers characterized a structure's size and shape by
isolating it from the remaining anatomy, for example the nasal diverticula
(Dormer, 1974; Schenkkan, 1977; Alcuri, 1980; Gurevich, 1980). It has been
difficult to measure sizes and shapes of structures relative to one
another. Even the most careful hand dissections suffer from slumping
tissue and changing shapes of organs. Studying morphology from serial
sections was an advancement over traditional dissection. The first
descriptions of delphinid serial section anatomy (Boice et al., 1964;
Hosokawa and Kamiya, 1965; Green et al., 1980), produced gross sections
ranging from 1.0 cm. to 6.0 cm. in thickness. Defining anatomic and

geometric relationships with precision is essential to forming a working model of the delphinid sound generation system. My preliminary study of anatomy shows that modern medical imaging techniques make such a model attainable. These techniques avoid most artifacts inherent to conventional dissection and permit precise measurement of geometries and densities in an intact dolphin head. This paper includes a brief description of the methods used, three examples of results, and some ideas that have come from them.

MATERIALS AND METHODS

X-ray computed tomography (CT or CAT) measures X-ray absorption and produces serial section images or tomograms. I have combined serial section tomograms with high resolution three-dimensional computer graphics to make a three-dimensional interactive computer model.

Cross-sectional tomograms or "slices" were generated using a General Electric model 9800 CT scanner.[1]

A tomogram is an image along a plane. X-ray CT uses an array of detectors to record many crisscrossed pathways of X-ray attenuation (a measure of electron density) to compute a tomogram. At points where paths cross, a Hounsfield or CT number, corresponding to the number of X-ray photons absorbed, is computed and stored in a (320 x 320) matrix. Each two-dimensional slice represents more than 100,000 points or pieces of information. Thus, each tomogram is composed of thousands of shaded points, called picture units or pixels, displayed on a high resolution graphics screen as one of 16 shades ranging from black to white. These tomograms are then stored as several successive two-dimensional matrices. Once this information is stored digitally it can be processed to form various types of images which enhance different features (see Mackay, 1984).

Two analytic programs have proven especially illuminating. It is possible to select any plane through the entire subject and to compute a derived tomogram through that plane. This method can yield an entirely new view of the subject in any plane, although some smearing may occur in accordance with the averaging and spacing of sections in the original scanning plane (Mackay, 1984). This technique has been valuable for illuminating the morphology of the posterior core of the dolphin's melon (see RESULTS). Another CT program allows the operator to display density gradients in any tomogram (Fig. 1). Placing markers (asterisk and square) on the tomogram denotes the interval along which a density gradient is to be measured. In the example given in this paper, a sampling width of three pixels was then chosen along the entire interval, and a graph was produced which displays the continuous change in CT number (and thus tissue density) along the interval.

One hundred and thirty transverse tomograms or "slices" were generated from the tip of the rostrum to the occiput for each of two postmortem male specimens of the spinner dolphin, Stenella Longirostris. Throughout the first 235 mm the X-ray beam was 5 mm wide, and the tomograms were representations averaged over that interval. Over the next 69 mm the region of greatest interest (through the posterior melon and nasal air sacs), the X-ray beam was tapered to yield 1.5 mm interval averages. The remaining posterior region of the head was imaged in 3 mm interval averages.

---

[1]Scanning protocol was 140 kV and 70 mA with 3 - second averages in the 1.5 mm region

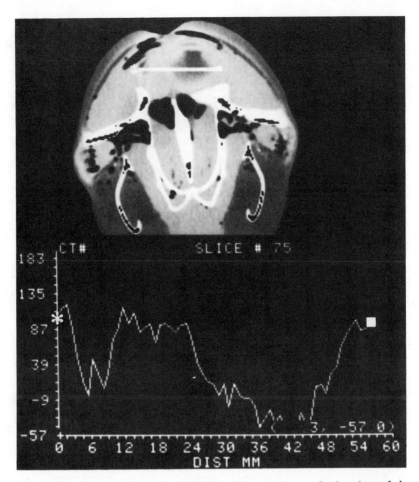

Figure 1 - Transverse tomogram (#75) from the region of the dorsal bursae.
The solid white horizontal line across the middle of this 1.5 mm
"slice" denotes a sampling interval for the measurement of a
density gradient. Bones are seen as bright white muscle as
light gray, fat as dark gray, and air as black. The graph in
the bottom half of the figure displays density (as CT #) against
distance along the interval (in mm) from the ✱ to the ■.

Each tomogram for the first specimen was then projected onto a digitalizing
tablet at the Quantitative Morphology Laboratory. The structures of
interest in each tomogram were traced into computer memory by means of a
scribe (Livingston, et al. 1976). I magnified tomograms for tracing, so
that the effect of any deviation from chosen boundaries during tracing
would be minimized. These two-dimensional contours were then stacked to
reconstruct a three-dimensional model. Three-dimensional images of
structures were displayed as a series of computer generated topographic
lines (Fig. 2). Thus, many combinations of structures can be made visible
in three dimensions on a high resolution graphics screen. Images can be
rotated, viewed from different angles, or magnified. Volumes, centroids,
and angles between any three points can also be determined.

Preliminary analysis revealed structures, geometric relationships, and
variations in tissue density that were previously undescribed.

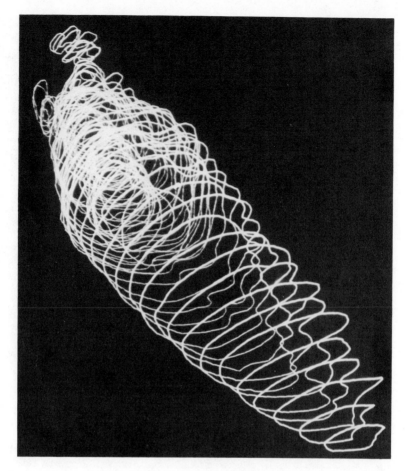

Figure 2 -   Three - dimensional computer graphic reconstruction of the
             melon in the dolphin's forehead. This structure focuses sound
             and is situated in front of the acoustically reflective air
             sac system. The inner set of circles represents the lower
             density core which splits and emerges from the posterior shell
             as two lateral branches ending near the dorsal bursae. Note
             that the lines are closer together after the tomogram
             thickness changes from 5 mm to 1.5 mm.

RESULTS

    The melon, which gives the characteristic bulbous shape to the dolphin's
forehead, is located immediately in front of the air-filled nasal
diverticula, and has been shown to focus sound (Norris and Harvey, 1974;
Litchfield et al., 1979; Varanasi et al., 1982) by using the sound velocity
topographies involving unique acoustic fats (Varanasi and Malins, 1971 and
1972; Litchfield et al., 1973).

    A complex of bifurcations was discovered in the posterior core of the
melon in both spinner dolphin specimens (Fig. 3a and 3b). The first split
occurs laterally just before the core emerges from the posterior boundary
of the melon's shell (Figs. 3a and 2). Posterior to the first disjunction,
each branch (ramus) emerges from the shell and gives off two small bulbous
dorsal projections and then ends bluntly near the nasal plug lip (Fig. 3b).
The lateral bifurcation of the core and part of the anterior dorsal
projection of the left ramus are shown in Fig. 1.

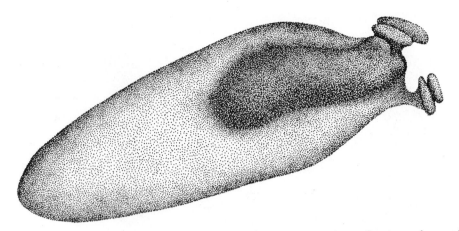

Figure 3a - This 3/4 view of the melon emphasizes the relative sizes of
the melon's branches (with their bursae), and illustrates
how the left side appears to be rotated approximately 50°
from the midsagittal plane.

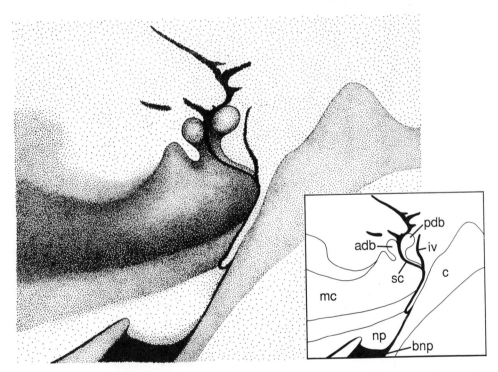

Figure 3b - Diagram of a generalized parasagittal plane through the right
side of the dolphin's head. Anterior is to the left of the
figure, while air spaces are shown solid black. the
spiracular cavity (sc) arises from the right naris (bnp) at
the bottom of the figure. It continues dorsally and gives off
the inferior vestibule (iv) before passing between the
anterior (adb) and posterior (pdb) dorsal bursae (mc = melon
core and c = cranium np = nasal plug).

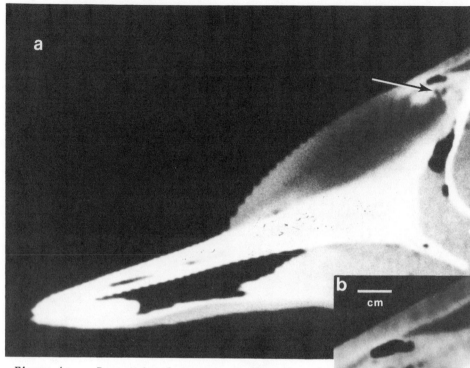

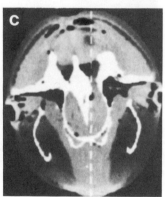

Figure 4a - Parasagittal section through the melon's major axis (the right side) of a male spinner dolphin. The melon is the dark gray elliptical structure composing the animal's forehead. Near the posterior end of this structure two small bulbous projections (arrow) resemble the ears of a rabbit. These dorsal bursae may function as sound transducers for the dolphin's sonar system.

Figure 4b - Magnification of the right dorsal bursae of the melon from Figure 4a (scale = 1 cm).

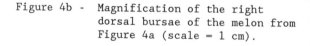

Figure 4c - Representative section through the region of greatest interest in which the dotted line indicates the parasagittal plane of Figs. 4a and 4b.

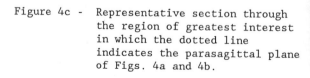

A density gradient was measured horizontally across the center of a transverse tomogram (slice #75). The graph at the bottom of the figure displays CT number[2] (a measure of density) against distance along the interval. The first two depressions of the graph, beginning at 6 mm, indicate lower density tissue in the left lateral ramus. The next major depression (beginning at 24 mm), indicates sampling across the large right ramus, which is of lower density than the left. The view presented in Figs. 4a and 4b is a CT reformation representing a parasagittal section bisecting the major (right) axis of the melon and viewing it from the left.

Magnification of the posterior melon from Fig. 4a is shown in Fig. 4b. It shows the configuration of the right dorsal projections, which, like the left, are embedded in the tissue above the nasal plus immediately antero-medial to the nasal plug lip. Each set of bursae superficially resembles a pair of "rabbit ears." Here I will call them the "dorsal bursae." The melon's core has left and right branches which invade both nasal plug muscles, as has been reported by other authors (Norris, 1968; Norris, 1969; Mead, 1975), but this is the first report of the dorsal bursae. The right bursae (anterior and posterior) project in the parasagittal plane and emerge from the core's major right branch. Bursae also arise from the left branch although in a paraxial plane rotated approximately 50o to the left of the midsagittal plane (Fig. 3a).

DISCUSSION

Dolphins are the "quintessential sound users in the ocean" (Dormer, 1974), and it is well known that many types of sounds can be recorded from wild dolphin schools almost continuously (except when resting). This is especially true at night (Norris et al., 1985). Au (1980) summarized a series of experiments in which dolphins produced single click intensities up to 230 dB re: 1 μPa at 1 yd. These intensities approach the finite limit of sound in water, so that additional energy put into the production of sound is increasingly dissipated as heat. Norris and Møhl (1983) have marshaled evidence suggesting prey debilitation by odontocetes using high intensity sounds. These observations suggest that dolphins are dependent on their ability to generate sounds and that these generators must be extremely reliable, resilient, and sufficiently strong to withstand the stress of frequent production of very loud sounds.

The location of the bursae suggests that they are probably involved in the generation of the sound beam. Two lines of evidence point to this hypothesis.

I. EVIDENCE FROM ANATOMIC STUDIES

The anatomic site of sound generation has been the focus of controversy for many years. The regions around the nasal plug lips have been implicated by several investigations. Cineradiographic evidence collected by Norris et al. (1971) and by Dormer (1974, 1979), show chirp and click production to be concurrent with the passage of air in this same region. Diercks et al. (1971) used acoustic triangulation and Mackay and Liaw

---

[2] 2CT number is a synonym for Hounsfield (H) number. This unit is actually a measure of electron density and is named after Dr. Geoffery Hounsfield, one of two men credited with the development of the first CT scanner. One Hounsfield unit is equivalent to "0.1% of the attenuation coefficient of water and the scale zero is water. Then air has H of .1000 and bone about +1000" (Mackay 1984).

(1981) used ultrasound to show that sounds were made concurrent with movements near the nasal plugs and lips. The electromyography and pressure events studies by Ridgway et al. (1980) and Amundin and Andersen (1983) also lend support to this suggestion. The electromyography and pressure event studies by Ridgway et al. (1980) and Amundin and Andersen (1983) also lend support to this suggestion.

In the same region, the melon terminates in the curious dorsal bursae and the blunt posterior end of the lateral rami that invade the nasal plugs near each lip. The bursae extend as channels of low density tissue located above, and in front of, the tissues of the lips, which are stout and invaded with connective tissue. Although the actual sound generation mechanism remains to be elucidated, these structures appear to be involved in the production of intense sound on a continuing basis. The bursae are connected directly to the melon's core by a low density tissue channel, and so they could serve as a transducer and/or conductor of sound. Further, vibrations could come not only from the lip itself, but also from the walls of the bursa(e).

II. EVIDENCE FROM SOUND GENERATION EXPERIMENTS

Dolphin sounds are generated or initiated using air and the airway. Evidence for a pneumatic source was suggested by Evans and Prescott (1962) when they forced air through the head of a postmortem _Stenella_ sp. and produced clicks and whistles similar to those made by live dolphins. Proof for a pneumatic source came from two studies (Ridgway et al., 1980, Amundin and Andersen, 1983), both of which found that pressure increases in the bony nares just prior to the onset of sound emission. Both teams recorded electromyographic events in muscles of the forehead during phonation, and increasing pressure in the nasal diverticula after phonation with concomitant decreasing pressure in the bony nares. Ridgway and his colleagues also found that when the narial catheter was opened to the atmosphere the animals ability to produce sound was impaired. It is important to note that the degree of impairment increased as the tube was opened wider and that during the animal's attempts to make sound air rushed from its open end. Collectively, the live animal studies found an absence of movement and air pressure events in the larynx during nearly all phonation.

The source of energy to ensonify the lip and bursae may originate with the passage of pressurized air bubbles past the lips or between the bursae (by way of the spiracular cavity), a process that may be analogous to the production of glottal pulses in the human larynx (Miller, 1959; Rosenberg, 1970; Altes et al., 1975). This line of reasoning is strengthened considerably by the work of Amundin and Andersen (pers. comm.). Dolphins and porpoises produce pulsed signals which emphasize two general frequency bands; pulses can be separated into high and low frequency components (Lilly, 1962; Dubrovskiy et al. 1971; Møhl and Andersen, 1973; Kamminga and Wiersma, 1981). Amundin and Andersen showed that when their porpoise inhaled a helium-oxygen mixture, the low frequency component was shifted up by almost one octave while the high frequency component was unchanged. They point out that this frequency shift indicates that low frequency sound must have been propagated in a gas, in this case respiratory air. At the same time, the high frequency component must originate in tissue: it was not affected by the higher sound propagation velocity of the inhaled mixture. As expected, sound pressure levels were significantly higher for the tissue component than for the gas dependent portion.

The exact mechanism of sound production remains to be demonstrated, but the lips and bursae are probably involved in the generation or propagation of dolphin sounds.

REFERENCES

Alcuri, G., 1980, "The role of cranial structures in odontocete sonar signal emission," in: Animal Sonar Systems, J.F. Fish, ed., Plenum Press, New York.

Altes. R. A. and W. E. Evans, 1975, "Cetacean Echolocation Signals and a New Model for the Human Glottal pulse." J. Acoustic. Soc. Am., 57(5): 1221.

Amundin. M. and S.H. Andersen, 1983, "Bony nares air pressure and nasal plug muscle activity during click production in the harbour porpoise, Phocoena phocoena, and the bottlenosed dolphin, Tursiops truncatus," J.' Exp. Biol., 105: 275.

Au. W.W.L., 1980, "Echolocation signals of the Atlantic bottlenose dolphin (Tursiops truncatus) in open waters," in: Animal Sonar Systems, J.F. Fish, ed., Plenum Press, New York.

Boice, R.C., M.L. Swift, and J.C. Roberts, 1964, "Cross sectional anatomy of the dolphin," Norsk Hvalfangst.Tidende (Norwegian Whaling Gazette). 53(7): 177.

Diercks, K.T., R.T. Trochta, C.F. Greenlaw, and W.E. Evans, 1971, "Recording and analysis of dolphin echolocation signals," J. Acoust. Soc. Am., 49(6): 1729.

Dormer, K.J., 1974, The mechanism of sound production and measurement of sound processing in Delphinid Cetaceans, University of California, Los Angeles. Ph.D. Dissertation.

Dormer, K.J., 1979, "Mechanism of sound production and air recycling in delphinids: cineradiographic evidence," J.' Acoustic. Soc. Am., 65: 229.

Dubrovskiy, N.A., P.S. Krasnov, and A.A. Titov, 1970. "On the emission of echolocation signals by the Azov Sea harbor porpoise," Akusticheskii Zhurnal. 16(4): 521. Translated from Akusticheskii Zhurnal.

Evans, W.E. and J.H. Prescott, 1962, "Observations of the sound capabilities of the bottlenose porpoise: a study of whistles and clicks." Zoologica, 47(3): 121.

Evans, W.E., W.W. Sutherland, and R.G. Beil, 1964, "The directional characteristics of delphinid sounds," in: Marine Bio-Acoustics, W.N. Tavolga, ed., Pergamon Press, New York.

Fraser. F.C. and P.E. Purves, 1960, "Hearing in cetaceans. Evolution of the accessory air sacs and the structure and function of the outer and middle ear in recent cetaceans." Brit. Mus. (Nat. Hist.), Bull. Zool., 7(1): 1.

Green, R.F., S.H. Ridgway, and W.E. Evans, 1980, "Functional and descriptive anatomy of the bottlenosed dolphin nasolaryngeal system with special reference to the musculature associated with sound production," in Animal Sonar Systems, J.F. Fish, ed., Plenum Press, New York.

Gurevich, V.S., 1980, "A reconstructing technique for the nasal air sacs system in toothed whales," in: Animal Sonar Systems, J.F. Fish, ed., Plenum Press, New York.

Hosokawa, H. and T. Kamiya, 1965, "Sections of the dolphin's head (Stenella coeruleoalba)," Sci. Repts. Whales Res. Inst., 19: 105.

Kamminga, C. and H. Wiersma, 1981, "Investigations on Cetacean Sonar II. Acoustical Similarities and Differences in Odontocete Sonar Signals," *Aquatic Mammals.* 8(2): 41.

Lawrence, B. and W.E. Schevill, 1956, "The functional anatomy of the delphinid nose," *Bull. Mus. Como. Zool.* (Harvard), 114(4): 103.

Lawrence, B. and W.E. Schevill, 1965, "Gular musculature in delphinids." *Bull. Mus. Como. Zool.* (Harvard), 133(1): 1.

Lilly, John C.. 1962, "Vocal behavior of the bottlenose dolphin," *Proc. of the Am. Phil. Soc.*, 106(6): 520.

Litchfield, C., C. Karol, and A.J. Greenberg, 1973, "Compositional topography of melon lipids in the Atlantic bottlenose dolphin (*Tursiops truncatus*): implications for echolocation," *Marine Biology*, 23: 165.

Litchfield, C., R. Karol, M.E. Mullen, J.P. Dilger, and B. Luthi, 1979, "Physical factors influencing refraction of the echolocative sound beam in delphinid cetaceans," *Marine Biology*, 52: 285.

Livingston, R.B., K.R. Wilson, B. Atkinson, G.L. Tribble, D.M. Rempel, J.S. MacGregor, R.E. Mills, T.C. Ege, and S.D. Pakan, 1976, "Quantitative cinemorphology of the human brain displayed by means of mobile computer graphics," *Trans. Amer. Neurol. Ass.*, 101: 99.

Mackay, R.S. and C. Liaw, 1981, "Dolphin vocalization mechanisms," *Science.* 212(8): 676.

Mackay, R.S., 1984, *Medical imazes and displavs: comparisons of nuclear magnetic resonance. ultrasound, X-rays, and other modalities*, John Wiley & Sons, New York.

Mead, J.G., 1975, "Anatomy of the external nasal passages and facial complex in the Delphinidae (Mammalia: Cetacea)," *Smithsonian Contribu. tions to Zoology.* no. 207: 1.

Miller, R.L., 1959, "Nature of the Vocal Cord Wave," *J.' Acoustic. Soc. Am.*, 31(6): 667.

Møhl, B. and S. Andersen, 1973, "Echolocation: high frequency component in the click of the harbour porpoise (*Phocoena phocoena* L.)," *J. Acoustic. Soc. Am.*, 54(5): 1368.

Norris, K.S., 1964, "Some problems of echolocation in cetaceans," in: *Marine Bio.Acoustics.* W.N. Tavolga, ed., Pergamon Press, New York.

Norris, K.S., 1968, "The evolution of acoustic mechanisms in odontocete cetaceans," in: *Evolution and Environment*, E.T. Drake, ed., Yale University Press, New Haven.

Norris, K.S., 1969, "The echolocation of marine mammals," in: *The Biology of Marine Mammals*, H.T. Andersen, ed., Academic Press, New York.

Norris, K.S., K.J. Dormer, J. Pegg, and G.T. Liese, 1971, "The mechanism of sound production and air recycling in porpoises: a preliminary report," in: *Proc. VIII Conf. Biol. Sonar Diving Mammals*, Menlo Park, California.

Norris, K.S. and G.W. Harvey, 1974, "Sound transmission in the porpoise head," J. Acoustic. Soc. Am., 56(2): 659.

Norris, K.S. and B. Møhl, 1983, "Can odontocetes debilitate prey with sound?," The American Naturalist, 122(1): 85.

Norris, K.S., B. Wursig, R.S. Wells, M. Wursig, S. M. Brownlee, C. Johnson, and J. Solow, 1985, "The behavior of the Hawaiian spinner dolphin, Stenella longirostris," Administrative report to the National Marine Fisheries Service. vol. LJ.85.06C, Southwest Fisheries Center P.O. Box 271, La Jolla, CA 92038.

Purves, P.E., 1966, "Anatomical and experimental observations on the cetacean sonar system," in: Proc. Sym. Bionic Models Animal Sonar System, R.G. Busnel, ed., Frascati, Italy.

Ridgway, S.H., D.A. Carder, R.F. Green, A.S. Gaunt, S.L.L. Gaunt, and W.E. Evans, 1980, "Electromyographic and pressure events in the nasolaryngeal system of dolphins during sound production." in Animal Sonar Systems, J.F. Fish, ed., Plenum Press, New York.

Rosenberg, A. E., 1971, "Effect of Glottal Pulse Shape on the Quality of Natural Vowels," J. Acoustic. Soc. Am., 49(2-2): 583.

Schenkkan, E.J., 1973, "On the comparative anatomy and function of the nasal tract in odontocetes (Mammalia, Cetacea)," Bijdragen tot de Dierkunde. 43(2): 127.

Schenkkan, E.J., 1977, "Notes on the nasal tract complex of the boutu, Inia geoffrensis (De Blainville, 1817)(Cetacea, Platanistidae)," Bijdragen tot de Dierkunde, 46(2): 275.

Varanasi, U. and D.C. Malins, 1971, "Unique lipids of the porpoise (Tursiops gilli): differences in triacylglycerols and wax esters of acoustic (mandibular and melon) and blubber tissues," Biochem. Biophys. Acta., 231: 415.

Varanasi. U. and D.C. Malins, 1972, "Triacylglycerols characteristic of porpoise acoustic tissues: molecular structure of diisovaleroylglycerides," Science, 176: 926.

Varanasi, U., D. Markey, and D. C. Malins, 1982, "Role of isovaleroyl lipids in channeling of sound in the porpoise melon," Chemistry and Physics of Lipids, 31: 237.

WHALE HEADS, MAGNETIC RESONANCE IMAGES, RAY DIAGRAMS AND TINY BUBBLES

R. Stuart Mackay

Boston University and San Francisco State University
San Francisco, California 94132

It has become common to make diagrams of the heads of toothed whales showing the paths of sound waves; this is more suitable in some cases than others. We will mention some limitations, give some observations made since the last of these meetings, and conclude with methods for further study.

Eyeglasses a millionth of a centimeter across would not be taken very seriously, nor would people discuss their own reflection in a mirror this small. (One could not even reliably count such objects under a light microscope, to say nothing of observing or denoting their shape, when they are less than a wavelength across.) But some investigators do a similar thing when they show ray paths of sounds undergoing specular reflections from air sacs in a dolphin head or focussing of sound by a lens for frequencies where a wavelength is comparable to or larger than the structure. In soft tissue, the wavelength of sound is 1/3 meter for a frequency of 4.5 thousand cycles per second, and it is 2 cm even at 75 kHz.

It is tempting to say that sound, from its source (perhaps the nasal plugs), travels upward where it is reflected forward from the underside of the vestibular sac in the head of a dolphin. At high frequency this could be true since the extent of the surface of the sac is then several wavelengths. But at low frequency the sound engulfs the sac which pulsates and acts as a second source radiating in all directions. Any small object different from its surroundings would reradiate similarly (Figure 1 gives well known examples following a Huygen's principle construction), but a pocket of gas does so especially strongly for easily visualized reasons. Air sacs reradiate more than a similar solid, as if they were larger than their real size without an increase in geometrical detail (Mackay and Rubissow, 1978). This figure also suggests why for objects very small relative to a wavelength, the shape or distribution of acoustic properties matters little, e.g., presence on one side of a small opacity such as the skull, in the dolphin case with very low frequencies.

The melon might be considered a lens to beam sound forward. But at low frequencies it could neither focus well because of diffraction effects nor could it gather much energy. For any sonar system the transverse or azimuthal resolving angle is determined by the size of the aperture through which the "world is viewed" as measured in wavelengths. This applies to transmission or reception or both, and for pulsed sounds such as are used by dolphins or for frequency swept sounds as used by bats. (The range resolu-

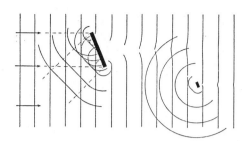

Fig. 1. The reflection and diffraction of waves
by small objects comparable in size to a
wavelength. The origin of the law of
reflection angles is seen, but note the
extension beyond the edges of the mirror
of some energy. (From Jenkins and
White, 1937)

tion depends instead on the speed of sound divided by the spread of frequencies employed or the bandwidth; this is speed times pulse duration for pulsed systems.) Humans best detect sounds above 25 kHz when coupled to their lower jaws, and it is presumed dolphins use their jaws for hearing; aperture limitations also apply to this case, even with large amplitude sounds and nonlinear effects.

There are three possible cases or regimes relating to sound production (and hearing or reception). The frequencies may be low, medium or high so that the wavelengths are large with respect to anatomical structures, comparable in size, and small in size with respect to the geometrical dimensions of the subject. (Also, conclusions are modified when sources or receivers are very close to the subject.) If the frequency is very low then there will be no directionality or even preferred directions of transmission (or reception), and one has a rather omnidirectional situation. Figure 1 is relevant (both to optics and acoustics), and one has diffraction and scattering dominant. An idea of lenses and mirrors of such size is misleading, and relevant concepts are like why the sky appears blue and the sunset orange.

For very high frequencies or very sharp clicks the ray diagrams may suffice. For the intermediate range of frequencies the situation is more complex, and wave equations must be combined to determine the response at any point in space. Thus, an initial structure vibrates and sends out a wave perhaps with differing strengths in different directions. If a small gas pocket (not dissolved gas) is struck it will pulsate in response and send out sound in all directions, with an initial phase that depends on its separation from the primary source. At each point in various directions the new wave combines with the original to give a calculable sum that depends on the distances from both sources, primary and secondary, i.e., the two waves combine differently at every point depending on the two phases and amplitudes. The wave radiated by the gas pocket will be in phase with the segment of pressure wave that strikes it except near the resonance frequency of the bubble, where it can appear to lead or lag. Near the bubble it does not appear as a point source. Such computations are common in radio antenna theory. Even when the wavelength approaches structure size and we expect a directed flow of sound, the beam will not be narrow. One can add the contributions at any point from several secondary sources combined with the primary source, and this result can be compared with experimentally determined distributions from animals to see if the assumptions of mechanism could be correct.

In Figure 2 is seen ultrasonic images of tiny bubbles involved with

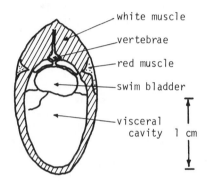

- white muscle
- vertebrae
- red muscle
- swim bladder
- visceral cavity

1 cm

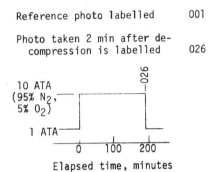

Reference photo labelled   001

Photo taken 2 min after decompression is labelled   026

10 ATA
(95% N_2,
5% O_2)

1 ATA

0   100   200

Elapsed time, minutes

Fig. 2. Bubbles of decompression sickness in a goldfish are seen in ultra-sonic images, with pressure cycle and anatomy indicated. Bubbles much less than a wavelength of sound exert a clear effect on sound waves.

decompression sickness in a goldfish. It is seen that the presence of little gas pockets has a not marginal effect, with each a significant source of reradiation though buried in tissue and much smaller than a wavelength; it is not a resonant effect so exact size is not critical. With a sound wavelength of 200 um one can localize and measure bubbles ranging in size down to 1 um. A real-time ultrasound motion picture of this will be shown. See Mackay 1984; Mackay and Rubissow, 1978.

This gives us confidence in the interpretation of the equations, and it also suggests a performance possibility for dolphins probably unavailable to humans. Bubbles forming in fat in the head might interfere with their hearing and warn of impending decompression sickness problems if they did not then swim with caution (Mackay, 1982). Furthermore, in any hypothesis about acoustic stunning of prey, an important aspect may be that small bubbles maintained near the ocean surface (e.g., Thorpe, 1982) bathe the lateral line of fishes, and circulate in their gills and in the syphons of squid; intense sounds could cause damage due to bubble pulsation (Mackay, 1982), and brief impulses of moderate intensity can produce effects (Miller and Williams, 1983). If bubbles do matter then laboratory experiments must take into account differences between ocean and fresh water; water squirted into a basin of the former produces noticeably more fine bubbles.

Similarly, if the frequency of sound of a small sperm whale is such that

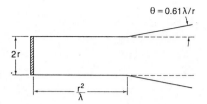

$$\theta = 0.61\lambda/r$$

Fig. 3. A vibrator radiates sound into roughly this
pattern. The near field is not very uniform.
This has application both to limits on ray
diagrams and to generating head-exploring beams.
(From Mackay, 1984)

the wavelength is roughly the length of the spermaceti organ, which is longer
than the "junk" is high, then the distribution of material across the layers
in the junk matters little, and though the junk probably can guide and couple
sound, it need not function as a row of lenses. If some sound travels in one
medium and some in an adjacent different medium, or there is variable
material thickness across a beam, then the velocity of the sound will be
intermediate.

There is another limitation. For an extended source or one with
several parts an approximation is a vibrating piston where the "far field"
starts at about a distance of source area/wavelength, or size squared
divided by wavelength (Figure 3), and inside that distance geometrical optics
considerations can yield misleading results. Some writings show the far
field starting at diameter-squared/wavelength rather than radius squared/
wavelength; depending on purpose, there is justification for the lesser
distance. In any case, the near field is criss-crossed with interference
patterns within an overall progression of sound, and ray diagrams do not
there adequately represent detail of interest. A small vibrator (dipole) has
a near field extending out about one wavelength (Mackay, 1984) within which
radial outward rays are overwhelmed by patterns attached to the source. A
separate point is that a radiator only a wavelength across tends to be rather
inefficient and is often used with a horn, or etc.

Sound can move in small spaces however. There can not be a propagated wave in a region less than half a wavelength across. Stranding of
whales in shallow water has been attributed to this confusing their sonar.
But vibrations obviously move through narrow tubes of stethoscopes, speaking
tubes, airplane movie sound systems, narrow organ pipes, string toy tele-
phones between tin cans, etc., even though the "perfect" reflections at the
walls of larger tubes is not involved.

Since the previous of these meetings at which new ultrasonic methods
of head exploration were described (Mackay, 1980), they have received
application (Mackay and Liaw, 1981). Ultrasonic beams (generated as in
Figure 3) were projected into the heads of dolphins and reflections from
moving surfaces returned with a shift in frequency (Figure 4) to indicate
what did and did not move during vocalization. An analysis of the method
has been given (Mackay, 1984). In some observations by this method on
Delphinus delphis and on Tursiops truncatus during phonation, it appeared the
larynx was not involved in sound production. The vestibular sac acts as a
reservoir to circulate air past the nasal plugs which vibrate. More details
are in Mackay and Liaw, 1981. For comparison with bats, Tursiops truncatus
produced a frequency sweep before any sound of specified constant frequency
(FM-CF), not for processing utility but probably simply to adjust the
frequency (Mackay, 1981) as a human singer might do.

Helium breathing and pressure changes are useful tools for studying

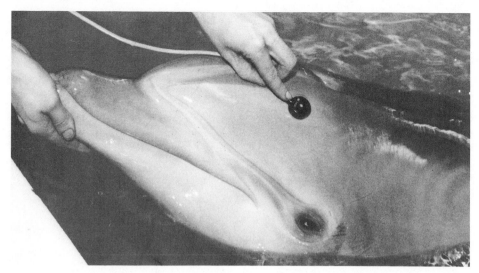

Fig. 4. A 2MHz motion-sensing Doppler ultrasound probe being applied to the
head of <u>Tursiops truncatus</u> to determine what moves during vocalization. Air sac vibrations can be observed directly, for example.

vocalization. Composition changes can distinguish a tissue from a gas
resonator. In humans, whistles are generated by a different mechanism from
other sounds, the former being quenched by an increase in ambient pressure,
and the same is probably true of dolphins (Mackay and Liaw, 1981). (It is
interesting that the deep-diving sperm whales have not been observed to
whistle.) The velocity of sound in an idealized gas is independent of
pressure (Mackay, 1981), though upward modifications of the human voice can
be noticed starting at 30m depth.

Some dolphins both whistle and buzz; human divers whistle with difficulty in air compressed to a depth corresponding to over about 20 m but can
easily speak. I had great difficulty whistling after a breath of sulfur
hexafluoride (with no impurities) which has a density at one atmosphere about
that of air compressed to a depth where a diver is unable to whistle, thereby
confirming the role of gas density under pressure change, relative to
adjacent structure stiffness, in the pressure quenching of a biological
whistle. A mixture of half air and half SF_6 did not prevent whistling, and
yet lowered the apparent frequency of the voice but not of the whistle;
helium also did not noticeably shift whistle frequency, all as expected
(Mackay and Liaw, 1981).

The data for making ray diagrams or other computations must come
from some form of anatomical measurements. The property being sensed
and mapped in any imaging process can be quantitated. There is some advantage to observing by acoustic properties when the result is to be used to
interpret acoustic phenomena. There may also be an advantage to incorporating a size determination by some other modality with which rays do not bend
due to refraction in going from one organ to the next. Viewing a subject
section from two perpendicular directions with a pulsed ultrasonic scan
should allow estimate of size and of sound velocity of different parts; this
is done in a "compound B scan." In a plane through a dolphin head one could
instead pass a family of either parallel or non parallel (e.g., diverging)
rays of sound in many angular orientations, and measure the passage time of
each; from this can be calculated the sound velocity at every point in the
plane (amounting to a form of ultrasonic computed tomography). An assumption
probably being made is that velocity is independent of direction, which is an

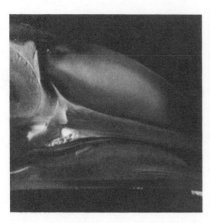
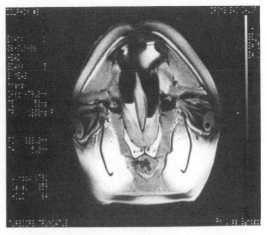

Fig. 5. Magnetic resonance images of intact head of <u>Tursiops</u> <u>truncatus</u>. No movement of subject or apparatus is needed during or between views. Note the brain, larynx and air sacs, and composition variation in melon, and in the cross section, the eyes and their muscles.

approximation since, e.g., both velocity and absorption of sound differ along and across a muscle bundle; absorption differs more than velocity. Such equipment does exist for transmission time-of-flight measurements using either a swept frequency or pulses to determine the fastest path signal in a situation where there can be reverberation, reflections and refraction.

Refraction and reflection of sound are determined by acoustic velocity and impedance, which depend on overall material density. Since x-rays strongly interact with electrons, x-ray images are maps of electron density, and in neutral molecules the number of electrons equals the number of heavier mass-contributing protons (though there are differences in the number of neutrons per proton in the atoms present) making such images dependent on density. But, acoustic parameters also depend on elasticity (e.g. consider velocity in air vs. water), and so x-ray pictures are somewhat better for measuring geometry than for quantitating acoustic parameters.

The layout of organs can be seen also in images invoking display of small chemical differences. In Fig. 5 is seen a sagital and a transverse section image of an intact dolphin head, made by nuclear magnetic reso-

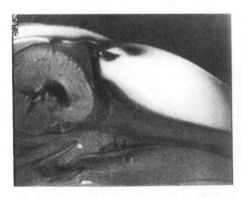
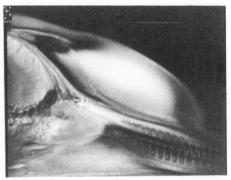

Fig. 6. Midline section of <u>Tursiops</u> <u>truncatus</u> formed in one minute (on left). The right picture is one of a series of 10 mm thick section images collected all at once, and being to the left of the midline, the teeth are seen.

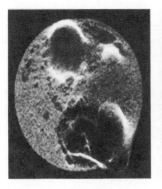

Fig. 7. Proton magnetic resonance images. Left: two chitons in which magnetite in the radula disrupts the image by forming a dark and light pattern and can even distort the pattern so the contents appear outside the container. Right: section of intact brain of Tursiops truncatus.

nance imaging, in which is depicted the distribution of mobile protons, the image also being weighted by the proton relaxation time to indicate properties of the vicinity of the protons. No mechanical motion of subject or apparatus was needed during or between exposures.

Though projections like ordinary x-ray images can be made by nuclear magnetic resonance, the present computed tomographic images are truly as if a succession of slices had been made and photographed (e.g., Mackay, 1984). In the ones shown, structure is averaged through a "slice thickness" of 5.0 mm for the cross section image and 10.0 mm for the longitudinal sections. Since these are not a projection of a 3-dimensional object down onto a 2-dimensional picture, the shape of the outer surface of an organ does not influence shading. Thus, for example, the pattern seen in the melon represents a partially stepless gradation in composition.

In evaluating acoustic structures, caution must be used with dead specimens (Mackay, 1966; 1984). This dolphin was not alive, but the structures seen here seem mostly unaltered by post mortem changes; however, the eye lenses appear slumped in position suggesting mechanical and acoustic changes in some tissues. Magnetic resonance seems without hazard (the author's auditory nerves will be shown, well resolved).

The brain, larynx, air passages and sacs, and melon are seen. The pictures are the responses of mobile hydrogen nuclei, the images being weighted to indicate relaxation times or variations in the environments of the protons. (Technically, displaying the amplitudes corresponding to different anatomical locations during either the first or second spin echo shows the distribution of protons as weighted by a combination of spin-lattice and spin-spin relaxation times; a short time to spin echo minimizes T_2 dependence and a long sequence repetition interval minimizes T_1 differences, both being roughly a second: See Mackay, 1984). Shading is seen in the melon in both T_1 and T_2 images. The transverse image, being near the blow hole, also shows the eyes and their nerves and muscles. A 5 per second display of a series of simultaneously collected sections can be inspected back and forth "through the subject" to get a feeling for 3 dimensional anatomy, or the subject made to appear to rotate about any axis under operator control in real time. The two pictures of Fig. 5 are different displays of overlapping data, all collected at once. The left image of Fig. 6 was produced in one minute of data collection. The white spots are disruptions of the image, possibly due to magnetic material such as magnetite. These images are some that were produced in a collaborative effort with Ted Cranford and Michael Reid. It should be noted that the views of Fig. 6 could not be directly produced by a CAT scanner (x-ray computed tomography), though these might be synthesized from a large number of thin cross sections.

An interesting and potentially confusing aspect of such images is the speckling at the top of the brain in Fig. 7. It is probably due to a distribution of small quantities of magnetite since it resembles what is sometimes seen after bone grinding or in mascara artifacts. To the left for comparison is seen the magnetic resonance image of two chitons (Mopalia muscosa) in a cup of water; some chitons have magnetic properties (Tomlinson, 1959). The radula of each of these two contains large amounts of magnetite which is ferromagnetic.

References

Jenkins, F.A. and White, H.E., 1937, Fundamentals of Physical Optics. McGraw-Hill, New York.

Mackay, R.S., 1966, Studying dolphins with ultrasound and telemetry. in: Whales, Dolphins and Porpoises, K. Norris, ed., U.C. Press, Berkeley, pp 378, 450.

Mackay, R.S., 1980, Dolphin air sac motion measurements during vocalization by two noninvasive ultrasonic methods. in: Animal Sonar Systems pp 933-940, R.G. Busnel and J. Fish, eds., Plenum Press, New York.

Mackay, R.S., 1981, Dolphin interaction with acoustically controlled systems: aspects of frequency control , learning and non-food rewards. Cetology 41:1.

Mackay, R.S., 1982, Dolphins and the bends. Science 216: 650.

Mackay, R.S., 1984, Medical Images and Displays. Wiley and Sons, New York.

Mackay, R.S. and Liaw, H.M., 1981, Dolphin vocalization mechanisms. Science 212: 676. See also cover caption pg. 593.

Mackay, R.S. and Rubissow, G., 1978, Decompression studies using ultrasonic imaging of bubbles. IEEE Trans, on Biomedical Engineering 25: 537.

Miller, D. and Williams, A., 1983, Further investigations of ATP release from human erythorocytes exposed to ultrasonically activated gas-filled pores. Ultrasound in Medicine and Biology 9:297.

Thorpe, S., 1982, On the clouds of bubbles formed by breaking wind-waves in deep water, and their role in air-sea gas transfer. Phil. Trans. R. Soc. Lond. A 304:155.

Tomlinson, J., 1959, Magnetic propeties of chiton radulae. The Veliger 2(2):36.

INDIVIDUAL VARIATION IN VOCAL TRACT RESONANCE MAY ASSIST OILBIRDS IN RECOGNIZING ECHOES OF THEIR OWN SONAR CLICKS

Roderick A. Suthers and Dwight H. Hector

School of Medicine and Department of Biology
Indiana University
Bloomington, IN 47405

INTRODUCTION

Oilbirds, Steatornis caripensis, live in colonies in caves, within which they navigate by echolocation (Griffin 1953, Snow 1961). The sonar "clicks" of these nocturnal birds typically last about 40 to 80 ms and are emitted at repetition rates up to about 12/s. Most of the acoustic energy lies beween 1 and 15 kHz. Sound during some clicks is continuous but other clicks, here referred to as double clicks, are divided by a silent period lasting about 20-30 ms into two brief bursts of sound.

The sonar clicks are produced by the oilbird's bronchial syrinx which is composed of two semi-syrinxes located, one in each primary bronchus, a short distance from the caudal end of the trachea (Suthers and Hector, 1985). A puzzling feature of this syrinx is the fact that the length of the primary bronchus between the right semi-syrinx and the trachea is shorter than that between the left semi-syrinx and the trachea, conferring a noticeable bilateral asymmetry on the oilbird's vocal tract (Fig. 1). Examination of a number of birds further reveals that each individual differs in the absolute and relative length of the cranial portion of the two primary bronchi.

In this paper we summarize evidence that this random variation in vocal tract dimensions between individuals alters their vocal tract resonances, and gives the clicks of each oilbird different formant frequencies which may assist it in identifying echoes of its own sonar clicks from those of other individuals flying nearby.

METHODS

Oilbirds were collected from colonies in Trinidad, West Indies and taken to the Field Research Station of the Asa Wright Nature Center for study. Sonar clicks of normal intact birds were recorded while the bird was held in front of a calibrated 1/2 inch (12.7 mm) microphone (Brüel and Kjaer model 4133), the amplified output of which was delivered to an instrumentation tape recorder (Racal model Store 4DS). The frequency response of this recording system was flat between 100 Hz and 40 kHz.

The length of the cranial portion of each bronchus, from the carina to the point where tips of the two cartilagenous semi-rings bordering the anterior and posterior margins of the external tympaniform membranes meet, was measured with vernier calipers in live or fresh specimens. The mean of several measurements on each bronchus was used to calculate its quarter wavelength fundamental resonant frequency (F_1) using the formula for a stopped tube:

$$F_1 = \frac{1}{4L} \, 20.1 \sqrt{T_A}$$

where L = the length of the bronchus in meters and T_A is the absolute temperature of the air in the primary bronchus, which we assume to be $310^{\circ}K$ ($37^{\circ}C$) (Roederer, 1974). No attempt was made to apply an end correction since the acoustic coupling between the bronchus and trachea is not known.

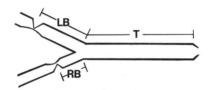

FIG. 1.   Schematic diagram of oilbird's vocal tract showing asymmetry of bronchial syrinx. Location of each semi-syrinx is indicated by V-shaped external tympaniform membrane opposite curved internal tympaniform membrane. T, trachea; LB, cranial portion of left primary bronchus; RB, cranial portion of right primary bronchus.

RESULTS

The oilbird's vocal tract can be considered to consist of three rigid tubes, each having a fixed length, i.e., the trachea and two smaller diameter primary bronchi. The sonar clicks of oilbirds often contain three principle formants (Fig. 2). The lowest of these, F_1 agrees well with the predicted one quarter wave fundamental resonance of the trachea acting as a tube open at one end. Tracheal length is about 8 or 9 cm in adult birds and the frequency of F_1 is usually about 1 kHz.

The frequency of F_2 and F_3 vary between individual birds in a way that is predicted by the quarter wavelength fundamental resonance of the cranial portion of each primary bronchus considered as a tube open at its junction with the trachea, where an abrupt increase in the cross-sectional area produces a mismatch in acoustic impedance, but closed at the syrinx. These two formants are thus determined by the distance of each semi-syrinx from the trachea. Since, in the birds we have examined, this distance is always greatest for the left semi-syrinx, the data suggest that F_2 arises from the cranial portion of the left primary bronchus and F_3 arises from the cranial portion of the right primary bronchus.

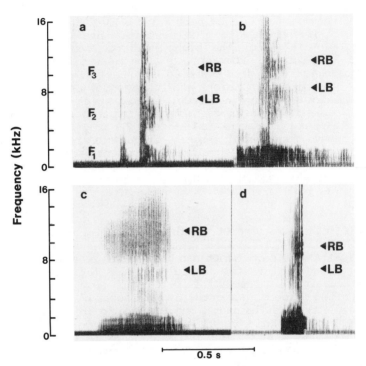

FIG. 2.   Sonar clicks of four different oilbirds showing formants F_1, F_2 and F_3. Predicted fundamental resonances and observed formant frequencies, respectively, in kHz are: Bird a, LB 7.3 and 6.0, RB 10.6 and 10.2; bird b, LB 8.5 and 7.8, RB 11.7 and 10.9; bird c, LB 7.0 and 6.2, RB 11.1 and 10.1; bird d, LB 7.1 and 6.3, RB 9.9 and 8.8. Arrows indicate predicted quarter-wave resonant frequency of cranial portion of left (LB) and right (RB) primary bronchi. Sonagrams were made on a Sonagraph model 6061B (Kay Electric) with an effective filter width of 1200 Hz.

This conclusion is supported by the effect on these respective formants of plugging one primary bronchus at its tracheal end so that all sound is generated by the contralateral semi-syrinx (Suthers, submitted). In sonar clicks, as in protest squawks, F_3 disappears when the right bronchus is plugged and F_2 disappears when the left bronchus is plugged. These two formants thus originate from the bilateral asymmetry of the oilbird's syrinx. Variation from bird to bird in the degree of this asymmetry gives each individual its own unique formant frequencies.

Formant frequencies characteristic of sonar clicks from four intact oilbirds with differing vocal tract dimensions are shown in Fig. 2, together with the predicted frequency of the fundamental quarter-wave resonance of the left and right cranial primary bronchi. The second member of the double click of bird a and b (Fig. 2) appear to ring the vocal tract giving rise to prominent formants visible as damped oscillations following the high intensity second member of the broadband double click. Formants are also present in the long duration continuous click from bird c and in the short single click of bird d (Fig. 2). Bronchial dimensions and fundamental resonant frequencies have been calculated in nine birds, each of which has a unique set of resonant frequencies. In these birds the length of the cranial portion of the left primary bronchus ranged from 8.9 to 14.6 mm, corresponding to quarter-wave fundamental resonant frequencies between about 10 kHz and 6 kHz, respectively. Similar measurements on the right primary bronchus varied from 6.8 to 9.9 mm, yielding quarter-wave fundamental resonances between 12.6 kHz and 8.7 kHz, respectively.

DISCUSSION

The formant structure of the oilbird's social vocalizations, such as agonistic squawks, arises from the resonant properties of its asymmetrical vocal tract, which can be modelled as a series of 3 tubes— i.e., the trachea and the two cranial portions of the primary bronchi (Suthers, submitted). The data reported here show that these formants are also often evident in short wide band sonar clicks. The fact that many clicks lack clear formants, suggests that oilbirds have some degree of control over their vocal tract resonance. In many social vocalizations the first formant is about twice the frequency recorded in these clicks suggesting the trachea may alternatively resonate as an unstopped tube.

The observed bronchial formant frequencies are consistently below those predicted. This difference may be due to the fact that we have arbitrarily picked an inappropriate length measurement on the primary bronchus, or to our ignorance of acoustic coupling between tubes, or to the effect of small diameter tubes on the velocity of sound. Whatever the reason(s), it is the consistency of the relationsip between predicted and observed formant frequencies, not their absolute values, that is important to the hypothesis we advance here.

The variable asymmetry of the oilbird's bronchial syrinx may have evolved to facilitate individual recognition based on the differing vocal tract filter functions of each bird. Our data suggest that variation in the length of the cranial portion of the primary bronchus can impose individually unique sets of formants on sonar clicks which could aid a bird in distinguishing the echoes of its own clicks from those of other birds. Further experiments are needed to determine if these potential acoustic cues are actually used by the birds.

This research was supported by NSF grants BNS 82-10256 and BNS 85-17099 to R.A.S.

REFERENCES

Griffin, D. R.   1953.   Acoustic orientation in the oilbird, Steatornis.
    Proc. Natl. Acad. Sci. U.S.A.   39: 884.

Roederer, J. G. 1974.   "Introduction to the Physics and Psychophysics of
    Music." Springer Verlag, New York.

Snow, D. W.   1961.   The natural history of the oilbird, Steatornis
    caripensis, in Trinidad W. I.   I. General behavior and breeding
    habits. Zoologica 46: 27.

Suthers, R. A.   1986.   Avian vocal tract resonance: Structural variation
    produces individually unique vocalizations in oilbirds. Submitted
    for publication.

Suthers, R. A. and Hector, D. H.   1985.   The physiology of vocalization
    by the echolocating oilbird, Steatornis caripensis.   J. Comp.
    Physiol. A 156: 243.

MIDBRAIN AREAS AS CANDIDATES FOR AUDIO-VOCAL INTERFACE IN ECHOLOCATING BATS

Gerd Schuller and Susanne Radtke-Schuller

Zoologisches Institut der Universität München
Luisenstr. 14, D-8000 München 2, F.R.G.

INTRODUCTION

In bats the auditory system and the vocalization system have a close functional relation to each other during acoustic communication as well as during echolocation. In many species the spectral parameters of the emitted echolocation sounds are adapted to the special echolocation tasks during a behavioral sequence. In order to control the parameters of the echolocation sounds appropriately the relevant information has to be transmitted from the auditory to the motor system.

The Doppler shift compensation behavior of CF-FM bats (Rhinolophus and Pteronotus) is a good model for the investigation of audio-vocal interaction. During this behavior, the bats vary only a single parameter of the emitted echolocation sounds, i.e. the emitted frequency, in order to cancel the Doppler-induced frequency shifts in the returning echoes.

Much data is available for the processing of the relevant frequencies (frequencies at and above the resting frequency of the bat) at different levels of the auditory system. The information on the efferent neuronal pathways controlling vocalization and on the neuronal connections mediating auditory information to the vocalization system are, in contrast, very scarce.

The efferent vocalization system has been studied recently at the level of the motor nucleus of the larynx, the Ncl. ambiguus, by HRP injections into physiologically defined areas of this nucleus and the adjacent formatio reticularis (1). Following this study, many brain areas at different brain levels showed retrograde labeling, but their specific involvement in the control of vocalization remained unclear. In order to investigate their participation in the control of the echolocation calls, electrical microstimulation was used and tracer injections (HRP, WGA) yielded their connectional pattern within the efferent motor system. In addition, possible sources of auditory input into the motor system could be revealed.

MATERIAL AND METHODS

Eighteen rufous bats (<u>Rhinolophus rouxi</u>) from Sri Lanka were used in this study. The electrical stimulation experiments were conducted in a stereotaxic device allowing a reconstruction of stimulation sites with a precision of 100-200 μm in all three dimensions (2). The electrical stimuli consisted of 15 msec long trains of 15 pulses of negative polarity with a duration of 0.1 msec and were applied through insulated tungsten electrodes with tip diameters between 2 and 20 μm.

The elicited vocalizations were picked up with a condenser microphone, analyzed in frequency and intensity and stored on tape together with the electrical stimuli and the signal monitoring the respiration cycle. In parallel the animals were observed with a TV-camera and ear and facial movements were recorded on video tape.

Midbrain structures were systematically scanned between the level of the rostral superior colliculus and the caudal half of the inferior colliculus in a dense grid (100-200 μm) of stimulation coordinates. Only the very dorsolateral parts of the midbrain and the areas adjacent to the midsagittal plane have not been probed with the same density. Locations were termed "specific" for eliciting echolocation sounds under the conditions, that
1) the threshold for triggering vocalizations was smaller than 20 μA, lying typically even below 10 μA;
2) the elicited vocalization corresponded to the natural echolocation sounds in respect to frequency, intensity and duration;
3) no other movements were elicited besides the vocalizations except ear or noseleaf movements which belong intimately to the pattern of sound emission;
4) the latency of vocalizations relative to the electrical stimulus was stable within 10 to 20 msec and smaller than 100 msec and
5) the vocalizations did not occur as a secondary reaction to stimulus induced general arousal.

Horseradish peroxidase (HRP) or wheat germ agglutinin (WGA) (3-4 μAmin and 4-5 1μAmin, respectively) was injected iontophoretically into the so defined "vocalization-specific" foci. The brains were histologically processed following the Mesulam TMB-protocol or a modified DAB-protocol (Adams) and every second section was counterstained with neutral red or the cytochrome oxidase reaction (Radtke-Schuller, in prep.). Data from stimulation experiments and anatomical information were reconstructed on a common data base.

RESULTS

<u>Type of elicited vocalizations.</u> When vocalizations could be elicited after the criteria (1,3,4,5) from above, in most cases the spectral pattern and duration were identical to those of natural echolocation sounds as uttered by the bat in the resting position. Only in pontine areas and the overlying fibers could vocalizations be elicited with extremely short duration and frequencies of the constant frequency portion lower than the resting frequency of the bat.
The latencies between electrical stimulation and onset of the vocalization ranged typically between 20 msec and 6o msec but in some locations it was consistently around 80 msec. With increasing stimulation intensity above threshold the latency

stabilized (variations below 10 to 20 msec) as long as there was no strong discrepancy between respiratory cycle and stimulation rate. The stimulation rate was found to be an important factor for consistently eliciting vocalization and most often was optimum at 7 Hz, which is about twice the spontaneous respiration rate.

Besides the amplitude of the emitted sound and in some locations the number of emitted vocalizations, no other parameter of the vocalization could be systematically influenced by changing the parameters of electrical stimulation. The resting frequency of the echolocation sounds could not be manipulated in the midbrain structures we have probed with electrical stimulation.

Correlation with ear and noseleaf movements. Stimulus induced ear movements could be elicited at many more brain locations than vocalizations. Movements of either one ear or coordinated movements of both ears occurred. At stimulation sites specific for triggering of vocalizations, the ear movements and vocalizations had very similar thresholds and were in strict temporal coordination to the emitted echolocation sounds. Nose leaf movements could be evoked either unilaterally or bilaterally and most often accompanied the vocalizations in close synchrony.

Correlation with respiration. The respiratory cycle was synchronized at many stimulation sites by stimulation currents smaller than the threshold currents needed to evoke vocalization. On the other hand, synchronized respiration could also occur as a secondary effect to the eliciting of vocalization as part of the entire pattern of sound emission.

Brain sites of specific triggering of vocalizations.
Figure 1 indicates the areas of lowest thresholds for evoking species-specific vocalizations. The crosses indicate stimulation sites with thresholds below 10 µA and the circles mark places where the thresholds lay between 10 and 20 µA. In the shaded ranges the vocalizations were optimally elicited at lower or the same thresholds as the other constituents of the pattern of sound emission (respiration, ears, noseleaf). These loci specific for eliciting echolocation sounds are:
1) the deep and intermediate layers of the superior colliculus (SC) adjacent to the griseum centrale (CG) and dorsal to the Ncl. cuneiformis (CUN),
2) the Ncl. mesencephalicus profundus (NMP) in the dorsolateral part of the reticular formation and
3) the Ncl. tegmentalis pedunculopontinus (NTPP), a cell aggregation medial and anteromedial to the rostral part of the Ncl. lemniscus lateralis dorsalis.

Injections of HRP and WGA. Figure 2 summarizes the interconnections of the superior colliculus (SC), Ncl. mesencephalicus profundus (NMP) and Ncl. tegmentalis pedunculopontinus (NTPP) within the efferent vocalization system and their inputs from auditory nuclei. None of these three areas showed a direct anatomical projection to the motor nucleus of the laryngeal nerves, the Ncl. ambiguus. The intermediate and deep layers of the SC have clear reciprocal connections with the NMP and with the CUN, where no vocalizations could be electrically elicited. The superior colliculus also receives inputs from the Ncl. tegmentalis pedunculopontinus (NTPP).

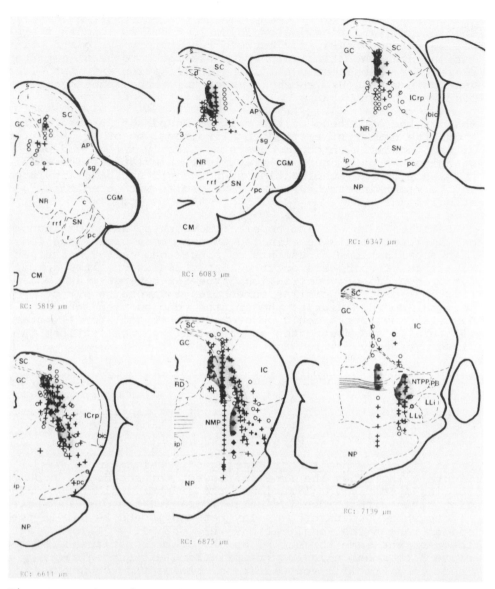

Fig. 1. Brain stimulation sites for species-specific vocalization in the bat, <u>Rhinolophus</u> <u>rouxi</u>. +:thresholds 10 µA,
o: thresholds 10-20 µA, shaded: vocalization optimally evoked.
GC: griseum centrale, SC: superior colliculus, AP: pretectal
area, NR: Ncl. ruber, SN: subst. nigra, pc: cerebral peduncle,
rrf: retrorubral field, IC: inferior colliculus, NMP: Ncl.
mesencephalicus profundus, LL: lemniscus lateralis, PB: Ncl.
parabigeminalis, bic: brachium of IC, CGM: corpus geniculatum
mediale, NTPP: Ncl.tegmentalis pedunculopontinus (NTPP).

WGA-injections in the NMP provided evidence for an anterograde
connection to the NTPP and for a reciprocal connection with
the Ncl. cuneiformis, as well as of a projection to the
Ncl. facialis. The link to the laryngeal motor nucleus was

demonstrated by tracer injections into the Ncl. cuneiformis which displayed strong anterograde transport to the Ncl. ambiguus, thus confirming the retrograde results of Rübsamen and Schweizer (1). Besides the connections of SC, NMP, NTPP with the Ncl. ambiguus via the Ncl. cuneiformis there are at least 3 more indirect connections mediated by a) the reticular formation dorsolateral to the Ncl. facialis, b) the Ncl. retrofacialis immediately rostral to the Ncl. ambiguus and c) the dorsal portions of the Ncl. parabrachialis.
Direct input from auditory brain areas (inferior colliculus, auditory cortex) reach the SC and the NTPP. In addition, indirect pathways to these nuclei and to the CUN may transmit auditory information to the efferent vocalization system via pontine and cerebellar nuclei.

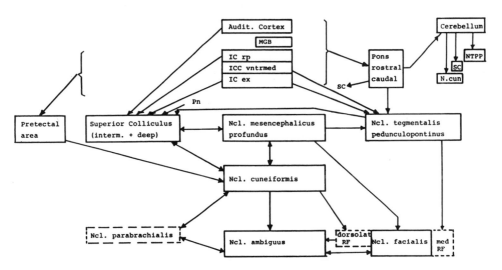

Fig. 2. Block diagram summarizing the projections of the "vocalization-specific" nuclei as revealed by HRP- and WGA-injections and connections to nuclei within the auditory system. Abbrev.: see Fig. 1 and rp: rostral pole, ex: external, RF: formatio reticularis, MGB: medial geniculate body.

CONCLUSIONS

Three foci for specific triggering of vocalization could be identified in the midbrain of the bat Rhinolophus rouxi. These areas are only indirectly coupled to the laryngeal motor nucleus. They are embedded in a complicated functional net coordinating the behavioral pattern of echolocation sound emission comprising the simultaneous control of respiration, the larynx and the facial movements. There also exists a number of direct and indirect connections of these premotor midbrain areas with auditory brain centers, which may convey auditory modulation of motor activity. Thus, the audio-vocal interface appears as a distributed, multipath connection between the hearing and the vocalization system, so that no unidirectional or hierarchical control of the vocal efferences can be expected.

Acknowledgement

Supported by Deutsche Forschungsgemeinschaft SFB 204/TP 10

References

(1) Rübsamen, R.,Schweizer, H.; J.Comp.Physiol.159 (1986, in press); (2) Schuller, G.,Radtke-Schuller, S., Betz, M.; J. Neurosc. Meth. (1986, in press))

THE SOUNDS OF SPERM WHALE CALVES

William A. Watkins and Karen E. Moore
Woods Hole Oceanographic Institution
Woods Hole, MA 02543

Christopher W. Clark
Rockefeller University, Tyrrel Rd.
Millbrook, NY 12545

Marilyn E. Dahlheim
National Marine Mammal Laboratory
7600 Sand Point Way, NE
Seattle, WA 98115

ABSTRACT

The development of the click sounds of sperm whales (Physeter catodon) has been investigated through comparisons of these vocalizations from calves of different sizes. The observations include sounds from four small stranded calves held for short periods in aquaria at Miami, Florida, and Seattle, Washington, and in a bay on Long Island, New York. These vocalizations were compared with those of larger calves encountered at sea. All the calves produced typical sperm whale sounds -- clicks with broadband spectra, often produced in short series. Vocalizations from the smaller calves sometimes included slightly noisy, tonal components -- similar noisy click sounds were also occasionally heard in the presence of calves at sea, and they were interpreted as the result of improperly formed clicks. The smallest calves produced few temporally repetitive click patterns, but the larger (older) ones produced sequences with stereotyped "coda"-like temporal patterns. The use of such patterned click sequences increased with the apparent age of the calves. In the larger calves, the sounds appeared more organized in sequences similar to the communicative signals of adults. None of the calves appeared to use their sounds in ways that were related to echolocation.

INTRODUCTION

The vocal behavior of sperm whales (Physeter catodon) has been studied in considerable detail, beginning with the earliest identifications of their underwater vocalizations (Worthington and Schevill, 1957; Schevill and Watkins, 1962; Backus and Schevill, 1966; Watkins and Wartzok, 1985). Sperm whale vocalizations have been shown to be short pulsed sounds (clicks) produced singly and at a variety of repetition rates. When recorded at sea from nearby animals, these underwater sounds have consistently been broad-spectrum pulses with variable energy at frequencies from about 50 Hz to 20 kHz. At a distance, underwater sounds are often modified by multipath sound transmission, reverberation, and other propagation char- acteristics of the water, so that the sperm whale sounds generally have ap- peared to have different bandwidths and emphases at particular portions of their spectra. However, because the temporal patterns of their vocaliza- tions are not as affected, these have been consistently distinguishable, and they would appear to be the important communicative acoustic feature of the sounds to these whales (Watkins, 1980).

Typical sperm whale vocalizations are broadband pulses with variable spectral emphases. They include relatively isolated clicks, long series of clicks at regular rates of 1 to 4 per sec produced when whales are spread out underwater (Watkins and Schevill, 1976, 1977a), short sequences of clicks produced in temporally repetitive patterns termed "codas" (Watkins and Schevill, 1977b), and rapid series of clicks lasting for 0.5 to 5 sec or more, usually produced at constant rates (up to 83 per sec have been recorded). With rare exceptions, we have heard only such pulsed sequences (Watkins and Wartzok, 1985) in more than 4000 hours of listening to sperm whales at sea.

In this paper, we have attempted to assess the development of the click sounds of sperm whales through comparisons of the vocalizations from calves of different sizes. We have encountered sperm whale calves on numerous occasions at sea, both with small groups of one to three adults and with large groups that included several calves and many more adults. Because of this close association with adults, the sounds of small calves at sea usually have been difficult to distinguish, but larger, more independent calves have been well recorded. In addition to such observations of calves at sea, the sounds of four stranded sperm whale calves, held in captivity for short periods, have been recorded and studied. The vocalizations from calves of different sizes (ages) are compared with the usual adult repertoire to assess the development of vocalization in sperm whales.

METHODS

The open-sea observations and recordings of vocalizations from sperm whale calves that were considered during this study were made during numer- ous cruises (Woods Hole Oceanographic Institution) in the western North Atlantic off the continental shelf as far as Bermuda, and in the southeast Caribbean. The sounds were recorded with single hydrophones as well as multiple-hydrophone arrays for acoustic tracking (cf. Watkins and Schevill, 1976). The smaller calves generally remained close to adults so that the calf sounds were not easy to distinguish, but larger (7 to 8 m) calves often appeared to move about independently. On occasion, larger calves closely approached our hydrophone cables and ships, facilitating good re- cordings and observations.

Comparisons were made between these sounds and vocalizations from four captive sperm whale calves that had stranded. These calves were held for short periods in aquaria and a bay while efforts were made to try to re- store them to health.

A male sperm whale calf, of 4 m, was found near Fort Lauderdale, Florida in July 1964, and it was kept alive for a few days at the Miami Seaquarium. Underwater listening (Watkins) on 1 and 2 August 1964 revealed only isolated clicks from this calf.

A female sperm whale calf of approximately 4 m, stranded 2 km north of Sands Key, Florida, on 26 August 1974. This calf was also taken to the Miami Seaquarium, where it survived 5 days. A four-min recording of the calf's underwater sounds was made by Warren Zeiller on 30 August, during attempts to feed the whale by staff veterinarian, Jesse White.

A female sperm whale of 4.5 m, with part of its umbilical cord still attached, stranded near Florence, Oregon, on 18 September 1979. It was brought to the Seattle (Washington) Aquarium, where it survived 4 days. Underwater recordings (Dahlheim) were made of sounds from this calf on 19, 20, and, 21 September. Mark Anderson (Friday Harbor Museum) also made recordings which were sent to us for comparison.

A male sperm whale calf of 7 m, stranded on the south shore of Long Island, New York, on 16 April 1981. It was taken to a boat basin on Fire Island, where it remained for 10 days while being treated, and then it was released. Although this calf was larger, it could still have been nursing (Best, Canham and MacLeod, 1984). Underwater vocalizations were recorded from this whale (Clark) on 17 April.

The acoustic equipment used for the open ocean and captive observations varied, but in each case the equipment response allowed recording of the general spectra of the sperm whale sounds. Acoustic analyses were made with a sound spectrograph (Kay Elemetrics 7029A), and a variety of oscillographs, and computer systems for assessments of frequency as well as amplitude and time relationships.

RESULTS

The sounds from all of the sperm whale calves recorded at sea and in captivity were generally alike in overall spectra, and they often had variable emphases from one click to the next. These vocalizations were consistently similar to those of adults in spectra and duration of clicks, and when produced in series, they had the same ranges of variation as adults in temporal repetition and numbers of clicks. In both calves and adult sperm whales, clicks even within series generally varied with no obvious relationship to size (or age), except perhaps for relative maximum intensities.

The young captive calves produced relatively low level clicks, while the clicks from older calves at sea were often higher level by 6 to 10 dB. Although the highest click levels we have recorded have been from adult sperm whales (approximately 185 dB broadband source level, re. 1 μPa at 1 m, calculated from many different array recordings), the larger calves produced clicks that were sometimes only 4 to 6 dB less intense.

In our sample of calf sounds, the only significant vocal feature that changed with calf size was the apparent development and stability of patterned sequences of clicks. The larger (older) calves exhibited the ability to produce apparently controlled, repetitive click sequences, while in the smaller calves the sounds were variable and apparently less structured.

The 1964 Miami calf produced only a few widely spaced clicks. These were short, low level, broadband, impulsive sounds that were characteristic of sperm whales. There were no click-series or repeated click sequences.

The 1974 Miami calf produced similar broadband clicks, but in short sequences of 4 to 30 clicks, rather than as isolated, single clicks. Click repetition rates were variable to 83 click per sec. Many of these discrete sequences had 4 to 10 clicks (Fig. 1), produced at repetition rates which varied within and between series. Some click sequences consistently had a longer interval before the final click, or other temporal click repetition patterns, much like the adult coda patterns. The sounds were relatively low level, although they usually could be heard well above tank noise.

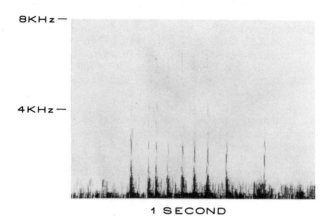

8KHz —

4KHz —

1 SECOND

Fig. 1. Spectrogram of a click sequence showing temporal patterning of click repetition by the 1974 Miami sperm whale calf. The filter bandwidth (resolution) of the analysis was 300 Hz.

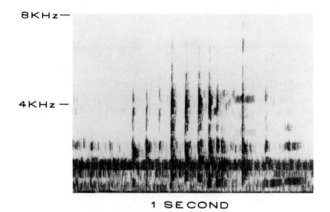

8KHz —

4KHz —

1 SECOND

Fig. 2. Spectrogram of a click sequence by the 1979 Seattle calf. The analysis filter bandwidth was 300 Hz.

The 1979 Seattle calf produced broadband clicks, singly as well as in sequences with varying numbers of clicks and a variety of repetition rates. Both wide bandwidth clicks and clicks with emphases at the lower frequencies were recorded, and some clicks were preceded or accompanied by low-level, tonal, noise components. Observers sometimes noted air escaping from the blowhole during sound production. The escape of air and added noise components to the clicks (the background of Fig. 2) may have been due to incompletely formed vocalizations (also occasionally heard in the presence of calves at sea). Discrete sequences from this calf had variable rates of click repetition, and they usually lasted between 0.5 and 1 sec, with many sequences having longer intervals before the final one or two clicks. The patterns of click repetition were similar to the types of coda patterns heard from adults, but there was little stereotypy in the click sequences. Clicks from the Seattle calf were generally low level, but distinguishable above tank noise when the calf was within 3 or 4 m of the hydrophone.

The 1981 Long Island calf produced clicks in patterns that were quite similar to those of adult whales, single clicks, slow series with 1- to 4-sec intervals between clicks, and coda-like sequences within about 1 sec consisting of 3 to 23 clicks (two sets had 2 clicks). Ten of 103 of these click sequences also contained lower-frequency, somewhat tonal, noise components with the clicks, somewhat like those of the Seattle calf. Discrete sequences of clicks usually had durations of less than one second. Although many of these click series often appeared coda-like, the temporal click patterns and number of clicks were variable, 24% had 4 clicks, and 9.7% had 10 clicks. There were 36 sequences (35%) with 9, 10, 11, or 12 clicks and very similar temporal patterns, even though the number of clicks varied, i.e. the first 3 and the last intervals were longer than intervening ones (Fig. 3). These patterns were sometimes repeated several times, indicating that this calf apparently had the ability to produce repetitive temporally patterned vocalizations. Click levels within sequences varied by about 12 dB, but overall the click levels of the Long Island calf were the highest of the captive calves.

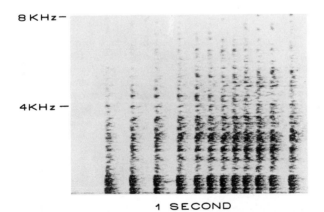

8 KHz—

4 KHz —

1 SECOND

Fig. 3. Spectrogram of one of several underwater click sequences with similar temporal patterning by the 1981 Long Island calf. The analyzing filter was 45 Hz to show the relative click spectra.

The temporal click sequences of Figs. 1, 2, and 3 show somewhat similar coda-like click organization. The 1974 Miami calf and the 1979 Seattle calf produced only a few such patterned sequences, but the 1981 Long Island calf had a variety of click patterns with some sequences repeated several times. Although none of the small calves appeared to have the stereotyped click sequences that have been typical of the identity codas used by adult sperm whales, as in Fig. 4 and Fig. 5, there appeared to be a progression in the ability to produce such sequences with the size of the calves.

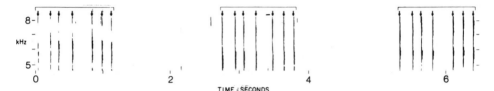

Fig. 4. Spectrogram of a typical sequence of codas from an adult sperm whale (from Fig. 3 of Watkins and Schevill, 1977b). Such temporally patterned click sequences are individually distinctive and appear to be used for identification during particular behaviors. Analyzing filter was 150 Hz.

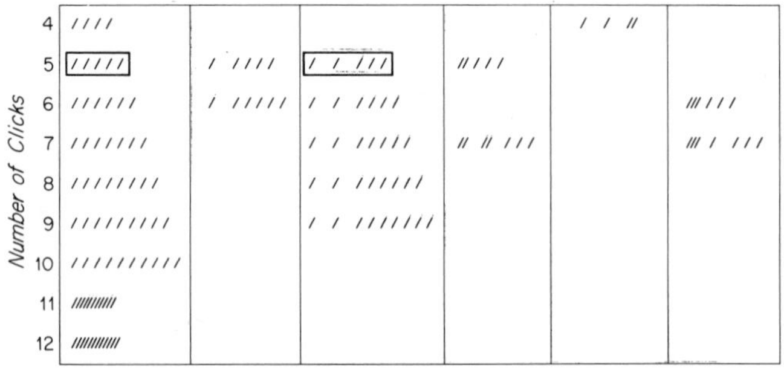

Fig. 5. Underwater coda patterns produced by a group of adult sperm whales from the southeast Caribbean (Nov. 1981) are shown schematically -- each slanted line represents one click, and relative click intervals are indicated by the spacing between lines. The codas are arranged vertically by the number of clicks and horizontally by the relative pattern of intervals between clicks to show the similarities in coda formation. The two "shared" codas (boxed) were produced by numerous whales, while the others appeared to be distinctive "identity" codas given repetitively by individual whales.

None of the calves have appeared to use their sounds in ways that were related to echolocation.

DISCUSSION

To summarize these results, the general progression in sizes of the
sperm whale calves that we recorded included new-born calves of 4 m in
Miami (July and August), a slightly larger 4.5-m calf in Seattle (September, though with umbilical cord still attached), a 7-m calf in Long Island,
and several young whales of 7-8 m recorded at sea. This progression in age
paralleled an apparent progression in development of patterned click
sounds, and in production of a more adult repertoire of sounds, including
single clicks, slow sequences of regular clicking, fast series of clicks,
and short repetitive sequences with more stereotyped temporal patterns of
click repetition (codas). Even the smallest calves produced broadband
clicks that were typical of sperm whales. Like those of adults, these
clicks varied in spectra within series as well as in successive sequences.
Although not fully organized in the repetitive coda patterns heard from
adults at sea (Figs. 4 and 5), the click sequences were produced in
similarly organized temporal patterns by the younger calves (Figs. 1, 2,
and 3), and they became more stereotyped and repetitive only in older
calves.

Although this data set on sperm whale calf sounds is small, it has
provided the only opportunity, so far, for making even preliminary assessments of the development of vocalization in sperm whales. The sounds of
calves at sea were recorded during apparently normal sperm whale interactions. However, the sounds from the captive calves were from stressed
animals away from their normal environment -- all had stranded on beaches
or in shallow water for some period, they were starved and in poor health
(most died within a few days of the recordings), and doubtless they were
affected by human handling, by their isolation, by noisy situations, and by
other conditions associated with their captivity. Therefore, it is the
similarities between their vocalizations and those from sperm whales at
sea, as well as the apparent progression in development of patterned
sounds, that are important, rather than any differences. The sounds
recorded from the captive calves are considered as representative of their
minimum acoustic capabilities.

The significant result of these comparisons of calf sounds is the
apparent development of repetitive temporal patterns of clicks, similar to
the coda signals used by adults. The calves could produce recognizable
sperm whale sounds soon after birth, but the communicative coda-like click
patterns developed in complexity and stereotypy with age. Similarly, in
Tursiops truncatus.it has long been known that sounds are produced soon
after birth (McBride and Kritzler, 1951), but the communicative whistle
sequences increase in complexity and stereotypy with age of the calves
(Caldwell and Caldwell, 1979).

Older calves produced click sequences that appeared to be more organized
into coda-like sequences, including a variety of temporal patterns. In
adult sperm whales at sea, each individual produced distinctive coda
patterns (Fig. 4) that were repeated again and again during particular behaviors (Watkins and Schevill, 1977b). Other codas were produced by numerous (perhaps all) whales in a local population (Watkins, Moore, and
Tyack, 1985). Some click sequences from the Long Island calf (particularly
the 9-to 12-click patterns) were produced in similar temporally repetitive
patterns, perhaps indicating that the number of clicks in a coda may not be
as important for conveying information as the temporal pattern of click
repetition. The relationship between the variety of coda patterns from
adult whales is illustrated in Fig. 5, and a comparison is also made
between the click patterns and the numbers of clicks in individually
distinctive codas and "shared" codas (boxed). Apparently, the sperm whale
calves that we have recorded could all produce such short, patterned click

sequences (compare the five Figs.), even if they had not yet developed the stereotyped patterns needed for the coda communication of adults.

From our numerous close encounters with sperm whales at sea, we know that adults produce clicks at a variety of spectra, levels, and repetition rates. This also has been true of the young sperm whales which have passed closely or approached us at sea. It was also true of the captive calves, so that the variations that we noted may have been a result of the same natural variability, emphasized by the relatively small sample size.

The steady, slow clicking (at rates of 2 to 4 per sec) by the Long Island calf was reminiscent of the most common click series heard from sperm whales at sea, occurring often in series of variable duration from many submerged sperm whales during apparent foraging dives (Watkins, Moore, and Tyack 1985). These sounds have appeared to be used to maintain communicative contact between whales that were spread out underwater (Watkins and Schevill, 1977a), and again, it was only the larger, more experienced captive calf that produced such sounds.

The lack of any indication of echolocation function for the clicks produced by both captive and free-ranging calves was in agreement with our previous assessments of the uses of click sounds by sperm whales (Watkins, 1980; Watkins and Wartzok, 1985).

ACKNOWLEDGEMENTS

We have appreciated the aid of many people at each of the facilities in which the different calves were held. Particularly, we thank Jesse White, Larry Tsunoda, Jane Small, and Jerry Joyce. Warren Zeiller (Miami) and Mark Anderson (Seattle) made recordings that helpfully contributed to these comparisons. We thank Amelia Giordano of Palermo, Sicily, and Peter Tyack of Woods Hole for their contribution to the acoustic analyses. We also thank William E. Schevill and Peter Tyack for their comments on the manuscript. Fig. 4 is reproduced by permission of the Journal of the Acoustical Society of America, from Watkins and Schevill 1977b. This is Contribution Number 6203 from the Woods Hole Oceanographic Institution.

LITERATURE

Backus, R. H., and Schevill, W. E., 1966, Physeter clicks. Pp. 510-527, in: "Whales, Dolphins, and Porpoises," K. S. Norris, ed., University of California Press, Berkeley.

Best, P.B., Canham, P.A.S., and MacLeod, N., 1984, Patterns of reproduction in sperm whales, Physeter macrocephalus, Rep. Int. Whal. Commn., (Special Issue 6): 51-79.

Caldwell, M. C., and Caldwell, D. K., 1979, The whistle of the Atlantic bottlenosed dolphin (Tursiops truncatus)--ontogeny. Pp. 369-401, in: "Behavior of Marine Animals, Current Perspectives in Research, Vol. 3: Cetaceans", H. E. Winn and B. L. Olla, eds., Plenum Press, New York.

McBride, A. F., and Kritzler, H., 1951, Observations on pregnancy, parturition, and post-natal behavior in the bottlenose dolphin, Jour. Mamm., 32:251-266.

Schevill, W. E., and Watkins, W. A., 1962, "Whale and porpoise voices, a phonograph record," Woods Hole Oceanographic Institution, 24 pp., phonograph record.

Watkins, W. A., 1980, Acoustics and the behavior of sperm whales. Pp. 283-290, in: "Animal Sonar Systems," R.-G. Busnel and J. F. Fish, eds., Plenum Press, New York.

Watkins, W. A., Moore, K. E., and Tyack, P., 1985, Investigation of sperm whale acoustic behaviors in the Southeast Caribbean, _Cetology_ 49:1-15.

Watkins, W. A., and Schevill, W. E., 1976, Biological sound-source location by computer analysis of underwater array data, _Deep-Sea Research_ 23:175-180.

Watkins, W. A., and Schevill, W. E., 1977a, Spatial distribution of _Physeter catodon_ (sperm whales) underwater, _Deep-Sea Research_ 24:693-699.

Watkins, W. A., and Schevill, W. E., 1977b, Sperm whale codas, _Jour. Acoust. Soc. Am._, 62:1485-1490, phonograph disc.

Watkins, W. A., and Wartzok, D., 1985, Sensory biophysics of marine mammals, _Marine Mammal Sci._, 1:219-260.

Worthington, L.V., and Schevill, W. E., 1957, Underwater sounds heard from sperm whales, _Nature_, 180:291.

APPARENT SONAR CLICKS FROM A CAPTIVE BOTTLENOSED DOLPHIN,

Tursiops truncatus, WHEN 2, 7 AND 38 WEEKS OLD

Morten Lindhard

Dept. of Zoophysiology
University of Arhus
8000 Arhus, Denmark

SUMMARY

A two-week-old dolphin produced clicks with peak energy as high as 60 kHz. At 7 weeks some clicks from the same dolphin peaked between 10 and 20 kHz. At 38 weeks it produced clicks with peak energy as high as 120 kHz (-10 dB BW 100 kHz), similar clicks were recorded from its mother and another adult female in the same pool. There was a 1/1 ratio between the -10 dB bandwidths and the respective spectral peak frequencies. Click duration was less than 50 μsec. Observations of the dolphins orientation relative to the hydrophones indicated that the selected clicks were recorded close to the beam axis. The click train from the two-week-old dolphin had click intervals from 70 to 2.5 msec. The intervals changed in a cyclic pattern with a decreasing mean interval as the dolphin approached the hydrophone. The possible implications of these observations for the click producing mechanism are discussed.

Abbreviations

RL = Received Sound Level peak-to-peak in dB re 1 μPa.
    (This convention implies that a continuous sine wave
    with the same peak-to-peak value as the click will have
    a 9 dB lower reading on an RMS indicating device)
SL = Source Level = RL 1 m from the source (blowhole)
BW = -10 dB bandwidth

MATERIALS AND METHODS

The dolphin Venus was born in a 40x20x4 m kidney shaped concrete pool, which it shared with its mother and five other dolphins at all times. There were no underwater viewing windows and no special underwater lights. The dolphins were not allowed to be isolated from each other. These constraints made sound source identification difficult at best.

Recordings were done with B&K 8100 and 8103 hydrophones, B&K amplifiers (2607, 2425, 2619 & 2608), Krohn-Hite filters (3322 & 3550) and instrumentation recorders (Lyrec TR47 and RACAL Store 7D) in various combinations in the order hydrophone, amplifier, filter, recorder. The 3 dB flat response limits were 1-100 kHz for the recording at 2 weeks and 1-140 kHz for the last two recordings at 7 and 38 weeks.

Commentaries describing dolphin movements were recorded to help identify the source of the sounds. In week 2 and 7 recordings were made with a single hydrophone. To increase the likelihood of correct source identification, the sound recordings in week 38 were done with a linear array of 3 B&K 8103 hydrophones spaced 0.5 m. Video recordings were made of the pool area (4x4 m) around the array. The commentaries recorded in parallel on the video and the instrumentation recorder served to synchronize the recordings.

During analysis the recordings were screened for overload distortions using the RACAL fast peak meter and a storage oscilloscope. Usable click trains with an identifiable source were digitized at a sample rate of 470 kHz and a resolution of 8 bit. A 288 μsec time series (=128 samples) was chosen as standard to describe each click and to calculate its power spectrum using a FFT algorithm. This gave spectra with 3.7 kHz resolution and more than 140 kHz bandwidth. The click intervals were measured on a storage oscilloscope.

RESULTS

The main problem in this study was to identify the source as the dolphins could not be isolated. The click train in fig. 1 is most likely from the two-week-old dolphin, Venus. The other dolphins were swimming silently in a circle and breathing in synchrony. Venus left the group and approached the hydrophone, slowed down and came to within 0.5 m from it, while the click train was heard. It sounded as coming from an approaching source very close to the hydrophone. The click intervals varied between 70 and 10 msec through the first 80 clicks (part 1). The intervals then decreased rapidly to about 2 msec for the last 2-300 clicks (part 2). Similar short click intervals are only resorted from echolocating dolphins less than 0.4 m from the target, (Au, 1980). The max RL was 115 dB. The first 80 clicks have spectra that peak at 65 kHz with 50 kHz BW, fig. 1 part 1. The clicks in part 2 peak at 20-40 kHz with 40 kHz BW. The examples in fig. 1 are selected to illustrate the stability and the variance in the spectra.

In week 7 the click train in fig. 2 was recorded while Venus alone approached the hydrophone to within 1 m. Her mother was only 10 m away, but she had her head out of the water and is thus not believed to be the source. The other dolphins were more than 25 m away. The clicks sounded as coming from a close and approaching source. The minimum click interval was 6 msec. 80 clicks were emitted within 1.5 sec, then the intervals approached 1 sec. The max. RL was 109 dB. The clicks peaked at 10-20 kHz with 15-20 kHz BW as seen in fig. 2.

In week 38 the clicks in fig. 3 were recorded while Venus was stationary at the end of the linear array approx. 0.5 m from the nearest hydrophone and pointing her rostrum towards the array. The different arrival times of each click to the 3 hydrophones indicated the source to be in her direction on the array axis. The distance to the source could not be calculated, but RLs were about 10 dB lower on the nearest hydrophone compared to the two more distant ones. This can be explained by the nearfield effect as described by Au (Au, et al., 1978), only if Venus was the source. The average click interval was 16 msec for the 7 click long click train. The max SL was 153 dB. The spectra shifts from peaks at 110-130 kHz to a peak at 50 kHz.

Analysis of 3 click trains from two adult dolphins including Venus' mother showed clicks peaking at 120-130 kHz similar to the ones in fig. 3. The directional effects on click RL, waveshape and spectrum as revealed by

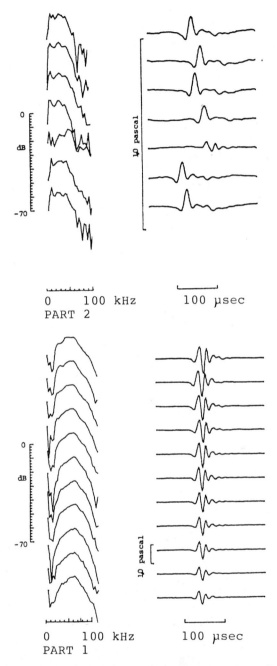

0

dB

-70

0   100 kHz
PART 2

10 pascal

100 µsec

0   100 kHz
PART 1

0

dB

-70

10 pascal

100 µsec

Fig. 1 Click spectra adjacent to their respective time series.
        Recorded from a 2-week-old <u>T. truncatus</u>. The click train
        could be divided into two parts. Representative clicks are
        chosen from each part.

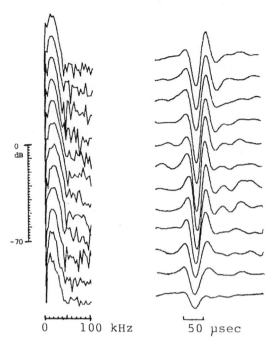

Fig. 2 Click spectra adjacent to their respective time series. From a 7 -week-old T. truncatus.

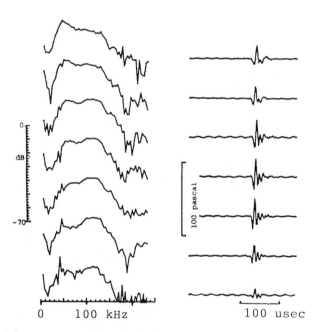

Fig. 3 Click spectra adjacent to their respective time series. From a 38-week-old T. truncatus.

112

the array and video recordings were within the limits given by Au et al. (Au, et al, 1978).

DISCUSSION

Venus seems to be the youngest dolphin from which 60 kHz clicks have been recorded. 120 kHz clicks are reported from a 60-day-old T. truncatus, (Ridgway, et al., 1983). The lack of knowledge about the hydrophone position relative to the click beam axis makes it impossible to rule out that the clicks in the first two click trains had in fact higher dominant frequencies at axis than recorded. Yet when comparing the wave forms to those found at various angles of axis by Au (Au, et al., 1978), then the clicks peaking at 60 kHz in fig. 1 part 1, at 20 kHz in fig. 2 and at 120 kHz in fig. 3 all seem to be recorded close to the beam axis. Likewise clicks with relatively lower frequencies like the 40 kHz clicks in fig. 1 part 2 and the 50 kHz clicks in fig. 3 have distorted waveforms as if recorded at a 5-20° angle to the beam axis. Thus it may be assumed that the beam axis clicks peaked at 60, 20 and 120 kHz in the 3 click trains respectively. If the peak frequency was determined by a physical structure setting up relaxation oscillations in the nasal plugs as suggested by Evans (Evans, et al. 1973) one would expect an inverse correlation between size and frequency. If however the click dominant frequency is determined by filtering of a white noise, analogous to the filtering of white noise forming human speech, then the peak frequency will be uncorrelated to the size of the dolphin. The first system is at variance with my observations, the latter is not. However the limited number of observations and the uncertainty concerning the site of recording relative to the beam axis make any conclusion concerning production mechanism preliminary awaiting more and better data from trained dolphins.

ACKNOWLEDGEMENTS

This study could not have been done without the kind cooperation of Kolmarden Djurpark, Sweden. Recordings in week 2 and 7 were done by Mats & Birgitta Amundin with assistance from Søren Andersen and Dan Dorschel. The analysis was done with assistance from Bertel Møhl. Technical assistance was supplied by the Danish Natural Science Research Council.

REFERENCES

1 Au, W.W.L., "Echolocation Signals of the Atlantic Bottlenosed dolphin (Tursiops truncatus) in Open Waters" in: "Animal Sonar Systems", R.-G. Busnel, ed., Plenum Press, New York (1980).
2 Au, W.W.L., Floyd, R.W., Haun, J.E., "Propagation of Atlantic Bottlenosed Dolphin Echolocation Signals", J. Acoust. Soc. Am. 64:411 (1978).
3 Evans, W.E., Maderson, P.F.A., "Mechanism of Sound Production in Delphinid Cetaceans: A Review and some Anatomical Considerations", Am. Zool., 13:1205 (1973).
4 Ridgway, S.H., Carder, D.A., "Apparent Echolocation by a Sixty-day-old Bottlenosed Dolphin, Tursiops truncatus" J. Acoust. Soc. Am., 74(S1):S74 (1983).

ONTOGENY OF VOCAL SIGNALS IN THE BIG BROWN BAT, *EPTESICUS FUSCUS*

Cynthia F. Moss

Universität Tübingen
Lehrstuhl Zoophysiologie
D-7400 Tübingen, F.R. Germany

INTRODUCTION

The production of vocal signals by infant bats is important for communication and may play a role in the development of sonar signals used for echolocation. Recordings of infant vocalizations from a variety of species show many similarities, even across different families of bats (e.g. Brown, 1976; Gould, 1971, 1975a, 1979; Matsumura, 1979; Brown and Grinnell, 1980; Brown, Brown and Grinnell, 1983). Young infant bats often emit multiple-harmonic sounds that are lower in frequency than the sounds of conspecific adults. In many species, the infant vocal repertoire includes sounds with relatively constant frequency components which are typically identified as isolation sounds. These sounds often promote approach and retrieval of an infant by its mother (e.g. Davis, Barbour and Hassell, 1968; Gould, 1971, 1975a; Brown, 1976; Thomson, Fenton, and Barclay, 1985).

A detailed study of vocal ontogeny in bats permits careful assessment of the relation between communication and echolocation sounds and can also provide information concerning vocal production, and how it changes over the course of development. It may be particularly valuable to study vocal ontogeny in a species for which sonar capacity has been carefully examined. The big brown bat (*Eptesicus fuscus*) has been the subject of extensive psychophysical studies on hearing and echolocation (e.g. Simmons and Vernon, 1971; Simmons, 1973; Kick, 1982; Kick and Simmons, 1984). Infant vocalizations have also been described (Gould, 1971); however, the published work on vocal ontogeny in *Eptesicus fuscus* omits a developmental series of spectrograms, thus precluding close examination of the structure of communication and echolocation sounds and how they change with age. With the goal of evaluating these factors, this study attempts a detailed, comprehensive spectrographic analysis of vocal development in *Eptesicus fuscus*.

METHODS

Pregnant female *Eptesicus fuscus* were collected from maternity colonies in Rhode Island and Massachusetts during the late spring of 1986. They gave birth in the laboratory between June 11 and July 4, 1986 and nursed their young until 4 weeks postnatal.

All recordings of vocalizations were made in a foam-padded sound-attenuating chamber. The sounds of 7 infant bats were recorded longitudinally during the first 6 weeks of life. Sounds were recorded from the infants while in contact with the mother, while temporarily separated from her, and then again when reunited with her. The mother's vocalizations were also recorded. In addition, vocalizations were recorded from females that did not give birth this year, from those whose infants did not survive this year, and from males. All vocalizations were recorded with a Racal tape recorder, located outside the chamber. During recording, the speed of the tape recorder was set at 60 inches per second. Sounds were later played back at 1/64 the original recording speed and analyzed using a Nicolet Spectrum Analyzer (VA-500A).

RESULTS

At 6 hours postnatal, *Eptesicus fuscus* emit a variety of sounds that differ in frequency content and duration. Spectrograms show that newborns emit both relatively constant frequency (CF) and frequency modulated (FM) calls, a finding which contrasts with Gould's report (1971; 1975a) that *Eptesicus* infants only emit CF sounds until 3-5 days. The frequency components are lower than those of adult echolocation calls: CF calls are often about 15 kHz and FM calls typically sweep from about 20-25 kHz down to 10 kHz, whereas the adult echolocation call of *Eptesicus* sweeps from about 60 to 25 kHz. Sound duration is also longer in infant calls than the 2 ms sonar sounds of adults. The duration of infant CF calls is highly variable (20-125 ms) and often exceeds that observed for infant FM calls (20-30 ms).

Infants at one day of age show the same pattern of vocalizations; however, calls are occassionally shorter than those recorded from newborns. Between 5 and 7 days postnatal, bats produce sounds that are similar to those of newborns, although they now also emit sounds of shorter duration. These shorter sounds are less than 10 ms and contain multiple harmonics. The fundamental frequency is somewhat lower than the longer duration sounds (8-15 kHz) and shows shallow FM.

Figure 1 summarizes the vocal patterns observed in bats from 6 hours to 7 days postnatal. Presented on the left are spectrograms of sounds emitted by infants in contact with their mothers and on the right, those emitted by infants separated from their mothers. This figure illustrates that, in general, infants in close proximity of their mothers tend to produce longer duration sounds with less FM than those completely isolated from their mothers. It also shows a tendency for sound duration to decrease with age. There is, however, a great deal of individual variability in the spectral and temporal structure of infant sounds, and those recorded in different contexts cannot be unambiguously categorized. In spite of the individual variability, mothers apparently use infant calls (along with olfactory cues) to identify their own young (Gould, 1971). Mothers leave their young behind in the roost when they forage. Upon return, they retrieve and selectively nurse their own infants (Davis et al., 1968).

Spectrograms of vocalizations by bats from 7 days to adult are presented in Figure 2. Between 7 and 14 days postnatal, the starting frequency of some FM calls rises and the duration of these calls shows a marked decrease (Fig 2A). Bats at this age also appear less distressed when separated from their mothers and often do not vocalize when isolated from

116

or reunited with them. It is interesting to note that in the wild, mothers begin leaving the roost for longer periods of time when their infants are older (Davis et al., 1968). By 14 days, FM calls closely resembling the adult sonar signals are present; these calls are 2-5 ms in duration and sweep from approximately 40 to 20 kHz.

At 21 days, young *Eptesicus* are capable of flight (Davis et al., 1968; Gould, 1971; present study), and their sonar signals are similar to those of adults, although somewhat lower in frequency (sweeping from 45-20 kHz) and longer in duration (up to 5 ms). Between 21 and 28 days, the sonar signals become higher in frequency and shorter in duration.

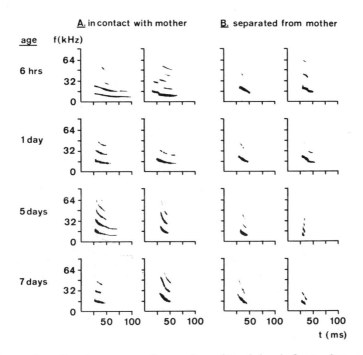

Figure 1.   Spectrograms of sounds emitted by infants between 6 hours and 7 days postnatal; in contact with mother (left) and separated from mother (right).

Twenty-eight-day-old bats are capable of foraging on their own (Gould, 1971), and their sonar sounds are virtually indistinguishable from those of adults (Fig 2A).

Figure 2B shows that infant and juvenile bats, as well as adults, emit the short, low frequency pulses first recorded from 5-day-old bats. The short duration and shallow FM of these sounds in younger bats first suggested that they may be precursors of adult sonar signals; however, the fact that these pulses show little change with age indicates that they do not evolve into adult sonar sounds.

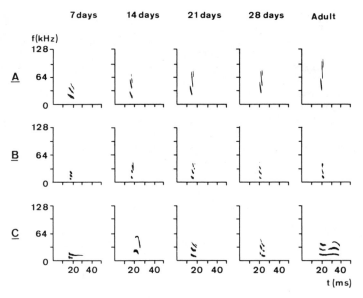

Figure 2. Spectrograms of sounds emitted by bats from 7 days to adult. A. FM sounds. B. Short low frequency pulses C. Communication sounds.

In Figure 2C, spectrograms of various other calls emitted by *Eptesicus* are presented. CF calls are not only recorded from infant bats but also from adults. The adult CF sound shown in this figure was emitted by a mother separated from her infant. A variety of other distress calls are also emitted by adults (not shown), many closely resembling the long, shallow FM and CF sounds produced by infants. Such adult distress calls are not unique to a mother separated from her infant. Similar vocalizations were also recorded from other adult *Eptesicus*, including females that did not give birth this year, those whose offspring did not survive this year, and males.

At intermediate ages (14-28 days), a variety of sounds were also recorded that could not be readily classified as FM sonar sounds or short, low frequency pulses. Shown for a 14-day-old bat is what appears to be an anomalous adult sonar sound, with an almost CF portion followed by a downward FM sweep. The structure of this sound may reflect lack of vocal control, or it may represent a communication signal. The spectrograms presented for 21- and 28-day-old bats are relatively low in frequency and resemble calls of younger infants. Perhaps they also serve a communicative function.

DISCUSSION

The major developmental trend in vocalizations by *Eptesicus* is a rise in sound frequency, accompanied by a reduction in sound duration. Although general patterns of vocalization do change with age, virtually all sound types emitted by infant *Eptesicus* are also emitted by adults. Adult calls differing from the typical echolocation sound are often produced by distressed or agitated animals and thus probably function as communication signals.

The observation that adult *Eptesicus* emit calls resembling those more typical of infants has two important implications: First, there may be some overlap of sounds used for communication and those used for echolocation. This implies that sounds emitted by young *Eptesicus* may serve a dual function. Indeed, the infant bat may inadvertantly begin using echoes of its own signals while engaged in social communication. Second, the calls of a young infant *Eptesicus* do not disappear from its repertoire as it matures, suggesting that no single infant sound evolves into the adult echolocation call. Rather, production of a group of sounds may stimulate the maturation of the vocal apparatus which then permits production of a brief FM sweep characteristic of adult sonar calls.

The results of this study may be useful to further evaluate the mechanisms of vocal production and their development. In *Eptesicus* the frequency of brief ultrasonic pulses is controlled by contraction and relaxation of the cricothyroid muscle surrounding the larynx. When the cricothyroid muscle contracts, it increases tension on the laryngeal membranes. Relaxation occurs during phonation. It appears that the magnitude of muscle contraction determines the frequency at which the laryngeal membranes vibrate, and gradual relaxation produces the downward FM sweep. The cricothyroid muscle is innervated by the motor branch of the superior laryngeal nerve (SLN), and section of this nerve results in a dramatic drop in the fundamental frequency of vocalizations and elimination of frequency modulation (Novick and Griffin, 1961; Suthers and Fattu, 1973, 1982).

Since infant bats emit sounds that are lower in frequency and show less depth of frequency modulation than adults, it is interesting to speculate that this may reflect immaturity of the larynx, its muscles and perhaps also its innervation. Gould (1975b) reports that the motor branch of the SLN is unmyelinated in premature *Eptesicus* and thickly myelinated in subadults. Data for intermediate ages are not reported, but it may be that postnatal myelinization of the SLN permits greater contraction of the cricothyroid muscle and hence higher frequency phonation and deeper frequency modulation. Clearly, further data on developmental changes in the vocal apparatus of bats are needed to determine the mechanisms involved in vocal ontogeny.

ACKNOWLEDGEMENTS

I thank J. Simmons and J. Szewsczak for their assistance in collecting the animals and I. Kaipf for preparing the figures. I would also like to express my appreciation to U. Schnitzler for his generous support. This work was funded by a NATO postdoctoral fellowship awarded to C. Moss and by the Deutsche Forschungsgemeinschaft (SFB 307).

REFERENCES

Brown, P., 1976, Vocal communication in the pallid bat, *Antrozous pallidus*, *Z. Tierpsychol.*, *41*: 34-54.
Brown, P. E., and Grinnell, A. D., 1980, Echolocation ontogeny in bats. *in* "Animal sonar systems," R. G. Busnel and J. F. Fish, eds., Plenum press, New York, pp. 355-377.
Brown, P. E., Brown, T. W., and Grinnell, A. D., 1983, Echolocation behavior, development, and vocal communication in the lesser bulldog bat, *Noctilio albiventris*, *Behav. Ecol. Sociobiol. 13*: 287-298.

Davis, W. H., Barbour, R. W., and Hassell, M. D., 1968, Colonial behavior of *Eptesicus fuscus*, *J. Mammalogy*, *49*: 44-50.

Gould, E., 1971, Studies of maternal-infant communication and development of vocalizations in the bats *Myotis* and *Eptesicus*, *Communications in behavioral biology*, Part A, 5, No. 5, pp. 263-313.

Gould, E., 1975a, Neonatal vocalizations in bats of eight genera, *J. Mammalogy*, *56*: 15-29.

Gould, E., 1975b, Experimental studies of the ontogeny of ultrasonic vocalizations in bats, *Develop. Psychobiol.*, *8*: 333-346.

Gould, E., 1979, Neonatal vocalizations of ten species of Malaysian bats (Megachiroptera and Microchiroptera), *Amer. Zool.*, *19*: 481-491.

Kick, S. A., 1982, Target-detection by the echolocating bat, *Eptesicus fuscus*, *J. Comp. Physiol.*, *145*: 431-435.

Kick, S. A., and Simmons, J. A., 1984, Automatic gain control in the bat's sonar receiver and the neuroethology of echolocation, *J. Neuroscience*, 4: 2725-2737.

Matsumura, S., 1979, Mother-infant communication in a horseshoe bat (*Rhinolophus ferrumequinum nippon*): Development of vocalization, *J. Mammalogy*, *60*: 76-84.

Novick, A., and Griffin, D. R., 1961, Laryngeal mechanisms in bats for the production of orientation sounds, *J. Exp. Zool.*, *148*: 125-146.

Simmons, J. A., 1973, The resolution of target range by echolocating bats, *J. Acoust. Soc. Am.*, *54*: 157-173.

Simmons, J. A., and Vernon, J. A., 1971, Echolocation: Discrimination of targets by the bat, *Eptesicus fuscus*, *J. Exp. Zool.*, *176*: 315-328.

Suthers, R. A., and Fattu, J. M., 1973, Mechanisms of sound production by echolocating bats, *Amer. Zool.*, *13*: 1215-1226.

Suthers, R. A., and Fattu, J. M., 1982, Selective laryngeal neurotomy and the control of phonation by the echolocating bat, *Eptesicus*, *J. Comp. Physiol.*, *145*: 529-537.

Thomson, C. E., Fenton, M. B., and Barclay, R. M. R., 1985, The role of infant isolation calls in mother-infant reunions in the little brown bat, *Myotis lucifugus* (Chiroptera: Vespertilionidae), *Canadian J. Zool.*, *63*: 1982-1988.

OBSERVATIONS ON THE DEVELOPMENT OF ECHOLOCATION IN YOUNG BOTTLENOSE DOLPHINS

Diana Reiss

Department of Speech and Communication
San Francisco State University
San Francisco, California 94132

With the birth of two male bottlenose dolphins (Tursiops truncatus) with a 72 hour period into our research colony we began a two year program of systematic observation and recording of the ontogeny of their vocal and non-vocal behavior and communication. While pursuing this program we were fortunate to obtain instances of apparent echolocation by the young animals during the fourth postnatal week. Previously, the earliest instance of dolphin echolocation was reported in a sixty-day old bottlenose dolphin by Carder and Ridgway (1983). They observed head scanning movements concurrent with high energy pulses with peak frequencies ranging from 33 to 120 kHz with 3-dB bandwidths of 28-81 kHz. The appearance of sonar-like pulses by our animals at so early an age prompted us to review the acoustic and behavioral data we had compiled for the period extending from birth to postnatal day 40 when the infant and adult pulses were indistinguishable to 16 kHz, the limits of our recording and analysis equipment (see Methods). Thus, we can report here on a wide range of behavioral development concurrent with the ontogeny of echolocation as it occurred within a captive but social milieu. Additionally, we can suggest and provide examples of vocalizations which may play a role as precursors to normal echolocation.

METHODS

The colony of animals observed was composed of a mature male Pacific bottlenose dolphin, approximately 30 years old, two female Atlantic bottlenose dolphins, approximately 20 and 9 years old, and their two male offspring, Pan, born on July 30 and Delphi, born August 2, 1983. The mature male fathered both infants. The group resided in a 7 foot deep, 57,000 gallon, kidney-shaped pool filled with treated bay water at Marine World Africa USA in Redwood City, California.

Audio and video recordings and ad lib notation using time and event based sampling methods were utilized to document behavioral and vocal development. The data presented in this paper is based on daily observations and recordings from birth through postnatal day 40.

Dolphin vocalizations were recorded with input from a Finley-Hill EM 8 hydrophone with a flat frequency response ($\pm$ 2dB) to 22 kHz onto an Ampex ATR 700 tape recorder at a tape speed of 19 cm (7.5 inches) per second. At this operating speed the Ampex recorder had a flat frequency response from

40 Hz to 22 kHz. A dual channel recording method permitted underwater vocalizations to be recorded simultaneously with a detailed behavioral narrative.

Vocalizations were analyzed and sonagrams were produced on a 7029-A Kay Sona-Graph (spectrum analyzer) calibrated from 160 Hz to 16 kHz with an effective filter bandwidth of 600 Hz. In order to identify the vocalizing dolphin we selected samples in which air bubbles were emitted from only one dolphin's blowhole that correlated with vocal production or chose instances in which an individual would vocalize directly at the hydrophone from within a 1 meter distance. This latter method also served to minimize errors in analysis since the frequency content of signals can depend on the angle of the dolphin's head to the transducer (Wood and Evans, 1980).

VOCAL AND BEHAVIORAL DEVELOPMENT

The behavioral development during the first 40 postnatal days can best be described as three general and slightly overlapping periods. By comparing these periods it is possible to follow distinct vocal development as it correlates with sensori-motor development and abilities toward the use of echolocation. A detailed summary of this data appears in Table 1.

1. The first period, from birth through the 5th postnatal day was characterized by almost constant mother-infant contact in close tandem formation swimming and brief infant departures. The precocial nature of the dolphin infant and the high degree of active maternal care has been qualitatively described by several investigators (McBride and Hebb, 1948; McBride and Kritzler, 1951; Tavolga and Essapian, 1957; Caldwell and Caldwell, 1977). Immediately following birth there is an intense period of reciprocal stimulation between mother and infant. During this postpartum period the neonate exhibits the thigmotaxic response and spends the first hours in not learning to follow but rather whom to follow. Mothers use vocal, tactile and visual stimuli to elicit and direct neonatal activities. Vocal exchanges of whistled contact or isolation calls occur frequently when mother-infant pairs are separated and usually continue until they are reunited (McBride and Kritzler, 1951). By days 2-4 the infants avoid tank walls with less maternal guidance and often reapproach their mothers without maternal vocal signalling.

Whistles, often tremulous and showing relatively little frequency modulation and whistle-squawks (Caldwell and Caldwell, 1967, 1979) were emitted by the infants immediately after birth and appear to function as contact-isolation calls. The whistles were sometimes produced with such intensity, that they became rich in harmonics and could be categorized as squeals. Whistle-squawks (Fig. 1) were emitted in more emotional situations such as when the young dolphins were separated from their mothers for prolonged times or when they would swim into the tank wall. The pulsed or 'squawk' component of these signals generally occurred at the beginning or end and simultaneously with the whistle suggesting that at this early an age, the dolphin is capable of producing two types of vocalizations at the same time. The burst pulse squawks were lower in frequency, below 4 kHz, than their whistles which varied between the 3-10 kHz range.

2. The second period extended from postnatal days 5 through 14 and was marked by more rapid and frequent departures by the young dolphins from their mothers. During days 9 through 12 rapid development of many sensori-motor and coordinated behaviors were first observed such as porpoising, breaching, tail slapping, and quick, sharp directional changes during infant departures and subsequent mother-infant chases. Deeper and longer duration dives to the tank bottom suggested more respiratory control. On

Table 1. Chronological Development of Vocal and Non-Vocal Behavior
(based on first observations)

Age in Days		Behavior	Vocal Behavior
Pan	Delphi		

1	1	mother-infant tandem swims
		nursing (3 hrs. postpartum)
		short departures from mother
		dependence on maternal vocal-physical
		  guidance for orientation
		beak-genital orientation & contact
		  by mother to infant
			whistles: tremulous with
			  little frequency modula-
			  tion
			whistle-squawks: narrow
			  band whistles with a
			  pulsed component
			(these signals appear to
			  be contact, isoloation calls)

2	2	better avoidance of tank walls	
4	2	rapid departures from mother mothers follow-chase infants infant approaches mother occasional open mouth swimming	
5	2	aunting (1 mother with both infants) brief approaches by infants to tankmates increased departures from mothers	
8	5	visual orientation in air tail slaps (as infant departs mother)	

9	6	porpoising during departures-chases
		infants learn to stop
			squawks: low frequency
			  burst-pulsed sounds
			longer duration, low fre-
			  quency pulses
			(occur in rapid departures)

12	9	breaching during departures-chases avoidance of walls without maternal aid dives to tank bottom (7 ft.) rapid changes in swimming direction (alone)	
14	10	increased exploration of tank visual orientation & mouthing of fish/objects	
18	15	erections increased open mouth orientation & mouthing of objects & tankmates increased departures, chases, beak-genital swims	
19	16		shorter duration, low fre- quency pulses

27	24	higher porpoising
		more active toy play
			apparent open mouth echo-
			  location on hydrophone &
			  during toy play

29	26	increased open mouth orientations and mouthing of objects & tankmates	
30	27	head to head orientations between infants and tankmates	

38	35	more closed mouth chasing and toy play	apparent closed mouth echo-
			  location on hydrophone,
			  in during chases & toy
			  play

days 10-14 they exhibited greater interest in exploring their environment, underwater and surrounding the tank. Visual-oral coordination developed as visual inspection, often with an open mouth, was frequently followed by the mouthing of objects such as dead fish and toy balls. During these orientations no head scanning or infant pulsed emissions were recorded. On post-natal day 6 for Delphi and 9 for Pan, both infants began to emit a variety of pulsed sounds with open mouths. These sounds were produced just as the infants initiated a rapid, often energetic departure from the mother or porpoised or breached during a chase. They emitted a series of 20-40 msec pulses that often graded into higher repetition rate burst pulse or squawk type vocalization (Fig. 2 and 3). These pulses have a click like component at their onset but a longer duration 20-40 msec high energy component bet-ween 2-3 kHz. They could be described as long duration tonal clicks. They sound similar to adult echolocation signal but lower in frequency.

3. The third period, from postnatal days 15 through 40, was typified by more social-sexual interactions between mother-infant pairs (Reiss, 1985). During days 15-18, we first observed erections in the young, now confirmed to be male dolphins, within the contexts of mother-infant beak-genital swimming and chases. During the third week, Pan and Delphi showed even more interest in exploring areas and objects in their environment. They were still spending 90% of their time however, in close mother-infant tandem swims.

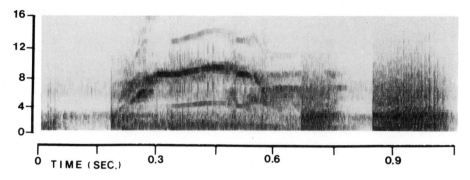

Figure 1. Example of a whistle-squawk emitted by Pan during postnatal
week 1.

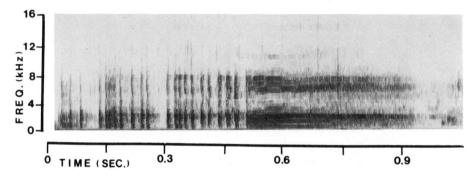

Figure 2. Example of long duration pulses grading into a higher repetition rate pulses emitted by Delphi as he rapidly departs
from his mother in week 2.

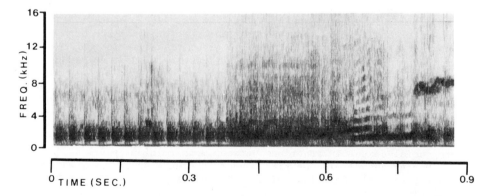

Figure 3. Example of low frequency, long duration pulses grading into
a squawk-type vocalization emitted by Delphi as he rapidly
departs from his mother in week 2.

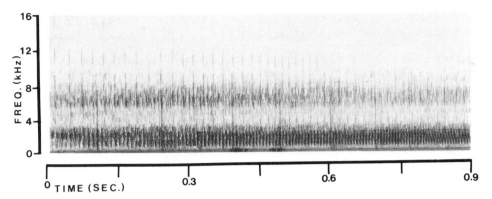

Figure 4. Example of shorter duration low frequency clicks emitted by Pan on his 19th postnatal day (week 3). Pulses were produced as the dolphin oriented to the hydrophone and his mother with an open mouth from a distance less than 1 meter.

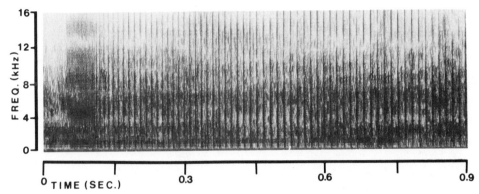

Figure 5. Pan emits echolocation-type clicks concurrent with open mouth head orientations and scanning movements at the hydrophone from less than a 1 meter distance in week 4.

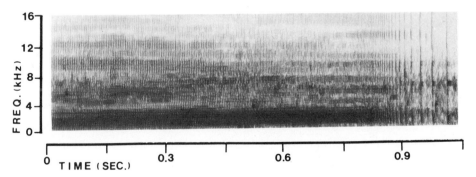

Figure 6. Delphi approached the hydrophone from a 1 meter distance and emits closed mouth clicks with concurrent head scanning movements on postnatal day 37 (week 6). Click repetition rate decreased just as he passed the transducer.

A sequence of sonar-like development occurred during days 16 through 38. On the same day, Delphi's 16th and Pan's 19th postnatal day, both males began to emit shorter duration 1 msec clicks that resembled adult click durations but were lower in frequency with peak frequencies below 4 kHz (Fig. 4). These clicks were also produced with the dolphins' mouths open as they oriented to the hydrophone and other objects. Head scanning movements were now observed concurrent with click production. During days 24-27 wider band higher frequency clicks were recorded as the young animals continued to orient to the hydrophone and other objects with open mouths. The signals looked like our examples of adult echolocation to 16 kHz. By days 35-38 both Pan and Delphi were producing adult-like clicks to 16 kHz with closed mouth approaches and orientations to the hydrophone and other objects and areas. Head scanning was observed at these times. Repetition rate and peak frequency varied with activity (Fig. 5 and 6).

DISCUSSION

In considering the pattern and sequence of the vocal and behavioral development in our two young dolphins, several new relationships and questions come to mind. First, in comparing our findings with the relatively rich literature on the ontogeny of bat echolocation (reviewed by Brown and Grinnell, 1980) certain similarities are striking. In both our dolphins and many species of bat, there is a developmental shift from open mouth to closed mouth pulse production. In the bats this is a shift from oral to nasal production and reflects the closing of the laryngo-nasal junction (Matsumura, 1979). Whether there is a similar anatomical basis for this shift in the dolphin and whether open to closed mouth production is indicative of oral to nasal production is unclear.

Both our dolphins and many bat species showed a progressive shift from production of longer to shorter duration pulses or clicks. Early pulsed sounds were lower in frequency and developed into higher frequency pulses. Adult-like echolocation signals developed in bats between the 25th and 45th postnatal day depending on species, which is similar to the beginning of adult-like pulses in our dolphins between the 24th and 40th days. The development of more adult-like echolocation in both bat and dolphin appears to correlate with their increased interest in the environment. In our dolphins there was also a visible pattern in their sensori-motor development and coordination progressing from visual, to oral-tactile, to acoustic (pulsed) investigations of the environment.

It is unclear whether young dolphins obtain, either intentionally or inadvertently, environmental information from their early pulsed sounds. The early squawks and longer duration pulses in the first and second weeks and in emotional or energetic contexts may be early precursors to the more developed echolocation signals. However, these early pulsed sounds, such as squawks, that may be initially reflexive, effort-related, or emotional signals remain within the adult repertoire. They may function as communicative and emotional signals as they are frequently observed in play-chases, aggressive, and confrontational contexts (Overstrom, 1983).

Finally, it is unclear what role the mother and conspecifics play in providing echolocation models and directives for young animals. We observed many ambiguous cases in which the mother and infant would simultaneously orient to objects and areas and two simultaneous pulsed emissions were evident, one adult type click train and one lower (2-3 KHz) frequency pulse train duringthe second and third weeks. This is a fertile area for future investigations.

ACKNOWLEDGMENTS

This research was supported by the Hugh J. Andersen Foundation and the
Marine World Research Foundation. I would like to thank Dr. S.H. Ridgway
for his suggestions, Stuart Firestein for his assistance in editing the manuscript, and Dr. R.G. Busnel for his encouragement. A special thanks to Bruce
Silverman and Denise Herzing for their assistance in data collection and to
the five dolphins.

REFERENCES

Brown, P.E. and Grinnell, A.D., 1980, Echolocation ontogeny in bats, in:
    "Animal Sonar Systems," R.-G. Busnel and J.F. Fish, eds., Plenum Press,
    New York.
Caldwell, M.C. and Caldwell, D.K., 1967, Intraspecific transfer of informa-
    tion via the pulsed sound in captive odontocete cetaceans, in: "Animal
    Sonar Systems: biology and bionics," R.-G. Busnel, ed., Imprimerie Louis-
    Jean, Gap (Hautes-Alpes).
Caldwell, M.C. and Caldwell, D.K., 1977, Social interactions and reproduction
    in the Atlantic bottlenosed dolphin, in: "Breeding Dolphins Present
    Status, Suggestions for the Future, S.H. Ridgway and K. Benirschke, eds.,
    Report No. MMC-76/07 Final Report to U.S. Marine Mammal Commission.
Caldwell, M.C. and Caldwell, D.K., 1979, The whistle of the Atlantic bottle-
    nosed dolphin Tursiops truncatus-ontogeny, in: "Behavior of Marine
    Animals, current perspectives in research, vol.3-cetaceans," H.E. Winn
    and Bl. Olla, eds., Plenum Press, New York.
Carder, D.A. and Ridgway, S.H., 1983, Apparent echolocation by a sixty-day
    old bottlenosed dolphin, Tusiops truncatus, JASA.,Supp. 1, 74:S74
Matsumura, S., 1979, Mother-infant communication in a horseshoe bat Rhino-
    lophus ferrumequinum nippon, I. J.Mamm. 60: 76-84
McBride, A.F. and Hebb, D.O., 1948, Behavior of the captive bottlenose dol-
    phin, Tursiops truncatus, J. Comp.Physiol. Psychol.,41:111-123
McBride, A.F. and Kritzler, H., 1951, Observations on pregnancy, parturition,
    and post-natal behavior in the bottlenose dolphin, J. Mammal., 32:251-266
Overstrom, N.A., 1983, Association between burst-pulse sounds and aggressive
    behavior in captive Atlantic bottlenosed dolphins Tursiops truncatus,
    Zoo Biol., 2:93-103
Reiss, D.L., 1985, Development of the social-sexual patterns through early
    mother-infant interactions in the bottlenose dolphin, Tursiops truncatus,
    Abstracts: Sixth biennial conf. on the biol. of marine mammals,
    Vancouver, B.C.
Tavolga, M.C. and Essapian, F.S., 1957, The behavior of the bottlenosed dol-
    phin Tursiops truncatus: mating, pregnancy, parturition and mother-
    infant behavior, Zoologica, 42(1):11-31
Wood, F.G. and Evans, W.E., 1980, Adaptiveness and ecology of echolocation in
    toothed whales, in:"Animal Sonar Systems," R.-G. Busnel and J.F. Fish,
    eds.,Plenum Press, New York.

THE SHORT-TIME-DURATION NARROW-BANDWIDTH CHARACTER

OF ODONTOCETE ECHOLOCATION SIGNALS

H. Wiersma [†]

Delft University of Technology
Mekelweg 4
Delft, Netherlands

ABSTRACT

Descriptions are presented of echolocation signals as observed in a number of Odontocete species. It is illustrated that all of these signals have a remarkably small bandwidth considering their time duration. In fact, these signals achieve the maximum concentration in the time/frequency space that is physically possible. This is demonstrated by comparing the signal waveforms with the result of a computation of the waveform that has the absolute minimum possible time duration and bandwidth for a given mean frequency and time delay. There appears to be a striking resemblance between the actual echolocation waveforms and the waveform computed with the same mean frequency and time delay. This holds not for one, but for all the species considered here and, in the case of bi-frequent sonar, not only for the high-frequncy component, but equally well for the low-frequency one. Moreover, it can also be shown that all these waveforms are approximately their own Fourier sine transforms. Waveform characterisation could, therefore, be very adequately based on one 'quantum' number only. Such waveforms yield the best possible signal-to-noise ratios when they have to be detected in broad background noise. In addition, for these signals of short duration and small bandwidth, a simple square-law detector closely approximates the performance of a matched-filter detector.

INTRODUCTION

Since the late forties and early fifties of this century, when the first observations were reported speculating on the possibility of the use of ultrasound for echolocation, it has become known that many Odontocetes produce ultrasound. The early descriptions of the sounds observed were purely subjective and the experiments performed did not actually prove that these sounds had been emitted with the intention of gaining information on the environment by the perception of their reflections. While this is what could be called echolocation in a strict sense, the production of ultrasound does not necessarily imply echolocation in this sense. In fact, for only a relatively small number of Odontocetes has the ability to echolocate

---

[†] Present address: Koninklijke/Shell Exploratie en Produktie
Laboratorium, Volmerlaan 6, Rijswijk, The Netherlands

been demonstrated conclusively. For the majority of species that are described below, no conclusive experiments or no experiments at all are reported that prove that the observed ultrasounds have an echolocational function. For some of the species, even the production of ultrasound has not been reported before. Hence, strictly speaking, we cannot freely use the term echolocation. We shall nevertheless continue to do so, since the ultrasounds that have now been observed in different species have many properties in common, making it very likely that these are indeed echolocation sounds.

The sound recordings that underlie the present work were made at various places over a period from 1977 until 1983. All recording was done in oceanaria or delphinaria, with the animals having been in captivity for various lengths of time. With respect to the acoustical behaviour, little seems to be known about the consequences and side-effects of the unnatural living circumstances in captivity. A few recordings could be made when a stranded and ill specimen of Lagenorhynchus albirostris was brought from the wild into a tank; here the only adaptation apparent was a lowering of the acoustical volume. Some of the recordings were obtained during well-controlled experiments, in which a dolphin had to perform a specified echolocation task, but a substantial part of tha data are from animals that were free-swimming in their tank or pool. These latter recordings thus represent spontaneous acoustic behaviour, which we assume to be echolocation behaviour. No details about the recordings or the animals are discussed here, since these were already reported elsewhere (KAMMINGA, 1979; KAMMINGA and WIERSMA, 1981; WIERSMA, 1982; KAMMINGA, WIERSMA and DUDOK VAN HEEL, 1983), or will be reported in the future (WIERSMA, diss.).

TIME DURATION AND BANDWIDTH

Although a variety of subjective classifications of the spectral width of echolocation signals can be found in the literature, these ranging from very narrow to very broad, it is probably more appropriate to refer to these kinds of pulsed sounds as short-time-duration narrow-band signals, the reason for this being the observation that echolocation signals usually have very short time durations in relation to their bandwidths. In fact, the product of duration and bandwidth, which is a dimensionless number, is usually very close to what is physically possible. This product is an important number that characterises the shape of a signal. It can, in fact, be interpreted as a measure of the number of degrees of freedom of a signal. Or, in other words, the lower it is, the fewer different waveforms exist with the same value for this product and conversely, the higher it is, the more variety is found among the set of all signals with the same product value.

The question now naturally arises whether there exist other waveforms with even smaller values for this product, and if so, how small can it get for a certain class of signals? If indeed within a certain class we can construct a sequence of waveforms with a monotonously decreasing value for the duration/bandwidth product, does this sequence then have a (unique) limit, and, if so, how does the limiting waveform compare with that of an echolocation signal of the same class? The fact that such a limit exists was already discovered in the beginning of this century; it was first demonstrated empirically and later shown to be a fundamental law of physics. It is generally referred to as the principle of uncertainty and is manifested whenever two quantities are each other's conjugate. Time and frequency are two such conjugated quantities, i.e. they are related to each other by a Fourier transformation. Uncertainty is here to be interpreted as the fundamental difficulty of measuring simultaneously the time of occurrence and the mean frequency of an echolocation signal with arbitrary accuracy. Various examples that illustrate this principle can be given in a number of fields, for example in quantum physics (position and momentum of a particle), in television technique (channel bandwidth and spot distribution on the screen), in sonar and radar technique (range and velocity ambiguity) and many more could be mentioned.

Unfortunately, the literature on uncertainty relations is not very exhaustive and the question of how small both duration and bandwidth can be for a given mean frequency and time delay has not been solved in general. (In this context by time delay we mean the interval time between the onset and the centroid of the signal's envelope, in the same way as the mean frequency is obtained from the envelope of the signal's spectrum. Duration and bandwidth are then defined as the second moment of the signal and spectrum envelopes relative to their centroids.) A special study therefore had to be made to investigate what waveform would yield the lowest possible value for the product of duration and bandwidth if the time delay and the mean frequency are specified.

METHODS

For any given signal, a scaling of the time axis corresponds with a reciprocal scaling of the frequency axis. Accordingly, quantities such as duration and bandwidth will change proportionally, the value of their product remaining the same. It is from this property that this product derives its importance. Since it is the shape of the signal that is of interest here and not so much the absolute time or frequency scale, we would like to apply a normalisation such that two signals with identical shape but different mean frequency, say, would not be considered different. This goal cannot be accomplished by fixing a certain choice for the time or frequency axis, nor by normalising all signals to the same mean frequency, nor by any other constraint in either the time or frequency domain only, if the possibility of proper waveform comparison is to be retained. The only way is to relate some signal quantity in the time domain to a corresponding quantity in the frequency domain. This can even be done such that relative time and frequency coordinates result that render signal representations in both domains perfectly symmetrical (WIERSMA, diss.). To facilitate interpretation, both the time and frequency axes are taken to be one-sided. The one-sidedness of time is a natural consequence of causality, while frequency is in itself non-negative. (Two-sided frequency spectra are often used, but this is for mathematical attractiveness only.) Consequently, the Fourier transform relationship between signal and spectrum reduces to a real-valued half-line sine transform. The signal and spectrum envelopes are obtained in the usual way, where, however, the Hilbert transform is to be replaced by the Kramers-Kronig transform (WIERSMA, diss.), because of the one-sidedness in both domains.

The question then is to find, for a given time delay and mean frequency, that waveform which attains a minimum in the product of duration and bandwidth. (This particular waveform will henceforth be referred to as the optimum waveform.) Since for this problem no closed-form solutions seem to be available, we have to rely on a numerical method. This means that we have to find the minimum of a non-linear objective funtion of inifitely many variables, which is constrained by non-linear equalities. Of course, it is computationally impossible to use the inifitely many variables that constitute the exact signal representation and, therefore, we somehow have to truncate that number. As a consequence, artificial bounds will be introduced on the possible solution space and not every pair of values for the time delay and mean frequency will yield a corresponding optimum waveform. In other words, with a limited representation accuracy, there are combinations of time delay and mean frequency for which no waveform exists. Again,

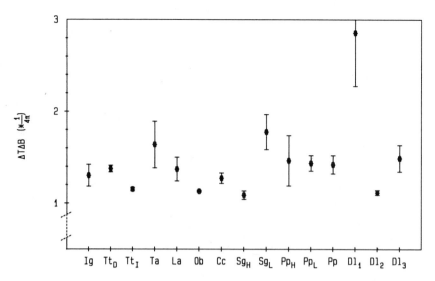

Figure 1. Time duration/bandwidth products (ΔTΔB) for
different species. Shown are averages (•)
with their standard deviations (|).

this situation is entirely brought about by the limitation in representation accuracy. The form of these bounds on the solution space can be obtained experimentally, but this is too lengthy to discuss here. The interested reader is refered to WIERSMA (diss.).

In order to solve this non-linear optimisation problem, we used a
second-order procedure for the unconstrained minimisation of a penalty
function constructed from the objective function (product of duration and
bandwidth) and the constraints (time delay, mean frequency and normalisation conditions). The mathematical and numerical aspects hereof are rather
involved and fall beyond the present scope. Details can be found in WIERSMA
(diss.). In addition, the computational burden of such a method is quite
heavy (WIERSMA, 1984). It is fair to say that these kinds of computations
have only become practical since the availability of present-day supercom-
puters.

RESULTS

In Figure 1 the values of the time duration/bandwidth product are
presented as obtained for different species. These are indicated horizontally, the abbreviations along the axis standing for the following species:

Ig	Inia geoffrensis
Tt_D and Tt_I	Two different specimens of Tursiops truncatus
Ta	Tursiops aduncus
La	Lagenorhynchus albirostris
Ob	Orcaella brevirostris
Cc	Cephalorhynchus commersonii
Sg_H	Sotalia guyanensis (HF component)
Sg_L	idem (LF component)
Pp_H	Phocoena phocoena (HF component)
Pp_L	idem (LF component)
Pp	Phocoena phocoena with only a HF component
Dl_1, Dl_2 and Dl_3	Different recordings of Delphinapterus leucas

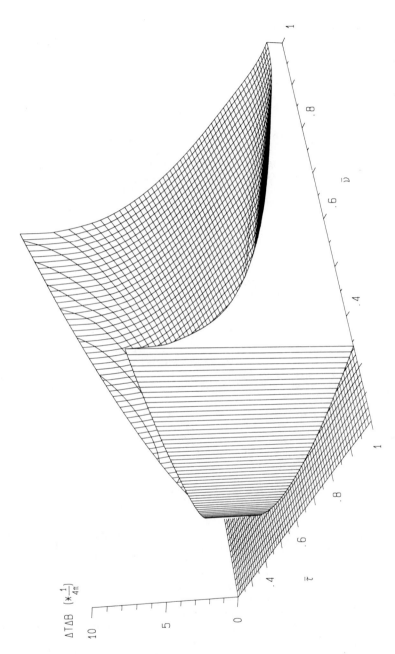

Figure 2. Minimum possible time duration/bandwidth products ($\Delta T \Delta B$) as a function of the relative mean frequency ($\bar{\nu}$) and the relative time delay ($\bar{\tau}$).

133

For each species, one or more complete pulse trains were analysed. Signals with two different frequency components were preprocessed (KAMMINGA and WIERSMA, 1981) in order to separate the high (HF) and low (LF) frequencies. These were then considered as individual signals themselves. The figure shows averages and standard deviations. The vertical units are such that the number 1 corresponds to the absolute physical lower bound that holds for the product of duration and bandwidth irrespective of what the mean frequency or time delay may be. Hence, no signal can ever have a lower value for the product than 1. It can also not have a value equal to it. No signal physically exists that can attain this lower bound, but it can be approximated as closely as desired, however, by letting at least either the time delay or the mean frequency become arbitrarily large. Although we now know that there is such a lower bound, we still do not know whether it is also actually the greatest lower bound possible.

It appears that the minimum value for the product of duration and bandwidth is a function of both the time delay and the mean frequency. For every pair of values for these we can find a waveform that has the lowest possible duration/bandwidth product among the set of all signals with the same time delay and mean frequency. For completeness we must add that the uniqueness of this waveform cannot be proven rigorously by means of iterative search methods as employed here, but that from many experiments it has become clear that such a waveform must be unique (WIERSMA, diss.).

Furthermore, because of the limited representation accuracy mentioned earlier, it is not possible to compute waveforms with either too large or too small a time delay or mean frequency. This is illustrated by Figure 2. Here the mimimum product of duration and bandwidth is plotted in a perspective view as a function of the time delay along one axis and the mean frequency along the other. Only a part of the time delay/mean frequency quarter plane is shown, that part for which the relative coordinates are less than 1 in magnitude. For the total number of representation variables chosen to compile this figure (in this case expansion coefficients relative to a suitable set of basis functions), there is a large area for small coordinates, where the product is show as zero. This, in fact, means that, in this region, no waveforms exist that are representable by the same number of variables (and using the same basis functions) and that have time delays and mean frequencies as indicated by the coordinate values. If the plotted region were to have been extended to large enough coordinate values, then we would have observed a similar effect for large time delays and mean frequencies. Theoretically, waveforms do exist everywhere in the interior of the entire quadrant plane that is bounded by the coordinate axes, but practically, it becomes exceedingly difficult to compute these for extremely small and extremely large values for the time delay and the mean frequency. But, it should be kept in mind that, by steadily decreasing the representation error (boundlessly increasing the number of variables), these zero regions can be made as small as desired.

It is recalled that the values along the coordinate axes in Figure 2 are all relative. To obtain real times and frequencies, these values should be multiplied by reciprocal numbers with, of course, reciprocal dimensions, for example, 0.1 ms and 10 kHz. The surface in Figure 2 is entirely symmetrical with respect to the imaginary line where the time delay is numerically equal to the mean frequency. This is not coincidence, since the scale normalisation was chosen deliberatedly such that time and frequency representations would become symmetrical. Had we used some other choice then the graph would have changed, thereby losing the symmetry, but it would, of course, still have represented the same information. The valley in between the steep edges becomes rather flat for coordinates larger than 1. It can be shown that the height of the surface rapidly decreases asymptotically to 1 unit on the vertical scale (representing the absolute lower bound) if

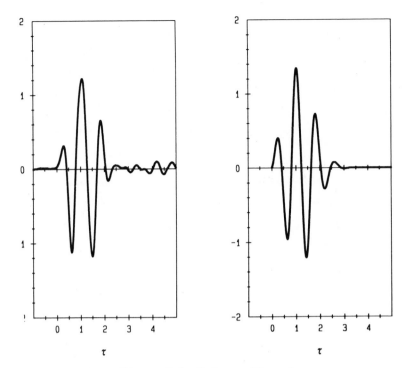

Figure 3.1. Inia geoffrensis.

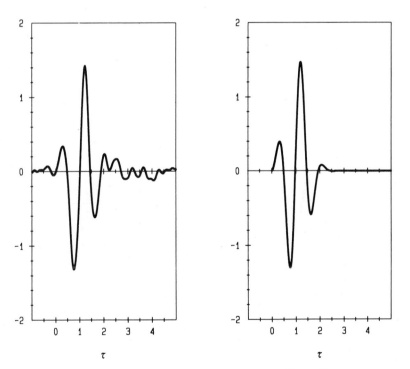

Figure 3.2. Tursiops truncatus (Doris).

Figure 3. Actual and computed echolocation signals with the same
time delay and mean frequency, plotted as a function of
relative time (τ).

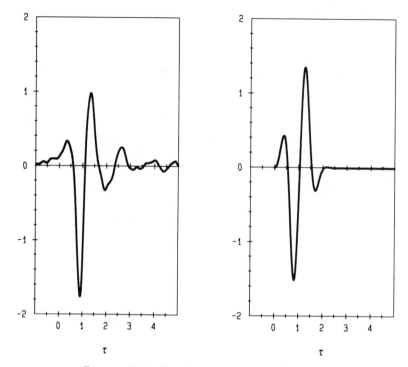

Figure 3.3. Tursiops truncatus (Ilias).

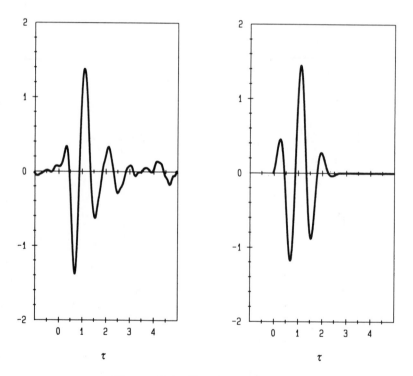

Figure 3.4. Tursiops aduncus.

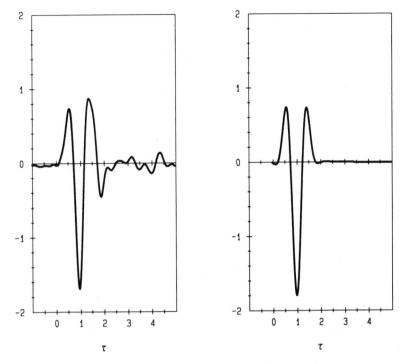

Figure 3.5. Lagenorhynchus albirostris.

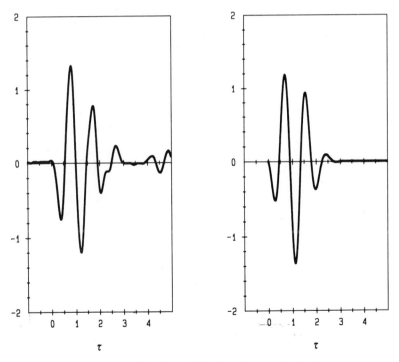

Figure 3.6. Orcaella brevirostris (a).

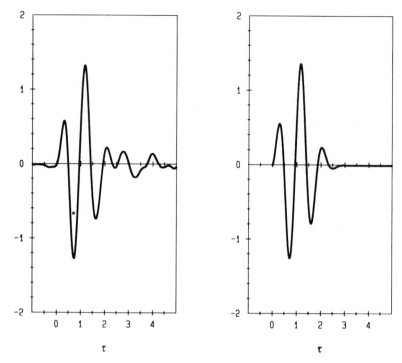

Figure 3.7. Orcaella brevirostris (b).

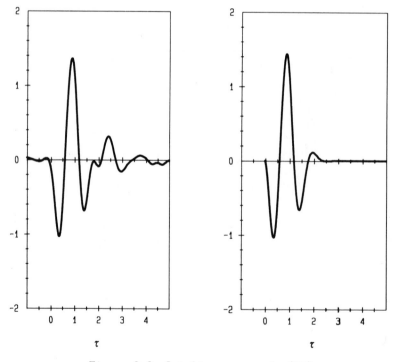

Figure 3.8. Sotalia guyanensis (HF).

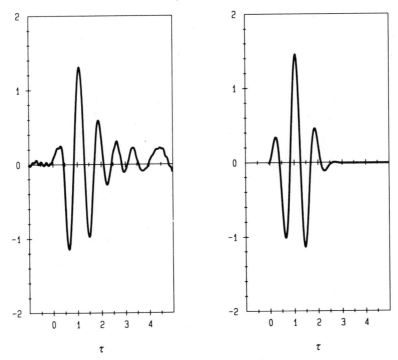

Figure 3.9. Sotalia guyanensis (LF).

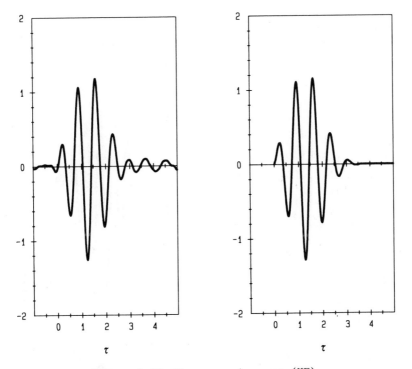

Figure 3.10. Phocoena phocoena (HF).

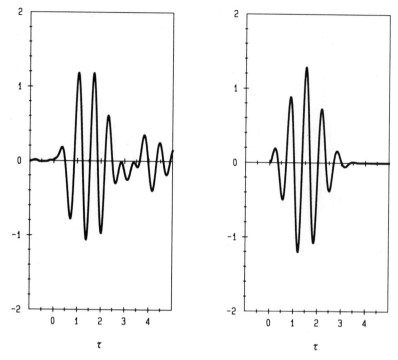

Figure 3.11. Phocoena phocoena (LF).

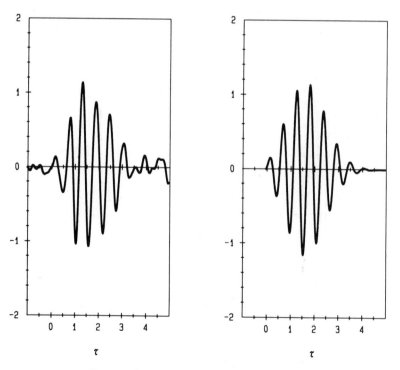

Figure 3.12. Delphinapterus leucas.

140

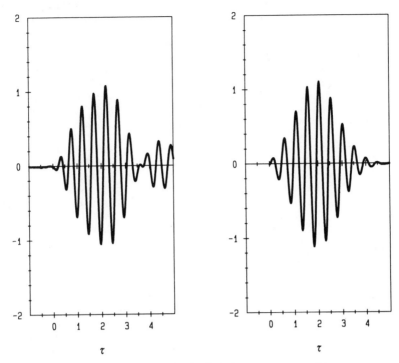

Figure 3.13. Cephalorhynchus commersonii.

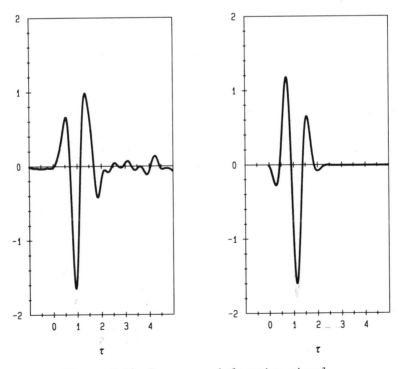

Figure 3.14. Improper echolocation signal.

either the time delay or the mean frequency becomes large. It will, however, always remain greater than and never exactly become 1 unit.

Typical examples of echolocation signals were selected for each of the species. The time delay and the mean frequency were estimated as well as possible from individual pulses within a pulse train. The main difficulty in determining the time delay is to detect the onset of the signal, which is usually submerged in the background noise. It is best estimated by backwards extrapolation of the first cycle of the signal. These two parameters, the time delay and the mean frequency, define a point in the coordinate plane of Figure 2 and, hence, uniquely determine an optimum waveform with that time delay and mean frequency. This waveform is presented in the right parts of Figures 3.1 to 3.13, together with the echolocation waveforms to which they pertain (left parts). The time scale is in relative units. The resemblance with the actual echolocation signals is apparent over the whole range of species considered. Note also the similarity between the individual HF and LF components of those species that employ bi-frequent sonar (KAMMINGA and WIERSMA, 1981; WIERSMA, 1982). Whether these two components both serve an echolocation function or not is something that can only be speculated upon, since so far no experiments are known that could answer such a question.

Figure 3.14 shows an example where no match was found between actual and computed waveform. A closer inspection of the actual signal reveals that negative half-cycle has a smaller apparent period time than the two positive half-cycles. This signal, therefore, is less sinusoidal than the others, and, consequently, has a duration/bandwidth product in excess of the theoretical minimum. This example was included only for illustration purposes and is probably not a properly recorded echolocation signal.

When all the values that were found for the time delay and the mean frequency of all the species are cross-plotted, then a remarkable relation appears between these two quantities themselves. Figure 4 shows what the result looks like. The time delay (horizontal) and mean frequency (vertical) are plotted by using a different symbol for each species or signal component. There is definitely a line-up along the axis of symmetry, which means that the numerical value for the time delay is very close to being equal to that of the mean frequency. Comparing this figure with Figure 2 shows that all the echolocation waveforms are in fact to be found right in the deepest part of the valley, where the minimum duration/bandwidth products are the smallest. Although not presented here, the waveforms that are found for other combinations of delay and mean frequency could very easily have been interpreted as echolocation sigals. For the same mean frequency, the time delay, and hence the time duration, can be much smaller, at the expense, however, of a more than proportionally increasing bandwidth. Yet no species known so far employs waveforms of this kind.

It can readily be proved that every waveform for which the time delay is equal to the mean frequency is its proper Fourier sine transform and conversely, that every signal which is its own Fourier sine transform will have equal time delays and mean frequencies. Given the relationship illustrated by Figure 4 we must conclude that, for whatever reason, the different species apparently all prefer to use waveforms that have time/frequency symmetry.

CONCLUSIONS AND DISCUSSION

As a consequence of the low time duration/bandwidth products that are used by the dolphins, the waveforms have a basic shape, a smooth envelope

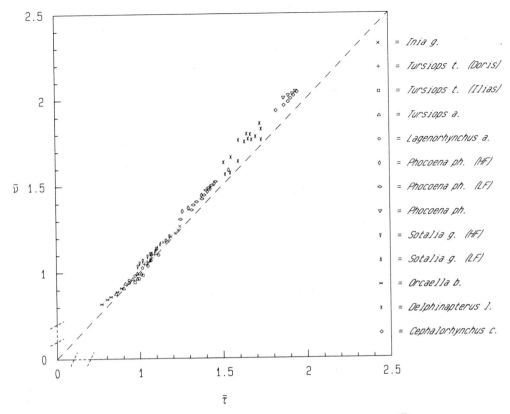

Figure 4. Mean frequencies ($\bar{\nu}$) and time delays ($\bar{\tau}$)
of echolocation waveforms for different
species.

and an almost sinusoidal structure. Sophisticated signal conceptions such as frequency-modulated waveforms are in direct conflict with this. No sign of any such modulation was ever observed while inspecting our entire data base of echolocation signals, except perhaps when out-of-phase reflections interfere with the original signal or when looking off centre in the emitted wavefield. For waveforms that contain no more than $1\frac{1}{2}$ to 2 or 3 cycles, it is very hard anyway to imagine what frequency modulation would be, even for extreme modulation depths, say well over 100%. But signals with more cycles, such as those of the Commerson's dolphin, do not exhibit what could be called frequency modulation. Figure 5 presents an analysis of a typical echolocation pulse of this type of dolphin. In the top part is shown the instantaneous frequency of the signal below, which actually consists of a main component followed by a number of reverberations that show up in the tail. The sharp peaks in the graph of the instantaneous frequency indicate the cross-over points, where one reverberation flows over into the other. A slight trough can be seen in the curve of the instantaneous frequency, where the instantaneous power in the waveform is maximum. It would be difficult, however, to interpret this as a deliberate frequency modulation that would serve a definite purpose.

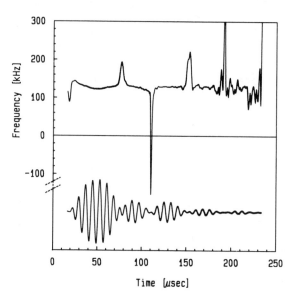

Figure 5. Echolocation signal of
Cephalorhynchus commer-
sonii. The top part shows
the instantaneous fre-
quency.

It is clear from Figures 3.1 to 3.13 that all the species considered here employ echolocation waveforms that have a very small – if not the smallest possible – extent in the combined time/frequency plane. Not only that, but these waveforms also have equal relative time delays and mean frequencies as evidenced by the data in Figure 4. What the reasons for this may be is something upon which we can only speculate. At this point we might call a few facts to the reader's attention.

Suppose that a sonar detector is to be designed that gives the best possible receiver characteristics in a general stationary and broadband background noise, i.e. noise that has a flat distribution in the combined time/frequency plane, then it can be shown that a waveform would be required that occupies the smallest possible area in this plane. This would be equivalent to requiring that the product of time duration and bandwidth be minimal. Hence, the dolphin echolocation waveforms maximise the performance of their sonar system in the absence of information a priori on the noise distribution.

The sonar waveforms could rightfully be called very elementary, considering that they have a very small number of degrees of freedom. In fact, given that they are optimum signals in the sense of having minimum duration/bandwidth products, they are entirely determined by only two parameters, the time delay and the mean frequency. As a consequence of this low intrinsic dimensionality, the structure of the sonar detector can likewise be expected to be simple. In fact, it can be shown that for such optimum waveforms a detector operating according to the square-law principle does not perform much worse than the much more sophisticated and, consequently, more complicated matched-filter detector. Conversely, if a waveform is to be designed for use with a square-law detector, then the best possible detection thresholds are achieved if and only if the duration/bandwidth products are as small as possible (URKOWITZ, 1967; WIERSMA, diss.). The smaller this product is, the smaller the difference in detection threshold is for the simplest (the square-law) and for the best (the matched-filter) detectors. Ultimately, for duration/bandwidth products that are close to the theoretical minimum, this difference in detection power becomes marginal. Therefore, dolphins would not gain very much by using matched-filter detection instead of energy detection. The latter only requires knowledge of the signal's duration and bandwidth, while for a matched filter complete knowledge of the signal waveform would be required.

The fact that over the whole range of species covered here, the relative time delays are numerically equal to the relative mean frequencies will enable a waveform classification to be based on one single 'quantum' number, the coordinate along the axis of symmetry in Figure 2. There is no reason a priori why delay and mean frequency should be equal. On the contrary, if the time duration were of primary importance to the dolphin, then he could have shortened the waveform considerably while keeping the mean frequency constant. The price that would have to be paid for this would be an increase in bandwidth that, according to Figure 2, would be more than proportional to the decrease in duration, since the surface steadily rises further away from the axis of symmetry.

ACKNOWLEDGEMENT

The author is greatly indebted to Mr. C. Kamminga not only for initiating, compiling and maintaining the data base of echolocation signals, but also for the many stimulating discussions and the pleasant cooperation. The author would also like to express his gratitude to Prof. Y. Boxma for his continuous support and for the opportunity to do this research as a member of his Group. Thanks are further owed to the Stichting Academisch Rekencentrum Amsterdam (SARA), Amsterdam, for their kind permission to use the CYBER 205 supercomputer. The author is also grateful to Mr. R. Llurba for his enthusiastic help in adapting and optimising the program code to make it run efficiently on this powerful computer. Without this patient and indefatigable electronic partner this entire research would not have been feasible.

REFERENCES

KAMMINGA, C., 1979, Remarks on dominant frequencies of Cetacean sonar, *Aq. Mamm.*, vol. 7, no. 3, pp. 93-100.
KAMMINGA, C., WIERSMA, H., 1981, Investigations on Cetacean sonar II; Acoustical similarities and differences in Odontocete sonar signals, *Aq. Mamm.*, vol. 8, no. 2, pp. 41-60.
KAMMINGA, C., WIERSMA, H., DUDOK VAN HEEL, W.H., 1983, Investigations on Cetacean sonar VI. Sonar in Orcaella brevirostris of the Mahakam River, East Kalimantan, Indonesia; First descriptions of acoustic behaviour, *Aq. Mamm.*, vol. 10, no. 3, pp. 83-95.
URKOWITZ, H., 1967, Energy detection of unknown deterministic signals, *Proc. I.E.E.E.*, vol. 55, no. 4, pp. 523-531.
WIERSMA, H., 1982, Investigations on Cetacean sonar IV. A comparison of wave shapes of Odontocete sonar signals, *Aq. Mamm.*, vol. 9, no. 2, pp. 57-66.
WIERSMA, H., 1984, Optimum wave forms in sonar of whales and dolphins, *Supercomputer*, vol. 4, pp. 22-30.
WIERSMA, H., Ph. D. Dissertation, Delft University of Technology, Delft, to appear.

SECTION II

AUDITORY SYSTEMS OF ECHOLOCATING ANIMALS

Section Organizer: Patrick W.B. Moore

ACOUSTICAL ASPECTS OF HEARING AND ECHOLOCATION IN BATS
   Anna Guppy and Roger B. Coles

PERCEPTION OF COMPLEX ECHOES BY AN ECHOLOCATING DOLPHIN
   Whitlow W. L. Au and Patrick W. B. Moore

MORPHOMETRIC ANALYSIS OF COCHLEAR STRUCTURES IN THE MUSTACHED BAT
Pteronotus parnellii parnellii
   O. W. Henson, Jr. and M. M. Henson

THE EFFERENT AUDITORY SYSTEM IN DOPPLER-SHIFT COMPENSATING BATS
   Allen L. Bishop and O. W. Henson, Jr.

SOME COMPARATIVE ASPECTS OF AUDITORY BRAINSTEM CYTOARCHITECTURE IN
ECHOLOCATING MAMMALS: SPECULATIONS ON THE MORPHOLOGICAL BASIS OF TIME-
DOMAIN SIGNAL PROCESSING
   John M. Zook, Myron S. Jacobs, Ilya Glezer and Peter J. Morgane

TEMPORAL ORDER DISCRIMINATION WITHIN THE DOLPHIN CRITICAL INTERVAL
   Richard A. Johnson, Patrick W. B. Moore, Mark W. Stoermer, Jeffrey L.
Pawloski and Leslie C. Anderson

DETECTION ABILITIES AND SIGNAL CHARACTERISTICS OF ECHOLOCATING FALSE KILLER
WHALES (Pseudorca crassidens)
   Jeanette Thomas, Mark Stoermer, Clark Bowers, Les Anderson, and Alan
Garver

BINAURAL NEURONS IN THE MUSTACHE BAT'S INFERIOR COLLICULUS: PHYSIOLOGY,
FUNCTIONAL ORGANIZATION, AND BEHAVIORAL IMPLICATIONS
   Jeffrey J. Wenstrup

ONTOGENY OF THE ECHOLOCATION SYSTEM IN RHINOLOPHOID CF-FM BATS: AUDITION
AND VOCALIZATION IN EARLY POSTNATAL DEVELOPMENT
   Rudolf Rubsamen

LIGHTMICROSCOPIC OBSERVATION OF COCHLEAR DEVELOPMENT IN HORSESHOE BATS
(Rhinolophus rouxi)
   Marianne Vater

ONLY ONE NUCLEUS IN THE BRAINSTEM PROJECTS TO THE COCHLEA IN HORSESHOE
BATS: THE NUCLEUS OLIVO-COCHLEARIS
   Joachim Ostwald and Andreas Aschoff

PARALLEL-HIERARCHICAL PROCESSING OF

BIOSONAR INFORMATION IN THE MUSTACHED BAT

Nobuo Suga

Department of Biology
Washington University
St. Louis, Missouri

INTRODUCTION

Neuroethologists working with bats may read original articles in their field, but may not necessarily get an overall view of the neural processing of complex biosonar signals. Researchers outside of bat neuroethology may have little time to read original articles and may hardly get an integrated view of the neural mechanisms of bat echolocation. Therefore, in this article, I review the present state of our understanding of the neural mechanisms for processing complex biosonar signals in a schematized way. Most of the description in this article are based upon a large amount of neuroethological data, but some require further studies. To the readers who are interested in this topic and want to see the "key" data, I recommend reading my review article entitled "The Extent to Which Biosonar Information is Represented in the Bat Auditory Cortex" (Suga, 1984).

BIOSONAR SIGNALS

For prey (flying insects) capture and orientation, the mustached bat Pteronotus parnellii from Panama and Jamaica emits orientation sounds (biosonar signals or pulses), each of which consists of a long constant-frequency (CF) component followed by a short frequency-modulated (FM) component. Since each orientation sound contains four harmonics (H_{1-4}), there are eight components that can be defined (CF_{1-4}, FM_{1-4}). In the emitted sound, the second harmonic (H_2) is always predominant and the frequency of CF_2 is about 61 kHz. The frequency of the CF component is different among subspecies and also among individuals of the same subspecies to some extent (Suga and Tsuzuki, 1985). It is also different between males and females (Suga et al., 1986). In FM_2, frequency sweeps down from 61 kHz to about 49 kHz (Fig. 1A). H_3 is 6-12dB weaker than H_2, while H_1 and H_4 are 18-36 and 12-24 dB weaker than H_2, respectively. Many species of moths are most sensitive to frequencies between 20 and 40 kHz (Suga, 1961; Fenton and Fullard, 1979). They show evasive behaviors when they hear the orientation sounds of bats (Roeder, 1962). Since H_1 (25-31 kHz) in the orientation sound of the mustached bat is suppressed, probably by antiresonance of the vocal tract, the bat can approach these moths closely before being detected (Suga and O'Neill, 1979).

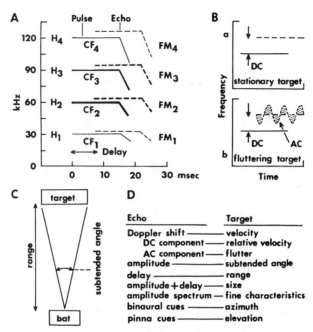

Fig. 1. Biosonar signals (orientation sounds) of the mustached bat,
Pteronotus parnellii, and the information carried by its
signals. A: Schematized sonagram of the orientation sound
(solid lines) and the Doppler-shifted echo (dashed lines).
The orientation sound is also called a pulse. The four
harmonics (H_{1-4}) of both the orientation sound and the echo
each contain a long CF component (CF_{1-4}) and a short FM
component (FM_{1-4}). Thickness of the lines indicates the
relative amplitude of each harmonic in the orientation sound.
H_2 is the strongest, followed by H_3, H_4, and H_1.
B: When the mustached bat flies toward or near a stationary
object, the frequency of the echo becomes higher than the
emitted sound by the Doppler effect (graph a). This steady
shift is called the DC component of the Doppler shift. When
the bat flies toward a fluttering target, for example, a
flying moth, the Doppler shift of the echo consists of a DC
component proportional to relative velocity and the periodic
frequency modulation (FM) proportional to the speed of wing
beat (graph b). This periodic FM is called the AC component
of the Doppler shift. The AC component is complicated because
the insect's four wings are moving in complex patterns and in
different phase relationships relative to the bat. The echo
from the fluttering target is also modulated in amplitude.
C: Target size is determined from both target range and
subtended angle.
D: Relationship between echo properties and target
properties. (Suga et al., 1983b).

150

Echoes eliciting behavioral responses in the mustached bat always overlap temporarily with the emitted sound. As a result, biosonar information must be extracted from a complex sound containing up to 16 components. The CF component is an ideal signal for target detection and the measurement of target velocity (relative movements and wing beats) because the reflected sound energy is highly concentrated at a particular frequency. The mustached bat uses the CF component for this purpose and performs a unique behavior called Doppler-shift compensation (Schnitzler, 1970; Henson et al., 1982). The short FM component, on the other hand, is suited for ranging, localizing and characterizing a target because of the distribution of its energy over many frequencies. Different parameters of echoes received by the bat carry different types of information about a target (Fig. 1D).

During target-directed flight, the duration of the orientation sound shortens from 30 to 7 msec and the emission rate increases from 5/sec to 100/sec. The shortening of the duration of the sound is mainly due to shortening of the CF component. The increase in emission rate increases the information carried by the FM component. Echo amplitude and delay from the emitted sound also change systematically during target-directed flight. The bat compensates the amplitude of emitted sounds to stabilize the amplitude of echoes (Kobler et al., 1985).

PARALLEL-HIERARCHICAL PROCESSING OF COMPLEX SOUNDS

The eight components (CF_{1-4} and FM_{1-4}) of the orientation sound of the mustached bat are all different from each other in frequency, so that they are analyzed in parallel at different regions of the basilar membrane (Fig. 2, bottom). Then the signals are coded and are sent into the brain by peripheral neurons. In the brain, the signals are sent up to the auditory cortex through many auditory nuclei. For simplicity, we may consider that there are 8 channels for processing of these signal elements: CF_1 channel, CF_2 channel, and so on. The CF_2 channel is very big compared with any other channel and is associated with extraordinarily sharply tuned local resonator in the cochlea for fine frequency analysis (Fig. 2). Therefore, the frequency-tuning curves of auditory nerve fibers tuned to 61 kHz, CF_2 frequency, have a quality factor of 210 and the slope of 1500-to-1900 dB/octave. Because of this extremely sharp tuning, they can code a 6 Hz frequency shift from 61 kHz. That is, they can easily code small frequency modulation which would be evoked by the wings of flying insects (Suga and Jen, 1977). The CF_4 channel is very small, if it exists at all.

In the CF_1, CF_2 and CF_3 channels (Fig. 2), frequency selectivity is increased and amplitude selectivity is added by inhibition to some neurons in the cochlear nucleus (Suga et al., 1975) and also to many neurons in higher levels (Suga and Manabe, 1982; O'Neill, 1985; Olsen, 1986). These neurons are thus tuned to particular frequencies and amplitudes. The extent to which neural sharpening occurs is different among groups of neurons tuned to different frequencies. The sharpening is most dramatic in the CF_2 channel. In a certain region of the medial geniculate body, a part of the CF_1 channel and a part of CF_2 or CF_3 channel are integrated, so that neurons in this region poorly respond to CF_1, CF_2 and CF_3 tones when delivered alone, but respond strongly when the CF_1 tone is delivered together with the CF_2 or CF_3 tone. A deviation of the best CF_2 or CF_3 frequency from the exact harmonic relationship with the best CF_1 frequency, i.e., an amount of Doppler shift is a critical parameter for their excitation. These CF/CF combination-sensitive neurons project to the CF/CF area of the auditory cortex (Olsen and Suga, 1983; Olsen, 1986). In the CF/CF area, two types of CF/CF neurons, CF_1/CF_2 and CF_1/CF_3, are separately clustered and form the frequency-versus-frequency coordinates in each cluster for

Parallel-Hierarchical Signal Processing (Tentative Scheme)

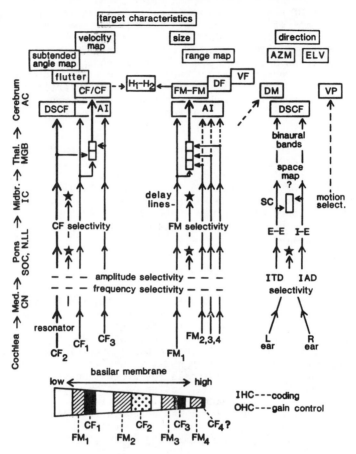

Fig. 2. Parallel-hierarchical processing of different types of biosonar
information carried by complex biosonar signals. The CF_{1-4} and
FM_{1-4} of the orientation sound are analyzed at different
portions of the basilar membrane in the cochlea (bottom). Inner
and outer hair cells (IHC and OHC) on the membrane are
respectively related to stimulus coding and gain control. These
signal elements are separately sent up to the auditory cortex
(AC) through several auditory nuclei (left margin): cochlear
nucleus (CN), superior olivary complex (SOC), nucleus of lateral
lemmiscus (N.LL), inferior colliculus (IC), and medial
geniculate body (MGB). During the ascent of the signals,
frequency, amplitude, CF and FM selectivities are added to some
neurons (arrows with a star). Each star indicates that the
addition of selectivity also takes place in the auditory nuclei
and cortex as well as in the nucleus where the arrow starts.
The CF_2 channel is disproportionately large and projects to the
DSCF (Doppler-shifted CF processing) area of the auditory
cortex. In certain portions of the MGB, two channnels
processing different signal elements (e.g., CF_1 and CF_2 or FM_1
and FM_2 channels) are integrated to produce "combination-
sensitive" neurons. CF/CF and FM-FM combination-sensitive
neurons respectively project to the CF/CF and FM-FM areas of the
auditory cortex, where target velocity or range information

the representation of Doppler shifts, i.e., target velocity information (Fig. 3; Suga et al., 1981, 1983a). CF/CF neurons show sharp "level-tolerant" frequency-tuning curves and are remarkably specialized to respond to particular frequency relationships of two CF tones (Suga and Tsuzuki, 1985). The signal processing in the CF channels is thus "parallel-hierarchical".

In the FM_1, FM_2, FM_3 and FM_4 channels (Fig. 2), frequency selectivity is increased and amplitude selectivity is added by inhibition to some neurons. Interestingly, FM selectivity is also added to some neurons by disinhibition, so that these "FM-specialized" neurons respond to FM sounds, but not to CF tones and noise bursts (Suga, 1965, 1969; O'Neill, 1985). In a certain region of the medial geniculate body, a part of the FM_1 channel and a part of FM_2 or FM_3 or FM_4 channels are integrated, so that neurons in this region poorly respond to these FM sounds when delivered alone, but respond strongly to the FM_1 sound combined with the FM_2 or FM_3 or FM_4 sound. The delay of the FM_2 or FM_3 or FM_4 from the FM_1, i.e., echo delay is a critical parameter for their facilitative responses. These FM-FM combination-sensitive neurons project to the FM-FM area of the auditory cortex (Olsen and Suga, 1983; Olsen, 1986). In the FM-FM area, three types of FM-FM neurons, FM_1-FM_2, FM_1-FM_3 and FM_1-FM_4, are separately clustered and form a time (echo-delay) axis in each cluster for the representation of target-range information (Fig. 3; Suga and O'Neill, 1979; O'Neill and Suga, 1982; Suga and Horikawa, 1986). Therefore, the signal processing in the FM channels is also parallel-hierarchical.

As described above, a part of one channel is integrated by a part of the other channel in the medial geniculate body. The remaining parts of these channels, which are not integrated, project to the auditory cortex which is not described above. For instance, a part of the CF_2 channel projects to the DSCF (Doppler-shifted CF processing) area of the auditory cortex which has the frequency-versus-amplitude coordinates to represent target-velocity information and subtended-target-angle information (Suga, 1977; Suga and Manabe, 1982). The DSCF area over-represents frequencies between the CF_2 resting frequency of the bat's own sound and 1.0 kHz above it. The CF_2 resting frequency differs among individual bats within a range between 59.7 and 63.5 kHz. In each bat, there is a unique match between the tonotopic representation and the CF_2 resting frequency of the bat's own sound. Since the orientation sound is sexually dimorphic, the tonotopic representation is also sexually dimorphic (Suga et al., 1986). The DSCF area can be divided into two subdivisions which predominantly contain either I-E or E-E neurons (Fig. 3; Manabe et al., 1978). About a half of DSCF neurons can clearly represent the AC component of a Doppler-shift through phase-locked discharges, but the other half cannot. In the CF/CF area, one-third of neurons can represent the AC component, but the remaining two-thirds cannot (Suga et al., 1983b). Therefore, there is a possibility that some of the channels described

(Fig. 2 continued.)
is systematically represented. Because of cortico-cortical connections, DF, VF and VA areas also consist of combination-sensitive neurons (center top). Target velocity and range information is thus processed in a parallel-hierarchical manner. The DSCF area has the frequency-versus-amplitude coordinates to represent velocity and subtended angle information of a target. The DSCF area consists of two subdivisions mainly containing I-E or E-E neurons (right column). Motion-sensitive neurons appear to be in the ventroposterior (VP) area of the auditory cortex (Suga, 1986).

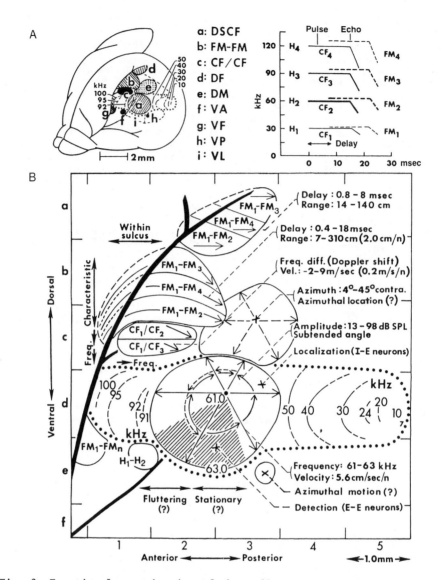

Fig. 3. Functional organization of the auditory cortex of the
mustached bat. A: Dorsolateral view of the left cerebral
hemisphere. The auditory cortex consists of several areas (a-
i). DSCF, FM-FM, CF/CF, DF, and DM areas (a,b,c,d, and e,
respectively) are specialized for the systematic
representation of biosonar information. The branches of the
median cerebral artery are shown by the branching lines. The
longest branch is on the sulcus. B: Graphic summary of the
functional organization of the auditory cortex. The tonotopic
representation of the primary auditory cortex and the
functional organization of the DSCF, FM-FM, CF/CF, DF, and DM
areas are indicated by lines and arrows. The DSCF area has
axes representing either target velocity (echo frequency: 61-
63 kHz) or subtended target angle (echo amplitude: 13-98 dB
SPL) and is divided into two subdivisions suitable for either
target detection

above consist of subchannels in terms of processing of another type of auditory information. Fig. 2 is only to explain the parallel-hierarchical processing of biosonar information which has been thus far explored.

Almost all frequencies found in the biosonar signals are projected not only to the areas which appear to be important to echolocation, but also to the areas which appear not to be important to echolocation. These areas are probably important to processing communication sounds. Except for the CF_2 channel which is specialized for processing biosonar information from the periphery through the auditory cortex, clear separation of biosonar-signal processing from non-biosonar-signal processing appears to take place first in the medial geniculate body.

The auditory cortex of the mustached bat is 0.9 mm thick and is about 14.2 mm^2 large, which is very large relative to the size of its brain (Fig. 3). The auditory cortex shows multiple cochleotopic (tonotopic) representation which is directly related to representation of different types of biosonar information. Figure 3 shows several functional areas explored electrophysiologically. In these areas, certain response properties of single neurons arranged orthogonally to the cortical surface are identical. In this sense, there is a columnar organization. Along the cortical surface, however, the response properties vary systematically and form axes for representation of particular types of biosonar information, as described above to some extent. Among the several functional areas, the CF/CF, FM-FM, DF, VF and VA areas consist of combination-sensitive neurons, so that these areas are particularly interesting in terms of neural mechanisms for processing complex sounds. [For further information of the auditory cortex of the mustached bat, see Suga (1984).]

The auditory cortex of the mustached bat consists of six layers and is expected to receive afferent fibers and to send out efferent fibers,

---

(Fig. 3 continued.)
(shaded) or target localization (unshaded). These subdivisions are occupied mainly by excitatory-excitatory (E-E) or inhibitory-excitatory (I-E) neurons, respectively. The FM-FM area consists of three major types of FM-FM combination-sensitive neurons (FM_1-FM_2, FM_1-FM_3 and FM_1-FM_4), which form separate clusters. Each cluster has an axis representing target ranges from 7 to 310 cm (echo delay: 0.4-18 msec). The dorsoventral axis of the FM-FM area probably represents fine target characteristics. The CF/CF area consists of two major types of CF/CF combination-sensitive neurons (CF_1/CF_2 and CF_1/CF_3), which aggregate in independent clusters. Each cluster has two frequency axes and represents target velocities from -2 to +9 m/sec (echo Doppler shift: -0.7 to +3.2 kHz for CF_2 and -1.1 to +4.8 kHz for CF_3). The DF area and a posterior part of the VA area receive nerve fibers from the FM-FM area. The DF area consists of the three types of FM-FM neurons, but the VA area contains only H_1-H_2 combination-sensitive neurons. The DF area projects to the VF area, which consists of the three types of FM-FM neurons. The DM area appears to have an azimuthal axis representing the azimuthal location of the target. In the VP area, motion-sensitive neurons have been found. The functional organization of the VF, VA and VP areas remains to be studied further (Suga, 1986).

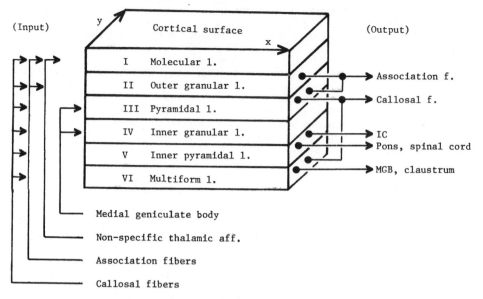

Fig. 4. The auditory cortex consists of six layers. It receives
afferent fibers (arrows on the left) from four different
regions of the brain for signal processing and sends out
efferent fibers (arrows on the right) to several different
regions. IC: inferior collicus; MGB: medial geniculate
body.

as that of cats and monkeys does (Fig. 4). Neurons located in the
superficial layers send out association fibers to communicate with
cerebral cortical areas on the same side or callosal fibers to
communicate with those on the other side. On the other hand, neurons
located in the deep layers send out descending fibers to subcortical
auditory nuclei or the motor system. We now know the flow of biosonar
information from the primary auditory cortex to other regions of the
brain to some extent, but we do not yet know about the signal processing
which may occur within each cortical column, consisting of these six
layers.

The FM-FM area representing target ranges from 7 cm to 310 cm
projects to the DF and VA areas of the cerebrum as well as other regions
of the brain (Fig. 3). The DF area consists of three clusters of FM-FM
neurons. In each cluster, a range (echo-delay) axis is formed, along
which target ranges from 7 cm to 140 cm are systematically represented
(Suga and Horikawa, 1986). The DF area thus represents only the shorter
half of the ranges represented in the FM-FM area. The DF area projects
to the VF area as well as other areas in the cerebrum, to which the FM-
FM area does not project. The VF area also consists of three clusters
of FM-FM neurons and appears to represent target ranges between 7 and 80
cm, the shorter half of the ranges represented in the DF area. We do
not yet know the functional significance of these multiple time axes.
One may hypothesize that these three different areas are related to
echolocation behavior at different distances to targets. The H_1-H_2
area, a part of the VA area contains combination-sensitive neurons which
are different from FM-FM and CF/CF neurons. They show facilitative
responses to the CF_2 and/or FM_2 of an echo when these are combined with
the CF_1 and/or FM_1 of the orientation sound.

Auditory information is sent not only to the association cortex from the auditory cortex, but also to the motor system. Both the FM-FM and CF/CF areas project to the pontine motor nuclei which in turn project to the cerebellum. In the cerebellar vermis, there are tiny clusters of FM-FM and CF-CF neurons. In addition to these, there is a cluster of noise-burst sensitive neurons. Different cerebellar lobules represent different harmonics of the orientation sound (Horikawa and Suga, 1986). There is no systematic representation of frequency in each lobule (Sun et al., 1983). Biosonar information is also sent to the vocal system. Some neurons in the periaqueductal gray and midbrain reticular formation, for instance, become active prior to vocalization and respond to acoustic stimuli delivered from a loudspeaker.

The projections of the CF/CF area thus far studied do not overlap with those of the FM-FM area. If overlapped, however, it may be in the H_1-H_2 area. All the data thus far obtained indicate that complex-acoustic signals are processed in a parallel-hierarchical way in the ascending auditory system and beyond the auditory cortex.

IMPORTANT PRINCIPLES FOR THE PROCESSING OF BIOLOGICALLY IMPORTANT SOUNDS

The data obtained from the auditory system of the mustached bat indicate not only the specialization of the bat's auditory system for echolocation, but also the neural mechanisms that are shared or probably shared with different types of animals. In the following, these mechanisms are listed as principles for the processing of biologically important sounds.

(1) The peripheral auditory system has evolved not only for the detection of biologically important sounds, but also for frequency analysis of the sounds to fulfil species-specific requirements. Therefore, the quality factor of a frequency-tuning curve can be higher for peripheral neurons tuned to frequencies of sounds most important to the species.

(2) Frequency tuning of central neurons can be sharpened by lateral inhibition which eliminates mainly the "skirt" of a frequency tuning curve. The more important the frequency analysis of a particular component of sounds, the more pronounced the neural sharpening for neurons tuned to that component.

(3) The central auditory system can create neural filters tuned to information-bearing parameters (IBPs) other than frequency. These IBP filters can also be sharpened by lateral inhibition. The IBP filters act as cross-correlators which correlate incoming signals with their filter properties: stored information.

(4) Complex sounds can be processed by IBP filters tuned to different combinations of signal elements.

(5) Different types of IBP filters are aggregated separately at particular locations of the central auditory system. In other words, the system contains functional subdivisions or areas specialized for processing particular types of auditory information essential to a species.

(6) In each subdivision or area, IBP filters are arranged along an axis or axes for the systematic representation of IBPs, that is, signal variation which has biological importance.

(7) The axis--population of neurons--representing an IBP is apportioned to the biological importance of the IBP.

(8) The functional organization of the auditory system can be quite different among different species, reflecting differences in the properties of the acoustic signals used by them and/or species-specific auditory behavior. Organization can also be different among individuals within the same species when the properties of their biologically important acoustic signals are different among the individuals.

SUMMARY

For the processing of different types of biosonar information carried by pulse-echo pairs, the auditory system "creates" arrays of neurons tuned to different values of information-bearing parameters (IBPs) in a parallel-hierarchical way. In the auditory cortex, these specialized neurons are systematically arranged to represent IBPs by locations of activated neurons. The different types of biosonar information represented separately in the auditory cortex are sent to other regions of the brain, including the motor system. The signal processing going on beyond the auditory cortex appears to be parallel-hierarchical.

ACKNOWEDGEMENT

This work on the auditory system of the mustached bat has been supported by a research grant from the U.S. Public Health Service, R01-NS17333 Javits neurosci. invest. award.

REFERENCE

Fenton, M. B., and Fullard, J. H., 1979, The influence of moth hearing on bat echolocation strategies, J. Comp. Physiol., 132: 77-86.

Henson, O. W. Jr., Henson, M. M., Kobler, J. B., and Pollack, G. D., 1980, The constant frequency component of the biosonar signals of the bat, Pteronotus parnellii, in: "Animal Sonar Systems," 913-916, Busnell, R. G., and Fish, J. F., eds., Plenum Press, New York.

Horikawa, J., and Suga, N., 1986, Biosonar signals and cerebellar auditory neurons of the mustached bat, J. Neurophysiol. 55:1247-1267.

Kobler, J. B., Wilson, B. S., Henson, O. W. Jr., and Bishop, A. L., 1985, Echo intensity compensation by echolocating bats, Hear Res., 20:99-108.

Manabe, T., Suga, N., and Ostwald, J., 1978, Aural representation in the Doppler-shifted-CF processing area of the primary auditory cortex of the mustached bat, Science 200:339-342.

Olsen, J. F., 1986, Functional organization of the medial geniculate body of the mustached bat, Ph.D. thesis of Washington Univ.

Olsen, J. F., and Suga, N., 1983, Combination-sensitive neurons in the auditory thalamus of the mustached bat, Ann. meet. of Soc. Neurosci. Abst., P.768.

O'Neill, W. E., 1985, Responses to pure tones and linear FM components of the CF-FM biosonar signal by single units in the inferior colliculus of the mustached bat, J. Comp. Physiol. 157:797-815.

O'Neill, W. E., and Suga, N., 1982, Encoding of target-range information and its representation in the auditory cortex of the mustached bat. J. Neurosci., 47:225-255.

Roeder, K. D., 1962, The behavior of free flying moths in the presence of artificial ultrasonic pulses, Animal Beh., 10:300-304.

Schnitzler, H. U., 1970, Echoortung bei der Fledermaus Chilonycteris rubiginosa., Z. vergl. Physiol., 68:25-38.

Suga, N., 1961, Functional organization of two tympanic neurons in Noctuid moths, Jap. J. Physiol., 11:666-677.

Suga, N., 1965, Analysis of frequency modulated sounds by neurons of echolocating bats, J. Physiol. (Lond.), 179:26-53.

Suga, N., 1969, Classification of inferior collicular neurons of bats in terms of responses to pure tones, FM sounds and noise bursts, J. Physiol., 200:555-574.

Suga, N., 1977, Amplitude-spectrum representation in the Doppler-shifted-CF processing area of the auditory cortex of the mustached bat, Science, 196:64-67.

Suga, N., 1984, The extent to which biosonar information is represented in the bat auditory cortex, in: "Dynamic Aspects of Neucortical Function," Edelman, G. M., Gall, W. E., and Cowan, W. M., eds., John Wiley & Sons, New York, 315-373.

Suga, N., 1986, Auditory neuroethology and speech processing: complex sound processing by combination-sensitive neurons, in: "Functions of the Auditory System," Edelman, G. M., Gall, W. E., and Cowan, W. M., eds., John Wiley & Sons. New York (in press).

Suga, N., and Horikawa, J., 1986, Multiple time axes for representation of echo delays in the auditory cortex of the mustached bat, J. Neurophysiol., 55:776-805.

Suga, N., and Jen, P. H. S., 1977, Further studies on the peripheral auditory system of "CF-FM" bats specialized for fine frequency analysis of Doppler-shifted echoes, J. Exp. Biol,. 69:207-232.

Suga, N., Kujirai, K., and O'Neill, W. E., 1981, How biosonar information is represented in the bat cerebral cortex, in: "Neuronal Mechanisms of Hearing," Syka, J., and Aitkin, L., eds., Plenum Press, New York, 197-219.

Suga, N., and Manabe, T., 1982, Neural basis of amplitude-spectrum representation in the auditory cortex of the mustached bat, J. Neurophysiol., 47:225-255.

Suga, N., Niwa, H., and Tanioguchi, I., 1983b, Representation of biosonar information in the auditory cortex of the mustached bat, with emphasis on representation of target velocity information, in: "Advances in Vertebrate Neuroethology," Ewert, P., ed., Plenum Press, New York, 829-867.

Suga, N., Niwa, H., Taniguchi, I., and Margoliash, D., 1986, The personalized auditory cortex of the mustached bat: adaptation for echolocation (in preparation).

Suga, N., and O'Neill, W. E., 1979, Neural axis representing target range in the auditory cortex of the mustached bat, Science, 206:351-353.

Suga, N., O'Neill, W. E., Kujirai, K., and Manabe, T., 1983a, Specialization of "combination-sensitive" neurons for processing of complex biosonar signals in the auditory cortex of the mustached bat, J. Neurophysiol., 49:1573-1626.

Suga, N., Simmons, J. A., and Jen, P. H. S., 1975, Peripheral specialization for fine analysis of Doppler-shifted echoes in "CF-FM" bat Pteronotus parnellii, J. Exp. Biol., 63:161-192.

Suga, N., and Tsuzuki, K., 1985, Inhibition and level-tolerant frequency tuning in the auditory cortex of the mustached bat, J. Neurophysiol., 53:1109-1145.

Sun, W., Jen, P. H. S., and Kamada,T., 1983, Mapping of the auditory area in the cerebellar vermis and hemispheres of the mustached bat, Pteronotus parnellii parnellii, Brain Res., 271:162-165.

DOLPHIN ECHOLOCATION AND AUDITION

Patrick W.B. Moore

Naval Ocean Systems Center
P.O. Box 997, Kailua, HI 96734

INTRODUCTION

Research on the auditory and echolocation performance of Tursiops truncatus since 1980 has proceeded slowly due to limited resources and the expense of conducting basic psychoacoustic research as compared to more general studies concerned with the natural history of the species. Echolocation is essentially a special extension and adaptation of the marine mammal hearing system coupled with an ability to generate sounds. Humans have the ability to judge room size based on reverberation from their own voice and some blind people use self generated sounds to detect reflective objects (Rice, 1966). Echolocation can be thought of as representing a highly refined acoustic ability on a broad acoustic sensory continuum.

Following the second Animal Sonar Symposium (Busnel and Fish, 1980), research began to more fully characterize the dolphin's hearing and echolocation detection capabilities. The basic state of affairs for any underwater sonar system, be it man-made or biological, is the requirement to detect a signal, usually an echo, in the surrounding noisy sea. For the dolphin this is accomplished in part by extensively adapted auditory neural systems about which the functional capabilities are not yet fully understood.

This paper reviews psychoacoustic data on Tursiops from experiments which have been conducted or are currently ongoing since the second Animal Sonar Symposium (Busnel and Fish, 1980).

CRITICAL INTERVAL

Vel'min and Dubrovskiy (1975, 1976) described a constant time interval (approximately 300 μsec) required by dolphins to perceive echo events. Johnson (1980, p517) discussed this concept at the Jersey meeting and asked "what is happening here?". His point was that the shortest pulse-echo interval available to the dolphin would be about 500 μsec (round trip time from pulse source to tip of the rostrum and back to the ear bones) and therefore any range determination would not be functional for such short intervals. However, if one considers the critical interval to be a function of the dolphin auditory system that is concerned with echo feature extraction and not related to range determination, then a 300 μsec time interval may represent a within-echo information extraction mechanism.

Moore et al. (1984) reviewed subsequent critical interval work and investigated the critical interval concept using a backward masking experiment performed in Kaneohe Bay, Hawaii. The problem was to determine if the 250 to 300 μsec critical interval would hold in backward masking during an active echolocation task. A backward masking function relating target detection to masked delay was determined for an actively echolocating dolphin in a target detection task. The masking noise was triggered by each outgoing echolocation click and was temporally adjusted between coincidence with the target echo to delays of up to 700 μsec. Indeed, the results indicated a backward masking threshold with a delay of 265 μsec which supported the Soviet notion that the 250 to 300 μsec delay was a "temporal resolving interval" in dolphin echolocation (Bel'kovich and Dubrovskiy, 1976). We pointed to time separation pitch (TSP), an old notion (Nordmark,1961, Pye,1967, R.A. Johnson, 1972, Floyd, 1980, C.S. Johnson 1980, Altes,1980, R.A. Johnson 1980, Diercks, 1980 and Hammer et al. 1980) as a possible explanation for certain divergent results (Vel'min and Dubrovskiy, 1975, 1976) reported and discussed by C.S. Johnson (1980) with TSP as a possible analysis mechanism for within-echo feature extraction. Updated information concerning this concept is discussed by Au (this volume) and R.A. Johnson (this volume).

Since 1980 echolocation research has been facilitated by microprocessor-based tools for echolocation signal acquisition and analysis. This has allowed greater access to basic questions concerning the acoustic behavior of actively echolocating dolphins. Au and Penner (1981) reported on the acoustic behavior of two dolphins in a masked echolocation detection task with five levels of masking noise presented in random 10-trial blocks. The study used an echolocation click detector to study the average number of clicks per trial, average response latency and the click interval distribution. Results showed that the number of clicks increased with increasing masking noise level until the noise prevented detection altogether. Beyond the masking point (82 - 87 dB re:1 μPa - spectrum level) the number of click emissions decreased and one animal even performed the behavioral motions of the task, including the present / absent response, without emitting echolocation signals on about 40% of the trials at the 87 dB masking level. A study of the acoustic behavior of the same two subjects indicated that the inter-click intervals were greater than the two-way transit time from the animal to the target and back, and each successive emission occurred after the reception of the proceeding signal's echo (Au et al., 1982). Average lag times of 7 to 9 ms from echo reception to click emission were reported. These lag times are shorter than those reported from earlier literature (15 to 20 ms, Morozov,1972), and probably resulted from well-practiced animals performing a simple detection task at a fixed range. This represents a type of "echolocation attention" which, until this study, had not been accurately measured and reported. This topic is discussed in more detail by Penner (this volume).

ECHOLOCATION ADAPTABILITY

Dolphins generally emit echolocation clicks at source levels sufficient to solve a particular task, and the emitted energy is controlled by a multitude of internal and external environmental conditions such as the ambient noise level, target range and target strength. Studies by Au and colleagues (1985) of Beluga (Delphinapterus leucas) emissions in open water indicate a highly adaptive animal which can shift both peak frequency and amplitude of its echolocation clicks to compensate for changes in the noise environment. Measures of Tursiops also have indicated that animals in different environments emit echolocation clicks with different source levels and frequency content. Au et al. (1980) reported echolocation signals with peak energies at frequencies over an octave higher (120-130 kHz) and source levels on the order of 30 dB greater (220 dB) that previously reported.

Schusterman et al. (1980) also demonstrated a binary on/off control of echolocation emission in the dolphin.

These observations lead to interesting questions about the degree of control the dolphin could exercise over it's echolocation emission system. Was the source level dependent on target attributes which determine echo-to-noise ratio? Could the emitted source level be controlled in ways other than just on and off? If a greater degree of control was possible was it mediated by higher order learning centers? Could the animal have voluntary and direct control over the emission level of it's click train in active target detection? Did a feedback loop exist which allowed learning, so that conditioning of the emitted source level can occur? Answers to these questions could go along way toward explaining how echolocation may have evolved and how it is developed in the new born animal and how it is used in assessing and learning target attributes.

We set out to discover if, in fact, dolphins could exert voluntary control over the emitted source level of their clicks. We additionally wanted to know if the animal could perform a target detection task with these controlled clicks. Our approach was to set up a simple detection task, one that the animal had to learn but was not so difficult that it would interfere or obligate the animal to a specific signal structure in order to perform the task. This allowed two observations: (1) the development of the "preferred" echolocation click for this task so that at best detection performance we could describe the baseline or "natural click" structure the animal used to solve this task, and (2) a description of the development, over time, of the animal's emitted source level as it gained experience with the task and target. These data formed a baseline from which to guide and compare the success of our attempted source level manipulations.

To control echolocation behavior it must be observed and reinforced in a timely manner. Clicks are emitted by the dolphin at very high repetition rates, sometimes hundreds per second, and have durations ranging from the low 60's to a few hundred microseconds, making the observation and measurement of the behavior a technological challenge. This problem was overcome with a click detector and peak frequency analyzer which we used to train and maintain the echolocation click emission behavior. The device used a calibrated hydrophone located in the center of the animals outgoing echolocation beam. The hydrophone (B&K 8103) detected each emitted click and fed the output to the computer which measured it's peak voltage and then measured the relative energy in the frequency range between 30 and 135 kHz. The frequency analysis consisted of the output of a bank of equally spaced 15 Khz wide filters which have their outputs fed to a eight channel analog to digital converter. The converter was in turn read by the computer for analysis and display.

After the animal had solved the detection task and was performing at 90% or better, we started training to establish control over outgoing emission level. Initially we used a computer program, activated at the start of the trial which turned on a tone from an underwater speaker. A tone on the right for increased level requirement and a different tone on the left for a decreasing level requirement. A monitor hydrophone detected each emitted click an fed it to the computer, which measured it's source level and compared it to the level criterion we entered for each trial. As the animal emitted clicks, a comparison was made between the criterion level and the measured emitted level. If the click was above (or below) the criterion the computer interrupted the trial by sounding the reinforcer tone while the animal was still emitting clicks. Slowly adjusting the criterion up or down in accordance with the right or left tones trained the animal to listen to the right and left tones as cues to produce high or low level click trains.

The click detector/analyzer was combined with a stationing device which kept the dolphin in a fixed position during signal emission. The stationing device was required to maintain the reference hydrophone in a fixed position within the animal's emitted beam. Au et al. (1986) detail the device and give results of measurements of the test animal's emission beam pattern.

The results indicated that operant control could be established over the emission amplitude of the dolphin. We developed a 23 dB difference in average emission amplitude using the tones to cue the animal for the required source level emission amplitude criterion required to receive reinforcement. We found that this amplitude control can be maintained even during a simple target detection task when both the target and the cuing tone are presented in a random sequence. The animal performed the detection task at 93% correct or better for seven 50-trial sessions. With emission level criterions of 190 dB or below versus 205 dB or above the animals average source level for the go-low condition was 184 dB and 207 dB for the go-high condition. The dolphin met the criterion 94% on average for the seven sessions (Moore and Patterson, 1983).

BASIC HEARING PARAMETERS

For biological sonar systems, the amount of energy which the animal actually attends to over a specific time or integration interval is unknown, whereas for man-made sonar systems the interval is a design parameter. Work to quantify biosonar target strength has examined the dolphin's ability to perceive complex or multi-highlight echoes. Masked detection thresholds have been determined for the dolphin using simulated or "phantom" echoes containing one or two highlights separated by varying amounts of time, 50 to 300 usec (Au et al.,1986). The results show that single element phantom echo thresholds in noise differ from double element thresholds by 3 dB as expected (doubling the energy equals a 3.0 dB increase). The 3 dB difference extends, in time, until the second element is separated from the first by about 290 µsec.; beyond 290 µsec the two element threshold is about the same as the single element threshold. These results show that the dolphin performs like an energy detector with a 290 µsec integration interval (see Au and Moore, this volume).

Attempts to quantify other basic hearing capabilities led to experiments to assess the critical bandwith for the dolphin. Our approach was to measure both the critical ratio and the critical band of a passive listening dolphin at frequencies above those previously tested (Johnson, 1968). Fletcher (1940) hypothesized that masking was not produced by all the noise in a broad band but was caused by only a small band centered about the stimulus frequency. He proposed that a measure of this band could be made by assuming the power in the signal to be detected would be equal to the power of the noise in the band when the signal was at the just-masked threshold. This procedure of determining the critical band by power ratios has become known as the critical ratio and is only an estimate of the actual critical band which would result from direct measurement. We calculated the critical ratio of the dolphin and directly measured the critical band using band narrowing techniques.

Johnson (1968) measured critical ratios for dolphins but not at the higher peak frequencies found in Kaneohe Bay echolocation signals. Moore and Au (1982,1983) conducted experiments to obtain critical ratio measures at 30, 60, 90, 100, 110, 120 and 140 kHz for the dolphin and then directly measured the critical band centered at 30, 60 and 120 kHz, using very sharp cut-off (greater that 100 dB/octave) bands of digital noise. The critical ratios were obtained using the tracking method of threshold determination at the 50% detection level (for a detailed description of this method see Moore and Schusterman, 1987). The critical bands were obtained by a modified method of constants using 8 bandwidths of noise for each testing frequency with Q_{3dB} values of 1.0, 1.3, 1.7, 2.4, 5.0, 10, 20 and 40. Estimates of the directly measured critical band were obtained by plotting the thresholds as a function of bandwidth and fitting two straight lines to the data to obtain the breakpoint where increased noise bandwidth ceased affecting threshold. The bandwidth was calculated at that point.

The low frequency critical ratio results correspond closely with Johnson's (1968) and support the validity of the higher frequency measures to at least 120 kHz. The results indicate a sharp increase in the critical ratio to 51 dB at 110 kHz, followed by a decline to 46 dB at 120 kHz. Direct measures of the critical band at 30, 60 and 120 kHz indicate that the relationship is different in dolphins from that measured in humans, the human critical band is about 2.5 times the critical ratio (Zwicker et al., 1957). For dolphins, the critical band was about 10 times (10 dB) wider than the critical ratio at 30 kHz, 8 times (9 dB) at 60 kHz and about equal at 120 kHz. Generally these results suggest that the pure tone frequency resolving ability of this animal is not optimal at the higher frequencies even though the peak emission energy has been observed in this range.

The Directivity Index is a measure of a sonar receiver's ability to discriminate a signal in isotropic noise. A **directionally** sensitive receiver will produce more output if it's pointed at the signal source than a **non-directionally** sensitive (or omnidirectional receiver). The directivity index is a complicated measure of the ratio of the noise power received by an omnidirectional receiver to that received by a directional receiver when both are exposed to the same isotropic noise field. To obtain this value for an animal requires considerable time and patience to gather enough signal thresholds from a dolphin to calculate this value. An experiment (Moore and Au, 1982 a & b; Au and Moore, 1984) obtained masked thresholds at several frequencies and at various vertical and horizontal positions about the dolphin and measured the vertical and horizontal receiving beam patterns at 30, 60 and 120 kHz. The results show that at the emitted peak frequency of 120 kHz, the dolphin's receiving beam was broader than the transmission beam in both the horizontal and vertical planes.

Using the beam pattern data, we calculated the directivity index assuming a linear transducer model. The calculations showed the behavioral data correlated well with an idealized 5.0-cm linear transducer in the vertical plane. Calculations for a two-element rectangular array also showed good correspondence with the horizontal beam pattern obtained from behavioral data. In general the dolphin showed a frequency dependant cone of sensitivity centered about the animal's midline, slightly elevated. The cone becomes narrower at higher frequencies.

A later remeasurement of the dolphin's **transmitting** beam by (Au et al., 1986) showed that the receiving and transmitting beams are aligned and the major axis of the transmitting beam is 5 degrees above midline, some 15 degrees lower that previously reported. The calculated transmitting directivity index in each plane was within 1 dB of previous measurements.

OUTLOOK

Since 1980, basic hearing and echolocation research has focused on only a few acoustic capabilities concerning dolphin echolocation. However, much remains to be learned, especially about the way dolphins process the return echoes. Particularly interesting questions include how the animal performs

feature extraction from a set of returning echoes and what salient echo
properties identify target characteristics such as shape, material, and
density.

Another question concerns the validity of techniques which apply passive
listening audiometric procedures to **all** aspects of dolphin echolocation
studies. One view of the dolphin auditory system (Vel'min and Dubroski,
1975, 1976, 1978) presumes two separate and distinct auditory subsystems.
This duplex theory of the dolphin hearing system involves both a passive and
an active subsystem. This concept infers that acoustic perception of "low
frequency" signals, used for conspecific identification, inter-specific
communication, general localization and tracking of sounds in space are
relegated to the passive subsystem. The active subsystem is concerned with
the emitted "high frequency" echolocation signals and the processing of the
echoes. If this view, that echolocating dolphins separate information flow
on the basis of frequency and signal source, has merit, then techniques
which use passive listening experiments to analyze "active" dolphin sonar
signal processing may be incomplete if not invalid. The duplex theory of
dolphin hearing does not rule out the continued use of traditional
psychoacoustic practices, but any particular application of a psychoacoustic
paradigm must be considered in the total context of the experimental
question being addressed. The phantom echo technique, as applied to studies
of dolphin echolocation which are described in this volume (Au et al., Au
and Moore) will likely provide a superior procedure to study echo feature
extraction by dolphins.

These observations and questions form a profuse core of future. Given
the opportunity, facilities and talent, we may yet better understand the
methods by which dolphins demonstrate such extraordinary sonar capabilities.

REFERENCES

Altes, R.A. 1980. Models for Echolocation. in: Busnel and Fish (eds.)
Animal Sonar Systems, Plenum Press, New York. pp 625-671

Au, W.W.L., and Penner, R.H. 1981. Target Detection in Noise by
Echolocating Atlantic Bottlenosed Dolphins. Jour. Acoust Soc. Am. 70, 687-
693.

Au, W.W.L., Penner, R.H., and Kadane, J. 1982. Acoustic Behavior of
Echolocating Atlantic Bottlenosed Dolphins. Jour. of the Acoust. Soc. Am.
71, 1269-1275.

Au, W.W.L. and Moore, P.W.B. 1984. Receiving Beam Patterns
and Directivity indices of the Atlantic Bottlenosed Dolphin Tursiops
truncatus. Jour. Acoust. Soc. Am. 75, 255-262.

Au, W.W.L, Carder, D.A., Penner, R.H. and Scronce, B.L. 1985.
Demonstration of Adaptation in Beluga Whale Echolocation Signals, Jour.
Acoust. Soc. Am. 77, 726-730.

Au, W.W.L, Moore, P.W.B. and Pawloski, D.A. 1986. The Perception of
Complex Echoes by an Echolocating Dolphin. Jour. Acoust. Soc. Am. S1 80,
S107.

Bel'kovich, V.M., and Dubrovskiy, N.A.. 1976. Sensory Basis of Cetacean
Orientation. Izdatel'stov Navaka, Leningrad, pp 82-87

Busnell, R.G. and Fish, J.A. 1980. "Animal Sonar Systems" Plenum Press,
New York.

Diercks, K.J. 1980. Signal Characteristics for Target Localization and Discrimination. in: Busnel and Fish (eds.) Animal Sonar Systems, Plenum Press, New York. pp 299-308

Fletcher, H. 1940. Auditory Patterns. Rev. Mod. Phys. 12, 47-65.

Floyd, R.W. 1980. Models of Cetacean Signal Processing. in: Busnel and Fish (eds.) Animal Sonar Systems, Plenum Press, New York. pp 615-623

Hammer, C. E. and Au. W.W.L. 1980. Porpoise Echo-Recognition: An Analysis of Controlling Target Characteristics. Jour. Acoust. Soc. Am. 68, 1285 - 1293.

Johnson, C.S. 1968. Masked Tonal Thresholds in the Bottlenosed Porpoise. Jour. Acoust. Soc. Am. 44, 965-967.

Johnson, C.S. 1980. Important Areas For Future Cetacean Auditory Study. in: Busnel and Fish (eds.) Animal Sonar Systems (pp.515-518) Plenum Press, New York.

Johnson, R.A. 1972. Energy Spectrum Analysis as a processing Mechanism for Echolocation. Ph.D. dissertation, University of Rochester, New York, pp 515-518

Johnson, R.A. 1980. Energy Spectrum Analysis in Echolocation. in: Busnel and Fish (eds.) Animal Sonar Systems, Plenum Press, New York. pp 673-693

Moore, P.W.B. and Au, W.W.L. 1982a. Masked Pure-Tone Thresholds of the Bottlenosed Dolphin (Tursiops truncatus) at Extended Frequencies. Jour. Acoust. Soc. Am. S1, S42.

Moore, P.W.B. and Au, W.W.L. 1982b. Directional Hearing in the Atlantic Bottlenosed Dolphin (Tursiops truncatus). Jour. Acoust. Soc. Am. S1, S42.

Moore, P.W.B. and Au, W.W.L. 1983. Critical Ratio and Bandwidth of the Atlantic Bottlenosed Dolphin. Jour. Acoust. Soc. Am. S1 74, p.S73.

Moore, P.W.B. and Patterson, S.A. 1983. Behavioral Control of Echolocation Source Level in the Dolphin (Tursiops truncatus). Fifth Biennial Conference on the Biology of Marine Mammals, New England Aquarium, Boston, Mass., p70.(Abstract)

Moore, P.W.B., Hall, R.W., Friedl, W.A. and Nachtigall, P.E. 1984. The Critical Interval in Dolphin Echolocation: What is it? Jour. Acoust. Soc. Am. 76, pp314-317.

Moore, P.W.B. and Schusterman,R.J. 1987. Audiometric Assessment of the Northern Fur Seals, Callorhinus ursinus. Marine Mammal Sci. 3, 1, pp31-53

Morozov, B.P., Akapiam, A.E., Burdin, V.I., Zaitseva, K.A., and Sokovykh, Y.A. 1972. Tracking Frequency of the Location Signals of Dolphins as a function of Distance to the Target. Biofizika, 17, pp139-145.

Nordmark, J. 1961. Perception of Distance in Animal Echolocation. Nature, 190, 363-364.

Pye, J.D. 1967. Discussion to: Theories of sonar Systems in Relation to Biological Organisms. in: R.G. Busnel (ed.) Animal Sonar Systems: Biology and Bionics. Laboratoire de Physiologie Acoustique, INRA-CNRZ, Jouy-en Josas, France. pp 1121-1136

Rice, C.E. 1966. The Human Sonar System. in: R.G. Busnel (ed.) Animal Sonar Systems: Biology and Bionics. Laboratoire de Physiologie Acoustique, INRA-CNRZ, Jouy-en Josas, France. pp719-755.

Schusterman, R.J., Kersting,D.A. and Au, W.W.L. 1980. Stimulus Control of Echolocation Pulses in Tursiops truncatus. in: R.G. Busnel (ed.) Animal Sonar Systems: Biology and Bionics. Laboratoire de Physiologie Acoustique, INRA-CNRZ, Jouy-en Josas, France. pp981-982

Vel'min, V.A., and Dubrovskiy, N.A. 1975. On the Analysis of Pulsed Sounds by Dolphins. Dokl. Akad. Nauk SSSR 225, pp470-473

Vel'min, V.A., and Dubrovskiy, N.A. 1976. The Critical Interval of Active Hearing in Dolphins. Sov. Phys. Acoust. 2, pp351-352

Vel'min, V.A., and Dubrovskiy, N.A. 1978. Auditory Perception by Bottlenosed Dolphin of Pulsed Signals. in: V. Ye. Sokolov (ed.) Marine Mammals: Results and Methods of Study. A.N. Severtsov Inst. Evol. Morphol. Anim. Ecol., Akad. Nauk SSSR, pp90-98

Zwicker, E., Flottop, G. and Stevens, S.S. 1957. Critical Bandwith in Loudness Summation. Jour. Acoust. Soc. Am. 29, pp548-557

PARALLEL AUDITORY PATHWAYS: I - STRUCTURE AND CONNECTIONS

John H. Casseday* and George D. Pollak#

*Departments of Surgery and Psychology
Duke University
Durham, NC   27710

#Department of Zoology
University of Texas
Austin, TX   78712

INTRODUCTION

The remarkable hypertrophy of the central auditory pathways in microchiropteran bats was recognized long before the bionsonar system of these animals was discovered. Some of the first experimental studies of the auditory pathways were conducted using the echolocating bat Rhinolophus (Poljak, 1926). In this historical context, it seems a paradox that after the discovery of echolocation, studies of the connections of central auditory pathways lagged far behind physiological and behavioral studies. It was not until the last sonar conference that the first studies to use modern neuro-anatomical techniques were reported (Schweizer,1980; Zook and Casseday,1980).

In this review of advances made since 1980 on defining the connections of the central auditory pathways in echolocating bats, we shall consider the anatomical basis for tonotopic organization, the organization of binaural and monaural pathways and the integration of these pathways at the inferior colliculi. We shall refer mainly to the literature on the anatomy of the bat's central auditory system. For reference to the broad literature on the neuroanatomy of auditory pathways in mammals, several recent reviews are available (Warr, 1982; Cant and Morest, 1984; Casseday and Covey, 1986).

It is useful to begin this review with tonotopic organization because the multiple tonotopy at each successive level of the pathway illustrates that the central auditory system consists of a number of parallel pathways. The connectional basis for tonotopy is of special significance for studies of bats which emit echolocation signals which consist of constant frequency (CF) and frequency modulated (FM) components. In the auditory centers of these bats an unusually high proportion of auditory neurons respond to frequencies of the CF part of the signal (Suga and Tsyzuki, 1985; see Suga, this volume). Tonotopic organization is no doubt the basis for processing certain features of sonar signals. For example, we shall suggest how the tonotopic connections in the lateral lemniscus of bats may form the basis for processing temporal features of the FM portion of the sonar signal.

We shall then describe pathways for monaural and binaural hearing. One of the fundamental principles which emerges from the study of the auditory pathways is that some pathways are designed to provide convergence of signals from the two ears, whereas others seem designed to provide separation of the inputs from the two ears. We shall refer to these two types of pathways as binaural and monaural respectively. This distinction is necessarily an over-simplification, because there is almost certainly more than one type of each pathway, but it does provide a useful system of classification.

Finally we shall turn to the question of how these pathways are inte-grated at the inferior colliculus. Although there is a clear distinction between binaural and monaural pathways at some levels, e.g., superior olive and lateral lemniscus, this distinction becomes hard to see at the inferior colliculus. Thus a major challenge in the study of the structure of auditory pathways is the question of how multiple pathways in the lower brain stem converge at the midbrain.

TONOTOPIC ORGANIZATION AND PARALLEL PATHWAYS

The auditory nerve

Echolocating bats emit biosonar calls that typically cover at least one and ususally several octaves (Simmons et al. 1975; Novick 1977). The basic conversion of frequency to place in the cochlea, whereby the power spectrum of echolocation sounds is transformed into a spatial pattern of activity along the cochlear partition, is described by Vater in this volume. An important point made in that chapter is that the cochleae of bats which emit long CF/FM calls do not afford equal treatment to each frequency. The length of cochlear partition devoted to the dominant frequency of the CF component of the echolocation call is greatly expanded and has numerous morphological and physiological specializations. These specializations impart exceptionally sharp tuning curves to the neurons which innervate the CF region of the cochlea.

Several studies have described the response properties of units in the peripheral parts of the central auditory system in mustache and horseshoe bats (Suga et al., 1975; 1976; Suga and Jen 1977; Suga and Tsuzuki, 1985), but in the absence of verified recording sites it is uncertain whether the neurons were located in the auditory nerve or cochlear nucleus. Neverthe-less, there is little doubt that the elegance of the frequency-to-place transformation in the cochlea is preserved in the population of fibers in the auditory nerve. The elongated region of basilar membrane devoted to the CF component in each species is reflected as a disproportionately large popul-ation of central auditory fibers tuned to 60 kHz in mustache bats or 80 kHz in horseshoe bats. Studies of the auditory nerve in these bats have revealed variations in the V shaped tuning curves that are typical of auditory nerve fibers in other mammals. The tuning curve can be described by a quality factor (Q_{10dB}), which reflects its sharpness or frequency selectivity and by the frequency to which it is most sensitive, "best frequency" (BF). Units with BFs above or below the dominant CF component of the orientation cry have small Q_{10dB} values that range from 3-18, much like the range in other mammals; however units tuned to the CF component have Q_{10dB} values as high as 300-400, considerably higher than in other mammals. The complete spectrum of frequencies to which the cochlea responds is, of course, re-represented in the auditory nerve, but the representation is greatly weighted in favor of units that are very sharply tuned to 60 kHz in mustache bats and to 80 kHz in horseshoe bats.

The cochlear nuclei

As it exits the cochlea, the auditory nerve is a single pathway, but as it enters the cochlear nuclei, its fibers bifurcate to form ascending and descending branches. This branching of the auditory nerve fibers forms the first division of the central auditory system into parallel pathways. The branching of fibers is the basis for three separate tonotopic organizations,

one in each main division of the cochlear nulcleus. The ascending branch
innervates the anteroventral cochlear nucleus (AVCN), and the descending
branch further divides to innervate two structures, the posteroventral (PVCN)
and dorsal cochlear nuclei (DCN).

Before considering tonotopic organization, we shall describe special
features in the cochlear nuclei of bats to show that there are differences
among the cochlear nuclei of the species studied in most detail, Rhinolophus
ferrumequinum, Rhinolophus rouxii and Pteronotus parnellii. In all these
species of bats AVCN is the largest division of the cochlear nucleus. The
anterior part of AVCN is characterized by a fairly homogeneous population of
spherical cells (Schweitzer, 1981; Zook and Casseday, 1982; Feng and Vater,
1985). Unlike similar neurons of other mammals, these spherical cells are
among the smallest cells in the cochlear nucleus. However, their connections
with the auditory nerve are probably similar to those of other mammals in
that they appear to receive very large endings (Feng and Vater, 1985). Thus,
as in other mammals, the input from the auditory nerve fibers to these
neurons consists of highly "secure" synapses. The posterior part of AVCN
contains a heterogeneous population of cells, including globular, multipolar,
and a few spherical cells. In addition to the usual type of multipolar
cells, the AVCN of Pteronotus contains a second population of multipolar
cells which are unusually large. These cells are located in the marginal
zone which extends along the medial edge of AVCN and over the dorsal portion
of the root of the auditory nerve. This area has not been identified in
Rhinolophus. Because the AVCN contains a variety of cell types, and because
it is the source of ascending pathways to several different cell groups, it
is important to determine whether these morphologically different cell types
also differ from one another in their projections and in their response
properties. Below we shall point out the significance of AVCN as the
principal source of projections that give rise to binaural pathways.

The cytoarchitecture of PVCN in bats appears to differ very little from
that of other mammals, but DCN in some bats has some unusual features. In
most mammals DCN has a laminar appearance, the chief characteristic of which
is a distinct layer of granular and fusiform cells below the outer edge of
DCN. In most bats this layer can be identified, although in many species it
is not well developed. In Rhinolophus rouxii, the ventral part of DCN is
laminated, but the dorsal part, which represents frequencies at and above the
CF, is not laminated and the division between the two parts is clear. This
division is not seen in other bats; for example, in Pteronotus it is difficult to distinguish a laminated part at all, and the entire DCN resembles
more the dorsal part of DCN of Rhinolophus than the ventral part. In
Eptesicus the lamination of DCN is apparent throughout. The significance of
the highly laminated part of DCN in Rhinolophus rouxii is not clear; it is
the largest part of DCN and represents frequencies below the CF in this bat.

The tonotopic organization of these multiple areas has been demonstrated
in detail in Rhinolophus rouxii by the method of tracing the connections of
physiologically defined parts of the cochlear nucleus (Feng and Vater, 1985).
Small injections of horseradish peroxidase (HRP) were placed in one division
of the cochlear nucleus, the tonotopic organization of which had previously
been mapped (Fig. 1). HRP is not only transported back to the cells of origin
in the spiral ganglion (see Vater, this volume) but is also transported
throughout all the other branches of the fiber into which it is incorporated.
For example, if HRP is injected in a region of DCN that responds to 78.8 kHz
as in Fig. 1a, branches that terminate in the 78.8 kHz regions of AVCN and
PVCN are also labeled (Fig. 1b). An example of transport from an injection
in an area responsive to 68.9 kHz is shown in Figs. 1c, 1d and 1e. By a
series of such experiments Feng and Vater were able to produce a detailed map
of the tonotopic arrangements within AVCN, PVCN and DCN, shown in Fig. 2.
The highest frequencies audible to the bat (i.e. at and above the resting

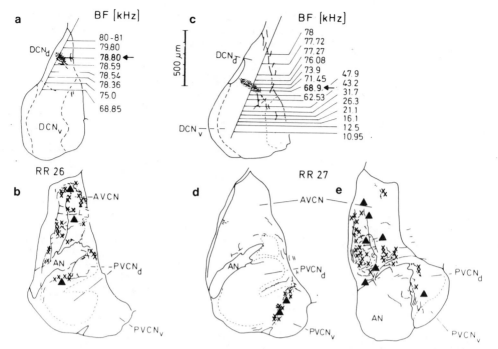

Fig. 1. Drawings to show the location of labeled cells and labeled fibers
after injections of HRP in DCN of two bats, RR 26 (a and b) and RR
27 (c, d and e). Following an injection centered at 78.8 kHz (a),
labeled cells (triangles) and labeled fibers (crosses) are seen in
the caudal part of AVCN and in the dorsal part of PVCNd (b). After
an injection in DCN centered at 68.9 kHz (c), labeled terminals and
labeled fibers are seen in ventral PVCNd (d) and in AVCN (e) in a
more rostral location than after the injection in a higher frequency
area shown in (b). Taken from Feng and Vater (1985).

ABBREVIATIONS USED IN FIGURES: ALD: anterolateral division of the inferior
colliculus; ALPO, anterolateral periolivary nucleus; AN: auditory nerve;
AV,a,d,m,p: anteroventral cochlear nucleus, anterior, dorsal, medial,
posterior divisions; AVCN: anteroventral cochlear nucleus; BC: brachium
conjunctivum; BIC: brachium of the inferior colliculus; BP: brachium pontis;
CB,CER: cerebellum; CG: central grey; CIC: commissure of inferior colliculus;
CP: cerebral peduncle; CU: cuneiform nucleus; DC: dorsal cortex of inferior
colliculus; DCN,d,v: dorsal cochlear nucleus, dorsal, ventral divisions;
DMPO: dorsomedial periolivary nucleus; DNLL: dorsal nucleus of lateral lemniscus; DPD: dorsal posterior division of inferior colliculus; DPO: dorsal
periolivary nucleus; GM,v: medial geniculate body, ventral; IC,c,e(ex,x),p:
inferior colliculus, central, external, pericentral; INLL: intermediate
nucleus of the lateral lemniscus; LNTB: lateral nucleus of the trapezoid
body; LSO: lateral superior olive; MBRF: midbrain reticular formation; MD:
medial division of inferior colliculus; MLF: medial longitudinal fasciculus;
MNTB: medial nucleus of trapezoid body; MSO: medial superior olive; nBIC:
nucleus of the brachium of inferior colliculus; Po: posterior thalamic
nuclear group; PV,a,c,l,m: posteroventral cochlear nucleus, anterior, caudal,
lateral, medial divisions; PVCN,d,v: posteroventral cochlear nucleus, dorsal,
ventral divisions; Pyr,Py: pyramidal tract; RB: restiform body; SC,d,i:
superior colliculus, deep layers, intermediate layers; SG: suprageniculate
nucleus; TB: trapezoid body; VII: seventh nerve; VIII: eighth nerve; VMPO:
ventromedial periolivary nucleus; VNLL,d,v: ventral nucleus of the lateral
lemniscus, dorsal, ventral divisions; VNTB: ventral nucleus of the trapezoid
body; VP: ventral posterior nucleus; VPO: ventral periolivary nucleus.

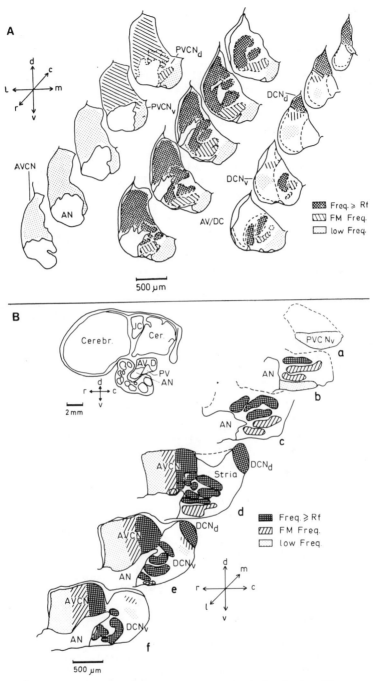

Fig. 2.   Schematic sections in two planes through the cochlear nucleus to
show patterns of connectivity related to tonotopic organization.
A) A series of sections in the frontal plane. B) A parasagittal
section of the head of Rhinolophus rouxi (left) and parasagittal
sections ("a" through "f") of the cochlear nuclei. Both views are
constructed from data such as shown in Fig. 1. Note the large areas
devoted to frequencies contained within the echolocation cry. (from
Feng and Vater, 1985)

frequency,) are located dorsally in the DCN, laterally in the PVCN and posteriorly in AVCN. Within each division there is a steady topographic progression from high to low frequency representation.

A very important aspect of these results is the finding that the connections have a planar organization. The projections to AVCN occupy sheets or slabs of neuropil that are parallel to the frontal plane and parallel to the isofrequency contours seen by electrophysiological recording. Feng and Vater (1985) proposed the hypothesis that such slabs are functional units. We shall return to this hypothesis when we consider the ascending projections from AVCN. For the present it is sufficient to point out that the slab hypothesis deserves further study. If a slab is a functional unit, then it must border one or more adjacent units, and we should be able to discover some functional as well as anatomical difference at the border between one slab and the next. Without knowing precisely what constitutes a functional or structural border, all we can say at present is that the system has a planar arrangement.

A second way of identifying the connectional basis for tonotopy is to examine cells labeled by retrograde transport after injections of HRP in physiologically defined areas of the inferior colliculus. High and low frequency areas of AVCN and DCN defined by this method in Eptesicus fuscus are shown in Figs. 3A and 3B.

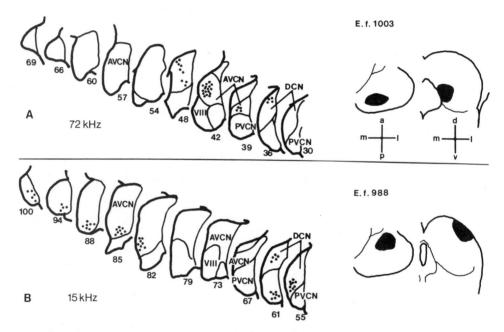

Fig. 3. Retrograde transport of HRP from the inferior colliculus to the cochlear nucleus. A) An injection in high frequency areas of the inferior colliculus labels cells in posterior AVCN as well as in dorsal DCN. Units at the injection site responded best to frequencies around 72 kHz. B) An injection in low frequency areas of the inferior colliculus labels cells in the most anterior and ventral part of AVCN as well as in ventral DCN and ventral PVCN. Units at the injection site responded best to frequencies around 15 kHz. The injection sites are shown in frontal sections (far right) and horizontal reconstructions (right) of the inferior colliculus.

Superior olivary complex

The structure of the superior olivary complex varies considerably from one species of bat to another. Probably the simplest organization and that most similar to other mammals is seen in Pteronotus, in which a lateral (LSO) and medial (MSO) superior olive can be identified in Nissl or fiber stains (Fig. 4). The MSO is of special interest because in large land dwelling mammals it is devoted largely to processing low frequency signals (Masterton et al. 1975), and its presence in bats has been questioned (Harrison and Irving 1966; Irving and Harrison 1967). As will be described later, the afferent and efferent connections of the structure labeled "MSO" in Pteronotus are consistent with the view that this structure is homologous with MSO of other mammals. In all bats that we have examined, an LSO can be identified, but in bats other than Pteronotus the cytoarchitecture of the structures medial to LSO is very complicated, and it is difficult to identify structures homologous with MSO of other mammals (Zook and Casseday, 1980).

Harnischfeger et al. (1985) described the tonotopic organization of MSO in the molossid bat, Molossus ater, as an orderly progression of best frequencies with low frequencies represented dorsally and high frequencies represented ventrally, similar to the tonotopic pattern of other mammals (Fig.5B). Recent studies in mustache bats also demonstrate an orderly tono-topic organization in MSO (Fig. 5A) as seen from projections to the inferior colliculus (Ross et al. 1985). In these studies the distribution of labeled cells in MSO was mapped after HRP was injected into isofrequency regions of the central nucleus of the inferior colliculus, identified with physiological techniques (Fig 6). These studies reveal the tonotopic pattern typical of mammals: low frequencies are located dorsally and high frequencies ventrally. In the case shown in Fig. 6 there is a large sheet of labeled cells within the ventral half of MSO following an injection of the 60 kHz region of the inferior colliculus.

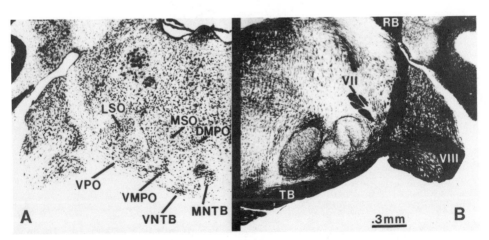

Fig. 4. Photomicrographs of the superior olivary complex of Pteronotus parnellii parnellii. A) Section through the left brain stem stained for cell bodies (cresylecht violet stain). B) Section through the right side of the brain stem, at the same level as shown in A, to show fibers (Heidenhain stain). The dense fiber plexus reveals the shape of MSO and LSO. (From Zook and Casseday, 1982a)

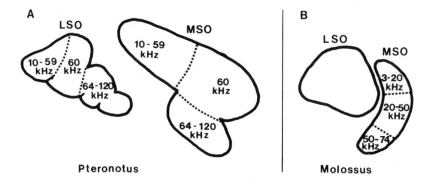

Fig. 5. tonotopic organization of superior olivary complex in two species
of bats. A) Tonotopic organization of MSO and LSO as seen by
retrograde transport from physiologically defined areas of
inferior colliculus of <u>Pteronotus</u>, based on data from L. Ross
(unpublished). B) Tonotopic organization of MSO in <u>Molosus ater</u>,
based on data from Harnischfeger et al. (1985).

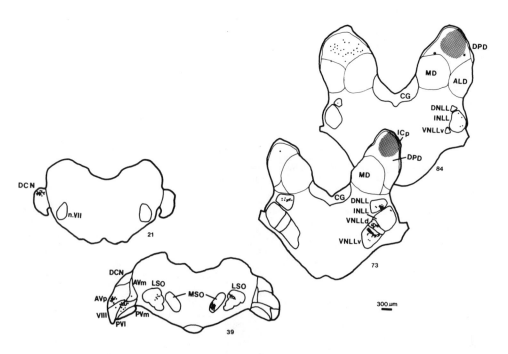

Fig. 6. Retrograde transport from the dorsoposterior (60 kHz) division of
the inferior colliculus of <u>Pteronotus</u>. The injection site (shaded
area) is entirely within the area responsive to frequencies around
60 kHz, as determined by recording responses of single units.
Large dots in the inferior colliculus indicate physiologically
defined borders of DPD. Small dots show labeled cells (from L. Ross,
unpublished).

The tonotopy in the bat's LSO is not as well understood as that of MSO. Harnischfeger et al. (1985) recorded from only 10 cells in LSO, a sample size too small to reveal frequency representation or tonotopic organization. Connectional studies in the mustache bat suggest that low frequencies are found in the lateral limb and high frequencies in the medial limb (Figs. 5A and 7). For example, the lateral limb receives projections from anterior AVCN (Fig. 7A), and the medial limb from posterior AVCN (Fig. 7B). This tonotopic pattern is similar to that of the cat (Tsuchitani and Boudreau 1966; Guinan et al., 1972). The frequency range represented in LSO and the fine structure of the tonotopic organization awaits direct physiological mapping studies.

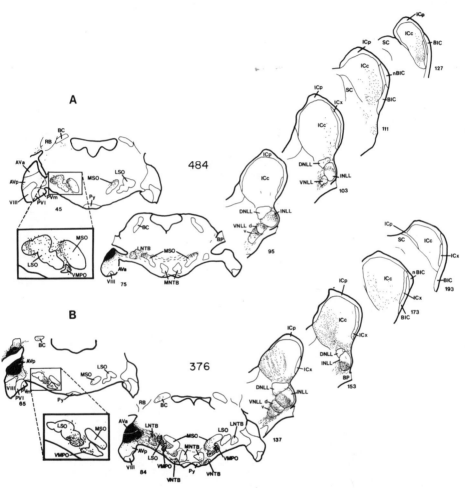

Fig. 7. Drawings to illustrate anterograde transport from AVCN to the superior olives, lateral lemniscus and inferior colliculus. A) An injection of [³H]-leucine involving low frequency areas, anterior AVCN. B) An injection involving high frequency areas, posterior AVCN. The projections from anterior AVCN distribute to lateral LSO, dorsal MSO, dorsal parts of INLL, VNLLd, VNLLv and lateral parts of the central nucleus of the inferior colliculus. The projections from high frequency areas distribute to the opposite poles of each of these nuclei.

Nuclei of the lateral lemniscus

The nuclei of the lateral lemniscus provide an excellent illustration of the utility of the comparative method in yielding clues about structure-function relationships. These nuclei consist of the ventral (VNLL) inter-mediate (INLL) and dorsal (DNLL) nuclei of the lateral lemniscus. The structure of the lateral lemniscus is very similar in all echolocating bats, even those in which the structure of the sonar signal differs markedly, such as Eptesicus and Pteronotus parnellii. This similarity is in contrast to pronounced differences in the appearance of the superior olivary complex in the two species. The similarity may have to do with common features in the sonar signals, so that our findings on the connections of the lateral lemniscus may be generalized to all echolocating bats. The VNLL and INLL are so highly differentiated in echolocating bats that the conclusion seems inescapable that these nuclei play an important role in echolocation. Not only are these nuclei much larger than they are in other mammals, each is clearly differentiated into several different subnuclei (Figs. 7,8 and 9). The most striking feature is a columnar arrangement of cells in one division of VNLL.

The first studies to show the connections of the nuclei of the lateral lemniscus in bats showed that the afferent projections diverge to separate targets which correspond to the main nuclei (Zook and Casseday, 1985), as shown in Figs. 7 and 8, and the efferent projections converge at the inferior colliculus (Schweizer, 1981; Zook and Casseday, 1982b). These observations suggest that each nucleus has a separate tonotopic organization. For example,

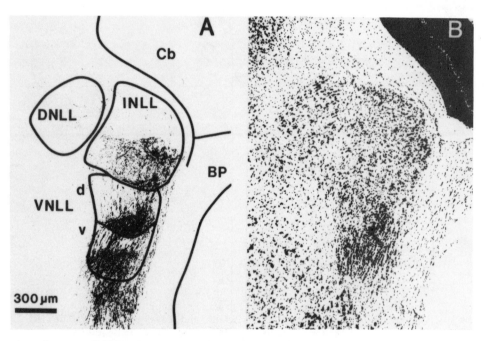

Fig. 8. Ascending projections from AVCN to the lateral lemniscus and cytoarchitecture of nuclei of the lateral lemniscus. A) Projections from the posterior part of AVCN diverge to three main targets, one in INLL and one in each part of VNLL. The photomicrograph is a negative image of the darkfield view of an autoradiograph to show transport of [^{3}H]-leucine from AVCN. B) The same section as shown in A) is stained to show the different divisions of the lateral lemniscus in this photomicrograph. (From Zook and Casseday, 1985).

in INLL low frequencies should be located dorsally and high frequencies ventrally (Fig. 7). However, the puzzling finding was that in the most highly structured nucleus, the columnar area of VNLL, no clear connectional basis for tonotopy was apparent.

This puzzle was resolved in a recent study on the lateral lemniscus of Eptesicus, in which we correlated the projections to and from the columnar area (VNLLc) with the tonotopy of the source of afferent projections and with the target of efferent projections, the inferior colliculus. The results of this study (Covey and Casseday, 1986) show a remarkably precise tonotopic arrangement that has important implications for the processing of temporal information, perhaps for encoding target range (Covey and Casseday 1986). The connections of this area are organized in thin sheets which are very precisely related to the tonotopic organization of the AVCN and the inferior colliculus. The ascending projections from AVCN diverge to several parts of the lateral lemniscus, but in the columnar area the projections are always arranged in a thin sheet which extends throughout the anterior-posterior dimension of the columnar area. The sheet of projections is perpendicular to the columns of cells and extends along one or two rows of cells. Retrograde transport from the inferior colliculus reveals a similar pattern; labeled cells extend as a horizontal sheet throughout one or two rows of cells in the columnar area. Figure 10 illustrates a row of labeled cells at the dorsal border of VNLLc.

There are three remarkable features of these projection sheets. First, the sheets are always approximately the same thickness over a large range in size of injections. Second, the positions of the sheets of anterograde

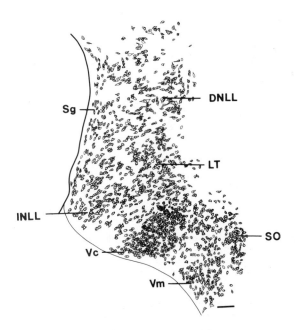

Fig. 9. Cytoarchitecture of the nuclei of the lateral lemniscus in
Eptesicus. The columnar area (Vc) is easily identified by its
small and densely packed cells and by the beaded or columnar
arrangement of the cells along the length of the ascending fibers.
Abbreviations not used in other figures: LT: lateral tegmentum;
Sg: sagulum, SO: superior olive; Vm: multipolar cell area of VNLL.

transport and the sheets of labeled cells are both precisely related to the tonotopic center of the injection site (Fig. 10). Finally, the entire range of frequencies audible to the bat is represented in this small slab of tissue, approximately 20–30 cells in the dorsoventral plane (Fig 10). Our interpretation of this compression is that only selected frequencies are processed in the columnar area, or that each sheet receives convergent input from a restricted range of frequencies, making it in effect a sort of neural comb filter. When one thinks of this "filter" in terms of processing an FM signal, the conclusion is that all the cells in one sheet must respond almost simultaneously to a small frequency band; the cells in an adjacent sheet must respond to a different frequency band, and these two bands must be separated by a gap. Thus, in response to an FM sweep, the combined output of all of the sheets in the columnar area might convert the smooth FM sweep into a series of temporally discrete responses that mark the time of occurrence of specific frequencies as they occur during the sweep.

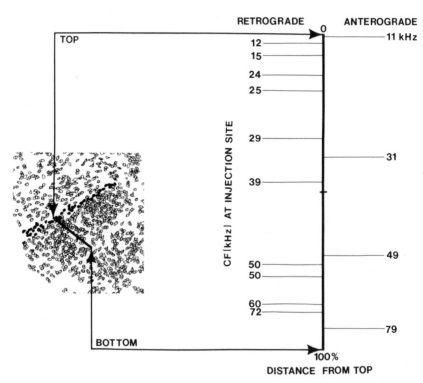

Fig. 10. Diagram to summarize the relationship between the tonotopy of the injection site and the location of anterograde or retrograde transport in the columnar area. Distance through the nucleus (left) is shown as per cent from top to bottom (right). Along this scale is shown the relative location of the sheets of retrograde or anterograde transport. The numerals indicate the best frequency in kHz of cells at the injection site. Cells shown in black depict the position of a sheet of cells labeled by an injection of HRP in a low frequency (11 kHz) region of the inferior colliculus (from Covey and Casseday, 1986).

Inferior colliculus

It is necessary to review the cytoarchitecture of the inferior colliculus, both for the present description of tonotopic organization and for our later purpose of examining converging pathways at the inferior colliculus. Figure 11 shows photomicrographs of Nissl stained frontal sections through the inferior colliculus of Pteronotus parnellii. Although a number of different areas in the inferior colliculus can be identified by cytoarchitectural analysis, it is clear that there are no borders that match in sharpness those of the auditory nuclei in the lower brain stem (e.g., Figs 4, 8 and 9). Nevertheless, the subdivisions of the central nucleus do have significance in terms of the ascending projections and the functions and organization of cells within the inferior colliculus.

We consider next the organization of cells within the central nucleus of the inferior colliculus as seen with Golgi methods. Fig. 12 shows the distribution and orientation of disc shaped cells in the inferior colliculus of Pteronotus. Using as a model the description of cell types in the cat (Morest and Oliver, 1984; Oliver and Morest, 1984), Zook et al. (1985) analyzed the orientation of the disc shaped cells in the inferior colliculus of Pteronotus. They divided the central nucleus into three main parts, anterolateral, medial, and dorsoposterior divisions, according to the orientation of the dendrites of cells within these divisions. These divisions correspond closely to those seen in Nissl stained material, i.e., the lateral, medial and dorsal divisions described by Zook and Casseday (1982a).

Most importantly, these different divisions can be related to frequency organization (Fig 13). The lateral division contains frequencies below the CF part of the echolocation pulse of the bat, that is, 10 to 59 kHz; the medial division contains the frequencies above the CF, that is, about 65 kHz

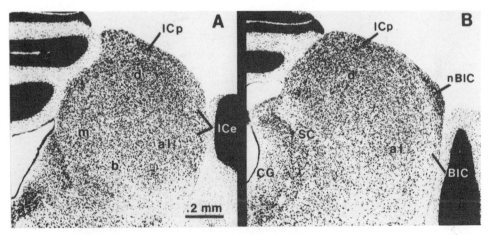

Fig. 11. Photomicrographs of the inferior colliculus of Pteronotus parnellii parnellii. The central nucleus can be divided into dorsal (d), anterolateral (al) and medial (m) parts on the basis of the relative distribution of different types of cells, although the borders between the divisions are not sharp. The clearest border is a cell sparse band (b) containing large cells between (m) and (al).

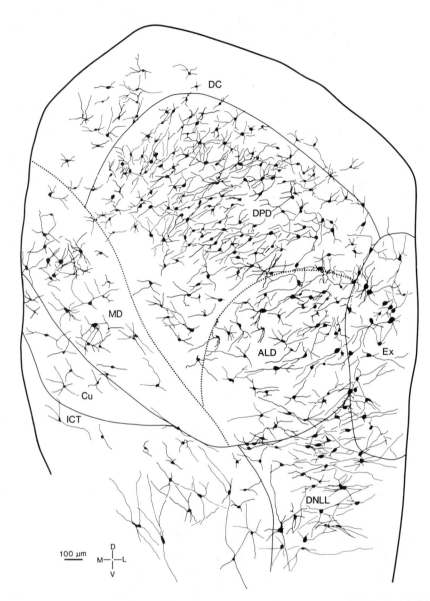

Fig. 12. Drawing of a frontal section of the inferior colliculus of
Pteronotus parnellii to show the orientation of small disc shaped
cells (Golgi impregnation). The area labeled DPD is characterized
by small stellate cells, the dendrites of which have a common
orientation. This area exclusively contains neurons which respond
best to the frequencies of the most intense harmonic of the CF
echolocation signal of Pteronotus. (From Zook et al., 1985)

and up. The dorsoposterior division contains frequencies just around the CF, as determined individually for each bat. These results suggest that in Pteronotus a cytoarchitecturally distinct region of the inferior colliculus is devoted to processing frequencies in and around the dominant harmonic of the CF pulse.

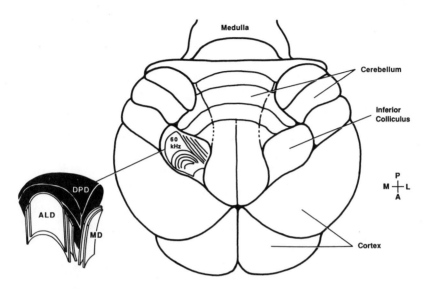

Fig. 13. Tonotopic organization of mustache bat's inferior colliculus. On the right a dorsal view of the bat's brain shows the isofrequency lines in the left colliculus, and the darkened region indicates the 60 kHz dorsoposterior division (DPD). A three dimensional view of the isofrequency sheets of the anterolateral division (ALD), the medial division (MD) and the DPD are shown on the left. (From Pollak et al. (1986).

To contrast the specializations just described in Pteronotus with the general mammalian plan of tonotopic organization in the inferior colliculus, we illustrate experiments on Eptesicus fuscus. These experiments show a direct relation between projections from AVCN and the tonotopic organization of the inferior colliculus. An injection of WGA–HRP in posterior AVCN, an area with cells responsive to frequencies about 79 kHz, as seen by the response of cells at the center or the injection, shows anterograde transport to ventral and medial, high frequency, areas of the inferior colliculus (Fig 14A). Similarly an injection in the anterior part of AVCN, in an area with cells responsive to 11 kHz, illustrates projections to low frequency areas of the central nucleus of the inferior colliculus (Fig 14B). Physiological mapping experiments reveal isofrequency contours in the central nucleus of the inferior colliculus as summarized in Fig. 15. The orientation of the bands of transport to the inferior colliculus corresponds closely with the orientation of isofrequency contours. The combined use of connectional and physiological methods can be utilized in future experiments to map the target area in the inferior colliculus of sonar signal frequencies transmitted from AVCN. Later we shall discuss further the banded nature of projections to the inferior colliculus.

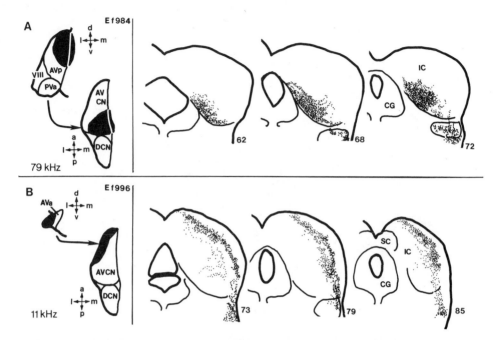

Fig. 14.   Anterograde transport to the inferior colliculus in Eptesicus.
A) Injection in high frequency area of AVCN B) An injection in low
frequency, anterior, area of AVCN. Frontal sections (upper left)
and horizontal reconstructions (arrow) show injection sites.

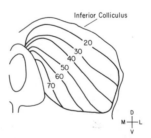

Fig. 15. Schematic representation of tonotopic organization in a transverse
section of the inferior colliculus of Eptesicus fuscus. The
isofrequency sheets are shown as lines running along the
dorsomedial to ventrolateral axis. The isofrequency sheets are
stacked in an orderly progression of frequencies.

Medial geniculate body

For our description of the tonotopic arrangement of the medial geniculate body it is first necessary to understand the cytoarchitectural organization of the auditory thalamus. Figure 16 shows the subdivisions that can be identified most clearly; these are very similar to those seen in other mammals (Morest, 1964; Casseday et al., 1976). The ventral division (GMv) consists of small densely packed cells and receives ascending input only from the central nucleus of the inferior colliculus. The posterior group (Po) contains a variety of cell types, the most striking of which is a group of large cells (arrowhead) homologous to the suprageniculate nucleus (SG) of other mammals.

We summarize a series of experiments in which injections of axonal tracers were placed in tonotopically identified regions of the inferior colliculus, medial geniculate body or auditory cortex. The main conclusion from these experiments is that only the ventral division of the medial geniculate body has topographic connections which relate to the tonotopic arrangement of the system. The posterior group, including the suprageniculate nucleus (SG), has no clear topography but projects to all of auditory cortex.

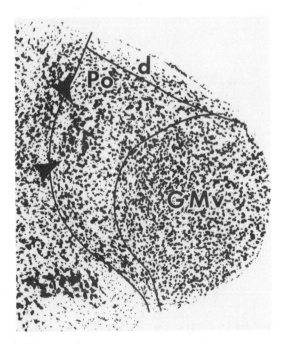

Fig. 16.   Photomicrograph of the medial geniculate body of Pteronotus parnellii parnellii. GMv contains densely packed small cells which project topographically to auditory cortex. Po is less densely packed with cells than GMv and the projections to auditory cortex are not topographic. The suprageniculate nucleus is a distinct group of large cells (arrowheads) within Po.

Figure 17 summarizes the location of labeled cells in the ventral division of
the medial geniculate body following small injections of HRP in the auditory
cortex. In each case the best frequency of cells was determined at the center
of the injection site, and in many cases the tonotopic organization of the
area surrounding the injection was also mapped. The point which relates to
biosonar is that a large area within GMv is connected to cortical units which
respond to the dominant harmonic of the CF echolocation pulse. This area
falls within a tonotopic sequence in which the lowest frequencies are located
anteriorly, laterally and dorsally, and the highest frequencies are located
posteriorly, medially and ventrally. The projections from Po and SG have
little if any topography and seem to project diffusely to the entire auditory
field. A further difference is in the cortical layer of termination. GMv
projects most densely to layers III-IV; Po and SG to layers I and VI. Whether
or not Po and SG are tonotopically organized remains to be determined.

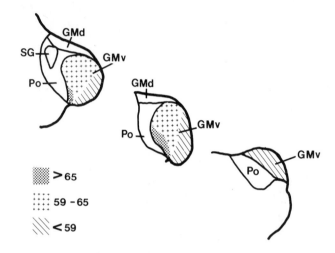

Fig. 17.   Summary diagram of relationship between location of labeled cells
            in the ventral division of the medial geniculate body and the
            tonotopic location of the HRP injection in auditory cortex. After
            injections in high frequency areas, labeled cells are located
            ventrally, medially and posteriorly in GMv. Injections in low
            frequency cortical areas yield labeled cells in a crescent shaped
            area which extends posteriorly from the ventral lateral edge to
            occupy the entire GMv anteriorly. Injections in areas responsive
            to frequencies at and around the CF yield labeled cells in the
            large area which comprises the remaining central part of GMv.

The central acoustic tract

        Although nearly all ascending pathways to the medial geniculate body
arise from the inferior colliculus, Cajal (1909) and Papez (1929) described
an auditory pathway that bypasses the inferior colliculus to reach SG in the
thalamus. The fibers of this pathway were thought to send collaterals to the
deep superior colliculus and the suprageniculate nucleus (see also Morest,
1965). Our results show that this pathway is prominent in Pteronotus. Figure
17 suggests that SG is the main thalamic target of the anterolateral
periolivary nucleus (ALPO), that GMv does not receive input from ALPO but
that the intermediate and deep layers of the superior colliculus do. Results
of retrograde transport from these targets confirm that ALPO is the source of
these projections. ALPO consists of a distinct group of large multipolar

186

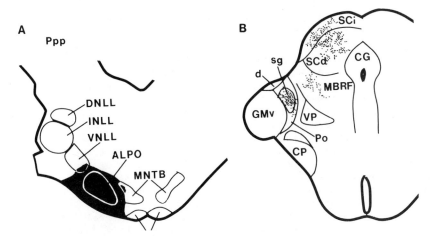

Fig. 18. Drawing to illustrate anterograde transport to the suprageniculate
nucleus and to the superior colliculus after an injection of WGA-
HRP in the anterolateral periolivary nucleus (ALPO). Other
experiments show that the areas adjacent to ALPO, involved in this
injection, do not contribute to the transport shown here (Data from
J. Kobler, S. Isbey and J. Casseday, unpublished).

cells which receive input from the contralateral cochlear nucleus. In short,
ALPO is the origin of a pathway whereby auditory information can reach the
suprageniculate nucleus without a synapse in the inferior colliculus. Thus it
is a "fast" route to auditory thalamus. Because of the relation of this
pathway to motor centers of the superior colliculus for head and pinna
movement the possibility that it may play a role in sonar performance
deserves attention.

BINAURAL AND MONAURAL PATHWAYS

Binaural pathways and the localization of echoes

Clearly, localization of sound is a very important part of echolocation,
and one in which the structures of the superior olivary complex must play an
important role. It is tempting to speculate that the variability seen in the
superior olivary complex of bats is related in some way to the differences in
the design of sonar pulses; the echoes from different types of pulses must
impose on the binaural system very different encoding requirements.

The bilateral connections to the superior olives were the first clues
that these structures might function in binaural hearing. It is now
generally accepted, for large land-dwelling mammals which are sensitive to
low-frequency signals, that cells in LSO process binaural intensity
differences, whereas cells in MSO process binaural time or phase differences
(see part II). If this is also true for bats, it is not surprising that LSO
is prominent in echolocating bats. By the same logic, it is surprising to
find a prominent MSO in Pteronotus. Thus, the connections become especially
important as a means of verifying that the structure which appears to be MSO
is really homologous to MSO of other mammals. Figures 7 and 19 illustrate this
and other points about the connections of the superior olivary complex.
First, LSO receives input only from the ipsilateral cochlear nucleus, whereas
MSO receives bilateral input. Second, most of this input to LSO and MSO
arises from AVCN (Fig.19); a small amount may come from PVCN, although this is
open to question. No input arises from DCN. In short, AVCN is the primary

origin of input to binaural centers which must be important for localizing echoes in space (see summary Fig. 20).

The ascending pathways from LSO and MSO have been examined in a number of experiments (Schweizer, 1981; Zook and Casseday, 1982b; 1986) and are summarized in Fig. 20. The results show that LSO projects bilaterally whereas MSO projects only ipsilaterally. Both project to the dorsal nucleus of the lateral lemniscus and to the inferior colliculus. The pattern of these projections can be seen by retrograde transport from the inferior colliculus (Fig. 6) and by anterograde transport from the superior olivary nuclei (see Casseday et al., this volume). We shall discuss further the projections to the inferior colliculus when we consider the problem of how to interpret convergent and divergent projections in terms of monaural and binaural pathways. Here we simply point out that the total target of projections from LSO and MSO does not occupy the entire central nucleus of the inferior colliculus but is restricted to approximately the ventral-lateral two-thirds of the central nucleus (see also Casseday et al., this volume).

In the lateral lemniscus only the dorsal nucleus (DNLL) is clearly connected with the binaural system via its input from MSO and LSO. This observation is important for echolocation because DNLL projects not only to the inferior colliculus but also to the deep layers of the superior colliculus (Covey et al., 1986). Thus DNLL is a major link, and the most direct source, of binaural input to motor centers responsible for turning the head and orienting the pinnae to sound sources.

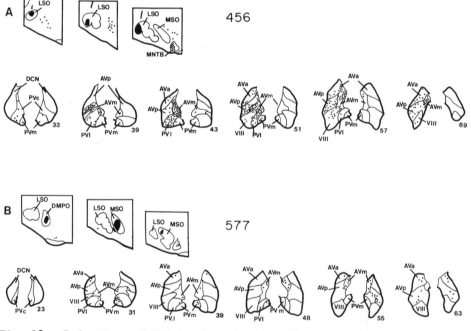

Fig. 19.   Injections of HRP in the superior olivary complex to show the location in the cochlear nucleus of cells which project to LSO and MSO. A) Results of an injection in LSO: labeled cells are located almost exclusively in the ipsilateral cochlear nucleus, mainly in AVCN; a large number of these cells are the large multipolar cells of the marginal division.   B) Results of an injection in MSO: labeled cells are located bilaterally and mainly in AVCN. In both cases some labeled cells are found in PVCN but none in DCN. (adapted from Zook and Casseday, 1986).

Monaural pathways and temporal analysis

As Fig. 21 shows, not all pathways from the cochlear nucleus project to the superior olivary complex. Considering the pathways from DCN first, it has already been shown that HRP injections in LSO or MSO do not label cells in DCN. This observation, combined with the results of anterograde transport from [^{3}H]-leucine injections in DCN, shows that fibers from DCN bypass the superior olivary complex and project directly to the contralateral inferior colliculus (Zook and Casseday 1985; 1986). Two additional monaural pathways to the inferior colliculus arise from the cochlear nucleus, one from AVCN and one from PVCN (Schweizer, 1981; Zook and Casseday, 1982b; 1985). Thus, each division of the cochlear nucleus sends a direct monaural pathway to the contralateral inferior colliculus. We shall return to this point later.

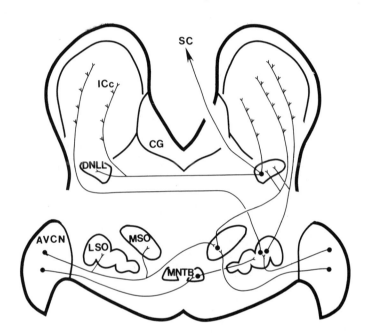

Fig. 20.   Summary diagram to show ascending binaural pathways. To simplify the picture only projections via the right superior olivary complex are shown. MSO receives direct bilateral projections from AVCN. LSO receives direct projections from the ipsilateral AVCN and indirect projections, via MNTB, from the contralateral AVCN. MSO projects to DNLL and inferior colliculus ipsilaterally. LSO projects to DNLL and inferior colliculus bilaterally. DNLL projects bilaterally to the inferior colliculus and also projects to the deep layers of the superior colliculus.

Experiments on Pteronotus, Rhinolophus and Eptesicus have shown that each nucleus of the lateral lemniscus sends a separate projection to the ipsilateral inferior colliculus (Schweizer 1981; Zook and Casseday, 1982b; Covey and Casseday, 1986). The ascending projections from the AVCN likewise diverge to three separate targets at the lateral lemniscus, one in INLL and

189

one in each of two divisions of VNLL (Figs. 7, 8). PVCN probably contributes
to these pathways as well, e.g. some octopus cells almost certainly project
to VNLL (Zook and Casseday 1985). Finally, MNTB projects to the ipsilateral
INLL. Figure 21 summarizes the main sources of pathways to the lateral
lemniscus. The major point is that these nuclei receive several
contralateral monaural inputs, some directly from the cochlear nuclei and one
indirectly via MNTB. This system obviously provides the opportunity for
different transformations, including different delays, of auditory stimuli.
The elaboration of the nuclei of the lateral lemniscus in echolocating bats
is a clear hint that this monaural system may perform frequency or temporal
analysis of echoes. Thus it is tempting to speculate that these pathways may
be the neural substrate for mechanisms, such as autocorrelation (Simmons,
1971; 1979), to analyze the temporal features of biosonar signals.

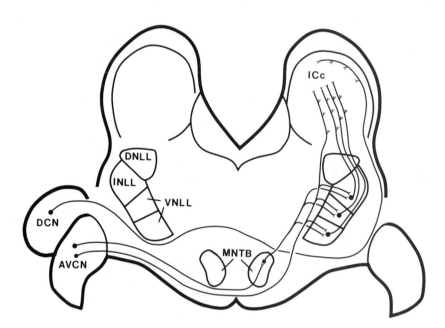

Fig. 21.   Schematic drawing to show some of the monaural pathways in the
           brainstem of <u>Pteronotus parnellii parnellii</u>. DCN projects directly
           to the contralateral inferior colliculus. AVCN projects directly
           and indirectly via relays in MNTB and in the nuclei of the lateral
           lemniscus. Pathways from PVCN are not shown because of
           insufficient knowledge about their course and terminations in the
           Mustache bat, but they do contribute to the monaural pathways.

     As an anatomical substrate for autocorrelation, one would expect to see
a number of pathways which have a common origin and project to common
targets; some of the pathways would arrive directly whereas others would
arrive only after a delay. Considering for the moment only the pathways from
AVCN, it is easy to see just such an arrangement, that is, different pathways
to the inferior colliculus involving one, two or three neurons. First, there
is a direct pathway from AVCN to the inferior colliculus. Second, at least

three separate pathways via the nuclei of the lateral lemniscus involve two neurons. Finally, the three neuron pathway from AVCN has a synapse in MNTB and another in the intermediate nucleus of the lateral lemniscus. Figure 20 depicts these pathways from AVCN. If we also included direct pathways from DCN and PVCN, additional sources of delays might be envisioned. However, the pathways via the nuclei of the lateral lemniscus are the ones that are especially elaborated in bats, and these are our candidates for processing neural delays that would be part of a mechanism for encoding temporal features of sonar signals.

CONVERGENCE AND INTEGRATION AT THE INFERIOR COLLICULUS

In the preceding sections we have presented evidence supporting the idea that there are a number of parallel auditory pathways in the lower brain stem. We have argued that there may be a major functional division into monaural and binaural systems. We suggest further that, in our opinion, most if not all of these anatomical subdivisions must represent functional subdivisions as well, many of which have not yet been revealed by physiological experiments. Viewed in this light, the inferior colliculus presents an especially interesting challenge for understanding how the ascending auditory system processes auditory signals. If the system divides the processing into separate tasks in the lower brain stem, then to some extent it merges the tasks again at the inferior colliculus, because many of the pathways that arise separately from the cochlear nuclei, superior olivary nuclei and nuclei of the lateral lemniscus converge again at the inferior colliculus. Our purpose in this section is to describe these pathways. We shall present evidence to show that there is convergence of many pathways at the inferior colliculus, that some separation exists between dorsal and ventral parts of the central nucleus and that the laminar distribution of ascending input to the central nucleus may be a clue to the organization of ascending pathways to the inferior colliculus.

In this examination of ascending projections to the inferior colliculus we shall try to relate the results to the analysis of the intrinsic organization of the inferior colliculus presented earlier. It must be kept in mind, however, that the relationship can be no more than a rough approximation because of differences in method. We shall first describe the targets of the cochlear nuclei in the inferior collciulus and then consider how these targets are related to the targets of the superior olivary nuclei and nuclei of the lateral lemniscus.

Targets of pathways from the cochlear nucleus

Figure 22 illustrates the topography of the AVCN target and also shows differences between the AVCN target and the DCN target. The anterior part of AVCN projects to the lateral part of the central nucleus (Fig. 22A). The medial division of the inferior colliculus is the target of the posterior part of AVCN as demonstrated by an injection of [^{3}H]-leucine in this part of the cochlear nucleus (Fig. 22B). These projections no doubt match the tonotopic arrangement of the inferior colliculus in the following sense. Assuming that AVCN of Pteronotus has a tonotopic organization like that of other bats studied (Feng and Vater,1985; Casseday and Covey,1985), the lowest frequencies are located in the most anterior and ventral parts of AVCN while the highest frequencies are located in the most dorsal and posterior parts of AVCN. The exception is that the frequencies near the main harmonic of the CF pulse (60 kHz) are located in the dorsoposterior division. That is, they fall out of the sequence in which the lowest frequencies are located anterior, dorsal and lateral whereas the highest are posterior, ventral and medial.

Figure 22C illustrates differences between the targets of DCN and AVCN. First, the injection in DCN results in a diffuse but widespread pattern of label in the inferior colliculus. There is no evidence of the banded pattern seen in the projections from AVCN. Second, although the DCN target overlaps the AVCN target, it includes the dorsal part of the inferior colliculus, extending into the pericentral areas. Incomplete evidence concerning projections from PVCN suggests that this system also has a diffuse projection to the inferior colliculus, although additional evidence is needed on this point.

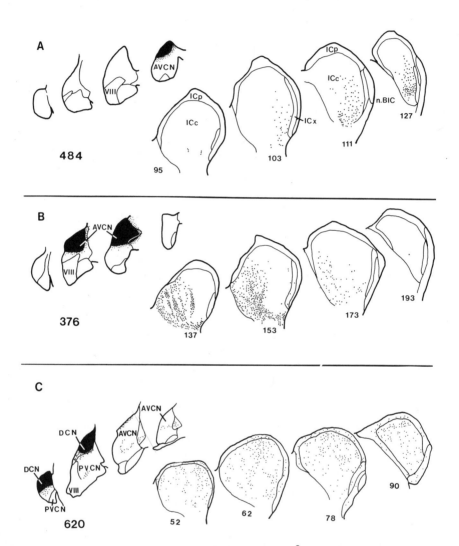

Fig. 22. Drawings of anterograde transport of [3H]-leucine in the inferior colliculus to show differences between the target of AVCN and DCN. A) transport to lateral and anterior (low frequency) areas of the central nucleus of the inferior colliculus following an injection in anterior AVCN. B) Transport to medial and posterior parts of the central nucleus following an injection in posterior AVCN. Note the absence of transport to most of the dorsal one-third of the central nucleus. C) Transport to inferior colliculus from an injection in DCN. Note that there evidence for transport from DCN to the dorsal parts of the inferior colliculus, that is to the areas that are not labeled after injections in AVCN.

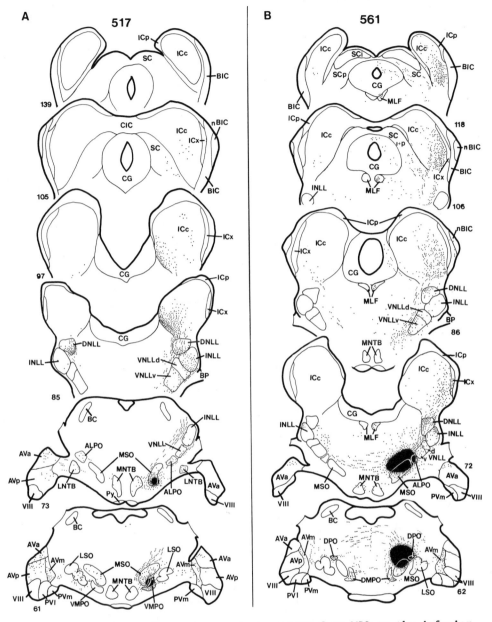

Fig. 23. Drawings to show anterograde transport from MSO to the inferior colliculus. A) An injection of [^{3}H]-leucine involving the ventral part of MSO yields labeling in the medial and posterior part of the inferior colliculus. B) An injection of [^{3}H]-leucine involving the dorsal part of MSO yields transport to the lateral part of the inferior colliculus. Note that the projections are virtually confined to the ventral two-thirds of the inferior colliculus in both cases.

Convergence of binaural and monaural pathways at the inferior colliculus

The next step in the analysis of pathways to the inferior colliculus is to illustrate the targets of the medial and lateral superior olives (MSO and LSO) and the target of the dorsal nucleus of the lateral lemniscus (DNLL). Figure 23 illustrates two points, the topography of projections from MSO in Pteronotus and the extent of the target of the entire MSO within the inferior colliculus. In one case (Fig. 23A) the ^{3}H-leucine injection is ventral, presumably in high frequency areas, and the projections are confined to the ventromedial part of the central nucleus of the inferior colliculus. The second case (Fig. 23B) shows that dorsal MSO, presumably the low frequency part, projects to the lateral part of the central nucleus. In neither case does the target extend into the most dorsomedial part of the central nucleus. The results from [^{3}H]-leucine injections in LSO of Pteronotus are similar. From a number of experiments to show anterograde transport from LSO and MSO and to show retrograde transport from the inferior colliculus, we conclude that the target of MSO and LSO does not extend into the dorsalmost part of the central nucleus. It remains to be determined how much of this target is included in the dorso-posterior area described by Zook et al.(1985). In Rhinolophus rouxii, the projections from LSO and MSO draw a clear border to separate the dorsomedial part of the central nucleus from the rest of the inferior colliculus (see Casseday et al., this volume).

The dorsal nucleus of the lateral lemniscus receives its main input from LSO and MSO and it in turn projects to the same target in the inferior colliculus as LSO and MSO. These projections, like those from LSO and MSO,

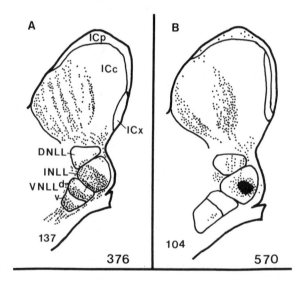

Fig. 24.   Drawings of anterograde transport of [^{3}H]-leucine to show convergence at the inferior colliculus of direct and indirect pathways from AVCN. A) Transport to the nuclei of the lateral lemniscus and inferior colliculus from the posterior AVCN. B) transport to inferior colliculus from an injection (blackened area) in INLL, mainly in the target of posterior AVCN. Note that the target of the indirect projections overlaps with the target of the direct projections. The orientation of the bands of projections is parallel to the orientation of the disc shaped cells in this part of the inferior colliculus.

are bilateral. The DNLL also projects to the deep layers of the superior colliculus, an area of the tectum which plays an essential role in initiating movements of head, pinnae and eyes toward objects in space. As we pointed out earlier, this pathway may serve some function in the orientation of pinnae and head position according to the location of echoes.

Turning to the monaural pathways, the main target of the ventral and intermediate nuclei of the lateral lemniscus is the ventral two-thirds of the central nucleus, i.e., the same target as that of AVCN, LSO, MSO and DNLL. As an example of the convergence of these pathways, Figure 24 shows the pattern of transport from posterior AVCN and from INLL. This figure helps illustrate important points concerning the banded pattern of projections to the inferior colliculus and the relationship between direct and indirect pathways from AVCN. The banded pattern of the projections seems to arise from several sources, not just the two shown here. This finding raises the question of whether the bands from one source overlap or interdigitate with the bands from another source. The answer would be a step in understanding how the various pathways interact with one another at the inferior colliculus. Figure 24 also illustrates that direct and indirect pathways from AVCN converge in the same general area of the inferior colliculus. That is, caudal AVCN projects to medial inferior colliculus and to ventral INLL (Fig. 23A); INLL in turn projects to the medial part of the inferior colliculus (Fig. 23B).

The AVCN system and convergence of pathways at the inferior colliculus

We have shown that the binaural pathways have a specific target in the inferior colliculus. This conclusion is not equivalent to saying that this part of the inferior colliculus is a "center" for sound localization or even for binaural processing, because this area of the inferior colliculus is also the target of a number of other pathways. The common element of this part of the inferior colliculus is that it is the target of pathways that have their origin, directly or indirectly, in AVCN. Therefore we refer to this set of pathways as the AVCN system. The system consists of the monaural pathways which are the direct pathways from AVCN and the indirect pathways from AVCN

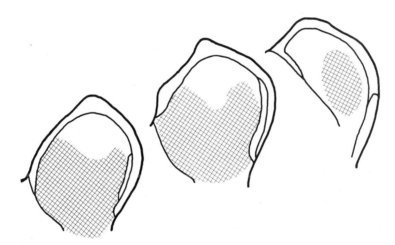

Fig. 25. Schematic drawing of the inferior colliculus to show the area which receives both direct and indirect projections from AVCN. The hatched area is the target of AVCN, as well as the target of those nuclei which receive their major input from AVCN: LSO, MSO, VNLL and INLL.

via the lateral lemniscus (Fig.25), it also includes the binaural pathways which consist of indirect pathways via LSO, MSO and DNLL. In short the AVCN system consists of several monaural and binaural pathways all of which converge in the same part of the inferior colliculus. An important question to answer in future experiments is whether these pathways converge at the level of single cells. The answer will be important for determining whether or not the monaural pathways play a role in sound localization. For example, does the direct pathway from AVCN terminate on the same cells as the binaural pathway from MSO? If so, then sound localization may require more complex integration than simply the timing and intensity processing done at MSO and LSO. A clue that some such interaction does occur at the cellular level is seen in the temporal pattern of responses of binaural cells at the inferior colliculus. Many, perhaps most, of these cells respond phasically to the onset of a stimulus, whereas cells in LSO or MSO respond tonically for the duration of the stimulus (see part II).

The banded pattern of projections is such a consistent feature of the ascending auditory pathways that it is difficult to ignore as a clue to the organization of the system. Recent evidence from intracellular injection of HRP suggests that individual axons of the lateral lemniscus have a terminal field which has dimensions similar to the bands of transport in the inferior colliculus shown in Fig. 24. The terminal arbor of such an axon occupies a sheet of neuropil in the inferior colliculus about 100 um wide, extending through much of the central nucleus in the anterior-posterior and dorsal-ventral dimensions. Such observations suggest the hypothesis that these sheets are functional units. Some of these axons must arise from structures which themselves have a slab or sheet-like organization, such as the cochlear nucleus (Feng and Vater, 1985) or nuclei of the lateral lemniscus (Fig. 10). These sheets are clearly related to tonotopy at each level. If we can determine how the sheets at one level are connectionally related to those at another, we should be able to determine whether tonotopic order at one level is a one-to-one reflection of the previous stage in the pathway or whether convergence and divergence in the system allows for comparison and separation of frequency representations. The latter possibilities would seem important requirements for analysis of some of the spectrally complex sounds of animal sonar.

CONCLUSIONS

The inferior colliculus is the target of a number of parallel auditory pathways. It has one tonotopic organization, which is the product of the separate tonotopic organizations of the pathways which ascend to it. Some of the pathways are primarily binaural and probably serve in localization of echoes and other sounds. Other pathways are primarily monaural and may play a role in temporal analysis such as pulse-echo delay. We have only a sketchy idea how these multiple auditory pathways converge at the inferior colliculus. This gap in our knowledge is the chief obstacle to understanding how the auditory pathways in the lower brain stem are related to auditory pathways in the thalamus and cortex. We will need to understand more about the transformations that take place at the inferior colliculus in order to suggest how these pathways might be represented in the thalamus and cortex.

ACKNOWLEDGMENTS

We thank E. Covey for help and discussions in the preparation of this manuscript. This research was supported by NIH grants NS 21748 (JHC), NS 13276 (GDP) and NSF grant BNS-8217357 (JHC).

REFERENCES (see part II)

PARALLEL AUDITORY PATHWAYS: II - FUNCTIONAL PROPERTIES

George D. Pollak* and John H. Casseday#

*Department of Zoology
University of Texas
Austin, TX 78712

#Department of Surgery and Psychology
Duke University
Durham, NC 27710

INTRODUCTION

In the previous chapter the wiring diagram of the auditory system was described and we pointed out the functional consequences of these connectional patterns, especially how they preserve the cochlear tonotopic organization. In this chapter we discuss the physiological characteristics of the brainstem auditory nuclei in echolocating bats, considering two major challenges bats face: 1) The evaluation of target range coded by the temporal interval between an emitted pulse and echo; and 2) The localization of a sound in space which is computed from interaural intensity disparities.

We will emphasize the pronounced divergence and convergence of these pathways as they ascend to the central nucleus of the inferior colliculus (ICc) where their afferents make terminal synapses. The properties of auditory neurons within the various levels of the system are heterogeneous and presumably reflect the different aspects of biosonar information processed within each region. We focus on the processing, and progressive signal transformations that occur within the AVCN and its parallel pathways innervating the lateral superior olive (LSO), medial superior olive (MSO), ventral nucleus of the lateral lemniscus (VNLL), and the ICc since these have been studied, and represent many of the principal brainstem nuclei in the primary auditory system. Diverging features will illustrate how the various parallel pathways process different aspects of biosonar information. We will then elaborate on signal processing and discuss some consequences of the converging inputs from the two ears for binaural processing in the MSO and LSO. The final section will consider hierarchical information processing, specifically the implications of the massive convergence of the projections from all lower brainstem centers upon the ICc for the encoding and representation of acoustic space.

We address these issues with the following caveat. The recent studies of the chiropteran auditory system have provided major insights into the detailed structure-function relationships which are correlated with the requirements of echolocation. A number of findings are especially intriguing, and are strongly suggestive of functional systems whose morphological underpinnings are seemingly well suited for certain demands. We emphasize these systems and point out the presumed functional implications, but we shall also show where the major gaps in our knowledge exist. In many cases, we shall buttress our arguments with references to homologous systems in cats, where these are known in greater detail. Our intent is to bring the recent structure-function relationships found in bats into the framework of mammalian auditory processing in general, pointing out those which seem to be emerging into functional systems, and at the same time suggesting experiments for future studies.

The Auditory Nerve

Echolocating bats emit biosonar calls that typically cover at least one, and usually several octaves (Simmons et al. 1975; Novick 1977). The basic coversion of frequency to place in the cochlea, whereby the power spectrum of orientation sounds is tranformed into a spatial pattern of activity along the cochlear partition, has been described by Vater in this volume. A noteworthy point made in that chapter is that the cochleae of long CF/FM bats do not afford equal treatment to each frequency. The length of cochlear partition devoted to the dominant frequency of the constant frequency (CF) component is greatly expanded and has numerous morphological and physiological specializations.

The more peripheral regions of the mustache and horseshoe bats' auditory system have been the subject of several reports (Suga et al. 1975, 1976; Suga and Jen 1977; Suga and Tsuzuki 1985) but the recording sites were never verified, leaving the uncertainty as to whether neurons in the auditory nerve or cochlear nucleus were recorded. Nevertheless, there is little doubt that the elegance of the cochlear frequency-to-place transform is preserved in the population of auditory nerve fibers. The elongated region of basilar membrane devoted to the CF component in each species is manifest as a disproportionately large population of auditory nerve fibers tuned to 60 kHz in mustache bats and 80 kHz in horseshoe bats. The physiological and morphological specializations in the CF region of the cochlea impart exceptionally sharp tuning curves to these neurons. The studies of the mustache and horseshoe bats' auditory nerve have shown that units have V shaped tuning curves, analogous to the auditory nerve tuning curves of other mammals that have been studied. Each tuning curve is described by a quality factor (Q_{10dB}), which reflects its sharpness or frequency selectivity, and by its best frequency (BF), or frequency to which it is most sensitive. Units having BFs above or below the dominant CF component of the orientation cry have small Q_{10dB} values that range from 3-18, whereas units tuned to the CF component have Q_{10dB} values as high as 300-400. The spectrum of frequencies to which the cochlea responds is re-represented in the auditory nerve, but the representation is greatly weighted in favor of units that are very sharply tuned to 60 kHz neurons in mustache bats and to 80 kHz neurons in horseshoe bats.

Anteroventral Cochlear Nucleus

Upon entering the brain, auditory nerve fibers bifurcate into ascending and descending branches (Casseday and Pollak in this volume).

The ascending branch innervates neurons in the anteroventral cochlear nucleus (AVCN), whereas the descending branch innervates the dorsal and posteroventral cochlear nuclei. Below we consider the physiological properties of AVCN neurons since their projections form the major pathways to the superior olives and nuclei of the lateral lemniscus.

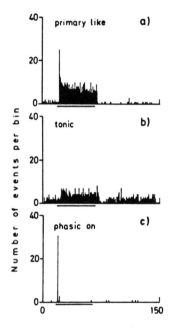

Figure 1. Peri-stimulus time histograms illustrating the major discharge patterns of neurons in the horseshoe bat's AVCN. Adapted from Feng and Vater (1985).

Discharge patterns of neurons localized in the AVCN have been described by Feng and Vater (1985) in horseshoe bats. The most common pattern, found in about half of the cells in this region, is a primary-like pattern, so called because of its similarity to the pattern evoked by tone bursts in auditory nerve fibers (Fig. 1). This pattern is characterized by an initially high discharge rate which then adapts to a lower rate that is maintained for the duration of the tone burst. Other AVCN neurons, however, respond with pure excitation, without the phasic component, or with a pure phasic-on pattern (Fig. 1).

It is tempting to suggest that the AVCN neurons discharging with a primary-like pattern are spherical and/or globular cells. These cell types dominate the AVCN, and in other animals are the recipients of calyciform and large bouton synaptic endings from auditory nerve fibers (Cant and Morest 1984). These synapses are some of the largest and most secure in the mammalian nervous system and are considered to

permit a faithful one-to-one transmission from pre- to postsynaptic
cell. However, the evidence for attributing a discharge pattern to a
particular neuronal type is only correlational and must be considered
to be tentative.

Medial Superior Olive

The MSO is a binaural nucleus receiving direct innervation from
the AVCN bilaterally (see Casseday and Pollak in this volume). The two
major types of binaural cells in the auditory system are E-E neurons,
defined as cells that can be driven by stimuli delivered to either ear
alone, and E-I cells, that receive excitation from only one ear and
inhibition from the other (Fig. 2). E-I neurons have traditionally
been considered important for the neural coding of sound location
because they compare the sound intensity at one ear with the intensity
at the other (Goldberg and Brown 1969; Goldberg 1975; Erulkar 1972; Jen

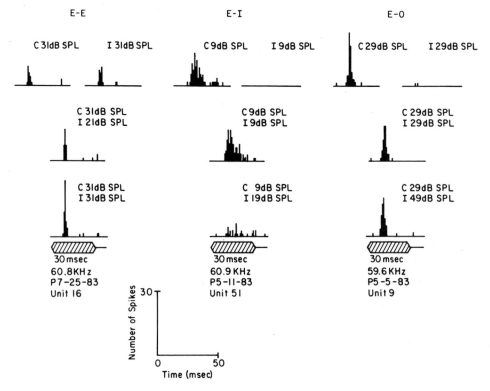

Figure 2. Peri-stimulus time histograms illustrating
major binaural types of neurons in the auditory system.
Binaural cell receiving excitatory input from both ears,
E-E cell, is on far left, an excitatory-inhibitory
cell, E-I, is in middle panel and a monaural E-O cell
is on far right. Stimulus intensity to contralateral ear
is indicated by C and stimulus to ipsilateral ear is
indicated by I.

1980; Schlegel 1977; Fuzessery and Pollak 1984, 1985). They are thus sensitive to interaural intensity disparities which are created by high frequencies due to acoustic shadowing and the directional properties of the ear (Grinnell and Grinnell 1965; Fuzessery and Pollak 1984, 1985; Harnischfeger et al 1985). E-E cells, on the other hand, are thought to be relatively insensitive to interaural intensity disparities, and are generally considered to play little or no role in sound localization (Goldberg and Brown 1969; Goldberg 1975). We shall comment upon the role of E-E neurons in sound localization in a later section.

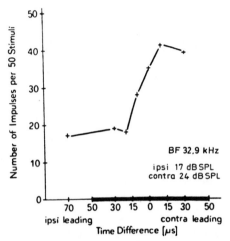

Figure 3. Impulse-count function for variations of interaural time differences in the microsecond range. The interaural intensity difference was fixed as indicated. Thickened part of abscissa indicates naturally occurring time difference. From Harnischfeger et al. (1985).

Harnischfeger et al. (1985) recorded from MSO neurons of Molossus ater. Stimulation of the excitatory ear evoked a tonic discharge pattern in 75% of MSO neurons. The aural types of most MSO neurons were E-E or E-I, and occurred in almost equal numbers, but a small number of monaural neurons were also found. The presence of monaural cells in the MSO is puzzling. The principal cells, and other types, in the cat's MSO receive binaural innervation. Thus, if the MSO in Molossus is homologous with the MSO in other mammals, it seems likely that the monaural units Harnischfeger et al. recorded were periolivary neurons located in close proximity to the MSO.

One of the surprising, and most important, findings was that several of the binaural MSO cells were sensitive to interaural timing differences of ± 500 microsec and two neurons were sensitive to timing differences as small as ± 50 microsec (Fig. 3). Interaural timing disparities in the range of ± 50 microsec are created by sounds located at different azimuthal positions around the bat's head. The ability of bats to utilize time disparities for sound localization has not previously been given much credence, but neurons exhibiting sensitivites to these small temporal disparities could be important for encoding sound location. These sensitivities to interaural time differences, however, were only evident with specific interaural intensity differences. That is, when the intensities at the two ears had an appropriate disparity, advancing or delaying the signal in one ear by a few microseconds resulted in a significant change in the neuron's discharge rate.

The time sensitive E-I MSO cells also exhibited marked changes in discharge rate to interaural intensity disparities (IIDs). In these cells, making the sound louder to the inhibitory ear, which is usually the contralateral ear, causes a profound depression of discharge rate. However, in the real world, a sound is louder in the contralateral ear only when the sound is closer to that ear, which also results in an earlier arrival time. The earlier arrival time by itself produces a depression of firing rate in these neurons. Furthermore, the change in firing rate that occurred with slight changes in the timing of the signal onset at one ear could be compensated for by changing the intensity to that ear. The explanation offered for this time-intensity trade involves the latency changes with intensity that are prevalent in sensory systems. Specifically, since firing latency shortens as signal intensity is increased, making the sound louder at one ear should effectively create a shortening in the discharge latency arriving from that ear. Thus delaying the sound to one ear, which let us say increases the firing rate, can be offset by increasing the loudness to the same ear, thereby shortening the firing latency evoked by that ear, and re-establishing the original binaural temporal pattern arriving at the neuron. The interpretation given to these results is that these intensity and time disparities work in a complementary fashion to change the discharge rate as the sound is moved within small increments around the bat's head.

It is noteworthy that intensity differences are far more potent for influencing discharge rates than are time differences. Indeed, for the neurons that were studied a change in timing to one ear of 8-50 microsec produced an effect equal to an intensity change of 1 db. In other words, if intensity and temporal effects were independent, the interaural intensity difference resulting from a change in azimuthal position of less than one degree, would have an equal effect to the temporal change that occurs when the azimuth of the sound is changed by 15-90 degrees. The very large time-intensity trading ratios suggest that intensity is the dominant cue for azimuthal location and not time.

The significance of these findings is fundamental because they address questions concerned with basic mechanisms of auditory processing. But as with all conceptually important studies they raise more questions than they answer. What, for example, is the significance of the sort of temporal sensitivity reported by Harnischfer et al. for sound localization? A feature requiring further clarification is whether the majority of these neurons display a sensitivity to temporal and intensity disparities that correspond to comparable regions of the sound field. Another question concerns the small number of time sensitive units recorded. Was this a consequence of sampling problems, or do those MSO units sensitive to such small temporal disparities constitute only a small subset of the population? It is questions such as these which were raised by Harnischfeger et al. that offer exciting possibilities for future experiments.

Lateral Superior Olive

In cats the LSO is a homogeneous nucleus in terms of its neuronal architecture (Cant 1984; Helfert and Schwartz 1984), discharge patterns (Tsuchitani and Johnson 1985), and binaural type, which are all E-I (Boudreau and Tsuchitani 1968; Tsuchitani 1977). Almost all neurons receive excitation from the ipsilateral ear, and respond with a chopper response pattern. Inhibitory inputs originate from the contralateral ear. Harnischfeger et al. (1985) recorded from ten LSO cells in Molossus and all of those cells had properties similar to those in the cat's LSO. The noteworthy feature, however, was that in contrast to MSO neurons, LSO neurons in Molossus are not sensitive to small time differences of $\pm$ 50 microsec or less. Rather they are sensitive predominantly to interaural intensity differences.

This finding suggests a fundamental difference in mechanism between the LSO and MSO that is consistent with traditional thinking. In cats and other animals, the LSO is thought to process only intensity disparities whereas the MSO processes interaural phase disparities (Erulkar 1972; Masterton and Diamond 1973; Masterton et al. 1975). But in these animals there is a pronounced difference in frequency representation in the two regions, where the LSO is predominantly high frequency (Tsuchitani and Boudreau 1966; Guinan et al. 1972) and MSO predominantly low frequency (Goldberg and Brown 1967; Guinan et al. 1972). Temporal sensitivity may not have been noted previously because high frequency neurons do not phase lock to the cycle-by-cycle action of a high frequency sinusoid, precluding any processing of interaural phase disparities.

Remarkably, the classical studies of the cat's LSO failed to test for sensitivity to onset disparities. Therefore, the possiblity exists that temporal processing may be occurring in the LSO, but in the millisecond rather than the microsecond range. The temporal mechanism could be one where a difference in the intensities at the two ears would generate a latency disparity impinging upon the LSO neuron. If this were the case, then the effect of a given intensity disparity should be compensated for by a temporal shift in the signal to one ear, but with a value equivalent to the intensity induced latency difference, which is in the millisecond rather than the microsecond range. Recent evidence from the cat's LSO is suggestive of such temporal processing, at least to certain degree (Caird and Klinke 1983). Thus the difference in temporal sensitivty between the LSO and MSO may be quantitative, rather than a qualitative difference in mechanism. Defining this distinction between LSO and MSO remains as one of the important questions in auditory physiology.

Nuclei of the Lateral Lemniscus

No published information is available documenting the discharge properties of neurons in any of the lateral lemniscal nuclei of bats. However, a number of correlations between the physiology and synaptology of AVCN neurons, and the connectivity with the columnar VNLL, are highly suggestive. The columnar VNLL is noteworthy because each row of cells receives projections from corresponding isofrequency regions of the AVCN. This region of the AVCN is dominated by small spherical and globular cells, cell types that likely receive calyciform or large bouton synapses. If the AVCN cells receiving calyciform synapses form the dorminant population projecting to the VNLL, then these same cells terminate with calyciform endings on the columnar neurons of the VNLL as well. A direct, monaural "calyciform pathway" form AVCN to VNLL would be well suited for precisely encoding the temporal features of an acoustic signal.

Inferior Colliculus

Temporal coding related to echo ranging and phasic-on neurons

One of the consistent transformations that occur between the lower brainstem auditory nuclei and the ICc is a change in discharge pattern, from predominantly tonic varieties in the lower regions to predominantly phasic-on patterns in the ICc (e.g., Suga 1971; Harnischfeger et al. 1985). In the section below we consider the monaural pathway from the AVCN leading to the columnar VNLL, and its projection upon the ICc. We suggest a relationship between the columnar VNLL and a variety of phasic-on ICc neurons, the constant latency neurons, and describe the presumptive role of these collicular neurons for temporal coding. Evidence for temporal coding by bats comes from behavioral studies concerned with target ranging (Simmons 1973). The signals from which range is extracted are brief, frequency modulated (FM) pulses that sweep down an octave or less in frequency. These experiments showed that the cue for range measurement is the time interval between the reception of a pulse and echo at the ear. By closely marking the time of occurrence of an emitted pulse and the time of occurrence of a returning echo, bats are able to determine target range with a precision great enough to detect changes as small as 75 microsec.

At the level of the inferior colliculus, there is a population of cells characterized by phasic-on discharge patterns and discharge latencies with only slight jitter (Pollak et al. 1977; Pollak 1981; Bodenhamer et al. 1979; Bodenhamer and Pollak 1981). These response properties enable these phasic constant latency responders (pCLRs) to mark accurately the time of occurrence of a pulse and echo such that the inter-response interval closely mirrors the actual pulse-echo interval (Fig. 4). When driven by a brief FM signal, pCLRs respond to only one frequency, or a very narrow band of frequencies, in the FM signal (Fig. 4), and not to the signal onset. When such a brief FM burst is presented repeatedly, the neuron's discharges lock onto the same frequency component, and the latencies fall into tight registration. The latency variation to repetitive FM signals is at most $\pm$ 250 sec, and the latency shortening with intensity is at most 1.75 msec, and usually less. Following from this is that when the duration of an FM signal is changed, but the starting and ending frequencies are kept constant, the discharge latency shifts in accordance with the frequency to which the cell is locked (Fig. 4) (Bodenhamer and Pollak, 1981). The frequency of an FM burst to which the discharges are locked is called the cell's excitatory frequency (EF).

Recall that a neuron's best frequency is thought to be a reflection of the place along the cochlear partition from which that neuron receives its primary innervation. Previous sections have amply demonstrated the orderly representation of BFs within each auditory nucleus. There is also an orderly progression of EFs along the dorsoventral axis of the ICc, where units with low EF's are located dorsally and those with high EFs are ventral, a pattern that closely follows the tonotopic progression of BFs in the inferior colliculus (Fig. 5) (Bodenhamer and Pollak 1981). Thus, the EF evoked by an FM

204

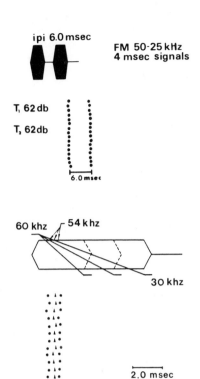

Figure 4. Top: Response of a constant latency unit to two
downward FM sweeps illustrating the precision with which these
units record the temporal interval between a pulse and echo.
The temporal, or interpulse interval (IPI) was 6.0 msec. Each
dot pair is action potentials evoked by the first and second
FM bursts. These bursts were presented repetitively. Lower:
Responses of a constant latency cell to 4.0, 6.0 and 8.0 msec
durations of a 60-30 kHz FM sweep. Each dot column represents
action potentials elicited by 16 stimulus presentations at
a given duration. The first column was produced in response
to the 4.0 ms duration signal, the second to the 6.0 ms duration
signal, and the third to the 8.0 ms signal. The dot columns
have been aligned directly beneath the FM sweeps to show their
temporal relationship to the location of the EF in each sweep.
The EF for this cell was approximately 54 kHz and its position
in the sweeps is indicated by arrows. Note lack of overlapping
dot columns even though latency shifts from one signal duration
to the next are quite small. Mean latency is given in ms below
each dot column. Stimulus intensity was 72 dB SPL. From Bodenhamer
and Pollak (1981).

signal is the equivalent of the neuron's BF determined with pure tones.

These units, then, are highly selective for both temporal and spectral features of the biosonar signals, and ensembles of these neurons should recreate the biosonar signal, in both time and frequency, across the tonotopically organized neural tissue (Fig. 5). Since the frequencies of an FM burst occur sequentially in time, and since these are further spread due to travel time in the cochlea, the neurons synchronized to high frequencies should be the first to discharge, followed by units having progressively lower EFs. Additionally, the pattern of activity should be spatially separated in the ICc due to tonotopy, where the most ventral neurons discharge first, followed sequentially in time by neurons in the more dorsal regions.

Target range is determined by evaluations of pulse-echo intervals. The pCLRs code for range with a precisely timed discharge to one frequency component of the pulse, and a precisely timed discharge to the same frequency component of the echo. The temporal separation between these discharges then is the representation of target range. This basic discharge pattern is repeated along the dorsoventral axis of the ICc, but for a different FM frequency component in each isofrequency sheet.

Although pCLRs have been studied in detail only in the ICc of Mexican free-tailed bats, neurons having similar features have been found in a variety of other species and seem to be common among bats (Suga 1970; Vater and Schlegel 1979; Pollak and Bodenhamer 1981; Pollak and Schuller 1981; O'Neill 1985). It should be noted, however, that while the majority of ICc units exhibit a much tighter synchronization of discharges to brief FM signals than they do to tone bursts, it is only a minority whose latency jitter is so small as to qualify them for the encoding of target range (O'Neill 1981). It is tempting indeed to suggest that the population of pCLRs are the midbrain neurons which receive innervation from the columnar division of the VNLL. This would certainly be a parsimonious explanation, since the VNLL forms only a small portion of the total projections to the ICc, and calyciform synapses are thought to convey precise temporal information which remain unaffected by intensity changes.

Nevertheless, we also emphasize the speculative nature of this proposal. Discharge characteristics of columnar VNLL neurons have yet to be studied, and it is by no means certain that neurons classified as pCLRs are monaural, or that they receive their innervation from the columnar VNLL. Clarifying the relation between the pCLRs and columnar VNLL is another challenge for future studies.

Temporal coding for binaural processing

Neurons sensitive to interaural time disparities are also found in the ICc of M. ater (Harnischfeger et al. 1985). Of the 95 ICc units studied in M. ater., eight were time sensitive. Seven were classified as E-I and all had a phasic-on discharge pattern. The characteristics of their binaural sensitivities to onset temporal disparities were similar to those observed in the MSO, and were discussed previously.

Organizational features related to convergence of inputs upon the ICc.

A major theme of the previous discussion is that lower brainstem regions are both tonotopically organized and have a constellation of defining physiological properties. The projections from each of these regions converge upon comparable isofrequency regions of the ICc to

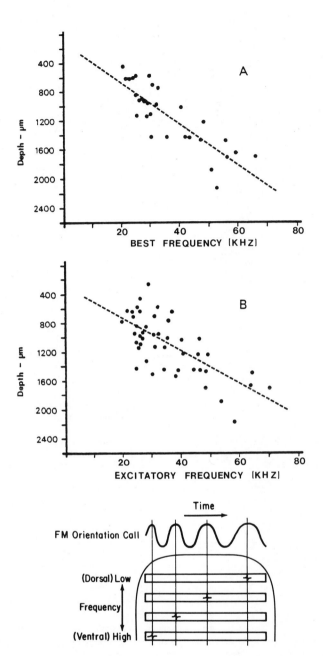

Figure 5. Upper panels: Plots of BF versus depth (A) and
of EF versus depth (B) in inferior colliculus of Mexican
free-tailed bat. In both cases there is an orderly arragement
of frequency with depth such that low EFs and BFs are found
dorsally, and high EFs and BFs are found ventrally. Dashed
lines in A and B are best fits for the data points. From
Bodenhamer and Pollak (1981). Lower panel: Schematic illustration
of discharges in time and space within the inferior colliculus,
elicited by a downward sweeping FM pulse. Further explanation
is provided in text.

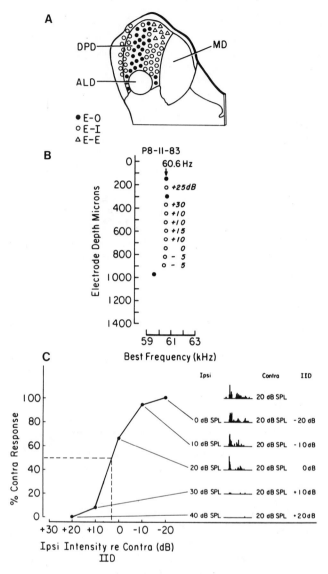

Figure 6. A: Schematic illustration of the topographic distribution
of aural response classes in a transverse section of the ICc. Symbols
are aural types recorded in the indicated regions of the DPD, and
in the adjacent external nucleus (not labeled). Responses recorded
from other divisions of the central nucleus are not shown. B: Aural
types and inhibitory thresholds of E-I unit clusters recorded in
one dorsoventral penetration through the DPD. Note the constant
BFs of the neurons in each penetration, and the regular progression
of inhibitory thresholds, shown to the right of each E-I cluster
recorded. C:Responses of a single E-I unit as a function of interaural
intensity disparity (IID). On the right are shown the peri-stimulus
time histograms of discharges evoked by the contralateral stimulus
(top histogram), and the inhibition produced as the intensity of
the ipsilateral (inhibitory) signal was increased. The best frequency
was 62 kHz and the threshold was 10 dB SPL. The inhibitory threshold
was +4 dB, as shown by the dashed lines in the IID function on the
left. From Pollak et al. (in press)

create a single tonotopic map. One of the questions raised previously concerned the specificity of connections between the columnar VNLL and the pCLRs in the ICc. This question can be cast in a more general sense to inquire into how the multiple projections from lower brainstem nuclei are represented within isofrequency sheets of the ICc. This problem has been studied by Wenstrup et al. (1985, 1986) in mustache bats using as a model the greatly hypertrophied dorsoposterior division (DPD), a 60 kHz isofrequency contour which receives a full complement of projections from all of the major lower auditory centers (Ross et al. 1985; Ross and Pollak, this volume).

The rationale was that since each lower region is either predominantly monaural or binaural, the degree to which monaural and binaural types are organized within the 60 kHz isofrequency DPD could be evaluated physiologically. The three aural types were classified as either monaural (E-O), or as E-E and E-I, the two major binaural types that were described previously. By mapping both single and multiunit responses, several zones in the DPD could be reliably identified, each having a predominant aural type (Fig. 6A). Monaural neurons are located along the dorsal and lateral parts of the DPD. E-E cells occur in two regions, one in the ventrolateral DPD and the other in the dorsomedial DPD. E-I neurons also have two zones: the main population is in the ventromedial region of the DPD, and a second population occurs along the very dorsolateral margin of the DPD, perhaps extending into the external nucleus of the inferior colliculus.

Inhibitory thresholds of E-I neurons are topographically organized within the DPD

The population of E-I neurons is of particular interest because they differ in their sensitivities to interaural intensity disparities (IIDs) (Wenstrup et al. 1985, 1986; Fuzessery and Pollak 1985; Fuzessery et al. 1985). Supra-threshold sounds delivered to the excitatory (contralateral) ear evoke a certain discharge rate that is unaffected by low intensity sounds presented simultaneously to the inhibitory (ipsilateral) ear. However, when the ipsilateral intensity reaches a certain level, and thus generates a particular IID, the discharge rate declines sharply, and even small increases in ipsilateral intensity will, in most cases, completely inhibit the cell (Fig. 6C). Thus each E-I neuron has a steep IID function and reaches a criterion inhibition at a specified IID that remains relatively constant over a wide range of intensities (Wenstrup et al. 1986). The criterion is the IID that produces a 50% reduction in the discharge rate evoked by the excitatory stimulus presented alone. We shall refer to this IID as the neuron's inhibitory threshold. An inhibitory threshold has a value of 0 dB if equally intense signals in the two ears elicit the criterion inhibition. An inhibitory threshold is assigned a positive value if the inhibition occurs when the ipsilateral signal is louder than the contralateral signal, and is assigned a negative value if the intensity at the ipsilateral ear is lower than the contralateral ear when the discharge rate is reduced by 50%. The inhibitory thresholds of E-I neurons in the DPD vary from +30 dB to -20 dB, encompassing much of the range of IIDs that the bat would experience.

The significant finding is that the inhibitory thresholds are topographically arranged within the ventromedial E-I region of the DPD (Fig. 6B) (Wenstrup et al. 1985, 1986). E-I neurons with high, positive inhibitory thresholds (i.e., neurons requiring a louder ipsilateral stimulus than contralateral stimulus to produce inhibition) are located in the dorsal E-I region. Subsequent E-I responses display a progressive shift to lower inhibitory thresholds. The most ventral E-I neurons have the lowest inhibitory thresholds; they are suppressed by ipsilateral sounds equal to or less intense than the contralateral sounds.

Spatially selective properties of 60 kHz E-I neurons

Binaural sensitivities to IIDs are suggestive of how a neuron will respond to sounds emanating from various spatial locations. To determine more precisely how binaural properties shape a neuron's receptive field, Fuzessery and his colleagues (Fuzessery and Pollak 1984, 1985; Fuzessery et al. 1985; Fuzessery (in press)) first determined how IIDs of 60 kHz sound vary around the bat's hemifield. They then evaluated the binaural properties of ICc neurons with speakers inserted into the ear canals, and subsequently determined the spatial properties of the same neurons with free-field stimulation delivered from speakers mounted on a hoop. With this battery of information, the quantitative aspects of binaural properties could be directly associated with the neuron's spatially selective properties.

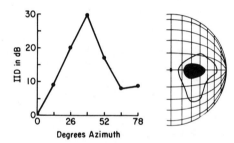

Figure 7. IIDs of 60 kHz sounds for azimuthal locations at 0° elevation. Right panel: Schematic illustration of IIDs of 60 kHz sounds at different azimuths and elevations. Blackened area is where all IIDs are 20 dB or greater, and gray are is where all IIDs are 10 dB or greater. All values were obtained from cochlear microphonic recordings. ·

The changes in IID values with azimuthal sound location are shown in Fig. 7 for 60 kHz in the mustache bat. With regard to azimuth, when the sound emanates from directly in front of the bat, at 0° elevation and 0° azimuth, equal sound intensities reach both ears, and an IID of 0 dB is generated. The largest IIDs originate at about 40° azimuth and 0°. elevation, where the sounds are about 30 dB louder in one ear than in the other ear. Within the azimuthal sound field from roughly 40° on either side of the midline, the range of IIDs created by the head and ears at 60 kHz is about 60 dB (+3 dB to -30 dB), and thus the IIDs change on the average by about 0.75 dB/degree. If recordings were made on the left side of the brain and the sound source located in the left hemifield at 40° azimuth and 0° elevation, the terminology would refer to this as a +30 dB IID. Figure 13B shows a schematic representation of the IIDs generated by 60 kHz sounds in both azimuth and elevation. Notice that an IID is not uniquely associated with one position in space, a feature that we shall address in a later section.

The binaural properties of three 60 kHz E-I units and their receptive fields are shown in Fig. 8. The first noteworthy point is that the spatial position at which each unit has its lowest threshold is the same among all 60 kHz E-I units, at about -40° azimuth and 0° elevation, and corresponds to the position in space where the largest IIDs are generated, i.e., the position in space at which the sound is always most intense in the excitatory ear and least intense in the inhibitory ear. The thresholds increase almost as circular rings away from this area of maximal sensitivity, as do the IIDs, attaining the highest thresholds, or even becoming totally unresponsive, to sounds presented from the ipsilateral sound field.

The second noteworthy point is that in each 60 kHz unit there is a position along the azimuth that demarcates the locations at which sounds can evoke discharges from locations at which sounds are ineffective in evoking discharges. That azimuthal position, and the IID associated with it, is different in each E-I cell and correlates closely with the neuron's inhibitory threshold. In 60 kHz units having low inhibitory thresholds, those demarcating loci occur along the midline, or even in the contralateral sound field, and sounds presented ipsilateral to those loci are incapable of eliciting discharges, even with intensities as high as 110 dB SPL. Units with high inhibitory thresholds require a more intense stimulation of the inhibitory ear for complete inhibition, and therefore the demarcating loci of these units are in the ipsilateral sound field.

E-I neurons create a representation of azimuth

The findings that an E-I neuron's inhibitory threshold determines where along the azimuth the cell becomes unresponsive to sound, coupled with the topographic representation of IID sensitivites among E-I cells, have implications for how the azimuthal position of a sound is represented in the mustache bat's inferior colliculus. Specifically, these data suggest that the value of an IID is represented within the DPD as a "border" separating a region of discharging cells from a region of inhibited cells (Fig. 9) (Wenstrup et al. 1986; Pollak et al. in press). Consider, for instance, the pattern of activity in the DPD on one side generated by a 60 kHz sound that is 15 dB louder in the ipsilateral ear than in the contraleteral ear. The IID is this case is +15 dB. Since neurons with low inhibitory thresholds are situated ventrally, the high relative intensity in the ipsilateral (inhibitory) ear will inhibit all the E-I neurons in ventral portions of the DPD. The same sound, however, will not be sufficiently loud to inhibit the E-I neurons in the more dorsal DPD, where neurons require a relatively more intense ipsilateral stimulus for inhibition. The topology of inhibitory thresholds and the steep IID functions of E-I neurons, then, can create a border betwen excited and inhibited neurons within the ventromedial DPD. The locus of the border, in turn, should shift with changing IID, and therefore should shift correspondingly with changing sound location, as shown in Fig. 9.

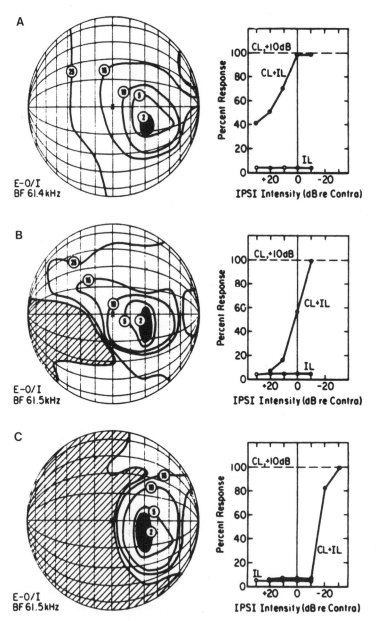

Figure 8. Spatial selectivity and IID functions for three 60 kHz E-I units. Blackened area on right is region of space from which lowest thresholds could be elicited. All thresholds in this region were within 2 dB. Isothreshold contours are drawn in increments of 5 dB. The uppermost unit (A) had a very high inhibitory threshold, and could not be completely inhibited regardless of the intensity to the ipsilateral ear. The inhibitory threshold of the unit in B was lower and was completely inhibited when the ipsilateral sound was about 10 dB more intense than the contralateral sound. Its spatial selectivity, on the left, shows that sounds in the ipsilateral field, indicated by the stripped area, were ineffective in firing the unit even with intensities of over 100 dB SPL. Similar arguments apply to unit C. From Fuzessery and Pollak (1985)

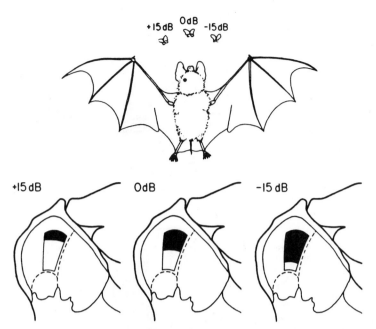

Figure 9. Schematic illustration of the relationship
between the IID value produced by a sound source at a
given location (moths at the top of figure) and the
pattern of activity in the ventromedial E-I region of
the left DPD, where IID sensitivity is topographically
organized. The activity in this region, indicated by
the blackened area, spreads ventrally as a sound source
moves from the ipsilateral to the contralateral sound
field. From Wenstrup et al. (1986).

E-E neurons code for elevation along the midline

E-E units are most sensitive to sounds presented close to or along
the vertical midline at an elevation determined by the directional
properties of the ears (Fig. 10) (Fuzessery and Pollak 1985; Fuzessery
in press). Their azimuthal selectivities are shaped by their binaural
properties that exhibit either a summation, or a facilitation of
discharges with binaural stimulation. The contralateral ear is always
dominant, having the lowest threshold and evoking the greatest
discharge rate. For 60 kHz, the ear is most sensitive, and generates
the greatest IIDs at $0°$ elevation, at about $-40°$ along the azimuth.
However, the position along the azimuth where those cells are maximally
sensitive, unlike E-I cells, is not so much a function of the ear
directionality, but rather is a direct consequence of the interplay
between the potency of the excitatory binaural inputs. A sound, for
example, presented from $40°$ contralateral will, due to the directional
properties of the ear, create the greatest intensity at the excitatory
ear. However, as the sound is moved along the azimuth the intensity at
the contralateral ear diminishes, but simultaneously, the intensity at
the ipsilateral ear increases. Since excitation of the ipsilateral ear
facilitates the response of the neuron, the net result is that the
response is stronger, and more sensitive, at azimuthal positions closer
to, or at the midline, than are responses evoked by sounds from the
more contralateral positions. In short, E-E neurons can be thought of
as midline units, being maximally sensitive to positions around $0°$
azimuth, at an elevation determined by the directional properties of
the ear.

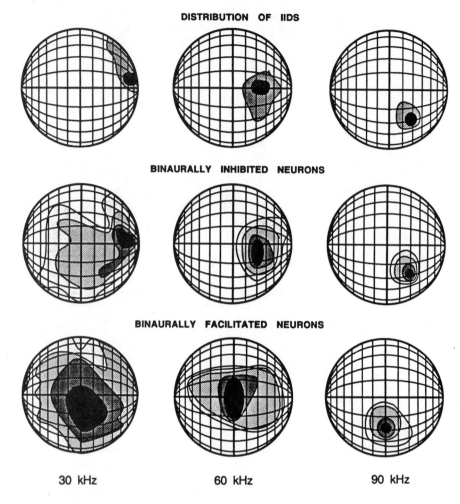

DISTRIBUTION OF IIDS

BINAURALLY INHIBITED NEURONS

BINAURALLY FACILITATED NEURONS

30 kHz   60 kHz   90 kHz

Figure 10. Interaural intensity disparities (IIDs)
generated by 30, 60 and 90 kHz sounds are shown in top
row. The blackened areas in each panel indicate the
spatial locations where IIDs are 20 db or greater, and
the gray areas indicate those locations where IIDs are
at least 10 dB. The panels in the middle row show the
spatial selectivity of three E-I units, one tuned to 30
kHz, one to 60 kHz and one to 90 kHz. The blackened
areas indicate the spatial locations where the lowest
thresholds were obtained. Isothreshold contours are
drawn for threshold inncrements of 5 dB. The panels in
the bottom row show the spatial selectivities of three
E-E units, tuned to each of the three harmonics.
Modified from Fuzessery (in press).

214

Spatial properties of neurons tuned to other frequencies

The chief difference among neurons tuned to other harmonics of the mustache bat's echolocation calls, at 30 and 90 kHz, is that their spatial properties are expressed in regions of space that differ from 60 kHz neurons, a consequence of the directional properties of the ear for those frequencies (Fig. 10) (Fuzessery and Pollak 1984, 1985; Fuzessery in press). The binaural processing of neurons tuned to those frequencies, however, are essentially the same as those described for 60 kHz neurons. Moreover, there is even evidence for an orderly representation of IIDs in the 90 kHz isofrequency laminae, further supporting the generality of a topology of inhibitory thresholds among E-I cells across isofrequency contours (Wenstrup et al., unpublished observations). In short, the sort of binaural processing found in the 60 kHz lamina appears to be representative of binaural processing within other isofrequency contours.

This feature is most readily appreciated by considering the spatial properties of E-I units tuned to 90 kHz, the third harmonic of the mustache bat's orientation calls. The maximal IIDs generated by 90 kHz occur at roughly 40° along the azimuth and -40° in elevation (Fig. 10, top panel). The 90 kHz E-I neurons, like those tuned to 60 kHz, are most sensitive to sounds presented from the same spatial location at which the maximal IIDs are generated. Additionally, the inhibitory thresholds of these neurons determine the azimuthal border defining the region in space from which 90 kHz sounds can evoke discharges from the region where sounds are incapable of evoking discharges (Fuzessery and Pollak 1985; Fuzessery et al., 1985). Since the population of 90 kHz E-I neurons have a variety of inhibitory thresholds, and those inhibitory thresholds appear to be topographically arranged within that contour, the particular IID will be encoded by a population of cells having a border separating the inhibited from the excited neurons, in a fashion similar to that shown for the 60 kHz lamina. The same argument can be applied to the 30 kHz cells, but in this case the maximal IID is generated from the very far lateral regions of space (Fig. 10, top panel).

The spatial behavior of E-E cells tuned to 30 and 90 kHz are likewise similar to those tuned to 60 kHz. The distinction is only in their elevational selectivity since their azimuthal sensitivity is for sounds around the midline, at 0° azimuth.

The representation of auditory space in the mustache bat's inferior colliculus

We can put this together and begin to see how the cues for azimuth and elevation are derived from the directional properties of the ears, in terms of the different IIDs they generate among frequencies at a particular location, and how these are represented across isofrequency contours in the bat's midbrain. Figure 11 shows a stylized illustration of the IIDs generated by 30, 60 and 90 kHz within the bat's hemifield, and below are shown the loci of borders that would be generated by a biosonar signal containing the three harmonics emanating from different regions of space. Consider first a sound emanating from 40° along the azimuth and 0° elevation. This position creates a maximal IID at 60 kHz, a lesser IID at 30 kHz and an IID close to 0 dB at 90 kHz. The borders created within each of the isofrequency contours by these IIDs are shown in Fig. 11 (bottom panel). Next consider the IIDs created by the same sound, but from a slightly

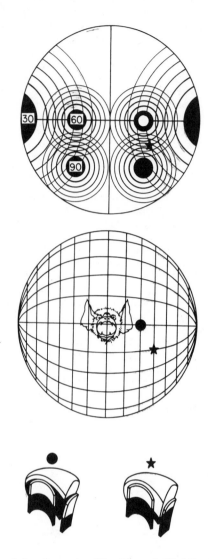

Fig. 11. Loci of borders in 30, 60 and 90 kHz E-I
regions generated by biosonar signals emanating from two
regions of space. The upper panel is schematic
representation of IIDs of 30, 60 and 90 kHz that occur
in bat's acoustic field. The blackened areas indicate
the regions in space where the maximum IID is generated
for each harmonic. The middle panel depicts the bat's
head and the spatial positions of the two sounds.
The lower panel shows the borders separating regions
of excited from regions of inhibited neurons in the 30, 60
and 90 kHz isofrequency contours.

different position in space, at about 45° azimuth and -20° elevation. In this case there is a decline in the 60 kHz IID, an increase in the 90 kHz IID, but the 30 kHz IID will be the same as it was when the sound emanated from the previous position. The constant IID at 30 kHz is a cruical point, and it occurs because for a given frequency an IID is not uniquely associated with one position in space, but rather can be generated from a variety of positions. It is for this reason that the accuracy of sound localization with one frequency is ambiguous (Stevens and Neuman 1936; Blauert 1969/70; Butler 1974; Musicant and Butler 1985; Fuzessery in press). However, spatial location, in both azimuth and elevation, is rendered unambiguous by the stimultaneous comparison of three IIDs, because their values in combination are uniquely associated with a spatial location.

This representation becomes ineffective along the vertical midline, at 0° azimuth, where the IIDs will be 0 dB at all frequencies and all elevations. The borders among the E-I populations will thus not change with elevation because the IIDs remain constant. The population of E-E neurons tuned to different frequencies becomes important in this regard (Fuzessery and Pollak 1985; Fuzessery in press). E-E units tuned to different frequencies exhibit selectivities for different elevations, and thus as the elevation along the vertical meridian shifts, the strength of responding also shifts for E-E units as a function of their frequency tuning. In combination, then, the two major binaural types provide a neural representation of sound located anywhere within the bat's hemifield.

A striking feature of the above scenario is its similarity, in principle, to the ideas about sound localization proposed previously by Pumphery (1948) and Grinnell and Grinnell (1965). What these recent studies provide are the details of how interaural disparities are encoded and how they are topologically represented in the acoustic midbrain.

SOME UNIFYING THEMES AND QUESTIONS FOR THE FUTURE

An important feature of the auditory system is that the information conveyed by the auditory nerve diverges progressively in the meduallary brainstem, where each nucleus seems to process information in a somewhat different, perhaps unique, fashion, extracting some feature or features from the incoming stream of discharges. All of this is then recombined in the ICc, both as a reconstitution of a single tonotopic map, and as a topological parcelling of the various acoustic attributes extracted previously in the lower centers. The clearest example of this is the orderly representation of IID sensitivities among 60 kHz E-I cells (Wenstrup et al. 1985, 1986) which are themselves topographically segregated from other aural types. We emphasize this does not require, or even suggest, that each cell type, or aural region, represents only one acoustic attribute. Quite to the contrary, it seems almost certain that each ensemble encodes multiple acoustic attributes, and some of these may be represented in several regions simultaneously.

With regard to binaural procesing, it is worth emphasizing that the auditory system operates on the cues generated by the physical dimensions of the head and ears. Stated differently, it is not space which the auditory system encodes, but IIDs, and the animal must then associate the pattern of IIDs with a spatial location. While seemingly obvious, this feature has a number of implications. In order to reconstruct acoustic space from the frequency dependent cues generated by the head and ears, it is mandatory that the intensity of a given frequency at one ear be compared with the intensity generated by the same frequency at the other ear. The same argument can be made for animals that utilize interaural time disparities. Binaural processing, thus, must be accomplished on a frequency-by-frequency basis, and the most efficient way of achieving this is by preserving the cochlear tonotopic map throughout the auditory pathway. Tonotopy, then, is a structural component, not only for pitch perception, but for sound localization as well. It is also noteworthy that a direct relationship between tonotopy and sound localization was originally proposed on the basis of human psycho physical studies by Butler in 1974.

We have stressed throughout this chapter that our understanding of the auditory system is far from complete. The binaural properties, for example, of lower brainstem nuclei can presently be only loosely correlated with binaural neurons in the ICc. To be sure, the basic aural types, E-E and E-I cells, are created in the MSO and LSO, but we can say little more about the similarities or differences of the binaural properties of neurons in these nuclei compared with similar types in the DNLL, about which we have almost no information, and the ICc.

Numerous questions of basic significance remain to be elucidated. We need to obtain a clearer and more detailed understanding of signal processing in lower brainstem regions and the sorts of tranformations that occur between these and the ICc, together with a more detailed picture of the microconnectivity among auditory brainstem nuclei. With current technologies, answers to these questions should be forthcoming in the near future, and with them will come much deeper insights into how the brain processes acoustic information.

ACKNOWLEDGEMENT

We thank Janet Young for the artwork, and Kristy Vick for typing the manuscript. Supported by NIH grant NS#21748 and grant NSF BNS#8217357 to J.H.C. and RO-1-2776 to G.D.P.

REFERENCES

Aitkin, L.M. (1976) Tonotopic organization at higher levels of the auditory pathway. Inter. Rev. Physiol. Neurophysiol. 10:249-279.

Blauert, J. (1969/70) Sound localization in the median plane. Acoustica 22:205-213.

Bodenhamer, R.D., Pollak, G.D. and Marsh, D. (1979) Coding of fine frequency information by echoranging neurons in the inferior colliculus of Mexican free-tailed bats. Brain Res. 171:530-535.

Bodenhamer, R.D. and Pollak, G.D. (1981) Time and frequency domain processing in the inferior colliculus of echolocating bats. Hearing Res. 5:317-335.

Boudreau, J.C., and C. Tsuchitani (1968) Binaural interaction in the cat superior olive s-segment. J. Neurophysiol. 31:442-454.

Butler, R.A. (1974) Does tonotopy subserve the perceived elevation of a sound? Federation Proc. 33:1920-1923.

Cant, N.B. (1984) The fine structure of the lateral superior olivary nucleus of the cat. J. Comp. Neurol. 227:63-77.

Cant, N.B. and D.K. Morest (1984) The structural basis for stimulus coding in the cochlear nucleus of the cat. In: Hearing Science. C. Berlin (Ed.), College-Hill Press, San Diego, California, pp. 423-460.

Caird, D. and R. Klinke (1983) Processing of binaural stimuli by cat olivary complex neurons. Exp. Brain Res. 52:385-399.

Cajal, S. Ramon y., 1909, "Le system nerveus de l'homme et des vertebres," Instituto Ramon y Cajal, Madrid.

Casseday, J.H., and Covey, E., 1986, Central auditory pathways in directional hearing in: "Directional Hearing," W.A. Yost and G. Gourevitch, eds., Springer Verlag, New York (in press).

Casseday, J.H., Diamond, I.T., and Harting, J.K., 1976, Auditory pathways to the cortex in Tupaia glis. J. Comp. Neurol., 166:303-340.

Covey, E., and Casseday, J.H., 1986, Connectional basis for frequency representation in the nuclei of the lateral lemniscus of the bat, Eptesicus fuscus. J. Neurosci., in press.

Erulkar, S.D. (1972) Comparative aspects of spatial localization of sound. Physiol. Rev. 52:237-360.

Feng, A.S. and M. Vater (1985) Functional organization of the cochlear nucleus of Rufous horseshoe bats: Frequencies and internal connections are arranged in slabs. J. Comp. Neurol. 235:529-553.

Fuzessery, Z.M. and G.D. Pollak (1984) Neural mechanisms of sound localization in an echolocating bat. Science 225:725-728.

Fuzessery, Z.M., J.J. Wenstrup, and G.D. Pollak (1985) A representation of horizontal sound location in the inferior colliculus of the mustache bat (Pteronotus p. parnellii). Hear. Res. 20:85-89.

Fuzessery, Z.M. and G.D. Pollak (1985) Determinants of sound location selectivity in the bat inferior colliculus: a combined dichotic and free-field stimulation study. J. Neurophysiol. 54:757-781.

Goldberg, J.M. and P.B. Brown (1968) Responses of binaural neurons of dog superior olivary complex to dichotic tonal stimuli: some physiological mechanisms of sound localization. J. Neurophysiol. 32:613-636.

Goldberg, J.M. (1975) Physiological studies of auditory nuclei of the pons. In: Handbook of Sensory Physiology, Vol. V/2, B.D. Keidel and W.D. Neff (Eds.), Springer Verlag, New York, pp. 109-144.

Grinnell, A.D. and V.S. Grinnell (1965) Neural correlates of vertical sound localization by echolocating bats. J. Physiol London 181:830-851.

Guinan, J.J., Jr., B.E. Norris, and S.S. Guinan (1972) Single units in superior olivary complex II. Locations of unit categories and tonotopic organization. Internat. J. Neurosci. 4:147-166.

Harnischfeger, G., G. Neuweiler, and P. Schlegel (1985) Interaural time and intensity coding in superior olivary complex and inferior colliculus of the bat, Molossus ater. J. Neurophysiol. 53:89-109.

Harrison, J.M. and R. Irving (1966) Visual and non-visual auditory systems in mammals. Science 154:738-743.

Helfert, F.H. and Schwartz, I.R. (1986) Morphological evidence for the existence of multiple neuronal classes in the cat lateral superior olivary nucleus. J. Comp. Neurol. 244:533-550.

Irving, R. and J.M. Harrison (1967) The superior olivary complex and audition: A comparative study. J. Comp. Neurol. 130:77-86.

Jen, P.H.S. (1980) Coding of directional information by single neurons in the S-segment of the FM bat, Myotis lucifugus. J. Exp. Biol. 87:203-216.

Masterson, B.R. and I.T. Diamond (1973) Hearing: Central neural mechanisms: In: Handbook of Perception, E. Carterette and M. Freedman (Eds.), Academic Press, New York.

Masterton, B.R., G.C. Thompson, J.K. Brunso-Bechtold, and M.J. Robarts (1975) Neuroanatomical basis of binaural phase-difference analysis for sound localization. A comparative study. J. Comp. Physiol. Psychol. 89:379-386.

Metzner, W. and S. Radkhe-Schuller (1985) The lateral lemniscus in Rufous horseshoe bat. Abstract in SIBRC, Aberdeen.

Morest, D.K., 1964, The neuronal architecture of the medial geniculate body of the cat. J. Anat. Lond. 98:611-630.

Morest, D.K., and Oliver, D.L., 1984, The neuronal architecture of the inferior colliculus in the cat: defining the functional anatomy of the auditory midbrain. J. Comp. Neurol. 222:209-236.

Musicant and Butler (1985) Influence of monaural spectral cues on binaural localization. J. Acoust. Soc. Am. 77:202-208.

Novick, A. (1977) Acoustic orientation. In: Biology of Bats, Vol. III. W.A. Wimsatt (Ed.) Academic Press, N.Y. pp. 74-273.

Oliver, D.L., and Morest, D.K., 1984, The central nucleus of the inferior colliculus in the cat. J. Comp. Neurol. 222:237-264.

O'Neill, W.E. (1985) Responses to pure tones and linear frequency components of CF-FM biosonar signal by single units in the inferior colliculus of the mustache bat. J. Comp. Physiol. 157:797-816.

O'Neill, W.E. (1986) The processing of temporal information in the auditory system of echolocating bats. In: Myotis, P.A. Racey (ed.), SIBRC, Averdeen Symposium (in press)

Papez, J.W., 1929, "Central acoustic tract in cat and man," Anat. Rec., 42:60.

Poljak, S., 1926, Untersuchungen am Oktavussystem der Saugetiere und an den mit diesem koordinierten motorischen Apparaten des Hirnstammes. J. f. Psychol. u. Neurol. 32:170-231.

Pollak, G.D. (1980) Organizational and encoding features of neurons in the inferior colliculus of bats. In: Animal Sonar System. Busnel, R.-G, and J. Fish, (Eds.) Plenum, N.Y. pp. 549-587.

Pollak, G.D., Marsh, D., Bodenhamer, R.D. and Souter A. (1977) Characteristics of phasic-on neurons in the inferior colliculus of bats with observation relating to mechanisms for echo-ranging. J. Neurophysiol. 40:926-942.

Pollak, G.D., D.S. Marsh, R. Bodenhamer, and A. Souther (1978) A single unit analysis of the inferior colliculus in unanesthetized bats: Response patterns and spike-count functions by constant-frequency and frequency-modulated sounds. J. Neurophysiol. 41:677-691.

Pollak, G.D. and Schuller, G. (1981) Tonotopic organization and encoding features of single units in the inferior colliculus of horseshoe bats: functional implications for prey identification. J. Neurophysiol. 45:208-226.

Pollak, G.D., Bodenhamer, R.D. and Zook, J.M (1983) Cochleotopic organization of mustache bat's inferior colliculus. In: Advances in Vertebrate Neuroethology. J.-P Ewert, R.R. Capranica and D.J. Ingle (eds). Plenum, N.Y. pp. 925-935.

Pollak, G.D., J.J. Wenstrup, and Z.M. Fuzessery (in press) Auditory processing in the mustache bat's inferior colliculus. Trends in Neurosci.

Pumphrey, D.J. (1947) The sense organs of birds. Ibis 90:171-199.

Ross, L.S., J.J. Wenstrup, and G.D. Pollak (1985) Anatomical projections to an isofrequency region: Basis for an organization of binaural response properties in mustache bats. Soc. Neurosci. Abst. 11, Part 1, pg. 734.

Ross. L.S., G.D. Pollak, and J.M. Zook (submitted) Ascending projections to an isofrequency region in the mustache bat's inferior colliculus. J. Comp. Neurol.

Ross, L.S., and G.D. Pollak (1986) Differential projections to monaural and binaural regions of one isofrequency contour in mustache bat's inferior colliculus. Soc. Neurosci. Abstr.

Schlegel, P. (1977) Directional coding by binaural brainstem units of the CF-FM bat, Rhinolophus ferrumequinum. J. Comp. Physiol. 118:327-357.

Schweizer, H. (1981) The connections of the inferior colliculus and the organization of the auditory brainstem nuclei in the greater horseshoe bat. J. Comp. Neurol. 201:25-49.

Schweizer, H., and Radtke, S., 1980, The auditory pathway of the greater horseshoe bat Rhinolophus ferrumequinum, in: Animal Sonar Systems, R.G. Busnel and J.E. Fish, eds., Plenum Press, New York, pp. 987-989.

Simmons, J.A. (1973) The resolution of target range by bats. J. Acoust. Soc. Am. 54:157-173.

Simmons, J.A, 1971, Echolocation in bats: signal processing of echoes for target range. Science. 171:925-928.

Simmons, J.A., 1979, Perception of echo pase information in bat sonar. Science. 204:1336-1338.

Simmons, J.A., Howell, D.J. and Suga, N. (1975) Information content of bat sonar echoes. Am. Sci. 63:204-215.

Stevens, S.S. and E.B. Newman (1936) The localization of actual sources of sound. Amer. J. Psych. 48:297-306.

Suga, N. (1964) Single unit activity in cochlear nucleus and inferior colliculus of echolocating bats. J. Physiol. (London) 172:449-474.

Suga, N. (1970) Echo-ranging neurons in the inferior colliculus of bats. Science 170:449-452.

Suga, N., J.A. Simmons and P.H.-S. Jen (1975) Peripheral specialization for fine analysis of Doppler-Shifted echoes in CF-FM bat, Pteronotus parnelli, J. Exp. Biol. 63:161-192.

Suga, N., G. Neuweiler, and P. Schlegel (1976) Peripheral auditory tuning for fine frequency analysis by CF-FM bat Rhinolopus ferrumequinum J. Comp. Physiol. 106:111-125.

Suga, N., and P.H.-S. Jen (1977) Further studies on the peripheral auditory system of CF-FM bats specialized for fine frequency analysis of Dopplershifted echoes. J. Exp. Biol. 69:207-232.

Suga, N. and K. Tsuzuki (1985) Inhibition and level-tolerant frequency tuning in the auditory cortex of the mustached bat. J. Neurophysiol. 53:1109-1145.

Tsuchitani, C. and J.C. Boudreau (1966) Single unit analysis of cat superior olive S-segment with tonal stimuli. J. Neurophysiol. 29:694-697.

Tsuchitani, C. (1977) Functional organization of lateral cell groups of cat superior olivary complex. J. Neurophysiol. 40:296-318.

Tsuchitani, C. and D.H. Johnson (1985) The effect of ipsilateral tone burst stimulus level on the discharge patterns of cat lateral superior olivary units. J. Acoust. Soc. Am. 77:1484-1496.

Vater, M. and D. Schlegel (1979) Comparative neurophysiology of the inferior colliculus of two molosid bats Molossus ater and Molossus molossus. II. single unit reponses to frequency modulated signals and signal and noise combination. J. Comp. Physiol. 131:147-160.

Warr, W.B., 1982, Parallel ascending pathways from the cochlear nucleus: neuroanatomical evidence of functional specialization: in: "Contributions to Sensory Physiology, "Vol. 7, W.D. Neff, eds., Academic Press, New York.

Wenstrup, J.J., L.S. Ross, and G.D. Pollak (1985) A functional organization of binaural reponses in the inferior colliculus. Hear. Res. 17:191-195.

Wenstrup, J.J, L.S. Ross, and G.D. Pollak (1986) Binaural reponse organization within a frequency-band representation of the inferior colliculus: Implications for sound localization. J. Neurosci. 6:962-973.

Zook, J.M., and Casseday, J.H., 1980, Ascending auditory pathways in the brain stem of the bat, Pteronotus parnellii, in: Animal Sonar

Systems, R.G. Busnel and J.E. Fish, eds., Plenum Press, New York, pp. 1005-1006.

Zook, J.M., and Casseday, J.H., 1980, Identification of auditory centers in lower brain stem of two species of echolocating bats: Evidence from injection of horseradish peroxidase into inferior colliculus. In D.E. Wilson and A.L. Gardner (eds): Proc. Fifth Int. Bat Res. Conference. Lubbock, TX: Texas Tech Press, pp. 51-60.

Zook, J.M. and J.H. Casseday (1982a) Cytoarchitecture of auditory system in lower brainstem of the mustache bat, Pteronotus parnellii. J. Comp. Neurol. 207:1-13.

Zook, J.M. and J.H. Casseday (1982b) Origin of ascending projections to inferior colliculus in the mustache bat, Pteronotus parnellii. J. Comp. Neurol. 207:14-28.

Zook, J.M., J.A. Winer, G.D. Pollak, and R.D. Bodenhamer (1985) Topology of the central nucleus of the mustache bat's inferior colliculus: Correlation of single unit properties and neuronal architecture. J. Comp. Neurol. 231:530-546.

Zook, J.M. and Casseday, J.H., 1985, Projections from the cochlear nuclei in the mustache bat, Pteronotus parnellii. J. Comp. Neurol. 237:307-324.

Zook, J.M., and Casseday, J.H., 1986, Convergence of ascending pathways at the inferior colliculus of the mustache bat, Pteronotus parnellii, submitted for publication.

COCHLEAR PHYSIOLOGY AND ANATOMY IN BATS

Marianne Vater

Zoologisches Institut
Luisenstr.14
8000 München 2 FRG

INTRODUCTION

The first steps in frequency analysis by the mammalian auditory system
are performed in the cochlea. Hair cells are tuned to a restricted frequency
range and the orderly pattern of frequency representation found at all
levels of the ascending auditory pathway is laid out in the frequency place
code along basilar membrane (BM) length, being produced by the gradient
in hydromechanical properties.

Research on the inner ear of numerous bat species has demonstrated a
variety of regional differentiations of cochlear structures which appear
to be correlated with the species specific representation of frequency information in the auditory system, which in turn relates to the type of sonar
signals used. This paper summarizes comparative data on physiological and
anatomical properties of the peripheral auditory system of bats emitting
either broadband or narrowband signals. It attempts to analyze the variations in basic cochlear design which ultimately generate the enhanced tuning
to a narrow frequency band in bats emitting long pure tone call components.

MIDDLE EAR

Knowledge about the characteristics of middle ear transmission is important in order to define the effective signal reaching the cochlea. Middle
ear morphology and function of middle ear muscles has been described in
several reports (e.g. Pteronotus: Pollak and Henson, 1973; Myotis: Suga
and Jen, 1975, Cabezudo et al., 1983). The transfer characteristics of the
tympanic membrane have been described in only two bat species. The eardrum
velocity vs. frequency curves in Rhinolophus (Wilson and Bruns, 1983a) can
now be compared with earlier data on Eptesicus (Manley et al., 1972). Data
from the two species of bats and from a nonecholocating mammal are given
in Fig.1. All response curves reach optimal efficiency in species specific
frequency ranges. There are clear adaptations for the ultrasonic range in
the two bats. The optimal frequency range is around 25 kHz for Eptesicus
and around 55 kHz for Rhinolophus. In the latter case, this is clearly
below the constant frequency (CF) component of the call (83 kHz); nevertheless, middle ear transmission is reasonable from 15 to 110 kHz.

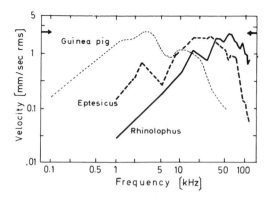

Fig. 1. Eardrum velocities at 100 dB SPL for R.ferrumequinum (Wilson
and Bruns, 1983a), E.pumilis (Manley et al., 1972) and Guinea
pig (Wilson and Johnstone, 1975). Arrows indicate the theoretically most efficient eardrum velocity (2.5 mm/sec rms). After
Wilson and Bruns, 1983a).

COCHLEAR PHYSIOLOGY

Cochlear potentials

     The properties of the cochlear potentials (cochlear microphonic, CM;
compound auditory nerve action potential, N1) have been described in many
earlier reports (long CF-FM bats: Pollak et al.,1972; Suga et al., 1975;
Suga and Jen, 1977; Schnitzler et al.,1976 ; FM-bats: Dalland et al., 1967;
Brown, 1973). This work has clearly demonstrated that the unique tuning
properties of the auditory system in long CF-FM bats are established at
the receptor level. Recent papers have emphasized a direct comparison of
horseshoe bats and mustache bats and inquired into the exact relation of
the CF-component of the echolocation signal to the sharply tuned region
of the audiogram (Henson et al., 1982; 1985a). As shown in Fig.2, in both
Rhinolophus and Pteronotus, the frequency of the CF-component emitted by
the nonflying bat (resting frequency, RF) is located at the steep low frequency flank of the sharply tuned region about 200 Hz below the most sensitive thresholds. In the nonflying bat, this probably prevents neurons within
the sensitivity maximum from being stimulated by the FM-component of the
call. During Doppler-shift compensation, the frequency of the emitted call
is lowered to keep the echo frequency within the sharply tuned region
(Schnitzler, 1968). The loud emitted pulse is thus displaced towards the
less sensitive frequency range and the sharply tuned region is held ready
to process the faint echoes.

     Although these relations are similar, horseshoe bats and mustache bats
differ in several aspects of the CM- and N1-responses (Henson et al.,1985a):
1. Rhinolophus provides a much less sensitive preparation for recording
CMs than Pteronotus. CM-thresholds are high and CM-amplitudes are low in
the sharply tuned region (Fig.2). 2. In Pteronotus the CM-amplitude reaches
an absolute maximum in the sharply tuned region whereas in Rhinolophus a
sharply tuned peak is mainly produced by a range of particularly low CM-amplitudes on the low frequency side. 3. Rapid phase changes in the CM occur
at the low frequency side of the sharply tuned region in Rhinolophus and
at the sharply tuned peak in Pteronotus. 4. In Pteronotus, stimuli in the
CF-frequency range generate CM-envelopes with slow rise times and long lasting ringing after stimulus end. Such pronounced ringing has not been found
in Rhinolophus. These data suggest that different mechanisms could lead
to enhanced tuning. However, the CM data alone are not conclusive, since
it is not clear whether the species differences are created by differences

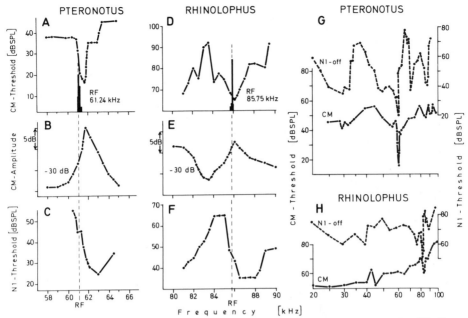

Fig. 2. Cochlear potentials in Pteronotus and Rhinolophus. A, D, CM-thresholds; B, E, CM-amplitudes; C, F, N1-on thresholds within the
sharply tuned region. The individual bat's RF is given by histograms and dashed line (after Henson et al., 1985a). G, H, N1-off
and CM-audiograms for the full frequency range (G: after Pollak
et al., 1979, curves from two different individuals; H: after
Schnitzler et al., 1976, curves from same individual).

in the geometry of the recording sites or whether they represent true differences in the BM-response properties reflected in the CM.

Ringing in the CM-response indicates the presence of a sharply tuned
resonator with low damping (Suga and Jen, 1977). Obviously, such a resonance provides high frequency resolution at the cost of temporal resolution. Under actual echolocation conditions both are important, but the problem might not be critical since echo intensity is usually low and furthermore
actively kept at low levels by the "echo intensity compensation" performed
by Pteronotus (Kobler et al.,1985).

Several reports demonstrate a physiological vulnerability of the responses within the sharply tuned region (Henson and Kobler, 1979; Pollak et al.,
1979, Henson et al., 1985a). Sensitivity can be impaired by factors such as
anaesthesia, temperature or recovery from surgery. Profound susceptibility
to loud pure tone exposure was observed in Pteronotus for the frequency
range around 61 kHz. (Pollak et al., 1979).

N1 off-audiograms

In Rhinolophus and Pteronotus, prominent N1-off responses occur.
These responses are more sharply tuned but less sensitive than the N1-on
and can be related to some extent to specialized response properties of
the CM (Schnitzler et al., 1976; Suga et al., 1975, Pollak et al., 1979).
The features of N1-off and CM within the 61 kHz region of Pteronotus have
been interpreted as due to a sharply tuned resonator and nonlinearity of
the cochlea.

Interestingly, the N1-off audiogram of Pteronotus (Fig.2G) exhibits a
series of distinct threshold changes at frequencies harmonically related

to the sharply tuned region at 61 kHz (Pollak et al., 1979), and additionally, in some recordings a sharply tuned minimum is found at 72 kHz, a
frequency range not related to any orientation call component. Enhanced
tuning at harmonically related frequencies is reflected in the Q 10dB values
of single units, reaching up to 60 for the first and third harmonic and
up to 400 for 61 kHz (Suga and Jen, 1977). In Rhinolophus, small threshold
and amplitude changes of CM and N1 off are seen at the first harmonic
(Fig.2H; Schnitzler et al., 1976).

Pollak et al. (1979) raised the interesting question of whether there
may be multiple specializations in the cochlea of Pteronotus corresponding
to the three or possibly four sharply tuned regions in the N1-off thesholds.
They point out that the sharp tuning at the first and third harmonics is
apparently not related to resonators at those frequencies since no ringing
can be observed in the CM.

Cochlear emissions

The mammalian cochlea not only acts as an acoustic receiver, it has
also been shown to emit faint sounds at characteristic frequencies (Kemp,
1979), which can be recorded with sensitive microphones in the external ear
canal. These otoacoustic emissions are believed to be generated by active
processes of the inner ear (outer hair cell feedback) and to play an important role in sound analysis at threshold level. In Pteronotus, strong otoacoustic emissions are produced at frequencies within the sharply tuned
region of the audiogram (Fig.3; Henson et al., 1985a; Kössl and Vater, 1985a).
The amplitude of tone evoked otoacoustic emissions is much higher than in
other mammals, reaching 70 dB SPL. Most importantly, they are clearly correlated with the indiviual specific CF-component (averaging about 700 Hz above
RF). This contrasts with data from humans, where large interindividual variations occur, although each ear has a constant set of emission frequencies.
The otoacoustic emissions in Pteronotus can be influenced by loud sound
exposure, cooling or anaesthesia, which strongly suggests the participation of active metabolic processes in their production. If outer hair cells
are involved as has been proposed for other mammals, the question remains
as to how feedback can be achieved through known cellular mechanisms at

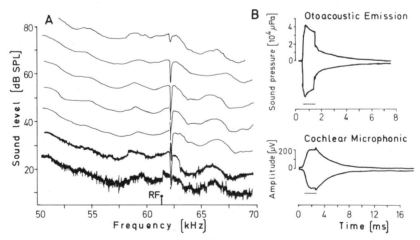

Fig. 3. Otoacoustic emissions in Pteronotus at 61.2 kHz revealed by
interference with continuous tone stimulus. B. Time course of
pulse-evoked otoacoustic emission. C. Time course of CM. Note
slow decay times after tone offset in B, C. (after Kössl and
Vater, 1985a).

such high frequencies. The high amplitude of the emissions in Pteronotus, and particularly their consistent presence at a specific frequency range, strongly suggest that hydromechanical specializations may play an additional role in their production by acting as reflection zones for cochlear waves.

COCHLEAR MORPHOLOGY

Basilar membrane

In bats, BM-length ranges from 6.9 mm (Myotis) to 16 mm (Rhinolophus) (Table 1). There is no simple relationship of BM-length and body weight among different taxonomic groups, since one of the largest species (Megaderma lyra) has the same BM-length as the small Hipposideros speoris. Although the correlation of cochlear size with body weight is more obvious within the same family (Pye, 1966), cochlear length seems to depend more on the type of echolocation signal used: the shortest cochlear ducts are apparently found in species emitting broadband calls (Myotis, Megaderma) whereas long CF-FM bats have the longest BMs.

There is considerable interspecies variation of BM-thickness and -width profiles which can be correlated with properties of the auditory system. These morphological parameters determine the stiffness of the BM and consequently the pattern of frequency representation along the longitudinal extent of the cochlear duct (Bekesy, 1960). Additionally, the specializations found in long CF-FM bats are thought to sharpen the mechanical frequency response (see last chapter). BM-data for different species are combined in Fig.4. In bats emitting FM-calls (Myotis: Ramprashad et al., 1979; Megaderma: Fiedler, 1983) and in species emitting pure tones or multiharmonic FM-sweeps depending

Table 1. Comparison of BM-length and cochlea innervation data in different species

Species	body-weight (g)	BM-length (mm)	number of IHC	number of OHC	HC/mm	number of spiral ganglion cells	References
R.ferrum-equinum	20	16.0				15953	Bruns & Schmieszek 1980
H.speoris	7	9.9	1400	4300	592	15200	Kraus, 1983
H.fulvus	7	8.8	1100	3800	556	13400	Kraus, 1983
P.parnellii	12	14.0					Kössl & Vater, 1985b
M.ater	37	14.6	2161	6512	594	31800	Fiedler, 1983
R.hardwickei	15	11.3	1600	5100	592	17900	Kraus, 1983
T.kachensis	50	14.4	2129	6331	587	23000	Fiedler, 1983
M.lyra	48	9.9	1267	4564	588	17500	Fiedler, 1983
M.lucifugus	5	6.9	700	2800	507	55300	Ramprashad et al., 1978, 1979
N.noctula		11.8	1410	4656	514		Burda & Ulehlova, 1983
E.serotinus		8.9	906	3372	482		Burda & Ulehlova, 1983
human	70000	34	3400	13400	494	30500	for references
cat	3300	22	2600	9900	543	49962	see Ramprashad
dolphin	118000	40	3451	12899	408	95000	et al., 1979

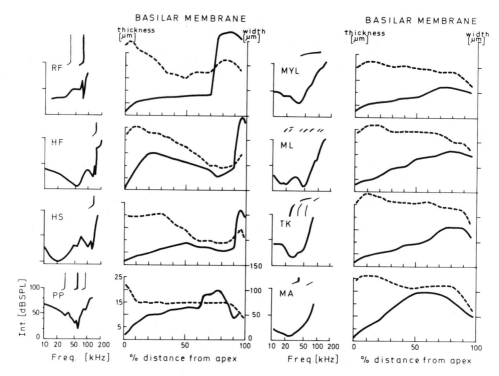

Fig. 4. Comparison of BM-thickness (continuous lines) and BM-width (broken
lines) on normalized BM-length scale for different bat species.
For each species, schematic audiograms and sonagrams are given.
HF Hipposideros fulvus; HS H.speoris, TK Taphozous kachensis;
MA Molossus ater; MYL Myotis lucifugus; ML Megaderma lyra; PP
Pteronotus parnellii; RF Rhinolophus ferrumequinum. Morphological
data after Bruns, 1976a (RF); Henson, 1978 (PP BM-width); Kössl
and Vater, 1985b (PP thickness); Kraus, 1983 (HF, HS); Fiedler,
1983 (ML, MA, TK); Ramprashad et al., 1979 (MYL). Physiological
data from Grinnell, 1963, 1970 (MYL, PP); Neuweiler, 1970 (RF);
Neuweiler et al., 1984 (HF, HS, TK, ML); Vater et al., 1979 (MA)

on the orientation situation (Taphozous: Fiedler, 1983; Rhinopoma: Kraus,
1983), a gradual increase in BM-width and decrease in BM-thickness towards
apex is observed, similar to nonecholocators. BM-morphology is more complex
in species with long or short CF-FM calls, where distinct regional differentiations in BM-thickness and -width are found. Rhinolophoidea (Rhinolophus
spec., Hipposideros spec.) exhibit a pronounced basal thickening of the BM,
which extends up to 4.5 mm distance from base in Rhinolophus (Bruns,1976a)
or up to 1 mm distance from base in Hipposideros (Kraus, 1983). The decrease
in thickness towards more apical regions is not gradual but quite abrupt.
Apical to this "discontinuity", the BM is again of constant thickness over
a species specific range. Only in the most apical regions, there is a gradual
decrease in BM-thickness. This peculiar thickness profile is paralleled
by a distinct change in BM-width. Two areas of relatively constant BM-width
can be distinguished, with the transition occuring at the discontinuity in
thickness. Pteronotus parnellii also has an irregular BM-thickness profile
(Kössl and Vater, 1985b). In contrast to Rhinolophus, BM in the most basal
region is not thickened. Instead, the region of maximally thickened BM is
located some distance from base. Thus, two "discontinuities" in thickness
are present, however of less magnitude than in Rhinolophus. Pteronotus also

differs in that there are no abrupt changes in BM-width parallel to the thickness profile. In scanning electron microscope, BM-width appears rather constant throughout the cochlea except for the extreme basal and apical ends (Henson, 1978). Molossus and Taphozous also posses nonuniform BM-thickness profiles (Fiedler, 1983), however the transitions in thickness are gradual and not accompanied by peculiar changes in BM-width, which increases regularly from base to apex.

The data on BM-thickness and -width might seem to imply that the geometry of vibrating structures is similar regardless of species or position along turns of the cochlea, differing only in magnitude. Fig. 5 demonstrates that this is clearly not the case. In most sections through the organ of Corti, particularly in the basal and middle turns, BM shows two distinct regions: an inner thickening (pars tecta, PT) and an outer thickening (pars pectinata, PP) separated by a thin zone. The prominent changes in thickness described so far relate to PP. Thickenings of PP are distinguishable in vestibular and tympanic sections. Thickenings of tympanic sections are found in most species, only differing in magnitude and shape. Pronounced vestibular thickenings have so far been reported only in the basal half turn in Rhinolophus (Bruns, 1976a, 1980), but are also present within the region of maximum BM-thickness in Pteronotus (Kössl; Henson pers. comm.). The shape of BM-thickenings is most unusual in Molossus. There is no distinction in PP and PT, rather the BM has the appearance of a solid plate (Fiedler, 1983). Another noteworthy feature is seen in the upper middle turns of Hipposideros fulvus (Kraus, 1983), where a second maximum of BM-thickness is located. There, the PP consists of a homogeneous lightly staining matrix encapsuled by darkly staining fiber strands.

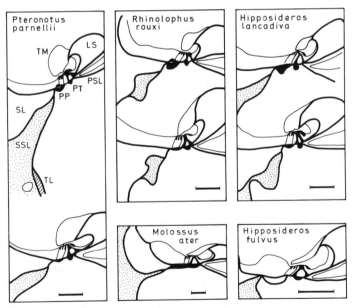

Fig. 5. Schematic crossections through the organ of Corti in different bat species. L limbus; PT pars tecta; PP pars pectinata; PSL primary spiral lamina; SL spiral ligament; SSL secondary spiral lamina; TL thick lining; TM tectorial membrane. For Pteronotus, Rhinolophus and Hipposideros, sections through thickened and "normal" BM-regions are shown. Graphs for Molossus and H. fulvus adapted from Fiedler, 1983 and Kraus, 1983. Calibration 100 um.

In an ultrastructural study, Bruns (1980) described the filament composition of the BM in Rhinolophus. The composition of the BM in the basal half
turn differs markedly from that in the apical regions, where filaments are
mainly oriented in a radial direction. The thickenings within the specialized
region are densely packed with filaments of complex orientation. The tympanic section of PP contains randomly oriented filaments in addition to radial
filaments. The vestibular section of PP is most remarkable since it is composed of a spirally oriented fiber bundle. This arrangement could provide a
pronounced longitudinal coupling within the BM of the basal half turn.

Attachment of BM

All microchiroptera possess a secondary spiral lamina (SSL) along the
entire cochlear duct formed by a bony structure within the tympanic side
of the spiral ligament (SL). Unique specializations of SSL have been reported within the basal turn of Rhinolophoidea (Bruns,1976a; 1980; Kraus, 1983).
They extend beyond the region of BM-thickenings. The SSL is bulky and pro-
trudes into scala tympani, being attached to the outer bony capsule only by
a thin bony lamella. In Hipposiderids its shape varies along the longitudinal
extent and it has been proposed that the SSL acts as a mechanical resonator
influencing the BM-response (Kraus, 1983).

The BM is attached indirectly at its outer margin to the otic capsule
via a fiber system embedded in the SL. In Rhinolophus and Pteronotus, the
size of the SL varies along the cochlear duct. The SL is considerably enlarged throughout the basal 8 mm of the cochlea in Rhinolophus (Bruns,
1976a, 1980), and in addition to the radial fiber bundles contains a group
of fibers which run spirally close to the BM-attachment. The SL is locally
enlarged in Pteronotus at the transition from the first to the second half
turn (Henson, 1978, Kössl and Vater, 1985b).

Based on ultrastructural and immunocytochemical evidence, Henson et
al. (1984, 1985b) proposed that the constituent elements of SL can create
radial tension to influence the vibration properties of the BM. Anchoring
cells located in the outer periphery of the SL attach its extracellular
fibers to the outer otic capsule. These cells are more abundant in Rhinolophus and Pteronotus than in mice and contain a set of proteins typically
found in contractile systems. They represent a possible site for active
control of the vibration characteristics; however, they are not innervated.

Hair cells and innervation

As in other mammals, the ratio of inner hair cells (IHCs) to outer
hair cells (OHCs) is about 1:3 up to 1:4 for all bat species. The absolute
number of receptor cells depends on the BM-length, but normalization to
BM-length shows that Myotis, Eptesicus and Nyctalus have the lowest hair
cell densities, close to that of humans, while all other bat species have
densities equal or greater than the cat (Tab.1). Hair cell density increases slightly from base to apex (Rhinolophus: Bruns and Schmieszek, 1980;
Myotis: Ramprashad et al., 1978; Nyctalus, Eptesicus: Burda and Ulehlova,
1983) with small regional maxima or minima corresponding to density changes in spiral ganglion cells (Bruns and Schmieszek, 1980).

Absolute numbers of spiral ganglion cells differ widely among species.
The FM-bat Myotis holds the absolute record with 55300 ganglion cells even
though it has the shortest BM (Ramprashad et al., 1978). The ganglion cells
are not enclosed within a spiral bony canal as in most other species, but
displaced into the internal auditory meatus. Ganglion cell distribution
along cochlear length is more or less uniform in species emitting FM-calls,
but pronounced regional variations are reported in Rhinolophoidea (Bruns
and Schmieszek, 1980; Kraus, 1983). These correspond to BM-specializations,

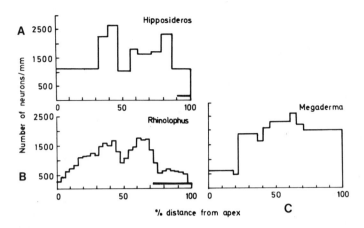

Fig.6. Comparison of spiral ganglion cell densities in three bat species.
(A, B, C). For Hipposideros and Rhinolophus, the extent of BM-
thickenings is shown by bar on abscissa. Hippossideros from Kraus,
1983; Rhinolophus from Bruns and Schmieszek, 1980); Megaderma
from Fiedler, 1983.

with the lowest number of ganglion cells found within the region of thickened BM (Fig.6).

The afferent innervation to the IHCs comprises 80-90% of the total
afferent supply of the cochlea of horseshoe bats (Bruns and Schmieszek,
1980), the remaining 10-20% contact OHCs; which is rather similar to data
from cat (90-95%, Spoendlin, 1973). Assuming a similar pattern of afferent
supply, Ramprashad et al. (1978) calculated the number of afferent terminals per IHC in Myotis as 70:1. This is much higher than in cat (20:1).
In horseshoe bats, pronounced regional variations in innervation density
are reported, ranging fom 8.3:1 in the basal half turn where BM is thickened to a maximum of 23.5:1 in the second half turn (Bruns and Schmieszek,
1980). No quantitative data are published for Pteronotus, but whole mount
preparations of the cochlea clearly demonstrate that regional differences
in innervation density are most pronounced in this species. Two clear maxima are observed, one in the hook region and a second even more prominent
one apical to the thickened BM-region, which receives only sparse innervation (Henson, 1973; Kössl and Vater, 1985b, Leake and Zook, 1985).

Data on the efferent innervation of the receptors were derived by ace-
tylcholinesterase histochemistry and electron microscopy. In FM-bats (Plecotus, Barbastella) efferent fibers supply both IHCs and OHCs (Firbas and
Welleschik, 1970). Among species of long CF-FM bats, striking differences
in the efferent nerve supply to the OHCs are reported. Each OHC in Pteronotus is contacted by one large efferent terminal (Bishop, et al., 1986).
The horseshoe bat cochlea completely lacks an efferent nerve supply to the
OHCs (Bruns and Schmieszek, 1980; Bishop, et al., 1986), while efferent
terminals are present in the region of IHCs. This finding raises the question of whether this system contributes to mechanisms which determine
threshold sensitivity and sharp tuning.

Stereocilia and tectorial membrane

Details regarding hair cell stereocilia and their contact with the
tectorial membrane are only known for horseshoe bats (Bruns and Goldbach,
1980). In cochlear regions apical to the discontinuity of BM-thickness and
-width, receptor cell arrangement, stereocilia or contact with tectorial
membrane resemble that described in other mammals. However, several specia-

lizations occur in the basal half turn. OHC-stereocilia are unusually short, the first row of OHCs is widely separated from the outer two rows; IHC stereocilia are twice as long as in other mammals or other turns, receptor surfaces of IHCs are unusually small and interreceptor distance is unusually wide.

The attachment of stereocilia to the tectorial membrane is peculiar in the basal half turn. The bottom side of the tectorial membrane has an unusual continuous zig zag elevation containing imprints of the innermost row of OHCs and lacks the Henson stripe above the IHC region, usually interpreted as an indicator of attachment of IHC stereocilia. Consequently, it has been proposed that in middle and apical turns of the horseshoe bat cochlea both IHC and OHC stereocilia are attached to the tectorial membrane and are excited by shearing motion, whereas in the basal half turn only the OHC stereocilia are sheared and the IHC stereocilia are moved by a subtectorial fluid motion component.

Morphological features unique to Pteronotus

Several specializations in cochlear morphology are unique to Pteronotus. The dimensions of the round window and cochlear aqueduct are exceedingly large and the scala vestibuli exhibits distinct volume changes within the basal turn (Henson et al., 1977, Kössl and Vater, 1985b). An extracellular substance covers the scala tympany in the basal half turn and part of the cochlear aqueduct (Pye, 1980; Jenkins et al., 1983). This "thick lining" disappears abruptly at the apical transition in BM-thickness (Kössl and Vater, 1985b). These findings suggest that specializations of perilymphatic spaces may also determine the acoustic properties of the inner ear.

FREQUENCY REPRESENTATION IN THE COCHLEA

Basilar membrane compliance estimates

It is generally accepted that within the mammalian cochlea, the pattern of frequency representation is largely determined by the compliance properties of the BM along its longitudinal extent. Although this parameter is of central importance, especially for cochlea models, it has not been measured directly in a bat cochlea. One indirect means of estimating the stiffness profile is based on morphological data, namely the ratio of BM-thickness to -width. Obviously, this does not take into account the factors outlined in previous chapters, namely the geometry of vibrating structures, the differential filament composition of BM-regions, the properties of attachment structures or possible active mechanisms. Fig.7 shows stiffness estimates for several bat species.

Bats with broadband echolocation signals are clearly set apart from species with narrowband signals. In Megaderma and Myotis, stiffness gradually decreases from base to apex and the two species are rather similar in absolute values and steepness of the curves. Only subtle irregularities are observed in Myotis and the pattern is close to that in nonecholocating mammals. Taphozous and Molossus exhibit more pronounced plateau areas with gradual transitions towards adjacent cochlear regions. In contrast, in Rhinolophoidea two cochlear regions of constant stiffness with species specific extent can be defined. A basally located plateau of increased stiffness (note the considerably higher values than in basal region of other bats with similar hearing range) is separated by a sharp transition from an apical plateau of less stiffness. A regular change in stiffness is only found in the most apical regions. Pteronotus exhibits a region of maximal stiffness at some distance from base, which is of less magnitude than in Rhinolophoidea and the transitions at its basal end are less abrupt. Within the most

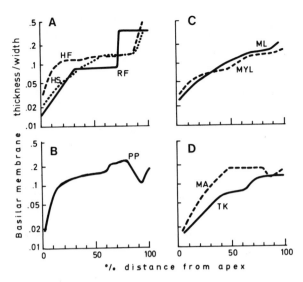

Fig. 7. BM-stiffness estimates for different bat species. Abbreviations and references as in Fig. 4.

basal part of the cochlea, a paradoxical change in stiffness is created. Apical to the stiffness maximum there follows a long stretch of BM with uniform properties.

Clearly, the pattern derived from the estimate of BM-compliance is species specific and could be expected to produce characteristic frequency place codes.

Frequency maps

The variety of regional differentiations in cochlear morphology found in different species of bats can be correlated with physiological proper- ties of the auditory system only if the frequency representation pattern on BM is known. Such frequency maps of the cochlea have been derived by different techniques: 1. The swollen nuclei technique (R. ferrumequinum; Bruns, 1976b), inducing morphological changes in the outer hair cell nuclei with intense pure tone stimulation. 2. The extracellular horseradish peroxi- dase (HRP) technique (R.rouxi: Vater et al., 1985; Pteronotus parnellii: Leake and Zook, 1985; Kössl and Vater, 1985b) analyzing the connectivity of cochlear spiral ganglion cells with physiologically characterized regions of cochlear nucleus, the central termination site of auditory nerve fibers.

Horseshoe bats: The cochlear frequency place maps for horseshoe bats are compared in Fig.8. According to the swollen nuclei map (Bruns, 1976b), the narrow frequency range from the bat's RF up to 3 kHz above RF is expanded on a 3 mm stretch of thickened BM just basal to the discontinuity, which corresponds exactly to the RF. The frequency range of unique filter proper- ties is therefore mapped directly onto the cochlear region where the main specializations of BM and hair cells are found, but innervation density is minimal. The mapping pattern for the low frequency range is regular and similar to other mammals. However it does not correspond to the stiffness estimations, which show a clear plateau within the second half turn.

These structure-function correlations are not supported by the HRP- map. HRP-injections into regions of cochlear nucleus sharply tuned to the CF-range label spiral ganglion cells apical to the BM-discontinuity, i.e.

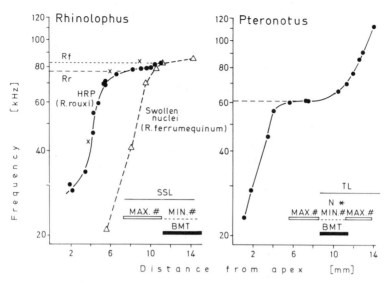

Fig. 8. Cochlear place frequency maps for Rhinolophus and Pteronotus:
The location of regional differentiations in cochlear morphology
is indicated above the abscissa: BMT: BM-thickenings; MAX., MIN.:
maxima or minima in innervation density; SSL: secondary spiral
lamina; TL: thick lining; N: narrowing of scala vestibuli. Rhinolophus rouxi: HRP-data from Vater et al., 1985; closed circles
represent data from Ceylonese bats with RF 77 kHz, crosses give
data from Indian bats with RF 84 kHz. Rhinolophus ferrumequinum:
swollen nuclei map from Bruns, 1976b; RF 83 kHz. Pteronotus parnellii: HRP-data from Kössl and Vater, 1985b; RF 61 kHz.

within the second half turn, where BM is not thickened and innervation density is maximal. The basal half turn was found to respond to frequencies
greater than 5 kHz above RF, but the exact frequency representation is
still unclear. Accordingly, the main morphological specializations of BM
and hair cells are located basal to the place of origin of sharply tuned
auditory nerve fibers. The HRP-map is in better agreement with stiffness
estimations and reports on central overrepresentation of the sharply tuned
frequency range (e.g. Feng and Vater, 1985). Furthermore, the mapping pattern
is independent of the absolute frequency range of the CF-component as demonstrated by comparison of the HRP-maps for Indian R.rouxi emitting 84 kHz
and Ceylonese R.rouxi emitting 77 kHz.

The causes for the pronounced discrepancy are unknown, but the shift
shows the same tendency as in other mammals, where acoustic overexposure
maps have been compared with physiological maps.

Mustache bats: The HRP-map for the cochlea of Pteronotus (Fig.8; Kössl
and Vater, 1985b, Leake and Zook, 1985) demonstrates that the sharply tuned
second harmonic CF-region (61 kHz) is located within the region of maximal
innervation density in the second half turn, apical to the stretch of thickened BM. Within the region of thickened BM where innervation density is low
and the spiral ligament is locally enlarged, label from injections into
66 kHz and 70 kHz areas was found. The frequency range from 70 to 111 kHz
corresponds to the region of maximal innervation density in the hook region,
which is terminated apically by the distinct narrowing of scala vestibuli.
Note the striking disagreement with stiffness estimates in this region.
There are no morphological specializations at the place of representation
of the first harmonic in the orientation call.

To summarize, in both species of long CF-FM bats, the HRP-technique demonstrates that the cochlear representation site of the sharply tuned centrally overrepresented CF-region ("acoustic fovea") corresponds to a maximum in innervation density and occupies a long stretch of non-thickened BM, which is located apical to main morphological specializations of BM. The frequency range above the sharply tuned area (<u>Rhinolophus</u>: 5 kHz above the CF-component up to the high frequency limit of the audiogram; <u>Pteronotus</u>: around 66 to 70 kHz) can not be correlated with an echolocation signal component. Nevertheless, it is expanded on the morphologically specialized BM, which in both species receives relatively sparse innervation. This raises the question of its functional role (see last chapter).

Basilar membrane measurements

Direct measurements of the mechanical response of the cochlear duct present considerable technical difficulties and the results are critically dependent on the physiological state of the preparation. So far, <u>R. ferrumequinum</u> is the only bat species in which vibration of cochlear structures has been measured (capacitive probe measurements, Wilson and Bruns, 1983b). The results did not provide clear evidence for a sharply tuned mechanical element which could directly account for the enhanced tuning characteristics at the CF-frequencies, nor was an indication of travelling wave reflections at the BM-discontinuity found. BM-vibration amplitudes at 100 dB SPL were exceedingly small as compared to other species (about 45 dB less response relative to stapes), especially at frequencies above 80 kHz. The most significant findings different from other species were 1) the occurance of sharp notches in BM-vibration amplitude at or just below the CF-frequency, which could either be created by cancellation effects or resonant absorption; 2) the divergence of phase of vibration of PT and PP for stimulus frequencies between 78 and 86 kHz, which appears to be related to a radial motion component in the region of thickened BM.

Considerations and models of tuning mechanisms

Several functional principles have been proposed to account for the enhancement of tuning characteristics within a narrow frequency range in horseshoe bats. These interpreted the role of the BM-thickenings either according to the swollen nuclei map or considered the HRP-map. These general considerations should prove applicable to interpretations of the cochlear specializations in other bat species as well.

<u>Resonators and guard ring principle</u> (Wilson, 1977): A two component model is proposed in which a population of high Q resonators loosely coupled to the BM influences BM-response at their resonance frequency. These "second mechanically tuned elements" are represented by structures in the organ of Corti or the tectorial membrane. Phase differences between PP and PT, which are exaggerated in the basal half turn of the horseshoe bat cochlea suggest a radial motion of these structures. This would produce strong hair cell excitation even at small BM-displacements. Additionally, a mechanism is proposed for maintaining high Q-values in a damped system (Guard ring principle). To avoid strong coupling between neighboring sequentially tuned resonantors (which would prevent high Q-values), it is necessary to match the rate of phase change of the driving function (BM-response) to the Q-value of the resonators. Such a mechanism could be common to all cochleae but not easily observed in acoustic organs where the Q-factor of the second resonator is too low to influence BM-response.

<u>Second mechanical filter</u> (Bruns, 1979, 1980): This model considers different modes of deflection of the BM and proposes that the mechanical configuration in the basal half turn consists of an independently vibrating outer

plate (PP, SL, SSL) which is only loosely coupled to the inner segment of
the BM and the outer bony wall. In contrast to nonspecialized cochleae,
where PP represents a flexing and bending element, both PT and PP of the
specialized region are considered inflexible. There, the only elastic element
is the thin narrow strand of BM between Deiters cells and outer pillars
allowing for out of phase motion of the two BM-compartments. A transversal
component of motion which would optimally shear the receptor organ is considered to be largest when PP and PT move in antiphase at 83 kHz and BM-amplitude is small. The independently moving outer plate is expected to create
radial fluid motion and quickly cancel deflections due to the longitudinal
traveling wave which would result in excitation confined to a narrow frequency band.

    Antiresonance and reflection (Miller, 1986): BM-amplitude and frequency
response of the horseshoe bat cochlea were calculated in a two-dimensional
mathematical model and compared with a cochlea lacking specializations.
The BM-specializations cause attenuation of BM-amplitude for all frequencies
with a local minimum at the discontinuity for a frequency of 83 kHz. They
also cause widely different BM-responses within a small frequency range
close to 83 kHz within a cochlear region apical to the discontinuity. Furthermore, significant reflection of cochlear waves occurs at frequencies
below 83 kHz. The condition at 83 kHz was decribed as antiresonance: sharp
notches are created instead of sharp response peaks. As a consequence, maximal neural excitation should occur at near minimal BM-motion.

    Acoustic interference filter (Duifhuis and Vater, 1986): The findings
of the HRP-technique raised the possibility that the enhanced tuning in
the horseshoe bat cochlea is produced by a mechanism based on wave interference. This was tested in a one-dimensional mathematical model. The abrupt
stiffness change of BM at the discontinuity represents a step in acoustic
impedance, which together with the impedance transition at the stapes makes
it likely that the basal part of the cochlea acts as an acoustic interference filter. Frequency dependent enhancement and cancellation effects will
produce multiple response maxima and minima (combfilter characteristic).
The response peak at the CF-frequency is sharp and sufficiently separated
from other response maxima occuring at multiple intervals.
In Rhinolophus, indications of multiple sensitivity maxima and minima as
predicted by this mechanism are observed in the N1-audiogram (Schnitzler
et al., 1976) or in the plots of tympanic membrane impedance (Wilson and
Bruns, 1983a).

    According to this model the functional role of the thickened BM-re-
gion is not seen in direct electromechanical transduction and relay of
sharply tuned information to the central auditory system. Rather, it is
interpreted to be a peripheral mechanical element designed to enhance the
tuning and sensitivity of the cochlear region apical to the discontinuity.

OPEN PROBLEMS

The persisting problems in research on the peripheral auditory system of
bats can be summarized as follows: There is a complete lack of direct
stiffness measurements of BM, which are essential for cochlea models.
There is a need for direct measurements of BM-vibration with sensitive techniques in physiologically intact cochleae. The significance of other specialized structures (e.g. tectorial membrane, SSL) or of subtle changes in
morphology for cochlear tuning is not yet quantitatively assessed. The functional importance of the species differences in the efferent supply of the
hair cells is unknown.

    It should be pointed out that in peripheral frequency analysis we are
dealing with a cooperative interaction among several mutually dependent

mechanisms, possibly with different frequency characteristics. In addition
to the still unresolved issues in describing the contribution of the
so-called "passive mechanics" in the different bat species, the possibility
of active mechanisms feeding back into the mechanical system of the mammalian cochlea has introduced a further set of problems.

REFERENCES

Bekesy, von G., 1960, "Experiments in hearing", McGraw Hill Book company.
    New York-Toronto-London.
Bishop, A. L., Henson, O. W. Jr., Henson, M. M., and Vater, M., 1986, Inner-
    vation of the outer hair cells in doppler-shift compensating bats, in:
    Abstracts of the ninth midwinter research meeting. D. J. Lim, ed.,
    Association for Research in Otolaryngology, p.28.
Brown, A. M., 1973, An investigation of the cochlear microphonic response
    of two species of echolocating bats: Rousettus aegyptiacus (Geoffroy)
    and Pipistrellus pipistrellus (Schreber), J. comp. Physiol. 83:407.
Bruns, V., 1976a, Peripheral auditory tuning for fine frequency analysis
    by the CF-FM bat, Rhinolophus ferrumequinum. I. Mechanical speciali-
    zations of the cochlea, J. comp. Physiol., 106:77.
Bruns, V., 1976b, Peripheral auditory tuning for fine frequency analysis
    by the CF-FM bat, Rhinolophus ferrumequinum. II. Frequency mapping
    in the cochlea, J. comp. Physiol., 106:87.
Bruns, V., 1979, Functional anatomy as an approach to frequency analysis
    in the mammlian cochlea, Verh. Deutsche Zool. Ges.,1979:141, Gustav
    Fischer Verlag, Stuttgart.
Bruns, V., 1980, Basilar membrane and its anchoring system in the cochlea
    of the greater horseshoe bat, Anat. Embryol., 161:29.
Bruns, V., and Schmieszek, E., 1980, Cochlear innervation in the greater
    Horseshoe bat: Demonstration of an acoustic fovea, Hearing Res.,3:27.
Bruns, V., and Goldbach, M., 1980, Hair cells and tectorial membrane in
    the cochlea of the greater horseshoe bat, Anat. Embryol., 161:51.
Burda, H., and Ulehlova, L., 1983, Cochlear hair-cell populations and limits
    of hearing in two vespertilionid bats, Nyctalus noctula and Eptesicus
    serotinus, J. Morph., 176:221.
Cabezudo, L., Antoli-Candela, F. Jr., and Slocker, J., 1983, Morphological
    study of the middle and inner ear of the bat: Myotis myotis, Adv. Oto-
    Rhino-Laryng., 31:28.
Dalland, J. J., Vernon, J. A., and Peterson, E. A., 1967, Hearing and coch-
    lear microphonic potentials in the bat, Eptesicus fuscus, J. Neurophy-
    siol., 30:697.
Duifhuis, H., and Vater, M.,1986, On the mechanics of the horseshoe bat coch-
    lea, in: Peripheral Auditory Mechanisms, J. W. Allen, J. L.Hall, A. Hub-
    bard, S. T. Neely, A. Tubis, eds., Springer Verlag Heidelberg New York.
Fiedler, J., 1983, Vergleichende Cochlea-Morphologie der Fledermausarten
    Molossus ater, Taphozous nudiventris kachensis und Megaderma lyra.
    Inaug. diss. Universität Frankfurt.
Feng, A. S., and Vater, M., 1985, Functional organization of the cochlear
    nucleus of Roufous horseshoe bats (Rhinolophus rouxi): Frequencies and
    internal connections are arranged in slabs, J. comp. Neurol., 235:529.
Firbas, W., and Welleschik, B., 1970, Über die Verteilung der Acetylcho-
    linesteraseaktivität im Cortischen Organ von Fledermäusen, Acta Oto-
    laryng., 70:329.
Grinnell, A. D., 1963, The neurophysiology of audition in bats: intensity
    and frequency parameters, J. Physiol., 167:38.
Grinnell, A. D., 1970, Comparative auditory neurophysiology of neotropical
    bats employing different echolocation signals, Z. vergl. Physiol., 68:
    117.
Henson, M. M., 1973, Unusual nerve-fiber distribution in the cochlea of the
    bat Pteronotus p. parnellii (Gray), J. Acoust. Soc. Amer., 53:1739.

Henson, M. M., 1978, The basilar membrane of the bat, Pteronotus p.parnellii, Amer. J. Anat., 153:143.
Henson, O. W. Jr., and Kobler, J. B., 1979, Temperature and its effect on the CM-audiogram of the bat, Pteronotus p. parnellii, Anatomical Rec., 193:744.
Henson, M. M., Henson, O. W. Jr., and Goldman, L. J., 1977, The perilymphatic spaces in the cochlea of the bat, Pteronotus p. parnellii (Gray), Anatomical Rec., 187:767.
Henson, O. W. Jr, Pollak, G. D., Kobler, J. B., Henson, M. M., and Goldman, L. J., 1982, Cochlear microphonic potentials elicited by biosonar signals in flying bats, Pteronotus p. parnellii, Hearing Res., 7:127.
Henson, M. M., Henson, O. W. Jr. and Jenkins, D. B., 1984, The attachment of the spiral ligament to the cochlear wall: Anchoring cells and the creation of tension, Hearing Res.,16:231.
Henson, O. W. Jr., Schuller, G., and Vater, M., 1985a, A comparative study of the physiological properties of the inner ear in Doppler shift compensating bats (Rhinolophus rouxi and Pteronotus parnellii), J.comp. Physiol.,157:587.
Henson, M. M., Burridge, K.,Fitzpatrick, D.,Jenkins, D. B., Pillsbury, H. C., and Henson, O. W. Jr., 1985b, Immunocytochemical localization of contractile and contraction associated proteins in the spiral ligament of the cochlea, Hearing Res.,20:207.
Jenkins, D. B., Henson, M. M., and Henson, O. W. Jr., 1983, Ultrastructure of the lining of the scala tympani of the bat, Pteronotus parnellii, Hearing Res., 11:23.
Kemp, D. T., 1979, The evoked cochlear mechanical response and the auditory microstructure - evidence for a new element in cochlear mechanics, Scand. Audiol. Suppl., 9:35.
Kobler, J. B., Wilson, B. S., Henson, O. W. Jr., and Bishop, A. L., 1985, Echointensity compensation by echolocating bats, Hearing Res., 20:99.
Kössl, M., and Vater, M., 1985a, Evoked acoustic emissions and cochlear microphonics in the mustache bat, Pteronotus parnellii, Hearing Res., 19:157.
Kössl, M., and Vater, M., 1985b, The cochlear frequency map of the bat, Pteronotus parnellii, J .comp. Physiol., 57:687.
Kraus, H. J., 1983, Vergleichende und funktionelle Cochlea-Morphologie der Fledermausarten Rhinopoma hardwickei, Hipposideros speoris und Hipposideros fulvus mit Hilfe einer computergestützten Rekonstruktionsmethode, Inaug.Diss. Universität Frankfurt.
Leake, P. A., and Zook, J. M., 1985, Demonstration of an acoustic fovea in the mustache bat, Pteronotus parnellii, in: Abstracts of the eighth midwinter meeting. D. J. Lim., ed., Association for Research in Otolaryngology:p.27.
Manley, G. A., Irvine, D. R. F., and Johnstone, B. M., 1972, Frequency response of bat tympanic membrane, Nature, 237:112.
Miller, C. E., 1986, Resonance and reflection in the cochlea: The case of the CF-FM bat, Rhinolophus ferrumequinum. in: Peripheral auditory mechanisms, J. W. Allen, J. L.Hall, A. Hubbard, S. T.Neely, A. Tubis, eds., Springer Verlag Heidelberg New York.
Neuweiler, G., 1970, Neurophysiologische Untersuchungen zum Echoortungssystem der Großen Hufeisennase, Rhinolophus ferrumequinum Schreber, 1774, Z. vergl. Physiol., 67:273.
Neuweiler, G., Singh, S., and Sripathi, K., 1984, Audiograms of a South Indian bat community, J. comp. Physiol.,154:133.
Pollak, G. D., and Henson, O. W. Jr., 1973, Specialized functional aspects of the middle ear muscles in the bat, Chilonycteris parnellii, J. comp. Physiol., 84:167.
Pollak, G. D., Henson, O. W. Jr. and Novick, A., 1972, Cochlear microphonic audiograms in the "pure tone" bat, Chilonycteris parnellii parnellii, Science, 176:66.
Pollak, G., Henson, O. W. Jr. and Johnson, R., 1979, Multiple specializa-

tions in the peripheral auditory system of the CF-FM bat, Pteronotus parnellii. J. comp. Physiol., 131:255.

Pye, A., 1966, The structure of the cochlea in chiroptera. I. Microchiroptera: Emballonuroidea and Rhinolophoidea, J. Morphol., 118:495.

Pye, A., 1980, The cochlea in Pteronotus parnellii, in: Animal Sonar Systems R.-G. Busnel, J. F. Fish, eds. Plenum Press, New York and London.

Ramprashad, F., Money, K. E., Landolt, J. P., and Laufer, J., 1978, A neuroanatomical study of the cochlea of the little brown bat (Myotis lucifugus), J. comp. Neurol., 178:347.

Ramprashad, F., Landolt, J. P., Money, K. E., Clark, D., and Laufer, J., 1979, A morphometric study of the cochlea of the little brown bat (Myotis lucifugus), J. Morph., 160:345.

Schnitzler, H.-U., 1968, Die Ultraschall-Ortungslaute der Hufeisenfledermäuse (Chiroptera-Rhinolophidae) in verschiedenen Orientierungssituationen. Z. vergl. Physiol., 57:376.

Schnitzler, H.-U., Suga, N., and Simmons, J. A., 1976, Peripheral auditory tuning for fine frequency analysis by the CF-FM bat, Rhinolophus ferrumequinum. J. comp. Physiol., 106:99.

Spoendlin, H., 1973, The innervation of the cochlear receptor, in: Basic Mechanisms in Hearing, A. R. Moller, ed., Academic Press New York and London.

Suga, N., and Jen, H.-S., 1975, Peripheral control of acoustic signals in the auditory system of echolocating bats. J. exp. Biol., 62:277.

Suga, N., and Jen, H.-S., 1977, Further studies on the peripheral auditory system of 'CF-FM' bats specialized for fine frequency analysis of Doppler shifted echoes. J. exp. Biol., 69:207.

Suga, N., Simmons, J. A., and Jen, P. H.-S., 1975, Peripheral specializations for fine frequency analysis of Doppler-shifted echoes in 'CF-FM' bat, Pteronotus parnellii, J. exp. Biol., 63:161.

Wilson, J. P., 1977, Towards a model for cochlear frequency analysis, in: Psychophysics and physiology of hearing, E. F. Evans, and J. P. Wilson, eds., Academic Press, London.

Wilson, J. P., and Johnstone, J. R., 1975, Basilar membrane and middle ear vibration in guinea pig measured by capacitive probe. J. Acoust. Soc. Am., 57:705.

Wilson, J. P., and Bruns, V., 1983a, Middle-ear mechanics in the CF-bat Rhinolophus ferrumequinum, Hearing Res. 10:1.

Wilson, J. P., and Bruns, V., 1983b, Basilar membrane tuning properties in the specialized cochlea of the CF-bat, Rhinolophus ferrumequinum, Hearing Res., 9:15.

Vater, M., Schlegel, P., and Zöller, H., 1979, Comparative auditory neurophysiology of the inferior colliculus of two molossid bats, Molossus ater and Molossus molossus. I. Gross evoked potentials and single unit responses to pure tones, J. comp. Physiol. 131:137.

Vater, M., Feng, A. S. and Betz, M., 1985, An HRP-study of the place frequency map of the horseshoe bat cochlea: Morphological correlates of the sharp tuning to a narrow frequency band. J. comp. Physiol.,157:671.

ASCENDING PATHWAYS TO THE INFERIOR COLLICULUS VIA THE SUPERIOR OLIVARY

COMPLEX IN THE RUFOUS HORSESHOE BAT, RHINOLOPHUS ROUXII

J.H. Casseday*#, E. Covey* and M. Vater+

Departments of Surgery* and Psychology#
Duke University
Durham, N.C. 27710

Zoologisches Institut der Universität München+
D-8000 München

In all species of mammals a number of parallel auditory pathways ascend to the inferior colliculus, but the relationships among these pathways at their targets are poorly understood. The pathways which project via the medial and lateral nuclei of the superior olivary complex (MSO and LSO) receive extensive bilateral input from the cochlear nuclei. For this reason they are probably the first stage in processing binaural cues for localizing sound in space. The projection area of MSO and LSO in the inferior colliculus can therefore be considered to be a structural definition of the binaural region of the inferior colliculus. In this study we sought to define the binaural region in the inferior colliculus of Rhinolophus rouxii by the method of anterograde axonal transport from physiologically defined areas within the superior olivary complex.

The superior olivary complex in Rhinolophus, like that of many echolocating bats, has a very unusual appearance. It has not been established whether the medial cell groups correspond to MSO, to superior paraolivary nucleus (SPN) of rodents, to some other periolivary group, or whether they are unique to bats. In the present experiments wheat germ agglutinin conjugated to horseradish peroxidase (WGA-HRP) was placed in physiologically defined areas of the superior olive. Because WGA-HRP is both an anterograde and retrograde tracer, it was possible to examine the patterns and sources of input to the nuclei of the superior olive as well as the pattern of ascending projections from the various component cell groups of the superior olivary complex to the inferior colliculus. In addition, we examined the binaural response properties of cells in and around the injection sites in the superior olive. With these methods it was possible to characterize the major cell groups in the superior olivary complex in terms of 1)excitation, inhibition, or lack of response to either ear, 2) bilateral or unilateral input from the cochlear nuclei, and 3) ipsilateral or bilateral ascending projections to the inferior colliculus.

METHODS

In 10 horseshoe bats (Rhinolophus rouxii), electrophysiological recordings were made in the superior olivary complex in order to characterize each subdivision according to the responses of its units to binaural stimuli, i.e., excitatory to both ears (EE), excitatory to ipsilateral, inhibitory to contralateral ear (EI), no response to ipsilateral, excitatory to the contralateral ear (OE), inhibitory to the ipsilateral ear and excitatory to the contralateral ear (IE), or some variation of these response types. Sinusoidal stimuli were presented, and the frequency to which the units responded at the lowest sound intensity was taken as a measure of the unit's best frequency. After determining the response properties of units at different locations, an electrode containing WGA-HRP was placed in an area selected on the basis of the results of the mapping experiment. Response properties of units were again recorded to insure that the electrode tip was in the desired location. Finally, an injection of WGA-HRP was made by electrophoresis. The animal was perfused 24 hours later, and brain sections were processed with tetramethylbenzidine to visualize anterograde and retrograde transport from the injection. A camera lucida was used to draw the location of labeled cells and to draw the trajectory and termination of fibers labeled by anterograde transport.

RESULTS AND DISCUSSION

The cases in Fig. 1 were chosen to illustrate the main points of the results. Two cases (Fig. 1A, 1B) illustrate the connections of LSO. In Fig. 1A (Case R710) a large injection includes all of the medial limb of LSO, an area in which high frequencies at and above the filter frequency are represented. Medially there is a small amount of spread to adjacent structures. There are ascending projections to the nuclei of the lateral lemniscus, especially to DNLL bilaterally. In the inferior colliculus, terminal label is present bilaterally throughout the ventrolateral two-thirds of the central nucleus, but is most dense ventromedially. The terminal fields on the two sides are approximately symmetrical and are organized as a series of fine bands parallel to the surface of the inferior colliculus. Cells labeled by retrograde transport are present throughout the medial two-thirds of the ipsilateral MNTB and in all divisions of the ipsilateral AVCN and PVCN. No labeled cells are present in DCN after this or any of the other injections in the superior olivary complex.

Fig. 1B (Case R907) illustrates the results of a smaller injection which fills the lateral and rostral part of the LSO, a region in which low frequencies, below the filter frequency, are represented. The ascending projections in this case are again bilateral, but in this case they are mainly to the ventrolateral part of the central nucleus of the inferior colliculus. Again, the terminal fields on the two sides are approximately symmetrical. This injection, taken together with the previous one, gives a good indication of the total target of LSO within the inferior colliculus. The combined projection field includes the entire ventrolateral two-thirds of the central nucleus on both sides.

In contrast to the above two cases, Fig. 1C shows an injection that fills the rostral and ventromedial part of the two large groups of cells medial to the LSO. In this case, virtually all of the ascending projections are ipsilateral. Although all divisions of the nuclei of the lateral lemniscus receive projections, those to INLL are particularly dense. Label in the central nucleus of the ipsilateral inferior colliculus is very dense ventrally, and it extends dorsally in finger-like bands about 2/3 of the way to the surface. Most of the labeled cells are in AVCN and PVCN contralateral to the injection. There are a few labeled cells in the ipsilateral cochlear nucleus.

Fig. 1D (Case R239) shows the results of an injection in MNTB. Antero-grade transport is present ipsilaterally only, but it extends no farther than the lateral lemniscus. There is dense label in INLL and in all of the major nuclei of the superior olivary complex, but none in DNLL or the inferior colliculus. Labeled cells are present only in contralateral AVCN and PVCN.

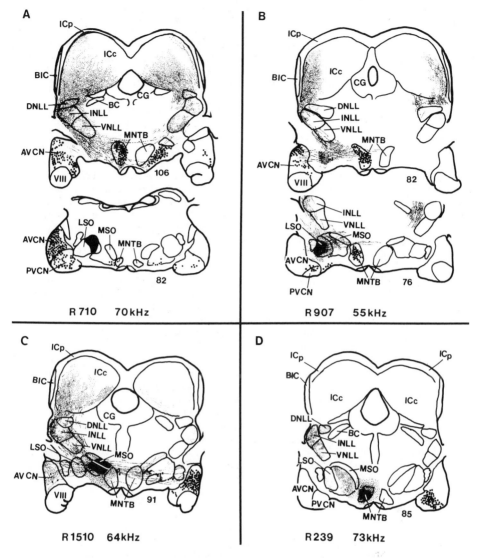

Fig. 1. Drawings to show different patterns of connections of cell groups
in the superior olivary complex of <u>Rhinolophus rouxii</u>. WGA-HRP
injection shown in black, labeled cells by large dots, anterograde
transport by stippling. (A) Injection in medial LSO shows anterograde transport to the ventromedial part of the inferior colliculus
bilaterally and labeled cells in the ipsilateral cochlear nucleus
and MNTB. (B) Injection in lateral LSO shows anterograde transport
as in (A), but to lateral part of inferior colliculus. (C) Injection in medial cell groups of superior olivary complex shows anterograde transport throughout the ventral two-thirds of the ipsilateral inferior colliculus. (D) Injection in MNTB.
Abbreviations: AVCN, anteroventral cochlear nucleus; BC, brachium
conjunctivum; BIC, brachium of inferior colliculus; CG, central
grey; DNLL, dorsal nucleus of lateral lemniscus; IC (c,p), inferior
colliculus (central, pericentral), INLL, intermediate nucleus of
lateral lemniscus; LSO, lateral superior olive; MNTB, medial nucleus of trapezoid body; MSO, medial superior olive; PVCN, posteroventral cochlear nucleus.

The pattern of ascending projections to the inferior colliculus can be summarized as follows. First, the target of the superior olivary complex, and thus, the binaural area, lies within the ventrolateral two-thirds of the central nucleus of the inferior colliculus (Fig. 1A). The dorsomedial part of the central nucleus receives no input from the superior olive. Second, LSO projects bilaterally (Fig. 1B) whereas the two major cell groups medial to LSO project only to the ipsilateral inferior colliculus (Fig. 1C). The target of LSO overlaps largely or entirely with that of the medial structures. Third, the medial nucleus of the trapezoid body (MNTB) does not project to the inferior colliculus but does project, ipsilaterally, to LSO, to the medial cell groups of the superior olive and to the ventral and intermediate nuclei of the lateral lemniscus (Fig. 1D).

The pattern of inputs to the nuclei of the superior olivary complex in Rhinolophus can be summarized as follows: LSO receives projections from the ipsilateral cochlear nucleus and the ipsilateral MNTB as it does in other mammals. The medial cell groups, unlike MSO, do not receive symmetrical bilateral input; instead the projections to these structures arise mostly from the contralateral cochlear nucleus. The input to MNTB is exclusively from the contralateral cochlear nucleus.

The significance of these projections with regard to functions of the inferior colliculus can be examined in the context of response properties of the cell groups of the superior olivary complex. LSO contains EI (82%, N=22) or EO (18%) units exclusively. Its main input is from cells in the ipsilateral cochlear nucleus and ipsilateral MNTB. The cochlear nucleus is, of course, the source of the ipsilateral excitation; the contralateral inhibition in LSO probably arises from MNTB which contains OE units (100%, N=13) and receives projections from the contralateral cochlear nucleus. The medial cell groups contain a mixture of OE (43%, N=46), EE (35%), IE (20%) and EO (2%) units, but no EI units. EE cells in this region tended to respond more vigorously and at lower thresholds to stimulation of the contralateral ear than to the ipsilateral ear. This, and the finding that the majority of cells in this region are type OE, is consistent with the finding that most projections to the medial cell groups of the superior olive in Rhinolophus are from the contralateral cochlear nucleus. More detailed mapping studies will be necessary in order to determine whether the cells having each response type are segregated into different areas.

The physiological and connectional evidence, taken together, suggest that the inferior colliculus receives the following classes of binaural inputs: EI from the ipsilateral LSO, IE from the contralateral LSO, and OE, EE and IE from the medial cell groups of the ipsilateral superior olive. In addition, MNTB and the nuclei of the superior olive must provide some form of input from the contralateral ear via the nuclei of the lateral lemniscus, but whether these inputs are excitatory or inhibitory remains to be determined.

ACKNOWLEDGMENTS

This research was supported by the Deutsche Forschungsgemeinschaft, NIH grant NS 21748 and NSF grant BNS-8217357.

PATTERN OF PROJECTIONS TO THE 60 KHZ ISOFREQUENCY REGION OF THE MUSTACHE BAT'S INFERIOR COLLICULUS

Linda S. Ross and
George D. Pollak

Department of Zoology
University of Texas
Austin, Texas

The central nucleus of the mustache bat's inferior colliculus (ICc) has three divisions, each representing a different portion of the cochlear partition (Pollak et al., 1983; Zook et al., 1985) (Fig. 1). The anterolateral division (ALD) represents low frequencies, from about 10 kHz to 59 kHz. The medial division (MD) represents high frequencies that range from about 64 kHz to over 120 kHz. The dorsoposterior division (DPD) is a specialized isofrequency region that occupies roughly a third of the volume of the inferior colliculus. DPD neurons are all sharply tuned to 60 kHz, the dominant frequency of the bat's echolocation calls (Pollak and Bodenhamer 1981; Zook et al., 1985).

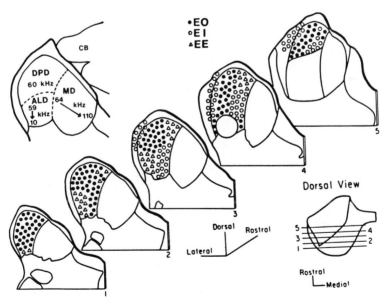

Fig. 1. Transverse section through the inferior colliculus showing the frequency representation in the three divisions (ALD, MD and DPD) and the topography of the four distinct binaural subregions of the DPD. See text for further explanation.

Additionally, neurons in this division are functionally organized such that neurons having similar binaural properties are grouped together and occupy specified loci (Wenstrup et al., 1985, 1986) (Fig. 1). Monaural neurons are localized in the dorsal and lateral portions of the DPD, and likewise, the two major types of binaural neurons, EE and EI, are confined to specified regions. EE cells occur in two regions, ventrolaterally and dorsomedially. The main EI population is located ventromedially with a second population found far laterally.

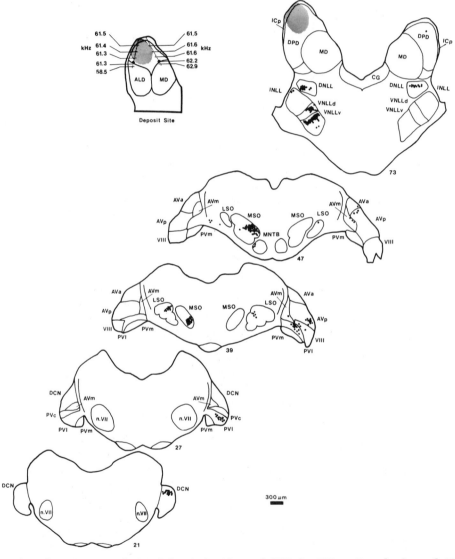

Fig. 2. Cells labelled by injection of HRP in DPD. Boundaries of DPD were first determined with neurophysiological mapping techniques. Notice that labelled cells are present in restricted regions of all lower auditory nuclei.

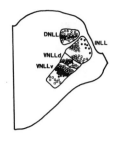

NUCLEI OF LATERAL LEMNISCUS

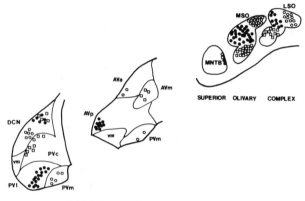

SUPERIOR OLIVARY COMPLEX

COCHLEAR NUCLEUS

Fig. 3. Neurons labelled by HRP injections in the low frequency anterolateral division (ALD) and high frequency medial division (MD) are shown together with cells labelled by HRP injections in the 60 kHz doroposterior division (DPD). Regions projecting to ALD are shown as open circles, regions projecting to MD are shown as squares, and regions projecting to DPD are shown as filled circles.

The first purpose of this study was to define the set of brainstem nuclei which project to the isofrequency DPD. Retrograde tracing techniques and neurophysiological recordings were combined by making large deposits of HRP after first defining its boundaries physiologically. In additional experiments, small deposits of HRP were placed in restricted regions of the low frequency ALD, or in the high frequency MD, thereby outlining the general tonotopic pattern of the bat's auditory brainstem.

The main finding is that all precollicular nuclei that project to the ICc also project to the DPD (Fig. 2). These include contralateral projections from all three divisions of the cochlear nucleus, ipsilateral projections from the medial superior olive (MSO), the ventral and intermediate nuclei of the lateral lemniscus (VNLL and INLL), and bilateral projections from the lateral superior olive (LSO) and the dorsal nucleus of the lateral lemniscus (DNLL).

Cells projecting to the DPD are spatially segregated from cells projecting to the MD and ALD, the other frequency representations of the inferior colliculus (Fig. 3). In general, the projections to the

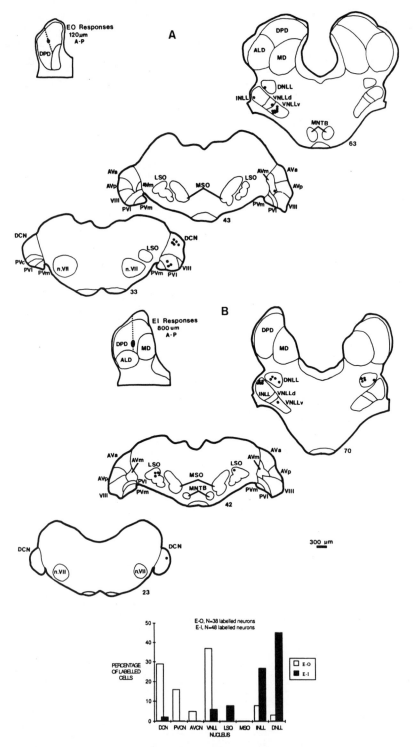

Fig. 4. Transverse sections through the brainstem illustrate the location of labelled cells following deposits of HRP in either the EO (monaural) (A) or EI subregions of the DPD (B). The bar graph shows the proportion of labelled cells in each of the brainstem auditory nuclei.

250

DPD were sandwiched between the cells projecting to the MD and ALD, reflecting a tonotopic organization similar to that seen in cats and other mammals (Guinan et al., 1972; Aitkin et al., 1984; Casseday and Pollak, this volume).

A second purpose of this study was to define and compare the set of brainstem nuclei which project to the various aural subregions of the DPD. We first documented the best frequencies and binaural type in electrode penetrations through the DPD and then made a focal deposit of HRP into a defined aural subregion. Two cases are illustrated here, one involving a deposit in the dorsal portion of the DPD, a region containing only monaural cells, and one describing a deposit in the ventromedial DPD, a region dominated by EI neurons.

The principal result is that the deposit in the dorsal region of the DPD produced labelling primarily in monaural nuclei of the brainstem (Fig. 4a). The largest number of labelled cells were in the VNLL, the dorsal cochlear nucleus (DCN), and the posteroventral cochlear nucleus (PVCN), with a much smaller number of labelled cells in the INLL and AVCN (Fig. 4b). In contrast, the injection in the ventromedial DPD produced labelling primarily in binaural nuclei. Of particular interest is that 80% of the labelled cells were in DNLL and INLL, with less than 10% occurring in the LSO and none in the MSO.

There are three major findings that emerge from these studies:
1) An isofrequency region of the inferior colliculus, the dorsoposterior division (DPD), receives afferents from the total set of auditory nuclei which project to the central nucleus of the inferior colliculus.
2) The projections to the dorsoposterior division arise from restricted loci in each of the lower auditory nuclei. Cells projecting to the DPD are spatially segregated from cells projecting to the MD and ALD, the other frequency representations of the inferior colliculus. This represents a point-to-point projection system which connects 60 kHz regions in the lower brainstem with the 60 kHz region of the central nucleus of the inferior colliculus.
3) Preliminary evidence suggests that two aural subregions of the DPD, the monaural region in the dorsal DPD, and the EI subregion in the ventromedial DPD, receive projections from different subsets of the nuclei that project to the DPD. The projections to the monaural, dorsal DPD originate primarily in monaural nuclei, whereas the binaural EI region receives projections primarily from a subset of the precollicular binaural nuclei.

REFERENCES

Aitkin, L.M., Irvine, D.R.F., and Webster, W.R., Central neural
    mechanisms of hearing, in: "Handbook of Physiology, Section I:
    The Nervous System, Vol. III, Sensory Processes, Part 2, "Ian
    Darian-Smith ed., Amer. Physiol. Soc., Bethesda, MD (1984).
Casseday and Pollak, this volume.
Guinan, J.J., Jr., Norris, B.E., and Guinan, S.S., 1972, Single
    auditory units in the superior olivary complex. II. Location
    of unit categories and tonotopic organization, Int. J.
    Neurosci., 4: 147-166.
Pollak, G.D., and Bodenhamer, R.D., 1981, Specialized characteristics
    of single units in inferior colliculus of mustache bat:
    Frequency representation, tuning, and discharge patterns,
    J. Neurophysiol., 46:605-620.

Pollak, G.D., Bodenhamer, R.D., and Zook, J.M., 1983, Cochleotopic
        organization of mustache bat's inferior colliculus, in:
        "Advances in Vertebrate Neuroethology," J.P. Ewert, R.R.
        Capranica, and D.J. Ingle, ed., Plenum, NY (1983).
Wenstrup, J.J., Ross, L.S., and Pollak. G.D., 1985, A functional
        organization of binaural reponses in the inferior colliculus,
        Hearing Res., 17:191-195.
Wenstrup, J.J., Ross, L.S., and Pollak, G.D., 1986, Binaural reponse
        organization within a frequency-band representation of the
        inferior colliculus: Implications for sound localization,
        J. Neuroscience, 6:962-973.
Zook, J.M., Winer, J.A., Pollak, G.D., and Bodenhamer, R.D., 1985,
        Topology of the central nucleus of the mustache bat's inferior
        colliculus: Correlation of single unit properties and neuronal
        architecture, J. Comp. Neurol., 231:530-546.

TARGET RANGE PROCESSING PATHWAYS IN THE

AUDITORY SYSTEM OF THE MUSTACHED BAT

William E. O'Neill, Robert D. Frisina, David M. Gooler, and
Martha L. Zettel

Center for Brain Research
University of Rochester School of Medicine and Dentistry
Rochester, New York 14642 USA

Echolocating bats determine target range from the time interval between emitted sonar pulses and returning echoes. Long CF-FM bats determine range using cues provided by the FM components of their signals (Simmons, 1973). The mustached bat emits pulses consisting of four harmonics each containing a long constant frequency (CF) component followed by a brief, descending FM. Suga and his colleagues have shown that there are neurons in at least two fields of auditory cortex (FM-FM and DF) which only respond to second (FM_2), third (FM_3), or fourth (FM_4) harmonic FM stimuli when preceded at a specific time interval by a first harmonic (FM_1) stimulus (O'Neill & Suga, 1982; Suga et al., 1983; Suga, 1984). Each of these "FM-FM" neurons is tuned to a specific, or best, time delay, and the cortical fields in which they are located are organized by increasing best delay ("chronotopic" axis) in the rostrocaudal direction. Thus, there are at least two neural axes representing target distance ("odotopic axes") in each hemisphere of the auditory cortex of the mustached bat.

Since each of these specialized cortical neurons responds to the temporal information contained in two different harmonic components in the emitted pulse (i.e., FM_1) and echo (e.g., FM_2), pathways separated within the tonotopically organized ascending (lemniscal) auditory system must converge at some point. Olsen & Suga (1983) have found delay-tuned neurons resembling the cortical type in the medial and dorsal divisions of the medial geniculate body (MGB). In contrast, O'Neill (1985) found no delay-tuned neurons at the level of the central nucleus of the inferior colliculus (ICC). The results reported here summarize an extensive series of experiments studying the response properties, locations and connectivity of neurons in the ICC which are tuned to frequencies associated with the FM_2 component, as well as preliminary data from ongoing work in the area representing FM_1. Taken together with the above-mentioned studies at higher levels, we have defined in some detail one pathway in the mustached bat auditory system whose function is to process target range information.

To delineate brainstem pathways carrying FM_2 information, we have employed single unit recordings, 2-deoxyglucose autoradiography, mapping of unit best frequencies (BF), tracing of connections using HRP, and various histological procedures including Golgi techniques. Single unit recordings (O'Neill, 1985) showed that neurons tuned to FM_2 frequencies generally had equal, and rarely, better sensitivity to FM_2 sweeps compared to pure tone

253

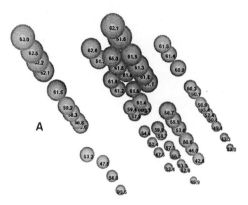

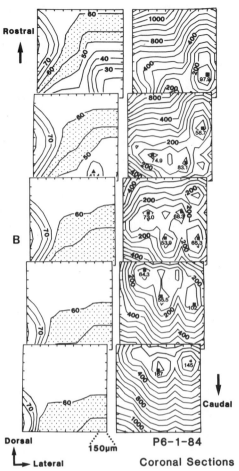

Fig. 1 Representation of frequency
in ICC. **A**: 3D reconstruction
of unit locations and BF's (in
kHz). Angle of view is caudo-
lateral (i.e., rotated by 130°
around horizontal axis).
B: Isofrequency contours within "slices" spaced 300 μm
apart in the coronal plane
(left row), paired with iso-
distance contours (right row;
distances expressed in μm
from the nearest actual unit
location). Shaded areas in
isofrequency contour slices
indicate region with units
tuned to FM_2.

Rostral

Dorsal
Lateral

150μm

Caudal

P6-1-84
Coronal Sections

stimuli. Stimuli mimicking CF-FM pulse-echo pairs did not reveal any
facilitation at different time delays like that seen in delay-tuned neurons
in cortex or MGB. The units simply respond to the FM_2 components of such
complex stimuli. Thus, at the level of the ICC, information from the FM_1
component does not reach neurons responding to FM_2 in order to produce
facilitation for particular time delays.

In order to define more accurately the ICC region that processes FM_2
information, we carried out extensive mapping studies of BF's of neurons in
the ICC, and made focal injections of HRP into the center of the region
where units were tuned to FM_2 frequencies (O'Neill & Frisina, submitted).
In each case, we reconstructed in detail the tonotopic organization of the
ICC. A three-dimensional view of the unit locations and BF's recorded in
one bat is shown in Fig. 1a, and isofrequency contours reconstructed from
these same data are shown in coronal sections through the IC in Fig. 1b.
Note that the shaded area representing the region of units tuned to FM_2
frequencies (50-60 kHz) occupies a large region of the ICC relative to
frequencies below 50 kHz and above 64 kHz. These findings are similar to
those of our 2-deoxyglucose studies which showed that FM_1 frequencies (25-
30 kHz) are represented in a somewhat smaller and more rostrally located
area in AL than FM_2 (Fig. 2; O'Neill et al, submitted). In general, the
isofrequency contours for frequencies below 60 kHz are oriented dorsoventrally in the anterior half of the ICC, so that the tonotopic axis runs
rostrocaudally. A large region located dorsal and caudal to the FM_2

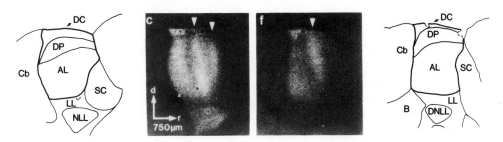

Fig. 2 2-deoxyglucose labeling showing repreentation of FM$_1$ and FM$_2$ in IC.
On the left is sagittal section through midbrain of bat exposed to
FM$_1$-FM$_2$ pairs (30.5-24.5 kHz, 62 dB SPL and 61.0-49.0 kHz, 59 dB
SPL, alternating delays of 3 and 8 msec), and on the right from a
bat exposed to FM$_2$ alone (60-48 kHz, 69 dB SPL). Arrowheads point
to regions in AL showing most uptake of 2-DG in each case. Note
that there are two areas in response to FM$_1$-FM$_2$. Sections are about
700 μm from lateral edge of IC. d, dorsal; r, rostral.

representation contains sharply-tuned reference frequency neurons with BF's
all near 60 kHz. Frequencies higher than 60 kHz are located ventromedially. These results are consistent with earlier studies (Pollak et
al., 1983; Zook et al., 1985).

Zook et al. (1985) have shown that there are three distinct divisions
of the ICC in mustached bats, namely, the dorsoposterior (DP), anterolateral (AL), and medial (M). DP contains reference frequency tuned
neurons, units tuned to lower frequencies are found in AL, and those tuned
above 60 kHz are represented in M. As can be seen in Fig. 3, our results
(O'Neill & Frisina, submitted) confirm these findings. The reconstructions

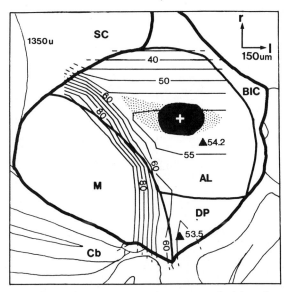

Fig. 3 Isofrequency contours overlaid onto
matching horizontal section in the
IC, showing correlation of frequency
representation and boundaries of the
divisions of the ICC. Cross shows
center of an HRP injection, the core
of which is indicated in black.

of isofrequency contours allow visualization of the frequency representation in any histological plane of section, one of which can then be superimposed on photomicrographs of brain sections cut in the same plane and
aligned according to the centers of HRP injection sites. We conclude from
these studies that there is an over-representation of the frequency band
between 50 and 60 kHz in the AL division of ICC, in accordance with its
importance as the most prominent FM echo component. The representation of
FM_1 is currently being investigated.

HRP injections carefully confined to the FM_2 representation in AL
(e.g., Fig. 3) have elucidated the connections to this area (Fig. 4;
Frisina & O'Neill, submitted). Ascending projections to this part of ICC
arose from brainstem and midbrain structures, in a pattern not unlike that
in other mammals. Weak connections were seen intrinsic to the IC between
the injection site and DP and M ipsilaterally, and to AL contralaterally.

More intriguing were the efferent connections of the FM_2 area. Weak
connections were seen to deep layers of the superior colliculus. A strong
descending connection was also seen to the lateral pontine nuclei, an area
which may be important in vocalization. Most interesting were the projections to MGB, which spare the ventral division and terminate almost
exclusively in a few discrete locations in the ipsilateral dorsal and
medial divisions. This is unusual since in cat the projection from ICC is
bilateral mainly to the tonotopically organized ventral division (Andersen
et al., 1980b). The terminations in the mustached bat MGB appear to be
located in areas where Olsen & Suga (1983) have discovered delay-tuned

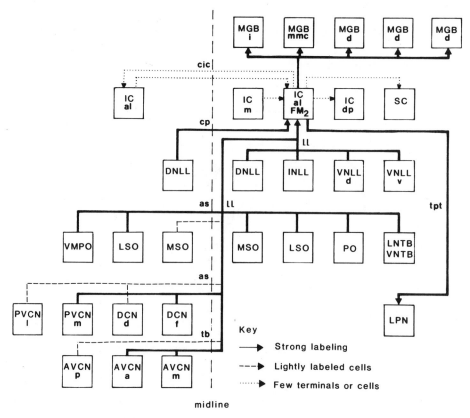

Fig. 4   Results of HRP injections into FM_2 region of ICC.

neurons similar to cortical FM-FM neurons (Olsen, pers. comm.). We have determined the cytoarchitectonic features of the MGB from Golgi-impregnated material, and compared to the cat, the medial division of the mustached bat MGB is more highly differentiated, with a much clearer segregation of different cell types (Zettel et al., in press). We are currently investigating the possibility that the FM_2 projection from ICC might preferentially innervate certain parts of this division.

These results imply that the response properties of delay-tuned FM_1-FM_2 neurons in MGB are the product of inputs redirected from their expected targets in the ventral division. Instead, the connections are found in divisions of the MGB which are known to have strong connections to cortical fields surrounding area AI (Andersen et al., 1980a). The FM-FM area of mustached bat cortex lies dorsal to a large tonotopically organized primary area homologous to AI in cat, and it contains a significant gap in the tonotopic axis for the FM_2 band (Suga & Manabe, 1982). The latter rather surprising observation now fits with the lack of FM_2 inputs to the ventral division of MGB, whose outputs are the main projection to AI in mammals. In the mustached bat the ventral division apparently lacks FM_2 representation. The role of the ICC in this instance has been to "re-wire" the circuits carrying FM_2 information away from the tonotopic thalamocortical pathway in order to provide inputs to the neurons of the medial and dorsal divisions which preferentially innervate FM-FM areas in cortex. These results corroborate the usefulness of studies in bats for demonstrating how the auditory system is functionally organized to subserve specific behavioral tasks.

ACKNOWLEDGEMENTS

Supported by BNS 83-11627 (NSF) to WEO, 2-F32-NS07343 (NINCDS) to RDF, and 5-T32-MH14577 (NIMH) to DMG.

Abbreviations: IC, inferior colliculus; ICC, central nucleus of IC; DP, M, AL, dorsoposterior, medial, and anterolateral divisions of ICC, resp.; SC, sup. colliculus; DC, dorsal cortex of IC; Cb, cerebellum; B, brainstem; BIC, brachium of IC; MGB, medial geniculate body; MGBi, interstitial nucleus of BIC; mmc, medial magnocellular nucleus of MGB; MGBd, dorsal nucleus of MGB; LL, lat. lemniscus; NLL, nucleus of LL; DNLL, INLL, VNLL, dorsal, intermediate and ventral nuclei of LL; LSO, lateral superior olive; MSO, medial superior olive; PO, periolivary nuclei; VMPO, ventromedial periolivary nuc.; LNTB, lateral and ventral nuc. of trapezoid body; PVCNl and m, lateral and medial ventral cochlear nucleus; DCNd, dorsal part of dorsal cochlear nuc.; DCNf, fusiform layer of DCN; AVCNp, a and m, posterior, anterior and medial anteroventral cochlear nuc.; LPN, lateral pontine nuclei; tpt, tectopontine tract; cic, commissure of IC; cp, comm. of Probst; as, acoustic stria; tb, trapezoid body.

REFERENCES

Andersen, R. A., Knight, P. L., and Merzenich, M. M., 1980a, The thalamocortical and corticothalamic connections of AI, AII and the anterior auditory field (AAF) in the cat: evidence for two largely segregated systems of connections, J. Comp. Neurol., 194: 663.

Andersen, R. A., Roth, G. L., Aitken, L. M., and Merzenich, M. M., 1980b, The efferent projections of the central nucleus and the pericentral nucleus of the inferior colliculus in the cat, J. Comp. Neurol., 194: 649.

Frisina, R. D., and O'Neill, W. E., 1986, Functional organization of mustached bat inferior colliculus. III. Connections of the FM_2 region, submitted.

Olsen, J. F., and Suga, N., 1983, Combination-sensitive neurons in the auditory thalamus of the mustached bat, Soc. Neurosci. Abstr., 13: 225.13.

O'Neill, W. E., 1985, Responses to pure tones and linear FM components of the CF-FM biosonar signal by single units in the inferior colliculus of the mustached bat, J. Comp. Physiol. A, 157: 797.

O'Neill, W. E., and Frisina, R. D., 1986, Functional organization of mustached bat inferior colliculus. II. Representation of FM_2 frequency band important for target ranging revealed by single unit recordings, submitted.

O'Neill, W. E., Gooler, D. M., and Frisina, R. D., 1986, Functional organization of mustached bat inferior colliculus. I. Representation of FM frequencies important for target ranging revealed by ^{14}C-2-deoxyglucose autoradiography, submitted.

O'Neill, W. E., and Suga, N., 1982, Encoding of target range and its representation in the auditory cortex of the mustached bat, J. Neurosci., 2: 17.

Pollak, G. D., Bodenhamer, R. D., and Zook, J. M., 1983, Cochleotopic organization of the mustache bat's inferior colliculus, in: "Advances in Vertebrate Neuroethology", J. P. Ewert, R. R. Capranica, and D. J. Ingle, eds., Plenum, New York.

Simmons, J. A., 1973, The resolution of target range by echolocating bats, J. Acoust. Soc. Amer., 54: 157.

Suga, N., 1984, Neural mechanisms of complex-sound processing for echolocation, Trends Neurosci., 7:20.

Suga, N., and Manabe, T., 1982, Neural basis of amplitude-spectrum representation in auditory cortex of the mustached bat, J. Neurophysiol., 47: 225.

Suga, N., O'Neill, W. E., Kujirai, K., and Manabe, T., 1983, Specificity of "combination sensitive" neurons for processing complex biosonar signals in the auditory cortex of the mustached bat, J. Neurophysiol., 49: 1573.

Zettel, M. L., Frisina, R. D., and O'Neill, W. E., 1986, Anatomical organization of the medial geniculate body of the mustached bat, Soc. Neurosci. Abstr. 16: in press.

Zook, J. M., Winer, J. A., Pollak, G. D., and Bodenhamer, R. D., 1985, Topology of the central nucleus of the mustache bat's inferior colliculus: correlation of single unit properties and neuronal architecture, J. Comp. Neurol., 231: 530.

PROCESSING OF PAIRED BIOSONAR SIGNALS IN THE CORTICES OF
RHINOLOPHUS ROUXI AND PTERONOTUS P. PARNELLII:
A COMPARATIVE NEUROPHYSIOLOGICAL AND NEUROANATOMICAL STUDY

G. Schuller, S. Radtke-Schuller, and W.E. O'Neill*

Zoologisches Institut, Universität München
8000 München 2, FRG

*Ctr. Brain Research, University Rochester, Med. Ctr.
Rochester, NY 14642, USA

Introduction

The old world horseshoe bat, Rhinolophus rouxi, and the new
world mustached bat, Pteronotus p. parnellii, belong phylo-
genetically to different bat families. On the other hand they
share similar types of echolocation calls, both using long
constant frequency (CF) pulses terminated by a downward frequency sweep (FM). The behavioural strategies of the two
species also show striking resemblences in that they both are
compensating for frequency shifts introduced into the echo by
the Doppler effect when flying, and they are known to hunt
for prey preferably in acoustically dense and cluttered
surroundings. Suga, O'Neill and coworkers have studied in
detail the auditory cortex of Pteronotus and especially the
processing of paired biosonar signals in this area (for
review: Suga, this volume). They found at least three cortical fields in which neurons could only be stimulated if the
two stimuli satisfied distinct spectral and temporal conditions (CF/CF-, FM/FM-fields and the dorsal fringe area).
These specialized cortical fields lie dorsal and dorsoante-
rior to the tonotopically organized primary auditory cortex.
Paired stimuli have not been used previously in the auditory
cortex of Rhinolophus and therefore no equivalent information is available in this species.

Methods

In both species identical physiological and anatomical procedures were applied. The stereotaxic technique developed for
Rhinolophus rouxi (Schuller et al., 1986) was adapted to
Pteronotus p. parnellii and thus allowed in both species
accurate reconstruction of recording sites in conjunction
with the atlas' of the brain (Radtke-Schuller, prep.). Single
units and occasionally multi units were recorded in oblique
penetrations relative to the cortical surface using insulated

tungsten microelectrodes (Micro Probe Inc.). The neuronal
responses to various stimuli including pure tones, pairs of
stimuli (CF, FM and combinations of those (CFM)) in different
temporal sequences and a variety of complex stimulus patterns
(frequency modulations, noise) were characterized. After physiological characterization of the cortical area, lesions
were made or HRP was iontophoretically injected for verification of the injection sites and for tracing the cortical
connections.

Results

The presentation of neurophysiological results will be concentrated on Rhinolophus rouxi and will be compared to published results from Pteronotus.
Recordings have been made along a 2000 μm distance (1500 μm
in Pteronotus) in rostrocaudal direction and covered laterally 2000 μm of the cortical surface starting at about 1000
μm lateral to the midline (Fig. 1, left).
The lower right graph shows in a projection on the flattened
cortical surface the location of neurons that preferentially
responded to pairs of CF, FM or combinations of these (CFM)
under appropriate spectral and temporal conditions for Rhinolophus. Data from five animals were pooled. Facilitated neural responses (247 neurons out of 446 neurons recorded) were
encountered in a band dorsal to the primary cortical area and
the anterior acoustic field. Facilitation to combinations of
CF-stimuli (CF/CF-neurons) occured in the most rostral part
of the field and this area extended ventrolaterally. In the
transitional zone between the FM- and CF-region neurons that
showed facilitated response to both types of stimulus pairs
were common (mixed type).
The stippled lines demarcate changes in the cyto- and myelo-
architectonic pattern of the cortex and show close coinci-

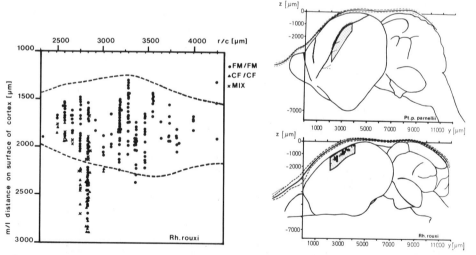

Fig. 1. Representation of recording areas in sagittal view in
Pteronotus p. parnellii (top) and Rhinolophus rouxi. The
lines above the brain show the profile lines used for stereotaxic orientation. Lower right: Projection of brain sites on
the flattened cortical surface of Rhinolophus rouxi, where
facilitation to stimulus pairs (CFM-CFM) was found.

dence with the borders of the physiologically-defined dorsal part of the dorsal auditory field.

Properties of FM-FM-neurons. The response properties of the neurons found in the FM-FM-region ranged from summation of the response to the individual components to extreme facilitation. The neuron represented in Fig. 2 shows facilitated response to the CF as well as to the FM of the stimulus pair (mixed type). Its FM-response is characteristic for most of the FM-neurons in that the facilitated response depends strongly on the delay between the two components (lower left graph: "best" delay for FM-response: 3-4 ms) and that the frequency of the lower and upper components of the pair are constrained within distinct frequency bands (lower and middle right). In about 25% of the recorded FM-FM-neurons (n=204) the lower component was optimum at exactly half the resting frequency of the bat in combination with the second component at the resting frequency. In the majority of neurons the ratio between higher and lower best frequency for facilitation was smaller than 2 so that the facilitating stimulus pair did not correspond to a typical pulse-echo pattern. In general the range of delay for facilitation decreased at lower "best delays" and the frequency tuning of the lower stimulus component was often broader than that of the upper component. These response features of FM-FM-neurons correspond closely to those of the same type of neurons found in Pteronotus (Suga and O'Neill, 1979, O'Neill and Suga 1982).

Topographic arrangement of "best delays". Fig. 3 displays the number of neurons with a specific best delay at different rostrocaudal levels of the FM-field. Besides a general increase of the best delay from rostral to caudal a strong overrepresentation of best delays between 2 and 4 ms (r/c: 2651-3531 um) is evident. O'Neill and Suga (1982) have found such overrepresentation of best delays between 2 and 8 msec in Pteronotus.

Properties of CF/CF-neurons. These neurons were facilitated by combinations of two CF-components at appropriate frequencies (similar to CF-response in fig. 2) and generally displayed the response properties as also found in Pteronotus (Suga et al.,1983). In Rhinolophus the facilitation was mostly best at zero delay and only few units exhibited very sharp tuning to delays. Other CF-neurons were no longer facilitated if the delay was greater than about 10 to 15 ms even though the stimuli were still overlapping. The best frequencies in a pair mostly did not have a harmonic relationship in Rhinolophus and the CF2-frequency often was below the resting frequency of the bat.

Mixed types and specifity of cortical layers. Not all neurons could strictly be attributed to either FM-FM or CF-CF neuron classes but showed enhanced response to CF-combinations as well as delay-dependent FM-responses (fig 2). The two combinations showed clearly different tuning for delay and frequency.
A trend to segregate response types by cortical layers was found, in that many penetrations showed neuronal properties in the following order: response to the lower component alone (CFM1 between 25 and 40 kHz) followed by facilitated responses to CFM1-CFM2 combinations, and finally at deep locations by responses to the second component alone.

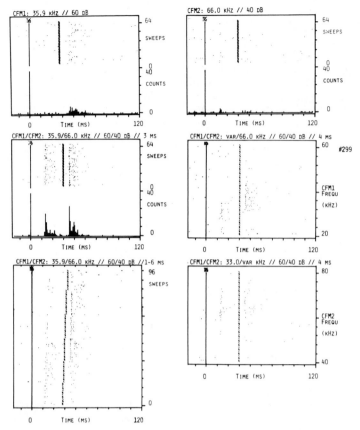

Fig. 2. Responses of mixed type neuron displayed as dot representations of spike discharges. Parameters are given above each dot display. Upper: response to isolated components (CFM1 and CFM2). Middle left: Facilitated response to a pair. Lower left: Differential temporal tuning to CF and FM (delay:1 - 6 ms). <u>Right</u> middle and low: Frequency tuning to CFM1 and CFM2 respectively (other component fixed).

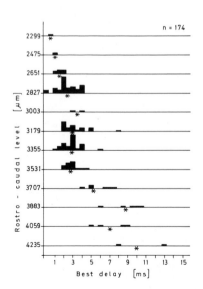

Fig. 3. Distribution of "best delays" at different rostro-caudal levels (r/c). The stars indicate the mean best delay for each r/c level. Note the clear overrepresentation of 2-4 ms between r/c and 2651 and 3531 μm.

262

<u>Other neuron types</u>   A variety of neurons was encountered
which displayed interesting response properties previously
not described in <u>Pteronotus</u>. a) In numerous neurons the
direction of the frequency sweep (FM) had little effect on
facilitation as long as the frequency bands were the same. b)
When presenting CF-stimuli combined with SFM pulses some neurons responded only to the combinations but not to either
component alone. The response was not synchronized to the
SFM-cycle, but occurred after the end of the stimulus pair
and depended on the modulation depth as well as the modulation frequencies of the SFM. c) An upward frequency sweep at
the beginning of the lower facilitating component, which is
also a component in the natural echolocation call, led to
suppression of the facilitated FM/FM-response in 5 neurons.
d) Active vocalization of the bat during the recording of
facilitated responses of neurons often led to complete
suppression of the response for several stimulus presentations.

Anatomical properties

In <u>Rhinolophus</u> <u>rouxi</u> the FM-FM sensitive cortical field borders the ventral rim of the cingulate area and represents the
most dorsal part of the auditory cortex. The FM-FM area
shares most cyto- and myeloarchitectonical features with the
rest of the dorsal field. The main differences are fewer
myelinated horizontal fibres in layer I and thinner layers
III/IV in <u>Rhinolophus</u>. In <u>Pteronotus</u> a corresponding dorsal
field can be defined which loses its characteristics as it
progresses dorsally into the sulcus. When compared on the
basis of general brain structures the FM-FM-field in <u>Rhino-
lophus</u> lies approximately 800 µm more rostral than that in
<u>Pteronotus</u>. This shift of location could be explained on the
assumption that the formation of the combination sensitive
area parallels the neuronal overrepresentation of the
species-specific CF-frequency which developed more rostrally
around 77 kHz in <u>Rhinolophus</u> and more caudally around 60 kHz
in <u>Pteronotus</u> along the tonotopic axis.
In both species the FM/FM area receives its main input from
rostral domains of the medial and dorsal medial geniculate
body (MGB) as well as from the rostrally adjacent thalamus.

Conclusions

The neurophysiological properties of combination sensitive
neurons in the dorsal auditory cortex are very similar in
both species of long CF-FM bats. Due to the harmonic composition of the echolocation call in <u>Pteronotus</u> neurons are
responding to a larger variety of pulse-echo combinations
than in <u>Rhinolophus</u> in which only one stimulus pair (the
first and second harmonic) leads to facilitated responses.
Seemingly fewer neurons in <u>Rhinolophus</u> responded to such
pairs nor could many response characteristics be easily associated with a behaviourally relevant situation. It was also
somewhat puzzling, that vocalization or an upward frequency
sweep at the beginning of the facilitating component could
lead to suppression of the response.
The cortical architecture shows close similarities even
though the location relative to general brain structures of
the areas sensitive to paired stimuli differs by almost a
millimeter. This rostro-caudal offset is most probably linked

to the neural overrepresentations at different frequencies along the tonotopic axis.

Acknowledgement

Supported by Deutsche Forschungsgemeinschaft SFB 204/TP 10.

References

O'Neill, W.E., Suga, N.: J.Neuroscience 2, 17-31 (1982); Schuller, G., Radtke-Schuller, S., Betz, M.: J.Neurosc.Meth. in press (1986); Suga, N., O'Neill, W.E.: Science 206, 351-353 (1979); Suga, N., O'Neill W.E.; Kujirai, K.; Manabe, T.:J.Neurophysiol. 49, 1573-1626 (1983);

CENTRAL CONTROL OF FREQUENCY IN BIOSONAR EMISSIONS OF THE MUSTACHED BAT

David M. Gooler and William E. O'Neill

Center for Brain Research, University of Rochester
School of Medicine
601 Elmwood Avenue, Rochester, NY 14642

Doppler-compensating, echolocating bats like the mustached bat (Pteronotus p. parnellii) present excellent models for studying how sensory feedback controls subsequent motor behavior, i.e., how auditory feedback may regulate vocalization. The mustached bat emits stereotyped orientation sounds consisting of a brief rising frequency modulation (FM) followed by a long constant frequency (CF) component and ending in a short downward sweeping FM. Doppler-shifts of the echo frequency result from the difference in the bat's velocity with respect to its surroundings. During "Doppler-shift compensation" (DSC), these bats actively compensate for Doppler-shifts in the echo by offsetting the frequency of the orientation sound in order to stabilize the echo at the "reference frequency" (about 100-150 Hz higher than the "resting frequency" emitted when the bat detects no Doppler-shift; Schnitzler, 1968, 1970). The variability in the CF frequency emitted during DSC as observed during obstacle avoidance tests is typically ± 10 to ± 110 Hz (Jen and Kamada, 1982). The precision with which mustached bats control the frequency of their emitted sounds, and consequently echo frequency, is regulated directly by feedback of information from echoes. The aims of this study were to define central vocal control regions, clarify their role in the production of echolocation sounds, and investigate the role of auditory processing in the control of vocal frequency.

A midline region of brain dorsal and anterior to the corpus callosum, presumably anterior cingulate cortex (ACg), has been explored for its role in the production of vocalization in 8 mustached bats. Pairs of constant current cathodal pulses (0.3 msec pulse duration, 100 Hz pulse repetition rate, 100 msec train duration, 2 Hz train repetition rate) were applied (via glass insulated platinum-iridium electrodes) to elicit vocalizations. Stimulus threshold current for vocalization was generally in the range of 20-30 µamps. Changes in stimulus intensity affected mainly the intensity of the emitted sound. Latency to response ranged from 70-275 msec. Accompanying motor patterns such as movement of the pinnae, the forming of the mouth into its characteristic cone shape, protrusion of the lips, and patterned, pulse-like exhalations appeared similar to those seen in bats that are echolocating voluntarily. No other gross body movements were seen.

Vocalizations elicited by microstimulation were virtually indistinguishable from spontaneously emitted sounds (Fig. 1). Frequency analysis (FFT; 200 Hz resolution) of the CF component of one of the spontaneously

A

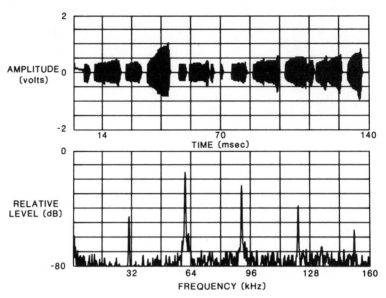

B

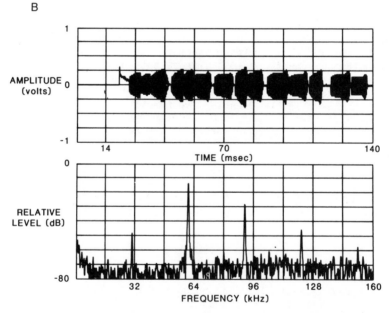

Fig. 1 Waveform analysis of mustached bat vocalizations.
(A) Echolocation sounds (spontaneous). At the top
are the envelopes of a number of sounds emitted
spontaneously. Frequency analysis (FFT; 200 Hz
resolution) of the CF component of one pulse
indicates 5 prominent harmonically related
frequencies. (B) Echolocation sounds emitted by the
same bat elicited by microstimulation of the ACg.
The envelopes of the waveforms and 5 harmonically
related frequencies are similar to those in (A).

emitted pulses in Fig. 1A indicates 5 prominent harmonically related frequencies: CF_1: 30.2 kHz, CF_2: 60.6 kHz, CF_3: 91.0 kHz, CF_4: 121.2 kHz and CF_5: 151.6 kHz. Five harmonically related frequencies identified in the elicited vocalizations (Fig. 1B) are similar to those in Fig. 1A: CF_1: 30.4 kHz, CF_2: 60.8 kHz, CF_3: 91.0 kHz, CF_4: 121.4 kHz and CF_5: 151.8 kHz (FFT of 10th pulse from left). The frequencies of CF_2 (second harmonic constant frequency component) emitted by the bat during stimulation of ACg were most often in the range from 57-62 kHz, typical of natural biosonar vocalizations. The electrically-elicited vocal pulses mimicked the pattern and the shape of the envelope of the spontaneously-emitted vocalizations. The durations of electrically-elicited vocalizations (8-32 msec per pulse, typically 12-24 msec) were well within the range of naturally emitted vocal pulses (7-40 msec). In contrast to earlier studies of the midbrain peri-aqueductal gray where one vocalization resulted for each stimulus train (Suga et al., 1973) 2 to 8 vocal pulses were emitted per train in ACg. An increase in the frequency of vocalizations over a number of consecutive pulses towards a steady state plateau is evident in both spontaneous and elicited emissions (Fig. 2). This pattern of increase in frequency is prominent in response to microstimulation just over threshold, while more intense stimuli produce vocalizations that more rapidly approach the steady state. In both spontaneous and elicited vocalizations the slopes of frequency vs vocal order are similar up to the frequency plateau. In this example, the steady state frequency for elicited vocalizations was lower than that for spontaneous vocalizations (usually at resting frequency).

In the naturally-occurring vocalizations four harmonics have been recognized with a fundamental frequency of about 30 kHz. Elicited echo-location sounds contain the same harmonic components. The CF_2 is dominant (maximum energy) in both types. In our data there is also evidence in both the natural and elicited vocalizations for a fifth harmonic, 30.0 dB

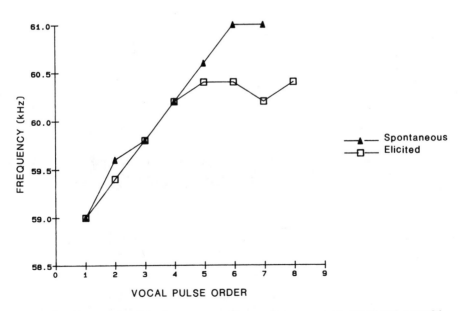

Fig. 2 Change in CF_2 frequency of spontaneous and elicited vocalizations as a function of the order of emitted pulses in a
train.

(+4.2 s.d.) less intense than CF$_2$. The relative intensities of the harmonics of electrically-elicited vocalizations are similar to those observed
for natural vocalizations: relative to CF$_2$, CF$_1$ is 34.4 dB (+4.4 s.d.)
less intense, CF$_3$ is 11.4 dB (+4.4 s.d.) less intense, CF$_4$ is 27.3 dB
(+5.9 s.d.) less intense. For example, the intensities of the CF
components of a spontaneously emitted vocal pulse shown in Fig. 1A are
(referred to CF$_2$): CF$_1$: -30.7 dB, CF$_3$: -9.2 dB, CF$_4$: -23.6 dB, CF$_5$:
-26.2 dB. The relative intensities of the CF components of a typical
elicited vocalization (Fig. 1B) are: CF$_1$: -34.4 dB, CF$_3$: -14.6 dB, CF$_4$:
-31.8 dB, and CF$_5$: -29.6 dB. (The values listed for relative intensity of
CF$_5$ are corrected from the attenuated values depicted in Figs. 1 and 3
because CF$_5$ was attenuated during recording). However, in some cases the
CF$_3$ and CF$_4$ of elicited sounds were as intense as the CF$_2$, an event which,
while not common, does occur in some natural vocalizations. For example,
in Fig. 3 both CF$_3$ and CF$_4$ are more intense than CF$_2$, while CF$_1$ and CF$_5$ are
nearly as intense as CF$_2$. This pattern might alternatively be interpreted
as selective attenuation of CF$_2$. Further study is needed to discern
whether the bat consistently alters the spectral pattern of the
vocalization under certain conditions and whether the ACg is organized for
different spectral patterns of multi-harmonic emissions.

Besides the biosonar signal, shorter latency, audible noise-bursts can
be elicited by microstimulation of ACg. The greatest energy falls near 12-
15 kHz, but it also extends into the ultrasonic frequencies. When this
sound is produced the oro-facial and pinnae movements are diminished in
amplitude. This sound is similar to that produced by bats in social interactions as they scramble about the roost.

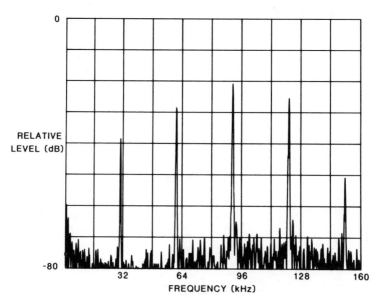

Fig. 3 Frequency analysis (FFT) of the constant frequency component of a vocalization elicited by
microstimulation. The frequencies of the emitted
pulse are: CF$_1$: 30.4 kHz, CF$_2$: 60.6 kHz, CF$_3$:
91.0 kHz, CF$_4$: 121.4 kHz and CF$_5$: 151.8 kHz. The
intensities of the components are referred to
CF$_2$: CF$_1$: -9.8 dB, CF$_3$: +8.0 db, CF$_4$: +3.7 dB,
CF$_5$: -8.0 dB. Note the difference in the
pattern of relative intensities as compared to
both A and B of Fig. 1.

A functional organization based on the type of vocalization produced by microstimulation appears to exist in ACg. The anterior portion is associated with echolocation sounds, while in the posterior section audible noise-bursts can be elicited. The two areas are continuous, and in the transition zone both types of vocalization can be elicited by the same stimulus train. Most interestingly, in contrast to the midbrain periaqueductal gray where electrically-elicited vocalizations are emitted near the resting frequency (Suga et al., 1974), the vocal frequencies elicited by microstimulation of anterior ACg increase along a rostro-caudal axis. While the frequencies of the vocalization range from 57 to 62 kHz (CF_2) along this axis, those from 59.5 to 61.5 kHz can be elicited from a disproportionately larger area of ACg in a way reminiscent of the tonotopically organized regions in auditory areas of the mustached bat brain. Thus, the ACg appears to be tonotopically organized for the frequencies normally emitted during Doppler-shift compensation.

We have examined the neural connections of ACg employing a double-barreled electrode (one side a micropipette and the other a glass insulated tungsten wire) in order to deposit HRP at the site of microstimulation. Preliminary identification of connections by degree of label includes: 1) Many cells labeled, ipsilateral: prefrontal cortex; frontal cortex, auditory cortex; amygdala; ventral anterior nucleus, anterior medial nucleus, nucleus reuniens and nucleus rhomboidalis of the thalamus; contralateral: anterior cingulate cortex; 2) Few cells, lightly labeled, ipsilateral: posterior cingulate cortex; pre-motor cortex; claustrum; medial lateral nucleus, medial ventral nucleus, interpeduncular nucleus; reticular formation of the thalamus and midbrain; dorsal raphe nucleus; contralateral: prefrontal cortex; frontal cortex; auditory cortex; 3) Moderately labeled terminals, ipsilateral: medial nucleus of the pons; 4) Lightly labeled terminals, ipsilateral: reticular formation of the thalamus and midbrain; tegmental reticular nucleus of the pons; contralateral: anterior cingulate cortex; tegmental reticular nucleus, and medial nucleus of the pons. The results thus far indicate a potential source for an auditory influence (via the auditory cortex) on an area controlling emitted frequency of echolocation pulses. They also demonstrate projections to other regions important for control of vocalization (i.e., midbrain reticular formation). Ongoing studies are aimed at eliciting neural responses of cells in ACg to acoustic stimulation in order to elucidate how echo information is integrated to control the frequency of the biosonar signal.

ACKNOWLEDGEMENT

Supported by PHS Grant NS21268 (NINCDS) to WEO.

REFERENCES

Jen, P.H.-J., and Kamada, T., 1982, Analysis of orientation signals emitted by the CF-FM bat, Pteronotus p. parnellii and the FM bat, Eptesicus fuscus during avoidance of moving and stationary obstacles, J. Comp. Physiol., 148:389.

Schnitzler, H.-U., 1968, Die Ultraschall-Ortungslaute der Hufeisen-Fledermause (Chiroptera-Rhinolophidae) in verschiedenen Orientierungssituationen, Z. vergl. Physiol., 57:376.

Schnitzler, H.-U., 1970, Echoortung bei der Fledermaus Chilonycteris rubiginosa, Z. vergl. Physiol., 68:25.

Suga, N., Schlegel, P., Shimozawa, T., and Simmons, J.A., 1973, Orientation sounds evoked from echolocating bats by electrical stimulation of the brain, J. Acoust. Soc. Amer., 54:793.

Suga, N., Simmons, J.A., and Shimozawa, T., 1974, Neurophysiological studies on echolocation systems in awake bats producing CF-FM orientation sounds, J. Exp. Biol., 61:379.

FRONTAL AUDITORY SPACE REPRESENTATION IN THE CEREBELLAR VERMIS OF ECHOLOCATING BATS

Philip H.-S. Jen and Xinde Sun

Division of Biological Sciences, University of Missouri
Columbia, Missouri, USA and Department of Biology, East
China Normal University, Shanghai, PRC

Since the classic work of Snider and Stowell 40 years ago, cerebellar neurons responding to acoustic stimuli have been reported in various animals including echolocating bats. The cerebellum has been suggested to play an important role in orienting an animal toward a sound source. As an echolocating bat essentially relies upon acoustic signal processing to sense its environment, its cerebellum must execute a highly coordinated sensorimotor integration in acoustic orientation. In particular, the cerebellum must integrate acoustic directional information for proper flight orientation. Judging the extremely successful echolocation performance of a bat, it apparently is able to detect and localize a target in order to make corrections to point at the target. Such an acoustic orientation requries the bat to be able to code accurately an echo source. In particular, its ability to code the echo source within its frontal auditory space is essential because of its forward beaming of the signal and its forward scanning position of its pinnae. To understand how the bat's cerebellum registers the echo source within its frontal auditory space, a study of the frontal auditory space representation in its cerebellum is conducted.

In this study, we used a free field stimulation condition to examine the spacial selectivity of neurons isolated from the cerebellar vermis. We also correlate the point of maximal sensitivity of each neuron with its recording site in order to study how the frontal auditory space is represented in the bat's cerebellar vermis.

A total of 168 electrode punctures were made on the exposed cerebellar vermis of 7 Nembutal-anesthetized (45-50 mg per kg body weight) FM bats, Eptesicus fuscus. A pure tone acoustic stimulus (4 msec in duration, 0.5 msec rise-decay times) was presented in front of the bat. Only 69 punctures encountered a total of 138 neurons that responding to the presented acoustic stimuli. By means of two micrometers, the coordinates of the point of electrode puncture relative to the cerebellar surface were registered. The recording site of each isolated neuron was determined from the scale of the electrode drive (Burleigh). The relative positions of the 69 electrode punctures are shown in Fig 1A. There were 39 punctures which isolated at least 2 neurons. Once a neuron was isolated and its best frequency was determined, the point of maximal sensitivity of each neuron was determined with the best frequency stimulus delivered from a selective position within the bat's frontal auditory space. At

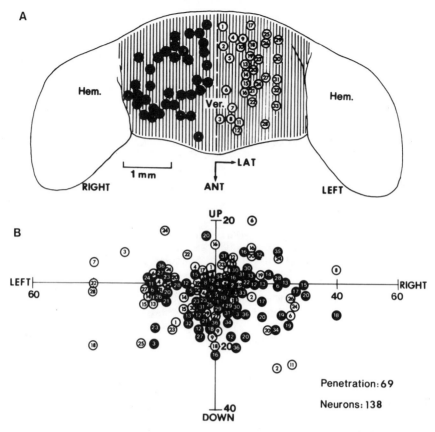

Figure 1   A. A composite map of the cerebellum showing positions of
            69 electrode punctures. For convenience, those 36
            punctures made on the right side of the cerebellum are
            indicated by filled circles. Those 33 punctures made on
            the left side of the cerebellum are indicated by
            unfilled circles. ANT:anterior, Hem:hemisphere,
            LAT:lateral, Ver:vermis.

            B. The locations of the point of maximal sensitivity
            of 138 neurons in the frontal auditory space of the bat.
            Circles with the same arabic numbers represent points of
            maximal sensitivity of those neurons that were
            sequentially isolated from the same electrode puncture.
            Note: most points of maximal sesitivity congregate in
            the central portion of the frontal auditory space.

the point of maximal sensitivity, the neuron had its lowest minimum
threshold. As shown in Fig 1B, all 138 points of maximal sensitivity
were located within 40^{0} lateral and 20^{0} up, 30^{0} down of the
frontal auditory space. However, the majority (112 neurons, 81%) of
them were located within 20^{0} lateral and 10^{0} up, 20^{0} down
indicating a highly directional sensitivity of the cerebellar neurons
to the frontal direction. There is no clear tendency that neurons
isolated form one side of the vermis were most sensitive to a
particular side of the frontal auditory space. Although a point of
maximal sensitivity could be determined within the frontal auditory

space for each isolated neuron, a systematic point-to-point representation is not apparent.

The auditory response area of each neuron was measured with 3 stimulus intensities (at 5, 10, and 15 dB above the best frequency minimum threshold respectively). Such a measurement was done by systematically moving the loudspeaker in azimuth and elevation to locations within the frontal auditory space where the neuron failed to respond. The size of the response area of a neuron generally expanded with the stimulus intensity but such an expansion was not even in all directions (Fig. 2). Although the degree of such an expansion in any particular direction was always unpredictable, the expansion would eventually cover a large part of the central portion of the frontal auditory space. Furthermore, the point of maximal sensitivity of a neuron is seldom the geometric center of its response area.

Our finding of an unsystematic point-to-point representation of the frontal auditory space in the bat's cerebellum concurs with our previous studies in the superior colliculus (Shimozawa et al 1984) and the inferior colliculus (Jen and Sun 1986) of the same bat. The fact that all cerebellar neurons were most sensitive to the central portion of the frontal auditory space suggests that the cerebellum can play an effective role in orienting the bat toward the echo source within its frontal gaze during insect hunting.

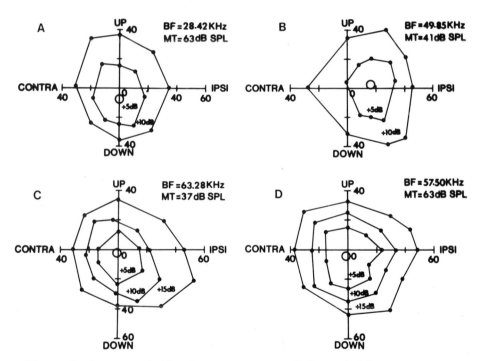

Figure 2 Respresentative response areas of 4 cerebellar neurons. Each unfilled circle represents the location of the pint of maximal sensitivity. Each solid line indicates the boundary of the response area measured at the indicated stimulus intensity above the best frequency minimum threshold of each neuron. Respectively, CONTRA and IPSI represent contralateral and ipsilateral to the recording site of the cerebellar neuron relative to the midline.

Acknowledgement

Supported by a NIH grant 1R01 NS20527 to Philip H.-S. Jen.

References

Shimozawa, T., Sun, X.D., and Jen, P.H.-S. (1984). Auditory space representation in the superior colliculus of the big brown bat, Eptesicus fuscus. Brain Res. 311, 289-296.

Jen, P.H.-S. and Sun, X.D. (1986). Auditory space representation in the inferior colliculus of the big bron bat, Eptesicus fuscus (submitted to Brain Res).

DIRECTIONAL EMISSION AND TIME PRECISION AS A FUNCTION OF TARGET ANGLE

IN THE ECHOLOCATING BAT CAROLLIA PERSPICILLATA

David J. Hartley[1] and Roderick A. Suthers[2]

School of Medicine[1, 2] and Department of Biology[2]
Indiana University
Bloomington, IN 47405

INTRODUCTION

Carollia perspicillata is a frugivorous bat which emits broadband, frequency-modulated pulses through the nostrils. Interference between the nostrils produces complex emission patterns containing amplitude minima at frequency-dependent angles (Fig. 1 and Hartley and Suthers, in preparation). Such patterns imply large, angle-dependent variations in the frequency structure of emitted pulses. The effect that this has on the potential time (target range) precision with which a point target can be located at angles that are off the main axis of emission can be investigated by calculating the autocorrelation function (ACF) of the emitted pulses at such angles. The potential time precision, which is reciprocally related to the effective bandwidth, can then be derived from the envelope of the ACF (Cahlander 1967, Skolnik 1980). This paper presents an analysis of the angle-dependency of time precision in two individual Carollia.

MATERIALS AND METHODS

Experiments were performed with the bats mounted in an echo-attenuating chamber. Brain-stimulation, following the method of Suga and Schlegel (1972), was used to elicit echolocation pulses. The emission pattern was obtained by sampling sound levels around the bat with a moveable microphone, the output of which was referenced to that of a stationary microphone placed directly in front of the bat. For each position of the moving microphone, the average spectrum of 10 pulses was obtained by FFT algorithm (Spectral Dynamics Corp. SD375). The average spectra were then transferred to a Digital Electronics Corp. PDP 11/23, where they were converted to autocorrelation functions by multiplication followed by an inverse FFT. The potential time precision of the pulses (τ) was then calculated as the time between −3 dB points of the envelope of the rectified ACF.

RESULTS

The results of this analysis are plotted in Fig. 2. The horizontal sections show that τ does not initially vary much with increasing off-axis angles in either bat. There is then an abrupt decrease in τ at about 50 degrees in bat 1 and 30 degrees in bat 2.

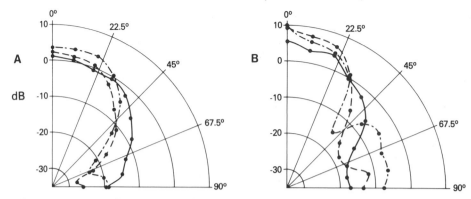

Figure 1. Horizontal sections through the emitted sound field at the level of the axis of the main lobe of the emission pattern. The main energy of <u>Carollia</u> pulses lies between 70 and 90 kHz (Hartley and Suthers, in preparation). The minima in the patterns are due to interference between the nostrils. (A) bat 1; (B) bat 2. Solid line: 70 kHz; dashed line: 80 kHz; dot-dash line: 90 kHz.

This decrease occurs at the angle at which emission pattern minima (Fig. 1) first occur for frequencies having significant energy in the emitted pulse, which in turn depends upon the nostril spacing (Hartley and Suthers, in preparation). In bat 2 this decrease is followed by an increase at larger off-axis angles. The S/N of the data for bat 1 did not allow correlation analysis of off-axis angles greater than 60 degrees. The vertical sections in Fig. 2 show that τ does not vary much in the vertical dimension around the main axis of emission.

To make sure that variations in pulse structure were not responsible for the changes in τ we performed an autocorrelation analysis on the reference microphone output for both bats. We found that these results were remarkably consistent, with variations of only +/-2 µs over the whole experiment.

DISCUSSION

In a directional system, emitted broadband pulses must show decreasing intensity of high frequency components, relative to low frequency components, at increasing off-axis angles. That is, the pulse bandwidth must decrease off-axis, and decreased bandwidth corresponds to an <u>increase</u> in τ. It is therefore surprising that τ is actually decreased at off-axis angles corresponding to the position of the minima in the emission pattern. Although this result appears to defy physical laws, a possible explanation lies in the method used to measure time precision. At low off-axis angles the ACF of the <u>Carollia</u> pulses has low sidelobe levels (ACF 1, Fig. 3). The-3 dB criterion can then be unambiguously applied to give the time precision of the function. At off-axis angles corresponding to the position of the first minima at frequencies of significant energy in the pulse, the ACF has sidelobes which are high but below the -3 dB level (ACF 2, Fig. 3). High sidelobes seem to correspond to a narrower central peak, perhaps due to the relationship that keeps the volume

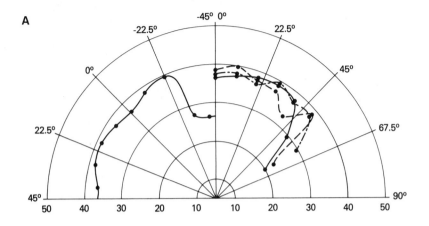

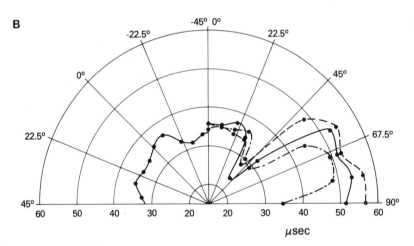

Figure 2. The time precision τ of the pulses plotted against angle. The right-hand side of each plot represents horizontal sections through the sound field at the level of the main axis of emission (solid line), 10 degrees below the main axis of emission (dashed line) and 10 degrees above the main axis of emission (dot-dash line). The left-hand side of each plot represents a vertical section through the sound field, at the horizontal position of the main axis of emission. Negative angles are those below the main axis of emission, positive angles thośe above the main axis of emission. (A) bat 1; (B) bat 2.

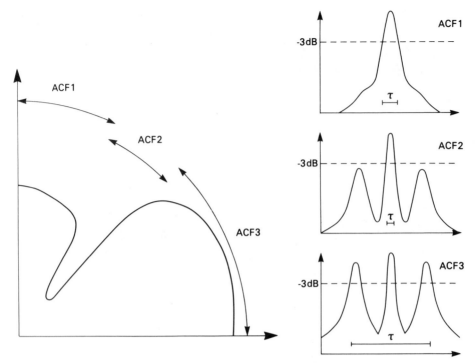

Figure 3.   The pattern of time precision versus horizontal angle
             obtained for bat 2 (left) and the type of autocorrelation
             function (ACF) obtained between the angles shown (right).
             Higher sidelobes seem to correspond to a narrower central
             peak of the ACF, and the time precision pattern can be
             partly explained by the failure of the −3 dB criterion to
             take sidelobes into account. The y-axis of the ACF's is
             normalized amplitude; the x-axis time delay. The bar at
             the bottom of each ACF plot shows the time precision (τ).

under an ambiguity diagram constant for a pulse of given energy, and
since only the central peak width is measured in this type of ACF, the
value obtained is lower than that for ACF 1. Finally, at higher off-
axis angles the sidelobes rise above the −3 dB level so that they are
included in the τ measurement (ACF 3, Fig. 3) giving a τ value much
higher than that obtained for ACF's for 1 and 2.

    The decrease in τ at off-axis angles may not then, by the above
arguments, represent a real increase in pulse bandwidth, but rather an
apparent increase in bandwidth due to the method of measurement of time
precision. Thus in pulses that give ACF sidelobes, care must be taken
in deducing an effective bandwidth from the ACF value.

    The result having the greatest biological importance is the
finding that the time (range) precision is fairly constant around the
main axis of emission in both horizontal and vertical dimensions.
Effectively, the bats have a fairly broad cone of emission within which
range precision is almost constant and resolution is good because
sidelobe levels are low. This may be useful in examining multiple
targets spread over a wide area—the type of cluttered environment that
Carollia probably encounters often. Space here does not permit a
discussion of the possible physical mechanisms behind this finding.

278

It should be pointed out that the terms "precision" and "resolution" are used in this text in the sense with which they apply to individual correlation functions calculated assuming a stationary target. The potential precision of the pulses when moving targets are taken into account cannot be estimated without full ambiguity diagrams, which have not been calculated at this time.

ACKNOWLEDGEMENT

Supported by NSF research grant BNS 82-17099.

REFERENCES

Cahlander, D. A., 1964, Echolocation with wide-band waveforms: Bat sonar signals, M.I.T. Lincoln Laboratory Technical Report 271.

Hartley, D. J. and Suthers, R. A., (in preparation), The sound emission pattern and the acoustical role of the noseleaf in Carollia perspicillata.

Sholnik, M. I., 1980, Introduction to Radar Systems, McGraw-Hill, Inc.

Suga, N. and Schlegel, P., 1972, Neural attenuation of responses to emitted sounds in echolocating bats, Science 177:82-84.

THE JAW-HEARING DOLPHIN: PRELIMINARY

BEHAVIORAL AND ACOUSTICAL EVIDENCE

Randall L. Brill

Chicago Zoological Society
Brookfield, Illinois

Parmly Hearing Institute
Loyola University of Chicago
Chicago, Illinois

INTRODUCTION

Two decades ago, Norris (1964, 1968) proposed that the lower jaw of the dolphin was the primary pathway to the tympanoperiotic bone for returning acoustic signals during echolocation. Unlike that of terrestrial mammals, the lower jaw in odontocetes is hollow and filled with a fatty material that extends beyond the pan bones to attach to the tympanic bulla. This material has been found to contain lipids which contribute to its ability to transmit sound (Varanasi and Malins, 1971; 1972). Electro-physiological studies have indicated that acoustical stimuli presented to the lower jaw evoke significant responses in the auditory system of the dolphin (Bullock et al., 1968; McCormick et al., 1970; 1980). Bullock et al. (1968) further reported that foam rubber or paper placed over the lower jaw to block acoustical stimuli significantly attenuated responses. Several investigators have considered the jaw-hearing hypothesis and its possible role, at least in part, in the findings of their acoustical experiments with dolphins as well (Renaud and Popper, 1975; Au, Floyd, and Haun, 1978; Au and Moore, 1984). Other than a limited attempt to hinder a dolphin's use of its lower jaw which remained inconclusive (Norris, 1974), what has lacked in the evaluation of this theory is behavioral and acous-tical evidence gained from a living animal actively echolocating under controlled conditions.

METHOD

A thirteen-year-old male Atlantic bottlenosed dolphin (Tursiops truncatus), one of five regularly performing dolphins housed in a common pool at the Chicago Zoological Park, was used as the subject for this experiment. Neither the dolphin's diet nor his performing schedule were altered during the course of the work reported.

The dolphin, Nemo, was conditioned to perform a discrimination task, by means of echolocation, and report his choices in a "Go/No-go" paradigm

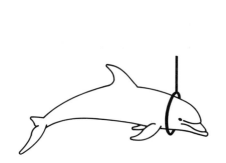

Fig. 1. Dolphin positioned in its
hoop station.

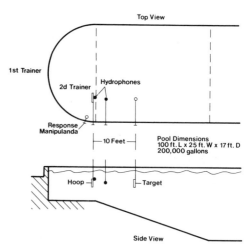

Fig. 2.   Positions of trainers and
equipment, in and around
pool, as they were for each
trial.

(Schusterman, 1980). Cued by a trainer, the dolphin would leave his
starting position, with a latex eyecup in place over his right eye, to
position his head and remain stationary in an 18" diameter, water-filled,
rubber hoop centered two feet below the surface of the water (Fig. 1).
Once in position, a second trainer standing in the water next to the hoop
placed a second eyecup over the dolphin's left eye after which one of two
targets, an aluminum cylinder, the "Go" target, or a sand-filled ring, the
"No-go" target, was lowered into the water at a distance of ten feet for
four seconds allowing the dolphin to echolocate on it. After the target
was withdrawn, the left eyecup was removed and the dolphin was free to
report his choice. Rather than look down range during target presentations, the second trainer looked back toward the first trainer for visual
indications of target presence and the appropriateness of the dolphin's
responses (Fig. 2). The order of target presentation was obtained from
Gellerman tables and each session conducted consisted of twenty trials
with both targets being presented an equal number of times.

To investigate the function of the lower jaw in performing the task
described, sessions were conducted in which ten trials, arbitrarily
selected out of the total twenty, additionally required the dolphin to
wear one of two rubber hoods, designed to cover the entire lower jaw from
the tip of the snout to the base of the pectoral fins along the line of
the gape of the mouth, held in place by rubber straps and small suction
cups (Fig. 3).

One of the hoods, used as a control, was constructed from 1/16"
non-foamed neoprene. The other hood, intended to block acoustical

Fig. 3. Dolphin fitted with neoprene hood. Small suction cups were
attached along jaw and on straps.

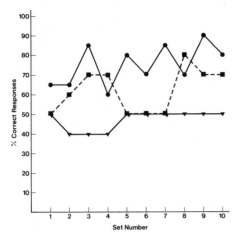

Fig. 4. Response rates for all 3 conditions over first 10 sets of experiment plotted as percent correct. (●) baseline=20 trials per point; (■) control=10 trials per point; (▼) experimental=10 trials per point.

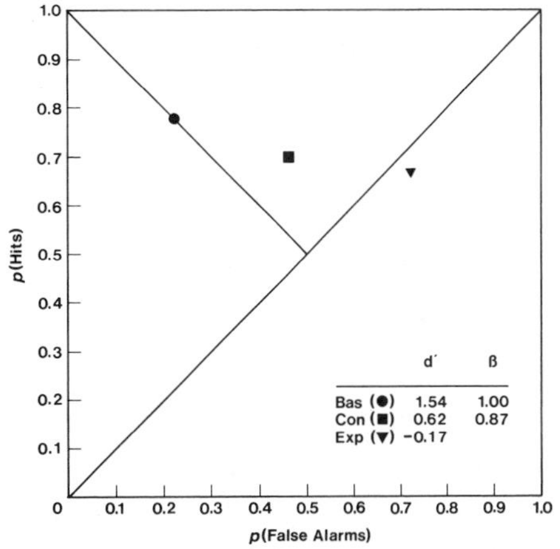

Fig. 5. ROC plot for the first 10 sets of the experiment. Each point represents 100 trials. (●) baseline, (■) control, (▼) experimental.

signals, was constructed from 3/16", closed-cell neoprene. Preliminary tests determined that the non-foamed and closed-cell neoprenes attenuated underwater acoustical signals of up to 115 kHz by 2 dB and 40 dB, respectively.

Due to the demands of the dolphin's performing schedule, sessions were conducted with some irregularity at a rate of one or two sessions per day. A baseline session, that conducted without the use of a hood, preceded each pair of hooded sessions, control and experimental, which were counterbalanced. Each sequence of three sessions constituted a set. The dolphin's correct responses were reinforced either with food, secondary reinforcers, or combinations of both in keeping with the training regimen familiar to the dolphin (Brill, 1981). Each incorrect response was followed by a brief time-out. During each session, the other dolphins in the group were occupied by a trainer at the opposite end of the pool.

The dolphin's outgoing acoustical signals were recorded during arbitrarily selected sessions. The signals from two hydrophones, an Edo-Western hydrophone attached to the underwater hoop directly over the dolphin's head and a Celesco LC-10 hydrophone positioned 1 m in front of the hoop and in line with its center were amplified, bandpass filtered between 3 kHz and 200 kHz, and recorded on an Ampex FR-1300 portable instrumentation tape recorder operating at 60 ips. To facilitate computer analysis, the recorded signals were later slowed by a factor of 16 (from 60 ips to 3-3/4 ips) and transferred to a Marantz PMD 360 portable stereo cassette tape recorder. The signals from the first hydrophone were used to trigger and synchronize the signals from the LC-10 during analysis so that Nemo's signals could be identified as opposed to occasional acoustical interference from the other dolphins residing in the pool.

RESULTS

Behavioral

The data reported are those collected over the first ten sets of the experiment since, at the time of this writing, the intended total of twenty sets has not been completed. Figure 4 summarizes the dolphin's performance in terms of per cent correct for all three conditions. Chi-square tests of association (Siegel, 1956) were applied to evaluate performance, or response rates, in terms of the proportions of correct responses to incorrect responses. To allow the use of N=100 in each condition, ten trials were selected out of each baseline session for inclusion in the evaluation with the only conditions being that they reflect an equal number of presentations of the two targets used in the task and approximately half of the errors observed within the session from which they were drawn.

The differences in response rates across all three conditions were significant (N=300, df=2, X^2=20.5, p<.001) as were the differences between the baseline and experimental conditions (N=200, df=1, X^2=19.2, p<.001) indicating that response rates were affected by the use of the hoods, the experimental hood in particular. The differences between the baseline and control conditions (N=200, df=1, X^2=5.4, p<.05) and the differences between the two hooded conditions (N=200, df=1, X^2=4, p<.05) were also significant.

To further evaluate the effect of the hoods on response rates, the same tests were applied to the differences observed between the trials using a hood and those not, within each of the hooded conditions. The differences in the proportions of correct responses to incorrect responses within the control condition were not significant (N=200, df=1, X^2=1.8, NS) while the differences within the experimental condition were significant (N=200, df=1, X^2=5.1, p<.05).

The data were further evaluated by the application of Signal Detection Theory (Green and Swets, 1966). Figure 5 displays the data in a receiver operating characteristic (ROC) format. The ordinate is the probability of hits, the abscissa is the probability of false alarms and values for sensitivity (d') and response bias (Beta) are shown. While values for performance fell from well above chance in the baseline condition to below chance in the experimental condition, the dolphin maintained a consistent response criterion and adhered to a conservative strategy across conditions.

Acoustical

Audio recordings of sessions in each of the three conditions have been and will continue to be recorded throughout the course of the experiment in the manner described above. At the time of this writing, signal analysis is in a preliminary stage consisting, for the most part, of digitizing the recorded signals and transferring them to diskettes. Acoustic signals recorded during the training for the experiment showed peak frequencies between 30 kHz and 60 kHz and individual clicks composed of three to seven cycles and 150 usec in duration. The frequencies above are in agreement with measurements reported by Diercks et al. (1971) for the same species in a concrete pool.

Based on what has been observed in the acoustic data recorded in the experiment thus far, it is anticipated that the most distinct differences in the dolphin's outgoing signals across conditions will be evident in the time domain with respect to repetition rates and interclick intervals. Figure 6 illustrates what thus far appears to be typical for baseline trials. These can be compared to Figure 7 which displays typical examples of experimental trials. Baseline signals appear to be characterized by a more consistent pattern across time. Click trains tend to be longer in duration, follow each other in close succession, and contain large numbers of clicks. Experimental signals, on the other hand, appear to be

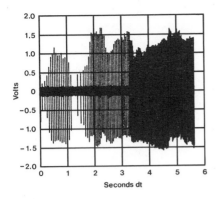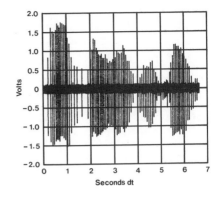

Fig. 6. Examples of echolocation signals recorded in the baseline
condition.

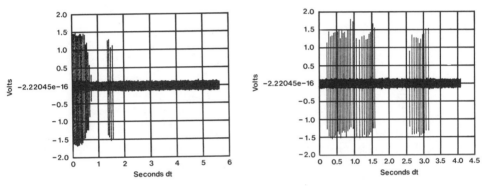

Fig. 7. Examples of echolocation signals recorded in the experimental
condition.

comprised of shorter click trains that are clearly separated in time and
contain fewer clicks with longer interclick intervals. It is conceivable
that the dolphin responded to signal attenuation at the lower jaw caused
by the experimental hood by employing a strategy of making more than one
attempt to detect the target within the time available to him.

DISCUSSION

In summary, the data reported support the hypothesis that the
dolphin's lower jaw acts as an acoustical pathway to the inner ear for
returning signals during echolocation. Attenuation of such signals at the
lower jaw significantly affected the performance of a living, actively
echolocating dolphin as would be predicted by the electrophysiological
evidence cited in the introduction and it is likely that signal analysis
will reveal differences in the dolphin's outgoing signals across the
conditions tested.

Novelty or discomfort associated with the use of the hoods could be
considered as an alternative explanation for the performance levels
observed. It is, however, unlikely. Since Nemo had never before been
used as a subject in a formal study, a great deal of caution was exercised
in the design and execution of this experiment so as to avoid any stress
or discomfort in requiring him to perform while both blindfolded and
hooded. It is reasonable to assume that the amount of time involved in
establishing the echolocation task and conditioning him to wear the hoods

would have adequately reduced any effects of novelty or discomfort before beginning the actual experiment.

Since it is probable that the experimental hood does not perfectly screen out all acoustic signals, it is likely that performance will improve over the second half of the experiment in that condition as it did in the baseline and control conditions. Signal analysis may also reveal possible strategies that the dolphin uses to compensate for the attenuation of incoming signals.

Additional behavioral data in the form of collateral behaviors observed during trials, signal analysis, and the completed experiment will be reported in a forthcoming manuscript.

ACKNOWLEDGMENTS

This work is being funded by a grant from the Chicago Zoological Society, assisted by the Parmly Hearing Institute, and conducted in partial fulfillment of the requirements for the degree of Doctor of Philosophy. Space does not allow me to recognize everyone who has been supportive of this effort but I am grateful to Dick Fay, Toby Dye, Ken Norris, and Bill Yost for their guidance; Sam Ridgway, NOSC, San Diego, for the loan of the Ampex FR-1300; Whitlow Au, NOSC, Hawaii, for testing the hood materials; Patrick Moore, NOSC, Hawaii, for his invaluable assistance and advice; Patrick Harder, PHI, for computer assistance; Brenda Woodhouse, Martha "Marty" Sevenich, Tim Sullivan, Janet Sustman, Ron Witt, and Ed Hausknecht, the Seven Seas Panorama training staff, for their help and patience with me; Ed Krajniak, curator of Seven Seas Panorama, for his friendship and understanding; the Chicago Zoological Society for providing a unique opportunity; my wife and sons for their support; and the porpoise in my life, Nemo.

REFERENCES

Au, W.W.L., Floyd, R.W., and Haun, J.E., 1978, Propagation of Atlantic bottlenose dolphin echolocation signals, J. Acoust. Soc. Am., 64:411-422.

Au, W.W.L., and Moore, P.W.B., 1984, Receiving beam patterns and directivity indices of the Atlantic bottlenose dolphin Tursiops truncatus, J. Acoust. Soc. Am., 75:255-262.

Brill, R.L., 1981, R.I.R. in use at the Brookfield Zoo: random and interrupted reinforcement redefined in perspective, in: "Proceedings of the Annual Conference of the International Marine Animal Trainers Association, 1981, Niagara Falls, New York," J. Barry and R.L. Brill, eds., New England Aquarium, Boston.

Bullock, T.H., Grinnell, A.D., Ikezono, E., Kameda, K., Katsuki, Y., Nomoto, M., Sato, O.,Suga, N., and, Yanigasawa, K., 1968, Electrophysiological studies of central auditory mechanisms in cetaceans, Z. Vergleich. Physiol., 59:117-156.

Diercks, K.J., Trochta, R.T., Greenlaw, C.F., and Evans, W.E., 1971, Recording and analysis of dolphin echolocation signals, J. Acoust. Soc. Am., 49:1729-1732.

Green, D.M., and Swets, J.A., 1966, "Signal Detection Theory and Psycho-physics," Robt. E. Krieger Publishing Co., Huntington, New York.

McCormick, J.G., Wever, E.G., Palin, J., and Ridgway, S.H., 1970, Sound conduction in the dolphin ear, J. Acoust. Soc. Am., 48:1418-1428.

McCormick, J.G., Wever, E.G., Ridgway, S.H., and Palin, J., 1980, Sound reception in the dolphin ear as it relates to echolocation, in: "Animal Sonar Systems," R.-G. Busnel and J.F. Fish, eds., Plenum Publishing Corp., New York.

Norris, K.S., 1964, Some problems of echolocation in cetaceans, in: "Marine Bioacoustics," W.N. Tavolga, ed., Pergamon Press, New York.

Norris, K.S., 1968, The evolution of acoustic mechanisms in odontoncete cetaceans, in: "Evolution and Environment," E.T. Drake, ed., Yale University Press, New Haven, Connecticut.

Norris, K.S., 1974, "The Porpoise Watcher," W.W. Norton and Co., Inc., New York.

Renaud, D.L., and Popper, A.N., 1975, Sound localization by the bottlenose porpoise Tursiops truncatus, J. Acoust. Soc. Am., 63:569-585.

Schusterman, R.J., 1980, Behavioral methodology in echolocation by marine mammals, in: "Animal Sonar Systems," R.-G. Busnel and J.F. Fish, eds., Plenum Publishing Corp., New York.

Siegel, S., 1956, "Nonparametric Statistics for the Behavioral Sciences," McGraw-Hill, New York.

Varanasi, U., and Malins, D.C., 1971, Unique lipids of the porpoise (Tursiops gilli): differences in triacylglycerols and wax esters of acoustic (mandibular and melon) and blubber tissues, Biochem. Biophys. Acta., 231:415-418.

Varanasi, U., and Malins, D.C., 1972, Triacylglycerols characteristic of porpoise acoustic tissues: molecular structures of diisovaleroyl-glycerides, Science, 176:926-928.

ACOUSTICAL ASPECTS OF HEARING AND ECHOLOCATION IN BATS

Anna Guppy and Roger B. Coles

Zoologisches Institut
Universität München
Luisenstrasse 14
8 München 2, West Germany

The structure of the external ear of Microchiropteran bats shows a
remarkable diversity, particularly in the shape and size of the pinna. There
is also a sharp contrast between the highly mobile pinnae of Rhinolophids
and Hipposiderids compared to the almost immobile pinnae of Megadermatids.
The presence of "large" ears (pinnae) is often associated with a gleaning
type of foraging behaviour, including the ability to detect and capture
prey by passive listening (Neuweiler, 1984). Within this context, the
Australian false vampire or ghost bat Macroderma gigas, has extremely large
pinnae, possibly the largest amongst the Microchiroptera and the acoustical
properties of the external ear have been examined. Similar data has been
obtained in Gould's long-eared bat Nyctophilus gouldi, which is also probably
a gleaning bat similar Antrozous pallidus (Bell, 1982).

In fresh specimens of M. gigas and N. gouldi, the tympanic membranes
were removed and replaced by a small calibrated microphone (Bruel and Kjaer
Type 4138) inserted through the bulla, leaving the external ear intact.
Under normal conditions in M. gigas, there is a rapid increase in the
amplification of sound pressure at the tympanic membrane, relative to the
free field, for frequencies above 3kHz. For speaker positions on the acoustic
axis of the ear, peak amplification of 20-30dB occurs between 5-12kHz,
gradually declining to less than 10dB above 40kHz. In comparison, the
pressure gain in the ear canal of N. gouldi rises rapidly above 5kHz to
peak values of 15-23dB between 7-22kHz. In both species, removal of the
pinna results in a loss of pressure in the ear canal and the gain of the
pinna can be estimated by the difference curve (Fig. 1). Pinna gain curves
resemble the gain characteristics of a finite-length acoustic horn (Beranek,
1954). Expected gain curves were calculated for a series of finite horns
as shown in Fig. 1 and the pinnae of M. gigas and N. gouldi are, to a first
approximation, similar to conical horns, at least in the low to mid frequency
range. The pinnae have similar cone angles ($\alpha=28°$), although they are
irregularly shaped and asymmetrical. The maximum pressure gain at the throat
of each pinna, as an efficient horn, would be expected to be 13dB, based
on the areal ratio of the pinna mouth and throat (entrance to the ear canal)
and is close to the observed values. The larger size of the pinna in M.
gigas (for dimensions see Fig.1 legend) produces a gain curve which rises
about an octave earlier than N. gouldi. However in both cases, pressure
gain starts to increase rapidly as the ratio of the circumference of the
pinna mouth to the wavelength approaches unity (ka=1; k=wavenumber $2\pi/\lambda$;
a=radius). The peak in pressure gain near 5kHz and 7kHz for M. gigas and

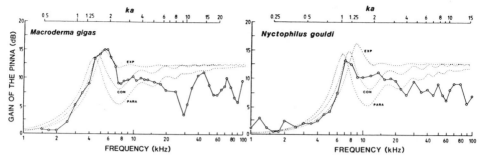

Fig 1. Average pinna gain curves (solid lines) for M. gigas (4 ears) and
N. gouldi (3 ears) based on the difference in sound pressure for the intact
and pinna-less external ear. Dotted curves are expected pressure gains of
finite-length paraboloidal (para), conical (con), and exponential (exp)
horns, calculated from the physical dimensions of the pinnae:-M. gigas,
radius of mouth = 1.7cm; radius of throat (at entrance to the ear canal)
= 0.4cm; average length = 2.4cm. N. gouldi, radius of mouth = 0.85cm; radius
of throat = 0.2cm; length = 1.3cm. Upper abscissa in each graph relates
the circumference of the pinna mouth to the wavelength or ka (where k =
wavenumber, $2\pi/\lambda$; a = radius).

N. gouldi respectively, suggest horn resonance as expected for a finite
horn (Fig. 1). The fundamental resonance peaks are sharper in individual
ears and are broadened by averaging.

     Normal directivity patterns based on pressure measurements at the
tympanic membrane are determined by the pinna and the pinna mouth experiences
sound diffraction similar to a single, approximately circular, aperture
(see also Coles & Guppy, 1986). Directionality becomes significant above
about 3.5kHz in M. gigas and 8kHz in N. gouldi and directivity patterns
develop an acoustic axis (Fig. 2a). The directivity characteristics of the
main lobe, e.g. -3dB acceptance angle (full angle 2θ) and semi-angle (first

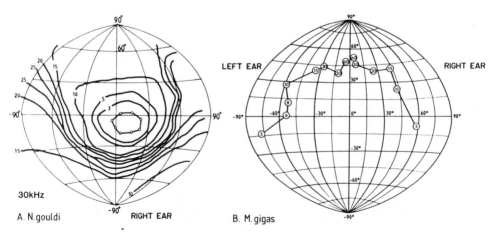

Fig 2 a) Directivity pattern for N. gouldi at 30kHz, based on sound pressure
in the right ear canal. Iso-pressure contours (solid lines) are plotted
on a zenithal projection of a hemisphere, in dB below the maximum pressure
at the acoustic axis (centre of 1dB contour, open circles).
     b) Movement of the acoustic axis as a function of frequency (kHz)
circles) for the left and right ears of a single M. gigas.

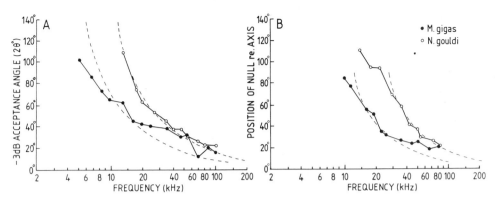

Fig 3a) Changes in the average (4 ears each species) acceptance angle (full angle=2θ°) of directivity patterns (see Fig. 2a) as a function of frequency, measured at -3dB relative to the sound pressure at the acoustic axis.
b) Angular separation between the first null and the acoustic axis (semi-angle) as a function of frequency. M. gigas = closed circles; N. gouldi = open circles. Dashed lines are expected curves based on sound diffraction by a circular aperture (in an infinite baffle) of a radius equal to the average radius of the pinna mouth in each species (M. gigas a=1.7cm; N. gouldi a=0.85cm).

null) are related to the diffraction limits imposed by the radius of the pinna mouth in relation to the sound wavelength (Fig. 3). The spatial location of the acoustic axis is frequency dependent in both species. For example in M. gigas (Fig. 2b) the axis shifts upwards by about 50° between 5-15kHz but remains relatively stationary at higher frequencies. In azimuth there is about 60° movement of the acoustic axis towards the midline between 5-20kHz and the acoustic axes of both ears are close to the midline between 35-60kHz. Since this bandwidth is used for sonar (Guppy et al., 1985), a sonar horizon can be defined by the position of the acoustic axes at ultrasonic frequencies and probably determines the head orientation in M. gigas when echolocating during horizontal flight. A sonar horizon can also be defined in N. gouldi for frequencies above 30kHz. In both species the frequency dependent spatial location of the acoustic axis can be related to diffraction effects, if the mouth of the pinna is treated as the oblique truncation of a (right) conical horn. Ear mobility is of interest in M. gigas since the pinnae are fused along the medial edge for half their length and gross movements are severely restricted. The pinnae of N. gouldi are also joined at the base, restricting movement, although in this species the pinnae can be concertinaed. Movement of the acoustic axis as a function of frequency may be important for sound localization in bats with restricted ear movements as shown for Pteronotus parnellii (Fuzessery & Pollak, 1984).

Acoustical data suggest that an advantage of large pinnae in bats maybe to extend the low frequency hearing range and maybe an important adaptation for gleaning, both by the amplification of sound in the external ear and improved directionality at low frequencies. The neural audiogram of M. gigas (Guppy & Coles in press) shows extremely high sensitivity to sound in the region 10-20kHz where thresholds approach -16dB SPL (re. 20μPa), similar to the audiogram of Megaderma lyra (Neuweiler et al. 1984; Schmidt et al. 1984). These data in Megadermatids strongly suggest that prey could be easily captured by listening, without the use of sonar, in direct analogy to non-echolocating volant predators such as Tyto alba. In M. gigas, the neural audiogram has a second, less sensitive hearing peak at 35-43kHz which corresponds to the main energy in the sonar pulse. In N. gouldi, the neural audiogram shows good sensitivity in the 8-14kHz bandwidth but thresholds do not exceed +5dB SPL, suggesting similar hearing sensitivity to Antrozous pallidus (Brown et al. 1984). The use of low frequencies for social

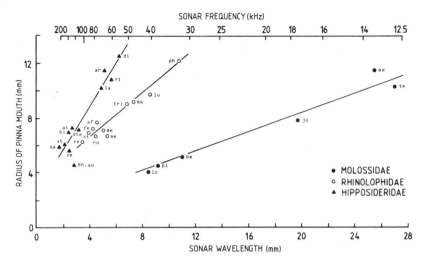

Fig 4. Average radius of the pinna mouth plotted as a function of the
wavelength of the dominant sonar frequency for 27 Hipposiderid (▲),
Rhinolophid (o), and Molossid (●) species. Sonar frequencies based on own
recordings, and Heller (1985), Roberts (1972), Gould (1980), Zbinden & Zingg
(1986). Solid lines are linear regression slopes, Molossids = 0.35 (r=0.98);
Rhinolophids = 0.81 (r=0.93); Hipposiderids = 1.57 (r=0.91). Key:- Molossids,
au-Tadarida australis; te-T.teniotis; jo-Chaerophon jobensis be-Mormopterus
beccarii; pl-M.planiceps; lo-M.loriae. Rhinolophus, ro-rouxi; ph-philipinensis;
tri-trifoliatus; ma-macrotis; af-affinis; fe-ferrumequinum; me-megaphyllus;
st-stheno; se-sedulus; re-refulgens; lu-luctus. Hipposideros, di-diadema
ar-armiger; ri-ridleyi; la-lankadiva; at-ater; bi-bicolor; ste-stenotis;
sa-sabanus; ce-cervinus; Rh.au-Rhinonicteris aurantius. H. ater is
represented by two specimens from separate regions of Australia.

communication is probably important for such species and a striking example
is the "chirp" call which is used extensively by M. gigas both in the day
roost and at night whilst foraging. The chirp has high energy in the audio
bandwidth from 6-15kHz (Guppy et al., 1985). Interestingly, the Vespertilionid Euderma maculatum has relatively enormous ears (height of pinna 34mm)
similar in size to Plecotus spp, but E.maculatum is not considered a gleaner
(Woodsworth et al., 1981). However E. maculatum uses sonar pulses with a
very low fundamental frequency around 10kHz which is suited to large pinnae.

In Microchiroptera, ear size is usually related to body size or forearm
length, e.g. Molossids (Freeman, 1981). Moreover, sonar frequencies are
related to forearm length as in Hipposiderids (Gould, 1980). Simmons (1982)
has pointed out that the height of the pinna bears a specific relationship
to the wavelengths used for sonar. An example of the relationship between
pinna size and sonar frequencies can be demonstrated in Molossids (Fig.
4). In each of five Australian Molossid species and the European T. teniotis
which vary significantly in body size (Freeman, 1981), the radius of the
pinna mouth is close to half a wavelength of the fundamental frequencies
used for sonar, typically expressed as a shallow single harmonic FM pulse
during cruising flight. Here the sonar wavelength is approximately equal
to the height of the pinna mouth. These frequencies are likely to be highly
amplified by the external ear and correspond to the lower limit for high
directionality (ka≈3) produced by sound diffraction at the pinna mouth.

In Hipposiderids and Rhinolophids the size of the pinna is correlated
to the wavelength of the CF component of the sonar call (Fig.4) but in these
bats the wavelengths are up to 3 times smaller than the radius of the pinna

mouth. Such a high ratio may well be at the upper limit for the use of the pinna as a diffraction device since sound reflection in the pinna may become critical. Hipposiderids and Rhinolophids may choose short wavelength harmonics to improve spatial and target resolution but such high frequencies do not optimize the pressure gain produced by the external ear, as seen in R. rouxi for example (Guppy, Coles & Schlegel in prep). These bats may still be attracted to prey by listening at lower frequencies and for social communication.

ACKNOWLEDGEMENTS

We thank N.H. Fletcher for deriving the horn equations and writing the computer programs to calculate expected horn gain, and C.R. Tidemann, B.Baker & W.R. Phillips for help with sonar recordings and collecting specimens. We also thank K.-G. Heller and M. Volleth for access to specimens of Rhinolophus & Hipposideros spp. from Malaysia, and M. Obrist for data on ear size and sonar recordings of E. maculatum.

REFERENCES

Bell, G.P. 1982. Behavioral and ecological aspects of gleaning by a desert insectivorous bat, Antrozous pallidus (Chiroptera: Vespertilionidae). Behav. Ecol. Sociobiol. 10: 217-223.
Beranek, L.L. 1954. Acoustics. McGraw-Hill, New York.
Brown, P.E., Narins, P.M. & Grinnell A.D. 1984. Low frequency auditory sensitivity of the pallid bat Antrozous pallidus. Am. Neurosci. Abst. 10: 1150.
Coles, R.B. & Guppy A. 1986. Biophysical aspects of directional hearing in the Tammar wallaby Macropus eugenii. J. exp. Biol. 121:371-394.
Fletcher, N.H. & Thwaites, S. 1979. Physical models for the analysis of acoustical systems in biology. Quart. Rev. Biophys. 12: 25-65.
Freeman, P.W. 1981. A multivariate study of the family Molossidae (Mammalia, Chiroptera). Fieldiana: Zoology New Series, No. 7, 175pp.
Fuzessery, Z.M. & Pollak, G.D. 1984. Neural mechanisms of sound localization in an echolocating bat. Science 225: 725-728.
Gould, E. 1980. Vocalizations of Malaysian bats (Microchiroptera and Megachiroptera). In : Busnel R.-G., Fish J.F. (ed) Animal Sonar Systems. NATO Advanced Study Institutes, Ser. A. Vol. 28. Plenum Press, New York, pp. 901-904.
Guppy, A. & Coles, R.B. Acoustical and neural aspects of hearing in the Australian gleaning bats Macroderma gigas and Nyctophilus gouldi. J. Comp. Physiol. In press.
Guppy, A., Coles, R.B. & Pettigrew, J.D. 1985. Echolocation and acoustic communication in the Australian ghost bat, Macroderma gigas (Microchiroptera: Megadermatidae). Aust. Mammal. 8: 299-308.
Heller, K.-G. 1985. The echolocation calls of 12 syntopic Rhinolophid species (Rhinolophus, Hipposideros). 7th Int. Bat Res. Conf., Aberdeen.
Neuweiler, G. 1984. Foraging, echolocation and audition in bats. Naturwissenschaften 71: 446-455.
Neuweiler, G., Singh, S. & Sripathi,K. 1984. Audiograms of a South Indian bat community. J. Comp. Physiol. 154: 133-142.
Roberts, L.H. 1972. Variable resonance in constant frequency bats. J. Zool., Lond. 166: 337-348.
Schmidt, S., Türke, B. & Vogler, B. 1984. Behavioural audiogram from the bat Megaderma lyra (Geoffrey, 1810; Microchiroptera). Myotis 22: 62-66.
Simmons, J. A. 1982. The external ears as receiving antennae in echolocating bats. J. Acoust. Soc. Am. 72: S41-42.
Woodsworth, G.C., Bell, G.P. & Fenton, M.B. 1981. Observations of the echolocation, feeding behaviour, and habitat use of Euderma

            maculatum (Chiroptera: Vespertilionidae) in south-central
            British Columbia. Can. J. Zool. 59: 1099-1102.
Zbinden, K. & Zingg, P.E. 1986. Search and hunting signals of echolocating
            European free-tailed bats, Tadarida teniotis, in southern
            Switzerland. Mammalia 50: 9-25.

THE PERCEPTION OF COMPLEX ECHOES

BY AN ECHOLOCATING DOLPHIN

Whitlow W. L. Au and Patrick W. B. Moore

Naval Ocean Systems Center
Kailua, Hawaii 96734

Dolphins echolocate with short transient broadband acoustic signals
having time-resolution constants between 12 and 15 μs (Au, 1980). Received
echoes are often complex and considerably longer in duration than the
transmitted signal, having many resolvable highlights or echo components.
Echo highlights come from signal reflection at external and internal
boundaries of a target and from the presence of different propagational
modes within a target. The purpose of this study was to investigate how
dolphins perceive and characterize the strength of complex echoes.

PROCEDURE

An experiment was conducted in which an echolocating <u>Tursiops
truncatus</u> was required to detect a variety of target echoes in noise.
Targets echoes were produced by a microprocessor-controlled electronic
target simulator which captured each signal emitted by the animal and
retransmitted the same signal back to the animal after an appropriate delay
to simulate a specific target range. The experimental configuration is
depicted in Fig. 1, showing a dolphin in a hoop station and three
hydrophones directly in front of the animal. An acoustic screen was
located between the hoop and the hydrophones. In the raised position the
screen blocked the animal's echolocation signals from the hydrophones.

The animal's echolocation signal was detected by the first hydrophone
which triggered an 8-bit A/D converter to digitize the signal received by
the second hydrophone. The A/D converter operated at a sample rate of 1
mHz, and digitized 128 points per signal. The digitized signal was then
stored in a static RAM (random access memory). The output of the first
hydrophone was also used to trigger an external delay generator and notify
the computer that a signal had been detected. After a delay corresponding
to a desired target range, the delay generator triggered the computer to
playback the signal stored in the RAM. The number of times the RAM played
back a stored signal, the time separation between and the amplitude of each
output signal was controlled by the computer. Masking noise was mixed with
the RAM signal and projected from the third transducer, located 2.4 m from
the hoop. The projector was driven by an equalization circuit which
flattened the transmit response of the transducer. The noise was projected
at a fixed level (64 dB re 1 μPa^2/Hz) and the level of the play back signal
was controlled with an adjustable attenuator.

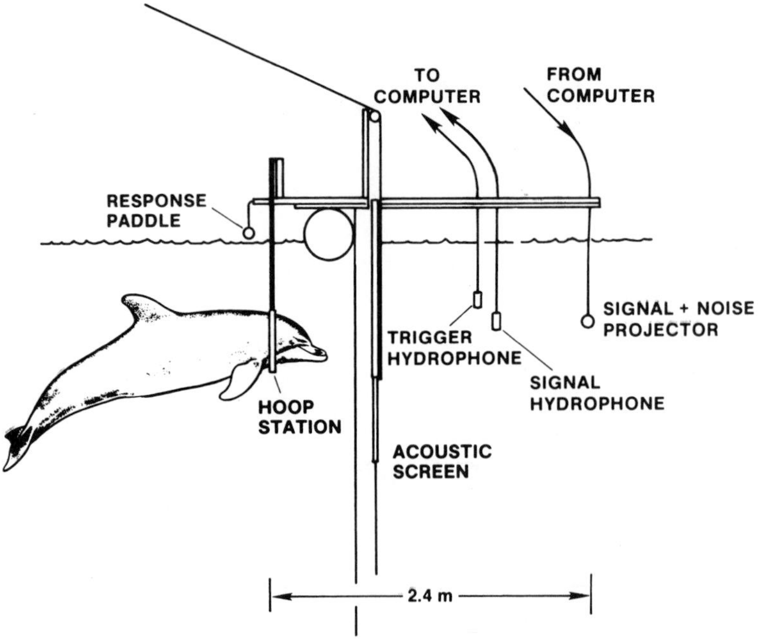
TO
COMPUTER
FROM
COMPUTER
RESPONSE
PADDLE
SIGNAL + NOISE
PROJECTOR
TRIGGER
HYDROPHONE
SIGNAL
HYDROPHONE
HOOP
STATION
ACOUSTIC
SCREEN
2.4 m

Fig. 1. Experimental configuration

A trial started when the animal swam into the hoop station, with the acoustic screen in the raised position. The masking noise was then turned on and the acoustic screen lowered, cueing the dolphin to begin echolocating. For target present trials, the animal was required to back out of the hoop and respond by striking a paddle. For target absent trials, the animal was required to remain in the hoop until a bridge tone was played. The dolphin received a fish reward for responding correctly on each trial.

A 2.5 x 3.8 cm polyurethane foam (internally weighted) cylinder was initally used to train the dolphin on the detection task. The target was located 20 m from the hoop. The dolphin was transferred without difficulties to the "phantom" target over several sessions after its introduction.

In the first experiment a doublet with two clicks separated by 200 μs was used as the simulated echo. A modified method of constants testing procedure was used with a session divided into six 10-trial blocks, each block having 5 target present and 5 absent trials. Five different attenuator settings in 3 db increments were used in each session. The first and last blocks of trials used the lowest attenuation. The settings for the other blocks were randomized. The purpose of this experiment was to determine the animal's detection performance using phantom echoes and to compare the results with previous results obtained with real targets.

The second experiment involved the determination of the dolphin's integration time for broadband transient-like signals. A staircase technique was used to determine the animal's threshold as the properties of the simulated echoes were manipulated. A minimum of 20 reversals from two consecutive sessions with the threshold attenuation for each session within 2 dB, was required for a threshold. The first signal was a single click. Then doublets with 7 different separation times (ΔT) varying from 50 to 600 μs were used. The specific separation time for each threshold measurement was randomized.

In the third experiment the dolphin's threshold was determined as a function of the number of clicks in each return echo. Thresholds were measured for simulated echoes with one, two and three clicks. The separation times between the two and three click echoes were all within the animal's integration time (discussed in next section).

RESULTS AND DISCUSSION

The results of the first experiment using a doublet as the simulated echo are shown in Fig. 2. The echo energy-to-noise (E/N) ratio was calculated using the maximum amplitude signal per trial, following the procedure of Au and Penner (1981). The horizontal axis represents the average of the maximum E/N per trial at each of the 5 attenuator settings. Also in the figure are the results of Au and Penner (1981) obtained with a 7.62-cm diam. sphere. The data came from ten sessions or 100 trials per attenuation setting.

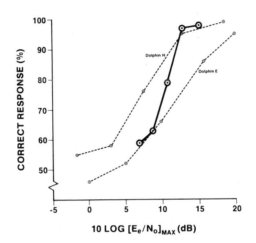

Fig. 2. Dolphin performance results with a doublet used as the simulated echo. The horizontal axis represents the average of the maximum echo energy-to-noise ratio per trial at each attenuator setting. The dash curves are the results of Au and Penner (1981) obtained with a 7.62-cm sphere.

The phantom target results were consistent with the real target results of Au and Penner (1981) despite differences in methodologies and animals. E/N at the 75% correct response level was 10 dB compared to 7.4 and 12.8 dB obtained with a real target. Differences in the slopes of the curves may be a result of differences in the E/N step sizes. Au and Penner varied the noise level in 5 dB increments. The noise level in this experiment was held constant and the signal-to-noise ratio was varied by attenuating the level of the simulated echoes in 2 dB increments.

Results of the second experiment to determine the dolphin's integration time are shown in Fig. 3. The threshold shifted approximately 3 dB towards a lower E/N per pulse as the stimulus changed from a single click to a double click. This shift is exactly what would be expected for an energy detector since there is 3 dB more energy in the two-click stimulus. The threshold E/N remained the same for ΔT up to 200 µs. As ΔT increased to 250 µs, the threshold E/N shifted towards the single click value, reaching the single click value for ΔT of 300 µs and greater. The solid curve in Fig. 3 is the response of an ideal energy detector with an integration time of 264 µs. This integration time best fitted the dolphin's data. A 264 µs integration time is considerably shorter than the tens of ms indicated by Johnson's (1968) data on pure-tone threshold as a function of duration. Apparently, dolphins may process broadband transient signals differently than narrowband pure-tone signals.

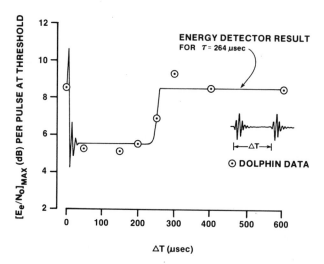

Fig. 3. Integration time experiment results. The ordinate is the average of the maximum E/N per trial involving a reversal, calculated for each click in the simulated echo. The abscissa represents the separation time between clicks in the double click signals. The data point at $\Delta T = 0$ represents the threshold for the single click echo. The solid curve is the response of an ideal energy detector with an integration time of 264 µs.

The integration time of 264 µs corresponds to the "critical interval" of approximately 260 µs measured for <u>Tursiops</u> (Vel'min and Dubrovskiy, 1976; Moore, et al., 1984). Vel'min and Dubrovskiy (1976) defined critical interval as a "critical time interval in which individual acoustic events merge into an acoustic whole." However, the results presented here suggest that Vel'min and Dubrovskiy's critical interval may merely be the integration time of an energy detector.

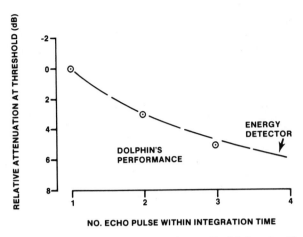

Fig. 4. Detection threshold versus number of clicks in the return echo.

Thresholds obtained with the staircase method for the single- and double-click stimuli were within 1.5 dB of the 75% correct response threshold obtained with the modified method of constants.

Results of the third experiment determining threshold shifts as a function of the number of clicks in the simulated echoes are shown in Fig. 4. The ordinate is the relative attenuation at threshold, which takes into account the signal attenuator and amplitude of the largest click per trial. Also shown in the figure is the response of an energy detector.

The dolphin's performance matched that expected for an energy detector. The two- and three-click stimuli had 3 and 4.8 dB more energy, respectively, than the one-click echo. The dolphins threshold shift was approximately the same as the increase in total energy in the echo.

REFERENCES

Au, W.W.L., and Penner, R.H., 1981, Target Detection in Noise by Echolocating Atlantic Bottlenose Dolphins, J. Acoust. Soc. Am., 70:687-693.

Johnson, C.S., 1968, Relation between Absolute threshold and Duration-of-Tone Pulses in the Bottlenosed Porpoise, J. Acoust. Soc. Am., 43: 757-763.

Moore, P.W.B., Hall, R.W., Friedl, W.A. and Nachtigall, P.E., 1984, The Critical Interval in Dolphin Echolocation: What is it?, J. Acoust. Soc. Am., 76: 314-317.

Vel'min, V.A., and Dubrovskiy, N.A., 1976, The Critical Interval of Active Hearing in Dolphins, Sov. Phys. Acoust., 2:351-352.

MORPHOMETRIC ANALYSIS OF COCHLEAR STRUCTURES IN THE MUSTACHED BAT,

PTERONOTUS PARNELLII PARNELLII

O.W. Henson, Jr. and M.M. Henson

Department of Anatomy
The University of North Carolina at Chapel Hill
Chapel Hill, N.C. 27514
USA

Mustached bats, Pteronotus parnellii, emit complex biosonar signals with a constant frequency (CF) component which is preceded and terminated by brief frequency modulated (FM) components. Each component is present in a series of at least four harmonics and the sense of hearing is very sharply tuned to the CF component frequencies of the second (ca 61kHz) and third (ca 91.5 kHz) harmonics (Pollak et al., 1979). The cochlea of P. parnellii is large and has a number of unusual anatomical and physiological properties (A. Pye, 1967; M. Henson, 1973; 1978; O. Henson et al., 1985; Kössl and Vater, 1985). The purpose of this paper is to report new observations on cochlear morphometry. These data will be related to the cochlear frequency map data of Kössl and Vater (1985) and Leake and Zook (1985), and to theories concerning cochlear micromechanics.

For these studies we used a variety of techniques. In order to obtain cross-sectional areas of different structures, cochleae were cut into segments approximately 2 mm in length; from these segments 2 µm serial sections were obtained and of these only those sections determined to be true transverse sections were measured. The outlines of structures were traced onto a data tablet and cross-sectional areas were calculated using a DEC Micro PDP-11/23 computer with custom software. To visualize the changing dimensions of structures, we obtained SEM micrographs and computer generated three-dimensional reconstructions from 60 and 30 µm serial sections of the entire cochlea.

The cochlea of P. parnellii has two and one-half turns, but 66% of its length lies in an enormous basal turn. Within this large basal turn the nerve fiber density varies (Fig. 1) and one can define: 1) a proximal densely innervated region (PDI); 2) a distal densely innervated region (DDI); and 3) a sparsely innervated region (SI) between them. Near the junction of the SI and DDI regions is an area where the curvature of the basilar membrane appears distorted and this is called the straight region (SR).

The PDI region is approximately 2.5 mm in length and is characterized by a series of large nerve trunks. According to the frequency map data of Kössl and Vater, the fibers in the most basal trunk process frequencies at least as high as 111 kHz while those near the junction of the PDI and SI regions have fibers which respond to frequencies near

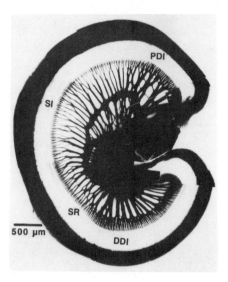

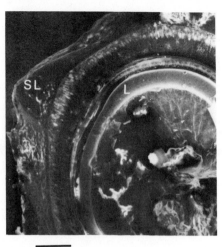

Fig. 1. Whole mount preparation
of basal turn of cochlea stained
with osmium tetroxide.

Fig. 2. SEM micrograph of the
enlarged spiral ligament (SL)
and limbus (L).

70 kHz. The part of this region with the thickest fibers occupies the
area near the SI region and these are probably the fibers which process
the third harmonic (92-70 kHz) components of the biosonar signals.

In the SI region the nerve trunks within the osseous spiral lamina
are about one-third the size of those in the densely innervated regions;
the SI region extends over an expanse of approximately 3.5 mm and the
frequency band represented extends from about 70 kHz to 62 kHz. The
available frequency maps for the cochlea of P. parnellii do not include
the SR region but it seems likely that it corresponds to an area of the
cochlea which processes sounds slightly higher (at least 100 Hz) than the
second harmonic CF resting frequency. This is a frequency band and an
area of the cochlea of special interest; it is here and in the adjacent
DDI regions that marked morphological specializations occur. Although it
cannot be stated with certainty, it seems most likely that the area
beginning as the straight region and extending into the DDI region is the
one associated with marked cochlear resonance, cochlear emissions, sudden
changes in sensitivity and cochlear microphonic phase (see O. Henson et
al., 1985; Kössl and Vater, 1985).

Near the middle of the straight region, the DDI region begins
abruptly; the spiral ganglion suddenly becomes larger and there are a
series of four to five thick nerve bundles radiating toward the edge of
the osseous spiral lamina. This is the area which appears to be concerned with the detection of the second harmonic resting frequency,
200-400 Hz below the resonance frequency of the ear. Here the 61 kHz,
second harmonic CF component echoes are processed and it is this area of
the cochlea that may represent what has been referred to as the acoustic
fovea (Leake and Zook, 1985). The units associated with this area are
very sharply tuned and very fine frequency resolution is achieved as evidenced by Doppler shift compensation. Within this DDI region the representation of small frequency bands is spread over an extensive length of
the cochlea.

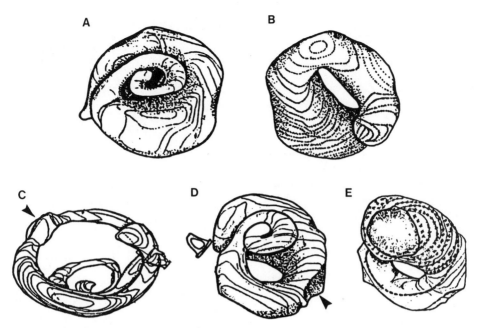

Fig. 3. Three dimensional reconstructions of the scala vestibuli (A,B), spiral ligament (C) and scala tympani (D,E). Arrows mark region of enlarged spiral ligament.

The second turn represents approximately 29% of the total length of the organ of Corti and the apical half turn approximately 5%. According to the data of Kössl and Vater (1985) the second and apical turns process a broad band of frequencies that include the initial and terminal FM sweeps of the second harmonic and the first harmonic CF (30.5 kHz) and FM (ca 30-23 kHz).

Computer reconstructions of the scala vestibuli are shown in Fig. 3 A,B. In the basal turn the cross-sectional diameter is large, but beginning with the second turn the scala is suddenly reduced in size and remains small to the end of the apical turn. In the basal turn the scala vestibuli has a sharp bend and the cross-sectional diameter is narrow at this point. This narrow portion, designated a "notch" by Kössl and Vater (1985), occurs about one-fourth of the way along the sparsely innervated region and it does not appear to be specifically associated with any special frequency region or with any other morphological discontinuities. The sudden narrowing of the scala vestibuli at the end of the basal turn occurs at the start of the second turn and near the junction of the area associated with the detection of second harmonic CF and FM signals. Computer reconstructions of the scala vestibuli in another species of Pteronotus (P. fuliginosa) have not shown any of the remarkable enlargements or sudden constrictions seen in P. parnellii. It should be noted that within the basal turn the cross-sectional area of the scala vestibuli reaches a maximum diameter in the second half turn and specifically in relation to the DDI region.

In contrast to the scala vestibuli, reconstructions of the scala tympani (Fig. 3 D,E) showed that it has a large volume in the most basal portion of the cochlea and then diminishes progressively from base to apex. It does not show any markedly enlarged or constricted areas associated with the DDI region.

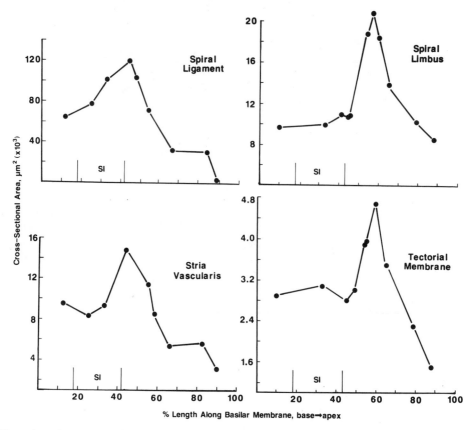

Fig. 4. Cross-sectional area measurements showing changes in dimensions of cochlear structures.

In most mammals the spiral ligament and stria vascularis show a progressive decrease in size from base to apex; in P. parnellii, however, there is a localized region where the size of these structures suddenly changes (Figs. 2,3C, 4). The region of maximum size is centered in the SR, and data suggest that the SR may in fact be created by the enlarged spiral ligament which distorts the normal curvature of the basilar membrane. The enlargement of the spiral ligament with its complement of tension-generating anchoring cells, would be expected to increase the stiffness of the basilar membrane in this area (M. Henson et al., 1985). One would thus expect that the compliance of the basilar membrane just apical to the SR would be much greater than in the SR itself. Whether the enlargement of the spiral ligament and stria vascularis in the SR relates to a corresponding enlargement of the cochlear duct or whether the stria creates a higher potassium ion level and different biochemical environment remains to be determined. One must ask if the high amplitude of the CM in Pteronotus at the resonance frequency might, at least in part, be associated with the enlarged stria vascularis.

The spiral limbus and associated tectorial membrane show a sudden increase in size at the junction of the SI and DDI regions(Figs. 2,4). The enlargements are just apical to the region of maximum size of the spiral ligament and stria vascularis (Fig. 4) and they correspond to the beginning of the region of the cochlea that processes the 61 kHz second harmonic CF, "reference frequency" signals. In the apical turn these structures are much reduced in size.

The basilar membrane has a relatively uniform width throughout most of the cochlea (M. Henson, 1978) and thus it is quite unlike that seen in most mammals where there is a progressive increase in width from base to apex. Measurements of the cross-sectional profiles have supported data obtained by Kössl and Vater (1985). It is now clear that the basilar membrane mass and stiffness are at a maximum in the straight region and that there is a sudden change at, or very near, the beginning of the DDI region.

Certainly the most interesting aspects of this study, in terms of micromechanics, are the abrupt changes in size, shape and mass of structures in and around the regions of the cochlea associated with very fine frequency resolution, resonance and sudden changes in sensitivity. The observed changes in the dimensions of the spiral limbus and tectorial membrane, which begin just prior to the DDI region, support the contentions of Zwislocki (1983) that the tectorial membrane mass may play a role in the sharpening of response properties of the mammalian cochlea.

ACKNOWLEDGEMENTS

Support for this work was supplied through NIH grants NS-12445 and NS-19031) and the Air Force Office of Scientific Research (85-0063).

REFERENCES

Henson, M.M. (1973) Unusual nerve-fiber distribution in the cochlea of the bat Pteronotus p. parnellii (Gray). J. Acoust. Soc., Amer. 53:1739-1740.

Henson, M.M. (1978) The basilar membrane of the bat, Pteronotus p. parnellii. Am. J. Anat. 153:143-158.

Henson, M.M., Burridge, K., Fitzpatrick, D.C., Jenkins, D.B., Pillsbury, H.C. and Henson, O.W., Jr. (1985) Immunocytochemical localization of contractile and contraction associated proteins in the spiral ligament of the cochlea. Hearing Research 20:207-214.

Henson, O.W., Jr., Schuller, G. and Vater, M. (1985) A comparative study of the physiological properties of the inner ear in Doppler shift compensating bats (Rhinolophus rouxi and Pteronotus parnellii). J. Comp. Physiol. A. 157:587-597.

Kössl, M. and Vater, M. (1985) The cochlear frequency map of the mustache bat, Pteronotus parnellii. J. Comp. Physiol. A. 157:687-697.

Leake, P.A. and Zook, J.M. (1985) Demonstration of an acoustic fovea in the mustache bat, Pteronotus parnellii. Abstracts of the Eighth Midwinter Research Meeting of the Association for Research in Otolaryngology. (D.J. Lim, ed). Clearwater Beach, Fla. pp. 27-28.

Pollak, G., Henson, O.W., Jr. and Johnson, R. (1979) Multiple specializations in the peripheral auditory system of the CF-FM bat, Pteronotus parnellii parnellii. J. Comp. Physiol. A. 131:255-266.

Pye, A. (1967) The structure of the cochlea in Chiroptera. III. Microchiroptera: Phyllostomoidea. J. Morph. 121:241-254.

Zwislocki, J.J. (1983) Some current concepts of cochlear mechanics. Audiology, 22:517-529.

THE EFFERENT AUDITORY SYSTEM IN DOPPLER-SHIFT COMPENSATING BATS

Allen L. Bishop and O.W. Henson, Jr.

Department of Anatomy
The University of North Carolina at Chapel Hill
Chapel Hill, North Carolina 27514
USA

Studies on the innervation of outer hair cells (OHCs) of common laboratory animals have repeatedly demonstrated a general pattern of efferent terminals throughout the cochlea (Smith and Sjöstrand, 1961; Iurato et al., 1978; Spoendlin, 1979). Many efferent endings typically contact each OHC in the basal turn of the cochlea and the number decreases in more apical regions, particularly on the outermost rows. The present study on the efferent innervation of outer hair cells in different species of Doppler-shift compensating bats has revealed marked interspecies differences as well as striking departures from the basic plan known to occur in common laboratory animals. Studies were carried out on the cochlea of the neotropical mormoopid, Pteronotus p. parnellii and on the Old World rhinolophids, Rhinolophus rouxi and Hipposideros lankadiva. The rhinolophids and P. parnellii have independently evolved sophisticated biosonar systems; although the frequency resolving properties of their ears are similar, marked structural and physiological differences exist (Bruns, 1980; Kössl and Vater, 1985; Henson et al., 1985).

Glutaraldehyde fixed cochleae were dissected into small (ca 1 mm) segments and prepared by standard methods for transmission electron microscopy. Serial sections through the OHC region were collected from representative portions of each turn, including the areas which process the constant frequency components of the biosonar signals (Kössl and Vater, 1985). In common laboratory animals synaptic contacts on OHCs often show no apparent systematic arrangement (Smith and Sjöstrand, 1961; Spoendlin, 1966). In P. parnellii, however, computer-assisted reconstruction of serial sections revealed a consistent pattern of afferent and efferent innervation throughout the cochlea (Figs. 1 and 2). A single, large (ca 3 μm) efferent ending contacted the central portion of the base of each hair cell and was surrounded by a ring of six or seven afferent terminals. This highly organized spatial relationship and the number of terminals contacting each OHC was uniform in all three rows of OHCs from base to apex. Only one of more than two hundred OHCs examined appeared to have two efferent endings and no OHC was found to lack an efferent terminal. In contrast, electron microscopic examination of the cochlea of Rhinolophus rouxi demonstrated a complete absence of upper tunnel radial fibers and efferent terminals on OHCs (Fig. 3). Similar, but less complete studies on Hipposideros lankadiva also showed an absence of upper tunnel radial fibers and efferent terminals on all three rows of OHCs. The lack of efferent innervation to OHCs in R. rouxi

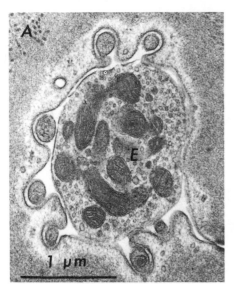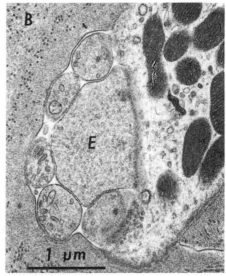

Fig. 1. Typical arrangement of nerve terminals at the base of an outer hair cell in P. parnellii. In A, the plane of section is beneath the base of an outer hair cell and shows a large efferent terminal (E) surrounded by afferent fibers. In B, a large efferent terminal (E) containing round, clear vesicles contacts the base of an outer hair cell.

supports the findings of Bruns and Schmieszek (1980) in the European horseshoe bat, Rhinolophus ferrumequinum. These data suggest that a lack of efferent innervation to the outer hair cells may be a feature characteristic of Rhinolophidae.

The origins of efferent neurons were determined in the brainstem of P. parnellii following intracochlear deposits of horseradish peroxidase (HRP). When HRP was introduced into one cochlea, many (up to 1585) efferent neurons were labeled and two separate cell populations could be identified; the cell bodies of 75% of the labeled cells were small and fusiform and they were localized ipsilaterally in a densely packed band, dorsal and medial to the lateral superior olivary nucleus. Labeled cells of the other type were also densely packed; they were larger, stellate and localized bilaterally in the dorsomedial periolivary nuclei (DMPO). Of this population, crossed neurons outnumbered the uncrossed group by a ratio of 2:1. The localization of efferent neurons was further demonstrated by the placement of HRP directly on the cut fibers of the crossed olivocochlear bundle in the floor of the fourth ventricle. In these studies, labeled neurons were found only in the DMPO of both sides. Previous studies of the efferent auditory system in other mammals such as the cat, have shown that fibers destined for inner hair cells (IHCs) of each cochlea arise from small neurons situated laterally in the ipsi- and contralateral superior olivary complex, while those to OHCs arise from more medial loci (Warr and Guinan, 1979). The situation in P. parnellii is different in that the efferent fibers associated with IHC afferents are supplied by a purely ipsilateral system of fibers, while bilaterally symmetrical loci account for the efferent innervation of OHCs. Evidence strongly suggests that the crossed olivocochlear bundle in P. parnellii is composed entirely of fibers that terminate on outer hair cells.

In recent years a number of studies have implicated the efferent

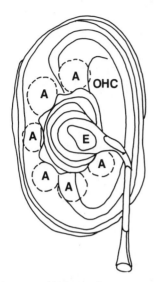

Fig. 2. Rendition of a computer-assisted reconstruction of nerve terminals contacting the base of an outer hair cell (OHC) of P. parnellii. Note the large efferent ending (E) surrounded by afferent terminals.

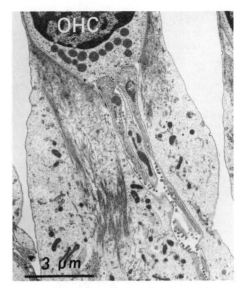

Fig. 3. Transmission electron micrograph from the cochlea of R. rouxi showing afferent nerve endings contacting the base of an outer hair cell (OHC).

innervation of OHCs in a mechanical regulation of cochlear transduction (Siegel and Kim, 1982; Guinan, 1986). An exceptional ability to resolve small frequency differences is represented by both P. parnellii and R. rouxi, yet, in one species the efferent system displays an unprece-dented level of uniformity in the distribution of efferent terminals to OHCs, while in the other species these fibers and terminals are conspi-cuously absent. These findings provide a unique opportunity to assess the functional significance of the efferent system in terms of the highly resonant basilar membrane properties in P. parnellii vs. the mechanically damped characteristics of the cochlea of rhinolophids (Henson et al., 1985).

ACKNOWLEDGEMENTS

We wish to thank Dr. Marianne Vater for her role in supplying coch-leae of Rhinolophus rouxi and Hipposideros lankadiva. Support for this work was supplied through NIH grants (NS-12445 and NS-19031) and a grant from the Air Force Office of Scientific Research (85-0063).

REFERENCES

Bruns, V. (1980 Basilar membrane and its anchoring system in the cochlea of the greater horseshoe bat. Anat. Embryol. 161:29-50.
Bruns, V. and Schmieszek, E. (1980) Cochlear innervation in the greater horseshoe bat: demonstration of an acoustic fovea. Hearing Res. 3:27-43.
Guinan, J.J., Jr. (1986) Effect of efferent neural activity on cochlear mechanics. Scand. Audiology (in press).

Henson, O.W., Jr., Schuller, G., and Vater, M. (1985) A comparative study of the physiological properties of the inner ear in Doppler shift compensating bats (Rhinolophus rouxi and Pteronotus p. parnellii). J. Comp. Physiol. A 157:587-597.

Iurato, S., Smith, C.A., Eldredge, D.H., Henderson, D., Carr, C., Ueno, Y., Cameron, S., and Richter, R. (1978) Distribution of the crossed olivocochlear bundle in the chinchilla's cochlea. J. Comp. Neurol. 182:57-76.

Kössl, M. and Vater, M. (1985) The cochlear frequency map of the mustache bat, Pteronotus p. parnellii. J. Comp. Physiol. A 157:687-697.

Siegel, J.H. and Kim, D.O. (1982) Efferent neural control of cochlear mechanics? Olivocochlear bundle stimulation affects cochlear bio-mechanical nonlinearity. Hearing Res. 6:171-182.

Smith, C.A. and Sjöstrand, F.A. (1961) Structure of the nerve endings on the external hair cells of the guinea pig cochlea as studied by serial sections. J. Ultrastraucture Res. 5:523-556.

Spoendlin, H. (1966) The organization of the cochlear receptor. Adv. Otorhinolaryngol. 13:1-226.

Spoendlin, H. (1979) Sensory neural organization of the cochlea. J. Laryngol. Otol. 93:853-877.

Warr, W.B. and Guinan, J.J., Jr. (1979) Efferent innervation of the organ of Corti: two separate systems. Brain Res. 173:152-155.

SOME COMPARATIVE ASPECTS OF AUDITORY BRAINSTEM CYTOARCHITECTURE IN ECHOLOCATING MAMMALS: SPECULATIONS ON THE MORPHOLOGICAL BASIS OF TIME-DOMAIN SIGNAL PROCESSING

John M. Zook [1], Myron S. Jacobs[2,5], Ilya Glezer[3,5] and
Peter J. Morgane[4,5]

[1]Dept. of Zoological & Biomedical Science and College of
Osteopathic Medicine, Ohio U., Athens, OH, 45701, USA
[2]Dept. of Pathology, NYU College of Dentistry, NY, NY, 10010
[3]Dept. of Anatomy, CUNY Medical School, NY, NY, 10031, USA
[4]Worcester Foundation for Experimental Biology, Shrewsbury
MA, 01545, USA
[5]Osborn Lab. Marine Sci., NY Aquarium, Brooklyn, NY, 11224, USA

INTRODUCTION

A number of studies have described the gross morphology and general
hypertrophy of the auditory brainstem nuclei in Cetacea and Chiroptera (See
Henson, 1970 for review; Zvorykin, 1959, 1963). However, there have been
only a few attempts to describe the cytoarchitecture of the auditory brainstem in detail (Osen & Jansen, 1965; Schweizer, 1981; Zook & Casseday, 1982)
or to make comparisons of brainstem auditory cytoarchitecture between echolocating species of bats and marine mammals. We will begin here to take a
closer look at the auditory brainstem of a number of microchiropteran and
odontocete cetacean species, initially focusing upon a few distinctive
cytoarchitectonic patterns found in three cell groups: the anteroventral
cochlear nucleus (AVCN), the medial nucleus of the trapezoid body (MNTB)
and the ventral nucleus of the lateral lemniscus (VNLL). In some species
of bat and dolphin, cells in these nuclei form orderly rows or columns that
are more uniformly aligned than is found in most nonecholocating mammals.
In the two dolphin species examined, rows of cells perpendicular or at an
angle to the fibers of the trapezoid body are present in the caudal part of
the AVCN and in the MNTB. Within the VNLL in some bat species there is a
distinctive alignment of cell soma, in columns rather than rows, parallel
to fiber bundles.

We wish to raise the possibility here of a functional as well as a
anatomical connection between these auditory nuclei; a collective role in
temporal feature detection that might also be present, but less elaborated,
in mammals where the cells are not as overtly aligned as will be shown here.
In cats, the rodent and the mustache bat these three cells groups are linked
anatomically by particularly large synaptic endings terminating on cell
soma. Cells of the caudal part of the AVCN receive synaptic and bulbs of
Held from the eighth nerve (Harrison and Irvine, 1965; Brawer and Morest,
1975). These cells are the main source of the even larger calyces of Held
which surround cell soma in the contralateral MNTB (one is shown in fig. B)
(Warr, 1972; Tolbert, et al., 1982). Part of the VNLL also receives large

calycine terminals from the contralateral AVCN (Adams, 1978; Zook & Casseday, 1985). This part of the VNLL is also the target of the principle cells of the MNTB on the same side (Glendenning, et al., 1981; Zook & Casseday, submitted). In some cases individual axons from the AVCN appear to divide, supplying branches to both the contralateral MNTB and the VNLL (Spirou, et al., 1985). Units in all three areas are characterized by large positive prepotentials and on-set response patterns which preserves the activity pattern of eighth nerve primary afferents (Guinan, et al., 1972; Adams, 1978). The specific combination of morphological and physiological features has led several observers to suggest that these cell groups might be particularly well suited to code and preserve the temporal relationships of an auditory stimulus (Morest, et al., 1973; Adams, 1978). We will discuss the distinctive spatial arrangement of cells into rows or columns, as well as the interconnections between these nuclei, in terms of their potential contribution to an encoding or analysis of temporal cues.

METHODS

Descriptions of chiropteran cytoarchitecture was based upon material from three bats each of the following species, Pteronotus parnellii, Antrozous pallidus, and Artibeus jamaisensis, and two bats of the species, Noctilio leporinus, and Phylostomus hastatus. The bat brains were all form-alin fixed, embedded in celloidin and sectioned at 20 um in either horizon-tal, transverse or sagittal planes. Sections were stained with cresylecht violet or thionin for cells; alternate sections were stained with the Heidenhain method for fibers as described previously (Zook & Casseday, 1982a). Descriptions of dolphin cytoarchitecture was based upon material from three bottlenose dolphins, Tursiops truncatus, and three dolphins of the species, Stenella coeruleoalba. Brains were formalin fixed, embedded in colloidin and sectioned at 35 um in one of three planes as described previously (Jacobs, et al., 1971). Alternate sections were stained by the Bielschowsky-Plien cresyl violet method or the Loyez-modified Weigert hematoxylin method. Photomicrographs at magnifications from 10x to 400x were taken with a Leitz photomacroscope or a Zeiss microscope.

RESULTS

The medial nucleus of the trapezoid body (MNTB) is a generally compact nucleus lying within the horizontal-running fibers of the trapezoid body near the medullary midline. In the bat and dolphin species examined, as in other mammals, the MNTB is characterized by oval or round principal cells with eccentric nuclei and darkly-staining Nissl granules; elongate and stellate cells are also present (Morest, 1968; Zook & Casseday, 1982). The MNTB of the dolphin species examined is distinguished by an unusual arrange-ment of cells along the medial side of the nucleus (fig. C). These cells are aligned into a row which angles across the horizontally-running fibers of the trapezoid body. From dorsal to ventral there is a gradual shift medially in the relative position of the stacked cells such that they conform to a slanted line with a regular slope of 1.6 in Stenella and 2.9 in Tursiops. The ventral-most MNTB curves again laterally. Sections cut in other planes indicate that the row is part of a curved three-dimensional sheet of cells spanning most if not all of the trapezoid body. it should be noted that only the medial-most principle cells of the MNTB follow this pattern. The rest of the MNTB has a more random or clustered appearance similar to the MNTB in most other mammals. While such a distinctive row of cells is not found in the MNTB of the bat species examined, some alignment of cells is present in the MNTB in many species such as Noctilio (Fig. F). It is notable that the medial border of the MNTB, in most mammals examined, tends to be more sharply defined than the other borders of this cell group. The unusually well-defined cell row of the dolphin MNTB may simply represent a more pronounced manifestation of this tendency.

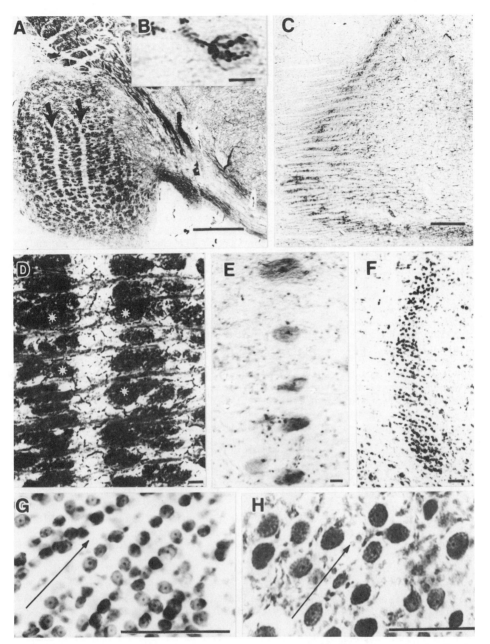

Figs. A-H. Photomicrographs of transverse sections through the brainstem.
Bar equals 1mm (A,C) 100um (FGH) or 10um (BDE). (A) Fiber-stained section
of the anteroventral cochlear nucleus (AVCN), dolphin, <u>Tursiops</u>. Arrows
mark unstained rows of cell soma in contrast to the dark-stained fiber fascicles. (B) Horseradish peroxidase-filled axon and a calyx of Held terminal
ending surrounding a cell soma in the medial nucleus of the trapezoid body
(MNTB), mustache bat, <u>Pteronotus</u>. (C) Nissl-stained section through the
MNTB, dolphin, <u>Stenella</u>. (D) Enlargement of fig. A. Note the horizontal-
running fibers of the trapezoid body. Stars indicate some of the fascicles
of the eighth nerve cut in cross-section. (E) Nissl-stained cell row from
the caudal AVCN of <u>Tursiops</u>. (F) Cell columns in the ventral nucleus of the
lateral lemniscus (<u>VNLL</u>) of the spear-nosed bat, <u>Phyllostomus</u>. Arrow
indicates course of lateral lemniscus fibers. (H) VNLL cells, <u>Tursiops</u>.

The second cell group of interest here, the caudal half of the AVCN, contains a wide variety of cells types identifiable in Nissl stained sections. Close to the eighth nerve bifurcation, globular cells with nissl-capped, large eccentric nuclei predominate. It is these cells which are thought to provide the main projection to the MNTB. Moving rostrally, there is a mixture of multipolar, elongate and small cells with an increasing proportion of small and medium spherical cells. These cell types are common to the AVCN of all bat and dolphin species examined. The dolphins possess an additional small number of giant cells as described by Osen & Jansen (1965). A typical giant cell is shown in fig. E (upper cell) along with several of the other cell types common to this region.

Cell soma are also condensed into rows within the caudal part of AVCN of Tursiops (fig. E). Very distinctive in fiber-stained sections (fig. A & D), these rows are almost exactly perpendicular to the fibers which exit the cochlear nucleus as the trapezoid body (horizontally- running fibers in fig. D). Rows of cell soma are visible but less precisely aligned in Stenella. Horizontal sections suggest again that these are sheets of cells extending rostrocaudally in the anteroventral cochlear nucleus but are only distinguishable in the caudal part of the nucleus. The pattern of dense fibers bundles in the ascending branch of the eighth nerve may physically dictate this columnar organization of soma (cross sections of these fiber bundles are marked by stars in fig. D). However this cell pattern is not prominent in other mammalian species, even where equally prominent fascicles of the ascending branch of the eighth nerve are found, as in the porpose Phocaena phocaena (Osen & Jansen, 1965).

The VNLL, like the MNTB, is a target of the ascending projections from the caudal part of the AVCN. In this nucleus in bats as well as some other mammalian species one finds a distinctive columnar organization of small to medium (8-15 um diameter), closely-packed, ovoidal and multipolar cells, often only in the most ventral part of the nucleus, such as shown in Fig. G. These columns are clearly defined by parallel fiber fascicles of the lateral lemniscus (arrows indicate fiber direction), but both columns and fascicle bundles disappear just above this area in the dorsal part of the ventral nucleus. A potentially homologous cell group is present in the marine mammals examined (Fig. H). Similar, although larger (15-25 um diameter), cells types are present, but there is only the suggestion of a columnar organization; cells are clumped together with only a slight orientation (arrow) relative to the fibers of the lateral lemniscus.

DISCUSSION

Here we have briefly described in three auditory brainstem nuclei two unusually well-ordered arrangements of cell soma: forming rows perpendicular to fiber tracts or in columns parallel to fiber tracts. Although they represent uncommon features, it is possible that these precise cell alignments have been guided by the prevailing course of the surrounding fiber tracts. In particular, the fascicles of the ascending branch of the eighth nerve or of the lateral lemniscus must contribute in some extent to the perceived cell order. However it is difficult to accept the course of fiber tracts as the entire organizing principle, otherwise one would expect cell rows to be more common in homologous regions in other mammals, especially those with similar proportions of cells to fibers as in Phocaena. In the case of the MNTB, it is highly unlikely that the course of fibers contributes to the alignment of cells as the cell rows are more perpendicular than parallel to the course of the trapezoid body fibers. In either case, the precision of cell alignment shown here is uncommon, and particularly notable as it appears at different levels of the brainstem pathway and within nuclei which may be physiologically and anatomically related.

Beginning with the dolphin MNTB, it is tempting to suggest that the cell rows in their slanted-line pattern might present an organizational elaboration to preserve or analyze temporal patterns. As a simple result of the difference in distance from AVCN to any two cells in the MNTB row, it is likely that two simultaneously generated signals from cells in the caudal AVCN would reach adjacent cells along the row of the contralateral MNTB with slightly different arrival times. Considering the entire row of MNTB cells, it seems reasonable to expect that such a systematic shift in axonal distance and travel time might translate into varying increments of arrival time delay. Such an anatomical delay line would be useful in a system of coincidence detection or operations similar to a correlation analysis of signals in the time domain. The dorsal-ventral tonotopic organization of the trapezoid body (Brownell, 1975) and evidence for a tonotopic order of units within the MNTB (Guinan, et al., 1972) raise the further possibility of operations in the frequency domain.

It is likely that most, if not all the cells in the rows found in the dolphin AVCN are contributing to the trapezoid body, and many may project to the MNTB or VNLL or both. The cell soma within a one AVCN row would appear to be equidistant from any one MNTB cell. If their targets were adjacent cells in the staggered cell-row of the contralateral MNTB, there would appear to be a precisely graded difference in travel distance, and thus in travel time. Projections from two different rows of the caudal AVCN would have set differences in distance to the MNTB imposed upon them. In either case the row arrangement could preserve, or induce, precise temporal relationships in the stream of information ascending via this auditory pathway.

Very little attention has been paid to the functional role of the VNLL in the auditory pathway. Any significance attributed to this cell group is pure speculation at this point. Nevertheless, in a hypothetical correlation system, one would not only need a delay line (from caudal AVCN to the MNTB), but a site to integrate the signal with the delayed signal. Since the VNLL is a target of both the caudal AVCN and the MNTB, it is tempting to point to the VNLL as a promising site for integration. To follow upon this speculation, one might then predict that the distinctive organization of cells in the VNLL and the pattern of ascending terminations within VNLL are optimally arranged to preserve or enhance both the temporal-delay gradient established and relayed from the MNTB and the original temporal relationships conducted directly from the AVCN. These speculations are raised in hopes of stimulating a theoretical framework for further explorations of the intriguing physiological and anatomical relationships between these nuclei of the auditory brainstem.

REFERENCE

Adams, J.C., 1978, Morphology and physiology in the ventral nucleus of the lateral lemniscus, Soc. Neurosci., 8:3.
Brawer, J.R., and Morest, D.K., 1975, Relations between auditory nerve endings and cell types in the cat's anteroventral cochlear nucleus seen with the golgi method and nomarski optics, J. Comp. Neur., 160:491.
Brownell, W.E., 1975, Organization of the cat trapezoid body and the discharge characteristics of its fibers, Brain Res., 94:413.
Brugge, J.F., and Geisler, C.D., 1978, Auditory mechanisms of the lower brainstem, Ann. Rev. Neurosci., 1:363.
Glendenning, K.K., Brunso-Bechtold, J.K., Thompson, G.C. and Masterton, R.B., 1981, Ascending auditory afferents to the nuclei of the lateral lemniscus, J. Comp. Neurol., 197-673.
Guinan, J.J., Jr., Norris, B.E., and Guinan, S.S., 1972, Single auditory units in the superior olivary complex II: Locations of unit categories and tonotopic organization, In J. Neurosci., 4:147.

Harrison, J.M., and Irving, R., 1964, Nucleus of the trapezoid body: dual afferent innervation, Science, 143:473.

Henson, O.W., Jr., 1970, The central nervous system, in: "Biology of Bats," W.A. Wimsatt, ed., Academic Press, N.Y.

Jacobs, M.S., Morgane, P.J., and McFarland, W.L., 1971, The anatomy of the bottlenose dolphin (Tursiops truncatus). Rhinic lobe (Rhinencephalon) I. The paleocortex, J. Comp. Neurol., 141:205.

Morest, D.K., 1968, The collateral system of the medial nucleus of the trapezoid body of the cat, its neuronal architecture and relation to the olivo-cochlear bundle, Brain Res., 9:288.

Morest, D.K., Kiang, N.Y.S., Kane, E.C., Guinan, J.J., Jr., and Godfrey D.A., 1973, Stimulus coding at caudal levels of the cat's auditory nervous system: II. Patterns of synaptic organization, in: "Basic Mechanisms in Hearing," A.R. Moller, ed., Academic Press, N.Y.

Osen, K.K. and Jansen, J., 1965, The cochlear nuclei in the common porpoise, Phocaena phocaena, J. Comp. Neurol., 125:223.

Schweizer, H., 1981, the Connections of the inferior colliculus and the organization of the brainstem auditory system in greater horseshoe bat (Rhinolophus ferrumeginum), J. Comp. Neurol., 201:25.

Spirou, G.A., Brownell, W.E., Zidanic, M., and Dulguerov, P., 1985, Low frequency response properties of trapezoid body fibers, Soc. Neurosci., 11:1052.

Tolbert, L.P., Morest, D.K., and Yurgelun-Todd, D.A., 1982, The neuronal architecture of the anteroventral cochlear nucleus of the cat in the region of the cochler nerve root, Neurosci. 7:3031.

Warr, W.B., 1971, Fiber degeneration following lesions in the multipolar and globular cell areas in the ventral cochlear nucleus of the cat, Brain Res., 40:247.

Warr, W.B., 11982, Parallel ascending pathways from the cochlear nucleus: Neuroanatomical evidence of functional specialization, Sens. Phys., 7:1.

Zook, J.M., and Casseday, J.H., 1982, Cytoarchitecture of auditory system in lower brainstem of the mustache bat, Pteronotus parnellii, J. Comp. Neurol., 207:1.

Zook, J.M., and Casseday, J.H., 1985, Projections from the cochlear nuclei in the mustache bat, Pteronotus parnellii, J. Comp. Neurol., 237:307.

Zook, Z.M., and Casseday, J.H., 1986, Convergence of ascending pathways at the inferior colliculus of the mustache bat, Pteronotus parnellii, submitted.

Zvorykin, V.P., 1959, Morphological basis of locative and supersonic abilities in bat. Arkh. Anat., Gistol. Embriol., 36:19.

Zvorykin, V.P., 1963, Morphological substrate of ultrasonic and locational capacities in the dolphin, Arkh. Anat., Gistol. Embriol., 45:3.

TEMPORAL ORDER DISCRIMINATION WITHIN THE DOLPHIN CRITICAL INTERVAL

Richard A. Johnson*, Patrick W. B. Moore**, Mark W. Stoermer**, Jeffrey L. Pawloski***, and Leslie C. Anderson**

*Western New Mexico University
Silver City, NM 88062

**Naval Ocean Systems Center
P O Box 997, Kailua, HI 96734

***SEACO, Inc.
146 Hekili St., Kailua, HI 96734

Human psychophysical experiments (Patterson and Green 1970, Ronken 1970, and others) have shown that human listeners can detect the difference between unequal amplitude click-pairs that arises from the order of the two clicks. The clicks of each pair are separated by a few milliseconds. Standard Fourier analysis indicates that this "time reversal" has no effect on the power spectrum. The apparent conclusion to be drawn is that human audition is phase sensitive, since only the phase spectrum can be different. As with many mathematical models, the conclusion is only as good as the assumptions upon which the model is based. The assumption which is in conflict here is that the waveform is known for all time, both past and future, since Fourier analysis employs integration with unbounded upper and lower time limits. The inability of (biological) systems to predict exactly ALL future details of a stimulus (or for that matter, to store ALL past details) inevitably leads to alternate mathematical formulations with less restrictive assumptions.

Such mathematical models useful for spectral analysis of non-stationary phenomena are generalized short-time power spectra and autocorrelation functions (Schroeder and Atal, 1962). By employing a limited integration time or a weighting function in the form of the impulse response of a low pass filter, a "running" power spectrum or autocorrelation function with time as a second variable depicts temporal variations. Such short-time spectra are not new to bioacoustics, having been the basis of sonagrams and "waterfall" spectrogram displays.

Our purpose was to create a "phase" experiment to determine if a dolphin can detect the difference in arrival order for appropriate click-pair stimuli. Explanations of dolphin echolocation performance have often included phase considerations (Johnson, 1968). Additionally, we wished to investigate the cues available to discriminate the stimuli.

PROCEDURE

The subject for the study was a female bottlenose dolphin (Tt-593) named Circe. The animal weighed 840 kgs and was fed a daily ration of mixed fish (smelt and herring) for a daily total intake of 35 kgs.

The animal was tested in a floating pen kept at the Naval Ocean Systems Center (NOSC), Kaneohe Bay, Hawaii. The pen was 6 x 12 m and moored in 5 to 6 m of water over a mud and silt bottom. On the pen was an instrument shelter which housed the signal generation equipment. Other apparatus consisted of a response manipulandum and the animal's stationing devices. A bite-plate tail-rest stationing combination, 1.0 m underwater, was used to position the animal and insure that the animal was maintained in a fixed listening location throughout the trial sequence and that her position was identical for each subsequent trial.

The stimuli to be discriminated consisted of a pair of single cycles from a 60 kHz sign wave (clicks) separated by 200 microsecs. The clicks were repeated every 20 msecs. This click-pair train was presented in two 1.5 second listening intervals separated by 0.5 seconds with no stimulus. The signals were generated by a Wavetek programmable signal generator (Model 154) which was controlled by a NOSC designed digital timer which managed the timing of the signal events. A Hewlett-Packard attenuator (model 305C) was used to control the relative amplitude of one click with respect to the other and was set 10 dB lower. The click-trains were amplified by a modified (for high-frequency response) David Hafler DH-220 calibrated amplifier and presented to the animal via a F-30 calibrated hydrophone provided by the Underwater Sound Reference Division of the Naval Research Laboratory. The output of the amplifier was continuously monitored by a Tektronix T912 oscilloscope. Measurements of the peak sound pressure level of the signals at the listening position of the animal was 178 dB re: 1 uPa.

A trial consisted of the animal stationing on a foam pad and, on hand cue, going to the bite plate station. When the experimenter visually verified that the animal had stationed correctly, a trial began with the presentation of the first interval of click-pairs followed by the second. In the first interval the larger click led the smaller. On 50% (randomized) of the trials, the second interval would be the same signal as the first interval. The other 50% of the trials contained a second interval with the time reversed stimulus (where the smaller click of the pair led the larger). The animal reported a discrimination between the two listening intervals using a go/no-go response paradigm. If the second interval was the same as the first, the animal was to remain at station for the full second interval plus 0.5 seconds and this constituted a no-go response. The go response was required if the animal detected a difference between the two intervals: it was to release station during the second interval and hit a response paddle before the trial ended.

The animal was initially trained to discriminate the two intervals by use of an amplitude cue: the second listening interval was made 18 dB louder on time reversed trials. The animal easily learned to discriminate and respond to the concept "stay if the same and go if louder". Since the second interval had both the time-reversed cue and the loudness cue, a fading process was used to slowly remove the loudness cue and leave only the time-reversed cue as the means by which to discriminate when the two intervals are different.

RESULTS

Initial results have demonstrated that the animal was making a discrimination between the two signal orders. The learning process was not an easy one. The first attempt to fade the loudness cue led to a performance breakdown. However, after retraining with the loudness cue and a second fading procedure, the animal has performed 630 trials (with time-reversal only being tested) with 75% correct discrimination performance. Detection performance based on the Theory of Signal Detection (TSD) indicates an average d' of 1.30 with a slight conservative bias (beta not calculated). Since the animal's performance has not reached an asymptotic value, testing continues and a 3.0 dB increase in overall level of the signals has been made in hopes of making the time reversal cue more salient for the animal. Experimentation along these same lines continues.

ANALYSIS OF FREQUENCY DOMAIN CUES

For humans, the cues reportedly used to discriminate the order of clicks are very similar to the cues used for time separation pitch. The qualitative sensation of a complex pitch suggests a frequency domain cue, although analysis in the time domain cannot be ruled out. Since humans cannot temporally separate the clicks of a pair, and a similar conclusion may be made for dolphins (Moore, et al. 1984, Au and Moore, this volume), we chose to concentrate on the frequency domain cues available.

We generated versions of similar stimuli consisting of single cycles of a 60 kHz sine wave (click) sampled at 2 microsecond intervals and placed appropriately in a 2048 point FFT analysis window. We could vary the individual click amplitude and the inter-click interval. The phase spectrum of an individual click reveals only the expected linear phase component with slope proportional to position (delay) in the FFT analysis window. For click-pairs, the phase spectrum of the "time-reversed" click pair is different from the "forward" pair, even when the linear phase components are equalized to the maximum extent possible. When the clicks have dissimilar amplitudes (Fig 1a), the phase spectrum shows a ripple (Fig 1b) not unlike that in the power spectrum (Johnson and Titlebaum 1976, Johnson 1980). It is the fine structure of the ripple which is different for the two temporal orders.

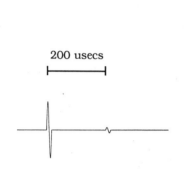

Figure 1a. Click-pair consisting of a large pulse and small pulse separated by 200 usecs.

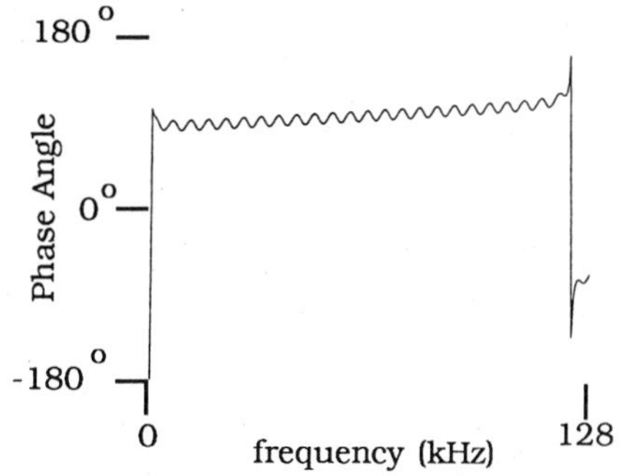

Figure 1b. Phase spectrum of click pair in Figure 1a with large linear component removed.

We applied the technique of short-time spectral analysis using a time scale sufficient to resolve the changes in the short-time spectra. For these click-pairs, the changes might occur within the time span of an individual click, so we investigated short-time spectra with a time resolution of 2 microseconds. Initially, our analysis involved just the short-time spectra over the course of arrival of the second click into the (rectangular) analysis window since all changes of interest occur during that period of time. When the click amplitudes are in a ratio of 10 to 1, the starting spectra for the two click pairs differ only by size (Fig 2a,b) but following the complete arrival of the second click, no matter which order they are in, the spectra are identical as ordinary Fourier analysis says they should be (fig 2c). The change in the spectra is more understandable when a "movie" is constructed from the short-time spectra (in a manner reminiscent of spectrum analysers that use an oscilloscopic display). It shows the time course of the spectral changes. Our first observation was that the ripple "interference" occurs almost immediately and thereafter the spectra settle to the final amplitude. We then used short-time logarithmic spectra because they behave in a similar manner and are more useful to understand the available cues in light of the dynamic range available to biological systems.

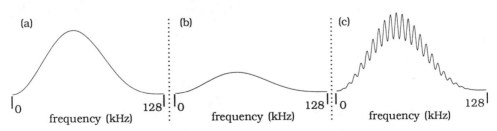

Figure 2. Power spectra of (a) large click, (b) small click, and (c) both clicks separated by 200 usecs (in either order of presentation).

After this initial investigation, we decided to pursue this short-time analysis. We felt that we had to overcome two shortcomings of the procedure: (1) the FFT rectangular analysis window and (2) the display of the short-time spectra.

With regard to the first short-coming, there are well known windowing techniques whose properties are documented widely (e.g. Harris, 1978). Since the use of an analysis window is equivalent to filtering, we felt that the window should reflect whatever information is known about dolphin audition. In particular, measurements of the critical interval (Moore, et al. 1984) or the integration time for the perception of complex echoes in dolphins (Au and Moore, this volume) suggest a filter with a rapid rise and exponentially decaying tail. The data are not exacting enough to be more explicit, so we arbitrarily and crudely approximated such behavior with the shape of the chi-square distribution function with the decay at 10% of the maximum value at 300 microseconds (fig 3). Similar windows might be defined, but we feel that there would be only negligible changes in the short-time spectra.

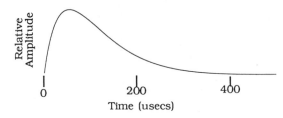

Figure 3. Chi-square window

The second shortcoming, the "movie" display, has been alternately presented as a sonagram or spectrogram. These are useful techniques but the sonagram lacks amplitude detail sufficient to study the rippled spectra and the spectrogram display overwhelms our visual system, particularly when applied to "real world" ripple (Beuter, 1980). Our solution was to enhance the sonagram technique using a color graphic display, where the amplitude in the spectrogram is represented with color.

We created color spectrograms for the theoretical click-pairs both with and without the new window function, using lograthmic spectra exclusively for the dynamic range available. Without the window, the sonagrams both evolve to an identical pattern following the occurence of the second click as in the initial analysis.

When the window is used to create color spectrograms (fig 4a,b) additional interesting features are observed. Here, color aids in identifying absolute amplitude as well as relative amplitude (ripple) and is largely lost in this black and white reproduction. First, the effects of rapid changes (clicks) are "delayed". Second, the "decay" in the spectrograms due to the window causes the spectrograms to be quite different from each other after the occurence of the second click. In particular, the ripple is far more pronounced when the second click is the smaller one. The window exaggerates the temporal order differences in the temporal course of the spectrogram.

Figure 4a. Spectrogram created using the Chi-square window with the large click arriving before the small click.

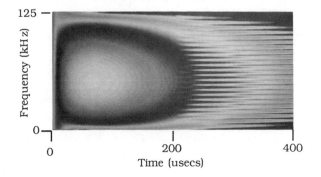

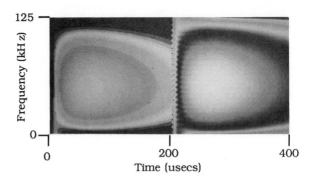

Figure 4b. Spectrogram created using the Chi-square window with the small click arriving before the large click

CONCLUSIONS

Our data suggest that a dolphin has the ability to discriminate the temporal order of click-pairs within the critical interval. Analysis of the available frequency domain cues for discriminating the temporal order reveal a difference in the classic phase spectra. However, the limitations of the mathematical foundations of that analysis lead to an alternate explanation based on the difference in short-time spectra, particularly when a plausible analysis window is employed. Since this cue involves rippled spectra, it supports the hypothesis that the analysis of rippled spectra may be an important function of dolphin audition. This hypothesis is similar to an hypothesis of bat echo evaluation (Beuter 1980). Finally, although these frequency domain cues offer an explanation of the discrimination, the possibility that analysis in the time domain might also explain this ability cannot yet be ruled out.

ACKNOWLEDGEMENTS

Much of this work was performed while the principle author was on an ASEE-Navy Summer Faculty Research Fellowship at the Naval Ocean Systems Center Hawaii Laboratory.

REFERENCES

Beuter, Karl J., 1980, A New Concept of Echo Evaluation in the Auditory System of Bats, in: *Animal Sonar Systems*, edited by R.-G. Busnel and J. F. Fish, Plenum, New York, pp. 747.

Harris, Frederic J., 1978, On the Use of Windows for Harmonic Analysis with the Discrete Fourier Transform, Proc. IEEE, 66:51.

Johnson, C. Scott, 1967, Discussion in: *Animal Sonar Systems*, edited by R.-G. Busnel, Laboratoire de Physiologie Acoustic, Jouy-en-Josas, France, pp. 384.

Johnson, Richard A. , 1980, Energy Spectrum Analysis in Echolocation, in: *Animal Sonar Systems*, edited by R.-G. Busnel and J. F. Fish, Plenum, New York, pp. 673.

Johnson, Richard A., and Titlebaum, E. L., 1976, Energy Spectrum Analysis: A Model of Echolocation Processing, J. Acous. Soc. Am., 60:484.

Moore, P. W. B., Hall, R. W., Friedl, W. A., and Nachtigall, P. E., 1984, The Critical Interval in Dolphin Echolocation: What Is It?, J. Acous. Soc. Am., 76:314.

Patterson, James H., and Green, David M., 1970, Detection of Transient Signals Having Identical Energy Spectra, J. Acous. Soc. Am., 48:894.

Ronken, Don A., 1970, Monaural Detection of a Phase Difference between Clicks, J. Acous. Soc. Am., 47:1091.

Schroeder, M. R. and Atal, B. S., 1962, Generalized Short-Time Spectra and Autocorrelation Functions, J. Acous. Soc. Am., 34:1679.

DETECTION ABILITIES AND SIGNAL CHARACTERISTICS OF ECHOLOCATING

FALSE KILLER WHALES (Pseudorca crassidens)

Jeanette Thomas[1]*, Mark Stoermer[1], Clark Bowers[2],
Les Anderson[2], and Alan Garver[3]

[1]Naval Ocean System Center, Kailua, HI   96734
[2]Naval Ocean System Center, San Diego, CA   92152
[3]Sea World Inc., San Diego, CA   92109

False killer whales (Pseudorca crassidens) are deep-diving, pelagic animals that inhabit tropical and temperate waters of the Pacific and Atlantic Oceans. Highly social animals, they have been seen in herds of more than 100 individuals. They feed on squid and large fish.

Busnel and Dziedzic (1968) reported pulses produced by a herd of Pseudorca in the Mediterranean Sea. A false killer whale at Sea Life Park in Hawaii was trained to acoustically retrieve rings while blindfolded (K. Pugh, pers. comm.). Watkins (1980) described characteristics of Pseudorca pulses recorded at sea. The ability to use pulses for echolocation and their acoustic characteristics have not been documented previously.

Our objectives were to: 1) document the ability of false killer whales to detect a target using pulse trains and 2) document the time, frequency, and amplitude characteristics of pulses used during echolocation.

METHODS

We conducted an experiment to demonstrate echolocation abilities in a subadult, male false killer whale (Bob) at Sea World in San Diego. During the tests, the whale was sent to station on an aluminum crook that supported the animal forward of the pectoral flippers and maintained his position at 1 m below the water surface (Fig.1). In front of the animal were three moveable arms constructed of PVC. The first arm supported an acoustic screen made of 6.4 mm-thick neoprene rubber, which limited his ability to detect the target while approaching the station. The second arm supported an opaque screen made of black mylar plastic, which is acoustically transparent. The target, a 7.62 cm diameter, hollow, water-filled stainless-steel sphere, was suspended with monofilament line from the last arm and could be positioned 1 m below the water or removed.

A go/no-go testing procedure was used. With the whale's head out of the water and attention on the trainer, the visual and acoustic screens were positioned in the water and the target was either lowered (target-present

*author conducted the target detection study as an employee of Hubbs Marine Research Institute at Sea World in San Diego

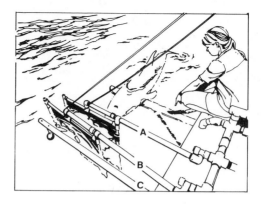

Figure 1. Station with: A. acoustic screen, B. visual screen, C. target.

trials) or not (target-absent trials). The animal was signaled to go to station. When he was resting properly on the crook, the acoustic screen was raised, signaling the whale to begin target detection. If the animal detected a target and backed-out of the station within five seconds (go trials), he received a fish reward. If the subject did not detect a target, he waited until the trainer called him back with a whistle at a randomly selected time (no-go trials) and then received a fish reward. Incorrect responses were not reinforced.

The whale's detection capabilities were tested at distances of: 1, 2, 4, and 6 m. For each distance, 100 trials were conducted in ten, 10-trial sessions. Equal numbers of target-present and target-absent trials were conducted with the order randomly selected. Tape recordings were made during the detection task to verify the presence of pulse trains.

To further examine the characteristics of echolocation signals, we recorded an adult female false killer whale (Makapu'u) at Sea Life Park in Hawaii. To roughly control her orientation relative to the hydrophone, we tossed a fish 1 meter in front of the hydrophone and narrated the distance and angular approach of the animal. All recordings were made on multiple channels of a Racal Store 4DS portable 4-channel instrumentation recorder with an H52 hydrophone (system frequency response 0.200 to 150 kHz $\pm$ 3dB).

We examined waveforms and spectra of pulses in ten hours of recordings using a Data Precision (model Data 6000) spectrum analyzer. Data were low-pass filtered at 250 kHz and sampled at 500 kHz with 16 bits of resolution. For each pulse, 256 sample s were taken. Discrete Fourier Transforms were performed, which resulted in 128 frequency bins each 1.95 kHz wide.

We selected four representative pulse trains (hereafter designated as Train A through Train D). Each of these trains was displayed using relative log plots (bandwidths at -20, -10, and -3 dB), normalized power spectra, and waveforms. Source levels in dB re 1 μPa were calculated from recorded peak-to-peak voltage levels with the assumption of spherical spreading loss.

RESULTS AND DISCUSSION

Detection performance varied between 90% and 95% correct for target ranges from 1 to 4 m (Fig. 2). At 6 m, the performance level was inconsistent. We lost access to the whale thereafter and are unsure whether he would have obtained high performance at this distance. He produced sounds when on station, with exaggerated head scans across the target. Scanning

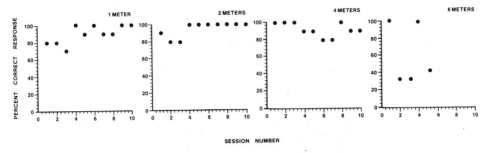

Figure 2. Performance data for target detection at four distances.

and signal production were more exaggerated during target-absent trials.
We conclude that the whale was detecting the metal sphere by echolocation.

Signals recorded outside the primary transmission beam of Tursiops (Au,
1980) have different waveforms, peak frequencies, and amplitudes. We
examined the characteristics of pulses produced by a false killer whale
scanning her head across a hydrophone (Train A). In Figure 3 the left
display of Train A is a relative log plot showing bandwidths of -20, -10,
and -3 dB for every other pulse. The right display is the normalized power
spectrum of every twentieth pulse. Waveforms of every fourth pulse in Train
A are shown at the top of Figure 4. Figure 5 illustrates the source levels
of all pulses in Train A.

We then analyzed pulse trains produced with a direct orientation
(Trains B and C). Bandwidths (-3dB) of single peaks ranged from 5 to 16 kHz
(Fig. 3, left displays of Trains B and C). The whale generally produced
pulses that had peak frequencies from 20-65 kHz (Fig. 3, right displays of
Trains B and C). Trains B and C (Fig. 4) show the more consistent waveforms
compared to the pulses in Train A. Data on source levels of pulses produced
with a direct orientation to the hydrophone (Fig 5., Trains B and C) show
more consistent levels than pulses recorded during head scanning (Train A).

Au (1980) showed that Tursiops has a directional transmission beampattern in both the horizontal and vertical planes and this also was found to
be true of Delphinapterus (Au et al., this volume). The transmission
beampattern of Pseudorca is unknown; however, our analysis of data collected
when the animal changed its head orientation implies a directional transmission beampattern.

On two recordings the animal produced a different type of pulse train;
a narrowband, low frequency component produced in conjunction with a narrowband, high frequency component. These signals were produced with a direct
orientation to the hydrophone. Train D in Figure 3 shows energy at two
distinct frequency ranges; 5-12 kHz and 40-50 kHz. Waveforms of Train D
(Fig. 4) show superimposed high and low frequency components that graded in
and out throughout the series. Source levels of pulses in Train D are quite
variable, but higher in amplitude than in other trains (Fig. 5).

Busnel and Dziedzic (1968) reported pulses from Pseudorca with "2 or 3
amplitude oscillations per pulse, 0.8 msec in duration, and peak frequencies
between 8-12 kHz." Although their recordings were limited to 30 kHz, their
data were comparable to our lowest frequency pulses. More commonly the peak
frequencies we recorded were between 17 and 56 kHz. Pseudorca also can
produce superimposed high and low frequency pulses. Belugas and harbor
porpoise exhibit similar superimposed high and low frequency pulses
(Kamminga and Wiersma, 1981).

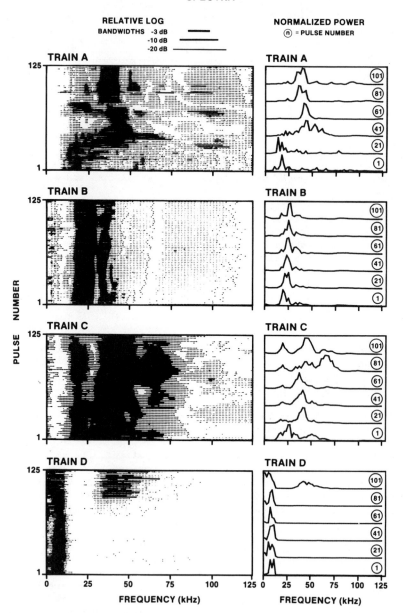

SPECTRA

RELATIVE LOG
BANDWIDTHS -3 dB
            -10 dB
            -20 dB

NORMALIZED POWER
Ⓝ = PULSE NUMBER

TRAIN A

TRAIN A

125

1

⑩⓵
⑧⓵
⑥⓵
④⓵
②⓵
①

TRAIN B

TRAIN B

125

1

⑩⓵
⑧⓵
⑥⓵
④⓵
②⓵
①

PULSE NUMBER

TRAIN C

TRAIN C

125

1

⑩⓵
⑧⓵
⑥⓵
④⓵
②⓵
①

TRAIN D

TRAIN D

125

1

⑩⓵
⑧⓵
⑥⓵
④⓵
②⓵
①

0   25   50   75   100   125   0   25   50   75   100   125

FREQUENCY (kHz)   **FREQUENCY (kHz)**

Figure 3.   Train A was produced while the animal scanned its head, Trains B
and C were produced with a direct orientation, and Train D has two-frequency
components. Relative log plots (showing bandwidths at -20, -10, and -3 dB)
are displayed in the left column.   Normalized power spectra are displayed in
the right column.

WAVEFORMS

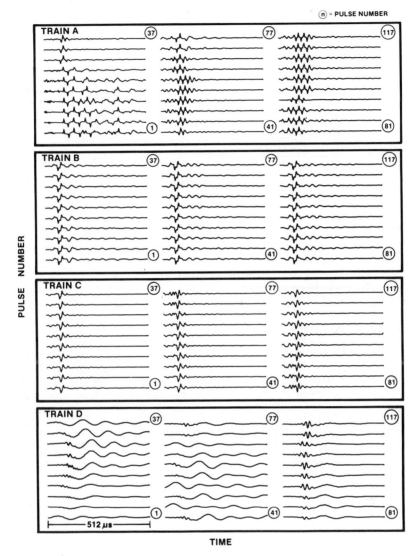

Figure 4. Waveforms from a pulse train produced while the animal scanned its head (Train A), from two trains with direct orientation (Trains B and C), and from a two-frequency component, direct train (Train D).

Our source levels are representative of the animal's abilities in a reverberant pool; they probably can selectively alter their source level depending on the target and environment. Fish and Turl (1976) and Wood and Evans (1980) summarized peak-to-peak source levels (re 1 μPa) from odontocetes. Our data are relatively low in amplitude compared to those

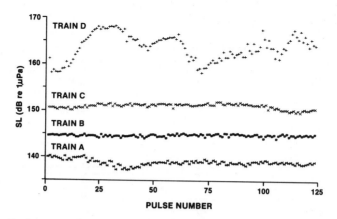

Figure 5. Peak-to-peak source levels for pulse trains with changing
orientation (Train A), for pulses in two trains with a direct orientation
(Trains B and C), and in a two-frequency component, direct train (Train D).

for Tursiops (175-228 dB) or Delphinapterus (160-180 dB), but similar in
magnitude to other pelagic species, such as Phocoena phocoena (112-149 dB),
Delphinus delphis (140 dB), or Stenella longirostris (108-115 dB).

Evans (1973) suggested that echolocation characteristics of odontocetes
are correlated with prey size. Because Pseudorca eat large prey compared to
the small fish in the diet of dolphins, their lower frequency echolocation
signals (i.e. larger target-size resolution) fit this suggestion. Watkins
(1980) suggested a correlation between body size and peak frequency, with
larger odontocetes producing lower frequency echolocation signals. Signals
from Pseudorca fit this generalized trend, being slightly higher in frequency than those of Orcinus and lower in frequency than those of small
odontocetes such as Tursiops, Inia, Phocoena, and Phocoenoides.

In conclusion, this captive false killer whales generally produced
midfrequency echolocation pulses, but seemed to have the ability to change
the peak frequency and bandwidth; sometimes producing superimposed, narrowband high and low frequency pulses. Source levels and pulse characteristics
associate Pseudorca echolocation signals with those from other large
toothed whales that consume large prey.

REFERENCES

Au, W.L., 1980, Echolocation signals of the Atlantic Bottlenose dolphin
    (Tursiops truncatus) in open waters. in: " Animal Sonar Systems,"
    Plenum Press. pp. 251-282.
Busnel, R.-G., and Dziedzic, A., 1968, Caracteristiques physiques des
    signaux acoustiques de Pseudorca crassidens OWEN (Cetacea Odonto
    cete). Mammalia, 32(1):1-5.
Evans, W.E., 1973, Echolocation by marine delphinids and one species of
    fresh-water dolphin, J. Acoust. Soc. Am., 54:191-199.
Fish, J., and Turl, C.W., 1976, Acoustic Source Levels of Four Species of
    Small Whales, Naval Undersea Center. Technical Report No. 547. 14 pp.
Kamminga, C., and Wiersma, H., 1981. Investigations on cetacean sonar II.
    Acoustical similarities and differences in odontocete sonar signals.
    Aquatic Mammals, 8(2):41-62.
Watkins, W., 1980, Click Sounds from Animals at Sea, in: "Animal Sonar
    Systems," R. -G. Busnel and J. Fish, eds., Plenum. pp. 291-297.
Wood, F.G., and Evans, W.E., 1980, Adaptiveness and ecology of echolocation
    in toothed whales. in: "Animal Sonar Systems," Plenum. pp. 381-426.

BINAURAL NEURONS IN THE MUSTACHE BAT'S INFERIOR COLLICULUS:

PHYSIOLOGY, FUNCTIONAL ORGANIZATION, AND BEHAVIORAL IMPLICATIONS

Jeffrey J. Wenstrup

Department of Physiology-Anatomy
University of California
Berkeley, CA, 94720, USA

This report describes how sound localization cues contained within a narrow band of the echo spectrum may be encoded in the auditory midbrain of the mustache bat (*Pteronotus parnellii*). Binaural neurons were studied in the 60 kHz region of the inferior colliculus (IC), the region in IC that analyzes echoes of the second harmonic, constant frequency (CF) component of the mustache bat's biosonar signal. In the enlarged 60 kHz region, binaural neurons are spatially segregated and available for detailed study. In particular, the responses of "EI" neurons were studied. These are excited by stimulation of the contralateral ear and inhibited by stimulation of the ipsilateral ear. Recordings of 60 kHz EI single units suggest what features of their response may encode interaural intensity differences (IIDs) and/or horizontal target location. These physiological studies suggest some directions for the behavioral study of echo localization.

EI neurons show a maximum response over a range of IIDs favoring the contralateral (excitatory) ear, a sharp cut-off in the response over a narrow range of IIDs, and a greatly reduced response at IIDs favoring the ipsilateral (inhibitory) ear (Fig. 1A). Among the population of 60 kHz EI neurons, cells differ in the particular IID at which the response is cut-off. For some, the response is suppressed at negative IIDs, i.e., when a sound is less intense in the ipsilateral ear than in the contralateral ear; the IID cut-off in other neurons occurs when the ipsilateral sound is more intense than the contralateral sound (Fig. 1A). Within the IC on one side of the brain, IID cut-off values are distributed between -15 dB and +30 dB, but most neurons (over 80%) display IID cut-offs between -5 and +25 dB (Wenstrup, Fuzessery, and Pollak, in preparation). Thus, a small change in the IID value will affect the firing rate of a large proportion of EI neurons if these changes occur within the IID range -5 dB to +25 dB.

Studies of the topographic distribution of IID cut-offs show a systematic shift in IID cut-offs within the segregated 60 kHz EI region (Wenstrup et al., 1986). Positive IID cut-offs (suppression only when the ipsilateral sound is more intense than the contralateral) are found dorsally, with a progression to lower or negative IID cut-offs at ventral locations (Fig. 1B). These data suggest that the IID value generated by an echo results in a particular topographic pattern of activity within the EI region.

IIDs provide a potent sound localization cue, particularly in the horizontal plane. How might the orderly representation of IID cut-offs encode

329

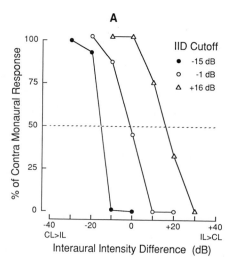 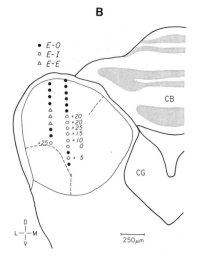

Fig. 1. A. Differences in IID cut-off among 60 kHz EI single units. Cut-off was defined as the particular IID required to suppress the contralateral response by 50%. Abbreviations: CL, contralateral; IL, Ipsilateral. B. Transverse section through IC showing systematic shift in IID cut-off within the 60 kHz EI region. Numbers to the right of open circles indicate the IID cut-offs of EI multi-unit responses.

the horizontal position of a target? Further studies sought to identify what features of an EI neuron's free-field response are influenced by the IID cut-off. These experiments compared the responses of single units to both closed- and free-field sounds (Fuzessery and Pollak, 1985; Wenstrup, Fuzessery, and Pollak, in preparation). Two features of a unit's spatial response are clearly not influenced by the IID cut-off. The most sensitive angle of a unit is defined as the location in the horizontal plane to which the unit responds at the lowest sound intensity. For the neurons in Fig. 2, the most sensitive angles are the same, 26° into the contralateral field, even though the IID cut-offs are quite different. The best angle is defined as the angular position in the horizontal plane from which sounds will evoke a near-maximum (75% or greater) response from the unit at the lowest intensity. Figure 2 shows that EI neurons with quite different IID cut-offs have the same best angle, located 26° into the contralateral field. The combined closed- and free-field studies show that the most sensitive angle and the best angle of 60 kHz EI neurons are uninfluenced by any binaural properties of these units, but rather are determined by the directional sensitivity of the external ear. It is thus unlikely that these response features can encode the location of a 60 kHz sound.

These studies demonstrated that the receptive field borders of EI units are strongly influenced by the binaural sensitivity to IIDs (Fig. 3). Neurons with IID cut-offs near 0 dB have receptive field borders located near the vertical midline in the free-field (Fig. 2, unit #2; Fig. 3). Neurons with higher (positive) cut-offs have receptive field borders located into the ipsilateral field (Fig. 2, unit #1; Fig. 3). It is thus proposed that a population of EI neurons, differing in IID cut-off, may encode horizontal location by the population of active neurons. For a sound positioned well into the ipsilateral field, only a few EI neurons will respond--those which are inhibited only by much louder ipsilateral sound. For a sound at the midline, many more EI neurons respond--only cells inhibited by less intense

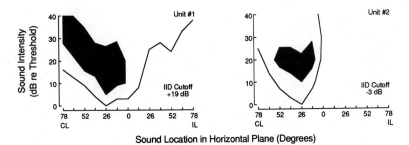

Fig. 2. Response profiles of 60 kHz EI neurons to sounds at different
horizontal locations. Data for two neurons are shown in the form
of threshold response contours (lines) and 75% maximum response
magnitude contours (blackened areas) as a function of both
horizontal sound position and sound intensity.

ipsilateral sound will be suppressed. Finally a sound well into contralateral field activates all EI neurons regardless of IID cut-off, since the
excitatory input is greatly favored over inhibitory input. Because IID cut-
offs are topographically arranged in the 60 kHz EI region, the position of a
sound in the horizontal plane is represented by the dorso-ventral extent of
activity within the 60 kHz EI region.

The neural mechanisms described above may not be exclusive to the 60
kHz region of the mustache bat's IC. Some data suggest that a similar
topographic organization of IID cut-offs exists among EI populations in
other isofrequency regions of the IC (Wenstrup et al., 1985). Thus, the
mustache bat's IC may analyze and systematically represent the IID values
within each spectral component of the biosonar echo. Some of the neural
mechanisms which underlie broadband echo localization may therefore be
understood by studying narrowband mechanisms like that described above.

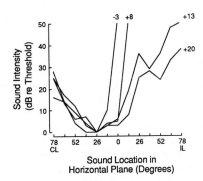

Fig. 3. Threshold response contours for
four 60 kHz EI units which differ
in IID cut-off (IID cut-offs in
upper right of figure).

BEHAVIORAL IMPLICATIONS

It is generally agreed that bats localize wideband signals more accurately than narrowband signals. However, many bats use relatively narrowband signals, particularly during the search phase of echolocation, and it
seems likely that they extract some information concerning target location
from these narrowband echoes. In both physiological and behavioral studies,
it seems important to distinguish the facets of echo localization that require wide signal bandwidth from those which need only one or a few spectral
components. Accordingly, this discussion focuses upon localization based
upon the narrowband neural mechanisms described above.

The above physiological studies principally sought to describe the
neural mechanisms underlying sound localization rather than the corresponding behavioral capabilities. The data consequently lack sufficient resolution for a precise estimate of behavioral acuity. Nonetheless, some rough
estimates can be made regarding the acuity of horizontal localization of a
60 kHz sound, the portions of the frontal sound field where acuity should be
greatest, and the limitations of the neural mechanisms proposed. Throughout
this discussion, I will consider only localization in the horizontal plane.

If the 60 kHz EI population encodes sound location, what factors shape
behavioral acuity? One factor is acoustic; the directional sensitivities
of the two ears determine where in the sound field IIDs are maximum and how
sharply the IID value changes with location. In the mustache bat, maximum
IID values for a 60 kHz sound are 25 to 30 dB; these maxima are located
near the horizontal midline, 26-52° to either side of the vertical midline
(Fuzessery and Pollak, 1985). The most rapid change in the IID occurs from
the midline to 26° on either side. Within this region the IID change averages about 0.85 dB per degree in the horizontal plane.

A second factor is of neural origin. EI neurons must be sensitive to
small changes in the IID value, i.e., they must have sufficiently sharp cut-
offs. For EI units with very sharp cut-offs, an IID change of 10 dB evokes
a change in firing rate from near-maximal to no response (e.g. the IID function on the left in Fig. 1A). If a 25% change in firing rate is used as a
criterion for detecting an IID change, such neurons could signal an IID
change of 2.5 dB. Based upon a 0.85 dB per degree change in the IID, individual neurons could thus signal a change in horizontal location of 3°. The
sharp directional sensitivity of the external ears improves the sharpness of
receptive field borders still further. In free-field studies, neurons with
less sharp IID cut-offs show a change in firing rate from maximal to no
response when a sound at a given intensity is moved by 13° in the horizontal
plane. Such data suggest that even neurons with less sharp cut-offs can
register changes in horizontal angle of roughly 3°.

A third factor influencing horizontal angle acuity is also neural. For
maximum acuity, this mechanism requires a distribution of IID cut-offs
throughout the IID range. Within the IC on one side of the brain, over 80%
of the neurons had IID cut-offs between -5 dB and +25 dB. These IID values
correspond to horizontal angles from about 6° in the contralateral sound
field to about 26° into the ipsilateral field. Together, both sides of the
IC will contain a distribution of topographically organized neurons with
cut-offs from 26° on one side to 26° on the other. Within this region,
covering the central portion of the frontal sound field, small changes in
horizontal position should influence the firing of a very large proportion
of 60 kHz EI neurons.

These considerations suggest that behavioral acuity should be best
within the central 50° of the frontal sound field, because the IID change
per degree is large, and because a large number of EI neurons have

reasonably sharp cut-offs at corresponding IID values. Bats should discriminate a difference in position of 3°, or possibly less. It is also suggested that acuity for horizontal localization within the central part of the frontal sound field may depend predominantly upon only a few spectral components of the echo. For example, IID contours at 60 kHz are very sharp, and there is a very large population of 60 kHz EI neurons available to analyze the IID. It is not clear how additional spectral components could improve this localization task. Another implication of the physiological results is that the EI population would show dramatic changes in activity in response to a target moving across the horizontal plane. Target movement could be encoded as a rapid change in the dorso-ventral extent of active 60 kHz neurons, perhaps within the duration of one echo.

Some difficulties become apparent for the localization of a narrowband 60 kHz signal. One problem arises from the distribution of IID values across the sound field in the horizontal plane. IID values decline from 40° to 90° in the lateral sound field. Thus, one IID value can apply to at least two positions in the horizontal plane, one in the medial sound field and one in the lateral field. With only a narrowband signal, horizontal localization by the mustache bat may suffer from a medial/lateral confusion. However, the directional nature of the sonar cry and the directional sensitivity of the ears may reduce such ambiguity.

Another difficulty concerns the manner in which the position of multiple sonar targets may be encoded within the topographically organized 60 kHz EI region. The inhibition evoked by a sound in the ipsilateral field will always be stronger than that by a sound in the contralateral field. Thus, the extent of active EI neurons, differing only in their cut-offs, would not reveal the presence of a second, contralaterally-located target. The solution to this problem probably requires diverse properties among EI neurons, other types of binaural neurons, and/or additional spectral components.

The above speculations are testable in behavioral studies. The ability to localize narrowband signals requires training bats to passively localize a narrowband sound source or to localize a filtered artificial echo. Such studies should test not only the acuity, but also the localization errors to which animals are subject under narrowband conditions. The role of signal bandwidth can then be examined by manipulating the bandwidth of the sounds or echoes to be localized.

ACKNOWLEDGEMENTS

I wish to thank my co-workers, Zoltan Fuzessery, George Pollak, and Linda Ross, who participated in many of the studies described here. Preparation of this manuscript was supported by NIH grant NS03377-02.

REFERENCES

Fuzessery, Z. M., and Pollak, G. D., 1985, Determinants of sound location selectivity in the bat inferior colliculus: a combined dichotic and free-field stimulation study, J. Neurophysiol., 54:757.

Wenstrup, J. J., Ross, L. S., and Pollak, G. D. 1985, Segregation of binaural responses in isofrequency sheets of the mustache bat inferior colliculus, Eighth Midwinter Meeting of the Association for Research in Otolaryngology (Abstracts), p. 20.

Wenstrup, J. J., Ross, L. S., and Pollak, G. D. 1986, Binaural response organization within a frequency-band representation of the inferior colliculus: implications for sound localization, J. Neurosci., 6:962.

ONTOGENY OF THE ECHOLOCATION SYSTEM IN RHINOLOPHOID CF-FM BATS:

AUDITION AND VOCALIZATION IN EARLY POSTNATAL DEVELOPMENT

Rudolf Ruebsamen

Lehrstuhl Allgemeine Zoologie
Ruhr-Universitaet Bochum, ND|6, Postfach 10 21 48
4630 Bochum, Federal Republic of Germany

The offsprings of the Old World bats Rhinolophus rouxi and Hipposideros speoris are born in a relatively immature state of development. During the first postnatal week these bats do not hear, but when separated from their mothers they produce multiharmonic isolation calls. These vocalizations differ conspicuously from the echolocation pulses of the adult, which in both species are higher pitched and are composed of an initial constant frequency component (CF) followed by a brief decreasing frequency modulation (FM), ranging about 15-20 kHz, emitted through the nose (nasal calls).

The isolation calls of young horseshoe bats (Rh. rouxi) are emitted through the mouth (buccal calls), and, in the course of the first to the third postnatal week, the vocalized frequencies undergo continuous increase (from 7 kHz up to 25 kHz for the fundamental frequency; Fig.1). With the onset of hearing in the second week, as indicated by the recording of evoked potentials in the inferior colliculus, the maximum sensitivity ranges between 15 and 25 kHz. During development, higher frequencies become successively incorporated into the hearing domain. At the beginning of the third week a narrowly tuned sensitivity maximum (auditory filter) emerges at frequencies between 55 and 60 kHz. Concurrently the same individuals start to emit nasal vocalizations resembling echolocation calls. These pulses are longer (30-50 ms vs 15-20 ms) and higher pitched, and show no harmonic resemblance to the isolation calls (Fig. 1). From the very first occurrence of the nasal calls their second harmonics are exactly matched to the auditory filter. In the third week the young bats first show persistent locomotive behaviour, when left behind inside the cave. If they fall from the ceiling of the cave they immediately orient themselves towards the nearest wall and start to climb it. This reaction probably requires echolocation, a suggestion that is undoubtedly true for four week old bats which have started to fly inside the cave. In the course of the third to the fifth postnatal week the auditory filter frequency and the frequency of echolocation pulses are jointly increased (Fig. 2), thereafter reaching the values typical for adult bats (73-77 kHz). Thus the echolocation system of young rhinolophids is functioning at lower frequencies than those manifested in adult bats.

The comparison of frequency shifts in the isolation and the echolocation calls to the development of audition (obtained from the same individuals) most plausibly leads to the suggestion of different reasons for changes in vocalized pitch: the frequency of the buccal calls (isolation calls) may be

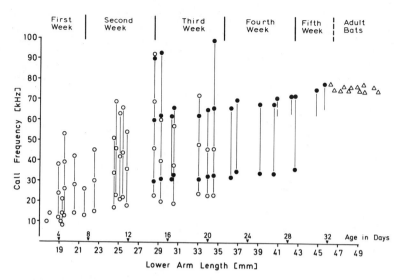

Fig.1 Shift in the vocalized call frequencies during postnatal development of horseshoe bats. Given on the abscissa are the forearm lengths and the ages of the bats. Open circles - frequency bands of isolation calls; filled circles - frequency bands of echolocation calls. Frequency bands found in the vocalizations of single bats are connected by vertical lines. Notice that during the third week both kind of calls are vocalized by the same individuals. Triangles - CF-components of echolocation pulses of adult rhinolophids.

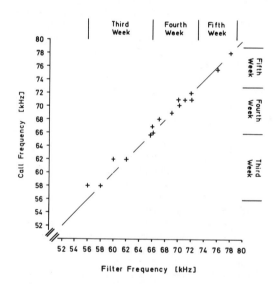

Fig.2 Comparison of the tuning of the high-frequency auditory filter with the frequency of echolocation calls in the course of the third to the fifth postnatal week. Each cross represents measurements for a single bat.

altered due to the maturation of the sound producing organs, whereas the nasal vocalizations (echolocation calls) might be produced under the frequency feedback control of the auditory system.

The postnatal development of audition raises the question of possible structural alterations of the nuclei of the central auditory system, especially of shifts in their internal frequency representations. This problem was tackled by stereotaxic recordings of auditory evoked multi-unit and single-unit activities in the inferior colliculus (IC) of young Hipposideros speoris, CF-FM bats which undergo postnatal development in vocalization (Habersetzer and Marimuthu, 1986) and audition comparable to that described for the rhinolophids.

In young H. speoris of different ages pure tone thresholds were obtained from systematic multi-unit recordings in the IC, which were integrated in stereotaxic frequency maps, covering the rostrocaudal and mediolateral extent of this midbrain nucleus. These maps were assigned, by the aid of constant current lesions, to three-dimensional morphometrical brain reconstructions yielded from serial brain sections.

At the end of the second postnatal week the audiogram of H. speoris is restricted to frequencies between 20 and 80 kHz and the minimum threshold values range between 5 and 15 dB SPL. Auditory responses can be evoked within the entire IC, but in terms of tuning properties there are two distinctly different areas: In the dorsal parts of the IC, 500-700 um in the rostrodorsal extent, and positioned like a cap on top of this ovoid-shaped nucleus, tonotopic organization can be determined with dorsoventrally increasing best frequencies from 20-80 kHz. Further ventrally, the orderly frequency representation abruptly ends and becomes replaced by areas which also display auditory excitability, but only at higher intensities with very broad tuning characteristics (mostly 40-60 kHz). This untuned and insensitive part of the IC initially covers more than two third of its total volume (Fig. 3).

In the course of development during the 3rd and the 4th weeks the tonotopically organized and more sensitive dorsal and dorsolateral regions are successively extended further ventrally, at the expense of the insensitive part of the nucleus. By this process higher frequencies are progressively incorporated into the tonotopic area. This development, additionally attended by further successive decrease of threshold values in the tuned area, is completed by the 5th postnatal week, when the entire IC shows coherent tonotopic organization (Fig. 4). But, in these young bats audition still does not cover the entire frequency range of the adult, and the frequency representation within the IC also differs from that of the adult. The general organization of the isofrequency contours is comparable to that of mature hipposiderids, and as in the adult the isofrequency contours fan out in the ventral and caudal parts of the nucleus causing an overrepresentation of a limited frequency band, 10-15 kHz in width. However, this filter area is tuned to frequencies 10-20 kHz below the filter frequencies of the adult.

The maturation of the tonotopic organization of the IC, which, as described above, follows a dorsoventral course, appears to proceed predominantly by repeated addings of sensitive IC-slabs successively tuned to higher frequencies, i.e. a given region within the nucleus does not change its tuning during this process. From the 5th to the 7th week (in the course of which the bats are volant and already can be found flying outside the cave) subsequent development is confined to the filter region, which in its entirety changes the tuning, i.e. a given region of the IC changes its best frequency. During this shift the frequency range established in the overrepresented filter region stays roughly constant, covering 10-15 kHz (see also Neuweiler et al., this volume). Differences in the tuning of single units confirms the particular position and significance of the filter region in the

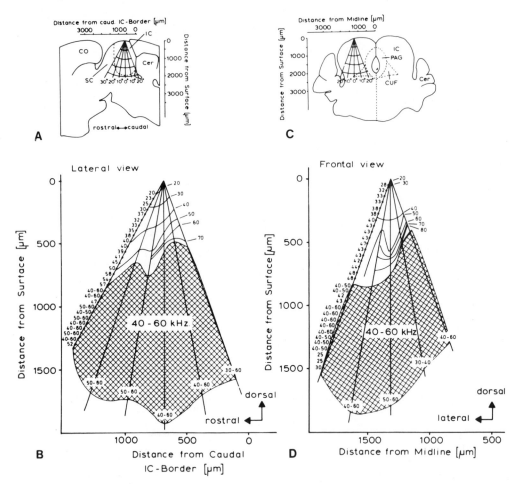

H. speoris 11

(FAL 26.3 mm)

Fig.3 Tonotopic organization of the IC in a two week old Hipposideros speoris
A.) Sketch of a parasagittal brain section demonstrating the borders of
the IC and the orientation of the six electrode penetrations by the aid
of which the stereotaxic frequency map in B is evaluated.
B.) Stereotaxic frequency map demonstrating the tonotopical organization of the IC in a lateral view. The electrode tracts are shown by
straight lines fanning from a single dorsal point. Representative for
the general recording procedure, the best frequencies are shown at the
leftmost penetration, measured at distances of 50 um. The solid curved
lines which cross the electrode tracts indicate isofrequency contours
in steps of 10 kHz, with the respective frequency values given on the
right side. The hatched area indicates locations of broadly tuned
recordings.
C.) Sketch of a frontal brain section demonstrating the borders of the
IC and the orientation of the five electrode penetrations by the aid of
which the stereotaxic frequency map in D is evaluated.
D.) Stereotaxic frequency map demonstrating the tonotopical organization of the IC in a frontal view. Design of the graph as in B.

H.speoris 13

(FAL 37.5mm)

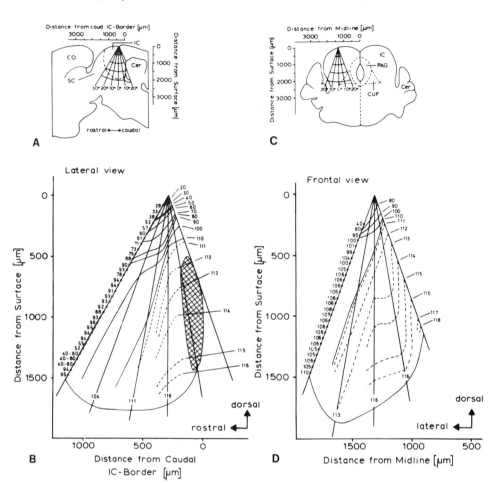

Fig.4 Tonotopic organization of the IC in a five week old Hipposideros
speoris.
A.) Sketch of a parasagittal brain section demonstrating the borders of
the IC and the orientation of the six electrode penetrations by the aid
of which the stereotaxic frequency map in B is evaluated.
B.) Stereotaxic frequency map demonstrating the tonotopical organization of the IC in a lateral view. Design of the graph as described in
B.
C.) Sketch of a frontal brain section demonstrating the borders of the
IC and the orientation of the five electrode penetrations by the aid of
which the stereotaxic frequency map in D is evaluated.
D.) Stereotaxic frequency map demonstrating the tonotopical organiza-
tion of the IC in a frontal view. Design of the graph as in B.

auditory system: with their first appearance, single units with best frequencies above 100 kHz are more sharply tuned than units with lower best frequencies.

It is shown here, for both rhinolophid and hipposiderid bats, that the echolocation system is established before audition develops to the adult state. Consequently it must be postulated that in the juvenile state the central nervous processing of auditory signals and the motor control system for vocalization must be functional. Moreover both subsystems must already be linked internally by an intermediate sensory-motor interface. The reason for the postnatal frequency shift in audition may be changes in the transfer function of the peripheral sensory structures, i.e. the middle ear and the inner ear. In rhinolophids for instance, a protracted maturation of the organ of Corti has been observed in the early postnatal development of these bats (Vater, this volume).

ACKNOWLEDGEMENT: This work was supported by the DFG as part of the SFB 114 and SFB 208, by the Joint Bat Research Project of the University of Kelaniya and of the University of Munich, and by the Indo-German Project of Animal Behaviour, Madurai Kamaraj University.

REFERENCES

Habersetzer, J., Marimuthu, G. (1986) Ontogeny of sounds in the echolocating bat Hipposideros speoris. J. Comp. Physiol. 158:147-157

Vater, M (1987) Lightmicroscopic observations on cochlear development in horseshoe bats. (This volume)

LIGHTMICROSCOPIC OBSERVATIONS ON COCHLEAR DEVELOPMENT

IN HORSESHOE BATS (Rhinolophus rouxi)

Marianne Vater

Zoologisches Institut
Luisenstr. 14
8000 München 2 FRG

INTRODUCTION

In horseshoe bats, the enhanced frequency selectivity within a narrow frequency range around the CF-component of the echolocation call is established in the cochlea (review: Neuweiler et al., 1980; Bruns, 1979; Vater, this volume). The frequency place maps of the inner ear of adult bats clearly demonstrate a correlation of the representation of the sharply tuned frequency range with the unusual thickness and width profile of the basilar membrane (BM) (Bruns, 1976, Vater et al., 1985), suggesting that the exceptional filter capacities of the auditory system are due to specialized hydromechanical mechanisms.

Several workers report that in young horseshoe bats, the hearing ability is restricted to the low frequency range and the hearing range progressively expands towards higher frequencies during postnatal development (Konstantinov, 1973; Rübsamen, 1986) parallel to a shift in frequency of the ultrasonic calls (Konstantinov, 1973; Matsumura, 1979), Rübsamen, 1986). Most interesting, already in the neurophysiological audiograms recorded from inferior colliculus of young bats, an indication of a sharply tuned region corresponding to the individuals CF-component is found, which however ranges about half an octave below the adult range (Rübsamen, 1986).

This study of cochlea morphology during development attempts to define the structural changes in the specialized hydromechanical system which parallel the pronounced changes in physiological characteristics of the auditory system.

MATERIAL AND METHODS

Cochleae of Rhinolophus rouxi of different developmental stages were obtained from specimens used in a physiological study of ontogeny of echolocation system in horseshoe bats (Rübsamen, 1986). Age encompassed bats of postnatal day 1-4 (reported to be deaf); and postnatal days 9-12; 13-16; 17-20; 21-24; 25-28; 29-32, during which the hearing characteristic develops. At least two cochleae of each stage were examined. Fixation and histology followed standard procedures used for lightmicroscopy. Frozen sections (42 μm) were used for analysis of BM-length and extent of morphological specializations; paraffin sections (14 μm) were used for measurement of

BM-thickness and -width; plastic sections (2 μm) were used for analysis
of cytological details. Additionally, the cochlear dimensions were analyzed
in a prenatal bat (unknown day of gestation).

The cochleae were three-dimensionally reconstructed according to the
procedure described in Vater et al.(1985) to obtain absolute BM-length
and longitudinal BM-thickness and -width profiles.

RESULTS

Gross morphological features

Ventral views of the skull and schematic crossections through the
cochlea of three age stages (prenatal; 1 week; adult) are given in Fig.1.
The size of the cochlea has already reached adult dimensions in the prenatal bat, although the skull itself is far from mature. The cochlea of
the prenatal bat is fully grown to the 7 half turns found in the adult
and crossections of the cochlea reveal that the characteristic BM-thickenings in the first half turn and the specializations of outer attachment
structures (secondary spiral lamina) in first and second half turn (arrows)
are already present.

In line with this evidence, a comparison of BM-length among different developmental stages shows that no pronounced or systematic change
in length is occuring. The cochlea from the prenatal bat is as long
(15.2 mm) as the cochlea of a specimen aged 21-24 days. The maximal length
difference of 9% between age 1-4 days (14.04 mm) and age 25-28 days
(15.4 mm) is within the variability caused by interindividual differences,
histological preservation and reconstruction technique. Maximal variation
is only 2.5% (i.e. 380 μm) among cochleae within the critical postnatal
time period, where pronounced changes in hearing characteristic occur (stage
9-12 to 25-28 days).

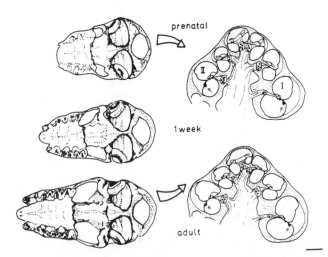

Fig. 1. Comparison of skull and cochlea size at different developmental
        stages in Rhinolophus rouxi demonstrating absence of growth in
        cochlear dimensions. left: ventral view of skulls; right: sche-
        matic midmodiolus sections through the cochlea, arrows indicate
        secondary spiral lamina in first and second half turn. Calibra-
        tion bars: left 2 mm; right 0.5 mm.

Histological aspects

The histological aspects of cochlear development during early postnatal life were closely examined in crossections of the organ of Corti.
Schematic drawings for first and second half turn are given for two age
stages in Fig.2. The description will focus on the main mechanical elements of the cochlear duct.

No differences are seen in shape and dimension of the BM among bats of
different age other than the presence of a huge spiral bloodvessel in the
pars tecta (PT) of the youngest specimen, which is closed at later stages.
Changes are however apparent in the filament composition of BM. Within
pars pectinata (PP), radial filament strands become more pronounced and
sharper delineated from the homogenous matrix with increasing age.

Prominent alterations take place in the constitution of the epithelial cell layer covering the tympanic side of BM (tympanic cover layer,
TL). This layer is progressively reduced with age. The reduction appears
to follow a baso-apical gradient with basal regions of the cochlea being
the first to loose the epithelial cover.

In all postnatal specimens, the tunnel of Corti is open throughout all
cochlea turns and size and inclination of pillar cells do not exhibit noticeable changes. In the youngest bat, the fibre material in the pillar shanks
is less differentiated than in later life.

Some observations on tectorial membrane (TM) are worth mentioning. In
the youngest specimen, TM is preserved in a close position above the organ
of Corti by extensions just lateral to the third row of outer hair cells
in all turns. In later development, the TM is shrunk away from the organ
of Corti in the first half turn, while histological procedure apparently
did not cause much shrinkage in more apical regions. This indicates a very
strong attachment of TM in early life and might also suggest differences
among cochlear turns in strength of its coupling to the receptor cells
in later life.

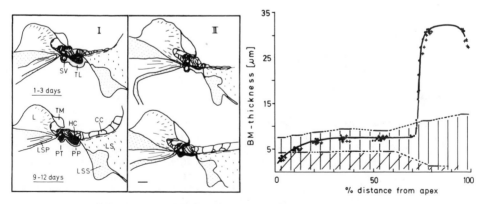

Fig. 2. left: Crossections through the organ of Corti in first half
turn (I) and second half turn (II) for two developmental
stages. CC Claudius cells; HC Henson cells; LS spiral ligament; LSP primary spiral lamina; LSS secondary spiral lamina;
PT pars tecta; PP pars pectinata; TM tectorial membrane; TL
tympanic cover layer; SV spiral vessel. Calibration bar 50 μm
Right: Thickness of basilar membrane (thick line and different
symbols) and thickness of tympanic cover layer (thin lines and
shaded areas) vs. BM-length given as percent distance from
apex. Further details see text.

343

Although the maturational state of hair cells can not be assessed with
the lightmicroscope, it is interesting to note the change in appearance of
outer hair cell nuclei with age. In the youngst bat, the nuclei are pale
and large throughout the cochlea. The size of the nuclei decreases in subsequent stages of age, again progressing from base to apex.

Quantitative measurements of BM-thickness and -width were made for the
postnatal cochleae representative for the time period of changing physiological properties (Rübsamen, 1986). In Fig.2, BM-thickness for two individuals of age 1-4 days (different symbols) is given together with the idealized BM-thickness curve (thick line) averaged across 7 cochleae encompassing the later stages of development and including one adult cochlea. It
is clearly shown that the cochleae of the youngest postnatal bats already
possess the BM-thickness profile characteristic for later stages of development. Also given in Fig.2 is the thickness of the tympanic cover layer for
developmental stage 1-4 (vertical shading) and 25-28 (oblique shading).
Note that the tympanic cover layer in the youngest bats is as thick as the
BM in second or more apical turns. It is considerably decreased in the
older specimen and nearly completely reduced in first half turn. Furthermore, BM-width was not found to change significantly with age. Comparison
of BM-width in radial sections through identical regions of first and second
half turn gave values of 108 μm and 90 μm for the youngest bat as compared
to 106 μm and 88 μm in the oldest bat respectively.

DISCUSSION

Cochlear morphology alone can certainly not account for the details
and exact time course of the change in physiological response of the auditory system in early postnatal life, since the establishment of adult hearing characteristics is a complex function of maturation processes of
middle ear, inner ear and central auditory pathway (review Romand, 1983).
Furthermore, the discussion of cochlear factors is necessarily limited
to morphological changes in hydromechanical parameters and has to leave
out the possible influences of immaturity of receptor cells and their
biochemical environment.

The cochlea of newborn horseshoe bats has already reached a remarkably
advanced level of differentiation and its cytological features are more
mature than those of newborn cats or rodents. Most important, the species
specific specialized BM-thickness and -width profile is established in
final dimensions at birth and morphological changes during early life are
restricted to structural details of the hydromechanical system like size
of the tympanic cover layer, filament composition of BM and pillar cells
and attachment of tectorial membrane. The change in physiological response
characteristics in early postnatal life of Rhinolophus can therefore not
be attributed to a simple mechanism of growth or gradual establishment
of gross mechanical specializations of the cochlea. Structural details
of BM and pillar cells of young bats might however indicate a less stiff
mechanical system, which additionally is loaded by the mass of the tympanic cover layer. Consequently, as discussed for other mammals (e.g. Kraus
and Kraus-Aulbach, 1981; Harris and Dallos, 1984) a lower BM-resonance
frequency than in the adult is expected.

In early postnatal life of Rhinolophus, similar to other species,
physiological responses are restricted to a lower frequency range than
in the adult, although the basal (i.e. high frequency) region of the cochlea
has reached maturity first. For other species, an age dependent shift in
cochlear frequency-place code has been proposed as a solution for this
paradoxon (Rubel et al., 1984; Harris and Dallos, 1984). A possible shift
in frequency-place code has interesting implications for Rhinolophus.

During development, the resonance frequency of the specialized mechanical system would then be matched to the young bats CF-components which are much lower than in the adult, gradually shifting upward with age. This attractive hypotheses needs to be tested in studies of single unit responses and the cochlear frequency place map during development.

ACKNOWLEDGEMENT

I want to express my gratitude to the members of the Sonderforschungsbereich for helpful criticisms and discussions. Thanks also to H. Kaupenjohann for technical assistence, to I. Patzak and S. Mommertz for their help in data analysis and to H. Tscharntke for photographical work. This project was supported by the SFB 204 "Gehör" München.

REFERENCES

Bruns, V., 1976, Peripheral auditory tuning for fine frequency analysis by the CF-FM bat, Rhinolophus ferrumequinum. II. Frequency mapping in the cochlea, J. comp. Physiol., 106:87.

Bruns, V., 1979, Functional anatomy as an approach to frequency analysis in the mammalian cochlea, Verh. Dtsch. Zool. Ges., 72:141.

Harris, D. M. and Dallos, P., 1984, Ontogenetic changes in frequency mapping of a mammalian ear, Science 225:741.

Kraus, H.-J., and Aulbach-Kraus, K., 1981, Morphological changes in the cochlea of the mouse after the onset of hearing, Hearing Res., 4:89.

Konstantinov, A. I., 1973, Development of echolocation in bats in postnatal ontogenis, Period. Biol., 75:13.

Matsumura, S., 1979, Mother-infant communication in a horseshoe bat (Rhinolophus ferrumequinum nippon): Development of vocalization. J. Mammolog., 60:76.

Neuweiler, G., Bruns, V., and Schuller, G., 1980, Ears adapted for the detection of motion, or how echolocating bats have exploited the capacities of the mammalian auditory system, J. Acoust. Soc. Amer.,68:741.

Romand, R., 1983, Development of the cochlea. in: "Development of auditory and vestibular systems," R. Romand, ed., Academic Press, New York.

Rübsamen, R., 1986, Ontogeny of the echolocation system in the rufus horseshoe bat, Rhinolophus rouxi. This volume.

Rubel, E. W., Lippe, W. R., and Ryals, B. M., 1984, Development of the place principle, Ann. Otol. Rhinol. Laryngol., 93:609.

Vater, M., 1986, Cochlear physiology and anatomy in bats. This volume.

Vater, M., Feng, A. S., and Betz, M., 1985, An HRP-study of the frequency-place map of the horseshoe bat cochlea: Morphological correlates of the sharp tuning to a narrow frequency band, J. comp. Physiol., 157:671.

ONLY ONE NUCLEUS IN THE BRAINSTEM PROJECTS TO THE

COCHLEA IN HORSESHOE BATS: THE NUCLEUS OLIVO-COCHLEARIS

Joachim Ostwald and Andreas Aschoff

Lehrbereich Zoophysiologie
Auf der Morgenstelle 28
D-7400 Tübingen, FRG

Investigations in the cat, rat, and guinea pig have shown that two separate populations of neurons form the efferent innervation of the cochlea (Warr and Guinan, 1979; White and Warr, 1983; Strutz and Bielenberg, 1984). Auditory nerve fibers at the base of inner hair cells (IHC) are innervated by small neurons that are located in the region of the lateral superior olive (LSO) predominantly or exclusively on the ipsilateral side. In rat and guinea pig, these neurons are found within the LSO, in cat they are located close to the LSO mainly in its dorsal hilus. Because of the position of these neurons in the olivary complex it was called lateral system. Outer hair cells (OHC) receive their efferent innervation from a group of large neurons in several periolivary nuclei. This medial system is organized bilaterally in all species studied so far.

One striking feature in the cochlea of Horseshoe Bats is the total lack of efferent innervation of OHC as has been shown in electron microscopic studies of Bruns and Schmieszek (1981) in Rhinolophus ferrumequinum and Bishop (1986; this volume) in Rhinolophus rouxi. We were interested whether this special feature of the cochlea is also reflected in the brainstem structures of the olivo-cochlear efferent system.

Experiments were performed in Rhinolophus rouxi. The fluorescent tracer Fast Blue (FB) was implanted into the cochlea through a small hole. After a survival time of up to seven days brains were fixated and cut in serial sections. FB is taken up by the terminals in the cochlea and transported retrogradely to the cell bodies in the superior olivary complex where it can be seen with the aid of a fluorescent microscope.

In contrast to all other species studied so far, in Rhinolophus rouxi only one group of neurons in the superior olivary complex was labeled. They comprise a separate T-shaped nucleus of small, densely packed cells between the medial (MSO) and lateral (LSO) superior olivary nucleus (Fig. 1) which we call nucleus olivo-cochlearis (NOC). Labeled cells were found exclusively ipsilateral to the implantation side. Outside the NOC, no neurons were labeled, neither in the periolivary region nor in the LSO. So there is no second system of olivo-cochlear efferents as it has been described in all other species. In sections stained for acetycholinesterase activity, these neurons in the NOC show up as a dense cluster of AChE-positive cells, which can be found in Rhinolophus rouxi as well as in Rhinolophus ferrumequinum.

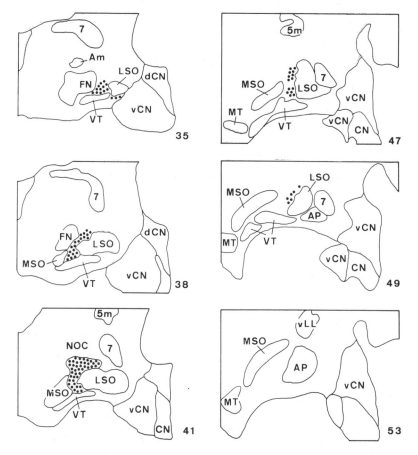

Fig. 1. Distribution of FB-labeled neurons in transverse sections of the brainstem of <u>Rhinolophus rouxi</u> after tracer implantation into the right cochlea. Appreviations: Am: ambigual nucleus; AP: anterior periolivary region; CN: cochlear nucleus; FN: facial nucleus; LSO: lateral superior olive; MSO: medial superior olive; MT: medial nucleus of the trapezoid body; NOC: nucleus olivo-cochlearis; vLL: ventral nucleus of the lateral lemniscus; VT: ventral region of the trapezoid body; 5m: motor nucleus of the trigeminal nerve; 7: descending root of the facial nerve.

Olivo-cochlear efferents in Horseshoe Bats share many features with the lateral efferent system of other species; neurons are small and located close to the LSO and they innervate auditory nerve fibers at the base of IHC. In <u>Rhinolophus</u> there seems to be a selective loss of the medial system of periolivary neurons innervating the OHC. This is not a general feature of the olivo-cochlear system of bats or of bats with a CF-FM echolocation sound since in all other bat species studied both, the lateral and medial systems, are present (Artibeus, Phylostomus, Rhinopoma, Tadarida: Aschoff and Ostwald, 1984; submitted; Pteronotus: Bishop, 1986).

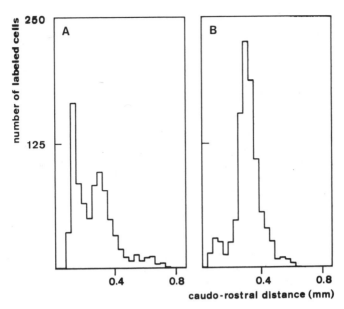

Fig. 2. Number of FB-labeled neurons in transverse sections of the brainstem of <u>Rhinolophus rouxi</u> after implantation of the tracer into the middle cochlear turn (A) or basal cochlear turn (B).

The functional significance of this loss of efferent innervation to the OHC remains unclear. Stimulation of OHC via efferent innervation seems to alter the mechanical tuning properties of the basilar membrane by changing the amount of its coupling to the tectorial membrane (reviewed in Dallos, 1985). In a mechanically highly specialized cochlea like that of rhinolophids (Bruns, 1976; Vater et al., 1985; reviewed in Vater, this volume) changing the tuning properties might not be of advantage.

In <u>Pteronotus parnelii</u> olivo-cochlear efferents of the lateral system are also concentrated in a densely packed nucleus between LSO and MSO (Bishop, 1986; this volume). It has been called interstitial nucleus by Zook and Casseday (1982). Since it serves the same function as the NOC in <u>Rhinolophus</u> we suggest to call it olivo-cochlear nucleus, too.

When the tracer was implanted into the basal or middle cochlear turn different rostro-caudal distributions of labeled neurons in the NOC were found. Neurons innervating the basal cochlear turn are found rostrally in the NOC, those innervating the middle turn are found in the caudal end of NOC, resulting in a topographical axis oriented from caudal to rostral. Fig. 2 illustrates examples of these two distributions. The bimodal distribution of labeled neurons is probably due to the spread of tracer from one cochlear turn to the next. As medial regions of the cochlea are already innervated by very caudal parts of NOC IHC in the apical regions of the Horseshoe Bats' cochlea seem to be less densely innervated than those in the basal region.

(Supported by Deutsche Forschungsgemeinschaft, SFB 307. We thank E. Friauf for critical discussion and I.Kaipf and H.Zillus for technical help.)

REFERENCES

Aschoff, A. and Ostwald, J., 1984, The origin of descending and ascending connections of the superior olivary complex in rat, guinea pig and bat. Neurosci.Lett., Suppl. 18,243

Aschoff, A. and Ostwald, J., submitted, Different origin of cochlear efferents in some bat species, rat and guinea pig. J.Comp.Neurol.

Bishop, A., 1986, The olivocochlear system in Doppler-shift compensating bats. Ph.D.Thesis. University of North Carolina, Chapel Hill

Bruns, V., 1976, Peripheral auditory tuning for fine frequency analysis by the CF-FM bat, Rhinolophus ferrumequinum. I. Mechanical specializations of the cochlea. J.Comp.Physiol. 106,77-86

Bruns, V. and Schmieszek, E., 1981, Cochlear innervation in the Greater Horseshoe Bat: demonstration of an acoustic fovea. Hear.Res. 3,27-43

Dallos, P., 1985, The role of outer hair cells in cochlear function, in: "Contemporary Sensory Neurobiology", Alan R. Liss, pp.207-230

Strutz, J. and Bielenberg, K., 1984, Efferent acoustic neurons within the lateral superior olivary nucleus of the guinea pig. Brain Res. 299,174-177

Vater, M., Feng, A.S., and Betz, M., 1985, An HRP-study of the frequency-place map of the horseshoe bat cochlea: Morphological correlates of the sharp tuning to a narrow frequency band. J.Comp.Physiol.A 157,671-686

Warr, W.B. and Guinan, J.J., 1979, Efferent innervation of the organ of corti: two separate systems. Brain Res. 173,152-155

White, J.S. and Warr, W.B., 1983, The dual origin of the olivo-cochlear bundle in the albino rat. J.Comp.Neurol. 219,203-214

Zook, J.M. and Casseday, J.H., 1982, Cytoarchitecture of auditory system in lower brainstem of the Mustache Bat, Pteronotus parnelii. J.Comp.Neurol. 207,1-13

SECTION III

PERFORMANCE OF ANIMAL SONAR SYSTEMS

Section Organizers: James Simmons and Hans-Ulrich Schnitzler

THE PERFORMANCE OF ECHOLOCATION: ACOUSTIC IMAGES PERCEIVED BY BATS
  James A. Simmons and Alan D. Grinnell

DESIGNING CRITICAL EXPERIMENTS ON DETECTION AND IN ECHOLOCATING BATS
  Dieter Menne

TARGET DISCRIMINATION AND TARGET CLASSIFICATION IN ECHOLOCATING BATS
  Joachim Ostwald, H. -Uli Schnitzler and Gerd Schuller

TARGET DETECTION BY ECHOLOCATING BATS
  Bertel Møhl

SONAR TARGET DETECTION AND RECOGINITION BY ODONTOCETES
  Whitlow W. L. Au

PREY INTERCEPTION: PREDICTIVE AND NONPREDICTIVE STRATEGIES
  W. Mitchell Masters

A MECHANISM FOR HORIZONTAL AND VERTICAL TARGET LOCALIZATION IN THE MUSTACHE
BAT Pteronotus p. parnelli
  Zoltan M. Fuzessery

ECHOES OF FLUTTERING INSECTS
  Rudi Kober

ENCODING OF NATURAL INSECT ECHOES AND SINUSIOIDALLY MODULATED STIMULI BY
NEURONS IN THE AUDITORY CORTEX OF THE GREATER HORSESHOE BAT, Rhinolophus
ferrumequinum
  Joachim Ostwald

DO SIGNAL CHARACTERISTICS DETERMINE A BAT'S ABILITY TO AVOID OBSTACLES?
  Albert S. Feng and Karen Tyrell

GREATER HORSESHOE BATS LEARN TO DISCRIMATE SIMULATED ECHOES OF INSECTS
FLUTTERING WITH DIFFERENT WINGBEAT RATES
  Gerhard von der Emde

PREDICTIVE TRACKING OF HORIZONTALLY MOVING TARGETS BY THE FISHING BAT,
NOCTILIO LEPORINUS
  Karen A. Campbell and Roderick A. Suthers

DISCRIMINATION OF TARGET SURFACE STRUCTURE IN THE ECHOLOCATING BAT,
Megaderma lyra
  Sabine Schmidt

A TIME WINDOW FOR DISTANCE INFORMATION PROCESSING IN THE BATS, Noctilio
albiventris and Rhinolophus rouxi
  Roald C. Roverud

THE PERFORMANCE OF ECHOLOCATION:

ACOUSTIC IMAGES PERCEIVED BY ECHOLOCATING BATS

James A. Simmons

Section of Neurobiology and
Department of Psychology
Brown University
Providence, RI   02912

Alan D. Grinnell

Jerry Lewis Center
University of California
Los Angeles, CA   90024

INTRODUCTION

   The performance of echolocation in bats is a measure of the
effectiveness of this system of orientation for representing significant
features of objects in perceptual images. Performance refers specifically
to the quality and content of these images--to the features of targets
which bats can perceive as distinctive and to the acuity of perception of
these features under different circumstances. A description of the
performance of echolocation is, in effect, a description of what bats can
"see" with their sonar.

   The primary data about the performance of echolocation necessarily come
from behavioral experiments which relate the features of targets or echoes
as stimuli to the perceptions that bats actually achieve. In the absence
of behavioral data, one can make inferences about performance from the
results of physiological experiments on the bat's auditory system or from
descriptions of the sonar signals themselves in the context of various
signal-processing strategies. However, physiological data are atomistic in
nature and do not convey a complete picture of the overall performance of a
perceptual system without the support of behavioral data. Furthermore,
descriptions of potential signal-processing operations require the support
of both behavioral and physiological data for their confirmation.
Conversely, it is also true that behavioral data can lead only to
inferences about the underlying mechanisms unless constrained by critical
physiological observations and a cohesive description of the signal-
processing operations involved. This chapter will review what is known at
present about the performance of echolocation in bats in relation to the
images that bats perceive and the mechanisms which produce these images.

ACOUSTIC IMAGES IN ECHOLOCATION

   Echolocation is fundamentally a mode of spatial perception. Bats can
detect, track, identify, and intercept airborne targets with an efficiency
that can only be accounted for if they perceive multidimensional spatial
images that explicitly portray the distance, direction, and velocity of
targets (Schnitzler and Henson, 1980; Simmons and Kick, 1983). Bats also
can distinguish between targets on the basis of size, shape, and movement,
so these qualities must be incorporated into the images as well (Schnitzler
et al., 1983; Simmons, in press; see Ostwald et al., this volume).

Although the content of the images perceived by bats is expressed in spatial terms, these images in fact are derived from acoustic information conveyed in echoes reaching the bat's ears. This acoustic information represents the effects of propagation, reflection, and scattering on the signals transmitted by the bat and broadcast into the immediate environment (Lawrence and Simmons, 1982a; Pye, 1980). To be reconstituted into images of targets, it must be transformed back into spatial information by the bat's sonar receiver.

The transformation of acoustic information in echoes into spatial information about targets is achieved by the bat's auditory system. The initial stages of the auditory representation of echoes are based on neural response properties that are much like those found in mammals generally (Suga, 1973). However, at higher levels of the bat's auditory system, neurons often exhibit responses driven more sharply by the spatial properties of targets than by individual acoustic parameters of sounds. To ensure the integrity of the eventual representation of spatial features of targets, some of the response properties found at more peripheral stages of the auditory pathways explicitly act to preserve the spatial information inherent in the acoustic features of echoes (Bodenhamer and Pollak, 1981; Grinnell, 1963a; Kick and Simmons, 1984; Lawrence and Simmons, 1982b; Neuweiler et al., 1980; Suga and Schlegel, 1972; Suga and Shimozawa, 1974). These are examples of signal-processing operations by which the bat's auditory system functions as a sonar receiver. Other auditory processes go beyond just preserving spatial information conveyed in neural responses to echoes so that they actually enhance the quality of this information (Simmons and Kick, 1984). Finally, neurons in the bat's auditory cortex systematically express spatial features of targets in terms of the distribution of activity over segments of tissue, in effect representing the images themselves (Feng et al., 1978; O'Neill and Suga, 1982; Suga, 1984; Suga and Horikawa, 1986; for example). Neural representations of spatial features of targets constitute the sonar display for echolocating bats, and the primary concern in seeking to understand the mechanisms of echolocation is to elucidate how these displays are produced and how they are read. The performance of echolocation is an important source of knowledge about these mechanisms because it allows us to distinguish between elements of the images that bats perceive and elements of the representation of these images in neural displays (Simmons, in press).

Acoustic information that arrives in echoes and is conveying spatial information about targets is labeled by its specific acoustic qualities, and the subsequent fate of these acoustic qualities serves as a tracer for determining how the corresponding spatial information itself is treated by the sonar receiver. The crucial initial distinctions between the two principal types of echolocation found in bats--FM and CF echolocation--were made in behavioral experiments that exemplified this strategy (Schnitzler, 1968, 1973; Schuller et al., 1974; Simmons, 1973, 1974), and major features of the neural mechanisms which underlie these two systems have been discovered by following the acoustic "label" (FM or CF responses) through progressively higher stages of the bat's auditory brain (Neuweiler et al., 1980; Suga, 1984). The same approach of letting behavioral data serve as a pathfinder for following information as it travels from echoes into images has recently revealed an unexpected gain-control system in echolocation (Kick and Simmons, 1984) and a range-gating mechanism (Roverud and Grinnell, 1985b). It has also provided a functional description of several aspects of image-processing in echolocation (Simmons and Kick, 1984). The summary of performance presented in this chapter explores by a similar approach the images perceived by bats and the representation of these images.

The acuity with which bats perceive the fundamental acoustic dimensions of echoes--frequency, time-of-occurrence, intensity--is the basis for their ability to perceive the corresponding spatial features of targets. Bats can locate targets in distance and direction and determine their velocity with an accuracy that depends upon perception of these acoustic dimensions. Localization of targets in distance and direction is achieved with an effective accuracy of roughly a centimeter in each direction for the practical purpose of seizing an airborne object (Trappe, 1982; Webster and Brazier, 1965; see Schnitzler and Henson, 1980). The perceptual capabilities of bats are mobilized in successive stages of the pursuit process that leads up to interception; detection, localization, tracking and identification of prey are articulated into a stereotyped maneuver that is quite complex (Kick and Simmons, 1984; Simmons and Kick, 1983; Trappe and Schnitzler, 1982). The perceptual anatomy of this pursuit process can be dissected in behavioral experiments that use targets or echoes as controlled stimuli (Simmons and Kick, 1983; Suthers and Wenstrup, in press). For example, the factors influencing detection of targets by bats range from those that are acoustic to those that are physiological, and ultimately perceptual, thus encompassing most aspects of information processing that have so far been identified experimentally (Simmons, in press; see Møhl, this volume). Beyond detection, the performance that bats would be expected to achieve at any point in the interception maneuver is conditioned by a host of acoustic and auditory factors which must be explicitly recognized and taken into consideration when experiments are conducted (Kick and Simmons, 1984; Menne and Hackbarth, 1986). By measuring the ability of bats to discriminate small changes in each dimension of targets or echoes, one can determine how sharp are the images produced through echolocation. From the influence of other targets or echoes on detection and discrimination, coupled with observations of neural responses within the bats auditory system to echoes, one can determine, further, the scale of the representations which underlie these sharp images.

Target Range and Echo Delay

Factors influencing target ranging. Echolocating bats perceive the distance to sonar targets from the time-delay of echoes after emissions (Simmons, 1973, 1979). The determination of echo delay occurs through a process initiated by reception of the outgoing sonar signal at the ears and terminated by reception of the echo (see below). Each echo thus provides an estimate of target range, and range estimates from successive echoes evidently are integrated together by the bat into what must amount to real-time acoustic-image motionpictures. The "frame rate," or the rate at which successive images are produced to update the nearly continuously-running image stream, is controlled by the repetition-rate of sonar emissions. Bats regulate the rate of emission of their echolocation sounds according to the distance to the target being tracked (Schnitzler and Henson, 1980) and according to the distance to background objects, which sets the approximate size of the space within which objects can be imaged (Novick, 1977; Simmons, in press). The basic acoustic strategy for detection and interception of prey consists of a continuum of types of sonar signals that extends from relatively narrowband signals appropriate for searching in completely open spaces to relatively broadband signals used when the target is being approached and seized (Simmons et al., 1979). Individual bats "work" along a segment of this continuum that appears to match the size of the space within which they fly. Some species are adaptable and seem to adjust their search behavior along the basic strategy's continuum according to how far away are background objects, while other species appear more rigid and seem fixed permanently on only part of the continuum regardless of whether they are flying in an open space or a space restricted by such objects as trees, vegetation, and the ground. This fixed behavior has previously been identified as a separate strategy for pursuing prey in the presence of obstacles (Simmons et al, 1979), but it really is a segment of the basic strategy, as shown by the behavior of species that switch from one to the other depending upon the presence of other targets in the background (Simmons, in press).

It has long been assumed that bats process echoes from any one emission during the silent interval that precedes the production of the next emission, and that the occurrence of this next emission probably blocks subsequent processing of echoes from the earlier emission. Regulation of the "depth of interest" or "depth of field" by the repetition-rate of emissions (Novick, 1977) amounts to a range "attention gate" (Schnitzler and Henson, 1980). Physiological evidence indicates that the neural mechanisms of target ranging are initiated by direct auditory reception of the outgoing emission (O'Neill and Suga, 1982; Suga and Margoliash, personal communication); subsequent emissions would initiate their own target-ranging cycles and would thus probably override the range-measuring process initiated by any previous emissions. The tracking of target range during pursuit by the repetition rate of emissions certainly suggests that the silent interval between emissions represents the maximum target-ranging window for the particular target being tracked, although it is conceivable that not all of the silent period between emissions is equally useful for different forms of echo analysis. For example, there is now reason to believe that echoes may be useful for target detection, and perhaps discrimination, at delays greater than are used for range measurement (Roverud and Grinnell, 1985b).

Detection of targets depends upon detection of echoes, and the maximum distance at which targets of different sizes can be detected appears limited ultimately by the bat's threshold of hearing (Kick, 1982). Eptesicus can detect insect-sized targets as far away as 3 to 5 m by receiving echoes with amplitudes of roughly 0 dB SPL (p. to p.), which is approximately the sensitivity of the bat's hearing. When searching for prey in relatively open spaces, bats emit longer duration signals having greater overall energy than when searching in more confined spaces (Schnitzler and Henson, 1980; Simmons et al., 1979). Under normal circumstances bats change the energy and the repetition-rate of their signals in tandem, so that the maximum operating range determined from the audibility of echoes is probably kept in reasonable coordination with the depth of field set by the repetition-rate. These two factors influencing target-ranging efficiency can be dissociated experimentally, however, as when noise is introduced to interfere with target detection or discrimination. In noise, some species of bats increase the energy of their individual signals by lengthening the CF component of their signals without changing the repetition-rate, while others increase the duty cycle of their sounds by increasing the repetition-rate without changing signal duration (Roverud and Grinnell, 1985a; Simmons, in press; Simmons et al., 1978). The lengthening of individual signals may prove to be a characteristic of FM/short-CF species, while increasing the repetition-rate may be a characteristic of short-CF/FM species.

In one short-CF/FM species, Noctilio albiventris, there is evidence that the "range window" can be much shorter than the "attention," or detection window (Roverud and Grinnell, 1985b). Distance discrimination, but not target detection or left-right position discrimination, is vulnerable to interference by artificial signals that are similar in character to the bat's own echolocation signals. Since these signals interfere when presented free-running at repetition-rates as low as 5 Hz, it is clear that they are not masking the bat's hearing of its echoes, but rather are introducing false distance information or interfering somehow with processing by the bat of its own echoes beyond the act of detection or hearing of their presence. It is possible to vary the parameters of the artificial sound and determine which features are essential for the interference effect (and hence, probably, are critical for determining range). These experiments have shown that the short CF component by itself, or the FM component by itself, have no interfering effect on distance discrimination. Instead, a combination of the beginning of the CF signal and an FM sweep of at least 10-11 kHz, beginning at approximately the frequency of the CF component, is needed for interference. Although earlier experiments, mostly neurophysiological, have indicated that the initial CF component of CF/FM sounds facilitates response to subsequent FM sweeps (Grinnell, 1963a, 1967, 1970, 1973; Grinnell and Hagiwara, 1972; Novick, 1977), these experiments go further in implying that the onset of the CF signal is needed to gate a range window that, under the conditions tested, in this species, lasts only approximately 27 msec. In this behavioral situation, with the bats emitting pairs of pulses every 100-120 msec, artificial CF/FM sounds interfere with distance discrimination only if the duration between the onset of the CF component and the FM sweep is between 2 and 27 msec, or if an artificial FM sweep alone is presented within the 27 msec window opened by the bat's own orientation sound (Roverud and Grinnell, 1985b; see Roverud, this volume, for illustration and more detailed discussion). These experiments are interpreted as indicating that the onset of a bat's own CF component opens a range window lasting approximately 27 msec, during which the time-intervals between the bat's emitted FM components and FM echoes are processed to produce distance information. During the remainder of the period between emissions, returning echoes may still be useful for detection and angular localization. Whether the duration of the range window is constant or varies under different conditions is not known; but it is clear that a 27 msec long CF-gated window cannot apply to the later stages of target pursuit, when the signals of this species lose their CF component and the intervals between successive emissions fall to well under 27 msec. Interestingly, similar behavioral techniques, when applied to the long-CF/FM bat, Rhinolophus rouixi, reveal a comparable CF-gated range window of much longer duration, appropriate to the longer R. rouxi pulses (see Roverud, this volume). Hence a CF-gated range window is likely to be a general feature of CF/FM bats. It seems probable that an analogous window exists in FM bats, but its features have not yet been worked out.

Separate processing mechanisms for range, as opposed to detection and determining the direction of targets, are implied also by the observation that there are differences in the number of echoes needed to obtain information of each type. Thus, Noctilio albiventris trained to make yes/no or right/left choices indicating whether or not a target is present and to the right or left of the mid-line, could make the decision with only the echoes of one pair of pulses (during a brief gap in masking noise). while even an easy distance discrimination requires the echoes from a minimum of two sets of emitted sounds. As the discrimination is made more difficult, the echoes of more emitted pulses have to be integrated (Roverud and Grinnell, 1985a).

Range Discrimination. The ability of a variety of species of bats to discriminate small differences in the distance to targets reveals that target range is a fundamental perceptual dimension of the acoustic images supplied by echolocation. Experiments on target ranging also provide an index of the sharpness of a target's image along the axis of range. Table 1 shows target-range discrimination "thresholds" obtained from seven species of bats representing five major families of Microchiroptera, all in substantially the same two-alternative forced-choice task. These data indicate that target-range discrimination thresholds for bats generally lie in the region of 1 to 2 cm, except for Rhinolophus, where acuity is 3 to 4 cm. A striking feature of these data is the apparent invariance of these thresholds at different absolute distances to the targets--from 30 to 240 cm in Eptesicus, for instance. This invariance may be illusory, however (see 360).

The acoustic dimension of echoes that conveys information about target range is the time-delay of echoes after emissions (Simmons, 1973; 1979; in press). Eptesicus and Phyllostomus can discriminate differences in the arrival-time of "echoes" with thresholds of 60-70 μsec which corresponds to the target range thresholds obtained with real targets (Simmons, 1971, 1973). The artificial echoes for these experiments were generated electronically by delaying the bat's emissions using a target simulator, and the two-choice discrimination procedure was in other respects the same as for real targets.

A crucial question is whether target-range discrimination thresholds provide a reasonable indication of the sharpness of the range-axis image of a single target. Discrimination experiments measure the capacity to resolve small changes along individual perceptual dimensions, and they have historically been the basis for specifying the "graininess" of perceived images. The term "resolution" in this context has the meaning of distinguishing the difference in distance between two closely-spaced targets to identify which is nearer. If the task is to distinguish a single target from a pair of closely-spaced targets, one is measuring a two-point threshold in psychophysical terms. "Resolution" has a different meaning when applied to this situation; the bat would be resolving two targets as different from one target (McCue, 1966). Some confusion has arisen over whether a range discrimination experiment measures range resolution (see Menne and Hackbarth, 1986; Schnitzler and Henson, 1980; Simmons, 1973). The answer depends upon whether one considers the experiment primarily a measurement of range by the bat or a measurement of the bat's ability to determine range. When the electronically-simulated echoes are presented successively, only from one "target" at a time (whichever target the bat is aiming its sonar signals at), Eptesicus discriminates echo delay differences as small as 60 μsec, the same threshold as is obtained when electronic echoes from both targets are presented simultaneously (Simmons and Lavender, 1976; Simmons, in press). The bat evidently achieves target-range or echo-delay discrimination by separately perceiving each target's range and then choosing the nearer target, rather than simply responding to the target that yields the earlier of two echoes that arrive together as a compound signal, which occurs when real targets are used. In other words, bats treat the target-ranging task as a spatial instead of an acoustic task; they deal with each target as a discrete source of echoes. The scanning movements of the bats head, along with the left-right spatial nature of the bat's response in the two-choice task, show that the range of each target is associated with its direction. It is thus likely that each target is separately displayed by the bat's sonar receiver as occupying a unique position in space. This conclusion implies that the underlying representation, which is the embodiment of the perceived image in the brain, is itself a spatial display (Simmons and Lavender, 1976). The results of target range or echo delay discrimination experiments thus yield at least a first approximation to the range-axis image of a target.

Whereas simultaneous or successive presentations of echoes do not change the threshold for delay discrimination in Eptesicus, a successive-presentation experiment with Noctilio albiventris, in which the bat was trained to respond positively to a target at a distance of 35 cm but not to one in the same direction at any other distance, yielded a threshold of 3 cm instead of the 1.3 cm obtained with simultaneous presentation (Roverud and Grinnell, 1985a). It is not surprising that the range-identification task, involving a yes-no choice between successive presentations of a target in one direction at different distances, separated by time-intervals of 30 seconds or more, would be more difficult than a left-right response based on comparison of echoes occurring within fractions of one second. However, when tested in a range-identification paradigm like that used for Noctilio, Eptesicus shows a threshold time difference of about 60 μsec. corresponding to the value obtained with simultaneous target presentation (Surlykke and Miller, 1985). Noctilio may be genuinely less accurate than Eptesicus at successive range judgments, but it seems more likely that the two successive-presentation procedures are not comparable.

Although it is evident from simultaneous and successive echo-delay discrimination experiments with Eptesicus that range discrimination results do indeed convey information about the image of a single target, ordinary range discrimination results do not portray the true sharpness of target-range images (Simmons and Stein, 1980). During individual experimental trials the bat moves its head from left to right to scan the targets, and from one trial to the next the bat occupies a slightly different position on the elevated platform from which it observes the targets. Variations in the position of the bat's head introduce unwanted variations in the distances to the targets. If the bat's head is thought of as the reference position for the judgement of target range or echo delay, the effect of variations in its position is to move the targets themselves to different distances from the bat (Simmons, 1969a, 1973). The discrimination performance of the bat is thus confounded with changes in the true distances to each target that are not considered when the raw results (percentage errors or correct responses) are graphed against the nominal distances to each target. The bat's performance must be corrected for variations in its head position by changing nominal target ranges to true target ranges after making measurements of the variability of the position of the head during experimental trials. The thresholds shown in Table 1 are thus an indication only that bats can perceive the range of a whole target to within a few centimeters in practical situations where movements are possible. After measuring head movements and correcting for the resulting variations in target range (which depend on the angular separation of the targets in the twochoice task), the bat's acuity for target-range discrimination appears much sharper (Simmons et al., 1975; Simmons and Stein, 1980).

Table 2 shows the acuity of target range discrimination by six species of bats after the movements of the bat's head are taken into account (Simmons et al., 1975). Head movements within and between trials have been measured directly for the first five species in Table 2; the sixth species (Pipistrellus) is included for reasons described below. Its head movements were approximated crudely from data for Eptesicus and Tadarida, keeping in mind that the target angle was 75° rather than the more usual 40°. The table shows the total range of corrected threshold values at an absolute range of 30 cm for the first four species, at 60 cm for Phyllostomus, and at 21-24 cm for Pipistrellus. These thresholds are considerably smaller than the uncorrected values from Table 1, indicating how significant is the head-movement artifact in the raw results. The corrected thresholds in Table 2 also show that these species differ from each other in range discrimination acuity by more than the uncorrected data suggest. These corrected thresholds are representative of the acuities at other absolute ranges, too, since the head-movement corrections are virtually identical at the different absolute ranges shown in Table 1. Consequently, the corrections do not affect the apparent invariance of range discrimination thresholds at different absolute ranges. Several previous discussions of target-ranging data were based on the raw threshold data shown in Table 1 (Schnitzler and Henson, 1980; Suthers and Wenstrup, in press) rather than the corrected threshold data shown in Table 2 (Simmons et al, 1975). The movements of the bat's head vary from species to species according to the size of the bat and the manner in which it rests on the observing platform

Table 1. Target Range Discrimination Thresholds

Species	Absolute Range	Range Difference Threshold	Angular Separation of Targets
Eptesicus fuscus[1]	30 cm	1.3 cm	40°
	60 cm	1.3 cm	40°
	240 cm	1.4 cm	40°
Myotis oxygnathus[2]	100 cm	0.8 cm	13°
	100 cm	2.3 cm	13°
Noctilio albiventris[3]	35 cm	1.3 cm	40°
Phyllostomus hastatus[1,4]	30 cm	1.3 cm	40°
	60 cm	1.2 cm	40°
	120 cm	1.2 cm	40°
Pipistrellus pipistrellus[5]	21-24 cm	1.5 cm	75°
Pteronotus gymnonotus[1]	30 cm	1.5 cm	40°
	60 cm	1.7 cm	40°
Rhinolophus ferrumequinum[1,2]	30 cm	2.8 cm	40°
	100 cm	4.1 cm	13°

([1]Simmons, 1973; [2]Ajrapetjantz and Konstantinov, 1974; [3]Roverud and Grinnell, 1985a; [4]Simmons, in press; [5]Surlykke and Miller, 1985)

Table 2. Acuity of Target Range Images

Species	Range Acuity	Equivalent Delay Acuity	Width of ACR Envelope
Rhinolophus ferrumequinum[1]	1.2-1.3 cm	70-75 μsec	67μsec (FM only)
Pteronotus gymnonotus[1]	0.7-0.8 cm	41-46 μsec	43 μsec
Tadarida brasiliensis[2]	0.6-0.7 cm	35-41 μsec	34 μsec
Eptesicus fuscus[1]	0.5-0.7 cm	29-41 μsec	33-38 μsec
Phyllostomus hastatus[1]	0.4-0.5 cm	23-29 μsec	29 μsec
Pipistrellus pipistrellus[3]	0.3 cm*	17 μsec*	14-28 μsec

([1]Simmons et al., 1975; [2]Simmons, in press; [3]Surlykke and Miller, 1985)

*estimated for 75° target angle using head-movement data from E. fuscus and T. brasiliensis during experimental trials. Furthermore, the actual thresholds for range discrimination vary more from species to species than Table 1 indicates as a consequence of the structure of the bat's sonar signals. The role of acoustic factors in determining the accuracy of target ranging is obscured when the uncorrected thresholds form the basis for discussion (see 360).

Echolocation Signals and Range Discrimination. To track the transformation of acoustic features of echoes into spatial features of images requires finding relationships between the structure of echolocation sounds and the perceptual images of targets that the bat derives from these sounds (see Simmons et al., 1975). Theories about the reception and processing of sonar echoes by bats either consider that the ultrasonic waveforms of emissions and echoes are transduced by the inner ear into patterns of nerve impulses conveyed along channels tuned to ultrasonic frequencies, or that the ultrasonic waveforms are demodulated into lower beat or envelope frequencies and eventually represented by patterns of nerve impulses that are presumably conveyed along channels tuned to these lower frequencies (see reviews by Novick, 1977; Pye, 1980; Schnitzler and Henson, 1980; Simmons, 1969a, in press; Stewart, 1979). Although of considerable interest, demodulation theories have played little part in the interpretation of the results of target-ranging experiments because physiological evidence indicates that information about the timing of echoes ascends the auditory system in channels tuned directly to ultrasonic frequencies (Bodenhamer and Pollak, 1981; Suga, 1970; see reviews by Pollak, 1980 and Simmons and Kick, 1984). Furthermore, even the earliest examinations of target-ranging data suggested that important features of the original ultrasonic waveforms themselves were in fact preserved to appear in the bat's perceptions (Simmons, 1969a). The problem underlying target ranging for the bat is to represent the time-of-occurrence of sonar echoes accurately enough to detect, locate, track, identify, and eventually capture prey. A fruitful interaction between experiments and theories has characterized efforts to learn the nature of this representation (see Altes, 1980, 1981; Menne and Hackbarth, 1986; Møhl, 1986; Pye, 1986; Schnitzler, 1984; Schnitzler and Henson, 1980; Simmons, 1969a, 1971, 1973, 1977, 1979, 1980, in press; Simmons and Stein, 1980).

Most species of echolocating bats transmit FM signals either as their entire sonar sound or as one component of a compound sound containing, in addition, a CF signal of variable duration (Grinnell, 1973; Novick, 1977; Pye, 1980; Schnitzler and Henson, 1980; Simmons et al, 1975). These FM signals have been suspected as target-ranging signals ever since they were discovered to be so widespread among bats (Griffin, 1958). The presence of FM signals in echolocation has led to several suggestions that the bat might exploit their particular properties for target ranging by processing them in an ideal, or optimal sonar receiver (Altes and Titlebaum, 1970; Cahlander, 1964; Glaser, 1974; McCue, 1966; Strother, 1961; van Bergeijk, 1964). The utility of FM signals for ranging depends upon the sonar receiver being "matched" to the signal in such a way that the FM sweep of echoes is compressed into a single, sharp pulse that nevertheless retains the bandwidth and frequency composition of the original waveform (Klauder, 1960; Klauder et al, 1960). The compressed pulse that results from complete matching of the receiver to the signal is the crosscorrelation function of echoes relative to emissions, and an ideal sonar receiver is one that replaces the echo with its corresponding crosscorrelation function and then estimates the target's location in range from the crosscorrelation function's location in time. The accuracy of target ranging depends upon the sharpness of the crosscorrelation function--which, in turn, is dependent upon the bandwidth and frequency composition of emissions and echoes and also upon the signal-to-noise ratio of echoes (Woodward, 1964). The crosscorrelation function between emissions and echoes (or a reasonable approximation to it) can be obtained in any of several ways--from a pulse-compression receiver using frequency-dependent delays (Klauder et al., 1960), from strict Fourier-transform computation in the time or the frequency domain (Floyd, 1980; Johnson, 1980), or from other representations of the waveforms than conventionally sampled time-varying signals or sampled spectra (Altes, 1980). Bats might conceivably use a correlation-equivalent sonar receiver based on the properties of some part of the auditory system. The earliest type of correlation-equivalent receiver to be proposed for bats--a pulse-compression receiver in which the cochlea provides frequency-dependent delays--was rejected on straightforward physiological grounds (McCue, 1969), but -t has remained a serious possibility that the necessary correlation-equivalent computations could be achieved within the nervous system.

It is important to distinguish between the use of crosscorrelation, or its equivalent in signal-processing terms, and the use of an ideal receiver (Menne and Hackbarth, 1986). Under ideal conditions the entire energy of the echo is brought to bear on maximizing the signal-to-noise ratio of echoes, and the echo itself is represented by a probability distribution along the time axis that is obtained by "reading" the crosscorrelation function according to a rule of thumb that the accuracy of target ranging is proportional to the width of the crosscorrelation function (its envelope or its central wave; see below) divided by the energy signal-to-noise ratio of echoes (Woodward, 1964). It is perfectly possible to use the crosscorrelation function to represent the time-of-occurrence of echoes without achieving the signal-to-noise-ratio criterion of ideal reception. The signal-to-noise ratio merely scales the display of the crosscorrelation function to produce the ideal probability distribution. From the perspective of understanding the processing of echoes in the bat's brain, it is the possible presence of a crosscorrelation operation, or its equivalent, rather than the effect of the signal-to-noise ratio, that leads to a description of the mechanisms which extract and display target range from echo delay. The display must then be read to control behavior, and it is convenient to think of the echo signal-to-noise ratio as exerting its limiting influence at this stage (see Woodward, 1964) instead of incorporating its effects into earlier stages such as those which actually compute the crosscorrelation function. Descriptions of pulse-compression receivers have emphasized that the characteristics of the signals and echoes directly contribute to the accuracy of target ranging; the usual index of the accuracy with which echo delay can be determined is the reciprocal of the signal's bandwidth, which corresponds to the width of the envelope of the crosscorrelation function (Klauder et al., 1960). Early pulse-compression theories were derived from radar applications where the radio frequencies transmitted were many times higher than the FM bandwidths. Under these conditions the wave structure of the crosscorrelation function would be so fine as to be useless; the envelope of the function instead serves to indicate the location of the function, and, hence, the echo, along the time axis (Woodward, 1964). The echolocation sounds of bats are at ultrasonic frequencies of roughly 10 to 200 kHz, not radio frequencies of hundreds to thousands of megaHertz, however, and their FM bandwidths typically are in the range of 10 to 100 kHz. Consequently, the transmitted frequencies and the bandwidths used by bats are similar in magnitude, and the fine wave structure of the crosscorrelation function for FM signals and echoes is similar in scale to the envelope of the function (Cahlander, 1964). Under the assumption that bats might process FM echoes in a correlation-equivalent receiver, the crosscorrelation functions for representative signals used by <u>Myotis</u> <u>lucifugus</u> during pursuit of airborne targets were computed to examine their properties. The widths of the envelopes of the crosscorrelation functions for these FM signals ranged from 85 to 125 μsec, corresponding to range accuracies of 1.5 to 2.2 cm (Cahlander, 1964).

The first target-range discrimination experiments were conducted with
Eptesicus and the resulting raw discrimination thresholds of 1.3 to 1.4 cm
(Table 1) were close enough to the range accuracies computed from the
signals of Myotis (which are similar in general features to the signals of
Eptesicus) to justify a closer examination of the crosscorrelation
functions for the signals of Eptesicus (Simmons, 1969a). Estimates of
range accuracy derived from the width of the crosscorrelation function's
envelope are an index only of the signal's contribution to target ranging;
the signal-to-noise ratio of echoes has to be taken into consideration,
too, for predicting the ultimate accuracy exhibited by the entire sonar
system (Cahlander, 1964; McCue, 1966; Woodward, 1964). The width of the
envelope of the crosscorrelation function thus provides only an informal
statement of the accuracy to be expected from the signals, not a formal
prediction of the performance to be expected from an ideal sonar receiver
(Simmons, 1969a). Nevertheless, this informal index has been in common use
to describe the important fact that features of the signals are manifested
in performance (Klauder, 1960; Klauder et al., 1960). Crosscorrelation
functions for the echolocation sounds of Eptesicus recorded during target-
ranging trials were therefore computed, and a prediction of the
discrimination performance to be expected from the bats was derived from
the envelopes of representative functions by assuming that the envelope
itself constituted the image of the target expressed in terms of echo delay
(Simmons, 1969a, 1973). The envelope only needed to be displaced in small
steps and the results averaged, thus mimicking the head-movement artifact,
to arrive at an estimate of the percentage of errors or of correct
responses likely to be achieved by the bats (Simmons and Stein, 1980). The
assumption that the height of the envelope of the crosscorrelation function
is directly related to the percentage of errors achieved by the bat is
merely an extension of the property that the crosscorrelation function's
location along the time axis is the best estimate of the location of the
corresponding echo. The likelihood that the target will be perceived at a
given distance is a monotonic function of the height of the envelope
(Cahlander, 1964). and, for the informal comparisons to be made first, the
effects of the signal-to-noise ratio of echoes were to be neglected
(Simmons, 1969a).

The percentage of errors achieved by Eptesicus in target-range
discrimination experiments was predicted accurately from the envelope of
crosscorrelation functions (Simmons, 1969a), so a search was undertaken to
find species of bats whose sonar signals were enough different than
Eptesicus that their crosscorrelation functions would, in turn, lead to
predictions of target-ranging performance that would be different from the
predictions obtained for Eptescus. The search was for bats that could be
trained in the target-range discrimination task and that emitted either a
single narrow FM sweep, leading to a broader envelope of the
crosscorrelation function, or narrower FM sweeps arranged in a harmonic
series, leading to prominent side-peaks in the envelope of the
crosscorrelation function due to the artificial lowering of signal-
periodicity to first harmonic frequencies (see Simmons and Stein, 1980).
Rhinolophus ferrumeguinum was chosen for its narrower transmitted
bandwidth, while Phyllostomus hastatus and Pteronotus gymnonotus
(=suapurenisis) were chosen for the harmonic composition of their signals,
particularly for the weakness or absence of the first harmonic. The
results of the target-range discrimination experiments with these
additional species could also be predicted from the structure of the sonar
signals, just as was true for Eptesicus. The percentage of errors achieved
by the bats corresponded closely to the envelopes of the crosscorrelation
functions for their respective signals after head movements were added to
these envelopes (Simmons, 1973). Not only did Rhinolophus perform with
less acuity to an extent related to the narrower bandwidth of its FM
signals, but Phyllostomus, and, in particular, Pteronotus exhibited

irregular target-range threshold curves with secondary peaks that
corresponded to the positions of side-peaks in the envelopes of the
crosscorrelation functions of their signals.

The collective results of the target-ranging experiments are
summarized in Table 2, which compares the acuity of target ranging by each
species with the width of the envelope of representative crosscorrelation
functions computed from the signals. These acuities have been corrected
for head movements (Simmons et al, 1975) to correspond to uncorrected
crosscorrelation-function envelopes rather than presenting raw
discrimination thresholds and corrected crosscorrelation function envelopes
(Simmons, 1973). Table 2 demonstrates in summary what is obvious from
discrimination curves graphed alongside predicted performance--that the
envelope of the crosscorrelation function provides a good description of
the image of a target along the axis of target range. (The data for
Pipistrellus are included in Table 2 because values for both discrimination
performance and the width of the envelope of the crosscorrelation function
for echoes are available (Surlykke and Miller, 1985). The head-movement
corrections for Pipistrellus have been crudely approximated for the
relatively large angular separation of the targets in this experiment from
the known head-movement data from other species.) The data in Table 2 have
been presented previously (Fig. 8 in Simmons et al, 1975) in a graph
comparing the bandwidth (-3 dB) of the energy spectra computed at the same
time as the crosscorrelation functions (Simmons, 1973) with both the width
of the envelope of the function and the actual target-ranging acuity
measured for each bat. This graph merely illustrates that the reciprocal
of the bandwidth of the signal is as good an index of target-ranging acuity
as the width of the envelope of the crosscorrelation function. The sound
spectrograms presented with this graph were drawn to show the limits (-6 dB
limits because "sonagrams" are amplitude rather than energy spectra) of the
FM sweeps used by each species within the confines of the energy spectra
derived from the correlation functions. The bandwidths used for the graph
are not "sonagram bandwidths" but "centralized RMS bandwidths" (Menne and
Hackbarth, 1986) approximated from energy spectra. The entire range of
frequencies present in the FM sweeps is greater than their strictly-defined
bandwidths (-3 dB from energy spectra or -6 dB from spectrograms) indicate.
Because the bandwidth is only approximately the reciprocal of the width of
the envelope of the crosscorrelation function for signals whose bandwidths
and average frequencies ("centralized RMS bandwidths" and "RMS bandwidths"
in the terms used by Menne and Hackbarth, 1986) are similar in magnitude
(see Simmons and Stein, 1980), the predictions of performance summarized in
Table 2 have been based directly on the crosscorrelation functions
themselves (Simmons, 1973; Simmons et al, 1975; 1979; see Pye, 1986; and
Zbinden, this volume). The strength of the relationship between the width
of the envelope of the crosscorrelation function and the acuity of target
ranging is expressed by the correlation coefficient (r) for the data from
the first five species in Table 2. For the 17 individual bats of these
species tested, r=0.97, which accounts for nearly all of the variance of
the data from the characteristics of the signals alone. This strong
relationship is concealed in the raw discrimination data by the head
movements of the bat.

The results summarized in Table 2 show that, under the broad
conditions of two-choice target-range discrimination experiments, the bat's
acuity is directly related to an identifiable acoustic feature of FM
echolocation signals. The width of the envelope of the crosscorrelation
function between sonar emissions and echoes, or the reciprocal of the
bandwidth of the signals, appears to determine the sharpness of the images
of targets along the axis of target range. Except for Rhinolophus, these
species emit either FM sounds or short-CF/FM sounds, and the
crosscorrelation function of the entire emission and echo adequately

predicts discrimination performance. For Rhinolophus the prediction is based on the FM signal alone. The index of acuity shown in Tables 1 and 2 is the discrimination threshold at 75% correct responses. This is a useful, though arbitrary, summary of the performance represented in more detail by the shape of the entire discrimination curve. The accuracy with which the entire discrimination curve is predicted by the envelope of the crosscorrelation function is a better indication of the acuity of range discrimination in relation to the sonar signals than is the threshold alone, because the much larger number of experimental measurements associated with the various points on the curve then contributes to the reliability of specifying the bat's performance.

The results of the target-range discrimination experiments solidly implicate the FM sonar signals of bats as vehicles for target ranging. Furthermore, interference experiments demonstrate that the FM sweeps at the end of the CF/FM echolocation sounds of Noctilio and Rhinolophus must be crucial for the range discrimination task (Roverud and Grinnell, 1985b; Roverud, this volume). These interference experiments raise the possibility that a neural template recognizes the FM sweeps of echoes, perhaps after being set up by the initial, direct reception of emissions and then "primed" by the onset of the CF component of echoes. In the case of Rhinolophus ferrumeguinum, the results indicate that the terminal FM component of the bat's compound long-CF/FM signal is used for target ranging. The echolocation sounds emitted by this species in the target ranging experiments are strongly "sweep-peaked" signals (see Pye, 1980), which suggests that the bat chooses this type of signal to emphasize FM components and the information they can convey when faced with a target-ranging task. Some recent observations of the variability of echolocation sounds emitted by Rhinolophus ferrumeguinum during insect pursuit show that the bat may narrow the range of its FM sweeps during the terminal stage of pursuit (Vogler and Neuweiler, 1983), although not all observations of pursuits yield this same conclusion (Griffin and Simmons, 1974; Schnitzler et al., 1985; Trappe and Schnitzler, 1982). Some of the signals emitted by Rhinolophus certainly do contain little or no FM sweep in comparison to the strength of the long-CF component, but recent studies of the sounds actually stimulating the inner ear of another long-CF/FM species, Pteronotus parnellii, call into question the validity of remote microphone recordings as a true indication of whether the sonar sounds and echoes received by a moving bat really are "sweep-peaked" or "sweep-decayed" (Henson et al., 1987). (Most species of echolocating bats narrow the frequency range of the individual FM sweeps in the sonar sounds they emit during the terminal stage (Schnitzler and Henson, 1980; Simmons et al., 1979), but these sounds also become weaker than the signals emitted at earlier stages of pursuit so it is difficult to determine how much the sweeps actually narrow. In contrast, Rhinopoma widens the frequency range in the sweep of each harmonic during the terminal stage (Habersetzer, 1981; Simmons et al., 1979, 1984).

The variability of the FM sweeps emitted by Rhinolophus enters into the interpretation of target-ranging experiments because the observed narrowing of the sweeps has suggested that the bat might determine target range from the CF signals instead (Vogler and Neuweiler, 1983). During the target-range discrimination experiments, no such loss or reduction of the FM sweeps was observed; if anything the signals were more consistently sweep-peaked than usual. Furthermore, the crosscorrelation function for the FM components specifically predicted the bat's performance. There is presently no evidence that Rhinolophus can use the onset of the CF components of echoes to determine target range, although the "range gate" observed in interference experiments with Noctilio and with Rhinolophus is activated by the onset of CF components (Roverud and Grinnell, 1985b; Roverud, this volume). The evidence cited in support of the idea that

<u>Rhinolophus</u> might use its CF components to determine range is the observation that binaural neurons in a different species of bat (<u>Molossus ater</u>) respond selectively to small interaural delays of CF stimuli (Harnischfeger, 1980; Harnischfeger et al., 1985). This species does not receive CF signals through very sharply-tuned channels in the auditory system as does <u>Rhinolophus</u>, however. The sharp tuning can be expected to distort and smear rapid envelope onsets. Furthermore, the onsets of the CF components emitted by <u>Rhinolophus</u> are very gradual in comparison with the temporal sensitivity of neurons in this other species, and they are often obscured by the upward FM sweep at the start of the sounds. More work is needed to learn whether <u>Rhinolophus</u> can determine target range from the CF components of echoes, and there is presently no evidence that identifies the role of the <u>initial</u> FM component.

<u>Fine acuity of echo-delay perception</u>. The fact that the envelope of the crosscorrelation function for echoes had to be shifted and averaged to take account of the movements of the bat's head during discrimination experiments raises the possibility that bats might be considerably more accurate at determining target range than the two-choice range discrimination experiment is capable of demonstrating (Simmons and Stein, 1980). A comparison of Tables 1 and 2 shows how the variability of the position of the bat's head conceals the importance of the signals themselves for determining the accuracy of target ranging. The crosscorrelation function has a fine structure consisting of several waves or peaks that lie beneath its envelope, and the bat could well be able to perceive this fine structure without it being revealed in the results of target-range discrimination experiments. The fine structure of crosscorrelation functions for echolocation sounds has dimensions of roughly 15 to 40 μsec between peaks, but the movements of the bat introduce variations as much as three times greater in the distances to the targets, which would smear the bat's performance to resemble the envelope rather than the fine structure of the crosscorrelation function (Simmons and Stein, 1980). When this possibility first came to be considered, two sets of data were available that might indicate whether bats could perceive more detail in the crosscorrelation function than just its envelope. An earlier experiment on discrimination of horizontal angles between targets (Peff and Simmons, 1972) yielded results that could be successfully modeled on the assumption that the time-difference between echoes arriving at the two ears provided an important cue for the bat to determine the horizontal direction of targets. In addition, earlier experiments on the ability of bats to discriminate the depth of holes in targets, which had been interpreted entirely as a discrimination of differences in echo spectra, yielded data that could be predicted from the crosscorrelation function of echoes. The significant feature of the re-examination of data from previous experiments was that the bat's performance was related to the fine structure, not the envelope, of the crosscorrelation function for echoes (Simmons, 1977; see Simmons and Stein, 1980, for illustration).

As a result of the surprising success of attempts to predict the results of horizontal-angle and hole-depth discrimination experiments from the fine structure of crosscorrelation functions, work was started on a new experimental procedure using jittered echoes to nullify most of the effects of the bat's head movements during discrimination experiments so that the true limits of echo-delay perception could be measured. In an echo-jitter discrimination task, <u>Eptesicus</u> can discriminate variations as small as 1 to 2 μsec in the time-of-arrival of echoes delivered sequentially with an electronic target simulator (Simmons, 1979). Due to the use of a sequential-presentation paradigm, this acuity is again based on the bat's ability to judge the absolute delay of individual echoes. It cannot be due to perception of the spectra of compound echoes such as those reflected back to the bat from targets with two surfaces at different distances

(Habersetzer and Vogler, 1983; Schmidt, this volume; Simmons, 1979; Simmons et al., 1974). It presently remains unclear how spectral and temporal cues contribute to the discrimination of complex targets. The primary objection to the possible use of temporal cues is that the time differences associated with range differences of fractions of a millimeter are only a few microseconds, which somehow seem implausibly small for bats to perceive. Yet, the results of the echo jitter experiments show unequivocally that bats can indeed perceive such small time differences, raising the strong possibility that temporal cues may contribute to the bat's ability to discriminate complex targets. Everyone assumes that bats can discriminate differences in the spectra of echoes; the difficulty is in proving it. Whereas the jitter technique effectively isolates temporal cues from spectral cues, experiments with complex targets do not perform the reverse of isolating spectral cues from temporal cues. The only cue that has been unambiguously demonstrated at present is temporal, and to provoke realization that the problem of distinguishing spectral from temporal cues is very serious, it has been proposed that bats use strictly temporal cues (Simmons, 1980). This strategy has succeeded in promoting further experiments to focus on aspects of discrimination performance that might be useful signs of the use of spectral cues by bats (Habersetzer and Vogler, 1983; Schmidt, this volume).

As two reflecting surfaces become closer together along the range axis, the echoes they return become increasingly more overlapped in time. Temporal overlap promotes interference leading to reinforcement and cancellation of various frequencies in the compound echo, so that one can think of a single composite echo having a complicated spectrum conveying information about the distance between the two glint surfaces, or one can think of two overlapping replicas if the incident sound, each having its own discrete time-of-occurrence. The capacity to recognize the existence of two replica components in the echo is crucial for using temporal information to perceive the distance between the glints. As the time separation of the glints becomes smaller, the role of spectral cues in perceiving glint separation can be expected to grow (Beuter, 1980). For large separations, only temporal cues can be used, and for infinitesimally small separations, only spectral cues presumably are available. For separations of moderate size, a mixture of temporal and spectral cues could be used to determine the distance between the glints. The question that needs answering concerns the size of "moderate" separations. Hole-depth discrimination and echo-jitter discrimination results are close enough to each other to indicate that temporal information may still be available to Eptesicus when glints are only 6-8 mm apart. Of course, spectral information is available, too, so what is really needed is an experiment to assess the relative contributions of time- and frequency-domain information to perception of complex targets.

During interception of flying insects by FM bats, the echoes that reach the bat's ears range in amplitude from about 0 dB SPL (p. to p.) at the maximum distance of detection to as much as 60 to 70 dB SPL at the moment of capture (Kick and Simmons, 1984). Except for periodic strong glints that occur when the wings of the insect achieve a favorable angle during the wing-beat cycle to return a strong specular reflection (see Schnitzler et al., 1983), the strength of echoes returning to the bat's ears is unlikely to rise much above 40 to 50 dB SPL during the crucial approach or tracking stages of pursuit. The gain-control action of the bat's middle-ear muscles also may effectively limit stimulation of the inner ear to no more than about 30 dB SPL for insect-sized targets. In the echo jitter discrimination experiments (Simmons, 1979), the amplitude of stimuli delivered to the bat was about 75 dB SPL (p. to p.), which is considerably higher than the amplitudes to be expected during pursuit. Additional jittered-echo experiments have explored the acuity of perception

of echo arrival-time for echoes at amplitudes close to the threshold of
detection to determine whether fine acuity occurs under these limiting
conditions as well as at higher echo amplitudes (Simmons, in press). The
original procedure was a discrimination of jittered from non-jittered or
"stationary" echoes presented in a two-choice task that delivered echoes
sequentially, one at a time. These further experiments use a procedure in
which bats detect echoes with varying amounts of jitter. Fig. 1
illustrates the performance of Eptesicus in both echo-jitter discrimination
and jittered-echo detection experiments. The curves demonstrate that the
bat is able to extract echo arrival-time accurately enough for jitters of
10 to 20 μsec to enhance detection performance even at the threshold (see
figure caption for explanation). It thus seems likely that a reasonable
degree of fine range acuity persists even when echoes are received at the
poor signal-to-noise ratios associated with detection thresholds. Such
acuity is substantially greater than the 1 to 2 cm range accuracy necessary
to successfully capture an airborne target (Schnitzler and Henson, 1980).
The possible role of fine range acuity for classification of target shapes
must be taken seriously in designing new experiments to explore the
significance of the results shown in Fig. 1.

The results of echo-delay jitter experiments demonstrate that
underlying the relatively gross disruptive effects of the bat's head
movements during discrimination trials is a very sharp image of the
location of a target along the range axis. As suspected from reexamination
of the results of horizontalangle and hole-depth discrimination
experiments, Eptesicus can indeed perceive the time-delay of echoes with an
acuity corresponding to the fine structure of the crosscorrelation function
for echoes (Simmons, 1977). The bat's head movements during trials had
obscured this finer acuity and made it seem that the images of targets
corresponded to the envelope of the crosscorrelation function rather than
its fine structure (Simmons and Stein, 1980). A particularly interesting
outcome of the echo-delay jitter experiment is that the bat's performance
can be predicted from the half-wave rectified crosscorrelation function;
the positive-going peaks in the fine structure of the function are
faithfully reproduced in the percentage of errors made by the bats at
certain jitter values (see Fig. 1; Simmons, 1979; in press). The evidence
that bats can perceive an image that is equivalent to the crosscorrelation
function for echoes is provocative for auditory theory. How does the
auditory system preserve and use the phase information at ultrasonic
frequencies necessary for computing a representation similar to the
crosscorrelation function, when it is known that the individual tuned
channels of the auditory system (consisting of a hair-cell and its
associated afferent neurons) extract the envelope of waveforms passing
through their tuning curves and smooth the envelope with a low-pass filter
having a cut-off frequency of only a few kiloHertz? The strict phase of
individual ultrasonic frequencies is destroyed by such a receiver. The
phase preservation problem was recognized from the results of the target-
range discrimination experiments: Perception even of the envelope of the
crosscorrelation function requires that the equivalent of phase be retained
and used until after the crosscorrelation function has been calculated
(Menne and Hackbarth, 1986). This is implicit in the terms "coherent" and
"semicoherent" used to describe receivers that operate respectively on the
fine structure of the crosscorrelation function or on its envelope. It was
proposed from the first target-ranging results (Table 2) that the secret of
phase preservation for ultrasonic FM echolocation signals and echoes must
reside in the unusual hybrid time- and frequency-domain representation that
the bat's auditory system uses to encode these waveforms (Simmons, 1973).
By representing emissions and echoes in terms of instantaneous frequency,
and by embodying the crosscorrelation operation in the comparison of neural
populations whose activity represents the instantaneous frequency pattern

368

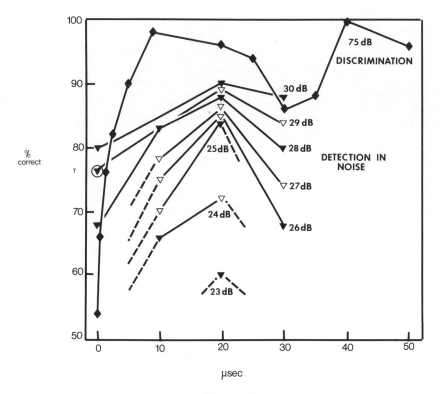

Figure 1

A graph showing the results of echo jitter discrimination and jittered-echo detection experiments with <u>Eptesicus</u> <u>fucus</u>. The uppermost curve (◆) shows the performance of a bat discriminating jittered echoes from "stationary" echoes, with the stimuli at amplitudes of about 75 dB SPL (p. to p.) (from. Simmons, 1979), The remaining curves (▽ and ▼) show the performance of a bat detecting echoes with varying amounts of jitter at amplitudes ranging from 23 to 30 dB SPL (p. to p.) (from Simmons, in press), All data-points show performance for 50 trials. The mean delay of echoes for jittered conditions is 2.9 μsec (corresponding to a target range of 50 cm), Under normal conditions (jitter of zero μsec) the threshold for echo detection at a delay of 2.9 μsec in <u>Eptesicus</u> is about 19 dB SPL (Kick and Simmons, 1984), Broadband random noise was mixed with the signals returned to the bat at a level sufficient to raise the bat's echo detection threshold to 29 dB SPL (circled data-point marked "T" on vertical axis, at left end of 29 dB contour curve), The family of contours shows that the presence of jitter 10, 20, or 30 μsec) alters the bat's detection performance in the region around threshold. At 29 dB SPL, for example, echoes that jitter by 20 μsec are more readily detected than echoes with no jitter at all or with a jitter of 30 μsec. The enhancement of detection performance with jitter demonstrates that the bat can determine the time-of-occurrence of echoes well enough to detect small time shifts in echoes even in the region of the threshold for echo detection. The results also confirm the observation from the uppermost curve that jitter values of about 30 μsec are less readily detected than smaller values of 10 or 20 μsec. The ambiguity about a jitter of 30 μsec evident in the bat's performance represents the average periodicity of the echoes reaching the bat's ears with respect to the periodicity of the original sonar emissions (Simmons, 1979; Simmons and Stein, 1980).

of FM sweeps in echoes, the auditory system could circumvent the nominal destruction of phase occurring at the receptor-neuron junction. The phase ascends the auditory system by being packaged internal to each spectrogram-like instantaneous-frequency image of the sound (for pictorial examples, see Simmons and Kick, 1984) by the fact that instantaneous frequency is the rate-of-change of phase. A sound theoretical basis for spectrogram correlation as a means to obtain crosscorrelation functions has been developed to describe the bat's sonar receiver (Altes, 1980, 1981; see Altes, this volume). One final point deserves comment: The bat behaves as though it perceives the half-wave rectified crosscorrelation function. For this particular waveform, the distinction between the function itself and its envelope is moot--the envelope of the half-wave rectified function is approximately the same as the half-wave rectified function itself. Bats appear to have chosen a representation for acoustic images along the range axis which could be the result either of coherent or semicoherent processing.

The bat's ability to perceive very sharp target-range images only a fraction of a millimeter across leads to an explanation of the apparent invariance of target-ranging acuity at different absolute ranges (see Table 1). The raw targetrange discrimination thresholds largely consist of the artifacts introduced into the bat's behavior by the effects of variations in the position of the bat's head during the experiments. Even the estimated acuities based on the envelopes of the crosscorrelation functions (Table 2) are substantially sharper than the raw thresholds, and the results of the jitter experiments indicate that the images are sharper than even the envelopes predict. The width of the central wave of the crosscorrelation function for _Eptesicus_ is a good approximation to the width of the image of the target (Simmons, 1979). In fact, the original range-discrimination experiments are better characterized as measurements of the bat's head movements than as measurements of the resolving power of the sonar system! The bat's acuity is sufficiently finer than the head-movement artifact that the bat's acuity could change from one absolute range to another and not be reflected as changing in the raw discrimination data. To ascertain whether the sharpness of range images changes or remains invariant at different absolute ranges requires repeating the jitter experiment at different distances, not repeating the range-discrimination experiment. There are many acoustic and perceptual factors that have to be taken into consideration in doing such an experiment (Kick and Simmons, 1984; Simmons, in press; see Møhl, this volume), an attempt at which is presently in progress.

For all of the species shown in Table 2, the sharpness of target-range images is better than the raw discrimination data show. For _Eptesicus_, the sharpness of the images corresponds to the width of the central wave of the crosscorrelation function (Simmons, 1979), but there is no reason to assume that the jitter experiment ought to yield this same result for all species of bats. _Eptesicus_, _Phyllostomus_, _Pteronotus_, and _Tadarida_ emit target-ranging signals containing several harmonic sweeps, which favorably suppresses the side-peaks in the crosscorrelation functions of echoes (Simmons and Stein, 1980). The lowest fundamental frequencies emitted by these bats are in the range of 15 to 25 kHz. Among FM bats as a whole, the lowest frequencies present in the FM sweeps of echolocation sounds are in the range of roughly 10 to 40 or 50 kHz, depending upon the size of the bat (the bat's "auditory" size, as indexed by the interaural distance or the length of the external ears (Simmons, in press) rather than the length of the body or the forearm, for example). In contrast, most long CF/FM bats emit FM sweeps that do not descent in frequency as low as 40 to 50 kHz; many do not descend below 100 kHz. _Pteronotus_ _parnellii_ is one well-known exception. Furthermore, the normal relationships between the length of the external ears or the interaural distance that hold for FM bats do not hold for long-CF/FM bats (Simmons, in press). The capacity of a spectrogram-correlation receiver to preserve phase information external to the instantaneous-frequency representation of the FM sweeps (that is, the phase of echoes relative to emissions) depends upon the temporal variability of the instantaneous-frequency measurement at least at one frequency being smaller than the period at that frequency (Altes, 1980, 1981). In physiological terms, this most likely means that the jitter in the timing of nerve impulses representing any one frequency must have a collective distribution that is narrower than this period (Simmons, 1979, 1980), and it is at the low-frequency end of the FM sweep for the lowest harmonic that the periods are longest. It thus is worth speculating that, if bats do indeed possess a coherent spectrogram-correlation receiver, then the critical echo-toemission phase reference is most likely passed along channels tuned to the low end of the FM sweep of the lowest harmonic. It also is worth speculating further whether FM bats may use such a phase reference because their lowest frequencies might have a period long enough to exceed some physiological limit on timing accuracy within a single frequency channel, whereas long-CF/FM bats might not have such a phase reference because their lowest frequencies still have periods that are shorter than this limit on timing accuracy. In such a case, other FM bats should perform as does Eptesicus in the echo-delay jitter experiment, while long-CF/FM bats should perceive only the envelope of the crosscorrelation function of echoes as the range image of a target.

Echo signal-to-noise ratios and target ranging. A formal test of the hypothesis that bats use an optimal, or ideal, correlation-equivalent receiver depends upon experiments in which the signal-to-noise ratio of echoes is manipulated (Menne and Hackbarth, 1986; Schnitzler and Henson, 1980; Simmons, 1969a). It is quite possible for the bat to perceive correlation-like images with a less-than-optimal, or suboptimal, receiver (Menne and Hackbarth, 1986; Simmons, 1977). This possibility was discussed at length at the Jersey Animal Sonar meeting in 1979 and at a Dahlem conference on auditory processing held earlier, in 1976. To evaluate the contribution of the bat's decision criteria associated with "reading" the crosscorrelation function to determine target range requires measurements of the bat's performance at relatively low echo signal-to-noise ratios where performance breaks down. The particular signal-to-noise ratio at the point of breakdown might indicate what kind of receiver the bat uses (Floyd, 1980; Menne and Hackbarth, 1986). Noise jamming experiments have been conducted with flying bats in the obstacle-avoidance task; Plecotus townsendii can detect wires that return echoes with an energy signal-to-noise ratio as low as roughly +5 to +15 dB (Griffin et al, 1963). Because two-choice detection or discrimination experiments keep the bat relatively stationary, the variability of echo signal-to-noise ratios can be reduced to provide a more accurate estimate of the relationship between the bat's performance in a well defined task and the signal-to-noise ratio of echoes.

Four series of experiments have examined the effects of reduced echo signal-to-noise ratios on target-range discrimination, echo-delay discrimination, and echo-delay jitter detection by Eptesicus fuscus. In addition, target-range discrimination with reduced echo signal-to-noise ratios was studied in Phyllostomus hastatus. The experimental procedure was to introduce broadband random noise into the discrimination task to attempt to interfere with the bat's performance. In the target-range and echo-delay discrimination tasks, the stimuli and the noise were at relatively high levels (planar targets reflect strong echoes and require correspondingly high levels of noise to obtain interference), but in the echo-delay jitter task, the stimuli were only about 10 dB above the bat's threshold for echo detection in quiet conditions and were thus more biologically realistic. The results of the noise-jamming experiments are summarized as follows (Simmons, in press):

Eptesicus and Phyllostomus can successfully discriminate differences of 2 to 4 cm in target range (or the equivalent in echo delay) as long as the energy at some of the frequencies in the echoes exceeds the power of the noise at those frequencies. Eptesicus can increase the total energy of its sonar emissions by 38 dB over the total energy in quiet conditions to adapt to noise, thus extending its ability to perform in noise to levels 35 to 38 dB higher than would be possible if it used the same signals in noise as in the quiet. Phyllostomus exhibits an adaptive range of only about 15 to 20 dB in the same experimental conditions. While Phyllostomus increases the energy at all frequencies by approximately the same amount (by increasing the amplitude and the duration of its FM sweeps), Eptesicus additionally alters the shape of its FM sweeps to add a shallow-sweeping, nearly CF signal of several milliseconds' duration to the end of each sound (Simmons et al., 1978). The shallow sweep is at 22 to 24 kHz in the first harmonic, and its presence concentrates most of the added energy of the sounds at 22 to 24 kHz and at 44 to 48 kHz. The energy signal-to-noise ratio for the echoes received by Phyllostomus when target-range discrimination breaks down is about +11 dB. For Eptesicus the signal-to-noise ratio for the entire echo is roughly +30 dB, but Eptesicus treats the shallow FM sweep added to its more usual broadband FM sweeps as a distinct signal for use under conditions of low echo signal-to-noise ratios. The energy signal-to-noise ratio of echoes of the shallow FM sweep at 22 to 24 kHz is about .5 dB when target-range discrimination breaks down. When trained to perform targe-range discrimination in alternation with echo-jitter discrimination, Eptesicus does not add the shallow FM sweep to its signals as noise is introduced. Under these conditions target-range discrimination again breaks down at an energy signal-to-noise ratio for echoes in the 25 kHz region of about +4 to +6 dB, even though the total energy of the signal is now considerably lower. These results suggest that at least near-ideal performance can be obtained by bats, but one must know how the bat parcels out the frequencies in its signals to identify what really is the "signal."

Target Velocity: Echo Doppler Shifts and Range Rate

Doppler Compensation. An echolocating bat can determine the approach velocity of a target either by observing the rate-of-change of the target's range from successive echoes or by observing the Doppler shift of individual echoes. Numerous behavioral experiments have demonstrated that bats which emit CF/FM echolocation sounds are sensitive to Doppler shifts of the CF components of echoes (Henson et al., 1987; Roverud and Grinnell, 1985c; Schuller, 1980; see Schnitzler and Henson, 1980). The auditory and neural mechanisms responsible for this sensitivity have been identified (Bruns and Schmieszek, 1980; Henson et al., 1985; Neuweiler et al., 1980; Schuller and Pollak, 1979; Suga, 1984); it appears to play a crucial role for the detection and and identification of prey, especially in cluttered environments (Bell and Fenton, 1987; Habersetzer et al., 1984; Henson et al., 1987; Pollak and Schuller, 1981; Schnitzler et al., 1983, 1985; see Schnitzler and Henson, 1980; Ostwald et al., this volume).

Most research on the perception of velocity by echolocation has focused upon the performance of bats in maintaining the stability of CF frequencies in echoes by tracking changes in echo frequency with compensatory changes in the frequency of emissions. The discovery of the Doppler compensation response (Schnitzler, 1968) provided the first glimpse of a wholly different system of echolocation than that served by broadband, usually FM signals. Many species of bats emit CF signals of varying durations in tandem with FM signals, and these CF signals are used to facilitate detection of targets at long range (Grinnell, 1973; Kick, 1982; Schnitzler and Henson, 1980; Simmons et al., 1975, 1978) and to distinguish among targets on the basis of differences in patterns of motion

(Henson et al., 1987; Neuweiler et al., 1980; Schnitzler and Henson, 1980). In flight, bats encounter echo Doppler shifts due to their own flight velocity and echo Doppler shifts due to the movement of targets. Bats that emit CF signals partially or completely compensate for the upward Doppler shift of echoes caused by their own forward movement by lowering the frequency of CF components in their emissions. In effect, they regulate the frequency of returning echoes to a reference frequency determined by the tuning of the ear. Doppler shifts that remain in echoes after compensation are attributable to the motion of the target, and it appears as though the Doppler compensation response is intended to facilitate perception of the target's own flight velocity and fluttering movements. The auditory systems of bats that emit long-CF/FM echolocation sounds are specialized to perceive small changes in the frequency of CF echoes around the reference frequency for the Doppler compensation response. In addition, they are specialized to detect rapid modulations of the amplitude and frequency of CF components in echoes caused by the rhythmic beating of the wings of insects (Henson et al., 1987; Neuweiler et al., 1980; Pollak and Schuller, 1981; Schnitzler et al., 1983; Suga, 1984). The Doppler compensation response has been observed in bats that emit both short and long CF components. Because it is a frequency tracking response, the performance of the subsystem of echolocation which processes Doppler shifts is usually expressed in terms of the variability of the frequency of CF components around a resting or a reference value. Table 3 summarizes the results of a variety of different types of experiments which measured either the stability of CF frequencies or the accuracy of frequency discrimination at the reference frequency for the Doppler compensation response. All of these species are CF/FM bats except for Eptesicus, which emits FM/short-CF signals instead. As has been predicted (Schnitzler and Henson, 1980; Simmons et al, 1975), the accuracy of frequency determination underlying the different observations shown in Table 3 is related to the duration of the CF components. The correlation coefficient (r) for the observed variability of CF frequency in relation to the duration of the signals is 0.62, which accounts for only about a third of the variance in the accuracy of frequency tracking or discrimination by these bats, however.

The species of bats shown in Table 3 are very diverse, and they are likely to use the CF or "almost CF" components of their sonar signals in different ways. The reactions of three of these species to externally-produced CF sounds that fall close to the frequency of the CF components of their echolocation sounds are quite different. Rhinolophus compensates for the upward Doppler shift in echoes by decreasing the frequency of subsequent transmissions (Schnitzer and Henson, 1980). The Doppler compensation response moves the frequency of the bat's emissions downward and away from the frequency of echoes by a large amount if the Doppler shift is great, and by a small amount if it is small. When the echo is only slightly higher than the emission, the bat does not change the frequency of its emissions at all. Rhinolophus appears to treat externally-produced CF sounds as echoes because it attempts to Doppler-compensate for them (Schnitzler, personal communication). Eptesicus emits FM sonar signals that sweep downward to a relatively shallow, nearly CF component at the end of the sound (Simmons et al, 1979). The frequency of the lowest part of the sweep normally is around 21-22 kHz, but, when externally delivered CF signals are in the range of 21-25 kHz, the bat raises the frequencies of its emissions to 22-25 kHz (Simmons, in press). The bat thus approximately matches the frequency of the external sounds with the frequency of the tail-end of its FM sweep. The range of frequencies in the sweep is thus kept entirely above the frequency of the external signals because the sweep terminates at the externally-introduced frequency. When raising the frequency of its sonar signals in response to external sounds, Eptesicus also increases the amplitude and the duration of its signals. These are characteristic parts of the bat's response to acoustic interference (Simmons et al., 1978), so it is likely that the bat also raises the frequencies of its sweeps to avoid interference. Noctilio albiventris emits CF/FM sonar sounds and behaves differently than Rhinolophus in the presence of CF/FM sounds introduced externally. Noctilio tracks the frequency of external sounds in the 72-76 kHz range by adjusting the frequency of echoes of its own sounds to fall approximately at the frequency of the external sounds. Thus, while Rhinolophus attempts to Doppler compensate for external sounds, Noctilio attempts to track the frequency of external sounds. This is what Eptesicus does, too, if the tail-end of the FM sweep is considered to be a CF component. Under natural conditions, other bats would be the most likely sources of CF sounds (Roverud and Grinnell, 1985c), and tracking the frequency of external sounds may be the bat's normal reaction to the presence of another bat. The tracking response may be a social signal to drive away other individuals by preempting the proper frequencies.

Noctilio leporinus emits relatively short-duration CF signals in contrast to the long CF signals emitted by Rhinolophus. Noctilio compensates for echo Doppler shifts and can perceive differences in the CF frequency of echoes that are adequate to determine a target's approach velocity with an accuracy of about 2 m/sec. This species can perceive target velocity itself with an accuracy of about 0.4 m/sec, however. Presumably Noctilio perceives target velocity from the progressive changes in distance occurring on successive echoes rather than from echo Doppler shifts (Wenstrup and Suthers, 1984). Rhinolophus and Pteronotus parnellii both are very sensitive to rapid amplitude and frequency modulations of CF echoes and attack prey on the basis of wing-beat cues carried by these modulations (Henson et al., 1987; Schnitzler and Henson, 1980; Schnitzler and Ostwald, 1983). The very long-duration CF components of their emissions have a high duty cycle and are ideal for gathering information about dynamic aspects of the target's appearance (Schnitzler et al., 1983). Noctilio and other species that emit CF signals that are shorter in duration probably cannot acquire detailed images of wing-beats because the individual wing-beat cycles last longer than the sonar sounds do. Some species, such as Hipposideros may overcome the limited duration of individual signals by increasing the repetition-rate rather than the duration as a means of increasing the duty cycle. It remains evident, however, that the short CF signals emitted by many species cannot serve the same function for perceiving wing-beats as long-CF signals and that there are qualitative differences between the images perceived by long-CF and short-CF bats (Schnitzler and Henson, 1980; Simmons et al., 1975). Accordingly, it is not surprising that the accuracy of CF frequency tracking is less closely related to signal duration (Table 3) than the accuracy of target ranging is to bandwidth (Table 2); the "CF signal" is used for different purposes by different bats, and it is quite possible that the accuracy of frequency tracking is less important to some species than to others.

Directionality of Echolocation and Localization of Targets

Directionality of Emissions and Hearing. The sonar signals which bats project into the environment to perceive objects are transmitted in a broad beam towards the animal's front. This beam is steered in different directions primarily by moving the head, which then steers the ears in tandem with the emissions. If the bat then also moves its pinnae, these movements would be superimposed on the joint movement of the ears with the head. There is little evidence presently available about the possibility that bats may additionally steer the emitted beam by moving the mouth or the nasal emitter. Both the mouth and the noseleaf of bats actively move during echolocation, so it would not be surprising to find

Table 3. Variability of CF Frequency, Accuracy
of Frequency Tracking, or Frequency Discrimination

Species	CF Frequency	Variability of Frequency (std. deviation)	Duration of CF Signals	Reciprocal of Duration	Actual Velocity Accuracy
Eptesicus fuscus[1]	22 kHz	750 Hz	1-2 msec	500-1000 Hz	5.8 m/sec
Hipposideros bicolor[2,3]	154 kHz	1,170 Hz	4-5 msec	200-250 Hz	1.3 m/sec
	154 kHz	460 Hz	4 msec	250 Hz	0.5 m/sec
Noctilio leporinus[4]	58 kHz	570-830 Hz*	3-6 msec	170-330 Hz	1.7-2.4 m/sec
Hipposideros speoris[2,3]	135 kHz	660 Hz	6-7 msec	140-170 Hz	0.8 m/sec
	133 kHz	400 Hz	3-6 msec	170-330 Hz	0.5 m/sec
Noctilio albiventris[5]	75 kHz	220 Hz	6-8 msec	120-170 Hz	0.5 m/sec
Asellia tridens[6]	120 kHz	100-200 Hz	5-7 msec	140-200 Hz	0.14-0.3 m/sec
Pteronotus parnellii[7,8]	57 kHz	ca 100 Hz	16-22 msec	45-62 Hz	0.3 m/sec
	61 kHz	ca 150-300 Hz	6-38 msec	26-160 Hz	0.4-0.8 m/sec
Rhinolophus rouxi[2]	83-85 kHz	130-230 Hz	50 msec	20 Hz	0.3-0.5 m/sec
Rhinolophus ferrumequinum[9-11]	82 kHz	60 Hz	20-40 msec	25-50 Hz	0.1 m/sec
	82 kHz	38 Hz	10-60 msec	16-100 Hz	0.08 m/sec
	82 kHz	50-200 Hz	10-60 msec	16-100 Hz	0.1-0.4 m/sec
	82 kHz	12 Hz*	90-100 msec	10-11 Hz	0.03 m/sec

([1]Simmons, in press; [2]Schuller, 1980; [3]Habersetzer, et al., 1984; [4]Wenstrup and Suthers, 1984; [5]Roverud and Grinnell, 1985c; [6]Gustafson and Schnitzler, 1979; [7]Schnitzler, 1970; [8]Henson, et al., 1987; [9]Simmons, 1974; [10]Schuller, et al., 1974; [11]Schnitzler and Fleiger, 1983)

* Discrimination threshold rather than standard deviation of frequency

that the beam can be shaped and aimed to some extent independently of the aim of the head, however. For <u>Myotis</u>, the emitted sonar sounds have a 10 dB beamwidth of about 60° in the horizontal plane and about 40 to 60° in the vertical plane (Shimozawa et al, 1974). For the big brown bat, <u>Rhinolophus</u>, the emitted sound has a 10 dB beamwidth of about 90° at 30 kHz (Simmons, 1969b). The horseshoe bat, <u>Rhinolophus</u>, emits its 83 kHz CF signals with a 10 dB beamwidth of about 60° in the horizontal plane and 90° in the vertical plane (Schnitzler and Grinnell, 1977).

The directionality that is characteristic of echolocation is already apparent from the fact that the sonar sounds of bats are beamed in a directional manner. This makes echolocation as a whole a directional process. The sensitivity of an echolocating bat to a target is determined by the intensity of the sounds being transmitted, by the reflective acoustic properties of the target, and by the propagation of sound to and from the target. A bat will be able to detect a target at a greater distance if it is located straight ahead than if it is off to one side because the sound reaching the target to produce an echo will be strongest straight ahead. The converse of directional sensitivity is also true--the bat is less sensitive to targets in directions other than straight ahead, so that echoes returning from objects not in the preferred direction will be relatively weaker. The importance of the directionality of sonar emissions for reducing the strength of echoes from extraneous targets, thus preventing interference, stands out as one of several aspects of echolocation related as much to avoiding interference as to obtaining good acoustic images of individual targets in the first place (Griffin, 1958; Grinnell, 1967; Grinnell and Grinnell, 1965; Henson, 1967; Shimozawa et al., 1974).

Not only are the bat's sonar emissions directional, but the reception of sounds is directional, too. The hearing sensitivity of bats is directed to the front, along approximately the same axis as the sounds are projected. In the horseshoe bat, <u>Rhinolophus</u>, the 10 dB directional receiving beam is 60° wide in the horizontal plane and 80° wide in the vertical plane, at the 83 kHz frequency of the bat's CF signals (Grinnell and Schnitzler, 1977). These receiving beamwidths were measured behaviorally. In bats that emit FM sonar sounds, the only data presently available on the directionality of hearing have been obtained using acoustical measurements or physiological rather than behavioral responses. As indicated by evoked potentials in the auditory midbrain, the 10 dB directional receiving beamwidth for <u>Myotis</u> is approximately 90° in the horizontal and vertical planes at the ultrasonic frequencies predominantly used for echolocation (Grinnell and Grinnell, 1965; Shimozawa et al, 1974). In <u>Eptesicus</u>, the receiving 10-dB beamwidth is about 75° in the horizontal plane (Simmons, 1987). In other species that use FM sounds, the directionality both of emissions and of hearing has not yet been determined.

Since both the sonar emissions and the auditory sensitivity of echolocating bats are directional, the overall sensitivity to a sonar target will be directional, too. The directional sensitivity to targets is compounded from the directionality of emissions and of hearing. For the horeseshoe bat, <u>Rhinolophus</u>, the 10 dB width of the beam of sensitivity to sonar targets is about 40° (Grinnell and Schnitzler, 1977). This refers to the directionality of echolocation at the 83 kHz frequency of the bat's CF signals. In the FM bat, <u>Mytois</u>, the combined directionality of emissions and reception results in a directional beam about 30 to 40° wide for sensitivity to targets. In <u>Eptesicus</u>, the combined directionality would yield a beam with a 10 dB width of 50° (In general, in bats the directionality of emissions and of reception are similar.) Insectivorous bats probably search over a conical zone about 120° wide when they are hunting for prey (Griffin et al., 1960; Schnitzler and Henson, 1980), so the directionality of detection in practice is quite broad, at least in FM bats.

Many species of bats (Eptesicus and Myotis, for example) keep their ears relatively stationary on the head, and the ears are moved when the aim of the head itself changes. If these bats make fine adjustments of the position of the ears, they have not yet been demonstrated. In several families of bats the external ears are rigidly constrained by their convoluted shape (Molossidae) or by a bridge of tissue which fuses them together and renders them immobile (Rhinopomatidae, Megadermatidae). The directional patterns of emission and of hearing are thus steered together when the bat changes the orientation of its head. Bats in the families Phyllostomidae and Noctilionidae often move their ears irregularly when they echolocate, but it is not known whether the directionality of hearing is redirected or merely changed in width by these movements. Horseshoe bats (Rhinolophidae, Hipposideridae), which emit long-CF/FM signals, move their ears in a repetitive, alternating manner that is synchronized to the production of echolocation sounds. The bat steers the echolocation system as a whole by moving its head, but the added movements of the ears scan up and down in the vertical direction with the directionality of hearing (Schnitzler and Henson, 1980).

Acuity of Localizationn of Targets. The accuracy with which a bat can determine the horizontal and vertical position of a sonar target is substantially sharper than the width of the directional beam of the sonar system for detecting sonar targets. The big brown bat, Eptesicus fuscus, can perceive a shift of 1.5° in the horizontal direction of a sonar target (Simmons et al., 1983) and a shift of 3° in the vertical direction of a target (Lawrence and Simmons, 1982b). also can track a target moving in the horizontal plane with an accuracy that appears to derive from an image of the target's position that is as sharp as 2° or thereabouts (Masters et al., 1985). To achieve an accuracy for target localization in the region of a degree or two using a sonar system with a directional sensitivity much broader than that requires echoes to be processed to extract from them the directional information they contain, beyond the processing required simply to detect targets in different directions. In effect, an acoustic image showing the target's position in relation to other targets and in relation to the orientation of the animal's body must be derived from the signals received at the two ears. Although the directionality of emissions and reception is likely to figure into the process of determining a target's location, it seems unlikely that the signal processing operations which produce acoustic images representing the target's location depend strictly upon the directional sensitivity of echolocation for their success. Directional sensitivity as such probably represents a compromise between the need to search for objects over a relatively broad front and the need to isolate individual objects located straight ahead from extraneous objects located elsewhere which produce interfering echoes. These two constraints are concerned with the detection of targets, whereas the acuity of localization refers to the sharpness of the acoustic images obtained after detection has occurred. The target's image occupies a small fraction of the zone of space being probed for targets.

The direction of a target could be perceived using three categories of information in echoes. The first, and simplest, category consists of the intensity of the echo as this is determined by the position of the target in the directional beam of the sonar system. To perceive a target's location solely from echo strength, the bat could move the beam up and down and from side to side around the target, observing the direction of aim which yields the strongest echoes. As far as is presently known, this is the method which Rhinolophus uses to determine the vertical position of a target (Grinnell and Schnitzler, 1977; see Schnitzler and Henson, 1980). Rhinolophus moves its ears in an alternating manner synchronized to the emission of sonar sounds. These movements sweep the directional receiving pattern for each ear vertically over the location of a target; the alternating movements of the ears have the effect of moving one ear's receiving beam upward past the target while the other ear's beam is moving downward. The emitted sound is beamed to the front, and the bat appears to probe along the vertical axis within the emitted beam to locate the target. Since the sonar signals of Rhinolophus contain long-duration CF components, there is a long enough interval of time during which each echo continues to return for ear-movements to pick up changes in echo intensity as the receiving beam-pattern sweeps over the target. Although the ears of Rhinolophus can also be moved significantly in a horizontal direction, and the bat's ears can be seen to "follow" passively generated sounds when the bat is hanging on a perch, there is apparently little or no horizontal movement of the ears during flight, so it is unlikely that intensity changes correlated with sweeping ear movements are used to perceive a target's horizontal position.

The second category of information in echoes that bats could use to localize targets includes differences in the amplitude, spectrum, or time-of-arrival of the echoes received at the two ears. Both ears must be unobstructed for echolocation to function effectively (Griffin, 1958; Schnitzler and Henson, 1980). When only one of the two ears of Rhinolophus is plugged, the bat experiences difficulty avoiding obstacles, but when both ears are plugged by about the same amount the bat is capable of avoiding obstacles virtually as effectively as if neither ear were plugged (Flieger and Schnitzler, 1973). Evidently the level of sound delivered to the ears must not undergo asymmetrical disruptions for the bat to localize targets. It is conventional to identify interaural intensity and arrival-time cues as the primary basis for horizontal sound localization (see Mills, 1972, for example), and these cues would be available to an echolocating bat (Schnitzler and Henson, 1980; Simmons, 1977; Simmons et al., 1983). To these cues must be added interaural differences in the spectrum of sounds (Fuzessery and Pollak, 1984; Grinnell and Grinnell, 1965). Isolation of binaural intensity, time, and spectral cues for localization depends upon experiments using earphones for delivery of controlled stimuli to the bat's ears, and such behavioral experiments have not been carried out. The heads of echolocating bats are small, and the interaural time differences that the bat would have to perceive are in the range of only a few microseconds if time cues are to be used for horizontal localization of targets. These have generally been judged too small for bats to perceive (Grinnell, 1963b; Schnitzler, 1973b; Peff and Simmons, 1972). However, an early suggestion that bats might in fact be able to perceive small interaural time differences (Simmons, 1977) is supported by the discovery that the echo delay acuity of Eptesicus is as small as about 1 μsec. The small size of the bat's head might be offset by the large bandwidth of sonar sounds. The acuity of horizontal-angle discrimination by Eptesicus is compatible with the use of interaural timing cues if the bat's echo-delay acuity also applies to perception of interaural time differences (Simmons et al., 1983). Neural responses recorded from the brainstem of bats show a surprising degree of sensitivity to interaural arrival-time differences (Harnischfeger et al., 1985; see Pollak, this volume). The interaural intensity differences available to bats are quite large (Grinnell, 1963b; Pollak, this volume) and can be presumed to contribute to horizontal localization. They may prove to be the primary source of directional information for many species. Species of bats may differ in the relative importance of interaural time and intensity cues for localization, too.

Echolocation sounds contain a broad range of frequencies, and echoes from targets arrive at the eardrums of bats with differences in their spectra imposed by the directional properties of the external ears (Grinnell and Grinnell, 1965; Shimozawa et al., 1974). In principle, a bat could locate a target by comparing the amplitudes of several different frequencies at the two ears (Fuzessery and Pollak, 1984; Grinnell and Grinnell, 1965), and neural responses recorded from the bat's auditory system suggest that spectral differences in echoes reaching the two ears are preserved and encoded in the auditory system's neural representation of echoes (Fuzessery and Pollak, 1984; Grinnell and Grinnell, 1965; Shimozawa et al., 1974). Individual auditory neurons exhibit directional sensitivity, but it is not clear whether these directional patterns arise from signal-processing apart from the directional sensitivity of the external ear itself (Jen, 1980; Jen et al., 1984; Poussin and Schlegel, 1984; Wong, 1984). Neural responses do exhibit binaural interactions and frequency-dependent directional sensitivity that are consistent with interaural intensity differences at different frequencies being an important cue for sound localization by bats (Fuzessery and Pollak, 1984; Wenstrup et al., 1985), but such observations do not rule out the possibility that interaural time differences are used, too.

The external ears modify the waveform and the spectrum of echoes reaching the bat's inner ears. The only direct evidence for a role of the external ears in localization of targets is the observation that disruption of the tragus profoundly degrades vertical localization by _Eptesicus_ (Lawrence and Simmons, 1982b). Such disruption does not affect horizontal localization, though (Simmons, in press). Removal of the tragus has only a slight effect on the directionality of the ear or on the frequency dependence of directionality (Grinnell and Grinnell, 1965; Simmons, in press), so it seems unlikely that localization of targets by _Eptesicus_ depends primarily upon directional spectral cues, although spectral cues may still be used as part of a multiple-cue system for localization. Removal of the tragus changes the time waveform of echoes by removing external-ear reverberations in the 40 to 60 μsec delay range. The dramatic disappearance both of sharp vertical localization as measured behaviorally and also of these reverberations when the tragus is disrupted from its normal position strongly implicates pinna-tragus reverberations as the principal acoustic cue for vertical localization by _Eptesicus_. The long-eared bat, _Plecotus_, becomes disoriented in flight when the pinnae are displaced medially by gluing their edges together, and this manipulation severely modifies the directionality of hearing (Griffin, 1958; Grinnell and Grinnell, 1965). It is not known how this affects the acuity of localization, though. Much work remains to be done to identify the contributions of the external ears to the acuity of localization as distinct from directional reception of sounds, particularly in terms of spectral and temporal cues generated by the tragus.

ACKNOWLEDGMENTS

Writing of this chapter was supported by NSF Grant No. BNS 83-02144, by ONR Contract No. N00014-86-K-0401, and by System Development Foundation Grant No. 57. Portions of the discussion of sound localization and the directionality of hearing and echolocation by bats are summarized from Simmons, 1987.

REFERENCES

Ajrapetjantz, E., and Konstantinov, A. I., 1974, "Echolocation in Nature," Nauka, Leningrad (English translation, Joint Publications Research Service, Arlington, VA).

Altes, R. A., 1980, Detection, estimation and classification with spectrograms, J. Acoust. Soc. Am., 67:1232-1246.

Altes, R. A., 1981, Echo phase perception in bat sonar? J. Acoust. Soc. Am., 69:505-508.

Altes, R. A., and Titlebaum, E. L., 1970, Bat signals as optimally Doppler tolerant waveforms, J. Acoust. Soc. Am., 48:1014-1020.

Bell, G. P., and Fenton, M. B., 1984, The use of Doppler-shifted echoes as a flutter detection and clutter rejection system: The echolocation and feeding behavior of Ecol. Sociobiol., 15:109-114.

Bergeijk, W. A. van, 1964, Sonic pulse compression in bats and people, J. Acoust. Soc. Am., 36:594-597.

Beuter, K. J., 1980, A new concept of echo evaluation in the auditory system of bats, in: R.-G. Busnel and J. F. Fish, eds., "Animal Sonar Systems", Plenum, New York, pp. 747761.

Bodenhamer, R. D., and Pollak, G. D., 1981, Time and frequency domain processing in the inferior colliculus of echolocating bats, Hearing Res., 5:317-355.

Bruns, V., and Schmieszek, E., 1980, Regional variation in the innervation pattern in the cochlea of the greater horseshoe bat, Hearing Res., 3:27-43.

Cahlander, D. A., 1964, Echolocation with wide-band waveforms: Bat sonar signals. MIT Lincoln Laboratory Report No. 271, Lexington, MA.

Feng, A. S., Simmons, J. A., and Kick, S. A., 1978, Echo detection and target-ranging neurons in the auditory system of the bat, Eptesicus fuscus, Science, 202:645-648.

Flieger, E., and Schnitzler, H.-U., 1973, Ortungsleitungen der Fledermaus Rhinolophus ferrumeguinum bei ein- und beidseitiger Ohrverstopfung, J. Comp. Physiol., 82:93-102.

Floyd, R. W., 1980, Models of cetacean signal processing, in: "Animal Sonar Systems," R.-G. Busnel and J. F. Fish, eds., Plenum, New York, pp. 615- 623.

Fuzessery, Z. M., and Pollak, G. D., 1984, Neural mechanisms of sound localization in an echolocating bat, Science, 225:725-728.

Glaser, W., 1974, Zur Hypothese des Optimalempfangs bei der Fledermaus-ortung, J. Comp. Physiol., 94:227-248.

Griffin, D. R., 1958, "Listening in the Dark." Yale University Press, New Haven, CT, (reprinted by Cornell Univ. Press, Ithaca, NY).

Griffin, D. R., Webster, F. A., and Michael, C. R., 1960, The echolocation of flying insects by bats, Anim. Behav., 8:141-154.

Griffin, D. R., McCue, J. J. G., and Grinnell, A. D., 1963, The resistance of bats to jamming, J. Exp. Zool.; 152:229-250.

Griffin, D. R., and Simmons, J. A., 1974, Echolocation of insects by horseshoe bats, Nature, 250:731-732.

Grinnell, A. D., 1963a, The neurophysiology of audition in bats: Temporal parameters, J. Physiol. (London), 167:67-96.

Grinnell, A. D., 1963b, The neurphysiology of audition in bats: Directional localization and binaural interaction, J. Physiol. (London), 167:97-113.

Grinnell, A. D., 1967, Mechanisms of overcoming interference in echolocating animals, in: "Animal Sonar Systems: Biology and Bionics," Vol. I., R.G. Busnel, ed., Laboratoire de physiologie Acoustique, Jouy-en-Josas, France, pp. 451-481.

Grinnell, A. D., 1970, Comparative auditory neurophysiology of neotropical bats employing different echolocation signals, Z. Vergl. Physiol., 68: 117-153.

Grinnell, A. D., 1973, Neural processing mechanisms in echolocating bats, correlated with differences in emitted sounds, J. Acoust. Soc. Am., 54:147-156.

Grinnell, A. D., and Grinnell, V. S., 1965, Neural correlates of vertical localization by echo-locating bats, J. Physiol. (London), 181:830-851.

Grinnell, A. D., and Hagiwara, S., 1972, Adaptations of the auditory
    nervous system for echolocation. Studies of New Guinea bats. Z. Vergl.
    Physiol., 76:41-81.
Grinnell, A. D., and Schnitzler, H.-U., 1977, Directional sensitivity of
    echolocation in the horseshoe bat, II. Behavioral directionality of
    hearing, J. Comp. Physiol., 116:63-76.
Gustafson, Y., and Schnitzler, H.-U., 1979, Echolocation and obstacle
    avoidance in the hipposiderid bat, Asellia tridens, J. Comp. Physiol.,
    131:161-167.
Habersetzer, J., 1981, Adaptive echolocation sounds in the bat, Rhinopoma
    hardwickei, J. Comp. Physiol., 144:559-566.
Habersetzer, J., and Vogler, B., 1983, Discrimination of surface-structured
    targets by the echolocating bat Myotis myotis during flight, J. Comp.
    Physiol., 152:275-282.
Habersetzer, J., Schuller, G., and Neuweiler, G., 1984, Foraging behavior
    and Doppler shift compensation in echolocating hipposiderid bats,
    Hipposiderso bicolor and Hipposideros speoris, J. Comp. Physiol.,
    155:559-567.
Harnischfeger, G., 1980, Brainstem units of echolocating bats code binaural
    time differences in the microsecond range, Naturwiss, 67:314-315.
Harnischfeger, G., Neuweiler, G., and Schlegel, P., 1985, Interaural time
    and intensity coding in superior olivary complex and inferior colliculus
    of the echolocating bat Molossus ater, J. Neurophysiol., 53:89-109.
Henson, O. W., Jr., 1967, The perception and analysis of biosonar signals
    by bats, in: "Animal Sonar Systems: Biology and Bionics," Vol. II, R.G.
    Busnel, ed., Laboratoire de physiologie acoustique, Jouy-en-Josas,
    France, pp. 949-1003.
Henson, O. W., Jr., Schuller G., and Vater, M., 1985, A comparative study
    of the physiological properties of the inner ear in Doppler shift
    compensating bats (Rhinolophus rouxi and Pteronotus p. penwllii, J.
    Comp. Physiol., 157:587-597.
Henson, O. W., Jr., Bishop, A. L., Keating, A. W., Kobler, J. B., Henson,
    M. M., Wilson, B. S., and Hansen, R. C., 1987, Biosonar imaging of
    insects by pteronotus p. parnellii, the mustached bat, Natl. Geog. Res.,
    3:82-101.
Jen, P. H-S., 1980, Coding of directional information by single neurones in
    the S-segment of the FM bat, Mytotis lucifugus, J. Exp. Biol., 87:203-
    216.
Jen, P. H.-S., Sun, Z., Kamada, T., Zhang, S., and Shimozawa, T., 1984,
    Auditory response properties and spatial response areas of superior
    collicular neurons of the FM bat, Eptesicus fuscus, J. Comp. Physiol.,
    154:407-413.
Johnson, R. A., 1980, Energy spectrum analysis in echolocation, in: "Animal
    Sonar Systems, R.-G. Busnel and J. F. Fish, eds., Plenum, New York, pp.
    673-693.
Kick, S. A., 1982, Target-detection by the echolocating bat, Eptesicus
    fuscus, J. Comp. Physiol., 45:431-435.
Kick, S. A., and Simmons, J. A., 1984, Automatic gain control in the bat's
    sonar receiver and the neuroethology of echolocation, J. Neurosci,
    4:2725-2737.
Klauder, J. R., 1960, The design of radar systems having both high range
    resolution and high velocity resolution, Bell System Tech. J., 39:809-
    820.
Klauder, J. R., Prince, A. C., Darlington, S., and Albersheim, W. J. 1960,
    The theory and design of chirp radars, Bell System Tech. J., 39:745-808.
Lawrence, B. D., and Simmons, J. A., 1982a, Measurements of atmospheric
    attenuation of ultrasonic frequencies and the significance for
    echolocation by bats, J. Acoust. Soc. Am., 71:484-490.
Lawrence, B. D., and Simmons, J. A., 1982b, Echolocation in bats: The
    external ear and perception of the vertical positions of targets,
    Science, 218:481-483.

Masters, W. M., Moffat, A. J. M., and Simmons, J. A., 1985, Sonar tracking of horizontally moving targets by the big brown bat, Eptesicus fuscus, Science, 228:1331-1333.

McCue, J. J. G., 1966, Aural pulse compression by bats and humans,J. Acoust. Soc. Am., 40:545-548.

McCue, J. J. G., 1969, Signal processing by the bat, Myotis lucifugus, J. Aud. Res., 9:100-107.

Menne, D., and Hackbarth, H., 1986, Accuracy of distance measurement in the bat Eptesicus fuscus: Theoretical aspects and computer simulations, J. Acoust. Soc. Am., 79:386-397.

Mills, A. W., 1972, Auditory localization, in: "Foundations of Modern Auditory Theory," J. V. Tobias, ed., Academic Press, New York, pp. 303-348.

Møhl, B., 1986, Detection by a pipistrelle bat, Acustica, 61:75-82.

Neuweiler, G., Bruns, V., and Schuller, G., 1980, Ears adapted for the detection of motion, or how echolocating bats have exploited the capacities of the mammalian auditory system, J. Acoust. Soc. Am., 68:741-753.

Novick, A., 1977, Acoustic orientation, in: "Biology of Bats," Vol. III, W. A. Wimsatt, ed., Academic Press, New York, pp. 73-287.

O'Neill, W. E., and Suga, N., 1982, Encoding of target-range information and its representation in the auditory cortex of the mustached bat, J. Neurosci., 47:225-255.

Peff, T. C., and Simmons, J. A., 1972, Horizontal-angle resolution by echolocating bats, J. Acoust. Soc. Am., 51:2063-2065.

Pollak, G. D., 1980, Organizational and encoding features of single neurons in the inferior colliculus of bats, in: "Animal Sonar Systems," R.-G. Busnel and J. F. Fish, eds., Plenum, New York, pp. 549-587.

Pollak, G. D., and Schuller, G., 1981, Tonotopic organization and encoding features of single units in inferior colliculus of horseshoe bats: Functional implications for prey identification, J. Neurophysiol., 45:208-226.

Poussin, C., and Schlegel, P., 1984, Directional sensitivity of auditory neurons in the superior colliculus of the bat, Eptesicus fuscus, using free field sound stimulation, J. Comp. Physiol., 154:253-261.

Pye, J. D., 1980, Echolocation signals and echoes in air, in: "Animal Sonar Systems," R.-G. Busnel and J. F. Fish, eds., Plenum, New York, pp. 309-353.

Pye, J. D., 1986, Sonar signals as clues to system performance, Acustica, 61:66-175.

Roverud, R. C. and Grinnell. A. D., 1985a, Discrimination performance and echolocation signal integration requirements for target detection and distance determination in the CF/FM bat, Noctilio albiventris, J. Comp. Physiol., 156:447-456.

Roverud, R. C. and Grinnell, A. D., 1985b, Echolocation sound features processed to provide distance information in the CF/FM bat, Noctilio albiventris: Evidence for a gated time window using both CF and FM components, J. Comp. Physiol., 156:457-469.

Roverud, R. C. and Grinnell, A. D., 1985c, Frequency tracking and Doppler-shift compensation in response to an artificial CF/FM echolocation sound in the CF/FM bat, Noctilio albiventris, J. Comp. Physiol., 156:471-475.

Schnitzler, H.-U., 1984, The performance of bat sonar systems, in: "Localization and Orientation in Biology and Engineering," Varju and Schnitzler, eds., Springer, Berlin, pp. 221-224.

Schnitzler, H.-U., and Flieger, E., 1983, Detection of oscillating target movements by echolocation in the greater horseshoe bat, J. Comp. Physiol., 153:385-391.

Schnitzler, H.-U., and Grinnell, A. D., 1977, Directional sensitivity of echolocation in the horseshoe bat, I. Directionality of sound emission, J. Comp. Physiol., 116:51-61.

Schnitzler, H.-U. Hackbarth, H., Heilmann, U., and Herbert H., 1985, Echolocation behavior of rufous horseshoe bats hunting for insects in the flycatcher-style, J. Comp. Physiol., 157:39-46.

Schnitzler, H.-U. and Henson, O. W., Jr., 1980, Performance of airborne animal sonar systems: I. Microchiroptera, in: "Animal Sonar Systems," R.G. Busnel and J. F. Fish, eds., Plenum, New York pp.109-181.

Schnitzler, H.U., Menne, D., Kober, R., and Keblich, K., 1983, The acoustical image of fluttering insects in echolocationg bats, in: "Neuroethology and Behavioral Physiology: Roots and Growing Points," F. Huber and H. Markl, eds., Berlin, Springer, pp. 235-250.

Schnitzler, H.-U., and Ostwald, J., 1983, Adaptations for the detection of fluttering insects by echolocation in horseshoe bats, in: "Advances in Vertebrate Neuroethology," J.P. Ewert, R.R. Capranica, and D.J. Ingle, eds., Plenum, New York, pp. 801-827.

Schuller, G., 1980, Hearing characteristics and Doppler-shift compensation in south Indian CF/FM bats. J. Comp. Physiol, 139: 349-356.

Schuller, G., Beuter, K., and Schnitzler, H.-U., 1974, Response to frequency shifted artifical echoes in the bat Rhinolophus ferrumeguinum, J. Comp. Physiol., 89:275-286.

Schuller, G., and Pollak, G. D., 1979, Disproprotionate frequency representation in the inferior colliculus of horseshoe bats: evidence for an "acoustic fovea," J. Comp. Physiol., 132:47-54.

Shimozawa, T., Suga, N., Hendler, P., and Schuetze, S., 1974, Directional sensitivity of echolocation system in bats producing frequency-modulated signals, J. Exp. Biol., 60:53-69.

Simmons, J. A., 1969a, Depth perception by sonar in the bat Eptesicus fuscus. Ph.D. Thesis, Princeton University.

Simmons, J. A., 1969b, Acoustic radiation patterns for the echolocating bats, Chilonycteris rubiginosa and Eptesicus fuscus, J. Acoust. Soc. Am., 46:1054-1056.

Simmons, J. A., 1971, Echolocation in bats: Signal processing of echoes for target range, Science, 171:925-928.

Simmons, J. A., 1973, The resolution of target range by echolocating bats, J. Acoust. Soc. Am., 54:157-173.

Simmons, J. A., 1974, Response of the Doppler echolocation system in the bat Rhinolophus ferrumeguinum, J. Acoust. Soc. Am., 56:672-682.

Simmons, J. A., 1977, Localization and identification of acoustic signals, with reference to echolocation, in: "Recognition of Complex Acoustic Signals," T. H. Bullock, ed., Berlin, Dahlem Konferenzen, pp. 239-277.

Simmons, J. A., 1979, Perception of echo phase information in bat sonar, Science, 204:1336-1338.

Simmons, J. A., 1980, The processing of sonar echoes by bats, in: "Animal Sonar Systems," R.G. Busnel and J. F. Fish, eds., Plenum, New York, pp. 695-714.

Simmons, J. A., 1987, Directional hearing and sound localization in echolocating animals, in: "Directional Hearing," W. A. Yost and G. Gourevitch, eds., Springer, New York.

Simmons. J. A. (in press), "The Sonar of Bats," Princeton University Press, Princeton, NJ.

Simmons, J. A., Fenton, M. B., and O'Farrell, M. J., 1979, Echolocation and pursuit of prey by bats, Science, 203:1621.

Simmons, J. A., Howell, D. J., and Suga, N., 1975, Information content of bat sonar echoes, Am. Sci., 63:204-215.

Simmons, J. A., and Kick, S. A., 1983, Interception of flying insects by bats, in' "Neuroethology and Behavioral Physiology: Roots and Growing Points," F. Huber and H. Markl, eds., Springer, New York, pp. 267-279.

Simmons, J. A. and Kick, S. A., 1984, Physiological mechanisms for spatial filtering and image enhancement in the sonar of bats, Ann. Rev. Physiol., 46:599-614.

Simmons, J. A., Kick, S. A., and Lawrence, B. D., 1979, Echolocation by mouse-tailed bats, Natl. Geog. Soc. Res. Rep., 1979 Projects, 681-691.

Simmons, J. A., Kick, S. A., Lawrence, B. D., Hale, C., Bard, C., and Escudie, B., 1983, Acuity of horizontal angle discrimination by the echolocating bat, Eptesicus fuscus, J. Comp. Physiol., 153:321-330.

Simmons, J. A., Kick, S. A., and Lawrence, B. D., 1984, Echolocation and hearing in the mouse-tailed bat, Rhinopoma hardwickei: Acoustic evolution of echolocation in bats, J. Comp. Physiol., 154:347-356.

Simmons, J. A., and Lavender, W. A., 1976, Representation of target range in the sonar receivers of echolocating bats, J. Acoust. Soc. Am., 60:S5.

Simmons, J. A., Lavender, W. A., and Lavender, B. A., 1978, Adaptation of echolocation to environmental noise by the Proceedings of the Fourth International Bat Research Conference, Kenya National Academy for Advancement of Arts and Science, Nairobi, Kenya, pp. 97-104.

Simmons, J. A., Lavender, W. A., Lavender, B. A., Doroshow, C. A., Kiefer, S.W., Livingston, R., Scallet, A. C., and Crowley, D. E., 1974, Target structure and echo spectral discrimination by echolocating bats, Science, 186:1130-1132.

Simmons, J. A., and Stein, R. A., 1980, Acoustic imaging in bat sonar: Echolocation signals and the evolution of echolocation, J. Comp. Physiol., 135:61-84.

Stewart, J. L., 1979, "The Bionic Ear." Covox Company, Eugene, OR.

Strother, G. K., 1961, Note on the possible use of ultrasonic pulse compression by bats, J. Acoust. Soc. Am., 33:696-697.

Suga, N., 1970, Echo-ranging neurons in the inferior colliculus of bats, Science, 170:449-452.

Suga, N., 1973, Feature extraction in the auditory system of bats, in' "Basic Mechanisms in Hearing," A. R. Moller, ed., Academic Press, New York, pp. 675-744.

Suga, N., 1984, The extent to which biosonar information is represented in the bat auditory cortex, in: "Dynamic Aspects of Neo-Cortical Function," G. M. Edelman, W. E. Gall, W. M. Cowan, eds., Wiley, New York: pp. 315-373.

Suga, N., and Horikawa, J., 1966, Multiple time axes for representation of echo delays in the auditory cortex of the mustached bat, J. Neurophysiol., 55:776-805.

Suga, N., and Shimozawa, T., 1974, Site of neural attenuation of responses to self-vocalized sounds in echolocating bats, Science, 183:1211-1213.

Suga, N., and Schlegel, P., 1972, Neural attenuation of responses to emitted sounds in echolocating bats, Science, 177:82-94.

Surlykke, A., and Miller, L. A., 1985, The influence of arctiid moth clicks on bat echolocation; jamming or warning? J. Comp. Physiol. 156:831-843.

Suthers, R. A., and Wenstrup, J. J., Behavioural discrimination studies involving prey capture by echolocating bats, in: "Recent Advances in the Study of Bats," M. B. Fenton, P. A. Racey, and I. M. V. Rayner, eds., Cambridge University Press, in press.

Trappe, M., 1982, Verhalten und Echoortung der Grossen Hufeisennase (Rhinolophus ferrumequinum) beim Insektenfang. Ph.D. Thesis, Universitat Marburg.

Trappe, M., and Schnitzler, H.-U., 1982, Doppler-shift compensation in insect-catching horseshoe bats, Naturwiss, 69:193-194.

Vogler, B., and Neuweiler, G., 1983, Echolocation in the noctule (Nyctalus noctula) and horseshoe bat (Rhinolophus ferrumeguinum), J. Comp. Physiol., 152:421-432.

Webster, F. A., and Brazier, O. G., 1965, Experimental studies on target detection, evaluation, and interception by echolocating bats, Technical Report No. AMRL-TR-65-172, Aerospace Medical Division, U.S.A.F. Systems Command, Tucson, AZ.

Wenstrup, J. J., Ross, L. S., and Pollak, G. D., 1985, A functional organization of binaural responses in the inferior colliculus, Hearing Res., 17:191-195.

Wenstrup, J. J., and Suthers, R. A., 1984, Echolocation of moving targets by the fish-catching bat, <u>Noctilio</u> <u>leporinus</u>, J. Comp. Physiol., 155:75-89.

Wong, D., 1984, Spatial tuning of auditory neurons in the superior colliculus of the echolocating bat, <u>Myotis</u> <u>luxidufua</u>, Hearing Res., 16:261-270.

Woodward, P. M., 1964, "Probability and Information Theory with Applications to Radar," 2nd Edition, Pergamon Press, New York.

DESIGNING CRITICAL EXPERIMENTS ON DETECTION AND ESTIMATION

IN ECHOLOCATING BATS

Dieter Menne[*]

Department of Zoology
National University of Singapore
10, Kent Ridge Crescent
Singapore 0511

INTRODUCTION

The similarity between sophisticated echolocation signals of bats and those of man-made sonar and radar has been a challenge to biologists and engineers since many years. The similarity in the structure of these signals suggested that their function might be analogous, and that signal processing in radar receivers and the brain of bats might be based on the same principles. The aim of many experimental studies has been not primarily to discover the limits of echolocation, but to find how these limits are imposed by characteristics of the signal and their processing in the nervous system. The most far-reaching goal of this approach has been to use behavioral data to predict possible receiver models that can be tested with electrophysiological methods.

The mathematical tools from the analysis and synthesis of radar receivers have been used to analyse bat signals and to design echolocation experiments. In this paper, I review some of the basic concepts of detection of signals and estimation of their parameters, and demonstrate these concepts with selected experiments from the literature and outlines of proposed experiments. Two major questions will be asked.

How can experiments in echolocation be designed to critically test theoretical receiver models?

What are the assumptions a system engineer has to make to translate the conditions and results of behavioral experiments into the formalized language of decision theory?

Certainly the most important theoretical contributions to signal processing in animal echolocation systems have been published by Altes (see Altes, 1984b, for a review). However, his work is mostly mathematical and consequently of limited accessibility to biologists who want to test the proposed models in behavioral experiments. This study is an attempt to

---

[*] On leave from: Lehrstuhl Zoophysiologie, Institut fuer Biologie III
D-7400 Tübingen, F.R. Germany

bring theory and experiment in bat echolocation research somewhat closer together.

NOISE AND FLUCTUATION

The responses of an animal in a detection task are normally not always correct, and estimated values of range, angular orientation or frequency jitter around some mean. This unpredictability may come from external sources or may be produced in the processing system itself. In the Theory of Signal Predictability (Peterson et al., 1954, Tanner and Birdsall, 1958, Green and Swets, 1966) it is modeled by Gaussian white noise added to the signal. To make sure that this assumption can be made, most experiments in modern psychoacoustics are designed so that external noise is the main factor limiting performance. When no external noise is present, the limits of performance depend on unpredictable processes in the hearing system itself; these can only be approximately represented as additive Gaussian noise ('internal noise') that is independent of the amplitude of the signal (Swets et al., 1959).

In the theoretical treatment of a system, noise is thought to result from statistical processes that are unpredictable in detail but quite stable in their overall characteristics, which can be quantified by power spectrum and amplitude distribution. When noise is added to a signal, the result is jitter in all the signal's parameters, and a theoretical analysis is necessary to find out how such additive noise affects the phase, amplitude and frequency of the signal. Beside noise, there exist other types of unpredictability, called fluctuations (Fig. 1). They act chiefly on one specific parameter of the signal and cannot be modeled by additive noise. As an example, consider slow amplitude fluctuations of the echo that affect the probability of target detection in radar (Swerling, 1960). These amplitude fluctuations may come from variable spreading loss (Michelsen and Larsen, 1983), changing target cross section or varying directivity of the ear. In bat echolocation the amplitude and frequency of the echo from a free flying insect will rapidly fluctuate as a result of the oscillating target cross section (Schnitzler et al., 1983). The fluctuating amplitude is a 'stray' or 'nuisance' parameter, as it carries unreliable information on, for example, range estimation. In radar theory a statistical model is sometimes used to get more independent of fluctuations of a parameter. The most radical statistical assumption for a stray parameter is that all its possible values are equally probable and therefore carry no information: in man-made ranging radar it is often assumed that the absolute phase is distributed with equal probability between 0 and 2π.

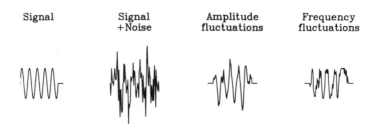

| Signal | Signal +Noise | Amplitude fluctuations | Frequency fluctuations |

Fig. 1. A short signal with additive noise and with fluctuations in amplitude and frequency.

Since many behavioral studies use noise to mask echolocation signals, in the next chapter I will discuss how such experiments can be designed to make them more accessible to theoretical interpretation. However, the intensity of external noise is very low in the bat's biotope, and it will be necessary to consider whether for bats there was really an evolutionary stress to develop a system that works well in additive noise. It may be that the echolocation system has more been adapted to produce reasonable results with fluctuating external parameters and in interfering clutter, and is <u>robust</u> enough to cope with unpredictabilities that cannot be simulated by additive noise. Most signal processing models discussed for bat echolocation are optimal only under very specific conditions and very sensitive to deviations from <u>a priori</u> assumptions. Robust techniques for signal processing (Kassam and Poor, 1985) trade optimal results under rigid conditions for acceptable results in a broad class of situations. They may provide alternative models more suitable for biological systems.

Noise Power, Signal Power and Signal Energy

A common experimental condition is one where the masking noise has a constant noise power density level N_0 over the whole band covered by the signal. The instantaneous power of an echolocation signal is not constant, but a function of time $N_s(t)$; the square root of it is the rms-(root-mean-square) amplitude $N_s(t)$. In an analog circuitry $N_s(t)$ and $A_s(t)$ can be calculated as shown in Fig. 2. This schematic of a rms-converter not only demonstrates a technical realization, but it can also serve as a model of certain features of the hearing system (see chapter on Detection and Discrimination). The time constant of a rms-converter is defined by the duration over which the running mean is taken; this time constant determines the degree of smoothing of the signal and critically influences the circuit's function. For theoretical applications and in digital signal processing, it is possible to define instantaneous power and amplitude without reference to a time constant (Rice, 1982), but most practical realizations use lowpass filters to implement the smoothing operation. Fig. 3 shows how the square of a signal is modified by the lowpass function. Both 'short' and 'long' signals are gated sinusoids with equal energy. Lowpass filtering is done by an exponentially weighted averaging over a sliding time window. A rapidly declining window (Fig. 3, center column) will average only over a few cycles of the sine wave; the output from such a filter follows the envelope of the squared signal. If a sliding window that is longer than the signal is used (right column in Fig. 3), the output is no longer similar to the envelope. Instead, the maximal amplitude is proportional to the <u>energy</u> of the signal. Since the energy of both the 'short' and the 'long' signal are equal, their maximal output amplitude is the same after lowpass-fil-

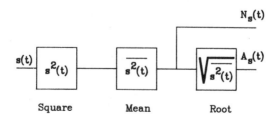

Fig. 2. Block diagram of a root-mean-square converter.

tering with a long window. In most commercially available rms-converters the time-constant of the averaging window is adjustable. With a short time-constant, the envelope or instantaneous amplitude can be recorded at the output; with a long time-constant, the maximal amplitude is proportional to the square-root of the total energy of the signal. For on-line determination of the signal energy, this maximal amplitude must be stored by an electronic maximum-hold circuit.

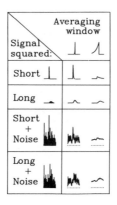

Fig. 3. The effect of lowpass filtering by time-weighted averaging on the square of a short and a long sinusoidal signal with equal energy. With a short averaging window (center column) the filter output approximates the envelope of the signal. With a long averaging window (right column) the maximum amplitude of both filters is proportional to the signal energy.

Signal-to-noise Ratio

The performance of an echolocation system depends on the signal-to-noise ratio (SNR), which can be defined in two different ways. The first is precisely called the 'signal-to-noise power ratio': it is the ratio of the instantaneous signal power to the instantaneous power of noise and is usually expressed in dB. To keep this ratio constant in an experiment with an active echolocation system, one could use an rms-converter with a short time constant and multiply its output signal by that of a noise generator (Fig. 4). This gated noise is then added to the signal and played back to the bat. Changes in the amplitude of a signal are immediately followed by proportional changes in noise amplitude, such that this circuit acts like a clamp for the signal-to-noise power ratio. This method has been used by Schnitzler and Grinnell (1977) and Grinnell and Schnitzler (1977) to measure the angular directivity of the echolocation system of Rhinolophus ferrumequinum.

The above definition of the SNR as a ratio of instantaneous powers is intuitively appealing and is often used to quantify the amplitude dynamics of electronic equipment. In communication theory, a different measure is used: the 'energy-ratio' (Woodward, 1955) compares the total energy E of the signal with the masking noise power density N_0

$$R = 2E/N_0.$$

This measure of the SNR is somewhat contradictory to intuition: it implies that in constant noise a short signal with high amplitude and a long signal with low amplitude can have the same SNR (=energy ratio), even if the latter will hardly be detectable on an oscilloscope (see Fig. 3, lower left). If the predicted performance of a receiver model only depends on the energy ratio and not explicitly on the instantaneous amplitude, this receiver must have the ability to concentrate all the energy of the signal in time, independent of its duration. The most straightforward implementation of this strategy is to accumulate energy in a time window that is supposed to contain signal and noise. This can be done in a rms-converter where the filter time constant is set longer than the expected duration of the signal. The output of a sliding integrator for both 'short' and 'long' signals in noise is shown in Fig. 3, lower right. As a result of the equal energy of the signals, the maximal amplitude in both cases is equal.

The energy integrator is one of the simplest models to detect a signal in noise. It is an optimal scheme for detection when the receiver knows nothing about the structure of the expected signal (Urkowitz, 1967). The process of energy concentration into a short time interval can be improved when the receiver has more detailed information on the signal to be detected. The optimum is achieved when the receiver makes use of all details of the signal in a crosscorrelator (= coherent receiver). The energy detector and the coherent receiver are therefore useful benchmarks to compare with the performance of a bat in a detection task.

Clamping the Energy Ratio

In an experiment with freely echolocating bats the energy of the echolocation calls is not fixed. To hold constant ('clamp') the value of the signal-to-noise ratio for one sound, one must measure its energy and adjust the power of noise in proportion to the sound energy. I have shown above how the energy of a signal can be determined, but its value is only available after the signal has ended. Since the signal must be masked by

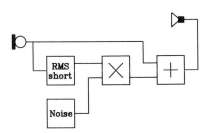

Fig. 4.   Block diagram of a circuit to clamp the signal-to-noise power ratio in a playback experiment with echolocating bats. 'RMS-short', root-mean-square converter with a short time constant that outputs the signal envelope; 'Noise', noise generator; 'x', multiplier; '+', adder.

noise from onset, the echolocation pulses must be delayed by at least their maximum duration. In a playback experiment this can be achieved by using a second loudspeaker (Fig. 5) or by an electronic delay line (Fig. 6). Another possibility is to hold the noise-power density constant and to use a gain-control circuit to clamp the replayed signal energy (Au and Moore, this volume).

No behavioral experiments with a clamped SNR have yet been performed with freely echolocating bats. It is thus not known if the bats, like an energy integrator or a crosscorrelation receiver, make use of the whole energy in the signal for a decision. With constant background noise, bats with such types of receivers could improve performance both by emitting louder calls or by extending the duration of their calls. An alternative model is one where bats use only those parts of their echolocation calls that clearly rise above the noise level. An intermediate possibility is that bats process several frequency-bands of their sweep separately and average the results of all bands (Menne, 'A matched filter bank for time delay estimation in bats', this volume).

An Experiment with Clamped Signal-to-noise Ratio

I will illustrate some of the concepts presented so far with the outline of an experiment. Assume that a bat has been trained to perform some arbitrary echolocation task in a playback experiment with virtual targets. The bat is tested under two conditions.

Condition A: the signal-to-noise power ratio is kept constant (Fig. 4).

Condition B: the energy ratio ($2E/N_0$) is kept constant (Fig. 5 or 6).

Two alternative classes of receiver models will be considered.

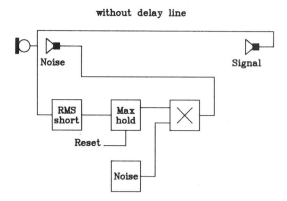

without delay line

Fig. 5. Circuit to clamp the SNR (energy ratio) in a play-back experiment with echolocating bats. The sound is picked up by a microphone close to the bat and replayed by a distant loudspeaker. A rms-converter with a long time constant and a maximum-hold circuit is used to calculate the square-root of the signal energy. This value controls the amplitude of noise that is played back to the bat from a second loudspeaker near the microphone. After every sound the maximum-hold circuit must be reset.

Receiver 1 can only evaluate the instantaneous power and does not integrate over the signal energy.

Receiver 2 uses all the energy of the signal. The energy integrator, semi-coherent receiver and coherent receiver are examples.

It is theoretically possible to predict the performance of both receivers in this task, but only if the type of echolocation sound is fixed and it is known how bats use the information in consecutive sounds for their decision. This means, that in practice a prediction is not possible. To overcome these limitations, one could adjust the noise level under both conditions so that the task becomes difficult (e.g. with 75% correct choices). If the bat can adapt its echolocation calls to the two situations, it should react in one of several ways.

With Receiver 1, short, high intensity sounds improve performance in Condition B, but do not so in Condition A. Therefore, in A the echolocation sounds should have lower peak intensity than in B.

With receiver 2, in Condition A there should be a tendency to increase the energy by lengthening the sound kept at moderate intensity. In this way the bat can increase signal energy without evoking a high noise level. In Condition B lengthening of the sound will not be effective, because the interfering noise level will increase proportional to the duration of the echolocation sound.

DETECTION AND ESTIMATION

'In radar, typical questions to be answered are: Is a target present? If it is present, what is its range? If it is moving, what is its velocity? Determining a target's presence or absence is the detection problem; measurement of the parameters of a target is the estimation problem' (DiFranco and Rubin, 1969, p. 219). These definitions are idealizations in radar theory and are initially useful. But are the definitions still applicable,

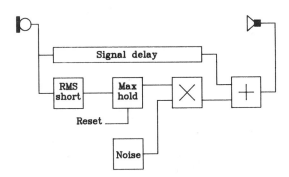

Fig. 6. Circuit to clamp the SNR (energy ratio) in a play-back experiment with echolocating bats. The principle is the same as in Fig. 5, but instead of a second loudspeaker a delay line is used to make sure that masking noise is present from the beginning of the echo.

when there are two, three or an unknown number of targets, an every night problem for bats hunting insects?

Passive Detection Experiments

The ideal experiment from the point of view of signal detection theory is one where a non-echolocating bat is placed into a soundproof room and trained to say 'yes' when a tone in noise is presented and 'no' when only noise is present. In cases where noise is not explicitly added one has to assume that ambient noise or processing noise in the hearing system limits performance. An experiment that comes close to this has been conducted by Sokolova (1973). Details of this work are not known to me, but her data reported in the review by Ayrapet'yants and Konstantinov (1974, p. 128) deserve mention. Sokolova measured the threshold of detection for tones of different length for Myotis oxygnathus. Her results are redrawn in Fig. 7 with the signal energy threshold on the ordinate scale; in the original graph the power was plotted. The energy of the signal at detection threshold remains constant up to 15ms and then rises with a slope of 10dB/decade. This is an experiment in time-intensity trading; the form of the trading curve is similar to that obtained with human observers where the change in slope occurs at about 150ms (Green et al., 1957; Watson and Gengel, 1969). In Sokolova's experiment, the signal energy at threshold is constant for durations shorter than 15ms. This can be interpreted as the bat integrating all energy in the signal over a time window of about 15ms, which is long compared with the echolocation signal (0.1-5ms). The result should not be interpreted as meaning that the hearing system always averages over such a long time and is unable so detect shorter details in amplitude structure. In human psychophysics an integration time longer than 100ms is found in this type of detection task, while the time constant of the human auditory system for the detection of changes in amplitude is shorter by a factor of almost 100 (Green, 1985).

ROC-curves and Theory of Signal Detectability

A major problem in the interpretation of all detection experiments is their sensitivity to bias of the animal in the test, that is to its internal threshold to say 'yes' when it thinks there was a signal. The Theory of Signal Detectability predicts that this bias can be manipulated by varying the payoff-matrix or the a-priori probability of an event (Green and Swets, 1966). The number of 'hits' and 'false alarms' for several values of bias can be plotted against each other, resulting in the 'receiver operating characteristics' (ROC). Schusterman et al. (1975) have carried out such ex-

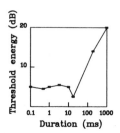

Fig. 7. Threshold energy of detection of a 40-kHz tone for Myotis oxygnathus. The absolute calibration of the ordinate is arbitrary. Data from Sokolova (1973), redrawn after Fig. 41 from Ayrapet'yants and Konstantinov (1974).

periments with a California sea lion (<u>Zalophus californianus</u>) and a porpoise (<u>Tursiops truncatus</u>): both animals had to detect the presence of a tone in ambient noise. If the SNR in such experiments were known, the data could be compared with theoretical predicted ROC-curves, for example a crosscorrelator or an energy detector (Tanner and Birdsall, 1958).

A passive signal detection experiment in noise to obtain ROC-curves has not yet been performed with bats. It seems to be ideal for comparison with theory:

1. It is an experiment in passive listening, where the SNR can be determined exactly. The bat does not have to use any complex processing mechanisms to get rid of echoes, signals from other directions of space, etc.

2. The signal can be presented only once or a known number of times, thus the bat is prevented from repeating the sampling until it is sure about a decision.

3. By fitting a predicted ROC-curve to the experimental data, in theory one should be able to infer the type of information processing the bat uses.

On the other hand, there are strong arguments that the information to be gained with this type of experiments is limited.

1. It is an experiment in passive listening, but we want to learn about the bat's active echolocation system. It may be that in experiments where no echolocation sounds are used the bat behaves like any other mammal and does not switch on its specific echolocation receiver.

2) To measure a point on a ROC-curve we must assume that an observer maintains a constant internal threshold during a number of trials. This is difficult for human observers and probably impossible for bats. Moreover, even with ideal observers the number of experiments necessary to differentiate between two different receiver types using the ROC can be forbiddingly high (Green and Swets, 1966, Appendix III).

Some years ago theoreticians urged experimental researchers to measure the ROC-curves of bats performing detection tasks to investigate the underlying mechanisms of echolocation. I believe that this approach is not worth the effort, as other experiments can be designed with much more power to discriminate critically between different receiver types.

Sound-triggered detection experiments

An improvement in experimental design over the passive listening task is used by Møhl (1985, 1986 and in this volume), who triggers the playback of a stored standard echolocation call after the bat has emitted a sound. Since the energy of the playback signal is fixed, the SNR is known, even without the clamping technique described above. The number of echolocation sounds bats used in Møhl's design was not limited, so some assumptions had to be made about the bat's strategy to average information in consecutive sounds. Møhl combined his method with a staircase procedure and obtained very stable estimates of auditory thresholds. It is not absolutely certain that bats make full use of their echolocation abilities when presented with 'canned' sounds, but it can be expected that this situation is closer to one of natural echolocation than passive listening. An experiment by Heilmann-Rudolf (1984) with a somewhat different aim may shed light on this problem. Heilmann-Rudolf has measured frequency discrimination in <u>Rhinolophus ferrumequinum</u> using passive listening to tone pulses and listening pulses triggered by echolocation calls. For most frequencies, she found similar thresholds for both conditions; in passive listening Rhinolophus refused to accept the task when stimuli were near its reference frequency (Schuller et al., 1974), while they showed consistent behavior in all cases when the sounds were triggered by their own echolocation calls.

2AFC-procedure – Detection or Discrimination?

Instead of the yes-no (YN) procedure, the two-alternative forced-choice (2AFC) method is commonly used in detection experiments. It is characterized by a symmetrical design, where a bat can clearly separate two directions in space. Signal and noise come from one of these directions and noise alone comes from the other, and the bat has to select that side containing the signal. The 2AFC-method is much less sensitive to variations in bias of the animal than the YN-method. In theory it is known how the results from YN and 2AFC procedures can be compared (Tanner and Birdsall, 1958), but experiments using both methods do not always lead to consistent results (Swets, 1959; Schulman and Mitchell, 1965). It should be noted that the decision strategy of a bat in a 2AFC task is different from that in a yes-no task. In a 2AFC-experiment, the animal has been conditioned to know that one side always contains a signal. It does not simply listen to the signal from one direction and then decide whether a signal was present or not, but it scans back and forth between directions and makes a decision after estimating, which of the two following events is more likely:

1. Signal+noise was left <u>and</u> noise alone was right;
2. Noise alone was left <u>and</u> signal+noise was right.

In the discussion of experimental results, predictions from YN-task have often directly been used erroneously for comparison with more complex 2AFC-procedures. On the other hand, there is a close relation between the 2AFC-procedure to detect a signal in noise and the <u>discrimination</u> procedure between two different signals. The alternatives in a discrimination task are:

1. Signal I + noise was left <u>and</u> Signal II + noise was right;
2. Signal II + noise was left <u>and</u> Signal I + noise was right.

For a theoretical analysis, it is useful to regard the 2AFC-detection procedure as a special case of a discrimination procedure where one signal is zero.

Kick (1982) and Kick and Simmons (1984) have used a 2AFC-procedure to measure the lowest echo intensity <u>Eptesicus fuscus</u> can detect; the bats had to detect their own echolocation sound, reflected from a ball (Kick, 1982) or simulated in a playback experiment (Kick and Simmons, 1984). Ambient noise was very low, and internal noise must have limited performance. The total energy necessary for a decision could only be estimated as the bats were allowed to vary freely the intensity of the returning echo and the number of echolocation pulses used. The decision strategy in the ball experiment probably comes quite close to that described above for a 2AFC-detection situation. The situation was more complex in the experiment with the virtual target, because passive reflection of echolocation sounds from the surface of the loudspeaker introduced additional clutter. Here, the bat had to test these alternative hypotheses:

1. virtual echo + clutter + noise from left <u>and</u>
                clutter + noise from right;

2.   clutter + noise from left <u>and</u>
    virtual echo + clutter + noise from right.

There are several likely hypotheses concerning a bat's strategy in this experiment. It may simply compare the energy of returning echoes from both sides and select the side with the higher echo intensity. A somewhat more sophisticated decision scheme would incorporate the different 'colors' of the echoes returning from different target types. Both these methods could also be used by mammals without a specialized echolocation system. Bats should be able to evaluate the spatial separation of the virtual target and loudspeaker by their different echo arrival times. If they can clearly separate both echoes, their detection performance should not deteriorate in clutter. But if they use the energy of returning sounds alone they should be strongly affected by additional reflecting targets. Both observations have been reported (Kick and Simmons, 1984; Møhl, 1985, 1986).

This example was chosen to demonstrate how difficult it is to plan a 'pure' detection experiment in active echolocation, one that can be interpreted without reference to a procedure that sorts complex echoes arriving from all directions in space and possessing different delays and various Doppler shifts. Numerous studies on the ability of bats to detect small objects in space, be they wires or insects, mostly give estimates of just detectable echo intensities returning from these objects, but they rarely report what additional echoes arrive at the bat's ear. If a bat has to fly through a frame of fine wires, the echo intensity from the frame may be higher than that of the wires; yet the observed reaction, an increase in duty cycle of the echolocation pulses, is quite certainly a response to the wires, as they impose a possibly dangerous obstacle to the bat. I doubt that the strategy of a bat hunting insects is 'First try to detect if any echo returns at all, averaged over all directions; then switch on your sonar display to find out where the echo comes from and if it is worth a diving maneuver.' Bats that hunt in highly cluttered surroundings are most unlikely to use this strategy; prior to detection and estimation they must form an image of the world to resolve different targets in range, azimuth, elevation and structure (Simmons, this volume). Detection by echolocation then is a process of making a yes-no decision based on a sonar display of the world.

RESOLUTION, CLASSIFICATION AND ESTIMATION

As noted in the previous chapter, prior to detection of a target and the final estimation of its position, an image of space is needed. In such an image, ideally every small cubicle in space would be represented by the intensity of the echo returning from it. The smaller the cubicles are, the more details the image contains, the better two nearby objects are resolved (Menne, 1985). In practice, however, signal power returning from one resolution cell will spill over to adjacent cells, so that the picture is blurred; the number of resolved points in space is lower than the number of cells. To detect an echo coming from some part of space, the summed echo intensity in a group of adjacent resolution cells would have to be higher than that of the background noise alone. To estimate range or direction, one has to find that resolution cell reporting the highest echo amplitude.

Resolution describes not only how clearly two points in space are separable, but in a more general sense it can define how distinctly patterns are resolved and how well classifications work (Altes, 1980). As an example, assume that bats use the frequency pattern of an insect's echo to classify different groups, and that the image of the echo is represented by the firing intensity of three neurons (Fig. 8). A good resolution in 'pattern space' means that the firing patterns in response to several individuals of insect classes A and B are clearly distinct as in the left part of

Fig. 8; with a bad pattern resolution, the firing pattern of the two insect classes will considerably overlap, resulting in a higher number of misclassifications (Fig. 8, right part).

Classification is a process intermediate between detection and estimation: in all three cases, one looks at a certain volume of the three-dimensional space or an abstract pattern space. The questions answered are:

for <u>detection</u>, which event is more likely, signal+noise or noise alone?;

for <u>classification</u>, to which of several discrete classes does the signal most probably belong?;

for <u>estimation</u>, which value on a continuous scale is the most probable?

Absolute Estimation and Bias

In an <u>estimation</u> process a system returns a numeric quantity, like an angle, range or velocity of a target. A good example is the estimation of horizontal angles by <u>Eptesicus fuscus</u> (Masters et al., 1985), where the bat's estimated angle is indicated by the direction of the head tracking a target. There are two types of error made by the bat in this situation: one is a systematic and predictable misreading of the target angle. In the model of Masters et al. this error is attributed to a non-unity gain in a feedback loop and is called <u>bias</u> of the system. [The definition of bias in an estimation task has nothing to do with the bias in detection experiments, which is a measure of the internal threshold for a decision.] The other error, a statistical and unpredictable jitter around the estimated angle, is ascribed to the influence of noise or fluctuations. Their contributions to the error can be reduced by averaging over repeated measurements, while bias as a deterministic phenomenon will remain constant even when many samples are taken.

To quantify how accurately bats can estimate a parameter, the contribution of bias plus noise is relevant, and the word 'accuracy' is used as a

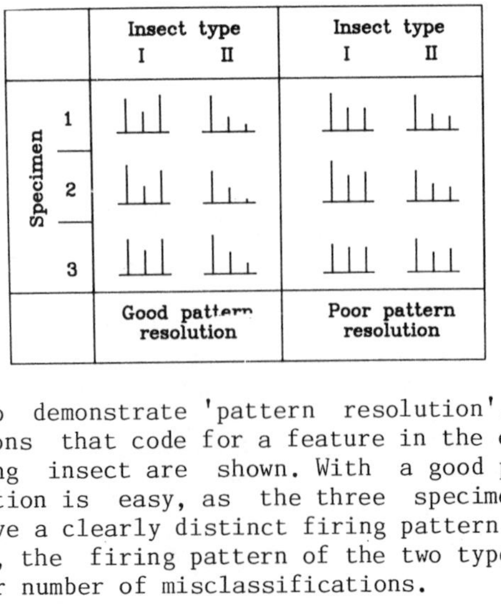

	Insect type I II	Insect type I II
1	⊥⊥ ⊥⊥	⊥⊥ ⊥⊥
2	⊥⊥ ⊥⊥	⊥⊥ ⊥⊥
3	⊥⊥ ⊥⊥	⊥⊥ ⊥⊥
	Good pattern resolution	**Poor pattern resolution**

Fig. 8. Example to demonstrate 'pattern resolution'. Firing rates of three neurons that code for a feature in the echo returning from a fluttering insect are shown. With a good pattern resolution, classification is easy, as the three specimen of insect type I and II have a clearly distinct firing pattern. With poor pattern resolution, the firing pattern of the two types overlap, leading to a larger number of misclassifications.

measure of the total error including noise and bias components. Following the definition in the IEEE Standard Dictionary of Electrical and Electronics Terms (1984), the word 'precision' refers to the reproducibility of a measuring process. This reproducibility is only influenced by statistical processes like noise and fluctuations. So if an angle of exactly $30°$ is to be estimated several times and the results are $25\pm0.1°$, then the precision of this measurement is very high, but the accuracy is low, since bias is the main source of error. In radar literature, 'accuracy' and 'precision' are sometimes used interchangeably; this is correct as long as there is no bias, but I believe that a clear distinction between the two terms is useful in the discussion of different strategies to dissect the system of signal processing in bats.

I will call the bat's task in the angular tracking experiment one of absolute estimation, because the measured orientation of the head is taken for the direction the bat estimates. There are not many studies of a bat's accuracy in absolute distance estimation. Photographic recordings by Webster (1963) and Trappe (1982) indicate an accuracy of localization of 1-3cm; these values should be taken as upper limits, because the bats were flying freely and constraints imposed by flight dynamics may have an impact on the achieved absolute accuracy.

Difference Estimation

Most ranging experiments in bats measure the threshold for relative estimation (e.g. Simmons, 1971,1973,1979; Roverud and Grinnell, 1985a). The bat has to decide which of two targets is closer or which target makes small jittering movements between consecutive sounds: the task of absolute range estimation is reduced to one of detection of a range difference. The lowest range difference thresholds are those obtained in Simmons' (1979) range jitter experiment; the bats were able to detect virtual target displacements of 0.22mm with 75% correct choices. One could assume that these values must be closely related to accuracy in absolute estimation, but analogous experiments in other fields of psychophysics show that this inference may not be correct. So humans are able to discriminate frequency differences in the order of 1% (Fastl, 1978), while the ability of those without 'absolute pitch' to estimate the absolute frequency of a tone is worse by orders of magnitudes (Bachem, 1937). In bat echolocation, the discrepancy between the observed absolute accuracy of several centimeter and the precision in the range difference experiment of 0.22mm may really be a feature of the system and not a consequence of the different measurement procedures alone.

An alternative method of measuring range difference estimation has been used in the careful study of Roverud and Grinnell (1985a). In one procedure they trained Noctilio albiventris to detect the closer of two simultaneously present targets; in the other procedure only one target was present and Noctilio was trained to go right when the target was 35cm away, and left when it was further than 35cm away. The authors named the first situation 'relative discrimination' and the second 'absolute discrimination'. They summarize their results, saying that 'the bat can determine an absolute distance around 35cm with an acuity of about 8.6%'. But this conclusion is not valid, as we do not know whether their bats really estimated the distance of 35cm correctly and without bias. The bats were, however, able to memorize the reinforced distance and to compare it with the distance of a target with an precision of about 8.6%, while the precision was 3.7% when two targets were present simultaneously.

This example demonstrates why the results from absolute and difference estimation procedures are difficult to compare: in range difference estimation the bias present in absolute range estimation can cancel out. Assume that there is a target at 30cm distance, but the bat always underestimates the distance by 20%; thus the error due to bias is 6cm. A target jittering by 0.5cm (between 30cm to 30.5cm) will produce an apparent jitter of 0.4cm. If the bat can detect this movement, its precision in a jitter experiment would be about 0.5cm, even if the accuracy of its absolute range estimation is worse. If some signal parameter is changed between two echolocation sounds, receiver types that use this parameter for range estimation will perceive a changing bias. This kind of bias manipulation can be used as a powerful tool to find parameters of a signal that the bat uses for distance estimation.

RECEIVER MODELS FOR RANGE ESTIMATION

While the results of studies on detection and estimation are certainly interesting in themselves, the primary rationale for these experiments has been to predict the receiver structure in the hearing system of bats. Because most theoretical work has been done on detection of signals and their use in range estimation, I will concentrate on this subject.

Three approaches have been used to make inferences from behavioral data on underlying receiver models:

1. theoretical analysis of the fine structure of echolocation calls;

2. comparison of accuracy of range estimation and accuracy of different theoretical receiver types; and

3. testing sensitivities of the echolocation system by manipulating signals that should elicit bias or should change the detection probability in some receiver types.

Analysis of echolocation calls

Strother (1961) noted that the time-frequency structure of bat echolocation signals resemble those of pulse-compression radar. He conjectured that bats could use a processing mechanism for these signals similar to that used in technical systems. Research on the fine-analysis of echolocation calls was triggered by a paper by Altes and Titlebaum (1970); the result of their investigation was that the echolocation call of Myotis lucifugus was optimally Doppler tolerant when processed in a crosscorrelation receiver. Their line of reasoning has been carried further by Beuter (1976); Escudié, Hellion, Munier and Simmons (1976); Mamode and Escudié (1985); Flandrin, Cros and Mange (1985). As some terms used in these publications are often confused by experimental researchers, I will explain them here.

Range-Doppler coupling. A signal shows range-Doppler coupling if it gives a systematically wrong (=biased) reading of the distance when the target moves relative to the bat. Range-Doppler coupling can sometimes be corrected with sophisticated signals (Altes and Skinner, 1977) or by using the constant-frequency part of a signal to obtain an independent measurement of relative velocity.

Doppler sensitivity. A signal is Doppler-sensitive, if the output of a cross-correlator matched to the emitted signal is considerably lowered when the echo is Doppler-shifted. With a Doppler-sensitive signal, the number of correct decisions in a detection task will decrease when the target moves relative to the receiver. Doppler-shifted echoes will also lead to a larger error (=lower precision) of time delay estimation in the presence of noise. An example for a Doppler-sensitive signal is a tone of a single frequency switched on for a specific time. The crosscorrelation receiver in this case is well approximated by a bandpass filter tuned to the emitted frequency. If the returning echo is Doppler-shifted, the output amplitude will be attenuated by filtering. Note that this Dopper-sensitive signal shows no range-Doppler coupling, as the arrival time of the signal will not be biased by a frequency shift. Range-Doppler coupling and Doppler sensitivity are independent effects; their influence on ranging accuracy can be read from the <u>ambiguity diagram</u>.

Doppler tolerance. A signal showing little Doppler sensitivity is called Doppler tolerant. This definition is used by Altes and in this paper. For Mamode and Escudié (1985), a Doppler-tolerant signal must also exhibit low range-Doppler coupling. The definition used by Altes refers to the <u>precision</u> of distance measurement and is more useful for theoretical work, while for Mamode and Escudié Doppler-tolerance describes the system's <u>accuracy</u> in a practical application, including bias and statistical error.

A theoretical analysis of echolocation sounds indicates that many bats produce signals with a frequency structure that is maximally Doppler tolerant and has low range-Doppler coupling when processed in a crosscorrelation receiver. It was concluded, that such sophisticated signals can only evolve if there is a selective advantage to use them. Their existence was taken as an indication that bats have a crosscorrelation receiver. This argument is not very strong, as the problem of Doppler sensitivity only occurs in receivers like the crosscorrelation receiver that are very 'narrow-minded', responding only to echoes that closely match the signal they are designed for. Suboptimal receivers do not make use of all details of the signal for detection and estimation, but they give reasonable response to a larger class of signals and therefore are implicitly Doppler-tolerant. Up to now there is no independent proof for the existence of a crosscorrelation receiver in bats (Schnitzler et al., 1985).

An experiment that would reveal whether bats are Doppler-sensitive is difficult to design. In theory, a Doppler-sensitive receiver should have a higher detection threshold for frequency-shifted signals than for an unshifted ones. I believe that this cannot be found, because the detection threshold for a considerably stronger signal manipulation is not raised (Møhl, this volume). It can be argued that in Møhl's detection experiment the bats did not switch on their full echolocation system. However, if in some future experiment no degradation in performance for Doppler-shifted signals can be found, two opposing interpretations are possible:

1. the bats have a crosscorrelation receiver, but their signal is Doppler-tolerant;

2. the bats have a much less sophisticated, more robust receiver, for which Doppler tolerance to <u>any</u> signal is a minor problem.

An outline of an experiments to test the degree of range-Doppler coupling is given in the chapter on 'Qualitative Features and the Manipulation of Bias'.

Detection and Discrimination of Range Differences

Behavioral experiments by Simmons (1979) demonstrated that <u>Eptesicus fuscus</u> has a precision of range difference measurement in the order of $1\mu s$. Following Simmons' conclusions these experiments were generally accepted as a proof for the existence of a crosscorrelation receiver. One of his major arguments was that the accuracy of time estimation is limited by the width of the autocorrelation function of the signal; this clearly contradicts

Woodward's (1955) relation, which indicated that the statistical error of time measurement goes to zero at high SNR (Schnitzler and Henson, 1980). While the order of magnitude of a bat's accuracy has been experimentally confirmed (Menne et al., in preparation), Menne and Hackbarth (1986) showed that Simmons's experiments are not sufficient to proof that bats have a crosscorrelation receiver.

The chief problem in the interpretation of Simmons' experiment is the unknown SNR. Since no external noise was explicitly added, we must assume that both microphone noise and internal noise limited performance. There-fore the effective noise intensity can only roughly be estimated. But let us assume that the experiments could be made with the external SNR clamped to a value where internal noise is negligible (Fig. 5, 6). When the bat is allowed to make an arbitrary number of echolocation sounds until it gives a response, the effective SNR for a group of signals must be calculated. There is a high degree of uncertainty how a bat uses the information from several sounds to make his decision (Floyd, 1980). Does it use a sequential decision scheme (Wald, 1947; Marcus and Swerling, 1962)? Does it select one signal in a sequence that 'sounds best'? Does it average over several signals (Swets et al., 1959)? These problems could be avoided if only one echolocation signal were allowed for each decision, but it could be difficult to train a bat to show consistent behavior in this situation. To use the bat's precision of time measurement in deciding between alternative receiver models, we cannot simply select the most likely estimate of effective SNR. Instead, it is necessary to make a worst-case analysis with many echolocation sounds and different hypotheses on the mechanisms a bat may use for evaluation of groups of signals. I presume that such a worst-case analysis would reveal that in no experiment where bats use their own echolocation calls the effective SNR is known better than ±6dB. This uncertainty is too high to allow a decision which theoretical receiver models do not contradict experimental results (Hackbarth, 1984).

The aim of theoretical analysis should not be to make quantitative predictions of behavioral results, but rather to describe strategies of in-formation processing. As an example, assume that a bat's performance is 6dB worse than receiver model A and only 1dB worse than model B. In this case, it is still possible that the bat uses the strategy A, even though the pro-cessing efficiency of its echolocation system is 6dB worse than that of an ideal receiver of type A (Tanner and Birdsall, 1958). Probing an imperfect receiver with noise therefore gives only limited information about the sig-nal processing involved. In the next chapter, I will discuss methods that can do this by testing qualitative features of a system.

An experimental paradigm that has not yet been used in echolocation research consists of testing the sensitivity of a system to random fluctua-tions of one parameter (Fig. 1). Fay and Passow (1982) have used such a paradigm in a fascinating way to study neuronal timing jitter in the gold-fish auditory system. In echolocation research one could measure how the detection of a jittering target is changed when an additional random jitter component is present. Another example is testing whether the envelope of an echolocation signal is important for range difference estimation. Here one could multiply the envelope of the echolocation calls with low-pass Gaus-sian noise to induce random fluctuations in the envelope while leaving the frequency structure almost constant.

Qualitative Features and the Manipulation of Bias

As discussed in the previous chapters, the estimation of effective noise in an active system is difficult and the discriminative power of such experiments is low. Therefore I believe that critical experiments should be designed in such a way that their interpretation is largely independent of the noise intensity in the system, that is, one should try to find <u>qualita-</u><u>tive</u> features of a system. The primary method for doing this is the manipu-lation of parameters in the returning echoes to detect insensitivities or 'blind spots' in a receiver.

Let me give two idealized examples to illustrate the concept of 'blind spots'. Assume that a bat has a receiver that does make no use of the frequency structure of the echolocation call, but only of its envelope. In such a bat the precision of its range estimation in a playback experiment is not changed by separating the envelope of the sound and filling it with another frequency (e.g. Gaussian noise or a constant frequency tone). As the second example, assume the contrary, that is the bat uses only the frequency structure of the call and not its envelope. In this case, a clip-ping of the sound to a constant amplitude does not change the precision of estimation. The receiver in the first example is blind to changes in the carrier frequency, while that in the second is insensitive to envelope manipulations. [A much less trivial case, that of phase sensitivity, will be discussed in the next chapter.] Blind spots in a receiver may result from an inability of the nervous system to use a specific part of informa-tion in the echo. More interesting is the hypothesis, that the use of cer-tain parameters of the echo could have proved unsuccessful, because they carried unreliable information for the localization of targets. The exis-tence of a blind spot then is an indication that a component of the echo is treated as stray parameter by the processing system.

The experiments of Møhl (1986) are an example of a search for 'blind spots'. Møhl assessed the detection thresholds of pre-stored echolocation calls played back in forward and in reversed order: the SNR at threshold was the same for both orders. Even though this is an experiment on detec-tion in noise, it tests a qualitative feature of the receiver. The value of the SNR at threshold is not directly interpreted, but only used to compare two different conditions. It is certainly one of the most confusing results in recent research that bats seem to be insensitive to a time reversal of the signal, and I am not sure whether in a similar experiment where the bat has to estimate range difference the same qualitative result would be ob-tained. Nevertheless, I believe Møhl's experiment to be a critical one, be-cause it gives a yes-no answer and does not allow bargaining over some dB of signal-to-noise ratio.

Numerous other manipulations of the signal can be imagined, such as filtering frequency components or altering the overall intensity, the enve-lope and of the frequency structure. Most of these manipulations can un-doubtedly be detected by bats, but knowing their detectability does not considerably improve our insight into the mechanisms of range estimation. A more specific question would be, how does a certain frequency band of the echolocation signal contribute to range measurement? Or, how does manipula-tion of the frequency structure bias time delay estimation?

An promising approach is that of Roverud and Grinnell (1985b). To discover, which component of the cf-fm echolocation signal of <u>Noctilio al-</u><u>biventris</u> was used for distance estimation, they produced artificial clut-ter that simulated different parts of the echolocation call and then they measured changes in the precision of range difference estimation when in-terference was present. The important methodological development of their design was to quantify performance using an estimation task, in which the bat had to use its echolocation system.

The most powerful paradigm in which to measure bias with psychophys-ical methods is to trade one parameter for another. The classical example is the trading of interaural time delay for intensity in binaural lateral-ization (Moushegian and Jeffress, 1959; Whithworth and Jeffress, 1961).

Within specific limits, human observers use both arrival time of a signal at the two ears and the interaural intensity difference for the lateralization of a sound source. In free field hearing, these two parameters are always coupled as a result of shading effects of the head. By using earphones, time delay and intensity can be uncoupled, and for a given time delay it is possible to find an opposing intensity difference that produces a forward directional sensation.

As an example of how this method can be used in echolocation research, assume that we want to measure whether bats are prone to range-Doppler coupling. Their range determination should be biased by shifting the frequency of the returning echo. To test this hypothesis, we could train a bat in a playback experiment with virtual echoes; in contrast to Simmons' (1979) experiment, both targets would be jittering, and the bat would have to detect the target having the larger time jitter. We would then replace the time delay jitter of one of the targets by varying degrees of electronically produced Doppler jitter. If the bat range estimation is biased by Doppler shifts, then it should be possible to find a Doppler jitter that the bat cannot distinguish from pure time delay jitter. An analogous experiment with Eptesicus fuscus in which constant phase shifts were traded against time delay has been carried out by Menne et al.(in preparation).

PHASE SENSITIVITY

Do bats use the phase of their echolocation signal to increase the accuracy of range estimation? This question has resulted in a controversy (Simmons, 1979, Altes 1981, Hackbarth, 1984, 1986, Menne and Hackbarth, 1986) that concentrated on the discrepancy between electrophysiological results, which show no evidence for phase-locking neurons at high frequencies, and behavioral experiments, which seemed to indicate that phase is used for range determination. How much information on the signal phase is lost and how much is retained in neurons that do not trigger in fixed phase relation to the input signal? Altes (1980, 1981) has shown that from a spectrogram alone the signal can be reconstructed accurately with only one unknown parameter, the absolute phase. His spectrogram definition uses the analytic envelope that contains many high frequency components, and it is questionable if this corresponds well enough to processing in the peripheral hearing system. Furthermore, the ability to completely reconstruct a signal does not imply that this process also works when noise is present. However, I will show that reconstruction processes similar to that described by Altes are used in the hearing system and I will give indications on their limits.

As an example consider a frequency sweep presented to a human observer starting at 15kHz and sweeping down to 10kHz within one second (Fig. 9); both frequencies are far above the limit where phase locking normally occurs (Anderson, et al., 1971), so we can be sure that something like a place mechanism is responsible for perception of change in frequency (Zwicker, 1970). We can easily discriminate this tone from a time reversed tone sweeping from 10kHz to 15kHz. This can be done despite the fact that the power spectra of the two tones are identical and they differ only by their phase. Yet nobody would attribute this discrimination to phase perception. In frequency analysis, the hearing system does not average over long tones; but instead, it takes a sequence of short-time spectra to form a spectrogram, which is different in the signal and its time reversed version. Signal manipulations that occur within the time interval of one short-time spectrum are difficult to recognize; in humans, this limit is at about 2.5ms (Green, 1985). Thus, if in our example we reduce the duration of the sweep to about 1ms, we will probably not be able to discriminate the signal and its time reverse. For frequency analysis of such short tones, the hearing system takes the average over the whole signal. In the signal of one second duration, a large part of the phase information is recoverable from the spectrogram. In the signal of one millisecond, this information is smeared out to such an extent that the hearing system cannot recover it easily, even if in absence of noise a mathematical 'deconvolution' of phase information is still possible (Altes, 1981).

Let me now present a more interesting example in which two signals differ only in their phase and not in their spectra. Ronken (1970) asked human observers to discriminate a signal consisting of a pair of very short clicks with different amplitude from the time-reversed signal. When the interval between the clicks was long enough, the observer heard 'loud-soft' or 'soft-loud'. At a temporal separation of 1-2ms, the two clicks could not be separated, but the two alternatives presented could still be recognized by their different tonal color. Such short clicks are very artificial stimuli for humans, but an analogous experiment with an active echolocation system simulates a natural situation (Fig. 10), which I call 'Ronken paradigm for active systems'. One of the alternatives simulates a strongly reflecting point target before a weaker one; the other alternative the reverse arrangement of targets in space. In this experiment, the echolocation signal is 'filtered' by the spatial arrangement of targets. This filtering can be simulated by adding a signal and its delayed copy with different amplitude weighting. By this delay-and-add operation all the frequencies of the echolocation call are weighted with the same comb-like amplitude transfer function (Bilsen, 1966); only the phase transfer functions in both cases are different (Fig. 10). When the time delay between the two echo components is not to long, the short-time spectra will also be clearly distinct, therefore the Ronken paradigm can be used to measure the temporal resolution of bats. They undoubtedly will be able to discriminate the two conditions when the targets are not too close, but I am reluctant to call this ability of the bat 'phase-sensitivity'. As a result of the phase shift, frequency groups in the echolocation sound are rearranged in time. It is this frequency-scrambling process that will make both spectrograms distinctly different.

The degree of frequency-scrambling produced by a filtering operation can be read from a plot of the group delay. The concept of group delay can be illustrated best with the example given at the begin of this chapter (Fig. 9). To characterize the difference between a signal sweeping from 15kHz to 10kHz and its time inverse, one could plot the time delay necessary to transform the first signal into the second one for each signal frequency. The two signals in Fig. 9 have been plotted so that for the 10kHz-component no time shift is required and thus the group delay is zero; the 15kHz-component must be shifted by 2 seconds. The vertical position of this

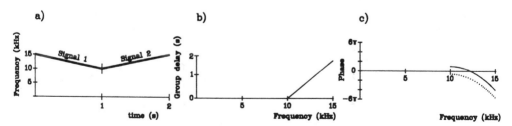

Fig. 9. (a) Simplified spectrogram of a tone sweeping from 15kHz down to 10kHz (signal 1) and the time-reversed signal (signal 2). (b) Plot of the group delay to transform signal 1 to signal 2. (c) Solid curve: phase transfer function for the transformation of signal 1 to signal 2. Dotted curve: another phase transfer function with the same group delay.

group delay plot is somewhat arbitrary; if the two signals in the upper part of Fig. 9 were selected with a larger gap between them, the group delay plot would shift in the vertical direction. But the span of the group delays, the difference between the lowest and the highest value, is always 2 seconds. This time describes the maximal time shifts necessary to transform a frequency group of one signal into the same frequency group of the other signal. In the case presented here, this time shift is much larger than the resolution time constant of the auditory system and the two signals should be easily discriminable. If one reduces the sweep time to 1ms, the group delay plot would have a span of 2ms, indicating that these two signals with the same spectrum could be much more difficult to discriminate. [Note that if the group delay graph in Fig. 9 is chosen with half the slope, the original signal is compressed to a very short time interval. A filter with this group delay acts like an 'optimal receiver' for this signal (Di Franco and Rubin, 1968, p. 161).]

For this pair of linear frequency sweeps, the group delay function could be found approximately by inspection, but for complex signals this is usually not possible. Instead, for all pairs of signals it is possible to calculate the phase shift as a function of frequency (Tribolet, 1979), and from this phase response function $\varphi(f)$, the group delay t_g as a function of frequency can be calculated using the relation:

$$ t_g(f) = - \frac{1}{2\pi} \frac{d\varphi(f)}{df} $$

Ronken paradigm for active systems

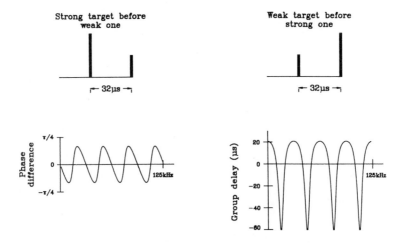

Fig. 10. A bat is trained to discriminate the echo from a strong target before a weak one and the reverse situation. Since echo components interfere, the returning signal is a comb-filtered version of the emitted signal. The amplitude transfer function of the comb-filter is equal in both conditions, but the phase transfer function is different. A plot of the group delay indicates how much frequency components of spectrograms are time-shifted relative to each other.

406

In other words: The group delay is the frequency derivative of the phase response function. Using this relation, the group delay function for the Ronken paradigm has been determined (Fig. 10, lower right). The range of group delays is from -60µs to 20µs; so when we compare the spectrograms of the same echolocation signal in both alternatives, frequency components are shifted up to 80µs relative to each other. If a bat can discriminate the two conditions, this means that it is able to resolve time shifts in the order of 80µs.

To obtain the phase function, one has to integrate the group delay:

$$(f) = -2\pi \int t_g(f) \, df \; + \; C$$

The value of the integration constant C cannot be directly derived from the spectrogram-like plot in Fig. 9a. It is this constant that describes the absolute phase relation between the two signals. In the example it was assumed that one signal was an exact, time reversed copy of the other, so at 10kHz there is a phase shift by π. If the signal are no time reversed copies of each other, this constant will be different and the phase function shifted vertically (dotted curve in Fig. 9c).

I have given examples how the hearing system can discriminate signal pairs when their short time spectrograms are distinctly different. The degree of difference between two signals with equal power spectrum can be read from the group delay plot. If the phase shift between two signals is constant for all frequencies, then the group delay will be zero. As a consequence, the spectrograms of the two signals will be equal and can be discriminated only if some mechanism exist that is sensitive to the phase shift itself. I therefore propose that the hearing system should only be considered 'phase sensitive' in a strict sense when the animal can discriminate between signal pairs with constant relative phase shift.

A special case of this phase sensitivity would be shown by an animal that is able to discriminate between a signal and the amplitude-inversed signal, corresponding to a 180 degrees phase shift (Altes, 1981). As a part of a more elaborate design to test setup testing the use of phase information in distance estimation, Menne et al. (in preparation) trained Eptesicus fuscus to discriminate between one virtual echo that is phase shifted by +45 degrees and another with a phase jittering between +45 degrees and -45 degrees between the sounds. The phase shifts were produced by digital filters, so that the frequency transfer functions for phase shifts could be made absolutely identical. Five bats performed this task after much training with 87%±3% correct choices. When phase jitter was traded against delay jitter, no value of time jitter could be found where phase jitter was not detectable. An interpretation of these results is that bats can detect phase shifts, but do not use this information for range estimation. The mechanism that enables a bat to sense this signal manipulation is unclear.

CONCLUSIONS

I have tried to show how theoretical analysis can supply the tools for the design of exeriments to discover the limits of a model, but I am quite sceptical if it is possible to find one receiver model that can explain even the limited complex of range estimation. Perhaps we will have to accept that several contradictory models of echolocation coexist, each describing some experimental results and hopefully some mechanisms of the echolocation system. It may be that bat echolocation and technical sonar will appear more and more dissimilar, as we learn about them. When an old sonar design gets inconsistent after too many updates, technicians always have the chance to start from scratch to realize consistent new concepts; evolution, however, cannot close for renovation.

ACKNOWLEDGEMENTS

I thank Dr. J. Counsilman, National University of Singapore, and Dr. C. Menne for their commenting on the manuscript. The author was supported by a 'Langzeitdozentur' of the German Academic Exchange Service (DAAD).

REFERENCES

Altes, R.A., 1980, Detection, estimation and classification with spectrograms, J. Acoust. Soc. Am. 67:1232.

Altes, R.A., 1981, Echo phase perception in bat sonar?, J. Acoust. Soc. Am. 69:505.

Altes, R.A., 1984a, Texture analysis with spectrograms, IEEE Trans. Sonics Ultrasonics, SU-31:407.

Altes, R.A., 1984b, Echolocation as seen from the viewpoint of radar/sonar theory, in: "Localization and Orientation in Biology and Engineering,", D. Varju and H.-U. Schnitzler, ed., Springer, Berlin.

Altes, R.A., Skinner, P.P., 1977, Sonar velocity resolution with a linear-period-modulated pulse. J. Acoust. Soc. Am. 61:1019-1030.

Altes, R.A., Titlebaum, E.L., 1970, Bat signals as optimally Doppler tolerant waveforms, J. Acoust. Soc. Am. 48:1014.

Anderson, D.J., Rose, J.E., Hind, J.E., Brugge, J.F., 1971, Temporal position of discharges in single nerve fibers within the cycle of a sine-wave stimulus: Frequency and intensity effects, J. Acoust. Soc. Am. 49:1131.

Ayrapet'yants, E.S., Konstantinov, A.I., 1974, Echolocation in nature, Nauka, Leningrad. English Translation: Joint Publications Research, Arlington.

Bachem, A., 1937, Various types of absolute pitch, J. Acoust. Soc. Am. 9:146.

Beuter, K.J., 1976, Systemtheoretische Untersuchungen zur Echoortung der Fledermäuse. Thesis, Universität Tübingen, F.R.G.

Beuter, K.J., 1980, A new concept of echo evaluation in the auditory system of bats, in: "Animal sonar systems", R.G. Busnel, J. Fish eds., Plenum Press, New York.

Bilsen, F.A., 1966, Repetition pitch: Monaural interaction of a sound with the repetition of the same, but phase shifted sound, Acustica 17:295.

Di Franco, J.V., Rubin, W.L., 1968, "Radar detection," Prentice Hall, Englewood Cliffs, N.J.

Escudié, B., Hellion, A., Munier, J., Simmons, J.A., 1976, Etude théorique des performances des signaux sonar diversité de certaines chauve-souris à l'aide du traitement du signal, Rev. Acoust. 38:216.

Fastl, H., (1978), Frequency discrimination for pulsed versus modulated tones, J. Acoust. Soc. Am. 63:275.

Fay, R.R., Passow, B., 1982, Temporal discrimination in the goldfish. J. Acoust. Soc. Am. 72:753.

Flandrin, P., Cros, P., Mange, G., 1985, Sensitivity of Doppler tolerance to the structure of bat-like sonar signals, in: "Air-borne sonar systems," Conference report Lyon, February 1985.

Floyd, R.W., 1980, Models fo Cetacean signal processing, in: "Animal Sonar Systems," R.G. Busnel, J.F. Fish eds., Plenum Press, New York.

Green, D.M., 1985, Temporal factors in psychoacoustics, in: "Time resolution in auditory systems," A. Michelsen, ed., Springer, Berlin.

Green, D.M, Birdsall, T., Tanner, W.P., 1957, Signal detection as a function of signal intensity and duration, J. Acoust. Soc. Am. 29:523.

Green, D.M., Swets, J.A., 1966, Signal detection theory and psychophysics, Wiley, New York.

Grinnell, A.D., Schnitzler, H.-U., 1977, Directional sensitivity of echolocation in the horseshoe bat, Rhinolophus ferrumequinum. II. Behavioral directivity of hearing, J. Comp. Physiol. 116:63.

Hackbarth, H., 1984, Systemtheoretische Interpretation neuerer Verhaltens- und neurophysiologischer Experimente zur Echoortung der Fledermäuse, Thesis Universität Tübingen, F.R.G.

Heilmann-Rudolf, U., 1984, Das Frequenzunterscheidungsvermögen bei der Grossen Hufeisennase, Rhinolophus ferrumequinum, Thesis Universität Tübingen, F.R.G.

IEEE Standard Dictionary of Electrical and Electronics Terms (3.ed), 1984, Wiley, New York.

Kassam, S.A., Poor, H.V., 1985, Robust techniques for signal processing: A survey, Proc. IEEE 73:433.

Kick, S.A., 1982, Target-detection by the echolocating bat, Eptesicus fuscus, J. Comp. Physiol. 145:431.

Kick, S., Simmons, J., 1984, Automatic gain control in the bat's sonar receiver and the neuroethology of echolocation, J. Neuroscience 4:2725.

Mamode, M., Escudié, B., 1985, Tolérance à l'effet Doppler et signaux optimaux. Signaux sonar émis par les chauve-souris, in: "Air-borne sonar systems," Conference report Lyon February 1985.

Marcus, M.B., Swerling, P., 1962, Sequential detection in radar with multiple resolution elements, IRE Trans. Inform. Theory, IT-8:237.

Masters, W.M., Moffat, A.J.M., Simmons, J.A., 1985, Sonar tracking of horizontally moving targets by the Big Brown Bat Eptesicus fuscus, Science 228:1331.

Moushegian, G., Jeffress, L.A., 1959, Role of interaural time and intensity differences in the lateralization of low-frequency tones, J. Acoust. Soc. Am. 31:1441.

Menne, D., 1985, Theoretical limits of time resolution in narrow band neurons, in: "Time resolution in auditory systems," A. Michelsen, ed., Springer, Berlin.

Menne, D., Hackbarth, H., Accuracy of distance measurement in the bat Eptesicus fuscus: Theoretical aspects and computer simulations, J. Acoust. Soc. Am. 79:386.

Michelsen, A., Larsen, O.N., 1983, Strategies for acoustic communication in complex environment, in: "Neuroethology and Behavioral Physiology," F. Huber and H. Markl, eds., Springer, Berlin.

Mohl, B., 1985, Detection by a Pipistrelle bat of normal and reversed replica of its sonar pulses, in: "Air-borne sonar systems," Conference report Lyon February 1985.

Mohl, B., 1986, Detection by a Pipistrelle bat of normal and reversed replica of its sonar pulses, Acustica, 60:??.

O'Neill, W.E., Suga, N., 1982, Encoding fo target-range information and its representation in the auditory cortex of the mustached bat. J. Neurosci. 2:17.

Peterson, W.W., Birdsall, T.G., and Fox, W.C., 1954, The theory of signal detectability, IRE Trans. Inform. Theory, PGIT-4:171.

Rice, S.O., 1982, Envelopes of narrow-band signals, Proc. IEEE 70:692.

Ronken, D.A., 1970, Monaural detection of a phase difference between clicks, J. Acoust. Soc. Am. 47:1091.

Roverud, R.C., Grinnell, A.D., 1985a, Discrimination performance and echolocation signal integration requirements for target detection and distance determination in the CF/FM bat, Noctilio albiventris, J. Comp. Physiol. 156:447.

Roverud, R.C., Grinnell, A.D., 1985b, Echolocation sound features processed to provide distance information in the CF/FM bat, Noctilio albiven-

tris: evidence for a gated time window utilizing both CF and FM components, J. Comp. Physiol. 156:457.

Schnitzler, H.-U., Henson, D.W., 1980, Performance of airborne animal sonar system: I. Microchiroptera, in: "Animal Sonar Systems," R.G. Busnel, J.F. Fish eds., Plenum Press, New York.

Schnitzler, H.-U., Grinnell, A.D., 1977, Directional sensitivity of echolocation in the horseshoe bat, _Rhinolophus ferrumequinum._ I. Directionality of sound emission, J. Comp. Physiol. 116:51.

Schnitzler, H.-U., Menne, D., Hackbarth, H., 1985, Range determination by measuring time delay in echolocating bats, in: "Time resolution in auditory systems," A. Michelsen, ed., Springer, Berlin.

Schnitzler, H.-U., Menne, D., Kober, R., Heblich, K., 1983, The acoustical image of fluttering insects in echolocating bats, in: "Neuroethology and Behavioral Physiology," F. Huber and H. Markl, eds., Springer, Berlin.

Schuller, G., Beuter, K., Schnitzler, H.-U., 1974, Response to frequency shifted artificial echoes in the bat _Rhinolophus ferrumequinum_, J. Comp. Physiol. 132:47.

Schulman, A.I., Mitchell, R.R., 1965, Operating characteristics from yes-no and forced-choice procedures. J. Acoust. Soc. Am. 40:473.

Schusterman, R.J., Barrett, B., Moore, P., 1975, Detection of underwater signals by a California sea lion and a bottlenose dolphin: variation in the payoff matrix, J. Acoust. Soc. Am. 57:1526.

Simmons, J.A., 1971, Echolocation in bats: Signal processing of echoes for target range, Science 171:925.

Simmons, J.A., 1973, The resolution of target range by echolocating bats, J. Acoust. Soc. Am. 54:157.

Simmons, J.A., 1979, Perception of echo phase information in bat sonar, Science 204:1336.

Simmons, J.A., Kick, S.A., Physiological mechanisms for spatial filtering an image enhancement in the sonar system of bats, Ann. Rev. Physiol. 46:599.

Sokolova, N.N., 1973, Perception and analysis of reflected ultrasonic signals of short duration by bats after extirpation of the auditory cortex, in: "Vopr. Sravn. Fiziol. Analizatorov", No 3, Izd-vo LGU, 154-165.

Strother, G.K., 1961, Note on the possible use of ultrasonic pulse compression by bats. J. Acoust. Soc. Am. 33:696.

Swerling, P., 1960, Probability of detection for fluctuating targets, IRE Trans. Inform. Theory, IT-6:271.

Swets, J.A., 1959, Indices of signal detectability obtained with various psychophysical procedures, J. Acoust. Soc. Am. 31:511.

Swets, J.A., 1973, The relative operating characteristic in psychology, Science 182:990.

Swets, J.A., Shipley, E.F, McKey, M.J., Green, D.M, 1959, Multiple observations of signals in noise, J. Acoust. Soc. Am. 31:514.

Tanner, W.P. Jr., Birdsall, T.G., 1958, Definitions of d' and η as psychophysical measures, J. Acoust. Soc. Am. 30:922.

Trappe, M., 1982, Verhalten und Echoortung der Grossen Hufeisennase (_Rhinolophus ferrumequinum_) beim Insektenfang. Thesis Universität Tübingen, F.R.G.

Tribolet, J.M., 1977, A new phase unwrapping algorithm, IEEE Trans. Acoust., Speech, and Signal Proc., ASSP-25:170.

Urkowitz, H., 1967, Energy detection of unknown deterministic signals, Proc. IEEE 55:523.

Watson, C.S., Gengel, R.W., 1969, Signal duration and signal frequency in relation to auditory sensitivity, J. Acoust. Soc. Am. 46:989.

Webster, F.A., 1963, Active energy radiating systems: the bat and ultrasonic principles II; acoustical control of airborne interceptions by bats. Proc. Int. Cong. Tech. and Blindness. A.F.B., New York 1:49.

Woodward, P.M., 1955, "Probability and Information Theory with Applications to Radar," McGraw-Hill, New York.

Wald, A., 1947, "Sequential Analysis", Wiley, New York.

Whithworth, R.H., Jeffress, L.A., 1961, Time vs intensity in the localization of tones, J. Acoust. Soc. Am. 33:925.

Zwicker, E., 1970, Masking and psychological excitation as consequences of the ear's frequency analysis, in: "Frequency Analysis and Periodicity Detection in Hearing", R. Plomb, G.F. Smoorenberg, eds., Sijthoff, Leiden.

TARGET DISCRIMINATION AND TARGET CLASSIFICATION IN

ECHOLOCATING BATS

J. Ostwald, H.-U. Schnitzler, and G. Schuller

Lehrbereich Zoophysiologie
Institut für Biologie III
Auf der Morgenstelle 28, D-7400 Tübingen, FRG

Zoologisches Institut
Luisenstr. 14, D-8000 München, FRG

INTRODUCTION

Geometrical and textural properties of reflecting targets result in a specific filtering of the echolocation sounds used by the various bat species. They are therefore represented in the temporal and spectral structure of the echoes. These echo features can be used as cues for target recognition, classification and discrimination. Target recognition and classification are perceptual processes in the bat's nervous system that imply the extraction of characteristic echo cues and a matching process with a stored template. In the recognition task this template has to be exactly (within the limits of the sensory and processing system) matched in the echo. For target classification the bat has to produce a more generalized template where echo cues are represented within a certain bandwidth for each parameter. The combination of several of these cues acts as a general template for matching. In contrast to the recognition task it is not necessary that the bat has ever perceived a specific combination of echo cues in order to classify the target.

In a two alternative forced choice procedure the bat has to recognize or classify one or both targets in order to discriminate the two targets and perform a correct response.

In order to find out how neuronal systems of echolocating bats can effectively use various parameters of the echoes for the recognition or classification of range-extended targets information from three different fields of study has to be used:
a) The spectro-temporal analysis of echoes by physical measurements and theoretical considerations gives insight into which echo parameters characterize the filter properties of a target. This shows which cues could be used by the bat.
b) Systematic variations of these cues in behavioral experiments in the laboratory demonstrate whether bats can use this kind of information. Field studies indicate which parameters seem to be most important for bats under natural conditions.
c) Electrophysiological recordings of neuronal responses to simple and more and more natural auditory stimuli give hints about the processing of these echo cues.

Only the crossfertilization of these different aspects of target echo processing will shed light on which echo parameters are used by the animal and how they can be analyzed. Information from each field can stimulate new approaches in the others.

ECHO PARAMETERS CHARACTERIZING A TARGET

In the following we will concentrate on echo parameters that encode geometrical and structural properties of range-extended targets such as size, shape, surface structure, and material and rhythmical fluctuations of target shape and their processing in the auditory system of bats. The analysis of target range, direction and relative velocity are discussed in other reviews in this volume.

Parameters encoding the size of spheres and cylinders

The amplitude of an echo is related to the size of a target. For simple targets like spheres and cylinders the echo intensity depends on the circumference of the target. When the circumference is less than one wavelength the echolocation pulses are scattered by the target (Rayleigh scattering region). For a constant frequency the echo intensity is inversely proportional to r^4 in a cylinder and r^6 in a sphere; which means that the echo intensity decreases very fast with decreasing radius. At a constant radius the echo intensity is proportional to f^3 in cylinders and f^4 in spheres, which means that in small targets high frequencies cause stronger echoes than low frequencies.

If the ratio of circumference to wavelength is between 1 and 10 echo intensity changes periodically with frequency (resonance region). If the wavelength is much shorter than the circumference all frequencies are reflected with the same intensity (optical region) (Skolnik, 1970).

The simple example of a spherical target shows that echo intensity is not merely a function of target size but also of signal frequency if target circumference is close to or smaller than signal wavelength.

Most of the signal energy is not reflected but scattered when a target has a circumference of less than about 30 mm at 10 kHz, 6 mm at 50 kHz and 3 mm at 100 kHz for echolocation in air. For a frequency independent reflection the target circumference has to be above 300 mm at 10 kHz, over 60 mm at 50 kHz and above 30 mm at 100 kHz. This condition is rarely met in the echolocation situation of bats.

Parameters encoding the geometrical structure of range-extended targets

The natural targets of bats are far more complex than spheres and cylinders. A target extended in three dimensions often contains discontinuities acting as separate points of scattering. The total echo of such a target is composed of interfering partial echoes from scattering points differing in depth and angular position. The information about geometrical target structure is therefore encoded in the complex spectral composition and temporal structure of the total echo field. If we assume that a receiver is unable to resolve the angular position of the individual scattering points the echo can be described in a first approximation as a multi-wavefront signal simulating the depth differences of the individual scattering points by corresponding time delays between the wavefronts.

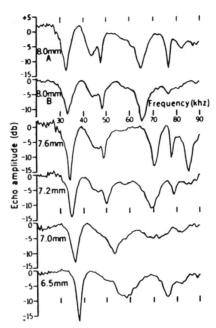

Fig. 1. Echo spectra of plates with holes of different depth using a linear FM-sweep. The difference to the spectrum of a plate without holes is plotted (after Simmons et al., 1974).

The echoes of complex range-extended targets are composed of multiple wavefronts which interfere with each other and therefore produce modulations in the spectral composition when a bat uses broadband echolocation sounds.

The interference pattern of two-wavefront echoes has been described by Simmons et al. (1974) and Beuter (1980). Such two-wavefront echoes are characterized by a pattern of periodic minima in the spectrum. The frequency separation between these minima is inversely proportional to the time delay between the wavefronts. Fig. 1 shows the power spectra of plates with holes of different depths (different time delay between the wavefronts) to demonstrate this effect (Simmons et al., 1974).

Parameters encoding roughness and elastic properties of the surface material

Another characteristic of a target is its surface roughness. Rough targets scatter the incident signals diffusely whereas smooth targets produce more regular scatter patterns. The fine structure of a rough target surface can be regarded as an aggregation of independent scattering grains if grain size is small compared to wavelength. In this case the same frequency dependencies can be found as has been described for targets in the Rayleigh scattering region. Rough surfaces additionally produce many multi-wavefront echoes with different delays thus causing interference patterns which depend on signal wavelength and the dimensions of surface particles (Skolnik, 1970).

If the acoustic impedance of a target is high (compared to air) all frequencies are reflected in the same way. In materials with low impedance, however, elastic vibrations and absortions in the target result in highly frequency dependent reflections (Hickling, 1962; 1967).

415

Tipula spec.

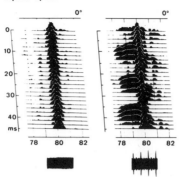

Fig. 2. Real-time spectrograms and oscillograms of the echoes of a non-flying (left) and a flying <u>Tipula</u> (right) (after Schnitzler et al., 1983).

Echo parameters of non fluctuating natural targets

Insects are important targets for many bats. If we assume that they consist of a little sphere (head), a cylinder (body) and two or four flat plates (wings) we can make some rough assumptions concerning the echo structure. The circumferenee of head and body usually ranges from 3 to 30 mm so at the frequencies used by bats the target is never in the optical region. For some size/frequency combinations the target is in the resonance region, for others - especially small insects - it is in the Rayleigh scattering region. Thus when "illuminating" an insect with a broadband echolocation signal there will be a size dependent alteration of the echo's spectral composition. Additionally the wings act as little "acoustical mirrors" which produce an optimal reflection or glint only when they are perpendicular to the impinging sound waves. All insects are extended targets with multiple scatter points which means that interfering multi-wavefront echoes produce additional spectral modifications. Special structures on the body surface like hairs will also cause changes of echo-composition.

If insects do not move their wings or if short echolocation sounds are used insect echoes are characterized by size dependent and multi-wavefront dependent spectro-temporal modulations of a broadband echolocation signal. The variability of insect echoes is further increased by the fact that small changes of the angular aspect of the target will expose different scattering areas thus producing a different echo spectrum. Systematic studies of the structure of echoes from nonflying insects have not yet been made. However it is clear that with an increase in bandwidth more spectral information can be conveyed.

Targets other than insects may also have an important meaning for bats especially if they must separate an insect echo from that of other targets which also consist of many scattering points in different depth and angular positions.

Up to now we do not know which echo parameters characterize an insect echo independent of its angular orientation and how such an echo differs from that of overlapping background echoes produced for instance by leaves and other targets.

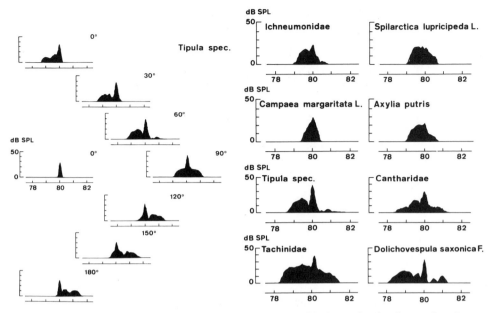

Fig. 3. left: Time averaged spectra of a non-flying <u>Tipula</u> from the front and from flying <u>Tipula</u> oriented at different directions. right: Time averaged spectra of different insect species. The flying insects were oriented at an angle of 60 degrees. (after Schnitzler et al., 1983).

Echo parameters of fluctuating natural targets

It has already been stated that a change in the angular aspect of a target may lead to distinct changes in echo amplitude and spectrum. In targets with rhythmical fluctuations, like insects beating their wings, the temporal variations of echo parameters encode the nature of target movements.

Bats with short FM-echolocation sounds may use amplitude variation to increase their chance to detect a flying insect (Griffin, 1958; Roeder, 1963). The long CF-FM echolocation sounds of Rhinolophids, Hipposiderids and <u>Pteronotus parnellii</u> can encode more information on wing movements of insects as they cover one or more wing beat cycles during a single echo which could be used to recognize insect echoes (Pye, 1967; Schnitzler, 1970; Schuller, 1972; Goldmann and Henson, 1977).

In order to look for possible echo cues contained in the CF-part of echoes from fluttering prey different species of insects have been mounted in the acoustical beam of an ultrasonic loudspeaker which transmitted 80 kHz. Returning echoes were modulated in amplitude and frequency in the rhythm of wingbeat. These modulations encode the wingbeat frequency in their temporal structure and contain information about the species used and the angular orientation of the insect in their spectral composition (Schnitzler, 1978; Schnitzler and Henson, 1980; Schnitzler et al., 1983; Schuller, 1984; Schnitzler, 1986; Kober, this volume). The most prominent features of echoes from fluttering insects are the "glints" which contain strong amplitude peaks produced when the reflecting wings that act as "acoustic mirrors" are perpendicular to the impinging sound waves. Because the wings move in the moment of glint production these amplitude glints are accompanied by broadenings in the echo spectrum due to Doppler shifts (frequency glints, Fig. 2).

The exact time of the glint in the wing beat cycle depends on the angular orientation of the insect. The movement of the wing towards or away from the loudspeaker during the glint results in different Doppler shifts of the echolocation frequency for different orientations. Because of these Doppler shifts spectral broadenings are caused in the echoes which extend more to the frequency range below the carrier frequency at 0 degree and more to the range above it at 180 degrees. Between these two extremes a continuous transition occurs. At 90 degrees the broadening extends more or less symmetrically to both sides (Fig. 3).

The echo spectra also contain species specific information. The repetition rate of glints encodes the wingbeat frequency, while the width of the sidebands depends on wingbeat frequency and size of the wings. The body echo is determined by its size and shape. Further species specific information is encoded in the fine structures of the sidebands (Fig. 3; see also Kober, this volume).

BEHAVIORAL STUDIES

Several behavioral studies in the laboratory and observations in the field have looked for the ability of bats to discriminate between various more or less complex targets.

Discrimination and classification of non fluctuating targets

Most of the experiments in which bats had to discriminate targets with different shape and size were made before 1980 and are reviewed in Schnitzler and Henson (1980). From these experiments it can be concluded that bats learn to discriminate a wide variety of targets especially when the overall echo intensity is different by more than 1-4 dB. However, as in most cases no detailed echo analysis has been made we do not know which echo parameters the bats could have used for discrimination.

Only in one group of experiments a single echo parameter was systematically changed. In these experiments the bats had to discriminate a two-wavefront control target (plate with holes of fixed depth) from a two-wavefront target with variable delay (plate with holes of variable depth). For discrimination the difference in depth had to be 0.6 - 0.9 mm in Eptesicus fuscus (wavefront delay difference 3.5 - 5.2 μs; Simmons et al., 1974), 0.8 - 1 mm (4.6 - 6 μs) in Myotis myotis (Habersetzer and Vogler, 1983), and 1.4 mm (8 μs) in Megaderma lyra (Vogler and Leimer, unpublished data as cited in Habersetzer and Vogler, 1983). In an electronically produced two-wavefront phantom target a delay difference of only 1.29 μs (corresponding to 0.22 mm) was necessary in Megaderma lyra (Schmidt, this volume). All authors assume that the bats use the spectral changes that are caused by the interference patterns of the two wavefronts (Fig. 1) for discrimination.

We want to point out that the results of the experiments with real targets (plates with holes) may not be comparable to the experiments with phantom targets produced by electronically simulated two wavefront echoes as can be seen by the large threshold differences found in Megaderma lyra. Real targets produce unpredictable changes in echo spectrum probably due to interference and resonance phenomena caused by the holes or by target edges and material. In order to find out which echo cues are used by the bat do discriminate between targets only electronically manipulated phantom targets should be used in further experiments, because only these target allow an independent control of echo parameters.

All the experiments described so far only tested the ability of bats to learn a discrimination task. In such a situation the bat has to recognize a known target and distiguish it from a different one offered either simultaneously or one after another within a short time intervall. Once the bat has matched the echo structure with it's internal template and recognized the positive target in many cases it does not even check the other one before making a decision. No experiments have been performed however to test the bats' ability to generalize target echo features and perform a classification of targets.

A true classification task has to be solved by gleaning bats that take insects from surfaces like leaves or branches of trees. If these bats recognize the insects due to echoparameter patterns they would be able to separate the class of insect echoes from the class of background clutter. Unfortunately nobody has made critical experiments which exclude the possibility that insects are classified due to other cues for instance rustling noises that they make when moving or their odor.

Discrimination and classification of fluctuating targets

When using CF-FM echolocation sounds the amplitude and frequency glint pattern of echoes from fluttering insects encode wingbeat rate, wing length, type of wing movement, angular orientation and species. The body echo gives additional information on the structure and size of insects.

Several observations indicate that Rhinolophids, Hiposiderids and the mormoopid bat Pteronotus parnellii use the modulations of their CF-FM signals to recognize insect echoes even in dense background clutter and to classify their prey.

Bats producing CF-FM signals only pursue fluttering prey. In the laboratory Rhinolophus ferrumequinum caught fluttering moths not only when they were flying in the middle of the room but also close to walls, or even when sitting on the ground. Moths which did not move their wings were ignored (Schnitzler and Henson, 1980; Trappe, 1982). Hipposideros ruber (Bell and Fenton, 1984) and Pteronotus parnellii (Goldmann and Henson, 1977) pursued fluttering prey whether flying or sitting, while insects not flapping their wings were ignored. The pursuits were aborted when targets stopped fluttering. A rotating propeller that produces amplitude and frequency glints in the echoes that are similar to those of a fluttering insect (v.d.Emde and Schnitzler, in press) attracts Pteronotus parnellii (Goldmann and Henson, 1977) and Rhinolophus ferrumequinum (Heblich, personal communication) in the laboratory and Hipposideros ruber (Bell and Fenton, 1984) and Pteronotus parnellii (Schnitzler, unpublished) in the field. These observations show that CF-FM bats use fluttering target information to recognize insects and discriminate insect echoes from unmodulated background clutter.

The detection of fluttering targets changes the echolocation behavior of bats with CF-FM sounds. In the laboratory Rhinolophus ferrumequinum that have located a fluttering insect increase the duration of their echolocation sounds thereby also increasing their duty cycle.

In insect hunting rhinolophids and in Pteronotus parnellii especially long sounds producing a high duty cycle are emitted just prior to the approach and terminal phase indicating the pursuit of an insect. Rhinolophus ferrumequinum (Trappe und Schnitzler, 1982) and Rhinolophus rouxi (Schnitzler et al., 1985) reach duty cycles of up to 75% in this phase whereas normal search flight is characterized by duty cycles of about 55%.

An increase of duty cycle was also found when bats were trained to locate a sinusoidally oscillating target or a glint producing propeller. In this situation Rhinolophus ferrumequinum increased the duty cycle from 55% to 75% by lengthening their sounds when the target was oscillating (Schnitzler and Flieger, 1983). This reaction was strongest at oscillation frequencies between 40 and 60 Hz, which is about the range of wingbeat frequencies of the preferred prey in captive bats (Trappe, 1982).

Hipposiderids achieve higher duty cycles by increasing the repetition rate of sound emision. For instance Hipposideros lankadiva reacts to a glint from a propeller with an increase of duty cycle from 16% to about 35%. Hipposideris speoris changes the duty cycle from about 20% to about 50% (v.d.Emde, 1984). Brown and Berry (personal communication) recorded long sequences with high repetition rate and duty cycle in a Hiposideros diadema when it left its perch to hunt for insects.

A high duty cycle not only allows the bats to detect echoes in flight and to separate such echoes from background clutter but also to evaluate fluttering target information. It was therefore suggested that CF-FM bats use temporal changes in the echoes for a more detailed evaluation of target properties (Schnitzler et al., 1983). This raises the question whether bats can read the wingbeat signatures in insect echoes and use it to classify their prey.

In the laboratory Rhinolophus ferrumequinum often selected one species of insect prey over another presumably on the basis of wingbeat rate (Trappe 1982). In Hiposideros ruber, however, Bell and Fenton (1984) found no evidence that the bats prefered one species of prey over any other; but they sometimes avoided larger moth. This data may indicate that there is a difference in the evaluation of fluttering target information in Hipposiderids and other CF-FM bats. Maybe Hipposiderids use fluttering target information only to separate insect echoes with periodic glints from background clutter whereas Rhinolophids and Pteronotus parnellii can evaluate additional echo information (like wingbeat frequency) for a more detailed analysis. Supporting evidence for the less developed evaluation of flutter information could be the much shorter duration of echolocation signals in Hipposiderids. If two glints are necessary within one echo to evaluate wingbeat frequency, sound duration has to be at least twice the wingbeat rate of the predominant prey. If so bats with shorter sounds could be adapted to hunt for insects with high wingbeat rates and vice versa. However, there is not enough comparative data to discuss this hypothesis.

The ability of bats to detect sinusoidal frequency modulations in echoes has been tested in discrimination experiments where bats were trained to distinguish an oscillating target from a similar motionless one. Rhinolophus ferrumequinum is more effective in detecting such modulations than Hipposiderid bats. For instance at an oscillation frequency of 40 Hz (typical wingbeat frequency for many moths that are prefered prey) Rhinolophus ferrumequinum needs an oscillation amplitude of 0.24 mm for discrimination (Schnitzler and Flieger, 1983), while in Hipposideros lankadiva the threshold amplitude is about 5 times higher, and Hipposideros speoris never learned the task (v.d.Emde and Schnitzler, in press). However, both Hipposiderids reacted strongly to a glint producing propeller. Rhinolophus ferrumequinum can also distinguish between simulated insect echoes which only differ slightly in wingbeat rate (v.d.Emde, this volume). Such selective reactions indicate that the bats can use the wingbeat rate of a fluttering target as an echo cue to classify their prey. Different bat species may differ in their ability to use the available information.

Echoes from fluttering targets also contain information on the angular orientation of an insect in the relative position of the sidebands of the frequency glints and species specific information in the fine structure of the echoes. We do not know yet whether this information is also evaluated by CF-FM bats.

The evaluation of fluttering target information allows CF-FM bats to hunt in areas where the insect echoes are obscured by a strong background clutter. The scattered observation on the feeding behavior in CF-FM bats shows that they use this chance and really hunt for insects close to or in bushes and trees, along walls, and near the ground. For instance Hipposideros caffer searches for insects in very thick bushes (Schnitzler, unpublished), Hipposideros speoris and Hipposideros bicolor enter into the foliage of trees (Neuweiler, 1983), and Pteronotus parnellii hunts under the canopy of large trees and between bushes (Schnitzler, unpublished). Some species of CF-FM bats also hunt for insects making short flights from a perch like a flycatcher: for instance Rhinolophus aethiops (Shortridge, 1934), Rhinolophus rouxi (Schnitzler et al., 1985), Rhinolophus hildebrandii (Fenton, personal communication), Hipposideros commersoni (Vaughan, 1977) and Hipposideros diadema (Brown and Berry, 1983). For bats which can classify their prey the flycatcher style of hunting is energy conserving because they can wait until an interesting insect comes within range.

Besides the CF-FM bats some FM bats (bats with mainly frequency modulated signals) also produce rather long sounds of almost constant frequency. But these signals are probably not used for the evaluation of fluttering target information but just to increase the prey detection distance (discussed in Schnitzler, 1986). This is supported by the fact that all these species immediately switch to short broadband signals when they have detected an insect.

ADAPTATIONS OF THE AUDITORY SYSTEM

We will focus on the auditory representation of echo cues that are caused by the reflective properties of range extended targets and not by their position and movement in space. But first we will briefly consider the influence of the type of echolocation sound on the recognition task.

Time invariant (non fluctuating) range extended targets are characterized by frequency dependent reflective properties. To constant frequency echolocation sounds all these targets appear uniform; their shape and size only producing intensity changes in the echoes. Target structure in depth additionally causes time smearing of the returning waveform. Harmonically structured CF-echolocation sounds may undergo differential attenuations of signal components that could yield some information about target structure. In order to use the full information of target structure that is contained in the spectral composition of echoes a broad-band echolocation sound is best suited.

Fluctuating targets offer some additional information in the time structure of their echoes. At least the rhythm of wing beat of insect prey is coded in the echo glint-pattern by amplitude peaks and specific spectral widenings that are caused by Doppler-shifts due to the movement of the wings. To test the temporal cues in this glint-pattern a long CF-signal, like it is used by Rhinolophids, Hipposiderids and Pteronotus parnellii, is especially well suited.

Analysis of range extended targets using broadband signals

To our knowledge up to now no experiments have been conducted where echo cues have been systematically changed to mimic natural echoes of range extended targets. How could such cues be represented in the auditory system?

Spectral cues

Studies of the mammalian auditory system with pure tone stimuli show that the spectral composition of a sound is well preserved in the cochlea and in higher stations by the spatio-temporal pattern of excitation. This is also true when FM-signals are used. The responses of single units to linear frequency-modulated signals have been studied in many stations of the auditory system of bats and these responses have been compared to those using pure tone stimuli (Bodenhamer and Pollak, 1981; O'Neill, 1985; Pollak et al., 1977; 1978; Suga, 1965a; b; c; 1968; 1969; Suga and Schlegel, 1973; Vater, 1981; Vater and Schlegel, 1979).

Most authors looked for adaptations to the processing of echolocation sounds in FM-bats by comparing pure-tone and FM thresholds. In the inferior colliculus (IC) and auditory cortex (AC) of Myotis lucifugus Suga (1965a; b; 1968; 1969) found units that exclusively responded to FM-sweeps. But they constituted only 3% of the sample in the IC (Suga and Schlegel, 1973). The responses to FM-sweeps depended on the "direction, range, speed and functional form" of the sweep.

Most units responded to both FM and CF signals of appropriate frequencies. In 21% of units recorded in the IC of Tadarida threshold was lower for FM-stimuli than for CF-stimuli (Pollak et al., 1978). In Molossus molossus 41% (Molossus ater, 27%) of IC units showed a lower threshold to FM-stimuli as long as short (2 ms) signals were used (Vater and Schlegel, 1979). Lower thresholds to FM-stimuli are not limited to FM-bats. In the IC of Pteronotus parnellii O'Neill (1985) found 8% of units with a lower threshold for FM-stimuli (58% had lower threshold for CF-stimuli). In most units, however, thresholds for CF- and FM-stimuli were very similar.

But comparison of thresholds does not clarify whether small spectral changes are coded in the response pattern of single units. Rather we need to know whether the spectral fine structure of an echo is still coded in the spatio-temporal response properties of the auditory system when FM-sounds are used instead of pure-tone stimuli.

In the IC of the Mexican Free Tailed Bat, Tadarida brasiliensis, Bodenhamer and Pollak (1981) found phasic constant latency responders (pCLRs). Neurons with this response type are especially well suited to address this question. They respond with a single action potential at a very stable latency when stimulated with an FM-sweep. By changing the sweep rate of the stimulus response time could be shifted (Fig. 4). Apparently these neurons responded to a restricted frequency component within the sweep that could be calculated from the various sweep latencies and was called excitatory frequency (EF). In most units EF coincided with the high frequency slope of the unit's pure tone tuning curve. As these slopes are very steep, EF changed only very little with intensity (Fig. 4). Similar to the tonotopical order of best frequencies, EFs are also systematically arranged in the IC with a frequency axis from dorsal (low EF) to ventral (high EF) (Bodenhamer and Pollak, 1981).

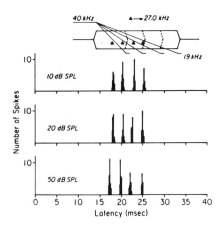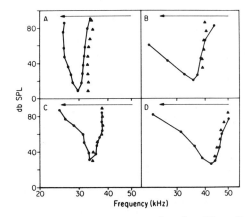

Fig. 4. left: Responses of a phasic constant latency neuron in the IC of Tadarida to 4 different durations of a 40-19 kHz FM-sweep plotted in one histogram. Discharge was elicited by the 27 kHz component of the stimulus (excitatory frequency; EF). Synchronization is very similar for different stimulus levels. right: Positon of EF in relation to the pure tone tuning curve of 4 different units (after Bodenhamer and Pollak, 1981).

Phasic-on neurons in the IC show monotonic spike count functions (Pollak et al., 1978; Vater and Schlegel, 1979) covering a range of up to 40 dB. Monotonic and nonmonotonic spike count functions were found in the IC of Rhinolophus ferrumequinum (Vater, 1981), Molossus ater and Molossus molossus (Vater and Schlegel, 1979) when using short FM-stimuli. In nonmonotonic units maximum discharges are elicited for intensities of 20-50 dB SPL (Vater and Schlegel, 1979) (Fig. 5).

This data indicate that the spectral information in the echoes is preserved in the spatio-temporal excitation of orderly arranged neurons in the auditory system also when FM-stimuli are used. It is completely open how much change in the response of a neuronal population is necessary for an animal to perceive a difference. Behavioral tests show that a difference in total echo intensity of 1-3 dB between two targets of identical shape can be learned by several bat species (Airapetyants and Konstantinov, 1974). In the two-wavefront experiments (Simmons et al., 1974) the frequency of the first two spectral gaps shifted over 10% (from 32.8 (47.1) kHz to 35.7 (52.8) kHz; Fig. 1). These differences should be coded by neurons with nonmonotonic spike count functions of a very limited range and with a Q 10-dB of about 20 which are found in FM-bats.

There is no information about how the auditory system "reads" the spectral information coded in the response properties of IC units. In order to code echo cues it is necessary to combine, by facilitation and inhibition, several of the spectral components in a specific way.

Two major problems are still open:
a) How can a bat recognize a specific spectral combination independent of overal echo intensity (termed "level tolerant tuning" by Suga)?
b) The waveform of echolocation sounds is not identical for each emission. Time structure and spectral composition of the sound changes. In some ways the receiver system has to be briefed about the sound structure used at that moment. In order to gain target filter characteristics the spectrum of the emitted sound has to be subtracted from the received echo. Currently there is no idea how the system performs that operation.

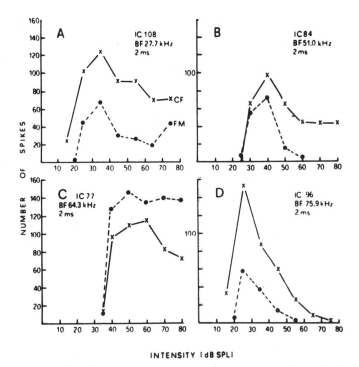

INTENSITY (dB SPL)

Fig. 5. Spike count functions of single units in the IC of <u>Rhinolophus</u> to pure tones at the units BF (crosses; solid lines) and FM-sweeps of 20 kHz centred at the units BF (circles; dashed lines) (after Vater, 1981).

Temporal cues

Miller (this volume) points out that spectral properties of FM-signals are also encoded in the time structure of its envelope. In the simulated two-wavefront targets of Beuter (1980) there was a strong amplitude modulation with gaps of about 0.5 ms distance corresponding to the frequency gaps in the spectrum. Can bats use this temporal information to gain insight into target structure?

Several experiments have been performed looking for the integration time of neurons in the auditory system of bats. The maximum discharge rate and the minimal threshold of a neuron is only reached when the stimulus duration is larger than a certain time. In a technical filter two stimuli occuring within this integration time cannot be processed separately.

Integration times of phasic-on units in the IC of <u>Tadarida</u> (FM-stimuli) were larger than 0.86 ms (Pollak et al., 1978), in <u>Rhinolophus</u> (CF-stimuli) they were smaller than 1 ms (Vater, 1977), only in <u>Myotis</u> integration times of below 0.5 ms were found (Sokolova cited in Airapetyants and Konstantinov, 1974). These times are well suited for echo-ranging but they do not enable the neurons to code fine temporal structure in a single echo. It is even more unlikely that in two-wavefront or more complex multi-wavefront echoes the time delay between the wavefronts is coded in the neuronal response as has been suggested by Simmons (1979).

Neurophysiological data support the processing of echo information in the spectral rather than the time domain when stationary targets are to be processed.

Analysis of fluctuating targets using long CF-signals

Echoes from fluctuating targets are characterized by time-dependent amplitude variations and changes in frequency due to Dopplershifts. These modulations can be random (e.g. in echoes from moving leaves) or they can be periodic as from insects beating their wings which are characterized by "amplitude glints" and "frequency glints" (Fig. 2).

Long CF-echolocation sounds act as carrier frequencies for these modulations. By Doppler-shift compensation the echo frequency is kept constant independent of the bat's own flight speed. Modulated echoes of fluttering targets therefore only occur in a small frequency band forming an "expectation window" (Schnitzler and Ostwald, 1983).

Basic properties of the auditory system in CF-FM bats

Neurons with best frequencies in the range relevant for CF processing are strongly overrepresented in the genus Rhinolophus and Pteronotus at all levels of the auditory system (Rhinolophus: Cochlear nucleus (CN): Suga et al. (1976), Feng and Vater (1985); Lateral lemniscus (LL): Metzner and Radtke-Schuller (submitted); Inferior Colliculus (IC): Schuller and Pollak (1979); Medial geniculate body (MGB): Engelstätter (1981); Auditory Cortex (AC): Ostwald (1984); Pteronotus: CN: Suga et al. (1975); IC: Pollak and Bodenhammer (1981); AC: Suga and Jen (1976)). Already on the most peripheral level, the inner ear, the small band of frequencies around the CF-frequency occupies a large part of the analyzing dimensions (see review of Vater, this volume). In Rhinolophus the expanded frequency range just above the resting frequency in conjunction with the Dopplershift compensation behaviour has been termed "acoustic fovea" in analogy to the visual system (Schuller and Pollak, 1979).

Most authors give information on the tuning characteristics in the form of best frequencies, Q(10dB)-values, minimum thresholds or tuning curves. Neurons in the small frequency band at and around the CF-frequency of the individual bat reach very high Q(10dB)-values (up to 400-500) in contrast to neurons with other best frequencies, which show the Q-values common to the mammalian auditory system (references as for 'overrepresentation', see above). This extremly high frequency resolution is of fundamental importance for separating adjacent frequency components and for the fine analysis of natural echoes and is found at all levels of the auditory system of CF-FM-bats.

Suga and Tsuzuki (1985) pointed out, that the Q(10dB)-value is only an appropriate characterization of the tuning properties of a neuron for intensities near minimal threshold. They demonstrated that in neurons in the auditory cortex of Pteronotus the tuning bandwidth is maintained over a wide intensity range (called level-tolerant-tuning). Thus the high frequency resolution capabilities are intensity-invariant and the contrast in frequency sharpens considerably.

Spike count functions yield information in which intensity range neurons are responding and the slope and form of the functions determine what response changes are provoked by changes in stimulus intensity. This neuronal property is important for the encoding of amplitude modulations ("amplitude glints") and is reported in most references cited above. Spike count functions have only been measured for the best frequency of the neuron by most authors. But a complete set of spike count functions within the entire response band of the cell would be necessary. The intensity changes in natural echoes rarely reach more than 30 dB and consequently result in rather moderate response changes compared to the activity

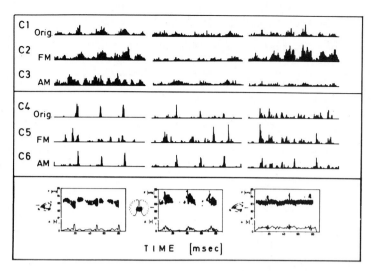

Fig. 6. Response of 2 neurons from the IC of <u>Rhinolophus ferrumequinum</u> to echoes of a flying insect from different directions. Response to the original echoes is compared to the response to their mainly frequency and mainly amplitude modulated components. Frequency and intensity of the stimuli are depicted below the columns (after Schuller, 1984).

differences in the same narrowly tuned neurons caused by the frequency variations commonly found (up to 1 - 2 kHz) in insect echoes.

Summarizing the basic properties of neurons found in the auditory system of CF-FM bats, it can be concluded that their frequency resolution is sufficient to resolve small frequency modulations in natural echoes and that amplitude modulations can also be traced to a limited extend. Whether the responses to natural echoes can be simply explained by a combination of basic neuronal response properties is questionable and the temporal processing of such information has to be looked at more closely.

Stimulation with natural echoes

Only in <u>Rhinolophus ferrumequinum</u> insect echoes or close simulations thereof have been used as stimuli during single unit recordings (IC: Schuller in: Neuweiler et al., 1980; Schuller, 1984; AC: Ostwald, 1980; this volume).

Schuller used natural insect echoes recorded from different aspects of the insect. Single units in the inferior colliculus (IC) showed clearly different responses when confronted with the echoes from different directions (Fig. 6). Discharges were phase-locked to rapid frequency or amplitude transitions ("glints"). Thus in all neurons that responded the information on wing beat frequency was preserved in the discharge pattern of the units. Which are the cues in the complex modulation patterns of insect echoes that are analyzed and reflected in the neuronal responses? Schuller (1984) extracted the mainly amplitude modulated component and the mainly frequency modulated component from the original echo signal and presented them separately. Units with best frequency in the range of echolocation calls showed generally, but not exclusively, a more vigorous response to the frequency transitions than to the amplitude transitions. The strongest response was obtained when a "frequency glint" coincided with an "amplitude glint" as can be seen in Fig. 6. On the other hand, the neurons not only extracted the fast transitions as a complex but also resolved the fine structure of the modulations. In order to establish the degree of correlation between neural responses and stimulus patterns more sophisticated methods like crosscorrelation of the response with stimulus parameters have to be applied.

Ostwald (1980; this volume) found similar discharge properties of neurons in the auditory cortex and his recordings also showed a superior synchronism of the response to the "frequency glints" than to the "amplitude glints" (see Fig. 1 in Ostwald, this volume). Neither in the AC nor in the IC neurons specialized to distinct target directions or different insect species were detected. The few recordings with natural echoes up to now have not revealed any clear hints of feature detections but only demonstrated that spectro-temporal changes may be apparent as distinct discharge patterns in the neuronal response.

Sinusoidally modulated stimuli

Sinusoidally frequency modulated (SFM) and amplitude modulated (SAM) stimuli were used by several authors to simulate in a first approximation the echoes from wing beating insects (Rhinolophus: CN: Vater, 1982; LL: Metzner and Radtke-Schuller, submitted; IC: Schuller, 1979a; Pollak and Schuller, 1981; Reimer, submitted; AC: Ostwald, 1980; this volume; Pteronotus: CN: Suga et al., 1975; IC: Bodenhammer and Pollak, 1983; AC: Suga et al., 1983). The modulation waveform of the stimulus is mirrored in the synchronism of the discharge pattern, which greatly depends on the amplitude of modulation, the modulation frequency, the carrier frequency and the overall intensity of the stimulus. Some response properties can be explained on the basis of the tuning characteristics of a neuron, but a few reject this interpretation.

The minimum depth of modulation which is reflected in a synchronized response has been determined for SFM as well as for SAM stimuli at the various brain levels. For SFM the lowest encoded depth was $\pm$ 20 Hz in the CN of Rhinolophus (Rh; Vater, 1982) and $\pm$ 6 Hz in the CN of Pteronotus (Pt; Suga and Jen, 1977), $\pm$ 10 to $\pm$ 20 Hz in the IC of Rh and Pt (Rh: Schuller, 1979; Pollak and Schuller, 1981; Pt: Bodenhammer and Pollak, 1983) and $\pm$ 150 Hz in Rh (Ostwald, 1980). Suga et al. (1983) report "best SFM depth" in the auditory cortex of Pteronotus as low as $\pm$ 3 Hz, but more typically between $\pm$ 97 and $\pm$ 970 Hz. Phase-locked discharge to such low modulation depths are only reached if the other modulation parameters (modulation frequency, carrier frequency, and intensity) are optimally set and it was only found in neurons that had best frequencies in the CF-band and exhibited extremely narrow tuning curves. As the frequency deviations introduced into the echoes by wing beating insects typically extent between 0.5 and 1 kHz and reach up to 2 kHz, this parameter can well be processed by SFM-sensitive neurons.

For the minimum degree of SAM modulation data are available for the CN, IC and AC of Rhinolophus (CN: 1 dB, Vater, 1982; IC: 1 dB, Schuller, 1979a; Reimer, subm.; AC: 7 dB, Ostwald, this volume). In Pteronotus no comparable data is available.

Both SFM and SAM cause different degrees of phase-locking in the response of neurons when the modulation frequency is systematically varied. Phase-locking only occurs in a certain range of modulation frequencies (Fig. 7). Highest modulation frequencies that cause synchronized discharge progressively decrease when ascending in the auditory processing level and are highest in CN-neurons with

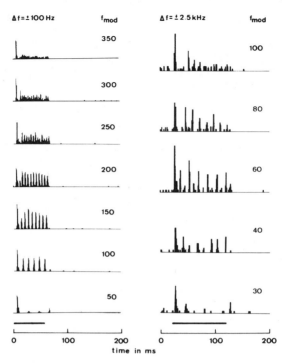

Fig. 7. Phase-locked discharge of a neuron from IC (left; after Schuller, 1979) and AC (right; Ostwald in prep.) of Rhinolophus to SFM stimuli with different modulation frequencies. Optimal discharge is found at higher modulation frequencies in the IC than in the AC.

synchronization up to 800 Hz in Rhinolopus (SFM; Vater, 1982) and up to 3 kHz in Pteronotus (SAM; Suga et al., 1975).

Maximum modulation frequencies vary progressively with depth in units of the different nuclei of the lateral lemniscus (LL) in Rhinolophus. In the ventral nucleus (vNLL) they were highest (800 Hz). A medium level (500 Hz) is found in the intermediate nucleus (iNLL) and lowest rates (250 Hz) in the dorsal nucleus (dNLL; Metzner and Radtke-Schuller, subm.). Synchronized responses were recorded up to modulation frequencies of 200 - 350 Hz in the IC of Rh by Schuller (1979) and Pollak and Schuller (1981) and up to 200 Hz in the IC of Pt (Bodenhamer and Pollak, 1983). In the auditory cortex of Rhinolophus and Pteronotus units differ in their upper modulation frequencies. Ostwald (1980) presents data indicating upper frequencies of 100 - 150 Hz in Rhinolophus, whereas Suga et al. (1983) show synchronized responses of AC-neurons up to at least 300 Hz in Pteronotus. More important than maximum modulation frequencies are the optimum rates at which synchronisation of the responses reaches a maximum. This optimum range was between 20 and 100 Hz in the IC of both species and between 40 and 70 Hz in the AC of Rhinolophus (Fig. 7). Some units in the CN as well as in the IC showed clear preference for certain modulation frequencies, but no such preference were discernible in the responses of cortical neurons in Pteronotus (Suga et al., 1983). In general the tuning of neurons to the modulation frequency is dependent on the modulation depth (see Fig. 2 in Ostwald, this volume). It has to be stressed at this point that only Reimer (subm.) has quantified the synchronisation of responses by calculating appropriate correlation coefficients.

Reimer prepared the most recent extensive investigation on the coding of SAM-stimuli in the inferior colliculus of Rhinolophus with special attention to the tuning to modulation frequencies. Synchronized activity could be elicited up to modulation frequencies of 400 Hz, but in 70% of the neurons the best modulation frequencies (BMF) lay at 100 Hz and below. BMF did not depend on the best frequency of the neuron nor on the intensity of the stimulus but generally changed with modulation depth. In the AC SAM with modulation frequencies between 40 and 70 Hz were most effective (Ostwald, this volume).

The tuning to distinct modulation frequencies in different neurons could constitute the neuronal basis for the classification of insects using wing beat frequency. However, neurons are tuned to particular modulation frequencies only in a broad sense.

If the carrier frequency of SFM stimuli is changed most dramatic effects in the response pattern occur especially in sharply tuned neurons. There are different neuron types in the NC (Vater, 1982), IC (Schuller, 1979; Pollak and Schuller, 1981; Bodenhamer and Pollak, 1983), and AC (Ostwald, 1980, this volume) that seem to emphasize the low or high frequency border of their tuning curve (asymmetrical neurons). Some response modifications can be explained by the characteristics of the exitatory pure tone tuning curve, but in many units, even on the level of the CN, synchronized discharge cannot be foreseen by looking at the properties of the neuron evaluated with pure-tone stimuli.

The attempt to derive the response properties to SFM-stimuli at different carrier frequencies on the basis of inhibitory fields in the vicinity was only partially successful (Bodenhammer and Pollak, 1983, Suga et al., 1983). SAM measurements were usually conducted at the units best frequency. In a few neurons Reimer (submitted) changed the carrier frequency of SAM stimuli and found only little effect as long as the carrier frequency stayed within the tuning area and intensity was appropriately selected.

In most neurons intensity thresholds for SFM are similar to the thresholds for pure tones and in single cases even smaller (Pollak and Schuller, 1981; Bodenhammer and Pollak, 1983). The responses to SFM are generally well synchronized within a wide range of intensities, although phase-related responses tend to fuse at high intensities in some neurons (CN: Vater, 1982; IC: Pollak and Schuller, 1981). CN as well as IC neurons often show preferences for distinct intensity ranges in which their overall activity as well as the synchronism of the response is optimum (Vater, 1982; Pollak and Schuller, 1981; Bodenhammer and Pollak, 1983). These prefered intensity ranges often change with carrier frequency.

SAM-responses are generally best when the intensity of the stimulus is within a range of 10 to 30 dB above the threshold of the neuron. Intensities higher than 50 dB above threshold most often blur the synchronism of the response. Both kinds of modulation are therefore well encoded and analyzed at intensities near and just above the threshold, thus providing the means for early detection and recognition of wing beating insects.

In natural echoes of fluttering insects frequency and amplitude modulations cannot be separated and have to be processed in parallel by the auditory system. Their relative phase depends on the direction of the insect. SFM and SAM with different phase shifts have been used in the CN (Vater, 1982) and in the IC (Reimer, subm.). Phase differences lead to strong modifications in the response pattern covering a range from optimum response to almost cancellation or changes in the synchronism. The combined modulations contain information of higher discrimination value than the separate components and might be a valuable cue encoded in the differential response of neurons.

The investigation of the response properties of neurons to modulated stimuli has revealed a complicated pattern of spectro-temporal analysis in which many though not all neurons throughout the auditory system show preferences for echo parameters that are present in the natural situation. Ascending in the auditory pathway the range of encoded echo properties is more and more limited. Modulated natural echoes will therefore cause a complicated spatio-temporal pattern of activities in the population of neurons covering the CF-frequency band. This pattern of excitation is not characterizing a specific target, but it will be modified by slight shifts in the carrier frequency or direction of the target.

Scheich (1977) has called such naturally occuring stimulus configurations that are preferentially processed by certain neurons "focal properties" (FP) which correspond to the "information bearing parameters" (IBP) of Suga et al. (1983). The question where in the central nervous system this information is used to classify targets and trigger an appropriate behavioral response still remains a matter of speculation.

Little information is available on those neurons that did not respond in a synchronized pattern to modulated stimuli. It would be interesting to know whether they display preferential responses to other echo parameters.

Influences of vocalization

In a critical review Symmes (1981) points out that the central nervous system of an animal is a highly variable structure in its response properties. This is especially important when using complex natural stimuli in higher centers. The situation as a whole for the animal has to be kept in mind. When studying the active orientation system of animals like echolocating bat this has to be concidered. In most experiments neurons in the auditory system of bats were tested in a passively receiving situation without considering that echolocation is an active process where echoes only occur after a prior vocalization. Recordings in the IC of vocalizing bats indicated that the vocalization itself might have an important influence on the processing of acoustic stimuli (Schuller 1979b, 1980). Some units only responded with a synchronized discharge to SFM stimuli, when the stimulus was presented short after the onset of vocalization. In this case the emitted sound acts as an acoustic stimulus to the ear. But obviously the process of vocalization itself additionally produces a neuronal conditioning as the influence of vocalization in these units could not be replaced by an artificial stimulus simulating the orientation sound. Suga et al. (1983) have shown, however, that for synchronized response to SFM stimuli in cortical neurons of Pteronotus (CF/CF-area) only the lower harmonic (H1) of the echolocation call has to be present simultaneously. Here the active process of vocalization does not seem to be necessary. As the data on this topic are scarce, the influence of vocalization on the auditory processing of complex stimuli in bats needs to be further investigated.

Acknowledgment

Supported by Deutsche Forschungsgemeinschaft, SFB 307 (Tübingen) and SFB 204 (München). We thank B. Fenton and A. Grinnell for critical discussion and I. Kaipf, T. Hoffmann, K. Rüttgers, and H. Zillus for technical help.

REFERENCES

Airapetiants, E.Sh., Konstantinov, A.I., 1974, Echolocation in nature. Nauka, Leningrad. English Translation: Joint Publication Research Service, Arlington.

Bell, G.P., Fenton, M.B., 1984, The use of Doppler shifted echoes as a clutter rejection system: the echolocation and feeding behavior of Hipposideros ruber (Chiroptera: Hiposideridae). Behav.Ecol. Sociobiol., 15:109-114.

Beuter, K.J., 1980, A new concept of echo evaluation in the auditory system of bats. in: "Animal sonar systems", Busnel, R.-G., Fish, J.F.(eds.) Plenum Press, pp.747-761.

Bodenhammer, R.D., Pollak, G.D., 1981, Time and frequency domain processing in the inferior colliculus of echolocating bats. Hear. Res., 5:317-335.

Bodenhammer, R.D., Pollak, G.D., 1983, Response characteristics of single units in the inferior colliculus of mustache bats to sinusoidally frequency modulated signals. J.Comp.Physiol., 153:67-80.

Brown, P.L, Berry, R.D., 1983, Echolocation behavior in a "flycatcher" bat, Hipposideros diadema. J.Acoust.Soc.Am. (Suppl.I), 74:532.

von der Emde, G., 1984, Detektion rhythmischer Zielbewegungen bei Fledermäusen aus der Familie der Hipposideriden. Diplomarbeit, Tübingen.

von der Emde, G., Schnitzler, H.U., in press, Fluttering target detection in Hipposiderid bats. J.Comp.Physiol.A.

Engelstätter, R., 1981, Hörphysiologische Untersuchungen an Neuronen der aufsteigenden Hörbahn der echoortenden Fledermaus Rhinolophus rouxi. Ph.D. Thesis, Frankfurt.

Feng, A.S., Vater, M., 1985, Functional organization of the cochlear nucleus of rufous horseshoe bats (Rhinolophus rouxi): frequencies and internal connections are arranged in slabs. J.Comp.Neurol., 235:529-555

Goldmann, L.J., Henson, O.W., 1977, Prey recognition and selection by the constant frequency bat, Pteronotus p. parnellii. Behav.Ecol. Sociobiol., 2:411-419.

Griffin, D.R., 1958, Listening in the dark. Yale University Press, New Haven.

Habersetzer, J., Vogler, B., 1983, Discrimination of surface-structured targets by the echolocating bat Myotis myotis during flight. J.Comp.Physiol., 152:275-282.

Hickling, R., 1962, Analysis of echoes from a solid elastic sphere in water. J.Acoust.Soc.Amer., 34:1582-1592.

Hickling, R., 1967, Echoes from spherical shells in air. J.Acoust. Soc.Amer., 42:388-390.

Metzner, W., Radtke-Schuller, S., submitted, The nuclei of the lateral lemniscus in the rufous horseshoe bat, Rhinolophus rouxi. A neurophysiological approach. J.Comp.Physiol.

Neuweiler, G., 1983, Echolocation and adaptivity to ecological constraints. in: "Neuroethology and behavioral physiology", Huber, F., Markl, H., (eds.) Springer, Heidelberg, New York, pp. 280-302.

Neuweiler, G., Bruns, V., Schuller, G., 1980, Ears adapted for the detection of motion, or how echolocating bats have exploited the capacities of the mammalian auditory system. J.Acoust.Soc.Amer., 68:741-753.

O'Neill, W.E., 1985, Responses to pure tones and linear FM components of the CF-FM biosonar signal by single units in the inferior colliculus of the mustached bat. J.Comp.Physiol., 157:797-815.

Ostwald, J., 1980, The functional organization of the auditory cortex in the CF-FM bat Rhinolophus ferrumequinum. In: "Animal sonar systems", Busnel, R.-G., Fish, J.F.(eds.) Plenum Press, pp.953-955.

Ostwald, J., 1984, Tonotopical organization and pure tone response characteristics of single units in the auditory cortex of the Greater Horseshoe Bat. J.Comp.Physiol., 155:821-834.

Pollak, G., Bodenhammer, R., Marsh, D., Souther, A., 1977, Recovery cycles of single neurons in the inferior colliculus of unanesthetized bats obtained with frequency-modulated and constant-frequency sounds, J.Comp.Physiol., 120:215-250.

Pollak, G., Marsh, D., Bodenhammer, R., Souther, A., 1978, A single-unit analysis of inferior colliculus in unanesthetized bats: response patterns and spike-count funktions generated by CF- and FM-sounds. J.Neurophys., 41:677-691.

Pollak, G.D., Bodenhammer, R.D., 1981, Specialized characteristics of single units in the inferior coliculus of the mustache bat: frequency representation, tuning and discharge patterns. J.Neurophysiol., 46:605-620.

Pollak, G.D., Schuller, G., 1981, Tonotopic organization and encoding features of single units in the inferior colliculus of horseshoe bats: Functional implications for prey identification. J.Neurophysiol., 45:208-226.

Pye, J.D., 1967, Discussion of the paper of Griffith. in: "Animal Sonar Systems", Busnel, R.G. (eds.) Jouy-en-Josas, pp: 1121-1136.

Reimer, K., submitted, Coding of sinusoidal amplitude modulated acoustic stimuli in the colliculus inferior of the rufous horseshoe bat, Rhinolophus rouxi. J.Comp.Physiol.

Roeder, K.D., 1963, Echoes of ultrasonic pulses from flying moths. Biol.Bull., 124:200-210.

Scheich, H., 1977, Central processing of complex sounds and feature analysis. In: "Recognition of complex acoustic signals", Bullock T.H. (ed.), Dahlem Konferenzen, Abakon Verlagsgesellschaft, Berlin, pp. 161-182.

Schnitzler, H.U., 1970, Echoortung bei der Fledermaus Chilonycteris rubiginosa, Z.vergl.Physiol., 68:25-39.

Schnitzler, H.U., 1978, Die Detektion von Bewegungen durch Echoortung bei Fledermäusen. Verh.Dtsch.Zool.Ges., pp. 16-33.

Schnitzler, H.U., Henson, O.W., 1980, Performance of airborne animal sonar systems: I. Microchiroptera. in: "Animal Sonar Systems", Busnel, R.-G., Fish, J.F.(eds.) Plenum Press, New York, pp: 235-250.

Schnitzler, H.U., Flieger, E., 1983, Detection of oscillating target movements by echolocation in the Greater Horseshoe Bat. J.Comp.Physiol., 153:385-391.

Schnitzler, H.U., Menne, D., Kober, R., Heblich, K., 1983, The acoustical image of fluttering insects in echolocating bats. in: "Neuroethology and behavioral physiology. Roots and growing points", Huber, F., Markl, H. (eds.) Springer, Heidelberg, New York, pp: 235-250.

Schnitzler, H.U., Ostwald, J., 1983, Adaptations for the detection of fluttering insects by echolocation in Horseshoe Bats, in: "Advances in vertebrate neuroethology", Ewert, J.-P., Capranica, R.R., Ingle, D.J. (eds.), Plenum Press, New York, pp. 801-827.

Schnitzler, H.U., Hackbarth, H., Heilmann, U., Herbert, H., 1985, Echolocation bahavior of rufous horseshoe bats hunting in the flycatcher style. J.Comp.Physiol.A, 157:39-46.

Schnitzler, H.-U., 1986, Echoes of fluttering insects - information for echolocating bats, in: "Recent advances in the study of bats", Fenton, M.B., Racey, P.A., and Rayner, I.M.V. (eds.), Cambridge University Press, pp. 226-234.

Schuller, G., 1972, Echoortung bei Rhinolophus ferrumequinum mit frequenzmodulierten Lauten. J.Comp. Physiol., 77:306-331.

Schuller, G., 1979a, Coding of small sinusoidal frequency and amplitude modulations in the inferior colliculus of 'CF-FM' bat, Rhinolophus ferrumequinum. Exp.Brain Res., 34:117-132.

Schuller, G., 1979b, Vocalization influences auditory processing in collicular neurons of the CF-FM-bat, Rhinolophus ferrumequinum. J.Comp.Physiol., 132:39-46.

Schuller, G., Pollak, G., 1979, Disproportionate frequency representation in the inferior colliculus of Doppler-compensating Greater Horseshoe Bats: Evidence for an acoustic fovea. J.Comp.Physiol., 132:47-54.

Schuller, G., 1980, Alterations of auditory responsiveness by the active emmission of echolocation sounds in the bat, Rhinolophus ferrumequinum. in: "Animal Sonar Systems", Busnel, R.-G., Fish, J.F. (eds.), Plenum Press, pp. 977-979.

Schuller, G., 1984, Natural ultrasonic echoes from wingbeating insects are encoded by collicular neurons in the CF-FM bat, Rhinolophus ferrumequinum. J.Comp.Physiol., 155:121-128.

Shortridge, G.C., 1934, The mammals of south west Africa. Heinemann, London.

Simmons J.A., Lavender, W.A., Lavender, B.A., Dorshow, C.F., Kiefer, S.W., Livingston, R., Scallet, A.C., Crowley, D.E., 1974, Target structure and echo spectral discrimination by echolocating bats. Science, 186:1130-1132.

Simmons, J.A., 1979, Perception of echo phase information in bat sonar, Science, 204:1336

Skolnik, M.I., 1970, "Radar Handbook", McGraw-Hill Book Company, New York.

Suga, N., 1965a, Analysis of frequency modulated tone pulses by auditory neurons of echolocating bats. J.Physiol., 179:26-53.

Suga, N., 1965b, Responses of cortical auditory neurons to frequency modulated sounds in echolocating bats. Nature, 206:890-891.

Suga, N., 1965c, Functional properties of auditory neurons in the cortex of echolocating bats. J.Physiol., 181:671-700.

Suga, N., 1968, Analysis of frequency-modulated and complex sounds by single auditory neurons of bats. J.Physiol., 198:51-80.

Suga, N., 1969, Classification of inferior collicular neurons of bats in terms of responses to pure tones, FM sounds and noise bursts. J.Physiol., 200:555-574.

Suga, N., Schlegel, P., 1973, Coding and processing in the nervous system of FM-signal-producing bats. J.Acoust.Soc.Amer., 54:174-190.

Suga, N., Simmons, J.A., Jen, P.H.S., 1975, Peripheral specialization for fine analysis of Doppler-shifted echoes in the auditory system of the CF-FM bat Pteronotus parnellii. J.Exp.Biol., 63:161-192.

Suga, N., Jen, P.H.-S., 1976, Disproportionate tonotopic representation for processing of CF-FM sonar signals in the mustache bat auditory cortex. Science 194:542-544.

Suga, N., Neuweiler, G., Möller, J., 1976, Peripheral auditory tuning for fine frequency analysis by the CF-FM bat, Rhinolophus ferrumequinum IV. Properties of peripheral auditory neurons. J.Comp.Physiol. 106:111-125.

Suga, N., Jen, P.H.-S., 1977, Further studies on the peripheral auditory system of 'CF-FM' bats specialized for fine frequency analysis of Doppler-shifted echoes. J.Exp.Biol., 69:207-232.

Suga, N., Niwa, H., Taniguchi, I., 1983, Representation of biosonar information in the auditory cortex of the mustached bat, with emphasis on representation of target velocity information. In: "Advances in vertebrate neuroethology", Ewert, J.-P., Capranica, R.R., Ingle, D.J. (eds.), Plenum Press, New York, pp. 829-867.

Suga, N., Tsuzuki, K., 1985, Inhibition and level-tolerant frequency tuning in the auditory cortex of the mustached bat. J.Neurophysiol., 53:1109-1145.

Symmes, D., 1981, On the use of natural stimuli in neurophysiological studies of audition. Hear.Res., 4:203-214.

Trappe, M., 1982, Verhalten und Echoortung der Großen Hufeisennase (Rhinolophus ferrumequinum) beim Insektenfang, Ph.D. Thesis, Marburg

Trappe, M., Schnitzler, H.U., 1982, Doppler-shift compensation in insect-catching Horseshoe bats. Naturwiss., 69:193-194.

Vater, 1977, Einzelzellantworten auf tonale Reize variabler Länge in der ungestörten und gestörten Reizsituation im Nucleus cochlearis von Rhinolophus ferrumequinum, Diplomarbeit, Frankfurt.

Vater, M., Schlegel, P., 1979, Comparative auditory neurophysiology of the inferior colliculus of 2 molossid bats, Molossus ater and Molossus molossus. II. Single unit responses to frequency modulated signals and signal and noise combinations. J.Comp.Physiol., 131:147-160.

Vater, M., 1981, Single unit responses to linear frequency modulations in the inferior colliculus of the Greater Horseshoe bat Rhinolophus ferrumequinum. J.Comp.Physiol., 141:249-264.

Vater, M., 1982, Single unit responses in cochlear nucleus of horseshoe bats to sinusoidal frequency and amplitude modulated signals. J.Comp.Physiol., 149:369-388.

Vaughan, T.A., 1977, Foraging behavior of the giant leaf-nosed bat Hipposideros commersoni. East Afr.Wildl., 15:237-250.

TARGET DETECTION BY ECHOLOCATING BATS

Bertel Møhl

Department of Zoophysiology
University of Aarhus
Denmark

INTRODUCTION

Echolocating bats form a highly diversified group. Their different
types of sonar signals have been proposed as a base for classification and
identification (e.g. Simmons and Stein, 1980, Ahlèn, 1981). The "design" of
the sonar pulses of a given bat species is believed to reflect adaptations
or trade-off's between various properties such as detection sensitivity,
ranging, clutter and noise rejection, and inconspicuousness. Operating at
the lowest signal to noise ratio, with parameter estimation being higher
order processes (Urick, 1983, Altes, 1984) detection can be argued to be the
fundamental process of a sonar. Measures of detection sensitivities – or
detection thresholds – therefore are informative characteristics of the
sonar of a given bat species.

Sensitivity can be stated in many ways, e.g. as the lowest echo sound
pressure, the lowest energy, or the largest detection range for a given
target. Whatever format is chosen, a broad set of conditions that influence
any detection threshold will have to be known to make the measure of sensitivity unambiguous. Such conditions are: signal properties, medium effects,
background (noise and/or clutter), directionality and other receiver properties like detection efficiency integration time, etc. The more important of
these conditions are related through the sonar equation (Urick, 1983):

$$(SL - 2TL + TS) - (No + BW - DI) = DT$$

(where SL is source intensity, TL one way transmission losses, TS target
strength, No noise spectrum density, BW receiver bandwidth, DI directivity
index, and DT detection threshold. The equation is in decibel form, implying 10 times the log ratio of each parameter to some standard).

The sonar equation can be regarded as a check list, not necessarily
complete, of parameters which have to be known for the specification of a
detection threshold. It further serves to emphasize that detection depends
on the signal-to-noise ratio (SNR), i.e. the difference between the signal
term (SL - 2TL + TS) and the noise term (No + BW - DI). Any detection
process is a question of identifying a signal against the background. Many
sources contribute to the background: ambient noise, clutter and reverberation, and noise sensu latu in the bat's hearing organs and central nervous
pathways. The always present background is a masker, and a detection experi-

ment may as well be regarded as a masking experiment. While this approach has been used extensively in dolphin sonar research (Au et al., 1974, Murchison, 1980, Au and Snyder, 1980, Au and Penner, 1981), only little or indirect use of the sonar equation concept is found within bat sonar work.

Isolating the detection process from parameter extraction may not be entirely satisfying. An echo stripped for clues about direction is hard to imagine. Also, some processes like the detection of fluttering or moving targets by CF-bats fall in the parameter extraction group although they are naturally thought of as detection processes. However, for parameters such as Doppler shifts to be discriminated, the echo first has to be detected, and the sonar equation is useful to evaluate the possibility for this detection to occur.

What follows is an attempt to describe the elements of the sonar equation in the context of bat echolocation and to summarize the results from behavioural experiments designed to investigate the sensitivity of the sonar receiver of bats.

THE ELEMENTS OF THE SONAR EQUATION

Source Level, SL

The source level is the sound pressure (or strictly speaking the sound intensity) measured in the acoustic axis and referred to a reference distance. In bat research a reference distance of 10 cm has gained universal acceptance since Griffin's book "Listening in the Dark" (1958).

The sound pressure of bat sonar signals is usually measured peak to peak with an oscilloscope and reported in dB SPL (i.e. referred to 20 micro-Pascal of -94 dB re. 1 Pa). In some respects, this practice is to be preferred to attempts to derive rms values, which are sensitive to pulse duration and effective integration time. However, peak measures are not unambiguous. A trivial source is the case where the nature of the reference, rms, peak (p), or peak-to-peak (pp), is not stated. Factory calibrations of the most commonly used 1/4 inch microphones are given in rms voltage for a standard rms sound pressure. Assume as an example the sensitivity of a microphone to be stated as 3 mV/Pa or -50 dB re. 1 V/Pa. The source level of a bat signal giving a reading from this microphone of 3 mV pp at 10 cm can then be given as a peak-to-peak reading of either 94 or 85 dB SPL, depending not on the reading method, but on the choice of the nature (pp or rms) of the reference. If, however, the same bat signal is compared with that of a sound level calibrator with a stated output of 94 dB SPL, the bat signal obviously is 9 dB below this reference, i.e. 85 dB SPL. In human auditory research this ambiguity is avoided by using the unit peSPL or "peak equivalent SPL", which is the rms sound pressure level of a continuous pure tone having the same amplitude as the transient (Stapells et al., 1982). This convention is followed here. To indicate peak measures referred to a rms standard, the term SPLpp is used.

Other causes of ambiguity in peak measures have to do with the often complex waveforms of bat signals. Should the signal be measured at the very maximum of the envelope, or should it be some kind of an average or "effective" peak? Reasonable arguments can be given for each of these strategies. And how should a multiple harmonic signal be dealt with? The fact that FM-bats sweep through a band of frequencies has led to peak values to be given for several frequencies within the signal (Kick, 1982).

In dolphin sonar research (Au and Snyder, 1980, Au et al., 1985) some of these problems are avoided by computing the energy flux density, E, of the signal:

$$E = \int_0^\infty p^2 (t) \, dt \quad (\text{unit: Pa}^2 \times s)$$

where p(t) is the time-varying sound pressure. This is the recommended SL-measure for short pulse sonars (Urick, 1983), and therefore would seem well suited for describing the SL of signals from FM-bats. Although such a measure is readily obtained from digitized time series, it is rarely used. However, energy units were already used by Griffin, McCue & Grinnell (1963). An argument for a more common use of energy units is that the auditory sensitivity of mammals, including dolphins (Johnson, 1968) and megachiropteran bats (Suthers and Summers, 1980) is determined by energy rather than sound pressure.

The above-mentioned questions of how to describe SL are of a theoretical nature and not at all serious as long as methodology is adequately described. The really perplexing problems arise when trying to obtain a meaningful SL from a bat species either in the field or in the lab. In the field, a major difficulty is to know if a bat signal is recorded in the acoustic axis. Also, determination of distance between the source and the microphone requires either transducer arrays (a technique better known from cetacean than microchiropteran acoustic research, see Watkins, 1976) or synchronized sound and optical recordings (Webster, 1967). Having solved such problems, the bats' behavioural variability is left to be measured.

In the laboratory, it is relatively easy to assure main beam recordings at known distances from bats trained for various sonar tasks (Simmons, 1969, 1971, Simmons et al., 1978, Kick, 1982, Kick and Simmons, 1984, Joermann, 1984, Roverud and Grinnell, 1985, Troest and Møhl, 1986). Variability, however, is generally high, which is to be expected since bats exert a certain amount of control over output amplitude. Search phase signals usually are more intense than signals from the terminal buzz (e.g. 12 dB for Nyctalus (Vogler and Neuweiler, 1983)). Increased output has been found in response to noise in masking experiments (6 to 10 dB for Eptesicus fuscus (Simmons, Lavender and Lavender, 1978), about 5 dB for Tadarida brasiliensis (Simmons et al., 1978), 25 dB in Rhinopoma (Schmidt and Joerman, 1985) and 4 to 9 dB for Eptesicus serotinus (Troest and Møhl, 1986)). In low noise ("no noise") detection experiments (Kick, 1982, Kick and Simmons, 1984) a typical output level of 109 dB SPL with variations only between 105 and 110 dB SPL was found in 4 E. fuscus. In similar experimental situations the inter trial, intra-specimen variation in Eptesicus serotinus could be up to about 20 dB (Troest and Møhl, 1986). In the latter experiment uncertainties of that magnitude were not acceptable and a technique developed to measure and log the peak level of each pulse. While uncertainties about actual peak levels can be reduced in this way it still leaves the experimenter with the task of deriving a representative, single value to describe the distribution of peak values recorded for a given trial. One solution is to report the average level of all pulses in a trial (Au and Snyder, 1980). Our approach has been to make the average of the 3 most intense pulses (Troest and Møhl, 1986).

The logging procedure also provides a handle, however weak, to another factor influencing the source level: the number of pulses used for a detection. Experience from man made sonars points to an improvement of DT (or SNR) of about 5 x log (number of pulses), provided the number is small (Urick, 1983). The problem is again to know the actual number of pulses used by the bat and preferably also their separation. In view of the difficulties in getting such numbers it may be justifiable to ignore these complications and regard the SL as determined by just one "typical" pulse.

An equally important factor for the estimation of the effective source level is the variability with regard to pulse duration. If pulse duration is shorter than the integration time of the ear, a mismatch occurs, which will raise the detection threshold. However, when energy measurements are performed, the effects of pulse duration are included in the measurements.

Transmission Losses, 2TL

The two way transmission loss is the sum of spreading losses (40 log r for spherical spreading, where r is the distance between target and bat, measured in the reference unit of distance), and losses due to absorption. Absorption values, which are frequency and humidity dependent, can be found in Griffin (1971), Lawrence and Simmons (1982). See also Surlykke, this volume. These dependencies complicate echo energy computations with a filtering operation. In many laboratory experiments distances are, however, so short that absorption is either neglected or included as part of the calibration of the system.

Whenever detection takes place within the near field of the target, the law of spherical spreading is not applicable (Urick, 1983). The extension of the near field depends on the geometry of the target, being small for spherical targets, but large for targets that cannot be treated as point sources of echoes. Wires used in obstacle avoidance experiments, plates, and even wings of large insects are examples of targets that can have a significant near field in relation to echolocating bats.

Target Strength, TS

The target strength is defined as 10 times the logarithm to the ratio of the returned to the incident intensity. The returned intensity is specified in a point at unit distance from the target's "acoustic center" in some direction; the source is assumed to be in the far-field. For a number of simple, geometric shapes TS-estimation formulas exist (e.g. Urick, 1983). Griffin, 1958, describes the calculations for echoes from wires and spheres. Recently, plates have served as targets in a number of biosonar experiments with bats. In the appendix, nomograms of TS for plates and spheres are given. For compatibility with the sonar equation dimensions and wavelengths should be measured in the same unit of distance as for the SL. TS-estimation formulas are valid only for certain relations between wavelength and target size. Spherical targets with diameters larger than the wavelength of the signal will reflect without spectral distortions. Planar targets, however, have a high pass-filtering transfer characteristic. Generally, estimated TS for complex signals such as multiharmonic FM-cries should be viewed as no more than crude approximations, useful in situations where calibrations cannot be undertaken.

The TS of biological targets is much too complex to be estimated theoretically. Instead, the specimen is ensonified with artificial pulses and the relative echo strength reported, e.g. Griffin, 1967, Schuller, 1984, Schnitzler et al., 1983). An important property of biological targets such as fluttering insects is a rhythmic, large variation in TS, in part due to specular reflections of the plane surfaces of properly oriented wing membranes (Schnitzler et al., 1983).

Microphones and loudspeakers, used in target simulators (Simmons, 1973, 1979, Kick and Simmons, 1984, Møhl, 1986, Troest and Møhl, 1986), have large TS-values that easily return echoes above detection threshold.

Noise Level, No

Due to its stochastic nature random noise usually is measured with a rms-reading device in a known bandwidth and expressed as noise power in a 1 Hz band at a given frequency, often referred to as spectrum level. For non-white noise spectra the entire spectrum should be described. No established practice with regard to terminology and units for this quantity is evident in the bat sonar literature. For compatibility with the transient form of the sonar equation a reference of 1 Pa x $Hz^{-1/2}$ is recommended. It follows from the nature of noise that a SNR for a given bat signal is meaningful only if the signal term is expressed in energy units (Griffin, McCue and Grinnell, 1963).

Noise is the sum of contributions from interfering sources which mask the detection of the wanted signal (reverberation and directionality effects are dealt with in separate sections below). It can be a very complex task to sort out the various sources, be it ambient noises such as wind and flow (flight) noise, rustling leaves, waterfall noise (including rain), and noises from biological sources, including fellow bats, or noises internal to the auditory system and central nervous auditory pathways. Not all such noises are stochastic and few have similar spectra, making assessment complicated. More importantly, they may not be measurable. Standard quarter inch microphones have a noisefloor of some 40 dB SPL in the .02 to 200 kHz range, or about -10 dB SPL x $Hz^{-1/2}$ at 63 kHz. Masking of FM-bats has been demonstrated at levels 10 dB below this value (Troest and Møhl, 1986).

Internal noise in the bat's auditory system can only be indirectly assessed, using the theoretical equality between DT and SNR and data from detection experiments where external noise can be excluded, such as Kick's (1982) or Dalland's (1965). Simmons et al. (1983) proposed that internal noise is the most important source. This conjecture was supported experimentally by Kick and Simmons (1984) by the finding of a range related change in sensitivity, interpreted as an automatic gain control (AGC), probably located in the bat's middle ear. Assuming this system not to contribute any noise of its own, it will operate quite differently in detection tasks limited by external noise and limited by internal noise. In the first case, SNR is not changed by the AGC and sensitivity (expressed as absolute echo level at threshold) should be independent of range. In the latter case, only the signal is attenuated by the AGC; the noise is unchanged. This leads to a decreased SNR with decreased range implying an increased sensitivity with range as actually found.

Noise Bandwidth, BW

With the mixed nature of the noise in mind, specification of its bandwidth should be expected to be a problem. It is if it is wanted per se, but only the part of the noise spectrum that coincides with the echo power spectrum has a masking effect. This is due to the filtering properties of the mammalian ear, known from sine wave stimulation as critical bands. For broad band transients like sonar cries of FM-bats - or dolphins - the auditory filter, and therefore effective BW, is perhaps better regarded as a single band-pass filter determined by the signal spectrum (Green, 1958). A prediction from this model is that CF-bats with their narrow band signals and highly tuned auditory filters would have an advantage in terms of detection range over FM-bats in situations limited by external noise.

For the coherent and the semicoherent receiver (Menne and Hackbarth, 1986) only the spectrum level of the noise is important. This is due to a special property of such receivers, in which processor gain can be regarded as a function of signal bandwidth (Glaser, 1974).

Clutter and Reverberation Level, RL

It is convenient to deal with the masking effects of the bat's outgoing cries and echoes from other objects than the target separately from other sources of interference. While the masking effect of other sources of interference can be reduced by increasing the output level, clutter and/or reverberation is a direct function of output level.

No clear distinction appears in the use of the terms clutter and reverberation. In the bat echolocation literature clutter is mainly used for interfering echoes from discrete objects. Such clutter depends on direction, aspect ratios, radial distance, etc., which are cumbersome to express properly within the sonar equation framework. Reverberation is a term well suited to describe the influence of homogeneously distributed sound scatters such as rain drops and insect swarms, possibly also certain kinds of foliage, i.e. scatters causing volume reverberation, which is quantifiable (Urick, 1983).

Directional Index, DI

This index is the ratio of the noise power entering a non (or omnidirectional ear in an isotropic noise field to the noise power entering an actual bat ear. The index has little if anything to do with directional discrimination abilities (Peff and Simmons, 1972, Simmons et al., 1983), and can be regarded as a modifier of the noise term for cases limited by external, isotropic noise. The directionality index varies with frequency and the morphology of the ear (species). Values in the order of 10 dB for Rhinolophus ferrumequinum and Myotis sp. can be inferred from Grinnell and Schnitzler, 1977, and from Shimozawa et al. (1974), respectively.

In a reverberation limited situation, DI is determined by the combined (or two-way) transmitting and receiving beam patterns (Urick, 1983). Transmitting beam patterns for Rhinolophus ferrumequinum (-6 dB at 24°) was measured by Schnitzler and Grinnell (1977). In FM-bats, the sound radiation pattern is a function of frequency (Strother and Mogus, 1970, Mogensen and Møhl, 1979). Combined beam patterns are given by Shimozawa et al. (1974) and Grinnell and Schnitzler (1977).

Direction Threshold, DT

This parameter is defined as 10 x the logarithm of the ratio of signal power to noise power required for detection to be performed at a given level of correctness, but the term is often used more loosely. In the sonar equation, DT is the equality term between the signal term (SL + 2TL + TS) and the noise term (No + BW - DI) at the point where the detector is just meeting its specifications. In this sense, DT is a dimensionless ratio (SNR), which for short duration pulse sonars of FM-bats requires the SL to be obtained in energy units (Griffin, McCue and Grinnell, 1963, Urick, 1983).

In many experimental situations the noise term is assumed to be negligible and the threshold, often referred to as absolute, is stated in sound pressure units as in audiometry. In audiometry, however, the signals used are at least as long as the integration time of the ear, i.e. several hundred ms. Signals of shorter duration require a higher intensity level to be detected, generally by 3 dB for each halving of duration. This implies that the ear is an energy integrator. In this case sound pressure measurements without envelope specifications are not sufficient to specify the threshold condition. Remarkably, energy integration time is not known for microchiropteran bats (Brown and Maloney, 1986). In the megachiropteran echolocating bat, Rousettus, integration time is in the order of 50 ms (Suthers and Summers, 1980). Porpoises, which employ even shorter pulses than do bats, have integration times in the order of 200 ms (Johnson, 1968), as have most other mammals including man. However, the mammalian ear has several other "integrators", which manifest themselves in different tasks such as gap detection, and which differ in integration time by several orders of magnitude (de Boer, 1985, Yost, 1980). Until further evidence on the energy integration time of bat ears is available, educated guesses on the magnitude of this property will have to be used. It appears that a long integration time, i.e. long with respect to the duration of the echoes of FM-cries, would be in accordance with the general finding in other mammals. See also Surlykke, this volume.

Since the noise term is part of the sum making up the detection threshold, it follows that DT itself is of a stochastical nature. Consequently, it is best described in statistical terms such as means, variances, etc. The sonar equation leads naturally to a description within the framework of theory of signal detection and the theory of ideal observers (Peterson, Birdsall and Fox, 1954, Swets, Tanner and Birdsall, 1961), where the task of the observer, knowing the costs of making mistakes, is to make a judgment as to whether a given observation is preferably reported to be caused by noise or by signal plus noise. A correct detection is called a hit. A detection reported in the absence of a signal is called a false alarm. The observation can be thought of as giving rise to a hypothetical neural activity distribution. How far the distribution caused by signal + noise differs from the distribution caused by noise alone is a monotonic function of SNR. The difference (symbolized d') is largely invariant over experimental techniques, subject motivation and bias, and can be found from the ratio of hits to false alarms from results of detection experiments. Also, the SNR at the output of the detector can be determined from this ratio, using the relation between d' and SNR.

This psychophysical approach has been used in porpoise sonar studies (Au and Snyder, 1980, Au and Turl, 1984) and marine mammal audiometry (Schusterman, Barrett and Moore, 1975), but has only recently been used in characterizing bat sonar sensitivity (Troest and Møhl, 1986).

The prevailing psychophysical procedure used in bat threshold determination is some variation of the method of constant stimuli (Guilford, 1954), combined with a two alternative, forced choice (2AFC) response technique, such as pioneered by J.A. Simmons for ranging experiments (Simmons, 1971, 1973). The 2AFC procedure appears to be readily "understood" by bats, but does not readily provide the statistical measures of the threshold determination; at least, such statistics are rarely reported. An exception is Schmidt et al. (1983), in which 2AFC-data are fitted to the accumulated gaussian distribution. In this way all the data points of the psychometric function can be used for the threshold determination – not just the pair bracketing the more or less arbitrarily chosen threshold criterion. Yes/No (Y/N) procedures have recently been used with Noctilio (Roverud and Grinnell, 1985), Pipistrellus (Møhl, 1986) and Eptesicus (Troest and Møhl, 1986). A rationale for using the Y/N technique is to get a measure of the ratio between hits and false alarms, as discussed above. Thresholds obtained with 2AFC procedures are generally lower than those obtained with the Y/N procedures (Swets et al., 1961).

EXPERIMENTS

The behavioural experiments designed to measure the sensitivity of the sonar system of bats can be grouped in three categories: obstacle avoidance with untrained animals, real target detection with trained animals, and simulated target detection, again with trained animals. The categories differ with respect to the number of elements of the sonar equation that can be measured rather than estimated. In this respect obstacle avoidance ranks lowest, target simulation highest. However, the methods involving training are time consuming; it may require several months of daily training for a bat to reach stable performance, although this time span varies considerably with the individual bat, the trainer, and the task.

1. Wire Avoidance Experiments

This is the classical technique in bat sonar experiments, used already by Hahn (1908), and refined by Griffin an co-workers (Griffin, 1958). A standard for wire avoidance experiments has been proposed by Schnitzler and Henson (1980). The technique allows for naive (untrained) bats to be tested in numbers fairly quickly and the method is well suited when sensitivity can be expressed in non-acoustical terms such as obstacle avoidance score (Jen and Kamada, 1982). With reference to the sonar equation, it appears that values with known accuracy on SL, TL, TS and No+DI are difficult to get, as is the false alarm rate (required to translate score differences into equivalent SNR differences). Estimates of DT in an artificial noise field for Plecotus obtained with this method thus range from -23 to 15 dB, dependent on assumptions and directionality, source level, bandwidth, etc. (Griffin, McCue and Grinnell, 1963).

Wire avoidance technique was used by Jen and Kamada (1982) to investigate the influence of obstacle movement on avoidance success in Pteronotus parnelli and Eptesicus fuscus. When the wires were moved at a rate of 10 cm/s, the average number of misses increased by about 10% for both species over the score found for stationary wires. Attention factors, rather than detectability factors, were suggested to account for this observation.

Joerman (1984) tested Desmodus rotundus in a wire avoidance experiment. 0.5 mm nylon threads were avoided in 87% of the flights at a mean response distance of 65 cm, as determined from the sound emission patterns. The echo level was calculated to be 49 dB SPLpp. Response distance decreased regularly from about 90 cm for 1 mm threads to 5 to 20 cm for .23 mm threads. In another experiment the bats had to dodge rods of various sizes, placed in various distances from a constriction in a flight tunnel. Reaction distances were found to increase with rod diameter and with the rods' axial separation from the constriction.

2. Real Target Detection, Conditioned Response

This technique requires the bats, trained to some level of reliability to react to the presence or position of a given target, using food as positive reinforcement, sometimes supplemented with negative reinforcement. Usually, a 2AFC paradigm is used. The distance between the bat and the target is known, and the sonar cries can be easily recorded from the axis. Only the target strength needs to be estimated (Kick, 1982), if not calibrated (Joerman, 1984). Ambient noise can be made negligible. Clutter is more difficult to avoid, but then the technique is well suited to study sensitivity to clutter (McCarthy and Jen, 1983, Joerman, 1984).

Real target detection appears to be the technique of choice for measuring absolute detection thresholds, i.e. thresholds not limited by factors in the environment. Kick (1982) used spherical targets at various distances and found detection thresholds for Eptesicus fuscus at computed echo levels of about 0 dB SPLpp. Such levels are remarkably low, considering the short duration (4-8 ms), low repetition rate (5-6 pps), and large bandwidth of the pulses (about 20 kHz). Compared with the auditory sensitivity found by Dalland (1965) for the same species for long-lasting, pure tone signals, the echo sensitivity is 3 to 4 orders of magnitude better in terms of energy. This result may be interpretable as evidence consistent with the hypothesis of coherent reception by bats (Strother, 1961, Simmons, 1979).

For vampire bats, Joerman (1984) found echo level thresholds between 35 and 42 dB SPLpp for millimeter sized, cylindrical targets at a range of 45 cm. The experimental design differed from that of Kick's in having restrictions in the sound field, implying the possibility of a clutter li-

442

mited threshold. The effect of clutter from various sized plates on detection performance was studied, but signal to clutter ratios not derived.

In an experiment involving Eptesicus fuscus detecting spherical (2 cm diameter) targets against background clutter, McCarthy and Jen (1983) found a 75% detection performance at a target distance of 74 cm. At this distance, the echo level exceeded the clutter level. Using Kick's (1982) data for SL at 30 kHz and the sonar equation, the echo level can be estimated to be about 40 dB SPLpp. Moving the target at slow speeds (4-10 cm/sec) increased detectability by up to 3 dB. The mechanism of this un-cluttering effect was not discussed.

3. Simulated Target Detection, Conditioned Response

For any given, real target TS is normally fixed. In order to determine the threshold region, one of the other parameters, usually distance, will have to be varied. This also influences the clutter conditions, absorption and the effects of the range-coupled, automatic gain control. Also, real targets may be sensed through other modalities, notably vision, which in critical experiments necessitates enucleation (Simmons, 1971, Simmons et al., 1983). With the simulated target technique (Simmons, 1973), such problems are avoided by generating a phantom target electro-acoustically. This allows for manipulation of TS and a number of other parameters without any change in geometry. The technique is also well suited for masking experiments, as noise can be added and projected from the same source as the "target", eliminating DI effects entirely. A disadvantage is that the required introduction of microphone and loudspeaker in front of the bat creates clutter that may be limiting detection.

3.a. Automatic Gain Control (AGC)

The simulated target technique has been used by Kick and Simmons (1984) to trace the detection threshold of Eptesicus fuscus for a range of echo delays from 1 to 6.4 ms. Sensitivity increased with delay at a regular rate of 11 dB per doubled distance, which is very close to perfect regulation of echo level due to spherical spreading losses. At the largest delay tested, the echo level at 75% detection was 8 dB SPLpp. This AGC-mechanism will have as a consequence that TS is represented directly by sensation level and is range tolerant at least for situations involving short range and spherical spreading losses. The mechanism is proposed to leave the bat free to observe echo amplitude changes that are caused by the target's own action, such as wing beats. The mechanism would also appear to be economical in terms of a hypothetical "library" of target characteristics in the bat's memory.

The results and nature of this experiment suggest that the bat has little or no control of the AGC-mechanism; it cannot switch it off. In some situations this could be a disadvantage to the bat. As noted by the authors, there appears to be a minimum TS which the bat can detect in the pursuit process. The bat's rising threshold for detecting earlier echoes will keep targets that are below threshold at 1 m below threshold at shorter ranges, too. A target, equivalent to a single, perfectly reflecting sphere as small as 1 mm in diameter, would thus never be detected.

A related effect of an obligate AGC-mechanism would seem to occur for detection of targets within the near-field. For wires, such as used in obstacle avoidance experiments, the echo intensity increases by only 9 dB per distance halved, while the AGC reduces the perceived level by 12 dB, leading to over-compensation. This system would be predicted to detect thin wires better at long than at short ranges. This prediction is not supported experimentally, since evidence from wire avoidance experiments show a reduced detection distance with reduced wire diameter (Joerman, 1984).

The AGC-experiment, however, may possibly be interpreted as a clutter limited experiment (Troest, personal communication). The rationale for this is that delay was varied by varying the distance to the loudspeaker. The clutter level of the loudspeaker also has a 40 x log r dependency. For a 2.4 diameter loudspeaker this clutter level will exceed that of the simulated target by about 30 dB at all ranges where the AGC is operative. Loudspeaker clutter was found to be a limiting factor at close range by the experimenters, and at other ranges by Møhl (1986) and Troest and Møhl (1986). A problem with this interpretation of the AGC-experiment is that it does not take into account delay related reduction of clutter effects, only the signal to clutter ratio.

3.b. Signal Fine Structure

The freedom to manipulate echo characteristics in target simulation has been used to address the question of the significance for detection of the down sweeping time/frequency structure of FM-cries (Møhl, 1986). A detection threshold of 35 dB peSPL in a loudspeaker clutter limited situation for a Pipistrellus pipistrellus was established for a phantom target, generated by playing back a pre-recorded cry each time the bat echolocated. Subsequently, the played back cry was time reversed, leaving energy, spectrum and duration unchanged. When the threshold was established again, no change was found. This indicates that detection primarily depends on signal energy, not on signal fine structure. The hypothesis of matched filter processing by FM-bats (Cahlander, 1967) is inconsistent with this result.

3.c. Masking

In target simulation experiments it is possible not only to control the signal term, but also the noise term, as required by sonar equation analysis. In particular, any possibility of improving SNR by directionality gain is eliminated, since the SNR is established before the signal is projected. Also, controlled noise levels far below what can be measured with currently available microphones can easily be established. A specified signal to noise ratio is required not only for detection experiments, but also for experiments on ranging (Menne and Hackbarth, 1986, Simmons, 1986). In the abovementioned pipistrelle experiment a masked detection threshold for a prerecorded signal was found at an E/No-ratio of 50 dB. In a different experiment (Troest and Møhl, 1986), the actual, emitted cry was played back directly after being mixed with noise to bats of the species Eptesicus serotinus. The E/No-ratio at threshold ranged from 36 to 49 dB (3 specimens). In both experiments the ratio of hits to false alarms was obtained and used to estimate the SNR at the output of the bat's receiver. This estimate of the SNR was only 1 to 2 dB. The discrepancy between the SNR estimates obtained this way and the SNR at the input is inconsistent with predictions based on the hypothesis of matched filter signal processing in bats, but in line with predictions on the performance of standard mammalian auditory processing of broad band signals in noise (Green, 1958, Scharf, 1970).

Contrary to the results of the reversed phantom experiment, which can be argued to be an experiment on passive rather than active sonar, Simmons (1986) found detectability to be influenced by the fine structure of the signal. In Simmons' experiment the probability of detection of a phantom target against a background of noise was measured as a function of jitter (small increments in delay in every second pulse, see Simmons, 1979). The results show a non-monotonic function, increasing from 76% correct at zero jitter to 89% at 20 microsecond jitter, followed by a decrease. As SNR was unchanged, the implication made from the unmasking effect of jitter was that

the bat perceives the phase of sonar echoes relative to the phase of its
sonar transmissions and suffers from a phase ambiguity effect at jitter
values representing small integer multiples of the average period of the
signal. The introduced jitter increased detection performance by an amount
equivalent to a 5 dB increase in signal-to-clutter ratio (SCR) for an ideal
observer in a comparable 2AFC task (Swets, 1959).

3.d. Clutter

In target simulator experiments the cluttering effect of the phantom
projecting loudspeaker is hard, if not impossible to avoid. While this may
complicate interpretations of the results, the observation as such may be of
some relevance for hypotheses about the working principles of the sonar receiver of FM-bats.

Kick and Simmons (1984) reported clutter effects when signal and clutter
were separated by 1 ms. The target distance was 17 cm, and the SCR can be
estimated to be about -30 dB. In the pipistrelle experiment (Møhl, 1986)
clutter led the signal by 3 ms; target range was 138 cm and the SCR was -20
dB. In the serotine experiment (Troest and Møhl, 1986), clutter was trailing
the signal by 2 ms at a target range of 55 cm and a SCR of -20 dB. Clutter
limitation was experimentally verified in both experiments. These observations do not suggest any pattern, but they do show that clutter rejection is
unlikely to be handled by range (or time) resolution, which is better by
about 3 orders of magnitude (Simmons, 1979). However, this cannot be interpreted to indicate that a FM-bat hunting within a metre of a cluttered background has little chance to detect insect prey. Such a bat has its directionality gain intact, and fluttering movements of the prey may highlight
its presence.

Simmons (1986) mapped the extent of sensitivity to clutter along the
target axis, using ring-shaped, cluttering objects and phantom targets. The
results show that the clutter interference zones have widths that are roughly
proportional to the range of the targets. This result again indicates that
clutter is not likely to be handled mainly by time resolution, which has
been found to be independent of range (Simmons, 1973).

CONCLUSIONS

In some areas knowledge about the target detection capabilities of echolocating bats has been expanded since the review by Schnitzler and Henson
(1980):

1. An absolute threshold for echoes at 0 dB SPLpp has been established
   for Eptesicus fuscus.
2. Sensitivity is found to be range dependent at ranges below 1 m, automatically compensating echo level for spherical spreading losses.
3. In masking experiments, FM-bats require a signal-to-noise ratio 40
   to 50 dB above that predicted for an ideal, coherent receiver.
4. Evidence on the significance of the fine structure of sonar cries
   from FM-bats may be conflicting. One experiment found no change in
   detectability when the echo was reversed in time. A different experiment found an increased detectability when the echo was jittered
   in time.
5. All evidence from FM-bats point to increased detection thresholds in
   the presence of clutter, even for signal to clutter separations of
   several ms. The sensitivity to clutter appears to be unrelated to
   range discrimination abilities. Detectability is better for situations with changing rather than stationary target/clutter patterns.

ACKNOWLEDGEMENT

I thank J.A. Simmons for providing advance information about experiments to be published in his book: "The Sonar of Bats".

REFERENCES

Ahlèn, I., 1981, Field identification of bats and survey methods based on sounds. Myotis, 18-19:128.

Altes, R.A., 1984, Echolocation as seen from the viewpoint of Radar/Sonar theory, in "Localization and Orientation in Biology and Engineering", D. Varju and H.-U. Schnitzler, eds., Springer-Verlag, Berlin.

Au, W.W.L., Carder, D.A., Penner, R.H., and Sconce, B.L., 1985, Demonstration of adaptation in beluga whale echolocation signals, J. Acoust. Soc. Am., 77:726.

Au, W.W.L., Floyd, R.W., Penner, R.H., and Murchison, A.E., 1974, Measurement of echolocation signals of the Atlantic Bottlenose dolphin, Tursiops truncatus (Montagu), in open waters, J. Acoust. Soc. Am., 56: 1280.

Au, W.W.L., and Penner, R.H., 1981, Target detection in noise by echolocating Atlantic Bottlenose dolphins, J. Acoust. Soc. Am., 70:687.

Au, W.W.L., and Snyder, K.J., 1980, Long-range target detection in open waters by an echolocating Atlantic Bottlenose dolphin (Tursiops truncatus), J. Acoust. Soc. Am., 68:1077.

Au, W.W.L., and Turl, C.W., 1984, Dolphin biosonar detection in clutter: Variation in the payoff matrix, J. Acoust. Soc. Am., 76:955.

de Boer, E., 1985, Auditory time constants: A paradox?, in "Time resolution in auditory systems", A. Michelsen ed., Springer, Heidelberg, New York.

Brown, C.H., and Maloney, C.G., 1986, Temporal integration in two species of Old World monkeys: Blue monkeys (Cercopithecus mitis) and grey-cheeked mangabeys (Cercocebus albigena), J. Acoust. Soc. Am., 79:1058.

Cahlander, D.A., 1967, Discussion of Batteau's paper, in "Animal Sonar Systems", R.G. Busnel ed, Laboratoire de Physiologie acoustique, Jouy-en-Josas.

Dalland, J.I., 1965, Hearing sensitivity in bats, Science, 150:1185.

Glaser, W., 1974, Zur Hypothese des Optimalempfangs bei der Fledermausortung, J. Comp. Physiol., 94:227.

Green, D.M., 1958, Detection of multiple component signals in noise, J. Acoust. Soc. Am., 30:904.

Griffin, D.R., 1958, "Listening in the dark", Yale University Press, New Haven.

Griffin, D.R., 1967, Discriminative echolocation by bats, in "Animal Sonar Systems", R.G. Busnel ed, Laboratoire de Physiologie acoustique, Jouy-en-Josas.

Griffin, D.R., 1971, The importance of atmospheric attenuation for the echolocation of bats (Chiroptera), Anim. Behav., 19:55.

Griffin, D.R., McCue, J.J.G., and Grinnell, A.D., 1963, The resistance of bats to jamming, J. Exp. Zool., 152:229.

Grinnell, A.D., and Schnitzler, H.-U., 1977, Directional sensitivity of echolocation in the Horseshoe bat, Rhinolophus ferrumequinum, J. Comp. Physiol., 116:63.

Guilford, J.P., 1954, "Psychometric Methods", McGraw-Hill, New York.

Hahn, W.L., 1908, Some habits and sensory adaptations of cave-inhabiting bats, Biol. Bull., 15:135.

Jen, P.H.-S., and Kamada, T., 1982, Analysis of orientation signals emitted by the CF-FM bat, Pteronotus p. parnellii and the FM bat, Eptesicus fuscus during avoidance of moving and stationary obstacles, J. Comp. Physiol., 148:389.

Joermann, G., 1984, Recognition of spatial parameters by echolocation in the vampire bat, Desmodus rotundus, J. Comp. Physiol., 155:67.

Johnson, C.S., 1968, Relationship between absolute threshold and duration-of-tonepulses in the Bottlenose porpoise, J. Acoust. Soc. Amer., 43: 757.

Kick, S.A., 1982, Target-detection by the echolocating bat, Eptesicus fuscus, J. Comp. Physiol., 145:431.

Kick, S.A., and Simmons, J.A., 1984, Automatic gain control in the bat's sonar receiver and the neuroethology of echolocation, J. Neuroscience, 4:2725.

Lawrence, B.D., and Simmons, J.A., 1982, Measurements of atmospheric attenuation at ultrasonic frequencies and the significane for echolocation by bats, J. Acoust. Soc. Am., 71:585.

McCarthy, J.K., and Jen, P.H.-S., 1983, Bats reject clutter interference for moving targets more successfully than for stationary ones, J. Comp. Physiol., 152:447.

Menne, D., and Hackbarth, H., 1986, Accuracy of distance measurement in the bat Eptesicus fuscus: Theoretical aspects and computer simulations, J. Acoust. Soc. Am., 79:386.

Mogensen, F., and Møhl, B., 1979, Sound radiation patterns in the frequency domain of cries from a vespertilionid bat, J. Comp. Physiol., 134:165.

Murchison, A.E., 1980, Detection range and range resolution of echolocating Bottlenose porpoise (Tursiops truncatus), in: "Animal Sonar Systems", R.G. Busnel and J.F. Fish eds., Plenum, New York.

Møhl, B., 1986, Detection by a pipistrelle bat of normal and reversed replica of its sonar pulses, Acustica, 60:??.

Peff, T.C., and Simmons, J.A., 1972, Horizontal-angle resolution by echolocating bats, J. Acoust. Soc. Am., 51:2063.

Peterson, W.W., Birdsall, T.G., and Fox, W.C., 1954, The theory of signal detectability, Trans. IRE Professional Group on Information Theory, PGIT-4, 171.

Roverud, R.C., and Grinnell, A.D., 1985, Discrimination performance and echolocation signal integration requirements for target detection and distance determination in the CF/FM bat, Noctilio albiventris, J. Comp. Physiol., 156:447.

Scharf, B., 1970, Critical bands, in "Foundations of Modern Auditory Theory", J.V. Tobias ed., Academic Press, New York, London.

Schmidt, S., Turke, B., and Vogler, B., 1983, Behavioural audiograms from the bat, Megaderma lyra, Myotis, 21/22:62.

Schmidt, U., and Joerman, G., 1985, The influence of acoustical interferences on echolocation in bats, in "Air-borne Animal Sonar Systems", B. Escudie and Y. Biraud eds., CNRS, Lyon.

Schnitzler, H.-U., and Grinnell, A.D., 1977, Directional sensitivity of echolocation in the Horseshoe bat, Rhinolophus ferrumequinum, J. Comp. Physiol., 116:51.

Schnitzler, H.-U., and Henson, O.W., 1980, Performance of airborne animal sonar systems, in: "Animal Sonar Systems", R.G. Busnel and J.F. Fish eds., Plenum, New York.

Schnitzler, H.-U., Menne, D., Kober, R., and Heblich, K., 1983, The acoustical image of fluttering insects in echolocating bats, in: "Neuroethology and behavioural Physiology", F. Huber and H. Markl eds., Springer, Heidelberg, New York.

Shimozawa, T., Suga, N., Hendler P., and Schuetze, S., 1974, Directional sensitivity of echolocation system in bats producing frequency-modulated signals, J. Exp. Biol., 60:53.

Schuller, G., 1984, Natural ultrasonic echoes from wingbeating insects are encoded by collicular neurons in the CF/FM bat, Rhinolophus ferrumequinum, J. Comp. Physiol., 155:121.

Schusterman, R.J., Barrett, R., and Moore, P.W.B., 1975, Detection of underwater signals by a California Sea Lion and a Bottlenose Porpoise: Variation in the payoff matrix, J. Acoust. Soc. Am., 57:1526.

Simmons, J.A., 1969, Acoustic radiation patterns for the echolocating bats Chilonycteris rubiginosa and Eptesicus fuscus, J. Acoust. Soc. Am., 46:1054.

Simmons, J.A., 1971, Echolocation in bats: signal processing of echoes for target range, Science, 171:925.

Simmons, J.A., 1973, The resolution of target range by echolocating bats, J. Acoust. Soc. Am., 54:157.

Simmons, J.A., 1979, Perception of echo phase information in bat sonar, Science, 204:1336.

Simmons, J.A., 1986, "The Sonar of Bats", Princeton University Press, in press.

Simmons, J.A., Lavender, W.A., and Lavender, B.A., 1978, Adaptation to acoustic interference by the echolocating bat, Eptesicus fuscus, Ann. E. Afr. Acad. Sci., FIBRC:97

Simmons, J.A., Lavender, W.A., Lavender, B.A., Childs, J.E., Hulebak, K., Rigden, M.R., Sherman, J., and Woolman, B., 1978, Echolocation by free-tailed bats (Tadarida), J. Comp. Physiol., 125:291.

Simmons, J.A., and Stein, R.A., 1980, Acoustic imaging in bat sonar: Echolocation signals and the evolution of echolocation, J. Comp. Physiol., 135:61.

Simmons, J.A., Kick, S.A., Lawrence, B.D., Hale, C., Bard, C., and Escudié, B., 1983, Acuity of horizontal angle discrimination by the echolocating bat, Eptesicus fuscus, J. Comp. Physiol., 153:321.

Strother, G.K., 1961, Note on the possible use of ultrasonic pulse compression by bats. J. Acoust. Soc. Am., 33:696.

Strother, G.K., and Mogus, M., 1970, Acoustical beam patterns for bats: Some theoretical considerations. J. Acoust. Soc. Am., 48:1430.

Stapells, D.R., Picton, T.W., and Smith, A.D., 1982, Normal hearing thresholds for clicks, J. Acoust. Soc. Am., 72:74.

Surlykke, A., 1986, Interaction between echolocating bats and their prey. This volume.

Suthers, R.A., and Summers, C.A., 1980, Behavioral audiogram and masked thresholds of the Megachiropteran echolocating bat, Rousettus, J. Comp. Physiol., 136:227.

Swets, J.A., 1959, Indices of signal detectability obtained with various psychophysical procedures, J. Acoust. Soc. Am., 31:511.

Swets, J.A., Tanner, W.P., and Birdsall, T.G., 1961, Decision processes in perception, Phychol. Rev., 68:301.

Troest, N., and Møhl, B., 1986, The detection of phantom targets in noise by serotine bats; negative evidence for the coherent receiver, J. Comp. Physiol., in press.

Urick, R.J., 1983, "Principles of Underwater Sound", McGraw-Hill Co., New York.

Vogler, B., and Neuweiler, G., 1983, Echolocation in the noctule (Nyctalus noctula) and horseshoe bat (Rhinolophus ferrumequinum), J. Comp. Physiol., 152:421.

Yost, W.A., 1980, Man as a mammal: Psychoacoustics, in: "Comparative Studies of Hearing in Vertebrates", A.N. Popper and R.R. Fay eds., Springer, Heidelberg, New York.

Watkins, W.A., 1976, Biological sound-source location by computer analysis of underwater array data. Deep Sea Research, 23:175.

Webster, F.A., 1967, Interception performance of ehcolocating bats in the presence of interference, in: "Animal Sonar Systems", R.G. Busnel ed., Laboratoire de Physiologie acoustique, Jouy-en-Josas.

APPENDIX

Target Strength Nomograms

Spheres and discs are frequently used as targets or clutter sources in
bat sonar experiments. The target strength (TS) of such objects can be estimated from fig. 2 and fig. 3, given that the objects are acoustically large.
This is determined from fig. 1. To be compatible with the other elements of
the sonar equation and conventions in bat sonar research, the unit of length
chosen is 10 cm. The choice of this unit determines the numerical value of
TS. The observation from fig. 3 that TS can attain positive values is a
consequence of this choice; it does not indicate that more sound is coming
back from the target than is incident upon it. Speed of sound is assumed to
be 3440 dm/s.
Abbreviations: a: radius; λ : wavelength; k: $2\Pi/\lambda$.

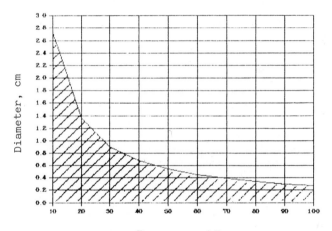

Frequency, kHz

Fig. 1. Relationship between diameter of an object and frequency for the condition 2 x Π x a/λ = 5. Object/
frequency combinations that fall outside the hatched
area are acoustically large, and their TS can be
predicted from fig. 2 or fig. 3.

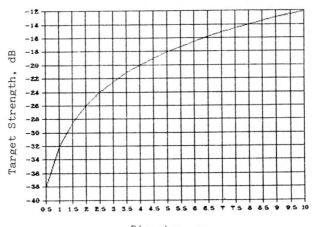

Diameter, cm

Fig. 2. TS of spheres as a function of diameter.

Basic formula: TS = 10 x log $(a^2/4)$.

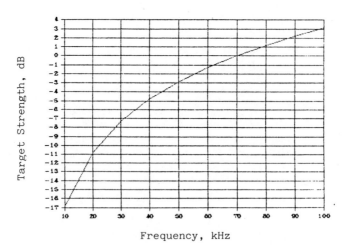

Frequency, kHz

Fig. 3. TS of 2.5 cm diameter disc as a function of frequency.

Basic formula: TS = 10 x log $((\Pi \times a^2/\lambda)^2)$.
TS-values for different sized discs can be found from:
$TS_q = TS_r + 10 \times \log (a_q/a_r)$, where r refers to the
2.5 cm disc, q to the different sized disc.

SONAR TARGET DETECTION AND RECOGNITION BY ODONTOCETES

Whitlow W.L. Au

Naval Ocean Systems Center
Kailua, Hawaii 96734

One of the most effective ways to probe an underwater environment for
purposes of navigation, obstacle avoidance, and prey detection is by sonar
(sound navigation and ranging). Many odontocete emit sounds and analyze
returning echoes to detect and recognize objects underwater. Prior to
1979, most of the experiments were performed in tanks, using mm sized targets and short ranges (see Murchison, 1980a). This review will consider
sonar detection and discrimination experiments performed in the open waters
of Kaneohe Bay, Oahu, Hawaii with the Atlantic bottlenose dolphin (Tursiops
truncatus) and beluga whale (Delphinapterus leucas).

Tursiops (Au, 1980) and Delphinapterus (Au et al., 1985) typically
emit echolocation signals with peak frequencies above 100 kHz in Kaneohe
Bay. Au et al. (1985) measured beluga whale echolocation signals with peak
frequencies between 100 and 120 kHz, an octave higher than the 40 to 60 kHz
measured in San Diego Bay. Examples of the ambient noise spectra in both
bays are shown in Fig. 1. A review of signals used by Tursiops in Kaneohe
Bay was given by Au (1980).

The target detection capability of any sonar system, man-made or biological, is limited by interfering noise and reverberation. The target
detection sensitivity of a sonar can be measured by a variety of equivalent
methods. The range of a target of known target strength can be increased
until the target can no longer be detected. A fixed target range can be
used and the size of the target can be reduced continuously until the target can no longer be detected. A fixed target range can be used with a
specific target and the amount of interfering noise or reverberation gradually increased until the target cannot be detected. Conversely, the
interfering noise or reverberation can be held constant and the target size
reduced until the detection threshold is reached. Whatever method is used,
measurement of source level, target strength and noise or reverberation
levels should also be made so that the echo-to-noise ratio (E/N), or the
echo-to-reverberation ratio (E/R) at threshold can be determined.

I. MAXIMUM DETECTION RANGE

Murchison (1980a, 1980b) discussed the results of a maximum range
detection experiment with two Tursiops in Kaneohe Bay. A 2.54-cm solid
steel sphere and a 7.62-cm diam. stainless steel water-filled sphere, were

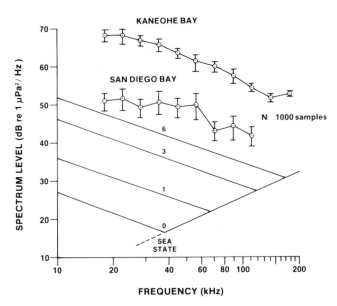

Fig. 1. Ambient noise of Kaneohe Bay and San Diego Bay measured in 1/3 octave bands. Deep water noise for different sea states are shown for comparison.

used as targets. The composite 50% correct detection thresholds were at ranges of 72 m and 77 m for the 2.54-cm and 7.62-cm spheres, respectively. However, a bottom ridge at approximately 73 m affected the animals' ability to detect the 7.62-cm target beyond 73 m. The animals' performance degraded rapidly when the target was in the vicinity of the ridge, suggesting that the animals were probably reverberation limited.

Au and Synder (1980) remeasured the maximum detection range of one of the dolphins (Sven) used by Murchison (1980a, 1980b) with a 7.62-cm diam. stainless steel water-filled sphere as the target. A different part of Kaneohe Bay, where the bottom was relatively flat and the water depth between 5.8 and 6.1 m, was used. In both studies, an overhead suspension system with a movable trolley and pulleys was used to vary the target range between two poles spaced 200 m apart. Targets were lowered or raise with a monofilament line extending back to the trainer's station.

Sven's target detection performances for the 2.54-cm (Murchison, 1980b) and 7.62-cm (Au and Synder, 1980) spheres are shown in Fig. 2 with correct detection and false alarm rates plotted as a function of target range. The 50% correct detection threshold for the 7.62-cm sphere occurred at 113 m, a considerably longer range than 76.6 m obtained by Murchison (1980a,1980b).

The dolphin's performance shown in Fig. 2 can be analyzed in terms of E_e/N_o by using the transient form of the sonar equation. Sven's echolocation signals were measured in the study of Au et al. (1974) for target ranges of 59 to 77 m. Following the procedure of Au and Penner (1981), the maximum source energy flux density (SE) per trial will be used to compute E_e/N_o. The maximum source level (SL) averaged over 12 trials at the 77 m

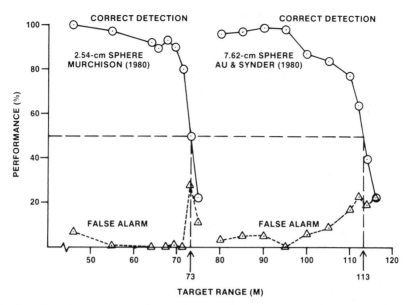

Fig.2.   <u>Tursiops</u> target detection performance as a function of range
        for two different spherical targets (From Murchison, 1980b;
        Au and Synder, 1980).

target range was 225 dB re 1 µPa.   The empirical equation given by Au (this
volume) indicates an SE_{max} of 166 dB re $(1 \ µPa)^2 s$.   Target strength measurement results on both targets (Au and Synder, 1980) using a simulated
dolphin signal, are shown in Fig. 3. From Fig. 1, the ambient noise level
at 120 kHz is approximately 54 dB re 1 $µPa^2/Hz$. The animal's performance
as a function of the $(E_e/N_o)_{max}$ is shown in Fig. 4. These results represent
the most conservative estimate of the dolphin's detection sensitivity. The
difference between average SE and maximum SE in a click train is typically
between 3 to 5 dB (Au, et al., 1974, 1982).   If the average SE was used, a
liberal estimate of the dolphin's detection sensitivity would be obtained.

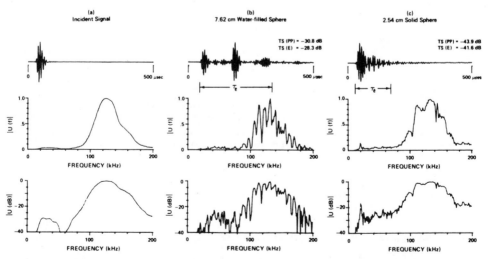

Fig. 3. Results of backscatter measurement, (a) simulated dolphin click
         (incident signal), (b) and (c) echoes from the 7.62 and 2.54-cm
         spheres, respectively (From Au and Synder, 1980)

The results depicted in Fig. 4 indicate that the animal's performance was consistent for the two studies. The 75% correct thresholds were at $E_e/N_o = 10.4$ dB for the 2.54-cm sphere results of Murchison (1980b) and 12.7 dB for the 7.62-cm sphere results of Au and Synder (1980). This difference of 2.3 dB is small considering the studies were done approximately two years apart.

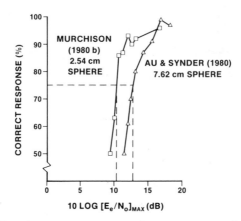

Fig. 4. _Tursiops_ target detection performance as a function of the echo-to-noise ratio for the range detection data of Figure 2.

II. TARGET DETECTION IN NOISE

Au and Penner (1981) performed a target detection in noise experiment with two _Tursiops_. The animals were required to station in a hoop and echolocate a 7.62-cm stainless steel water-filled sphere at a range of 16.5 m. A noise source with a flat spectrum between 40 and 160 kHz was located between the animal and the target, 4 m from the hoop. Masking noise levels between 67 and 87 dB re 1 $\mu Pa^2/Hz$ in 5 dB increments were randomly used in blocks of 10-trials for a 100 trial session. The dolphins' performance as a function of the masking noise level is shown in Fig. 5. Each point represents the average performance after 200 trials. Accuracy decreased monotonically as the noise level increased.

The dolphins' performance plotted as a function of E_e/N_o is shown in Fig. 6. The average value of the maximum source energy flux density per trial was used in the calculations. At two highest noise levels, both dolphins began to guess since Ehiku did not emit any detectable signals during 20% and 41% of the trials at 82 and 87 dB, respectively. Heptuna did not emit any signals for 14% of the trials at the 87 dB noise level. Therefore,

the average of the maximum signal per trial between 67 and 77 dB noise
levels were used to calculate E_e/N_o at 82 and 87 dB.

Au and Moore (this volume) performed a phantom target detection in
noise experiment with a <u>Tursiops</u>. A hydrophone 1.9 m in front of a stationing hoop was used to detect the projected signal which was digitized
and stored in RAM (random access memory). The stored signal was then
played back through a projector located 2.4 m from the hoop to simulate a
target. Masking noise at a fixed level was also played to the animal and
the intensity of the simulated echo was randomly varied in 10- trials
blocks with increments of 2 dB. For each click emitted by the animal, two
clicks separated by 200 μsec were played back to the animal at a time delay
representing a 20 m target range. The phantom target results are shown in
Fig. 7, along with results from the maximum range (Fig. 5), and the noise
studies (Fig 6).

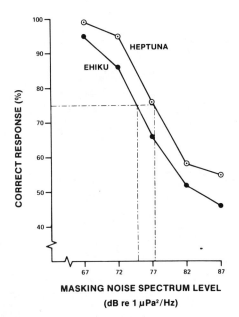

Fig. 5. <u>Tursiops</u> target detection performance as a function of the
masking noise level (From Au and Penner, 1981)

The 75% correct response thresholds for the different studies were
within 5.5 dB. The results agree extremely well despite the differences in
animals, time periods and experimental procedures. Murchison (1980b) and
Au and Synder (1980) varied target range in small increments in terms of
transmission loss. Au and Penner (1981) used a constant target range and
randomly varied the noise levels in relatively large increments of 5 dB. Au

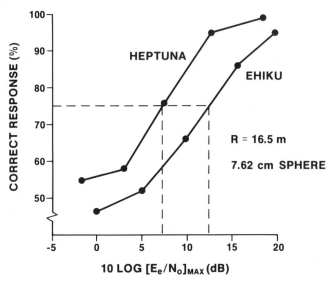

Fig. 6. <u>Tursiops</u> target detection in noise performance as a function
of the echo S/N (From Au and Penner, 1981).

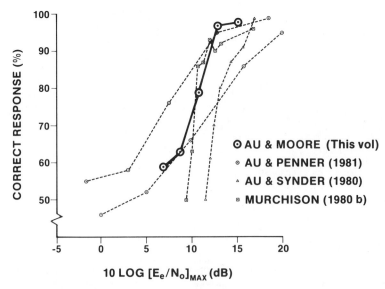

Fig. 7. <u>Tursiops</u> phantom target detection performance as a function
of the echo S/N (From Au and Moore, this volume).

and Moore (this volume) used a fixed phantom target range and noise level and randomly varied the amplitude of the echoes in 2 dB increments. The shallower slopes of the performance curves in the Au and Penner (1981) study were probably the result of using large noise increments. The curves for the other studies have similar slopes.

Turl et al. (1986) compared the target detection in noise capability of a beluga whale and a bottlenose dolphin. Sessions were alternated between the two animals, one in the morning and another in the afternoon. The same noise source used by Au and Penner (1981) was placed between the target and a hoop station, 5 m from the hoop. Detection performance was determine with a 7.62-cm diam. sphere at 16.5 and 40 m, and a 22.86-cm diam. sphere at 80 m. Both targets were water-filled stainless spheres.

The results of the target detection experiment as a function of the masking noise levels for the three target ranges are shown in Fig. 8 for the three target ranges. The data clearly indicated that the beluga whale could detect the targets considerably better than the Tursiops. The results plotted as a function of E_e/N_0 for the 7.62 cm sphere at the 16.5 and 40 m range are shown in Fig. 9. At the 75% correct response threshold 2beluga operated with an average of 11.5 dB higher noise than Tursiops. Even if we consider the best performance in noise of three Tursiops detecting a 7.62-cm sphere at a range of 16.5 m (Heptuna in the study of Au and Penner, 1981) the beluga would be approximately 7 dB better than Tursiops.

The reasons for the beluga's superior performance over Tursiops are not known. The maximum source levels per trial of both animals were approximately the same. Since the masking noise was emanating from a point source, a difference in the receiving directivity indices should not affect

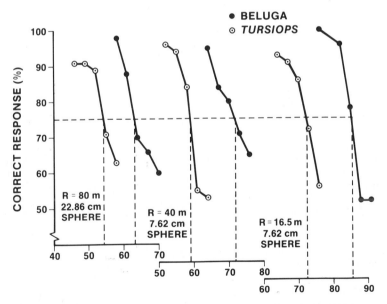

Fig. 8. Comparison of target detection performance in masking noise between Delphinapterus leucas and Tursiops truncatus for the same targets and experimental geometry.

457

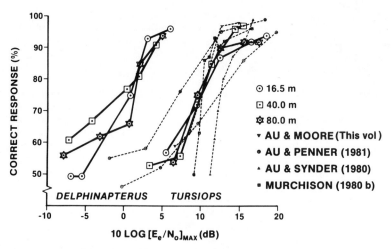

Fig. 9. Comparison of target detection performance as a function of
the echo S/N between <u>Delphinapterus leucas</u> and <u>Tursiops</u>
<u>truncatus</u> (From Turl et al. 1987).

the animals' performance. The beluga may have a narrower critical bandwidth than <u>Tursiops</u>, which would reduce the amount of masking noise received by the beluga. The beluga may also be able to process more signal
out of the noise than the bottlenose dolphin. Finally, the beluga does
have a narrow transmitting beam (Au et al., 1985) than <u>Tursiops</u> (Au, 1980)
and probably a narrower receiving beam. A narrower receiving beam may have
allowed the beluga to spatially separate the target and the noise source,
especially if the noise transducer was not perfectly aligned between the
animal and the target. However, the <u>Tursiops</u> results in Fig. 9 are consistent with those of Fig. 7, arguing against an alignment problem. The echo-
to-noise ratios at the 75% correct response level in Fig. 9 are within 1 dB
of the results from Au and Synder (1980), and Ehiku's results from Au and
Penner (1981).

III. TARGET DETECTION IN REVERBERATION

Reverberation differs from noise in how it interferes with the target
detection capability of a sonar system. It is caused by the interaction of
the sonar with the environment, and is the total contribution of unwanted
echoes scattered back to a sonar from objects and inhomogeneities in the
medium and on its boundaries. The spectral characteristics of reverberation are similar to those of the projected signal and its intensity is
directly proportional to the intensity of the projected signal. Therefore,
in a reverberation-limited situation, target detection cannot be improved
by increasing the intensity of the projected signal. In the presence of
reverberation, target detection also depends on the ability of the sonar
system to discriminate between the target of interest and false targets and
clutter that contribute to the reverberation.

Murchison (1980a,b) studied the effects of reverberation on the
ability of two <u>Tursiops</u> to detect a 6.35 cm solid steel sphere by measuring detection range as a function of target depth. The target depth was
varied from 1.2 to 6.3 m (lying on the bottom). Murchison found that the
animals' detection range decreased monotonically with target depth varying

from 74 m (target close to the surface) to 70 m (target on the bottom). His results also indicated that bottom reverberation affected the animals more than surface reverberation. Unfortunately, Murchison did not determine the echo-to-reverberation ratio (E/R) at the animals' threshold.

Au and Turl (1983) measured <u>Tursiops</u> ability to detect targets placed near a clutter screen, consisting of forty-eight 5.1-cm diameter cork balls spaced 15.2 cm apart in a 6X8 rectangular array. The cork balls were tied eight to a line and attached to a 1.9 x 1.9 m frame made of 3.2 cm diam. water filled PVC pipes. The clutter screen was placed 6 m from the animal's hoop station. Targets were hollow aluminum cylinders (3.81-cm diam. and 0.32-cm wall thickness), 10.0, 14.0 and 17.8 cm long. By using cylinders of the same diameter and wall thickness, but of different lengths, the echo amplitude could be varied without changing the echo structure, so echoes from each cylinder would "sound" the same.

Backscatter measurements of the targets and clutter screen were made with simulated dolphin signals to determine E/R based on the energy of the incident and reflected signals (E/R_E) and on the peak-to-peak values (E/R_{pp}). Results of five consecutive pings with the 10.0-cm long cylinder at a separation distance (ΔR) of 10.2 cm from the screen are shown in Fig. 10. Echoes from the cylinder were relatively similar from ping to ping, whereas echoes from the clutter screen varied slightly, being the results of complex constructive and destructive interferences of scattered signals from the balls and the frame.

The dolphin's detection results when the targets were located within the plane of the clutter screen are shown in Fig. 11 as a function of E/R. The linear least square lines fitted to the data indicate that the 50% detection threshold corresponded to E/R_E value of 0.25 dB and E/R_{pp} value

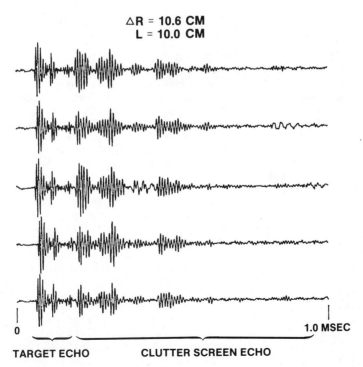

Fig. 10.   Echoes from the 10.0-cm-long cylinder and the clutter screen located 10.2 cm behind the target, for five consecutive pings (From Au and Turl, 1983).

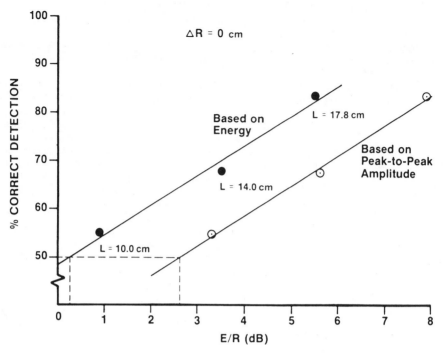

Fig. 11. Dolphin's target detection performance as a function of the
echo-to-reverberation ratio. The targets were within the
plane of the clutter screen. The closed circles refer to E/R_E
and the open circles to E/R_{pp} (From Au and Turl, 1983).

of 2.5 dB. The straight lines also indicate that the animal's target detection sensitivity is directly proportional to E/R.

   Au and Turl (1984) performed a followup experiment with the clutter
screen in which the response bias of the dolphin was manipulated by varying the symmetry of the payoff matrix. The payoff matrix (number of pieces
of fish reinforcement) was varied in terms of the ratio of correct detections to correct rejections in the following manner: 1:1, 1:4, 1:1, 4:1,
1:1, and 8:1. The targets were placed in the plane of the clutter screen
($\Delta R = 0$ cm). The results plotted in an ROC format are shown in Fig. 12.
Included in the figure are the ideal isosensitivity curves that best
matched the dolphin's performance excluding the results for the two smaller
targets at the 8:1 payoff.

   Changes in the animal's performance with changes in the payoff matrix
were relatively systematic and predictable. As the payoff increased from
1:1 to 4:1 and 8:1, the dolphin became progressively more liberal in reporting the presence of a target with a subsequent increase in the false
alarm rates and a decrease in the response bias, β. With the largest target, the animal became strongly biased towards the target-present response
as the payoff matrix shifted to 4:1 and 8:1. However, its detection sensitivity remained relatively constant as its response bias varied. With
the two smaller targets, the dolphin shifted from being conservative to
being relatively unbiased as the payoff matrix shifted from 1:1 to 4:1 and
8:1. The animal's sensitivity also remained relatively constant except at
the 8:1 payoff. Very little difference in the animal's performance occurred when the payoff matrix shifted from 1:1 to 1:4. The animal was already
conservative at the 1:1 payoff so that the shift to the 1:4 payoff did not
induce it to become more conservative.

IV TARGET RECOGNITION AND SHAPE DISCRIMINATION

Hammer and Au (1980) performed three experiments to investigate the target recognition and discrimination capability of an echolocating bottle nose dolphin. Two hollow aluminum cylinders, 3.81 cm and 7.62 cm in diameters and two coral rock cylinders of the same diameter, all 17.8 cm in length, were used as standard targets. The dolphin was required to echolocate the targets and respond to paddle A if one of the aluminum standards was present or paddle B if one of the coral rock standards was present. Target ranges of 6 and 16 m were used. After baseline performance exceeded 95% correct with the stand targets, probe sessions conducted to investigate the dolphin's ability to discriminate novel targets varying in structure and composition from the standards. All probe targets were cylinders, 17.8 cm in length. The first experiment involved a general discrimination in which four aluminum (two solid and two hollow) and four non-metal cylinders were used as probe targets. The dolphin reported all the probe targets as "B" (not A) even though several were aluminum and one of those was hollow with dimensions close to the larger aluminum standard. The results indicated that the dolphin recognized the salient features of the standard aluminum target echoes and excluded all the other targets from this class.

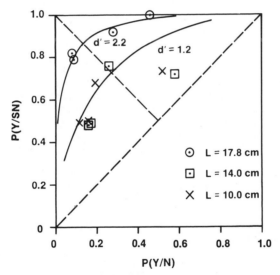

Fig. 12. Dolphin variable payoff results. Points towards the left are for the 1:4 and 1:1 payoff matrices. Points on the right are for the 8:1 payoff condition (From Au and Turl, 1984).

The second experiment studied the dolphin's ability to discriminate target wall thicknesses. Hollow aluminum probe targets with the same outer diameters but different wall thicknesses as the aluminum standards were used. The results showed that the dolphin could reliably discriminate wall thickness differences of 0.16 cm for the 3.81 cm O.D. targets, and 0.32 cm for the 7.62 cm O.D. targets.

461

In the third experiments, the dolphin's ability to discriminate material composition was tested using bronze, glass and stainless steel probes that had the same dimensions as the aluminum standards. The dolphin discriminated the bronze and steel targets from the aluminum but classified the glass probes with the aluminum standards.

Hammer and Au (1980) measured the target echoes with a simulated dolphin signal. The echoes were then processed with a matched filter using the projected signal as the reference, and the envelope of the matched-filter output was determined. They found a strong correlation between the dolphin's behavioral results and the envelope of the matched filter response. The arrival time differences between echo components seemed to be the dominant cue. The time differences may be perceived as a time separation pitch (TSP) by the dolphin.

In a follow-on study, Schusterman, et al. (1980) trained the same dolphin to discriminate between the aluminum and glass targets, using a two-alternative forced-choice paradigm. After 30 sessions, the dolphin could perfectly discriminate the 3.61 cm aluminum and glass cylinders. However, the animal was never able to discriminate between the 7.62-cm aluminum and glass cylinders.

Au et al.(1980) conducted an experiment to determine if an echolocating dolphin could discriminate between foam spheres and cylinders located 6 m from a hoop station. Three spheres and five cylinders of varying sizes but overlapping target strengths (Table 1) were used so that target strength differences would not be a cue. Two spheres and two cylinders were used in each 64 trial session. The dolphin was required to station in a hoop and the targets were 6 m from the hoop.

Results of the sphere/cylinder discrimination experiment are shown in Fig. 13. The dolphin was able to discriminate between spheres and cylinders with an accuracy of at least 94% correct. Au, et al. (1980) postulated that the major cue was the larger surface reflected component in the echoes from the spheres. However, when a "horsehair" mat was introduced in a session to absorb the surface reflected component of the target echoes, the dolphin still performed the task perfectly. Therefore, the dolphin probably performed the target discrimination based on differences in the echoes. Target strength for foam cylinders vary with log (frequency) and is constant with frequency for spheres (Urick, 1983).

Table 1. Dimension and measured target strength of foam targets used in the sphere (S) vs cylinder (C) discrimination experiment.

Target	Diameter	Length	Target Strength
S_1	10.2 cm	---	-32.1 dB
S_2	12.7 cm	---	-31.2 dB
S_3	15.2 cm	---	-28.7 dB
C_1	1.9 cm	4.9 cm	-31.4 dB
C_2	2.5 cm	3.8 cm	-32.3 dB
C_3	2.5 cm	5.1 cm	-28.7 dB
C_4	3.8 cm	3.8 cm	-30.1 dB
C_5	3.8 cm	5.1 cm	-27.6 dB

V DISCUSSION AND CONCLUSIONS

The sonar target detection sensitivity of <u>Tursiops</u> has been measured by determining: (a) The maximum detection range for two targets. (b) Target detection performance for a target at a fixed range in the presence of variable artificial masking noise. (c) Target detection performance for a variable sized target at a fixed range in the presence of artificial masking noise. The results of the various methods when considered in terms of E_e/N_o, were very consistent. E_e/N_o at the 75% correct response threshold varied from 7.2 to 12.4 dB. The shape of the performance curves as a function of the E_e/N_o was also similar except for the case in which the noise levels changed in 5 dB increments.

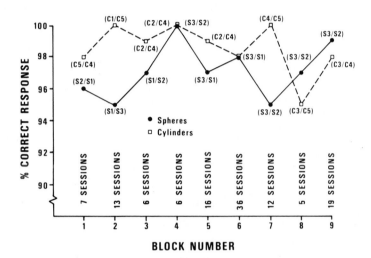

Fig. 13. <u>Tursiops</u> echolocation performance discriminating between foam spheres and cylinders (From Au et al., 1980).

The beluga whale was shown to possess a much keener ability to detect targets in noise than <u>Tursiops</u> by approximately 6 to 13 dB. This difference is very large and it is not obvious why it exists. The signal levels utilized by both species were similar ruling out the possibility that the gain was from the transmitting end. The most likely possibility for beluga's keener ability may be differences in the critical ratios of both species. The beluga whale's transmitting beam is narrower than <u>Tursiops'</u>, and its is highly likely that its receiving beam is also narrower. Therefore, it is possible but unlikely that the beluga may have been able to spatially resolve the noise from the target. More research with beluga whale is needed before the problem can be resolved.

<u>Tursiops</u> ability to detect targets in reverberation was found to be proportional to E/R. The echo-to-reverberation ratio at the 50% correct detection threshold was found to be 0.25 dB for E/R based on energy and 2.6 dB for E/R based on the peak-to-peak amplitude. The geometry used in the clutter screen experiment of Au and Turl (1983) represented a situation of a dolphin echolocating perpendicular to a bottom, with the clutter and

target echoes arriving at the animal at the same instance.

The response bias of an echolocating dolphin can be easily manipulated by varying the food reinforcement payoff matrix. Au and Turl (1984) showed that the dolphin's target detection sensitivity remained relatively constant as its response bias shifted. Their results clearly showed that the use of a balanced payoff matrix, a technique strongly favored in dolphin psychophysical experiments, will result in a conservative response bias.

The results of the target recognition and shape discrimination experiments clearly indicate that dolphins can discriminate and recognize various features of targets with their sonars. Target features such as structure (hollow or solid), wall thickness, material composition and shape are discriminable by an echolocating dolphin. The time separation between the primary and secondary echo components may provide TSP cues that dolphins use to make fine target discrimination. Discrimination between foam spheres and cylinders were probably based on differences in target strength with frequency.

REFERENCES

Albers, V.M., 1965, "Underwater Acoustics Handbook II", (The Pennsylvania State U. Press, University Park, PA).

Au, W.W.L., 1980, Echolocation Signals of the Atlantic Bottlenose Dolphin (Tursiops truncatus) in Open Waters, in "Animal Sonar Systems", R.G. Busnel and J.F. Fish, ed., Plenum, New York, pp. 251-282.

Au, W.W.L., Floyd, R.W., Penner, R.H., and Murchison, A.E., 1974, Measurement of Echolocation Signals of the Atlantic Bottlenose Dolphin, Tursiops truncatus Montagu, in Open Waters, J. Acoust. Soc. Am., 56: 1280- 1290.

Au, W.W.L., Schusterman, R.J., and Kersting, D.A., 1980, Sphere-cylinder Discrimination via Echolocation by Tursiops truncatus, in "Animal Sonar Systems," R.G. Busnel and J.F. Fish, ed., Plenum, New York, pp 859- 862.

Au, W.W.L., and Snyder, K.J., 1980, Long-Range Target Detection in Open Waters by an Echolocating Atlantic Bottlenose Dolphin (Tursiops truncatus), J. Acoust. Soc. Am., 68:1077-1084.

Au, W.W.L., and Penner, R.H., 1981, Target Detection in Noise by Echolocating Atlantic Bottlenose Dolphins, J. Acoust. Soc. Am., 70:687-693.

Au, W.W.L., Penner, R.H., and Kadane, J., 1982, Acoustic Behavior of Echolocating Atlantic Bottlenose Dolphins, J. Acoust. Soc. Am., 71: 1269-1275.

Au, W.W.L., Turl, C.W., 1983, Target Detection in Reverberation by an Echolocating Atlantic Bottlenose Dolphin Tursiops truncatus, J. Acoust. Soc. Am., 73:1676-1681.

Au, W.W.L., and Moore, P.W.B., 1984, Receiving Beam Patterns and Directivity Indices of the Atlantic Bottlenose Dolphin Tursiops truncatus, J. Acoust. Soc. Am., 75:255-262.

Au, W.W.L., and Turl, C.W. 1984, Dolphin Biosonar Detection in Clutter: Variation in the Payoff Matrix, J. Acoust. Soc. Am., 76:955-957.

Au, W.W.L., Carder, D.A., Penner, R.H., and Scronce, B.L., 1985, Demonstra-stration of Adaptation in Beluga Whale Echolocation Signals, J. Acoust. Soc. Am., 77:726-730.

Au, W.W.L., Penner, R.H., and Turl, C.W., 1986, Propagation of Beluga Whale Echolocation Signals (A), presented at Sixth Bien. Conf. Bio. Marine Mamm., Vancouver, B.C.

Hammer, C.E. Jr., and Au, W.W.L., 1980, Porpoise Echo-Recognition: An Analysis of Controlling Target Characteristics, J. Acoust. Soc. Am., 68:1285-1293.

Murchison, A.E., 1980a, Maximum Detection Range and Range Resolution in Echolocating Bottlenose Porpoises Tursiops truncatus (Montagu), Ph.D. Dissertation, University of Calif., Santa Cruz.

Murchison, A.E., 1980b, Detection Range and Range Resolution in Echoloca-ting Bottlenose Porpoise (Tursiops truncatus), in "Animal Sonar Systems," R.G. Busnel and J.F. Fish, ed., Plenum, New York, pp.43-70.

Schusterman, R.J., Kersting, D., and Au, W.W.L., 1980, Response Bias and Attention in Discriminative Echolocation by Tursiops truncatus, in "Animal Sonar Systems," R.G. Busnel and J.F. Fish, Plenum, New York, pp. 983-986.

Turl, C.W. and Penner, R.H., Au, W.W.L., 1987, Comparison of the Target Detection Capability of the Beluga Whale and Bottlenose Dolphin," J. Acoust. Soc. Am., 82:1487-1491.

Urick, R.J., 1983, "Principles of Underwater Sound", McGraw-Hill, New York.

PREY INTERCEPTION: PREDICTIVE AND NONPREDICTIVE STRATEGIES

W. Mitchell Masters

Department of Zoology
Ohio State University
Columbus, OH 43210

INTRODUCTION

Insectivorous bats use echolocation as their sole source of information to detect, locate, approach and capture flying insects. Although the sequence of behaviors they go through is relatively sterotyped (Webster, 1967b; Kick and Simmons, 1984), much remains unknown about how bats accomplish this incredible feat. I am particularly concerned here with what bats do with the positional information they obtain about a target by echolocation: how does a bat choose its flight path so as eventually to intercept the target?

Some work on this type of problem has been done on head-aim tracking. Once a bat decides to pursue a target it aims its head at it and keeps its head trained on the target until capture (Webster and Brazier, 1965). Masters et al. (1985) investigated whether a bat's head-aim tracking is predictive or nonpredictive. A predictive approach entails calculating where the target will be at some future time based on its past trajectory. A nonpredictive approach simply entails keeping up with where the target was on the last echolocation sound. With an irregularly moving target they found that in the laboratory head-aim tracking appears to be nonpredictive.

Given that head aim tracking may be nonpredictive, is target interception, a related aspect of prey capture, also nonpredictive? Since to some this idea may seem to contravene common sense it is worth asking whether a nonpredictive interception strategy is at all feasible. I believe it is, assuming bats satisfy three conditions. (1) They must tolerate some degree of error since lag error is unavoidable: the bat is always aiming where the target was, not where it is. Catching insects in the flight membranes could increase error tolerance. (2) The error must be kept fairly small. Bats could do this by frequent update of target position (high rate of sonar emission) and rapid processing of sonar information. (3) The bat must fly faster than its prey, otherwise it could capture only insects flying toward it on a near-collision course. Since these conditions seem to be fulfilled, we can at least entertain the possibility that bats may use a nonpredictive approach to target interception.

Modelling Interception

To understand interception better, I modelled the bat's approach to a target using three strategies involving increasing degrees of prediction.

I. <u>No prediction</u>. The simplest strategy, the bat merely flies towards the target position obtained from its last sonar emission.

II. <u>Predict one call ahead</u>. Here the bat uses straight-line extrapolation from the last two target positions (i.e., it measures velocity) to predict where the target will be on the next sonar emission and aims there.

III. <u>Predict path intersection</u>. In this strategy the bat extrapolates the target's future position from its position on the last two sonar calls and, knowing its own flight speed, determines the heading it must take to arrive at the intersect point at the same time as the target.

For these models to be at all realistic I had to make certain assumptions about the bat's flying speed, its minimum turning radius and its reaction time. Flying speed varies from about three to ten meters/second and can be controlled by the bat (Webster and Brazier, 1965); I used values of 4 to 6 m/s. I could find no information on turning radius, which must also depend on flying speed, so I chose values from 0.5 to 1.0 m as reasonable estimates. The bat's reaction time was estimated by Cahlander and Webster (1960) to be about 100 ms based on its observed tendency to miss a thrown target (see below) and by Webster (1967a) to be about 50 ms. A bat's head aim tends to lag target position by about 50 to 100 ms (Masters et al., 1985), a delay one might interpret as reaction time. Based on these observations I assumed a reaction time of 50 to 75 ms, that is, the bat was assumed not to act on information in a sonar signal until 50 to 75 ms after emission. Other assumptions were that the target was first detected at a distance of two meters, and that the rate of sonar emissions was similar to that given by <u>Myotis</u> approaching an 8 mm sphere (Webster and Brazier, 1965, Fig. 3b), namely, 80 ms between calls when the target was first detected at two meters decreasing linearly with bat-to target distance to a minimum of 5 ms between calls at a distance of 15 cm.

The situation examined in this paper is one described by Cahlander and Webster (1960) that seems at first glance to offer strong support for a predictive interception strategy. They reported that bats (presumably <u>Myotis</u>) accustomed to catching <u>Drosophila</u> in the laboratory at first missed mealworms tossed in the air, going through the catching motions at a point about 5 cm <u>above</u> the target's true position. With experience some bats learned to compensate and catch thrown mealworms. These bats, when then tested with a suspended mealworm, would go through the catching procedure 5 cm <u>below</u> this stationary target. Cahlander and Webster interpreted these results as suggesting that bats are sensitive to target velocity (but not acceleration) and use it to predict the interception point. Thus bats initially passed too high because they failed to allow for the downward acceleration of gravity. With practice they learned to compensate by aiming below the target and thus missed the stationary target.

RESULTS

Using the three strategies given above, I calculated the approach paths for a bat in level flight, assuming it first detected the target 2 m directly ahead. The target was assumed to be moving vertically with ballistic motion (i.e., acted on only by the acceleration of gravity). This corresponds to a mealworm thrown straight up in front of the bat.

Figure 1 shows the bat's approach paths for the three strategies. The first thing that is apparent is that with all strategies the bat tends to pass above the target; therefore, the fact that bats overfly the target is not, by itself, sufficient to prove they use a predictive strategy, contrary Cahlander and Webster's (1960) assumption. If the mealworm is thrown rather high above the bat's plane of flight (e.g. Fig. 1-I and 1-II, top paths), the bat will pass below the target; otherwise it will pass above it. Assuming that in the laboratory mealworms were not thrown high above the bat, either predictive or nonpredictive strategies would result in overshoot of the target, as observed by Cahlander and Webster.

The second point from Fig. 1 is that strategies I and II can be distinguished from III by the form of the flight path. For III the bat tends to make a marked upward turn early on due to the large upward velocity of the target at this time. Later, as the target decelerates, the bat is forced to make a downward turn to compensate. For I and II the bat's course is a gentle curve upward over the whole approach. It also is clear from Fig. 1 that it will be very difficult to distinguish between strategy I, where no prediction is involved, and strategy II, where there is a small amount of prediction, solely on approach paths since these are very similar.

By using different values for the bat's flying speed, turning radius and reaction time it is possible to see which of these parameters influence the bat's approach path. I found that for strategies I and II, flying speed is important but reaction time and turning radius are relatively unimportant. For instance, in the middle tracks of Fig. 1-I and 1-II the bat overshoots the target by about 4 cm; if its flight speed is reduced from 5 m/s to 4 m/s, the overshoots increases to about 14 cm, but reducing reaction time from 75 to 50 ms or reducing turning radius from 0.75 m to 0.5 m causes essentially no change in overshoot. For strategy III all three parameters are important. For example, in the middle track of Fig. 1-III the bat overshoots the target by about 10 cm; reducing the flight speed to 4 m/s increases the overshoot to about 15 cm, reducing turning radius to 0.5 m decreases it to about 7 cm, and reducing the reaction time to 50 ms decreases it to about 5 cm.

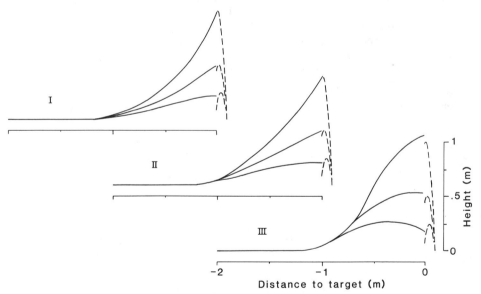

Fig. 1. Examples of approach paths using model strategies I, II and III. In each case, the bat (solid lines) is at -2 m when it detects the target (dashed lines) at the origin. The target is moving straight up but its path is drawn with a slight horizontal component for clarity. Model parameter values were flying speed 5 m/s, turning radius 0.75 m, and reaction time 75 ms. In (I) and (II) the bat's overshoot is 12, 4 and 0 cm, and in (III) 11, 10 and 7 cm, for targets reaching heights of 0.25, 0.5 and 1 m, respectively.

DISCUSSION

The major point I wish to make here is that a nonpredictive strategy is feasible -- or at least cannot be rejected out of hand -- and that it would be worthwhile looking at what bats actually do. The observation that some bats tend to pass above a thrown mealworm, which might seem prima facie evidence for a predictive strategy, is seen from models of interception not to be compelling. What is needed is to observe the form of the bat's approach path. With a ballistic target, strategies I and III, no prediction vs. maximal prediction, show obvious differences in shape. However, intermediate degrees of prediction are possible, the result being that no sharp boundary can be drawn between predictive and nonpredictive approaches. If a bat uses only a small amount of prediction, its trajectory will be nearly indistinguishable by most criteria from no prediction at all.

Cahlander and Webster's (1960) observations illustrate a potentially confounding variable, namely learning. The bats in their study learned to compensate for their initial overshoot, but whether they did so by lowering their aim point or by switching to another strategy is not clear. Analyzing the change in approach path with experience might answer this question.

In raising the possibility that bats might intercept insects by a nonpredictive strategy, I want to stress that I am not questioning whether bats can in fact predict where an insect seems to be headed. There is reasonably convincing anecdotal evidence that bats can tell where an insect is headed, at least in general terms, and anticipate potential problems such as flying into the ground or trees. In such cases bats have been known to alter their course so that their approach to the insect is essentially parallel to the obstacle (Webster, 1967a). But it seems to me that there is a distinction to be made between this behavior, where the bat chooses it direction of attack, and the rapid, last instant course adjustments necessary to effect capture. By analogy, shortly after a baseball is hit a good fielder can predict the general area where it will land and run there. He may then use a different strategy to catch the ball in his glove. Likewise a bat may anticipate the insect's approximate path and choose an appropriate approach direction but then follow a different strategy actually to catch the insect in its flight membranes, that is, during the tracking and terminal stages of pursuit described by Kick and Simmons (1984). This possibility is all the more likely given that insect flight paths are unlikely to be inherently predictable.

REFERENCES

Cahlander, D. A, and F. A. Webster, 1960, The bat's reaction time, MIT Lincoln Lab. Report No. 47G-0008.
Kick, S. A., and J. A. Simmons, 1984, Automatic gain control in the bat's sonar receiver and the neuroethology of echolocation, J. Neurosci., 4:2725.
Masters, W. M, A. J. M. Moffat, and J. A. Simmons, 1985, Sonar tracking of horizontally moving targets by the big brown bat Eptesicus fuscus, Science, 228:1331.
Webster, F. A., 1967a, Interception performance of echolocating bats in the presence of interference, in: "Animal Sonar Systems," R.-G. Busnel, ed., Lab. Physiol. Acoust., Jouy-en-Josas, France.
Webster, F. A., 1967b, Some acoustical differences between bats and men, in: "International Converence on Sensory Devices for the Blind," R. Dufton, ed., St. Dunstan's, London.
Webster, F. A., and O. G. Brazier, 1965, Experimental studies on target detection, evaluation and interception by echolocating bats, U.S. Air Force Tech. Documentation Report No. AMRL-TR-67-192.

A MECHANISM FOR HORIZONTAL AND VERTICAL TARGET LOCALIZATION

IN THE MUSTACHE BAT (Pteronotus p. parnellii)

Zoltan M. Fuzessery

Department of Neurophysiology
University of Wisconsin
Madison, WI 53706, USA

Considering their aerial existence, bats have an obvious need to extract both horizontal and vertical spatial information from the echoes of their emitted pulses. Behavioral studies indicate that bats can resolve the relative angular separations of pairs of targets to within 5° in both the horizontal and vertical planes (Peff and Simmons, 1971; Lawrence and Simmons, 1972; Airapetianz and Konstantinov, 1974; Simmons et al., 1983). Bats appear to be equally adept at resolving the absolute location of single targets. Observations of the prey-catching behavior of Myotis, for example, indicate this bat points its head towards an insect with an accuracy of also about 5° (Webster and Durlach, 1963; Webster and Brazier, 1968). The present contribution examines the acoustic cues available for resolving the azimuthal and elevational coordinates of a target, and suggests how this information might be encoded within the bat's central auditory system. The subject is the greater mustache bat (Pteronotus p. parnellii), a species which requires accurate spatial information because it hunts insects amid dense foliage.

Two assumptions are made, and while both seem likely, neither have been directly demonstrated. The first is that bats can resolve azimuthal and elevational target coordinates within the beam of its echolocation pulse, as opposed to simply detecting whether the target is present within the beam. Implicit in this assumption is that this spatial information can be extracted from a single echo. Such information would be of obvious value, allowing the bat to make immediate flight adjustments following initial detection of an insect, and to rapidly update the insect's location during pursuit. Webster (1967) noted that Myotis aims its head at an insect within 0.1 sec after inital detection. Assuming a search flight pulse rate of aboout 10 pulses/sec (e.g., Schnitzler, 1970), it is possible that Myotis makes this initial orientation on the basis of a single echo.

The second assumption is that most bats use the same spatial cues employed by passive sound locators, because whether a target is emitted or reflecting sound, similar spatial cues will be generated. While this is perhaps true for bats with relatively fixed pinnae (such as <u>Pteronotus</u>), other bats may create unique cues. The rhinolophids, for example, may obtain vertical information through rapid pinna "scanning" (Neuweiler, 1970; Schnitzler, 1973; Gorlinsky and Konstantinov, 1978).

Horizontal sound localization is thought be largely a binaural process, comparing time and/or intensity differences at the two ears (reviews, Erulkar, 1972; Aitkin et al., 1984). In bats, attenuating sound at one ear reduces obstacle-avoidance performance in most species tested (Griffin and Galambos, 1941; Flieger and Schnitzler, 1972). Such results strongly suggest the importance of interaural intensity differences (IIDs), but do not rule out the use of interaural time differences. Additional considerations also suggest the importance of IIDs: (1) the heads of bats are typically small and consequently generate very small interaural time differences, (2) their ears generate large IID values (review, Schnitzler and Henson, 1980, also Grinnell and Grinnell, 1965; Fuzessery and Pollak, 1984,1985), and (3) binaurally sensitive neurons are most sensitive to IIDs (Schlegel, 1977; Fuzessery and Pollak, 1985; Harnischfeger et al., 1985). It is likely that, at least while echolocating with high frequencies, bats rely primarily on IIDs to determine horizontal location. However, appropriate behavioral studies, involving dichotic stimulation paradigms, are needed to directly evaluate the relative importance of interaural time and intensity disparities.

Elevational information can be obtained from spectral cues. The directional selectivity of the external ears is frequency-dependent (Grinnell and Grinnell, 1965; Shaw, 1974; Calford and Pettigrew, 1984; Fuzessery and Pollak, 1985). If directionality shifts in the vertical plane, different frequencies will be maximally amplified by the ears at different elevations. Psychoacoustic studies reveal that humans are able to use the resultant spectral cues to determine sound source elevation (Roffler and Butler, 1968; Blauert, 1969/1970). This ability to associate frequency and space has not been demonstrated in other species, but it is unlikely that it is unique to humans. It is known that altering the structure of the external ears, and hence their acoustic properties, interfers with vertical, more than horizontal, localization in bats (Gorlinsky and Konstantinov, 1978; Lawrence and Simmons, 1982). One interpretation of these results is that modification of external ear structure eliminates the spectral cues used to resolve sound elevation.

Combining binaural intensity cues and spectral cues can provide both azimuthal and elevational information. Grinnell and Grinnell (1965) suggested that bats could resolve these coordinates though a binaural analysis of the echo power spectrum. Physiological studies of <u>Pteronotus</u> (Fuzessery and Pollak, 1985) support this idea. <u>Pteronotus</u> emits a long CF-FM pulse containing four harmonics at 30, 60, 90 and 120 kHz (Novick, 1963). The 60 and 90 kHz harmonics are the loudest. To reduce the system to the minimum number of essential elements, <u>Pteronotus</u> could obtain azimuthal and elevational information by the comparing IID values of the 60 and 90 kHz harmonics in the echo. The external ears of this bat are

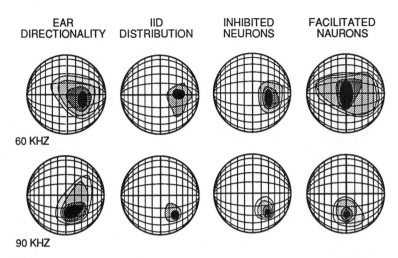

EAR DIRECTIONALITY   IID DISTRIBUTION   INHIBITED NEURONS   FACILITATED NAURONS

60 KHZ

90 KHZ

Fig. 1. Spheres represent the bat's frontal sound field.
Azimuthal axes are incremented by 13°, elevational axes by
20°. First column shows the directional selectivity of the
external ear, measured by cochlear microphonic threshold
responses to tones presented at various locations. Black area
shows where thresholds are within 2 dB of lowest recorded
value, with expanding circles representing the +5, +10 and +15
dB threshold contours. Second column shows the spatial
distribution of IIDs. Black area is where IID values <20 dB;
gray area, <10 dB. Third and fourth columns show,
respectively, the threshold sensitivities of binaurally
inhibited and facilitated neurons for sound location.
Threshold values as in first column.

highly directional at these two harmonics, and maximally
amplify these frequencies when they originate at different
elevations (Fig. 1, column 1). Calculations of IID values
show a similar trend (Fig. 1, column 2). The 60 kHz harmonic
is 20 dB louder in one ear only if the echo originates near
30° azimuthal and 0° elevation, while similar IID values at 90
kHz can only be generated by an echo originating 40° lower in
elevation. Thus a comparison of IID values at only two
frequencies could provide a means of resolving target azimuth
and elevation. It is important to note that these "difference
areas" are located well within the beam of the bat's
echolocation pulse, which drops to 50% of maximum amplitude at
about 20° off the vertical midline (Simmons, 1969).
Consequently, there is likely to be sufficient energy at both
harmonics to allow Pteronotus to compare IIDs across
frequency.

To determine how this information might be encoded within
the brain, neurons tuned to these frequencies and sensitivity
to IIDs were recorded in the inferior colliculus, and their
spatial selectivity was evaluated. Two general types of
binaurally sensitive neurons were found: binaurally inhibited
and binaurally facilitated neurons. Binaurally inhibited
neurons are excited by input to one ear and inhibited by input
to the other. They can be completely suppressed when a sound
is sufficiently louder at the inhibitory ear. Consequently,
these neurons are most responsive to sound locations
generating large IIDs at the excitatory ear (FIG. 1, column 3).

Binaurally facilitated neurons, on the other hand, are often maximally excited when intensities at the two ears are combined to generate specific IID values. Many of these neurons are maximally excited at IID values of 0 dB, which are normally created by sounds on the vertical midline of the frontal sound field. Two such neurons are shown in Figure 1, column 4; their frequency tuning determines their elevational selectivity along the vertical midline.

The result is that Pteronotus is provided with an array of neurons which differ systematically in their azimuthal and elevational selectivity. Figure 2 illustrates the +2, +5 and +10 dB threshold contours of these neurons, which serve to indicate the sound intensity they will experience as a function of target location. The important points are that, for any location within these contours, there will be a unique ratio of effective intensities among these neurons, and that, for any neuron, these intensities can vary as much as 20 dB. To oversimplify the model, it is assumed that effective intensity translates into a proportionate level of neuronal excitation. Any target location near the center of the bat's frontal sound field would therefore result in a unique level of excitation among these neurons. Relative levels of excitation could provide a mechanism for encoding two-dimensional space.

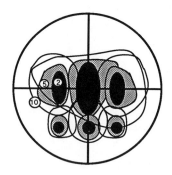

Fig. 2. The +2 dB (black areas), +5 and +10 dB threshold contours of the 60 and 90 kHz binaurally inhibited and facilitated neurons in Fig. 1, superimposed to show the extent to which the sound levels experienced by these neurons will differ as a function of sound location. The contours of the binaurally inhibited neurons were reproduced as mirror images to represent neurons that would be found on the opposite side of the inferior colliculus.

This mechanism assumes that Pteronotus can associate changes in the binaural power spectrum of the echo with target location. This could be tested by manipulating the power spectra of echoes and observing errors in vertical orientation – perhaps utilizing reflexive orientation behaviors. The model predicts, for example, that Pteronotus would associate an amplification of the 90 kHz harmonic with target locations in the lower part of its frontal sound field. At present, we know little about bats' ability to detect changes on power spectra, or to resolve IID values; it would therefore be premature to speculate on the spatial resolution potentially inherent in these cues.

The model also assumes that <u>Pteronotus</u> could distinguish changes in echo power spectra resulting from target location from those caused by other factors, such as the reflectant properties of the target and target distance. Atmospheric attenuation will have a great effect on the higher harmonic. Given that bats are capable of precise range resolution, it is possible that range information could be integrated with spectral analysis to provide appropriate compensation.

To summarize, bats create much of their acoustic environment, and they are no doubt sensitive to the information inherent in subtle changes in returning echoes. They appear to be able to extract diverse information from a single echo parameter. The echo power spectrum, for example, may provide information about target quality (Simmons et al., 1974; Habersetzer and Vogler, 1983), as well as target course and velocity (review, Schnitzler and Henson, 1980). Present results suggest that if <u>Pteronotus</u> were able to associate frequency and space, it could also extract elevational information.

REFERENCES

NOTE: References not cited below are found in:
Schnitzler, H-U., Henson, O. W. Jr , 1980, Performance of airborne animal sonar systems: echolocation in microchiropteran bats.In:Animal Sonar Systems, eds R. G. Busnel, J. F. Fish, eds., pp 109-182.New York: Plenum.

Aitkin, L. M., Irvine, D. R. F., Webster, W. R., 1984, Central neural mechanisms of hearing. In: Handbook of Physiology, Darian-Smith, I., ed., v.3, pt.2, pp.675-737, Bethesda, MD.
Blauert, J., 1969/1970, Sound localization in the median plane. Acustica 22:205-213.
Calford, M. B., Pettigrew, J. D., 1984, Frequency dependence of directional amplification at the cat's pinna. Hearing Res. 14:13-19.
Erulkar, S. D., 1972, Comparative aspects of spatial localization of sound.Physiol. Rev. 52:237-360.
Fuzessery, Z. M., Pollak, G. D., 1984, Neural correlates of sound localization in an echolocating bat. Science 225:725-728.
Fuzessery, Z. M., Pollak, G. D., 1985, Determinants of sound location selectivity in the bat inferior colliculus: a combined dichotic and free-field stimulation study. J. Neurophysiol. 54:757-781.
Griffin, D. R., Galambos, R., 1941, The sensory basis of obstacle avoidance in flying bats. J. Exp. Zool. 86:481-506.
Grinnell, A. D., Grinnell, V. S., 1965, Neural correlates of vertical localization by echo-locating bats. J. Physiol. 181:830-851.
Habersetzer, J., Vogler, B., 1983, Discrimination of surface-structured targets by the echolocating bat <u>Myotis myotis</u> during flight. J. Comp. Physiol. 152:275-282.
Harnischfeger, G., Neuweiler, G., Schlegel, P., 1985, Interaural time and intensity coding in superior olivary complex and inferior colliculus of the echolocating bat <u>Molossus ater</u>. J.Neurophysiol. 53:89-109.

Lawrence, B. D., Simmons, J. A., 1982, Echolocation in bats:
    the external ear and perception of the vertical
    positions of targets. Science 218:481-483.
Lawrence, B, D., Simmons, J. A., 1982, Measurements of
    atmospheric attenuation at ultrasonic frequencies and the
    significance for echolocating bats. J. Acoustic. Soc. Am.
    71:585-590.
Novick, A., 1971, Echolocation in bats: some aspects of pulse
    design. Am. Sci. 59:198-209.
Peff, T. C., Simmons, J. A., 1972, Horizontal-angle resolution
    by echolocating bats. J. Acoust. Soc. Am. 51:2063-2065.
Roffler, S. K., Butler, R. A., 1968, Localization of tonal
    stimuli in the vertical plane. J. Acoust. Soc. Am.
    43:1260-1266.
Schlegel, P., 1977, Directional coding by binaural brainstem
    units of the CF-FM bat, Rhinolophis ferrumequinum. J.
    Comp.Physiol. 118:327-352.
Schnitzler, H-U., Henson, O. W., Jr., 1980, Performance of
    airborne animal sonar systems: echolocation in
    microchiropteran bats.In: Animal Sonar Systems, Busnel, R.
    G., Fish, J. F., eds., pp109-182, Plenum, New York.
Shaw, E. A. G., 1965, Ear canal pressure generated by a free
    sound field. J. Acoust. Soc. Am. 39:465-470.
Simmons, J. A., 1969, Acoustic radiation patterns for the
    echolocating bats Chilonycteris rubiginosa and Eptesicus
    fuscus. J. Acoust. Soc. Am. 46:1054-1056.
Simmons, J. A., 1973, The resolution of target range by
    echolocating bats. J. Acoust. Soc. Am. 54:157-173.
Simmons, J. A., Lavender, W. A., Doroshow, C. A., Kiefer, S.
    W., Livingston, R., Scallet, A. C., 1974, Target structure
    and echo spectral discrimination by echolocating bats.
    Science 186:1130-1132.
Webster, F. A., 1967, Interception performance of echolocating
    bats in the presence of interference. In: Animal Sonar
    Systems, Vol. 1, Busnel, R. G., ed., pp 673-713, Lab de
    Physiologie acoustique: Jouy-de-Josas.
Webster, F. A., Brazier, O. B., 1965, Experimental studies on
    target detection, evaluation and interception by
    echolocating bats. Aerospace Medical Res. Lab., Wright-
    Patterson Air Force Base, Ohio.
Webster, F. A., Durlach, N. I., 1963, Echolocation systems of
    the bat. MIT Lincoln Lab. Rep. 41-G-3, Lexington, Mass.

476

ECHOES OF FLUTTERING INSECTS

Rudi Kober

Lehrbereich Zoophysiologie
University of Tübingen
Auf der Morgenstelle 28
D-7400 Tübingen, Federal Republic of Germany

INTRODUCTION

Bats with long cf-calls can not only detect their prey in strong background clutter but can also distinguish some insect species from others (Schnitzler and Henson, 1980; Trappe, 1982; von der Emde, this conference). What parameters do these bats use to classify their prey and - first of all - what are the echo parameters which allow this classification.

Schnitzler et al., 1983, showed that the echoes of flying insects are characterized by amplitude and frequency modulations following the rhythm of the wing beat. The echo amplitude does not change sinusoidally but shows short peak intensities or glints. A glint is produced each time the wing or parts of the wing are in a position perpendicular to the impinging sound waves (glint hypothesis). In the spectrograms the glints are represented as spectral broadenings, which encode the direction of wing movement at this moment. Wing movements towards the sound source yield positive Doppler shifts whereas wing movements away from the sound source yield negative Doppler shifts. Consequently, the sidebands in the spectrum are situated above and below the carrier frequency, respectively. Furthermore the wing movements results in an echo structure which is very much dependent on the angular orientation of the insect.

In the first part of this study the changes of the insect echo with different angular orientations will be presented. In the second part echo parameters that encode species specific information will be described.

MATERIAL AND METHODS

Insects from different orders were captured at night with a light trap in the vicintiy of Tuebingen, West-Germany. (One exception: Pachnoda spec. (Coleoptera) from West-Africa).

The insects were mounted in the acoustical beam of an ultrasonic loudspeaker transmitting 80 kHz (Machmerth et al., 1975) and the returning echoes were picked up with a Brüel and Kjaer 1/2" condensor microphone.

In the experiment concerning the angular dependance of the insect echo the loudspeaker-microphone setup was continuously turned around the flying insect in a vertical and a horizontal plane. By filtering out the carrier frequency echoes were obtained only from moving targets (e.g. insect wings) and not from stationary targets like the insect body or surrounding walls.

To determine the species specific echo parameters echoes were recorded from insects looking towards (0 degree) and away (180 degree) from the loudspeaker and from insects with their longitudinal axis perpendicular to the loudspeaker (90 degree).

RESULTS

Information on the angular position

At all angular orientations both the spectrograms and the oscillograms show a periodic glint pattern which encodes the wing beat frequency (Fig. 1). Mostly one or two prominent glints occur per wing beat. In comparison to the oscillograms the spectrograms encode more clearly the wing beat frequency. Sometimes the oscillograms show no periodicity whereas the frequency glints are still visible in the spectrograms.

Fig. 1 illustrates that in the horizontal plane comparable angular orientations of the insect to the left and right side produce similiar echoes whereas in the vertical plane opposite positions of the loudspeaker produce a mirror image of the spectrograms relative to the carrier frequency. In the horizontal plane a mirror image is also found at left and right side at positions with the angular orientation and 180 degree - .

In the horizontal and vertical planes the spectrograms change gradually in a systematic way when moving around the insect. In the spectrogram obtained from the 0 degree position most energy is situated below the carrier frequency. By turning the loudspeaker in a horizontal plane around the flying insect the energy is slowly moving to higher frequencies and is symmetrically ditributed at the 90 degree position. Further turning results in a spectrogram with most of its energy above the carrier frequency at the 180 degree position.

Species specific information

The most evident insect specific information in the echo is the wing beat frequency as it is encoded in the interval between glints (Fig. 2).

At a given wing beat frequency the width of the spectral broadenings is a cue for the wing length. The longer the wings the higher is the velocity of the wings and consequently the corresponding Doppler effect.

The amplitude glint shows a different angular dependence at different insect species. Except for beetles and flies all other insect species measured so far show a glint intensity at 90 degree that is higher than that at 0 degree or 180 degree. In two beetles and two flies the amplitude of the glint was essentially higher at 0 degree and 180 degree than at 90 degree. This must be caused by different wing movements. Thus information on type of wing movement is also contained in the echo.

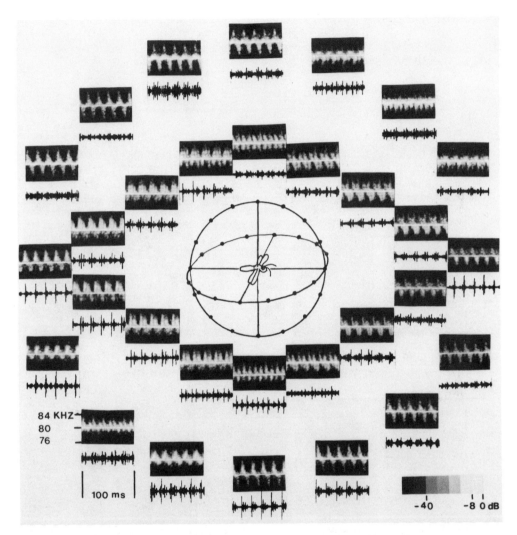

Fig. 1. Echoes (spectrograms and oscillograms) of Melolontha melolontha (Coleoptera). Echoes were recorded from 16 equally spaced (22.5 degree apart) positions on a vertical (arranged in the outer circle) and a horizontal circle (arranged in the inner circle) around the flying insect. Each echo corresponds to a dot on the orbits of the insect. Dots indicate the position of the loudspeaker-microphone setup.

Diptera have one pair of slender and stiff wings and spectral broadenings of short duration. In contrast Lepidoptera have two pairs of large wings twisting during wing beat and spectral broadenings of long duration.

DISCUSSION

According to the glint hypothesis (Schnitzler et al., 1983) spectrograms of echoes from opposite positions of the scattering wing should show a mirror image relative to the carrier frequency. For if one side produces a positive Doppler shift the other side should produce a negative Doppler shift. Since this study demonstrates this symmetry the glint hypothesis is confirmed.

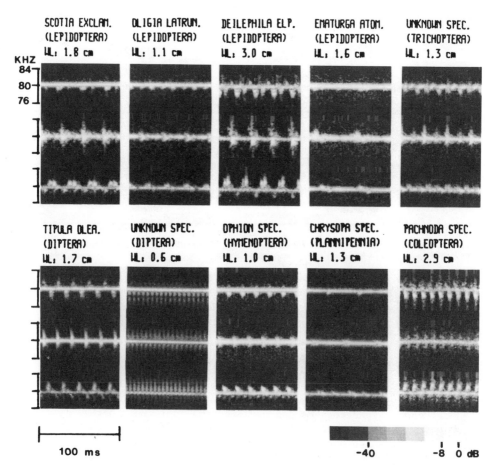

Fig. 2. Spectrograms of 10 insects (6 orders). Each insect was oriented at 0 degree (upper trace), 90 degree (middle trace) and 180 degree (lower trace) relative to the sound source. WL = winglength.

The present study provides evidence that the echo contains some information on the wing beat type. The results can be explained by the following hypothesis: Insects which move their wings mostly vertically (Lepidoptera-type) produce strong echoes at a position perpendicular to the impinging sound waves (90 degree) and echoes of lower intensity at 0 degree and 180 degree. In contrast insects which move their wings mostly horizontally (Coleoptera-type) produce strong echoes at a position looking towards or away from the sound source. It should be noted that an intermediate wing beat type also occurs (e.g. Ophion, Hymenoptera).

If one relates the wing structure of the Diptera and the Lepidoptera with their spectrograms the different durations of the broadenings can be explained: The slender and stiff wings of the Diptera act like one single scattering area. Once per wing beat they are in an optimal scattering position thus producing one single glint. In contrast the large wings of the Lepidoptera twisting during a wing beat act like an array of many small scattering areas thus producing many rapidly occuring glints.

In summary, the insect echo contains insect specific information on the wing beat frequeny, the wing length, the wing beat type and probably the wing structure. If bats use this information and moreover non-acoustic parameters from insects like occurence, seasonal appearence or behavior (e.g. reaction on ultrasonic sound), they should be able to identify their prey in a sufficient manner.

REFERENCES

Machmerth, H., Theiss, D., and Schnitzler H.-U., 1975, Konstruktion eines Luftultraschallgebers mit konstantem Frequenzgang im Bereich von 15 KHz bis 130 KHz., Acoustica 34, 81-85.
Schnitzler, H.-U., Henson, O. W., Jr., 1980, Performance of airborne animal sonar systems: I. Microchiroptera, in: Animal sonar systems. Nato advanced study institute series (A) 28, 109-181, R.-G. Busnel, and J. F. Fish, ed., Plenum Press, New York.
Schnitzler, H.-U., Menne, D., Kober, R., and Heblich, K., 1983, The acoustical image of fluttering insects in echolocating bats, in: Neurophysiology and behavioral physiology, F. Huber, and H. Markl, ed., Springer, Berlin Heidelberg New York.
Trappe, M., 1982, Verhalten und Echoortung der Grossen Hufeisennase beim Beutefang., Dissertation, University of Tübingen.

ENCODING OF NATURAL INSECT ECHOES AND SINUSOIDALLY MODULATED STIMULI BY NEURONS IN THE AUDITORY CORTEX OF THE GREATER HORSESHOE BAT, RHINOLOPHUS FERRUMEQUINUM

Joachim Ostwald

Lehrbereich Zoophysiologie
Auf der Morgenstelle 28
D-7400 Tübingen, FRG

Greater Horseshoe Bats use their long constant frequency (CF) echolocation sounds to distinguish between insect echoes and background clutter. Only insects flapping their wings are persued (Trappe and Schnitzler, 1982). Flying insects produce a species specific complex pattern of frequency and amplitude modulations in the rhythm of wingbeat (acoustical glints; Schnitzler et al, 1983) that is superimposed on the constant carrier frequency. Somehow these modulations seem to contain the cues that are used by the bats for prey classification and identification.

Transmitter and receiver system of Greater Horseshoe Bats are especially suited to encode those modulations. By Doppler-shift compensation the frequency of the CF-part of their echoes is kept constant in a small frequency band (Schnitzler, 1968; Trappe and Schnitzler, 1982). In the auditory system this frequency band is overrepresented forming an analysis window (reviewed in Schnitzler and Ostwald, 1983). Here units show extremely sharp tuning. In the auditory cortex (AC) of the Horseshoe Bat a large area (CF-area) is devoted to the processing of these frequencies (Ostwald, 1984). The detailed modulation structure of echoes from flying insects depends on the insect species and the relative angle of the insect (see Kober, this volume). In order to look for specialisations of the auditory system to the processing of those echoes single units in the CF-area of the AC of Greater Horseshoe Bats were tested for their responses to natural echoes of flying insects from different species and different directions. In Fig. 1 the response of two units to various stimuli is depicted. Both units responded with phase-locked discharges in the rhythm of wingbeat to most of the stimulus combinations when wing beat frequencies of the insects were 40 and 50 Hz respectively (center two columns). Units showed no phase-locking, however, when the echoes of a fly were presented that had a wing beat frequency of 180 Hz (right column) and only unit B synchronized to the echoes of Ematurga (wingbeat frequency: 20 Hz; left column). Amplitude glints in the echoes that are shown in the oscillograms in Fig. 1 were always combined with simultaneous widenings in echo frequency, but in some cases frequency modulations without pronounced amplitude glints were present (eg. in the echoes of Tipula, 180 degrees). Unit B showed a phase-locked discharge to this part of the stimulus. At least for some units in the auditory cortex frequency modulations in the echoes of flying insects seem to be more important cues than amplitude peaks.

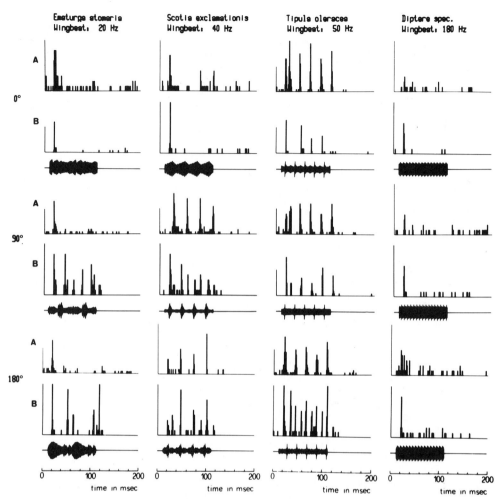

Fig. 1. Response pattern of two cortical neurons (A, B) to the natural echoes of 4 insect species with different wing beat rate (columns). Echoes were recorded from different angles (0 degree: frontal view). Oscillograms of the echoes are given below the PST-histogramms. Acoustic delay (2 ms) and latency of the unit (about 10 ms) is not corrected. Additionally to the amplitude peaks there were pronounced frequency modulations in the echoes which not allways corresponded to the amplitude peaks (carrier frequency = best frequency; 20 stimulus presentations; bin width: 1 ms; calibration bar: 10 spikes/bin)

During pursuit the carrier frequency of insect echoes varies depending on insect flight speed (Trappe and Schnitzler, 1982). Insect echoes were therefore presented with a carrier frequency that was systematically varied. Units showed phase-locked responses when the carrier frequency was within a range of several kHz around the units' best frequency (BF). In some units, however, synchronisation could only be elicited when the carrier frequency was below BF and the spectral widenings were crossing the low frequency slope of the unit's tuning curve.

In order to test which encoding mechanism could possibly cause the phase-locked response to complex modulations in the insect echoes the units' response to sinusoidal frequency (SFM) and amplitude (SAM)

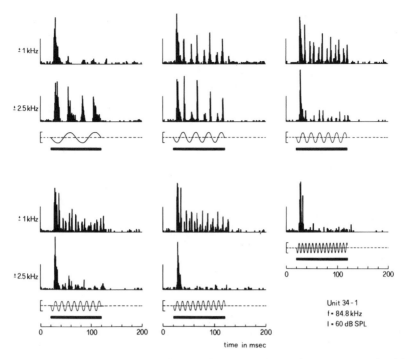

Fig. 2. Synchronization behavior of a neuron for SFM stimuli with different modulation frequencies (20, 40, 60, 80, 100, 150 Hz) and different modulation depths ($\pm$ 1 kHz, $\pm$ 2.5 kHz). Best phase-locked discharges for $\pm$ 1 kHz were found for 40 and 60 Hz modulation frequency. (Center frequency = best frequency; 50 stimulus presentations; bin width: 1 ms; calibration bar: 20 spikes/bin)

modulated stimuli was tested. 83% of the units showed a phase-locked response to SFM stimuli. The degree of synchronisation depended on the modulation depth of the stimulus. In most units minimal modulation depth had to be $\pm$ 250 Hz. In the lowest case a frequency variation of $\pm$ 150 Hz was necessary.

The amount of synchronisation was tested for modulation frequencies between 20 Hz and 200 Hz. Best phase-locked response was found in a frequency range between 40 and 70 Hz but it depended on the modulation depth used (Fig. 2). The upper limit of phase-locking was between 80 Hz and 150 Hz. In no unit recorded there was a synchronisation of discharge for modulation frequencies above 150 Hz.

When the center frequency of an SFM stimulus (corresponding to the carrier frequency in insect echoes) was changed relative to the units' best frequency two types of response characteristics could be distinguished. Symmetrical units showed phase-locked discharge as long as the stimulus touched the pure tone tuning curve within its cycle. In asymmetrical units discharge was only elicited when the stimulus swept through the low frequency slope of the tuning curve. Phase-locked responses to SAM stimuli are less common than to SFM stimuli. Only 30% of the units showed a phase-locked response even with modulation indices of up to 100%. To elicit synchronized response a minimal modulation index of 60% to 80% was necessary. Like with SFM stimuli modulation frequencies between 40 Hz and 70 Hz were most effective.

485

CONCLUSIONS

In the Greater Horseshoe Bat's auditory cortex the complex frequency and amplitude modulations in the echoes of flying insects are encoded in the neuronal responses. Like in the inferior colliculus (IC; Schuller, 1984) no units were found that selectively responded to a single insect species or a single echo pattern. When using SFM or SAM stimuli units in the AC only responded to a more limited range of modulation frequencies than in the cochlear nucleus (Vater, 1982) and the IC (Schuller, 1979; Pollak and Schuller, 1981). This more limited frequency range emphasizing 40 Hz to 70 Hz corresponds well to the wing beat frequency of nocturnal moths which are potential prey to horseshoe bats. The same holds true for the minimal modulation index necessary to elicit phase-locked responses. In SFM and SAM stimuli they are much smaller in lower levels of the auditory system than in the AC. Response properties of cortical neurons seem to be limited more to the behaviorally relevant stimulus range of prey recognition and do not cover the wide range like it is found in more peripheral stations of the auditory system.

ACKNOWLEDGEMENTS

Supported by DFG, SFB 307. I thank Dr. Menne for constructing the equipment to reproduce insect echoes from the computer and R. Kober for supplying tape recordings of insect echoes.

REFERENCES

Ostwald, J., 1984, Tonotopical organization and pure tone response characteristics of single units in the auditory cortex of the Greater Horseshoe Bat. J.Comp.Physiol. 155, 821-834

Pollak, G.D. and Schuller, G., 1981, Tonotopic organization and encoding features of single units in the inferior colliculus of horsehoe bats: Functional implications for prey identification. J.Neurophysiol. 45, 208-226

Schnitzler, H.-U., 1968, Die Ultraschall-Ortungslaute der Hufeisen-Fledermäuse (Chiroptera-Rhinolophidae) in verschiedenen Orientierungssituationen, Z.vergl.Physiol. 57, 376-408

Schnitzler, H.-U., Menne, D., Kober, R., and Heblich, K., 1983, The acoustical image of fluttering insects in echolocating bats, in: "Neuroethology and Behavioral Physiology", Huber, F. and Markl, H. (eds.), Springer Verlag, Berlin, pp. 235-250

Schnitzler, H.-U. and Ostwald, J., 1983, Adaptations for the detection of fluttering insects by echolocation in horseshoe bats, in: " Advances in Vertebrate Neuroethology", Ewert, J.-P., Capranica, R.R., and Ingle, D.I. (eds) Plenum Press, New York, pp. 801-827

Schuller, G., 1979, Coding of small sinusoidal frequency and amplitude modulations in the inferior colliculus of 'CF-FM' bat, Rhinolophus ferrumequinum, Exp.Brain Res. 34, 117-132

Schuller, G., 1984, Natural ultrasonic echoes from wingbeating insects are encoded by collicular neurons in the CF-FM bat, Rhinolophus ferrumequinum, J.Comp.Physiol. 155, 121-128

Trappe, M. and Schnitzler, H.-U., 1982, Doppler-shift compensation in insect-catching horseshoe bats, Naturwissenschaften 69, 193-194

Vater, M., 1982, Single unit responses in cochlear nucleus of horsehoe bats to sinusoidal frequency and amplitude modulated signals, J.Comp.Physiol. 149, 369-388

DO SIGNAL CHARACTERISTICS DETERMINE A BAT'S ABILITY TO AVOID OBSTACLES?

Albert S. Feng and Karen Tyrell

Department of Physiology and Biophysics
524 Burrill Hall
University of Illinois
Urbana, IL 61801 (USA)

Bats of the suborder Microchiroptera navigate and hunt for prey chiefly by means of acoustic perception of the echoes of their own sonar signals. Echolocating bats emit species-specific sonar signals. In a compressive medium such as air, different sounds have different propagation, reflective and refractive properties. As such, different sounds, along with the different analyzing capacity of the auditory system, dictate the performance of bats in various hunting situations.

Sonar signals of the Microchiroptera from several global regions have been analyzed in detail. Generally, they are either short, broad-band downward frequency-modulated (FM), or short or long constant-frequency (CF) signals (Simmons et al., 1975; Fenton, 1984). Most CF signals terminate with a brief downward FM component and are thus called CF/FM signals. The type of signal a bat uses has been shown to influence the bat's ability to resolve target range (Simmons, 1973). Bats emitting broad-band sonar signals, such as Eptesicus fuscus, typically can resolve target range better than those emitting narrow-band signals. This difference is attributed to the variation in the autocorrelation function of the two signal types.

Until recently, few studies have attempted to relate the echolocation performance of a bat other than target range resolution and localization with the signal the bat emits. The relationship between signal structure and target detection and discrimination of target features has not been systematically studied. This is surprising in view of the fact that target detection capability, which was among the first aspects of echolocation behavior studied, can be assessed using the wire-avoidance technique developed by Griffin many years ago. Jen and his colleagues recently studied the wire-avoidance behavior of a number of bat species and noted that CF/FM bats such as Pteronotus parnellii perform somewhat better than do FM bats such as Eptesicus fuscus and Myotis lucifugus (Jen et al., 1980; Jen and Kamada, 1982). However, it is not clear if the difference in behavioral performance is directly related to the type of signals these bats produce. Suthers and Thompson (1987) have compared the threshold for wire detection of various echolocating bats and birds, and found that the threshold is related to the wavelength of maximum energy in the signal. Sounds of shorter wavelength permit better detection thresholds, i.e., smaller minimum wire diameter, than do longer wavelengths. The comparative data therefore imply that the absolute frequency dictates the detection threshold. Another facet of echolocation behavior influenced by signal frequency is the target detection range. High-frequency signals decay more rapidly as they propagate from a source than do low-frequency signals. Thus there is a trade-off between the maximal target detection range and the size of a detectable target.

Additionally, it has been suggested (Fenton, 1984) that narrow-band shallow FM signals are well suited for detection of targets whereas broad-band steep FM signals are suited for gathering information regarding precise target characteristics as well as target range. Ground-gleaning Megaderma lyra generally produces low-intensity brief multiple-harmonic FM signals. The broad bandwidth is thought to be essential for discriminating fine prey features by means of spectral cues. However, for a fixed power-bandwidth function of a vocal apparatus, generated broad-band signals will be less intense than narrow-band signals. Therefore, echolocation with broad-band signals is restricted to a shorter range than it is with narrow-band signals. Here again, there is a trade-off between the various attributes of echolocation performance. Other species which forage in the open space tend to use long CF (or shallow FM) sonar signals while cruising to optimize long-range detection (see also Griffin and Thompson, 1982). During the approach phase, these same bats will instead emit brief FM sounds for determining precise target feature and range information.

In Malaysia, we used mist-nets to capture a variety of Microchiroptera. Mist-netting was done in both densely forested and open areas. The frequency of capture provides an indication of how well bats of Malaysia detect and avoid the mist-net, a behavioral act similar to wire-avoidance. We were impressed by the fact that while we were able to repeatedly capture some species (Table 1), others consistently avoided the mist-net. Inspection of the sonagrams (Figs. 1 and 2) shows that specimens captured included both CF/FM and FM bats. With one exception, all CF/FM bats captured emit signals with a CF frequency of 62 kHz or less. On the other hand, FM bats emitting high as well as low frequency signals were captured. Most notable are the flat-headed bats (Tylonycteris robustula and T. pachypus) which emit broad-band high-frequency FM sonar signals. Our records show that these bats were captured in large numbers. This is in contrast to the high-frequency CF/FM bats which were frequently observed flying close to our mist-nets without contacting them. Using a bat-detector, we noted that CF frequencies of these bats were above 78 kHz. Avoidance of the mist-net by CF/FM bats has also been observed in the field by other investigators (Grinnell, Menne, Neuweiler, Pettigrew, pers. comm.).

The difference in capture frequency between various species might simply be attributed to differences in local population density or flight behavior. While we cannot rule out these hypotheses, the fact flat-headed bats were captured in various different locations indicates that an alternative hypothesis should also be considered. In agreement with data from Jen's laboratory, our data suggest that high-frequency CF/FM bats, which were abundant where we netted, more effectively avoid obstacles than do FM bats emitting signals in the same frequency range. Perhaps factors affecting mist-net avoidance include sonar signal characteristics or the analytic capacity of the bat's nervous system. As far as the sonar signal is concerned, there may be several contributing factors: design of the sonar signal, signal duration, intensity, frequency bandwidth and absolute frequency. Most FM bats captured emit brief broad-band multiple harmonic signals with a bandwidth of 48-64 kHz whereas the bandwidth of CF/FM bats is consistently less than 11 kHz. While the FM and CF/FM bats captured emit signals in an overlapping fre-

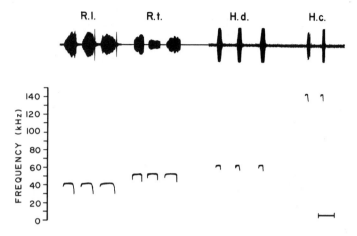

Fig. 1 Oscilloscope traces (top) and sonagrams (below) of echolocation signals of Malaysian CF/FM bats which were captured in standing mist-nets. The calibration time bar represents 50 msec. R.l. = Rhinolophus luctus, R.t. = Rhinolophus trifoliatus, H.d. = Hipposideros diadema. H.c. = Hipposideros cervinus.

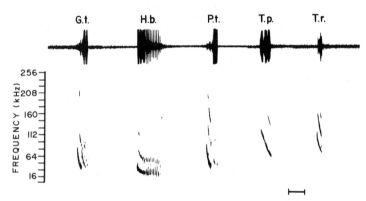

Fig. 2 Oscilloscope traces (top) and sonagrams (below) of echolocation signals of Malaysian FM bats which were captured in standing mist-nets. The calibration time bar represents 6.25 msec. G.t. = Glischropus tylopus, H.b. = Hesperopterus blanfordi, P.t. = Pipistrellus tenius, T.p. = Tylonycteris pachypus, T.r. = Tylonycteris robustula.

quency range, the most frequently captured bats were FM bats with the highest frequency signals (see Table 1). Both the broad frequency band-width and high absolute frequency of the Tylonycteris sonar signals should, according to existing hypotheses, enable these bats to resolve both range and target features accurately (Fenton, 1984; Simmons and Stein, 1980). The high frequency of the signal should also permit excellent target detection because short wavelength sounds are effec-tively reflected from small targets (Suthers and Thompson, 1986). Our data on Tylonycteris do not follow these predictions. There are several possible explanations for this disparity. One possibility is that these small bats may at times rely on spatial memory during flight in order to conserve the energy required by echolocation. Alternatively, broad-band high-frequency FM echolocation signals may be limited to a very short range due to low emission intensity as well as rapid attenuation of the sonar signals. The animals therefore may not have sufficient time to avoid capture once the net has been detected.

Is the high frequency of capture of Tylonycteris related to the type of signal they produce? If this is the case, is it because of the properties of the signal itself or the processing of signals by the ner-vous system? Or, are ecological factors and flight behavior primarily responsible for the difference in capture frequency of different bat spe-cies? We do not have conclusive answers to any of these questions. Similarly, we do not have satisfactory explanations for the low incidence of capture of high-frequency CF/FM bats. Our simple field observations point to the need for conducting more rigorous behavioral experiments to directly investigate these questions.

Table 1

Sonar signal characteristics of Malaysian Microchiroptera. Signals were obtained from individual bats housed in plastic mesh cages. Calls (spon-taneously emitted or vestibularly evoked) were taped using a Racal Store-4D tape recorder and subsequently analyzed on a Kay 7800 Digital Sona-Graph with a 6078A frequency translator. N is the number of bats of each species captured in standing mist-nets.

Species	(N)	Avg. bandwidth* (kHz)	Peak* or CF freq. (kHz)	Duration (ms)
Glischropus tylopus	2	49	83	1.3
Hesperopterus blandfordi	2	24	48	3.1
Pipistrellus tenuis	1	56	91	1.3
Tylonycteris pachypus	>12	64	128	2.5
Tylonycteris robustula	>12	48	120	1.4
Hipposideros cervinus	1	11	142	8.0
Hipposideros diadema	2	5	62	12.0
Rhinolophus luctus	2	10	42	40.0
Rhinolophus trifoliatus	3	11	51	25.0

* For FM bats, the average bandwidth and peak-frequency listed represent those for the dominant element of the echolocation pulses.

Acknowledgment. We wish to thank Prof. Yong Hoi-Sen and the staff of the Department of Zoology at the University of Malaya for their hospitality and assistance during our visit to Malaysia. The study was supported by a grant from the Research Board of the University of Illinois, National Institute of Health BRSG grant RR7030 and National Science Foundation grant BNS 85-11055.

LITERATURE

Fenton, M. B., 1984, Echolocation: Implications for ecology and evolution of bats, Quart. Rev. Biol., 59: 33-53.

Griffin, D. R., and Thompson, D., 1982, High altitude echolocation of insects by bats, Behav. Ecol. Sociobiol., 10: 303-306.

Jen, P. H.-S., and Kamada, T., 1982, Analysis of orientation signals emitted by the CF-FM bat, Pteronotus p. parnellii and the FM bat, Eptesicus fuscus during avoidance of moving and stationary obstacles, J. Comp. Physiol., 148: 389-398.

Jen, P. H.-S., Lee, Y. H., and Wieder, R. K., 1980, The avoidance of stationary and moving obstacles by little brown bats, Myotis lucifugus, In "Animal Sonar Systems", R. G. Busnel and J. F. Fish, eds., pp. 917-919, Plenum Press, New York.

Simmons, J. A., 1973, The resolution of target range by echolocating bats, J. Acous. Soc. Amer., 54: 157-173.

Simmons, J. A., Howell, D. J., and Suga, N., 1975, Information content of bat sonar echoes, Amer. Sci., 63: 204-215.

Simmons, J. A., and Stein, R. A., 1980, Acoustic imaging in bat sonar: Echolocation signals and the evolution of echolocation, J. Comp. Physiol., 135: 61-84.

Suthers, R. A., and Thompson, D. B., 1987, Sensitivity of echolocation by the oilbirds Steatornis caripensis: Relationship between detection thresholds of bats and birds (In preparation).

GREATER HORSESHOE BATS LEARN TO DISCRIMINATE SIMULATED

ECHOES OF INSECTS FLUTTERING WITH DIFFERENT WINGBEAT RATES

Gerhard von der Emde

Lehrbereich Zoophysiologie
University of Tübingen
7400 Tübingen

INTRODUCTION

If a cf-fm bat echolocates a fluttering insect, the cf-portion of the returning echo contains distinct amplitude and frequency modulations. A noteworthy feature of such an insect echo is the so called acoustical glint (Schnitzler et al. 1983), represented by a sudden amplitude peak and spectral broadening. A glint is produced each time an insect wing stands perpendicular to the impinging sound waves. Because there is usually one glint per wingbeat, the glint frequency encodes the wingbeat rate of the insect. In addition to information about wingbeat rate, the echoes of fluttering insects also contain information about the angular orientation of the insect, the insect's size and other information characterizing the insect species (Kober, this conference).

It has been shown that Rhinolophids (Schnitzler and Flieger 1983) using long cf-fm sounds and Hipposiderids (von der Emde and Schnitzler, in press) with short cf-fm calls are able to perceive the echo modulations caused by fluttering prey. This permits the detection of flying insects even against strong background clutter. Research findings indicate that several species using cf-fm echolocation calls hunt only fluttering prey, either flying or sitting on a substrate (e.g. Bell and Fenton, 1984; Goldman and Henson, 1977; Trappe, 1982).

The species-specific fluttering target information in the cf-echoes could also be used to learn more about the potential prey. Wingbeat rate for example is a constant quality of many insect species (Sotavalta, 1947). If the bats could extract wingbeat rate and maybe some additional information about the insect species from the perceived echo, this would allow them to select their prey (Schnitzler and Henson, 1980).

This study addresses the question if Rh. ferrumequinum, a bat with long cf-fm signals, is able to discriminate between insects with different wingbeat rates.

METHODS

Three Greater Horseshoe bats (Rhinolophus ferrumequinum) were trained in a two-alternative forced-choice procedure to discriminate between the echo of a flying insect fluttering at a rate of 50 Hz (positive stimulus) and the same insect fluttering at a lower rate (negative stimulus). Bats indicated selection of a stimulus by moving towards it along the arms of a U-shaped hanging stand. When the bats selected the positive stimulus they received a food reward. A variant of the method of limits was used to determine the discrimination threshold (75 % correct choices) for wingbeat rate.

The echo stimuli were produced in the following way: A fluttering insect (in this case a Tipula oleracea) was mounted in the acoustical beam (80 kHz) of a ultrasound loudspeaker and the returning echoes were recorded. The information encoding one wingbeat was stored on a programable chip (EPROM). This stored wingbeat could be used to modulate the bats own pulses, thus producing a typical insect echo. Different insect wingbeat rates were produced by changing the sampling rate of the read-out unit. These simulated insect echoes were played back to the bat as stimuli. Using this method, the carrier frequency, the relative sound pressure level, and the pulse duration of the individual bat-calls were preserved.

The modified echolocation calls were played back to the bat through one of two speakers, located 120 cm from the two arms of the hanging stand. The stimuli were alternated between the left and right speakers, and only one speaker was activated at a time. The location of the positive stimulus was varied in a pseudorandom order (Gellermann, 1933).

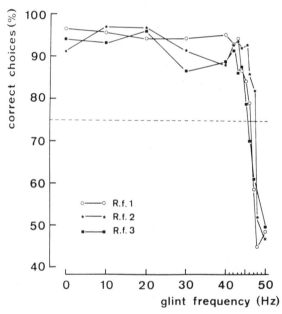

Fig. 1 Performance of three Rh. ferrumequinum discriminating between an insect echo with a wingbeat rate of 50 Hz (S+) and an insect echo with a lower wingbeat rate (S-). The abcissa shows the wingbeat rate (= glint frequency) of S-. Each point consists of at least 100 decisions by the bat.

RESULTS

The artificial insect echoes played back to the bats showed the typical amplitude and frequency structure of a cf-echo modulated by a fluttering insect. A change in insect wingbeat rate produced changes in the glint frequency, the spectral bandwidth and the glint duration of the echoes. With a lower wingbeat rate the glint frequency decreased, the spectral bandwidth became narrower, and the duration of the glints became longer.

After the bats had learned to fly to the middle of the hanging stand and to move to either side for a food reward, they learned very quickly to select the 50 Hz insect echo. Only a few Hz above threshold, the reaction time increased and the performance of the bats declined rapidly towards chance level. The 75% correct criterion was reached at a negative stimulus of 45.4, 46.2, and 47.2 Hz for the three bats (Figure 1).

After the threshold curves were established, the animals were tested in additional experiments with higher frequency stimuli. Either a 60 Hz or 70 Hz stimulus was paired with the previously positive 50 Hz stimulus. Selection of both the 50 Hz and the higher frequency stimulus was rewarded. In this situation two bats still mainly selected the insect fluttering with a wingbeat rate of 50 Hz, although their performance deteriorated in comparison with the earlier experiments. The third bat preferred the faster fluttering insect over the previously preferred 50 Hz stimulus (Table 1).

During all discrimination tasks the echolocation behavior of the bats changed when the fluttering phantom targets were switched on. The bats increased the duty-cycle of sound emission by up to 20%, even though different strategies for doing so were used. Two bats reduced the interpulse-intervals of their calls and started to emit many double-sounds, i.e. two calls following immediately after one another, followed by a longer interval. The duration of individual pulses remained constant or was slightly reduced (Figure 2a). The third bat increased its duty-cycle by emitting distinctly longer calls. The interpulse-intervals were only slightly reduced and this bat emitted only few double-pulses (Figure 2b).

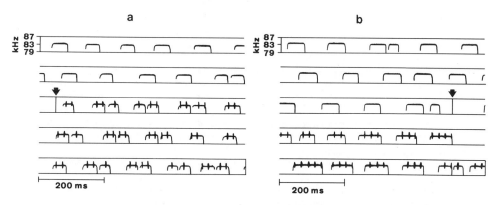

Fig. 2 Echolocation behavior of two Rh. ferrumequinum before and after the onset of the fluttering target echoes. Arrows = Onset of the stimulus. Before stimulus onset the sonagrams of the calls emitted by the bat, after stimulus onset the modified calls played back to the bat.

Table 1 Performance of the three bats using a S- with a higher wingbeat rate than 50 Hz. Numbers are percent responses to the 50 Hz stimulus (S+).

	negative stimulus	
	60 Hz	70 Hz
bat 1	26.0 %	23.6 %
bat 2	80.4 %	81.7 %
bat 3	66.7 %	77.3 %

DISCUSSION

At an insect wingbeat rate of 50 Hz, Rhinolophus ferrumequinum is able to perceive differences in wingbeat rate of only 3-5 Hz. What are the possible cues for the bats to solve the task of this experiment?
1) The bats might 'count' glints contained in the echoes, or measure the period between them. Because wingbeat rate is encoded by glint frequency, this would give them information about wingbeat rate.
2) They might measure the bandwidth of the spectral broadenings caused by the moving insect wings. The spectrum of the perceived echo is broader if the wingbeat rate is higher.
3) The bats might measure glint duration. With the method used in this experiment to produce the insect echoes, glint duration is a function of wingbeat rate. It is shorter if the wingbeat rate is higher.
Up to now it is not possible to determine which of the available cues were used by the bats in this experiment to discriminate between different wingbeat rates. Under natural conditions the measurement of wingbeat rate is probably achieved by the first method. The evaluation of the bandwidth of the spectral broadenings or the measurement of glint duration are not reliable factors for determining wingbeat rate, because they depend very much on wing size and the type of wingmovement of the insect.

The results of the experiments using a 60 and 70 Hz insect echoes reveal two different strategies of the bats to discriminate the stimuli: 1) Two bats learned to discriminate the characteristics of the 50 Hz positive stimulus. They continued to select it even when the alternative stimulus was a faster beating insect echo.
2) The third bat had learned the concept that it had to move towards the faster beating insect. When an insect beating its wings at a rate of 60 or 70 Hz was presented, it generalized this concept and selected these stimuli over the previously positive 50 Hz stimulus.

The results of this study show that Rh. ferrumequinum can use fluttering target information to dicriminate echoes differing in wingbeat rate. The evaluation of the insect echoes might give them information about the kind of prey they are echolocating. This might allow them to select specific insects and to avoid others - using wingbeat rate as a cue. Even though the insect echo of a cf-echolocation call contains information about several target features (angular orientation, insect size and so on) many echo parameters are very much dependent on the angular orientation of the insect (Kober, this conference). Wingbeat rate however is a constant, which does not change with angular orientation. If echolocating bats are able to accurately measure this parameter, they possess a very useful tool to classify their prey.

There is also some other species-specific information besides wingbeat rate encoded in the fine structure of the echoes, which Rh. ferrumequinum is able to extract. Experiments showed that this bats are able to discriminate between different insect species fluttering at exactly the same wingbeat rate. That this ability is independent of the angular orientation of the insect was shown by the following experiment: Rh. ferrumequinum was trained to select one insect species, which had a specific angular orientation relative to the bat. When this insect species was then presented at different angular orientations, the bats still preferred it over other insect species (fluttering at the same wingbeat rate). From this results follows that Rh. ferrumequinum is able to extract the species-specific information contained in the echoes, regardless of the angular orientation of the prey.

ACKOWLEDGEMENTS

I especially thank Dieter Menne, who designed and built the electronic apparatus to produce the synthetic insect echoes. I am also indebted to Cynthia Moss for her helpful comments and criticisms while preparing this manuscript. This research was supported by the 'Deutsche Forschungsgemeinschaft' (SFB 307).

REFERENCES

Bell GP, Fenton MB (1984) The use of Doppler-shifted echoes as a clutter rejection system: the echolocation and feeding behavior of Hipposideros ruber (Chiroptera: Hipposideridae). Behav Ecol Sociobiol 15:109-114

Emde G vd, Schnitzler HU (1986) Fluttering target detection in Hipposiderid bats. J Comp Physiol A (in press)

Gellermann LW (1933) Chance disorders of alternating stimuli in visual discrimination experiments. J Genet Psychol 42:205-208

Goldman LJ, Henson OW Jr (1977) Prey recognition and selection by the constant frequency bat, Pteronotus parnellii. Behav Ecol Sociobiol 2:411-419

Schnitzler HU, Flieger E (1983) Detection of oscillating target movements by echolocation in the greater horseshoe bat. J Comp Physiol A 135:385-392

Schnitzler HU, Henson OW Jr (1980) Performance of airborne animal sonar systems: I. Microchiroptera. In: Busnel RG, Fish JF (eds) Animal Sonar Systems. Plenum Press New York, Nato advanced study institude series (A) 28:109-181

Schnitzler HU, Menne D, Kober R, Heblich K (1983) The acoustical image of fluttering insects in echolocating bats. In: Huber F, Markl H (eds) Neurophysiology and behavioral Physiology. Springer, Berlin Heidelberg New York. pp 235-250

Sotavalta O (1947) The flight tone (wing stroke frequency) of insects. Acta Ent Fenn 4:5-117

Trappe M (1982) Verhalten und Echoortung der Grossen Hufeisennase beim Insektenfang. Dissertation, University of Tübingen

PREDICTIVE TRACKING OF HORIZONTALLY MOVING TARGETS BY THE

FISHING BAT, NOCTILIO LEPORINUS

Karen A. Campbell and Roderick A. Suthers

School of Medicine and Department of Biology
Indiana University
Bloomington, IN 47405 U.S.A.

Bats hunting by sonar do not receive continuous target information; rather, the echo from each sonar pulse provides an acoustic bulletin by which they update their current perception of target range and position. This suggests that a hunting bat might keep track of moving prey by two general techniques. It might simply fly toward the last known position of the target (a non-predictive tracking strategy), or it could somehow predict the target's trajectory on the basis of known target parameters such as velocity, acceleration and position at the time of the most recent echo (a predictive strategy).

There is evidence that some echolocating bats employ a non-predictive tracking strategy when monitoring the location of a moving target. Masters et al. (1985) trained Eptesicus fuscus to acoustically follow a small styrofoam ball suspended in front of the bat as it sat on a platform. Based on its head aim and sonar emissions with respect to target position, this bat did not appear to be predicting target trajectory, but was simply aiming its head at the target's last known position. A flying bat searching for and pursuing prey may or may not employ the same strategy.

We have investigated the tracking strategies used by the fish-catching bat, Noctilio leporinus. These neotropical bats have disproportionately large feet with sharp, laterally compressed claws which they dip into the water to gaff fish near the surface (Bloedel, 1955). When feeding, fishing bats fly above the water, using their biosonar to detect surface disturbances associated with fish swimming just below the surface (Suthers, 1965). The characteristic sonar emission of N. leporinus consists of a long (10 ms) narrowband portion at approximately 60 kHz followed by an FM sweep down to about 30 kHz (Suthers, 1965). N. leporinus is able to evaluate the relative velocity of a target moving along the axis of its flight path, apparently with the steep FM component of the pulse, using range-rate information available from echoes of successive sonar emissions (Wenstrup and Suthers, 1984).

In order to investigate this bat's ability to track moving prey, an adult male N. leporinus was trained to dip its feet at a moving vertical wire supporting a food reward which projected above the surface of a rectangular pool, 4.3 m long by 2.3 m wide by 0.3 m deep. For each trial, the bat left its perch at one end of the flight room and made a single flight over the length of the pool, dipping at a target moving at a constant speed in either direction along an axis perpendicular to its flight path. To remove any cues that a wake might provide regarding the position of a submerged target moving through the water, the area of the pool between the bat's perch and the target was covered with fiberboard, a smooth flat surface at which N. leporinus will also dip. A submerged target, therefore, "disappeared" beneath the surface of the fiberboard. As the bat flew towards the target during each trial, it intersected two light beams projecting across the pool. Each of these beams activated a relay controlling one of two flashes on a camera mounted at the end of the pool in front of the bat. Intersection of the first light beam also triggered the disappearance of the target, so these flashes produced a double-exposure during each trial showing the bat's position as the target disappeared and then the position of the target as the bat dipped its feet. In each daily session the bat flew 35 trials, of which approximately 5 disappeared. On any one trial, the target was either stationary or was moving auross the pool at a constant speed in either direction, from left to right or from right to left. On moving trials, the target travelled at one of two speeds, 0.66 m/s or 1.25 m/s. The combination of these three variables (disappearance or non-disappearance, direction of movement, and target speed) was determined randomly for each trial.

The measurements obtained from each photograph are shown in Figure 1. The point of target disappearance is indicated at "A", and the bat's position, taken as the midpoint between its feet, is shown at "B". "C" represents the distance the bat is displaced from the point at which the target disappeared. "D" represents the point where the bat dipped with respect to the position of the submerged target at that instant. A scale along the edge of the fiberboard permitted measurements to be taken directly from the photographs. With reference to either the point of disappearance or the position of the moving target, a positive measurement indicates that the bat dipped ahead of the reference point, whereas a negative measurement indicates that it dipped behind the point of reference. For Figure 1, then, C is positive, but D is negative, as the bat's dip lagged behind the position of the submerged target.

The data are summarized in Table 1. The bat's position relative to the stationary disappearing control target was determined by assigning a positive value to dips displaced to the right of the target, and a negative value to dips displaced to the left. The resulting mean, therefore, indicates that this bat had a tendency to dip 1.4 cm to the right of the stationary disappearing target, a distance which is probably not biologically significant, as the feet of N. leporinus span approximately 7 to 8 cm during a dip. This bat successfully caught stationary non-disappearing targets even when its dip was displaced by 3.5 to 4.0 cm, as measured to the midpoint between the feet.

In all moving trials, at all target speeds, N. leporinus consistently dipped ahead of the point of target disappearance, suggesting that it allowed for continued movement of the submerged target. The position of the dip with respect to the point of disappearance (Figure 1, C) was statistically different for slow (0.66 m/s) and fast (1.25 m/s) moving targets ($p < 0.05$), with the dip placed farther ahead of the point of disappearance when the target was moving at the higher speed. There was no statistical difference between the accuracy of dips at the stationary control targets and those at moving targets (Figure 1, D).

A series of control experiments were conducted to test the possibility that <u>Noctilio</u> located the submerged target by listening to sounds associated with target movement.   Information about the position of the target could potentially be conveyed by the sounds of surface turbulence created as the target moved through the water, or by the sound of the sled on which the target moved along its track.   In order to eliminate these possible acoustic cues, we lowered the water level below the level of the track, so there was no surface disturbance directly associated with the movement of the target.   We also mounted a horizontal extension on top of the sled, extending parallel to, but not touching, the track.   This apparatus was all still hidden from the bat by the fiberboard covering the pool.   The disappearing target, presented

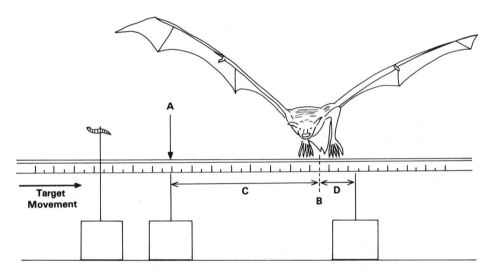

Figure 1.   Measurements taken from photographs of the bat
             dipping at a moving target. A – point of target
             disappearance; B – position of bat as measured to
             midpoint between feet; C – displacement of bat
             from point of target disappearance; D – bat's
             displacement from position of submerged target at
             time of dip.

on randomly determined trials, was mounted at the end of the extension, displaced 19.0 cm to the right of the sled. The non-disappearing target, presented in all other trials, was always mounted on the sled, so the bat had no opportunity to learn that a disappearing target was displaced a fixed distance from the sound of the moving sled. If <u>Noctilio</u> was utilizing sound generated by the moving target, it should <u>dip</u> at the position of the sled, rather than at the position of the disappeared target. There was no statistical difference between the position of the dips relative to the target under these conditions and that of dips obtained in previous trials (Table 1). These results clearly indicate that the bat was not using acoustic cues associated with target motion to determine the position of the target.

Table 1. Position of dips at disappearing targets [5]

Target Speed (m/s)	n		Bat's Position Relative to Disappearance (C) (cm)	Bat's Position Relative to Submerged Target (D) (cm)
Stationary[2]	19		+ 1.4 (4.7)	+ 1.4 (4.7)
0.66	21	R	+ 9.2 (0.8) *	+ 1.2 (1.2)
	17	L	+ 10.0 (0.7) *	+ 0.6 (0.7)
1.25	16	R	+ 15.6 (1.6) *	− 3.9 (1.5)
	18	L	+ 12.1 (1.2) *	− 4.9 (0.8)
Control for Use of Passive Acoustic Cues				
0.96	9	R	+ 13.2 (3.5) *	+ 1.1 (4.3)
	6	L	+ 12.4 (3.8) *	+ 0.9 (4.2)

[1]Sample mean (standard error); [2]Displacement to right and left of target assigned positive and negative values, respectively; n – number of trials; C and D refer to Figure 1. R – target moving left to right; L – target moving right to left; "+" – displacement in the direction of target movement; "−" – displacement opposite to direction of target movement. Asterisk indicates values statistically different from stationary control, $p < 0.001$.

The temporal pattern of sonar pulses emitted during trials in which the moving target disappeared differed from that preceding dips at a moving non-disappearing target. In the latter case, the bat increases its pulse repetition rate as it begins a flight over the pool, and decreases the initial narrowband component of its emissions such that it produces exclusively wideband pulses (Wenstrup and Suthers, 1984). The interpulse intervals (IPI), as measured from the end of one pulse to the beginning of the next, for sonar emissions during the 300 ms immediately preceding the bat's dip (Figure 2) were progressively reduced to a minimum of 4.5 ms when the target did not disappear until, in the terminal phase of a flight, the pulse repetition rate reached 185 pulses/s. During disappearing trials, the IPI was typically about 15 ms at the time the target disappeared, approximately 160 ms before the bat dipped its feet. The pulse repetition rate continued to increase to a maximum of 100 pulses/s for the next 80 ms, reaching a minimum IPI of 10 or 11 ms. During the final 80 ms before dipping at the position of the submerged target, the IPI rapidly increased, and the pulse repetition rate dropped to 30 pulses/s. The pulses emitted during the final 80 ms resembled those from the earlier approach phase of the trial, containing a prominent narrowband component. Despite the absence of a terminal buzz in these disappearing trials, N. leporinus executed an apparently normal, coordinated dip at the position of the submerged target.

A natural situation similar to this disappearing target experiment is the disappearance of a fish below the water surface, since N. leporinus is unable to echolocate submerged objects (Suthers, 1965). If

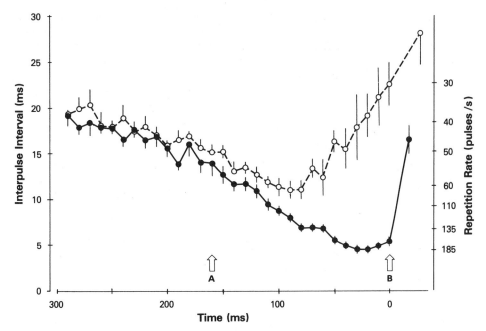

Figure 2. Interpulse intervals between sonar pulses emitted during final 300 ms prior to dipping at a moving target. Sonar emissions were recorded during 20 normal trials (solid dots) and 20 disappearing trials (open circles) at both target speeds, and pulses were pooled over 10 ms intervals. Error bars denote standard error of the sample mean. Scale of vertical axis is corrected for pulse duration. A - time of target disappearance in appropriate trials; B - time at which bat dipped (Time Zero).

a fish-catching bat has obtained sufficient information about the fish's movement and position before it disappears, the bat should be able to extrapolate the future position of the prey in order to catch it. Noctilio leporinus often catches fish from rippled, wavy, or choppy water surfaces. These experiments suggest that its success in accomplishing this task is improved by using a predictive tracking strategy. The experiments of Masters et al. (1985) were performed with stationary bats trained to follow a non-food item. It would be interesting to know whether Eptesicus continues to use a non-predictive strategy when actively pursuing insect prey, or whether under these conditions it too is able to predict the position of its moving target.

ACKNOWLEDGEMENTS

Supported by NSF grant BNS 82-17099 to RAS.

REFERENCES

Bloedel, P., 1955, Hunting methods of fish-catching bats, particularly Noctilio leporinus, J. Mammal. 36:390-399.
Masters, W.M., Moffat, A.M.J., and Simmons, J., 1985, Sonar tracking of horizontally moving targets by the big brown bat Eptesicus fuscus, Science, 228:1331-1333.
Suthers, R.A., 1965, Acoustic orientation by fish-catching bats, J. Exp. Zool., 158:319-348.

Wenstrup, J.J., and Suthers, R.A., 1984, Echolocation of moving targets by the fish catching bat, Noctilio leporinus, J. Comp. Physiol. A., 155:75-89.

DISCRIMINATION OF TARGET SURFACE STRUCTURE IN THE ECHOLOCATING BAT,

MEGADERMA LYRA

Sabine Schmidt

Zoologisches Institut der Universität München
Luisenstr.14, 8000 München 2, FRG

INTRODUCTION

Microchiropteran bats use echolocation for orientation in space and the pursuit of prey. For bats hunting near the ground or in dense foliage, a complex form of echo pattern recognition is implied due to background clutter. Texture information may be important in order to successfully track objects or identify palatable prey on the ground. The efficiency of echolocation for the discrimination of naturally occurring structures and the information processing involved are far from clear. A twofold approach is adopted here: (1) The performance in surface structure discrimination of the Indian False Vampire, Megaderma lyra, is established in a real target situation. (2) A phantom target experiment is presented where a well controlled "echo" input is provided to the auditory system of the bats, as will be crucial for theoretical analysis and as a starting point for more refined experiments.

(1) DISCRIMINATION PERFORMANCE USING OBJECTS WITH RANDOM SURFACE STRUCTURE

Material and Methods

Three M.lyra were successfully trained to fly towards two targets with different random surface structure offered in a forced two choice behavioral test. Target plates with a parabolic surface (ϕ 21.3 cm, focus about 1.48 m) were made of epoxide-resin. They were covered with one layer of quartz sand. Mean grain sizes (MGS) used were $\leqslant 0.4$ mm, 1.5 mm (range 1.0 mm - 2.0 mm), 2.5 mm (2.0 mm - 3.0 mm), 3.8 mm (3.0 mm - 4.7 mm) and 6.0 mm (4.8 mm - 8.0 mm), respectively. In all trials, targets of $\leqslant 0.4$ mm represented the rewarded reference. Experiments were performed in a sound-proof chamber (flight cage: 2.6 m x 1.6 m x 1.9 m) illuminated with a red darkroom light. Bats were trained to return to a fixed starting position at a distance of 1.48 m from the targets after completing each trial and decide for the rewarded reference target before takeoff (angular separation between target centers 32°). Correct choices were rewarded with mealworms. Two sets of targets each comprising all five grain size classes were mounted on two wheel mechanisms which permitted to present any sequence of plates. All test targets were paired with the references in a random order during one experimental session. For every series tested, data were pooled from several days (No. of trials n for any pair of plates n $\geqslant$ 30; i.e. 75% correct choices are

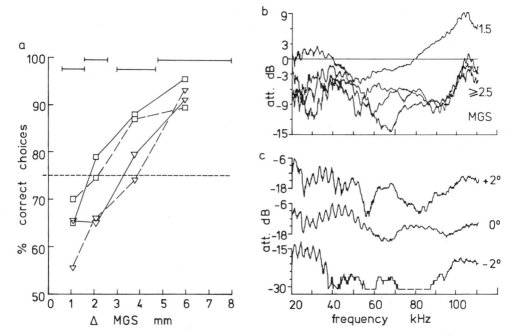

Fig. 1.   Performance of M. lyra in a texture discrimination task (two individuals, symbolized by triangles and rectangles, respectively) is
given in a. % correct choices are plotted as a function of the
differences in mean grain size (Δ MGS mm) between pairs of targets
offered. Solid lines connect values obtained in dim red light,
dashed lines those obtained from a control series made in complete
darkness. The horizontal dashed line marks 75% correct choices.
Horizontal bars indicate the range of scatter of the grain used.
In b, spectral attenuation (att. dB) relative to the reference
target is plotted for 0° deflection angle from plates of four
grain size classes. The horizontal line indicates 0 dB attenuation. Note that spectral attenuation for all frequencies is more
pronounced for MGS ≥2.5 mm than in the 1.5 mm target. For some
frequencies, this target produces even less attenuation than the
reference. In c, spectral attenuation of a 6 mm MGS target is
plotted relative to the emitted signal for +2°, 0° and -2°
yaw angles. The formation of sidelobes can be observed clearly.

significantly different from random choice with 1% error probability). Nine
series were completed for at least two animals with n ≥ 30 using different
target orientations (inverted 0° and 180° about the bat-target axis) and
configurations of the target sets.

     Energy spectra of echoes from each plate were determined at a distance
of 1.48 m within the 20 - 110 kHz range in 200 Hz steps. Pure tone pulses of
4 ms duration (1 ms rise/fall time) were emitted by an ultrasonic speaker
pointing at the center of the plates (peak amplitudes about 110 dB SPL).
Echoes were picked up by a B&K 1/4 inch microphone mounted under the speaker
as close as possible. No signal/echo overlap occurred. Maximum amplitudes of
the echo envelopes were stored in a LSI 11/23 computer (programs by M.Betz).
Echoes were recorded from plates at 0° position and 90° rolled, and yawed
in 2° steps (-6° to +6°) about the speaker-target axis. Spectral patterns
of the echoes were analysed either relative to the emitted signal or else
relative to the echo from the reference targets.

Results

The rewarded references (MGS ≤ 0.4 mm) are discriminated from the 6 mm
MGS targets in all series (81% to 100% correct choices). Targets of 3.8 mm
MGS are well above or close to threshold applying the 75% criterion (70% to
88% correct choices), whereas targets of 2.5 mm MGS are discriminated by one
animal in five series, only (70% to 84% correct choices for this bat, 60%
to 75% correct choices for the two others). Targets of 1.5 mm MGS are not
discriminated from the references (54% to 73% correct choices). One series
repeated in complete darkness indicates no difference in discrimination
performance (see Fig. 1a.).
In the real target situation, the bats discriminate grain size differences
of about 2 mm in randomly structured targets. As target presentation was
randomized and vision can be ruled out for the discrimination process this
performance can be exclusively attributed to audition.

Acoustical cues provided for target discrimination can be estimated
from the spectral attenuation patterns obtained for different targets.
Two features are prominent: (1) As can be expected for targets with smeared
texture, broadly tuned attenuation maxima appear. For a given deflection
angle relative to the speaker-target axis, their positions and amplitudes
vary with MGS. At 0° yaw, spectral attenuation measured for the rewarded
reference is between -3 dB and - 20 dB. Values measured rise from - 6 dB to
- 30 dB in 6 mm MGS targets. When summarized over the frequency range tested, increasing grain size produces a reduced echo intensity (see Fig. 1b.).
(2) Positions of the attenuation maxima on the frequency scale depend on
the angle of deflection. For targets with fine MGS (≤ 0.4 mm, 1.5 mm), echo
intensity is mainly concentrated in 0° direction and decreases gradually
within $\pm 4^{\circ}$ for the frequencies tested. For increased MGS, pronounced
sidelobes are measured varying in spectral content with the angle of deflection (see Fig. 1c.).

(2) DISCRIMINATION PERFORMANCE USING TWO-FRONT PHANTOM TARGETS

Material and Methods

The psychophysical procedure was designed in close analogy to the first
experiment. Three bats learned to compare phantom echoes from two loudspeakers in order to find out on which channel a rewarded reference target appeared. Sound emissions from the bats were picked up by a QMC microphone
13 cm below the head of the animal hanging at the starting position. They
were modified and played back through one of two ultrasonic speakers crea-
ting a phantom target at 1.34 m from the bat. Speakers used were flat within
± 2 dB between 30 kHz and 100 kHz; spectral amplitudes of the two speakers
did not differ more than 5 dB at any given frequency. The channel active at
a time was determined by electronically comparing the signal amplitudes from
two AKG microphones placed to the left and the right of the QMC microphone.
Phantom targets were situated 33 cm in front of the speakers. Given the
short echolocation call of M.lyra (500 µs duration), overlap between the
phantom echoes and any other echoes was avoided. To simulate two-front targets, signal input was continuously added to its delayed version and played
back to the bat. The internal delay could be varied in 1.294 µs steps. The
resulting interference patterns displayed sharply tuned attenuation peaks
exceeding 40 dB (see Fig. 2b). Maximum echo amplitudes of about 80 dB SPL
could be achieved. In the discrimination task, a rewarded reference with
t_{ref} = 7.77 µs internal delay was tested against targets with internal
delays t_{tst} ranging from 1.29 µs to 25.89 µs. Up to nine different test
delays were presented at random during one experimental session and data
were pooled for several days (No. of trials n per test target n $\geq$ 30). As
a control, a 7.77 µs internal delay was also offered among the unrewarded

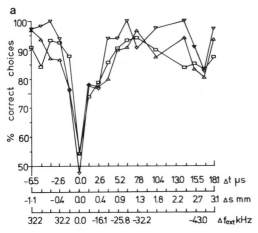 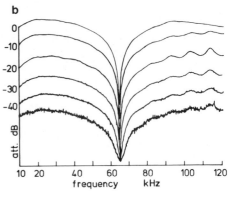

a

b

Fig. 2. In a, the performance of three M. lyra (symbolized by upright triangles, upside down triangles and rectangles) in choosing a rewarded reference with 7.77 μs internal delay is plotted as a function of delay differences (Δt μs) between this reference and 17 test targets. Data from each bat are linearly interpolated to make visible trends in the data points measured. The abscissa is rescaled twice: (1) in terms of depth differences (Δs mm) with $s = v_s \times t / 2$ ($v_s = 340$ m/s), interpreting the phantom echoes as coming from two planes, (2) by Δf_{ext} as frequency axis, since a difference in internal delays implies a shift in the first harmonic frequency extinctions given by $f_{ext} = 1/2t$. In b, the energy spectrum of a phantom echo (att dB) with $t_{ref} = 7.77$ μs is plotted relative to flat input for the dynamic range available.

test targets in all series. The internal delays of the targets offered were unknown to the experimenter during the sessions; therefore, any subjective bias can be excluded.

Results

The behavioral performance of the bats is given in Fig. 2a.. For identical internal delay ($t_{ref} = t_{tst} = 7.77$ μs) bats choose at random. Targets with internal delays of < 5.18 μs and > 11.65 μs are discriminated from the reference. Delays of 6.47 μs, 9.06 μs and 10.36 μs are close to threshold, i.e. the bats were able to discriminate delay differences of about |1.29| μs. When this performance is transformed into spatial depth differences of a real target, a resolution of |1.29| μs corresponds to a threshold value of 0.22 mm in depth discrimination. Although the bats were initially trained to decide between internal delay differences of 7.77 μs versus 15.53 μs, internal delays < 7.77 μs are discriminated from the reference without further training by all three animals. Two more features deserve comment: (1) There is a slight drop in performance around + 15.53 μs delay difference. For internal delays of 7.77 μs and 20.3 μs, the corresponding frequency extinctions at $v_s \times t_{ref}/2$ and $v_s \times 3t_{ref}/2$ are harmonically related. (2) The performance of the bats seems to deteriorate at smaller values of |Δt| for targets with internal delays < 7.77 μs than for those with internal delays > 7.77 μs. This asymmetry disappears when performance is read as a function of differences between the first harmonic frequency extinctions (Δf_{ext}).

510

CONCLUSIONS

The sonar information available to the bats in the two experiments described here can be estimated from the spectral reponse patterns of the targets offered. Smeared spectral patterns resulting from the complex properties of the real targets are opposed to sharply tuned, harmonically related notches in the spectra of the phantom targets. These target properties are reflected in a marked difference in discrimination performance when the depth discrimination threshold of about 2 mm in the real target experiment is compared to that of 0.22 mm in the phantom target approach. On the assumption that M. lyra makes use of its broadband sonar to exploit spectral cues in the echoes received, a functional hypothesis is provided for the information processing taking place in surface structure discrimination (see Beuter, 1980). Such a theoretical framework predicts several features which must arise in the performance of bats exposed to a two-front situation: (1) As the "depth" of a two front-target is proportional to its internal delay t and inversely proportional to the arising frequency extinctions f_{ext}, the number of spectral notches reduces and their spacing gets wider with decreasing depth. Therefore, the performance of the bats should improve - in the mm range - with decreasing depth of the reference. The threshold in depth resolution in the phantom target experiment of only 0.22 mm is obtained by using a reference of 1.3 mm depth. The threshold values found in other bat species with broadband echolocation calls of about 0.8 mm (Simmons et al., 1974, Habersetzer and Vogler, 1983) might result - apart from offering real targets with less distinct reflection properties - from using references of 8.0 mm and 4.0 mm depth. (2) Also, an asymmetry of the performance graph of the kind seen in the presented data can be predicted on account of the $f = 1/t$ relationship, if spectral information is processed by the bat. (3) Since the frequency extinctions occurring are harmonically related with $(2n - 1)$ x t for $n \in \mathbb{N}$, the reduced performance observed for a test target with $t_{tst} = 3t_{ref}$ can be conjectured.
A spectral hypothesis gives an easy explanation of the experimental results reported. Moreover, it has the anatomical and physiological evidence on its side, as frequency mapping and tonotopic organization are constituent features of the ascending auditory pathways, and the frequency discriminations required by the phantom target experiment are in the kHz range. On the other hand, a framework based on the processing of propagation delays per se has to cope with discriminated delay differences of 1.29 µs.

REFERENCES

Beuter, K. J., 1980, A new concept of echo evaluation in the auditory system
    of bats, in "Animal Sonar Systems", R.-G. Busnel and J. F. Fish, eds.,
    Plenum Press, New York.
Habersetzer, J., and Vogler, B., 1983, Discrimination of surface structured
    targets by the echolocating bat Myotis myotis during flight,
    J. Comp. Physiol., 152: 275.
Simmons, J. A., Lavender, W. A., Lavender, B. A., Doroshow, C. F., Kiefer,
    S. W., Livingston, R., Scallet, A. C., and Crowley, D. E., 1974,
    Target structure and echo spectral discrimination by echolocating
    bats, Science, 186: 1130.

A TIME WINDOW FOR DISTANCE INFORMATION PROCESSING

IN THE BATS, <u>NOCTILIO</u> <u>ALBIVENTRIS</u> AND <u>RHINOLOPHUS</u> <u>ROUXI</u>

Roald C. Roverud

Zoologisches Institut, Universität München
Luisenstrasse 14
D-8000 München 2, West Germany

Microchiropteran bats emit echolocation sounds that have structured patterns of frequency changes over time. Classes of frequency pattern have been observed among bat orientation pulses. Bat echolocation sounds generally consist of constant frequency (CF) and frequency modulated (FM) elements alone or in a combination of the two components. For example, in certain species (known as CF/FM bats), the echolocation sounds contain a CF component preceding the FM sweep; in some species the CF component is short (under 12 msec), while in others it is long (over 12 msec). The defined structures of echolocation sounds presumably reflex specific information processing requirements. One generally accepted requirement is that a broadband signal is necessary for accurate perception of target distance by neural measurement of the time interval between the emitted broadband event and a returning echo. Nevertheless, the essential elements and processing requirements of complex CF/FM echolocation sounds has mostly been a matter of conjecture. In this paper I describe experiments that demonstrate which structural elements of complex CF/FM echolocation sounds code target distance information and provide a mechanism regarding hcw this information is processed by the nervous system. These studies suggest that CF/FM bats use both the CF and FM components of their CF/FM echolocation sounds for the determination of target distance, with the onset of the CF component activating a gating mechanism that establishes a time window during which FM component pulse-echo pairs are processed for distance information.

In a psychophysical procedure bats of the species <u>Noctilio albiventris</u> (a short-CF/FM bat, Roverud and Grinnell 1985) and <u>Rhinolophus rouxi</u> (a long-CF/FM bat) were trained to discriminate a 5 cm and 8 cm difference in target distance respectively. Both species can discriminate these presented range differences with 90% or greater accuracy. During the discrimination trials <u>N. albiventris</u> emits 7-10 pairs of pulses each second. The first pulse is an 8 msec duration short CF signal that rises from about 71 to 75 kHz and then descends to 71 kHz. About 28 msec later the bat produces a short-CF/FM signal with an initial 6 msec CF component (again rising from about 71 to 75 kHz) and a terminal 2 msec FM sweep to about 57 kHz. The "CF" component of the calls of <u>N. albiventris</u> are actually shallow FM signals that seem to serve a narrowband function. During similar discrimination trials <u>R. rouxi</u> emits pulses at a repetition rate of about 10/sec. The bats usually emitted second harmonic FM/long-CF/FM pulses with an initial 2 msec upward FM sweep followed by an approximately 45 msec long CF component and terminating in a 2 msec downward FM sweep. The second harmonic CF frequency

513

varied between individuals from about 73-83 kHz, partly depending upon the natural population from which the individual originated. Individual bats, however, exhibited consistency in the frequency of their CF components. Both the initial upward and the terminating downward FM sweeps were about 10 kHz in bandwidth. Certin individuals occasionally emitted first harmonic FM/long-CF/FM pulses along with the more typical second harmonic pulses.

Artificial pulses, simulating the natural CF/FM echolocation sound of the bat and presented during the psychophysical trials, interfered with the ability of both N. albiventris (Roverud and Grinnell 1985) and R. rouxi to discriminate target distance. Interference was recognized when the performance of the bats in the distance discrimination trials was less than or equal to 75% correct responses. Systematic modification of the frequency and temporal structure of the artificial pulses resulted in orderly changes in the extent of interference. The artificial pulses were presented at an intensity of about 90 dB SPL and were free-running at a repetition rate of 10/sec. Presumably the disrupting effect of artificial CF/FM pulses is due to interference with the bat's processing of information from its own sounds by stimulating the same population of neurons that extract distance information from the bat's echolocation sounds. The specific artificial pulse structures required to achieve interference reflect the essential structural elements of natural echolocation sounds that code distance information and provide cues to neural mechanisms for processing this information.

The bat's auditory system processes specific structural elements of echolocation sounds for target distance determination. Although artificial pulses simulating the bat's natural CF/FM echolocation sounds severely interfered with distance discrimination, pulses simulating the natural CF component or the FM component did not affect the bat's performance (N. albiventris Roverud and Grinnell 1985; R. rouxi, this paper). In both N. albiventris and R. rouxi, artificial pulses must contain both CF and FM components to interfere. The temporal relationship between the two necessary components of the interfering artificial pulses, however, differs in the two species and depends on the echolocation sound structure of each species.

In the short-CF/FM bat, N. albiventris, interference occurred when the duration of the CF component of the CF/FM artificial pulse was between 2 and 30 msec, with maximum effect between 10 and 20 msec (Fig. 1a, Roverud and Grinnell 1985). It is not the duration of the entire CF component, however, but rather the time interval between the onset of the CF component and the following FM component that determines the degree of interference. As Fig. 1b shows, a brief CF signal 2-27 msec before a separate FM signal was as effective as an uninterrupted CF component of equal duration. When the artificial FM pulse was triggered by the bat's own emitted sounds, the interfering effectiveness depended on the temporal relationship between the onset of the bat's emitted CF/FM pulse and the artificial FM sweep (Fig. 1c). When the time delay between the onset of the artificial FM pulse was less than about 8 msec, a condition where the artificial pulse begins during the bat's own emission, interference did not occur. It appears that the bats process the artificial pulses differently if the onset of the sound occurs during pulse emission than if the sound begins after pulse emission. Nevertheless, an artificial FM sweep, which had no effect when presented at 10/sec free-running, was as effective at interfering with distance discrimination as a free-running FM pulse following a CF pulse when it was coupled with the bat's own CF/FM emission and came during the 8-27 msec period, with maximal effect again at 10-20 msec delay, after the onset of the bat's CF/FM pulse. The artificial CF component and the bat's emitted CF component seem to have a similar effect on the population of auditory neurons that process distance information; both alter the responsiveness of these neurons to FM sweeps over a restricted time interval. These results support the contention that the artificial pulses that interfere with distance discrimination act on the same population of neurons that extract distance information from the bat's own echolocation sounds.

R. rouxi emit long-CF/FM echolocation sounds with a CF component duration more than 7 times that of the short-CF/FM bat, N. albiventris. The temporal relationship between the CF and FM components of the artificial pulses required to achieve interference was correspondingly longer in

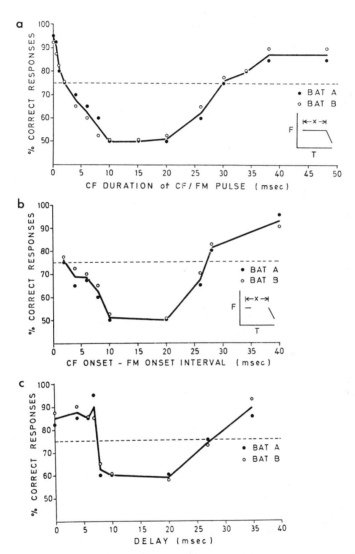

Fig. 1. Performance of two N. albiventris during the distance discrimination trials when presented with artificial (a) CF/FM (CF 75 kHz; FM 75-50 kHz, 2 msec) pulses with varying duration of the CF component, (b) CF plus FM combinations with a 2 msec CF (75 kHz) signal preceding a 2 msec FM sweep (75-50 kHz) at various intervals and (c) a FM (75-50 kHz, 2 msec) pulse at various delays after the onset of the bat's own CF/FM sound. The horizontal dashed line in this and the subsequent figure indicates the 75% correct response level. Cartoon inserts in this figure and Fig. 2 show the type of signal manipulation tested (From Roverud and Grinnell 1985).

R. rouxi than in N. albiventris. In R. rouxi, interference occurred when the duration of the CF component of the CF/FM artificial pulse was between 2 and 70 msec, with maximum effect between 10 and 60 msec (Fig. 2a). As was the case with N. albiventris, it is the time interval between the onset of the CF component and subsequent FM signals that determines the extent to which the artificial pulses will interfere. A discrete CF signal 2-67 msec before an unjoined FM signal had the same interfering effect as a continuous CF component of the same duration (Fig. 2b). Furthermore, an artificial FM signal, which again had no effect when presented at 10/sec free-running, once more was as effective as a discrete artificial FM signal following an artificial CF signal when it was coupled with the bat's own CF/FM emissions and

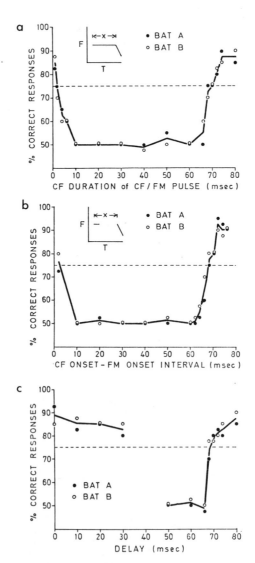

Fig. 2. Performance of two R. rouxi when presented with artificial (a) CF/FM (CF at resting frequency of individual bat, FM falls 20 kHz below resting frequency in 2 msec) pulses with varying duration of the CF component, (b) CF plus FM combinations with a 2 msec CF signal preceding a 2 msec FM sweep at various intervals and (c) a 2 msec FM pulse at various delays after the onset of the bat's own emissions.

came within about 67 msec after the onset of the bat's CF component
(Fig. 2c). Artificial FM pulses that overlap with the bat's pulse emissions
appear not to interfere.

These results suggest that in both Noctilio and Rhinolophus the essential echolocation sound elements used for the processing of distance information are the onset of the pulse CF component and the pulse and echo FM components. The onset of the CF component appears to activate a gating mechanism
that frames a processing time window by changing the responsiveness of a population of distance information processing neurons to FM signals during a
restricted time interval after the onset of the CF component. In
N. albiventris, neurons extracting distance information from FM signals do so
with greatest sensitivity 10-20 msec after the CF onset, with the entire
time window of sensitivity being from about 2-27 msec. The CF component of
R. rouxi is about 39 msec longer than that of N. albiventris and the time
window of R. rouxi is proportionally longer than that of N. albiventris.
In R. rouxi, the most criticial delays between CF onset and FM signals range
from about 10-60 msec, with the whole time window extending from about 2-67
msec. Thus in different species of bats the duration of the time window corresponds to the duration of the bat's echolocation sounds.

The gated time window regulates the processing of sensory information
from the environment and acts as a central filter. The gate mechanism serves
as a temporal filter since only FM signals that occur within a restricted
time window after the onset of the bat's CF component are processed for distance information. Because bats evaluate target distance from the time interval between the FM component of pulse emission and echo return, temporal
filtering of FM component echoes at different delays from the pulse emission
is equivalent to spatial filtering of targets at different distances. In
N. albiventris the gated time window would allow for accurate processing of
distance information from targets within a maximum response range of about
2.4-3.6 m, corresponding to echo FM components occurring 14-21 msec after the
emitted FM sweep (20-27 msec after the emitted CF onset). Since for each
species the duration of the time window is in proportion to the duration of
the CF component, the maximum pulse FM to echo FM processing interval is
about the same in R. rouxi as it is in N. albiventris. The time window of
R. rouxi would permit fine measurement of target distance within a maximum
response range of about 2.6-3.8 m, corresponding to echo FM components coming
15-22 msec after the emitted FM sweep (60-67 msec after the emitted CF
onset). Thus the maximum target range for accurate distance information
processing is about the same in both species.

In many bats the transition from the search stage to the approach stage
of target-directed flight is marked by range related changes in pulse emission at a distance of about 2-3 m. The time window seems to be adapted to
the bat's pulse design so that FM component echoes enter some criticial
region of the time window at a characteristic distance associated with the
initiation of the approach stage and the approach stage may be related to
processing of FM signals within the time window. A gated time window would
allow the bat to attend selectively to distance information from targets in
the near environment and to reduce interference from echoes from more
distant targets.

REFERENCES

Roverud, R. C., and Grinnell, A. D., 1985, Echolocation sound features pro-
cessed to provide distance information in the CF/FM bat,
Noctilio albiventris: evidence for a gated time window utilizing both
CF and FM components, J. Comp. Physiol. A, 156:457.

SECTION IV

NATURAL HISTORY OF ECHOLOCATION

Section Organizers: M. Brock Fenton and Gerhard Neuweiler

NATURAL HISTORY ASPECTS OF MARINE MAMMAL ECHOLOCATION: FEEDING STRATEGIES
AND HABITAT
  William E. Evans and Frank T. Awbrey

BEHAVIOUR AND FORAGING ECOLOGY OF ECHOLOCATING BATS
  Gerhard Neuweiler and M. Brock Fenton

INTERACTION BETWEEN ECHOLOCATING BATS AND THEIR PREY
  Annemarie Surlykke

LOUD IMPULSE SOUNDS IN ODONTOCETE PREDATION AND SOCIAL BEHAVIOR
  Ken Marten, Kenneth S. Norris, Patrick W. B. Moore and
  Kirsten A. Englund

HARMONIC STRUCTURE OF BAT ECHOLOCATION SIGNALS
  Karl Zbinden

ACOUSTICAL VS. VISUAL ORIENTATION IN NEOTROPICAL BATS
  Uwe Schmidt, G. Joermann and G. Rother

ECHOLOCATION STRATEGIES OF AERIAL INSECTIVOROUS BATS AND THEIR INFLUENCE ON
PREY SELECTION
  Robert M. R. Barclay

FORAGING BEHAVIOR, PREY SELECTION AND ECHOLOCATION IN PHYLLOSTOMINE BATS
(Phyllostomidae)
  Jacqueline J. Belwood

VARIATION IN FORAGING STRATEGIES IN FIVE SPECIES OF INSECTIVOROUS BATS -
IMPLICATIONS FOR ECHOLOCATION CALL DESIGN
  M. Brock Fenton

DETECTION OF PREY IN ECHOCLUTTERING ENVIRONMENTS
  Gerhard Neuweiler, A. Link, G. Marimuthu, R. Rübsamen

HOW THE BAT, PIPISTRELLUS KUHLI, HUNTS FOR INSECTS
  H. Ulrich Schnitzler, Elisabeth Kalko, Lee Miller and Annemarie Surylkke

THE COMMUNICATION ROLE OF ECHOLOCATION CALLS IN VESPERTILIONID BATS
  Jonathan P. Balcombe and M. Brock Fenton

AUDITORY INPUT TO THE DORSAL LONGITUDINAL FLIGHT MOTOR NEURONS OF A NOCTUID
MOTH
  Lee A. Miller and Bent M. Madsen

NATURAL HISTORY ASPECTS OF MARINE MAMMAL

ECHOLOCATION: FEEDING STRATEGIES AND HABITAT

William E. Evans

Frank T. Awbrey

Hubbs Marine Research Inst.
1700 South Shores Dr.
San Diego, CA 92109

Biology Department
San Diego State University
San Diego, CA 92182-0057

INTRODUCTION

The state of knowledge of echolocation in marine mammals and cetaceans in particular has been reviewed effectively in recent years (Wood and Evans, 1980; Watkins and Wartzok, 1985). Both of these reviews reveal that most efforts are still directed toward biophysics rather than the natural history, biological, and functional aspects. At the First Animal Sonar Conference in 1966, Donald Griffin and others asked researchers working with cetaceans how echolocation is used in navigation or feeding. Unfortunately 20 years later these two "obvious" uses of this extraordinary capability are still understood mainly by inference. The "Scylla" paradox described in Wood and Evans (1980), indicated that a dolphin deprived of vision can use directional hearing rather than active acoustic scanning to detect and capture moving, avoiding prey. This use of listening ("passive sonar") has been well studied in bats (Fiedler et al., 1980), but not considered, in print at least, for use by marine mammals. That is changing. Recent tests by Sonafrank, Elsner, and Wartzok (1983) demonstrated that a spotted seal could use vision and listening to find ice holes and to navigate.

Since the publication in 1980 of Animal Sonar Systems, considerably more information on the nature of Tursiops' echolocation capability has been added, e.g., width of the transmission beam, and ability to detect objects at distances and in varying noise conditions (Au et al., 1980, 1981, 1982, 1983, 1984, 1985). Experimental verification of echolocation and measurement or re-measurement of hearing has been accomplished for Pseudorca, Grampus, Globicephala, Platanista indi and Delphinapterus. For example, see Awbrey et al., 1985, Kamminga and Wiersma, 1981, Purves and Pilleri, 1983, Thomas et al., 1986 and this volume, and Zbinden, 1982. The kinds of narrow band ultrasonic signals produced by Commerson's dolphin and Dall's porpoise while taking food or avoiding objects suggest an echolocation capability by these species also (Evans, Awbrey and Hackbarth, 1986; Hackbarth, Awbrey and Evans, 1986). The nature of the signals recorded both in the field and in captivity for Commerson's dolphins suggests that echolocation in this species, if it exists, might be somewhat more specialized than in the other delphinids studied to date. The pulses have peak energy at 120-130 kHz, are 10-20 kHz wide and are considerably longer than those produced by the species discussed in Wood and Evans (1980)-- 200 to 500 microseconds compared with 15-75 microseconds.

Our main goal in this paper is to discuss those data that have become available since 1980 with special emphasis on how they have expanded our understanding of the natural history aspects of marine mammal echolocation. We will discuss what is known about prey species and feeding patterns of marine mammals known or suspected to be echolocators. As has been mentioned previously (Wood and Evans, 1980; Evans, 1973), marine mammals, including cetaceans, occupy all aquatic habitats from fresh water lakes and rivers to estuaries and the littoral and pelagic realms in all oceans from the equator to the polar ice edges. Some species are cosmopolitan and some are narrowly endemic. Some species are catholic feeders, others are highly specialized. Because we plan to relate pulse trains and behavior to foraging strategy and other behaviors, we will organize our discussion by groups, starting with generalists (both in feeding and habitat selection) and work our way through the specialists. Most of our discussion will concentrate on odontocete cetaceans. The only experimental evidence for echolocation in pinnipeds involves harbor seals (Phoca vitulina) (Renouf and Davis, 1982). As with other known underwater vocalizations of seals, sea lions, fur seals and walrus, the clicks emitted by the animal in this test were in the frequency range below 20kHz. Recently, however, the leopard seal (Hydrurga leptonyx) has been observed to emit ultrasonic signals in association with feeding (Thomas et al., 1982), so we also will speculate about echolocation in pinnipeds.

GENERALISTS IN PREY AND HABITAT SELECTION

The two best examples of cetacean generalists, as far as prey and habitat are concerned, are the bottlenose dolphin (Tursiops sp.) and the killer whale (Orcinus). Both of these genera are found far at sea (oceanic), in coastal waters, bays, fjords, the surf zone, and even occasionally up rivers in nearly fresh water. Tursiops is in all seas from approximately 40 degrees north to 40 degrees south. Orcinus also is reported from all seas (except Caspian and Black) from the ice edge in the antarctic to the ice edge in the arctic. Both are catholic feeders, taking a wide range of food types and sizes. The killer whale probably holds the record for range of recorded prey types, with the list ranging from herring to blue whales. Tursiops feeds on both fast-moving and stationary prey. Feeding strategies in both these species are also quite variable. Tursiops has been seen foraging as individuals as well as in small and large groups. Cooperative feeding has also been observed. When feeding on mullet, Tursiops sometimes "herds" the prey into shallow water and even onto the beach (Evans, personal observation). They feed both at night and during the day and have been observed to feed "silently" in the wild in extremely turbid water. Just as in the case of a single blind folded Tursiops capturing a free swimming fish without vocalizing (Wood and Evans, 1980) they may have used listening. The use of listening instead of active scanning for prey detection is not uncommon (bats, owls, wild dogs), and may be used more by cetaceans than was believed previously.

Tursiops may use pulse trains for social communication (Bastian, 1967). Because of this, determining the actual function of pulse trains recorded in the wild without the benefit of behavioral observation is essentially impossible. Most investigators have assumed that those pulses used for echolocation contain much more ultrasonic energy than those intended for social communication. Perhaps this assumption should be viewed with a little more skepticism. Assignment of sounds to mutually exclusive communication or echolocation categories may or may not match their uses by dolphins. Without good experimental data we are limited to educated guesses, which may well be wrong. If what data we do have are faulty, then we are on even shakier ground. For example, frequency content of recorded pulses is affected by location of the hydrophone relative to the head of the individual being recorded because a dolphin's sound beam becomes narrower as frequency increases. This is at least partly responsible for past reports that certain species' signals have no significant ultrasonic component. Orcinus was, and occasionally still is, listed as producing only low frequency pulses (Schevill and Watkins, 1966; Watkins and Wartzok, 1985). These conclusions could be due either to the limitations of the recording system used or to hydrophone location. Killer whales in the Pacific northwest, the antarctic and the controlled environment of a marine park produce pulses with peaks at frequencies up to 40 kHz and significant energy to 80 kHz (fig.1). Dall's porpoise, harbor porpoise, Chinese finless porpoise and river dolphins to mention a few, have been listed as using only low frequency pulses in echolocation (Evans, 1973; Watkins and Schevill, 1980). Now we know that the pulses produced by all of these species contain significant ultrasonic components (Awbrey et al, 1979; Herald et al., 1969; Kamminga et al., 1981; Zbinden, 1982; Pilleri et al., 1980). Convincing evidence of an echolocation capability exists now for 12 odontocete species. This is certainly more impressive than in 1980, but considering that we know that 9 mysticete, 43 odontocete, 16 pinniped and 2 sirenian species vocalize underwater (Watkins and Wartzok, 1985), the record is not all that good. Even less impressive is the lack of data on the functions of all of this vocal behavior. Collection of meaningful data on the functions of vocalization in general is difficult, but, for obvious reasons, verifying an echolocation capability and determining its functions under field conditions is impractical if not impossible for aquatic mammals.

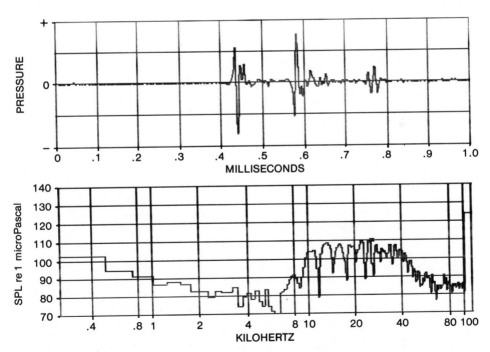

Fig. 1. Spectrum and waveform of a high frequency click from a British Columbia killer whale. The harmonic lobes spaced about 3 kHz apart are from the 297 microsec delay between the click and its strong reflection. Effective filter bandwidth was 250 Hz.

The literature contains many papers on various aspects of Tursiops' echolocation capability, but only three or four on feeding behavior. Just one of those incorporates observations about associated sounds. When feeding on menhaden in turbid water in the Gulf of Mexico, Tursiops circle the schools and engage in what appears to be cooperative (group) feeding. They also swim upside down when apparently feeding from the bottom (Leatherwood, 1975). If echolocation is being used, which is reasonable in conditions of low or no visibility, upside down swimming is adaptive given the emitted sound radiation pattern (Evans, 1973; Au, 1980). In the Eastern tropical Pacific, Tursiops were photographed and recorded as they fed at night on flying fish (Cypselurus sp.), which had been stunned when they ran into an anchored ship. Visibility was not a problem, as the area around the ship was well lighted, but every dolphin that pursued a fish produced pulse trains. The observers modified the acoustic characteristics of some of the fish by inserting empty gelatin capsules into their mouths or by popping their air bladders before throwing them back into the water for the dolphins to catch. These modifications caused the dolphins to turn away at the last instant and make another run before taking the fish. (Ljungblad, Leatherwood, and Awbrey, unpubl. data).

Hall and Johnson (1971), in association with a study of hearing, demonstrated that a blindfolded killer whale (Orcinus) was able to navigate easily around a 15 meter diameter pool, locate and retrieve an air filled ring. This is the extent of the experimental study of echolocation in this species. Recordings made while a small group of Orcinus was chasing, capturing and eating a small harbor porpoise revealed few or no vocalizations of any type; neither screams, whistles nor pulse trains of the type used in echolocation. In contrast, Awbrey has made recordings in the antarctic where killer whales were making long dives under the fast ice. Whales nearby, in the lead made by an icebreaker, were photographed with antarctic cod (Dissostachus mawsoni) in their mouths. The recordings include long pulse trains with both low frequency and high frequency clicks. Many of the pulses recorded had ultrasonic components. Conversations with divers who have worked in similar ice conditions support the assumption that visibility was limited. These pulses are not noticeably different in waveform from those produced by several animals that were lying on the surface less than 15 meters from the hydrophone, apparently interacting with each other rather than echolocating. Even though such observations are suggestive they do not allow for analyses that would delineate characteristics of the waveform, repetition rate or sound field. The questions raised by the use of a passive acoustic capability by the Tursiops Scylla (Wood and Evans, 1980) cannot be addressed at present. It would be interesting to repeat this experiment with the hydrophone on a killer whale searching for or capturing live prey. This type of definitive experiment can, however, only be accomplished with individuals held in captivity. Unfortunately, such work is becoming increasingly difficult because of burgeoning national and international regulation.

Apparent echolocation sounds can have other functions. Tursiops can be trained to discriminate targets by using echoes from pulses produced by a source other than itself (Scronce and Johnson, 1975). Given this capability, can an echolocating dolphin communicate information on presence, type and location of prey? K.S. Norris, in lectures, has speculated that the acoustic transparency of muscle and blubber, in water, might provide an echolocating Tursiops with information about stress, health, reproductive state or other internal characteristics of its conspecifics. Sounds may be important in mother-calf interactions, but the availability of cow-calf pairs of even the most common species (i.e. Tursiops) in a controlled environment conducive to observation and experimentation has been limited. The first study of mother-young relationship was conducted by Caldwell and Caldwell (1972). More research is being conducted by Dr. Peter Tyack, Woods

Hole Oceanographic Institution. These studies have concentrated mostly on Tursiops whistle vocalizations. Carder and Ridgway (1983) have suggested echolocation by a 60-day-old Tursiops. We recorded pulsed vocalizations by a two-week-old Orcinus during the first opportunity to watch a killer whale cow and her normal, nursing calf (September, 1985, Sea World, Orlando, Fla.). Now that success of captive cetacean breeding programs is improving, many more opportunities are becoming available for controlled studies of the ontogeny of echolocation and of a possible role for high frequency pulsed sounds in bonding. (see papers by D. Reiss and M. L. Schultz in this volume).

Some intriguing anecdotal evidence shows the direction some of those studies might take. In 1974, Evans participated in the training of a ten month old Tursiops male for an echolocation experiment. The initial phase involved training the subject to accept rubber suction cup blindfolds. This procedure turned out to be more difficult than expected. He would wear a cup on either eye but not both simultaneously. Only after several weeks would he accept having both eyes covered. Still more time passed before he would accept a fish from a hand when his vision was blocked, even though he had been hand fed for months. The development of the ability to navigate around the pool with both eyes covered took even longer. In this individual's case, echolocation obviously was a learned behavior, even though he had produced pulse trains from the first time he was recorded after being separated from his mother. This reluctance to accept blocking of the visual channel was also documented by K. S. Norris (1969) during training of a Tursiops cf gilli captured in Hawaiian coastal waters.

SPECIALISTS IN PREY AND HABITAT SELECTION

Marine mammal specialists are much more common than generalists. This is especially true for cetaceans. Many species are pelagic, others are exclusively littoral/riverine, others just riverine and still others live only in arctic or antarctic waters. Many are teuthophagus, others feed on benthic and epibenthic vertebrates and invertebrates and still others feed on mesopelagic fishes.

Most populations of long snouted or oceanic dolphins (e.g. Stenella, Delphinus) are found far at sea. Until recently little was known of their taxonomic status, let alone their behavior. The incidental take of signif- icant numbers of Delphinus and three Stenella species during purse seining for yellowfin tuna motivated a rather extensive as well as intensive study of their natural history. Behavioral studies, including foraging behavior, were emphasized. Radio telemetric studies and stomach content analyses revealed that these species feed predominantly at night on organisms associ- ated with the migrating acoustic deep scattering layer (Perrin et al., 1973; Evans, 1975, 1982; Norris and Dohl, 1980). Other than establishing that these species are very vocal, little or no work on echolocation and associated behavior has been done. Experimental studies with blindfolded Delphinus from the Black Sea, demonstrated that they can discriminate between complex planar targets that differ from each other only in the size of one element at least as well as Tursiops can (Bel'kovich et al., 1969). The pulses produced by both Stenella sp. and Delphinus tend to have multi- modal spectra, with strong peaks in the ultrasonic range. Recordings of Delphinus in the wild indicate diurnal differences in vocalization. During the day, whistles, click trains and squeals are heard. At night, click trains predominate. If these species use echolocation to locate prey, they must use it in a highly cluttered environment. The density of acoustic reflectors is such that man-made ultrasonic sonars cannot effectively detect targets of the size captured and consumed by Delphinus (20 cm length overall). Furthermore, most of the organisms in the layer, with the exception of squid, move so slowly that they do not offer good doppler cues. However, as many of the fish species that make up the scattering layer (myctophids, bathylagids) appear to be lethargic, they are probably easy to catch (Barham, 1970).

Norris and Dohl (1980) suggest that vocalizations serve to synchronize movement of large schools of Hawaiian spinner dolphins (Stenella longirostris). Moving schools of Delphinus also click synchronously. Often, ranks within a school move in cadence with pulse trains produced almost in unison.

Observations of the ecological aspects of echolocation are limited in this group since availability for study in captivity has been limited. Delphinus have produced calves in captivity, but few have lived longer than several months. Production of pulse trains during the day is limited and most of the contact between cow and calf appears to be maintained visually, although whistles are produced. Comprehensive studies of the ontogeny of echolocation and its possible use in maintaining contact between cow and calf or between individuals have not been conducted.

Four other species that are known to echolocate also are pelagic--the Pacific white-sided dolphin (Lagenorhynchus obliquidens), the false killer whale (Pseudorca crassidens), Risso's dolphin (Grampus griseus) and the Pacific pilot whale (Globicephala cf scammoni). The first two species feed during the day as well as at night, mainly on fast swimming, schooling fishes such as sardine, small tunas, and mackerel. The other two species are almost exclusively squid eaters. Data on pulses used by each of these species during echolocation are insufficient to test for differences in pulse waveform that might be related to prey selection. We know that Pacific white-sided dolphins can discriminate as well as Tursiops can between cylinders that differ from each other only in diameter (Evans, 1973). The major difference is that white-sided dolphins cannot turn the head from side to side because their cervical vertebrae are fused. Once the Pacific white-sided dolphin being tested committed itself to the right or left target, it could not examine the other target in the last 50-60 cm. Tursiops in exactly the same experimental set-up would stop short of the target for a last minute comparison. All of the aforementioned species share this cervical inflexibility. The false killer whale detected the presence or absence of a 7.5 cm diameter water filled steel sphere behind a visually opaque, acoustically transparent screen (Thomas, 1986). Evidence for echolocation in the other two species is less direct. W. E. Evans has seen a blindfolded pilot whale and Risso's dolphin retrieve objects (air filled rings) and concluded that they probably were echolocating.

Rivers represent an environment that is essentially at the other end of the environmental spectrum as far as echolocation is concerned. Ambient noise levels in rivers are significantly higher than in the pelagic environment (fig. 2). They are fresh water or brackish - so sound velocity is different. Rivers that dump into the ocean have significant thermoclines and haloclines at high and low tide. Most rivers have poor visibility, and carry suspended material that could represent false targets. Small cetaceans that feed, breed and calve in rivers have developed a number of adaptations, behavioral as well as physiological. Four species are considered to be exclusively "river dolphins" - The Amazon and Orinoco river dolphins (Inia geoffrensis), the La Plata river dolphin (Pontoporia blainvillei), the Indus and Ganges river dolphins, (Platanista sp.), and the Chinese lake dolphin or white flag dolphin, Lipotes vexillifer. Echolocation has been demonstrated experimentally in Inia and is inferred in Platanista minor (Herald et al., 1969; Penner and Murchison, 1970; Purves and Pilleri, 1969). Inia discriminated between wires and strands of nylon of different diameters that were presented behind a visually opaque, acoustically trans-

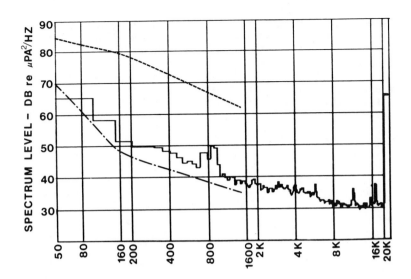

Fig. 2. Exponential average of ambient noise in the Snake R.,
Alaska, at slack tide with no wind, some small waves.
Inflatable boat was anchored in 5m of water approximately
100m from shore at first bend, about 6 km upstream from
mouth. Averaging time constant was 320 ms. SPL re 1
microPa was 85 dB. The small peaks around 800 Hz are
from waves slapping against the boat's hull and bottom.
Analysis filter bandwidth was 50 Hz. Interrupted line is
the average summer spectrum level in deep water of the
Beaufort Sea (Greene, 1981). Dashed line is the average
summer spectrum level in Baffin Bay (Leggat et al.,
1981).

parent plastic screen. Platanista is sometimes called the blind river
dolphin because the eye is poorly developed and vision is apparently limited
to differentiating between light and dark. The specimens studied at the
Steinhart aquarium produced pulse trains continuously as they swam around
the pool. They could locate fish equally well under conditions of light or
dark. Platanista pulses, although broad-band, have little or no energy
below 15 kHz. This is also true for Inia except that they appear to have
both a low frequency and ultrasonic component in their pulses (Herald et
al., 1969). An intriguing thing that these riverine species have in
common is that none are known to whistle. Of the small cetaceans that
venture far up rivers, only Tursiops and belukha whales, which are primarily
oceanic species, whistle. Phocoenids, platanistids and Cephalorhynchus are
also specialized and do not whistle. If these specialized species really
do not or even cannot whistle, does this mean that they produce their
echolocation signals differently than whistling species do, or does something about their ecology suppress whistling?

Those data that are available suggest that most of the river species
are somewhat catholic in their prey selection, perhaps feeding on whatever
is most plentiful. Both Inia and Platanista feed on benthic and epibenthic
fishes and crustacea. Platanista in the Indus river also feed on herring-
like schooling fishes.

Far too little is known about the ecology and natural history of these species. Unfortunately this lack of knowledge may have contributed to the significant decline in all river dolphin populations. The Chinese estimate that only a few hundred lake dolphins remain. Although this species apparently has a well developed echolocation capability, it seems to be poor at avoiding gill nets. This also appears to be true for the phocoenid species that is sympatric with it - Neophocoena asiaeorientalis (Chen et al., 1979).

Becoming entangled in monofilament gill nets is a prevalent problem for most phocoenids. This is particularly interesting because the waveform used by Dall's porpoise Phocoenoides dalli (fig. 3) would seem to have adequate resolution for detecting such things as nets (Evans, et al., in press). Commerson's dolphin, a delphinid that is convergent with Dall's porpoise in coloration and phonation (fig. 4), also is not known to whistle and is prone to entanglement in gill nets. Evans et al. speculate that these species may use their sonar mainly to detect rapidly moving objects and so, would disregard the nets as background clutter. Another possibility, of course, is that they cannot or do not use their specialized sonar to get general information about their environment and become entangled because they are usually silent when travelling. Either of these would explain why modifying a gill net to raise its target strength has done little to decrease net entanglement. Recent studies of the hearing of the harbor porpoise from the Azov Sea (Phocoena phocoena) by Voronov and Stosman (1986) demonstrates that this species' hearing is sharply tuned to the dominant frequency of its echolocation pulses (fig. 5). This is also true for Platanista indi and Inia (Purves and Pilleri, 1983). If other species that produce narrow band ultrasonic pulses have this kind of specialized hearing, the apparent low hearing sensitivity to most frequencies might contribute to entanglement.

Belukha whales spend considerable time in shallow water. They often are in water less than one meter deep and move long distances up rivers. Some of these rivers bear such heavy silt loads that underwater visibility is zero, yet we have watched belukhas chase and catch salmon there. Once, we attempted to record their vocalizations while several were chasing salmon in a small, quiet embayment just upstream from the mouth of the Snake River, Alaska. The recorder and hydrophone were flat (+3 dB) to 40 kHz, with measurable response to above 60 kHz, yet we recorded only an occasional buzz in spite of all the activity within just a few meters of our position. If the whales were echolocating, virtually all energy was above 60 kHz. We were able to record very little sound of any kind from the belukha whales that passed our stations. Most of the groups coming downriver had newborn calves and, usually, even though they passed within a few meters of our hydrophones, we heard nothing. Again, we cannot dismiss the possibility that the whales were relying on passive acoustic cues. In contrast, the large numbers of whales in shallow water at the river's mouth clicked and whistled continuously whenever we listened.

PINNIPEDS

In spite of considerable effort with a variety of species (Watkins and Wartzok, 1985), there is no really convincing evidence that pinnipeds can echolocate. This situation may be changing at last. Renouf and Davis (1982) reported that harbor seals can echolocate, but their work has been challenged on methodological grounds (Wartzok, et al., 1984) and needs replication with better controls. Polar pinnipeds, in particular, would be

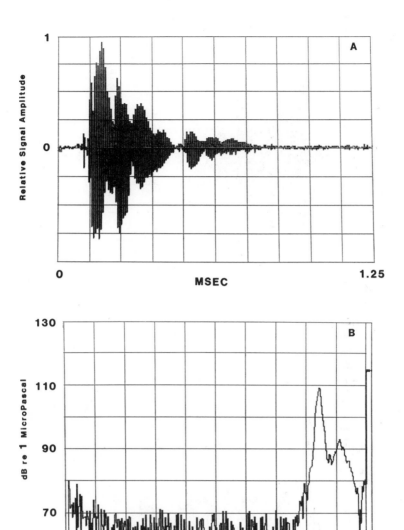

Fig. 3. Waveform (A) and energy density spectrum (B) of a pulse
typical of those produced by <u>Phocoenoides dalli</u> when
orienting toward an object such as a hydrophone. In this
case the hydrophone (a Bruel and Kjaer 8103) was approximately 2 meters deep. The recorder and analyzing system
are described in Evans, Awbrey and Hackbarth, 1986.

529

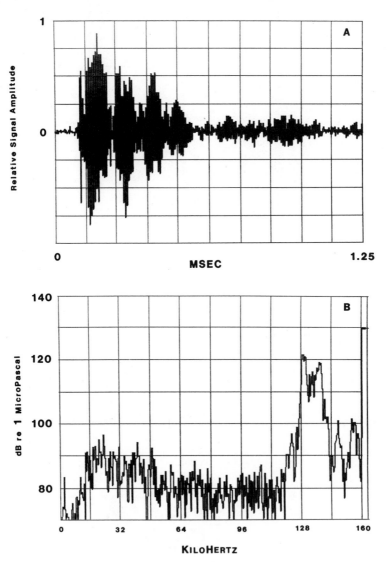

Fig. 4. Waveform (A) and energy density spectrum (B) of a pulse
typical of those produced by Cephalorhynchus commersonii
at Sea World in San Diego when orienting towards a fish.
The hydrophone (ITC 8095) was 2 meters deep. The recorder
and analyzing system are described in Evans, Awbrey and
Hackbarth, 1986.

expected to depend on echolocation because vision is useless under thick
ice in winter. At least one antarctic pinniped produces signals suitable
for high resolution echolocation (Thomas, et al., 1982). A female leopard
seal at Sea World, San Diego, emitted a variety of low-level ultrasonic
signals (fig. 6) while chasing live anchovies and small striped bass. The
wave forms of these signals are much more suggestive of echolocation signals
used by terrestrial echolocators (bats) than those of aquatic mammals.
Constant frequency-frequency modulated chirps 10-15 milliseconds long,

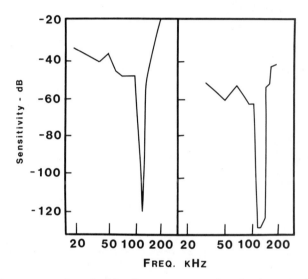

Fig. 5. Frequency thresholds for two Azov Sea harbor porpoises
        (Phocoena phocoena) determined using auditory evoked
        potentials measured from the brain stem. Hearing is most
        sensitive to frequencies between 100 and 140 kHz. The
        sensitivity falls off sharply above. (From Voronov and
        Stosman, 1986)

beginning at 130 kHz and then sweeping to 60kHz, other frequency modulated
signals of similar frequency and duration, frequency modulated buzzes, and
narrow band pulses (60-70kHz) were recorded. The unexpected variety of
signals this animal produced parallels the species' large repertoire within
the human hearing range and raises a multitude of questions. Was she
actually using echolocation? If so, did she use all the sounds she made?
Would such variety allow leopard seals to match their echolocation to
different kinds of prey? Finally, if this and other polar pinnipeds can
echolocate, how do their abilities compare with those of odontocetes?

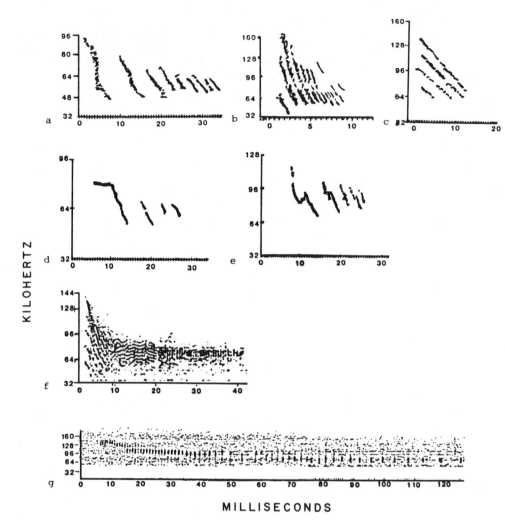

KILOHERTZ

MILLISECONDS

Fig. 6. Some of the high frequency sounds produced by a leopard
seal while chasing live anchovies or small striped bass
in a small pool. From Thomas, et al., 1982.

REFERENCES

Au, W. W. L., Penner, R. H., and Kadane, J., 1982, Acoustic behavior of
echolocating Atlantic bottlenose dolphins, J. Acoust. Soc. Am.,
71:1269-1275.

Au, W. W. L., Floyd, R. W., Penner, R. H., and Murchison, A. E., 1974,
Measurement of echolocation signals of the Atlantic bottlenose dolphin,
Tursiops truncatus, Montague, in open waters, J. Acoust. Soc. Am.,
56:1280-1290.

Au, W. W. L., and Turl, C. W., 1983, Target detection in reverberation by
an echolocating Atlantic bottlenose dolphin (Tursiops truncatus), J.
Acoust. Soc. Am., 73:1676-1681.

Au, W. W. L., and Martin, D. W., 1983, Insights into dolphin sonar discrimination capabilities from broadband sonar discrimination experiments with human subjects, J. Acoust. Soc. Am. Suppl. 1, 74:S73.

Awbrey, F. T., Norris, J. C., Hubbard, A. B., and Evans, W. E., 1979, The bioacoustics of the Dall's porpoise-salmon drift net interaction, Technical Rept. 79-120, Hubbs Sea-World Research Institute, San Diego.

Awbrey, F. T., Thomas, J. T., Evans, W. E., and Kastelein, R. A. 1986, Hearing threshold measurements and responses of belukha whales to playbacks of underwater drilling noise, in: "Underwater Drilling--Measurement of Sound Levels and Their Effects on Belukha Whales," API Publication No. 4438, American Petroleum Institute, 1220 L Street Northwest, Washington, D.C. 20005.

Barham, E. G., 1970, Deep-Sea Fishes: Lethargy and Vertical Distribution of Fishes and Cephalopods, in: "Proceedings of an International Symposium on Biological Sound Scattering in the Ocean," G. B. Farquhar, ed., U. S. Government Printing Office, Washington.

Bastian, J. R., 1967, The transmission of arbitrary environmental information between bottlenose dolphins, in: "Animal Sonar Systems: Biology and Bionics," R.-G. Busnel, ed., Laboratoire de Physiologie Acoustique, Joy-en-Josas, France.

Bel'kovich, V.M., Borisov, V. I., Gurevich, V.S. & Krushinskaya, N.L. 1969, The ability of echolocation in Delphinus delphis, (In Russian). Zoologicheskii Zhurnal, 48:876-884. (English translation JPRS 48780).

Caldwell, D. K., and Caldwell, M. C., 1972, Senses and Communications, in: "Mammals of the Sea, Biology and Medicine," S.H. Ridgway, ed., Chas C. Thomas, Springfield, Ill.

Caldwell, M. C., and Caldwell, D. K., 1968, Vocalization of naive captive dolphins in small groups, Science, 159:1121-1123.

Caldwell, M. C., and Caldwell, D. K., 1972, Behavior of Marine Mammals, in: "Mammals of the Sea, Biology and Medicine," S.H. Ridgway, ed., Chas C. Thomas, Springfield, Ill.

Carder, D. A. and Ridgway, S. H., 1983, Apparent echolocation by a 60 day old bottlenosed dolphin, J. Acoust. Soc. Am. Suppl. 1, 74:S74.

Chen, P., Peilin, L., Renjun, L., Kejie, L., 1979, Distribution, Ecology, Behavior and Conservation of the Dolphins of the Middle Reaches of Changjiang (Yangtze) River (Wuhan-Yueyang), Invest. Cetacea, 9:87-103.

Evans, W. E., 1973, Echolocation by marine delphinids and one species of fresh-water dolphin, J. Acoust. Soc. Am., 54:191-199.

Evans, W. E., 1982, Distribution and differentiation of stock of Delphinus delphis Linnaeus in the northeastern Pacific, Mammals in the Seas, FAO Fisheries series no. 5, 4:45.

Evans, W. E., Awbrey, F. T., and Hackbarth, H., In press, High frequency pulses produced by free ranging Commerson's dolphin (Cephalorhynchus commersonii) compared to those of phocoenids, IWC special report on Commerson's dolphin.

Fiedler, J., Habersetzer, J., and Vogler, B., 1980, in: "Animal Sonar Systems," R.-G Busnel and J. F. Fish, eds., Plenum Press, N.Y.

Greene, C. R., 1981, Underwater acoustic transmission loss and ambient noise in Arctic regions, in: N. M. Peterson, ed., "The Question of Sound from Icebreaker Operations: Proceedings of a Workshop," Arctic Pilot Project, Petro-Canada.

Hackbarth, H., Awbrey, F. T., and Evans, W. E., 1986, High frequency sounds in Commerson's dolphin, Cephalorhynchus commersonii, Tech. Rept. 86-193, Hubbs Sea-World Research Institute, San Diego.

Hall, J. D., and Johnson, C. S., 1971, Auditory threshold of a killer whale Orcinus orca Linneaus, J. Acoust. Soc. Am. 51, 515-517.

Herald, E. S., Brownell, R. L. Jr., Frye, F. L., Morris, E. J., Evans, W. E., and Scott, A. B., 1969, Blind river dolphin: first side-swimming cetacean, Science, 166:1408-1410.

Kamminga, C. and Wiersma, H., 1981, Investigations on cetacean sonar II. Acoustic similarities and differences in odontocete sonar signals, Aq. Mammals 8:41-62.

Leatherwood, S., 1975, Some observations of feeding behavior of bottle-
nosed dolphins (Tursiops truncatus) in the Northern Gulf of Mexico and
(Tursiops cf gilli) off Southern California, and Nayarit, Mexico, Mar.
Fish. Rev., 37(9):10-16.

Legatt, L. J., Merklinger, H. M., and Kennedy, J. L., 1981, LNG carrier
underwater noise study for Baffin Bay, in: N. M. Peterson, ed., "The
Question of Sound from Icebreaker Operations: Proceedings of a Workshop," Arctic Pilot Project, Petro-Canada.

Norris, K. S., 1969, The echolocation of marine mammals, in: "The Biology
of Marine Mammals", H. T. Andersen, ed., Academic Press, New York.

Norris, K. S., and Dohl, T. P., 1980, Behavior of the Hawaiian spinner
dolphin, Stenella longirostris, Fish. Bull., 77:821.

Perrin, W. F., Warner, R. R., Fiscus, C. H., and Holts, D. B., 1973,
Stomach contents of porpoise, Stenella spp., and yellowfin tuna, Thunnus
albacares, in mixed-species aggregations, Fish. Bull., 71:1077-1092.

Pilleri, G., Peixun, C.,Peilin, L., Renjun, L., and Kejie, L., 1980,
Distribution, ecology, behavior and conservation of the dolphins of the
middle reaches of Changjiang (Yangtze) River (Wuhan-Yueyang), Invest.
Cetacea, 10:87-103.

Penner, R.H. and Murchison, A.E., 1970, Experimentally demonstrated echolocation in the Amazon River porpoise Inia geofrensis (Blainville),
Proc. 7th Ann. Conf. Bio. Sonar and Diving Mam., 7:17-38.

Purves, P.E. and Pilleri, G.E., 1983, "Echolocation in whales and dolphins",
Academic Press, New York.

Renouf, D. and Davis, M. B., 1982, Evidence that seals may use echolocation,
Nature, 300:635.

Schevill, W. E. and Watkins, W. A., 1966, Sound structure and directionality in Orcinus (killer whale). Zoologica, 51:71-76.

Scronce, B.L., and Johnson C.S., 1975, Bistatic target detection by a
bottlenose porpoise, J. Acoust. Soc. Amer., 59:1001-1002.

Sonafrank, N., Elsner, R., and Wartzok, D., 1983, Under-ice navigation by
the spotted seal, Phoca largha, Abstract, Fifth Biennial Conf. on the
Biol. of Mar. Mammals, Boston, November 1983.

Thomas, J. A., Fisher, S. R., Evans, W. E., and Awbrey, F.T., 1982, Ultrasonic vocalizations of leopard seals (Hydrurga leptonyx), Antarctic
Journal, 17:186.

Thomas, J., Fisher, S. R., Yohe, E., Garver, A., Spafford, J., and Peterson,
J., 1986, Experimental verification of the echolocation abilities of a
false killer whale Pseudorca crassidens. Tech Rept. 86-197, Hubbs
Marine Research Institute, San Diego.

Voronov, V. H. and Stosman, I. T., 1986, Electrical responses of the stem
structures of the acoustic system of Phocoena phocoena to tonal stimuli,
in: "The Electrophysiology of the Sensory Systems of Marine Mammals",
V. E. Sokolov, ed., Nauk, Moscow (in Russian).

Wartzok, D., Schusterman, R.J., and Gailey-Phillips, J., 1984, Letter to
the editor, Nature, 308:753.

Watkins, W. A., and Schevill, W. E., 1980, Characteristic features of the
underwater sounds of Cephalorhynchus commersonii, J. Mamm., 61:738-
739.

Watkins, W. A., and Wartzok, D., 1985, Sensory biophysics of marine mammals,
Mar. Mamm. Sci., 1:219.

Wood, F. G. and W. E. Evans, 1980, Adaptiveness and ecology of echolocation
in toothed whales, pp. 381-425, in: "Animal Sonar Systems", R. G.
Busnel and J. F. Fish, eds., Plenum Press, New York.

Zbinden, K. 1982, "Das Sonarsystem der Zahnwale," Universitaet Bern,
Switzerland.

BEHAVIOUR AND FORAGING ECOLOGY OF ECHOLOCATING BATS

G. Neuweiler[1] and M. B. Fenton[2]

[1] Zoologisches Institut, Universitat Munchen, 14
Luisenstrabe, 8 Munchen 2, West Germany

[2] Department of Biology, York University, Downsview
Ontario, Canada M3J 1P3

INTRODUCTION

Since the 1979 Jersey Biosonar we have accumulated a great deal of
information about the ecology (review in Kunz 1982) and hunting behaviour of
echolocating bats in field and laboratory settings. These advances have been
paralleled by work on the hearing abilities of bats and efforts in several
geographic settings to place research on echolocating bats and their insect prey
in a broader perspective. The 1980 (Busnel and Fish) volume represents the
starting point for our review.

AUDITORY ADAPTATIONS TO ECHOLOCATION

Before we discuss how the designs of echolocation calls might or might not
be associated with specific ecological constraints we must consider adaptations of
the auditory system associated with echolocation. In the past the question of
auditory adaptations for echolocation rarely have been discussed and the available
data base is limited (Grinnell and Hagiwara 1972a, b; Jen and Suthers 1982;
Neuweiler 1984; Schmidt, Turke and Vogler 1984; Neuweiler, Singh and Sripathi
1984; Taniguchi 1985; Rubsamen, Neuweiler and Sripathi, in press). We can,
however, make some general points.

It is important to recognize that ultrasonic is not synonymous with
echolocation even though they commonly are equated (e.g., Pettigrew 1986).
Since sensitivity to ultrasonic sound is not a prerequisite for echolocation, why do
some animals echolocate while others do not and is there an adaptation central to
echolocation? Because the audiograms of nonecholocating Megachiroptera differ
little from those of many echolocating Microchiroptera, Grinnell and Hagiwara
(1972 a and b) suggested that the auditory adaptation to echolocation might be
found in the domain of time analysis. This was strongly supported by the discovery
of echoranging neurons in the auditory cortex of Pteronotus parnellii
(Mormoopidae) and has been further bolstered by other neurophysiological
evidence (reviewed in Neuweiler 1983). More recently, Roverud (this volume) has
shown the importance of time windows in distance discrimination by echolocation,
although the importance of time windows in other tasks (e.g., flutter target
detection or texture discrimination) remains to be determined.

If time windows improve auditory echo analysis for many tasks, the hypothesis that the sensory adaptations for echolocation occur in the auditory regions of the brain and not in the ears will become more attractive. This will mean that echolocation is a specific neuronal and not a cochlear achievement. If this hypothesis is correct, neuronal time windows will be a general adaptation to echolocation irrespective of ecological conditions. There is evidence, however, that the frequency band used for echoanalysis, absolute auditory sensitivity, and specific sensitivities to distinct types of auditory stimuli are tailored to different foraging strategies.

CALL DESIGN

By the Jersey meeting there was dissatisfaction with the terminology commonly used to describe the echolocation calls of bats (e.g., Gustafson and Schnitzler 1979, Pye 1980a; Habersetzer 1981). Although the terms Constant Frequency (CF), and Frequency Modulated (FM) can adequately describe the components of echolocation calls, they do not necessarily reflect the bats' use of the calls. Sometimes the terms do not accurately describe the calls, e.g., the "CF" of Noctilio spp. (Noctilionidae). The use of echolocation calls as the basis for preparing phylogenies of bats (e.g., Simmons and Stein 1980) has not been enthusiastically endorsed by other workers (e.g., Gustafson and Schnitzler 1979; Pye 1980a; Habersetzer 1981), although Simmons, Kick and Lawrence (1984) maintain that the echolocation calls of Rhinopoma hardwickei are relatively 'primitive' in design.

However, Simmons and Stein (1980) also showed how bats might collect data from calls of different design and made it easy to visualize how broader bandwidth calls could provide better resolution of target position. There has been considerable disagreement about the hypotheses of Simmons and Stein (1980) explaining how bats analyze acoustic data (Schnitzler, pers. comm.), but their presentation makes it easier to visualize how call design could influence the information available to the bat.

A summary of different basic call designs used by echolocating bats is presented in Table 1, while Table 2 shows associations between foraging strategies and echolocation call designs.

Flutter-Detection

An exception to the generalization about bandwidth is provided by species exploiting Doppler shifted echoes to collect information about background and fluttering targets (reviewed in Schnitzler and Ostwald 1983; Schnitzler, in press). In bats with calls dominated by CF components, notably P. parnellii and rhinolophid and hipposiderid bats, "acoustic foveas" - the specializations of the auditory system (summarized in Vater, in press) - permit the echolocating bat to exploit Doppler shifted echoes. Field studies of Rhinolophus rouxi (Rhinolophidae) (Schnitzler, Hackbarth, Heilmann and Herbert 1985; Neuweiler, Metzner, Heilmann, Rubsamen, Eckrich, and Costa in press) and Hipposideros caffer (Hipposideridae) (Bell and Fenton 1984) demonstrated how bats detected fluttering targets in a variety of situations, supporting laboratory work on Rhinolophus ferrumequinum (Schnitzler and Ostwald 1983).

Rhinolophids, hipposiderids and P. parnellii use the CF component as a detection signal, and in P. parnelii, neurophysiological studies have demonstrated that the terminal FM component serves for target ranging (O'Neill and Suga 1982) as it does in rhinolophids (O'Neill, Schuller and Radke-Schuller 1985). However, the CF component and the FM sweep are not processed completely independently. Roverud (this volume) has demonstrated that the onset of the CF triggers auditory mechanisms important for target range discrimination in R. rouxi. Biologists should

Table 1. Design of search phase echolocation calls and their clutter resistance and use in foraging

call design	bandwidth	duration in ms	clutter-resistance	families or species of bats known to use the calls
A- calls dominated by constant frequency, but ending in frequency modulated sweep (CF-FM).	narrow	10 – 50	high	Rhinolophidae, Hipposideridae and Mormoopidae (Pteronotus parnellii)
B- Shallow FM sweep (FM)	narrow	> 10	low	Rhinopomatidae some Emballonuridae, some Vespertilionidae and some Molossidae
C- Steep and shallow FM (FM) or sometimes short CF	broad	c. 10	high	some Vespertilionidae (Lasiurus borealis, L. cinereus, Myotis volans), and Noctilionidae
D- Steep FM	broad	< 5	high	many Vespertilionidae, some Emballonuridae
E- Short, low intensity and multiharmonic	broad	< 2	high	Nycteridae, Megadermatidae some Phyllostomidae, and some Vespertilionidae (may use echolocation only facultatively)

Table 2. Foraging strategies of animal-eating bats and associated echolocation calls

foraging strategy	call design (from Table 1)	species	source
1. fluttering prey, whether flying or sitting, whether from continuous flight, or flights from perches	A	R. ferrumequinum	Schnitzler & Ostwald 1983
		R. rouxi	Schnitzler et al. 1985
		R. hildebrandti	Fenton (this volume)
		H. caffer	Bell & Fenton 1984
		H. diadema	Brown & Berry 1983
		H. speoris	Habersetzer et al. 1984
		H. bicolor	"
		P. parnellii	Goldman & Henson 1977
2. pursues flying insects			
... 2.1 from continuous flight			
...... 2.1.a. detected at long (> 5 m) range	B, C	Rhinopoma	Habersetzer 1981; Simmons et al. 1984
		Taphozous mauritianus	Fenton et al. 1981
		T. georgianus	Fenton 1982
		T. kachensis	Neuweiler 1983
		many molossids	Fenton & Bell 1981; Fenton 1982
			Neuweiler et al. 1984
		Nyctalus noctula	Vogler & Neuweiler 1983
		Lasiurus cinereus	Barclay 1985; Belwood & Fullard 1984
...... 2.1.b. detected at short (< 5 m) range	D	Myotis volans	Fenton & Bell 1979
		M. lucifugus	Fenton & Bell 1979
		M. californicus	"
		Pipistrellus pipistrellus	Racey & Swift 1985
		P. dormeri	Neuweiler 1984
		P. mimus	"
		Lasionycteris noctivagans	Barclay 1985

... 2.2 from a perch	E	Cardioderma cor	Vaughan 1976
		Lavia frons	Vaughan & Vaughan 1986
3 moving, but non-flying prey			
3.a. often over water	D	Myotis lucifugus	Buchler 1980
		Nyctophilus balstoni	Fenton 1982
	C	Noctilio leporinus	Wenstrup & Suthers 1984
		N. albiventris	Roverud & Grinnell 1985;
			Brown et al. 1983
3.b. not necessarily over water	D	Myotis adversus	Thompson & Fenton 1982
	E	*Megaderma lyra	Fiedler 1979
		Nycteris thebaica	Fenton et al. 1983
		N. grandis	"
		*Antrozous pallidus	Bell 1982
		*Macrotus californicus	Bell 1984
4. stationary prey (often providing other cues)	D	Myotis auriculus	Fenton & Bell 1979
		M. myotis	Neuweiler 1983
	E	*Megaderma lyra	Fiedler 1979
		*Macrotus californicus	Bell 1984
		*Antrozous pallidus	Bell 1982
		Trachops cirrhosus	Barclay et al. 1981
		Nycteris thebaica	Fenton et al. 1983
		N. grandis	"

* often does not use echolocation during attacks

not treat the CF and FM components as separate entities. In some cases, CF-FM calls are preceeded by upwards FM sweeps (e.g., R. rouxi – Neuweiler et al. in press) and the significance of this call component has yet to be studied.

The rhinolophids, hipposiderids and P. parnellii tested to date respond only to fluttering or vibrating targets and Hipposideros bicolor also responds to targets moving on the ground (Link, Marimuthu and Neuweiler in press). In horseshoe bats, the CF component of the echolocation signals and the cochlear frequency fovea produce a specific and noise-proof flutter-detection system. In a rhinolophids' brain major parts of all auditory nuclei transfer this movement information. Because the CF component of hipposiderid calls is shorter than the rhinolophid CF (< 10 vs > 20 ms), their ability to detect flutter had been questioned. Comparative behavioural experiments have shown that Hipposideros speoris, and H. bicolor detect fluttering prey as well as rhinolophids (Link et al. 1986), while Hipposideros commersoni (Vaughan 1977) and H. caffer (Bell and Fenton 1984) also attack fluttering targets. Both H. bicolor and H. speoris have narrow filters tuned to the species-specific CF frequencies and at least in H. speoris, the CF frequency is vastly over-represented in the colliculus inferior (Link et al. 1986). It seems clear that hipposiderids also have an acoustic fovea which enables them to detect fluttering prey as well as rhinolophids or P. parnellii. The question "why do rhinolophids use longer CF signals than hipposiderids?" remains unanswered.

Long Narrowband Signals

Several types of bats produce search phase echolocation calls (Tables 2 and 3) dominated by shallow FM sweeps – narrow bandwidth over relatively long (> 10 ms) duration. Among these bats are Rhinopoma spp. (Rhinopomatidae) (Habersetzer 1981; Simmons et al. 1984), Taphozous spp. (Emballonuridae) (Fenton, Bell and Thomas 1980; Fenton 1982; Neuweiler 1984), Pipistrellus dormeri, Pipistrellus mimus (Vespertilionidae) (Neuweiler et al. 1984), and many molossids (Simmons et al. 1978; Fenton and Bell 1981). Other species use FM calls which include both steep and shallow FM components (e.g., the vespertilionids Myotis volans, Lasiurus borealis, Lasiurus cinereus, Nyctalus noctula –Fenton and Bell 1979; Fenton, Merriam and Holroyd 1983; Barclay 1985, Vogler and Neuweiler 1983, respectively). Shallow FM sweeps appear well designed for detecting targets, and if Schnitzler (in press) is correct, glints associated with changes in wing positions in flying insects are conspicuous when the duration of the echolocation call exceeds one wingbeat cycle of the prey. Glint detection should increase the bats' effective range of detection.

The shallow FM sweeps typical of many bats which forage in the open reflect a different adaptation than the narrowband calls associated with flutter-detection, primarily because of the way echoes are treated by the bats. Whereas flutter-detecting species have narrowly tuned filters to maximize effectiveness, these auditory specializations are lacking in the species using narrowband signals in open habitats. Typically shallow FM sweeps are lower in frequency (usually < 30 kHz) than the CF components of flutter-detectors (usually > 60 kHz).

Increased range may be an important factor favouring shallow FM sweeps. Sound energy travelling through air is reduced by the expansion of sound in space (geometric losses) and by atmospheric attenuation which is positively correlated with increasing frequency (Lawrence and Simmons 1982). For a 19mm diameter sphere at 10 m the echo will be attenuated by 93 dB for an emitted signal of 30 kHz, and by 145 dB for one of 120 kHz (Lawrence and Simmons 1982). This emphasizes the importance of lower frequencies in achieving long range detection and reflect the hunting behaviour of bats in a variety of settings (Fig. 1; Table 3).

Table 3. Bats using long, narrowband, multiharmonic search calls when hunting in the open.

family	species	harmonic with the most energy (kHz)	source
Rhinopomatidae	Rhinopoma hardwickei	35	Habersetzer 1981
Emballonuridae	Saccopteryx bilineata	45	Pye 1980b
	Coleura afra	28	Pye 1980b
	Taphozous peli?	30	Pye 1980b
	T. kachensis	24	Habersetzer 1983
	T. melanopogon	26	Habersetzer 1983
	T. georgianus	17	Fenton 1982
	T. mauritianus	26	Fenton et al. 1980
Phyllostomidae	Vampyrum spectrum	18	Bradbury 1970
Vespertilionidae	Vespertilio murinus	25	Ahlen 1981
	Nyctalus noctula	25	Pye 1980b
	Lasiurus cinereus	24	Barclay 1985
	Eptesicus fuscus	28	Fenton and Bell 1981
	E. pumilus	51	Fenton 1982
	Pipistrellus pipistrellus	50	Pye 1980b
	P. mimus	50	Habersetzer 1983
	?Myotis volans	46	Fenton and Bell 1981
	?M. dasycneme	35	Ahlen 1981
	Plecotus phyllotis	27	Simmons and O'Farrell 1977
Molossidae	Tadarida aegyptiaca	18	Fenton and Bell 1981
	T. macrotis	17	Fenton and Bell 1981
	T. ansorgei	18	Fenton and Bell 1981
	T. chapini	19	Fenton and Bell 1981
	T. fulminans	17	Fenton and Bell 1981
	T. beccarri?	28	Fenton 1982
	T. jobensis?	22	Fenton 1982
	T. midas	15	Fenton, this volume
	Otomops martiensseni	12	Fenton and Bell 1981

? before = narrowband sound not unequivocally proved

? after = identity of species not certain

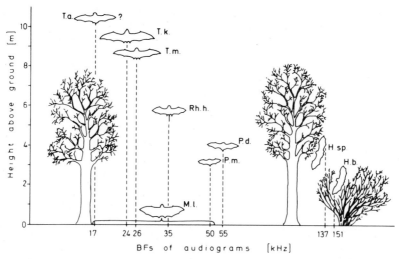

Fig. l. Altitude of foraging bats from southern India plotted against habitat features and best frequencies (BFs) of audiograms in kHz. The species include Tadarida aegyptiacus (T.a.), Taphozous kachensis (T.k.), Taphozous melanopogon (T.m.), Rhinopoma hardwickei (Rh.h.), Megaderma lyra (M.l.), Pipistrellus dormeri (P.d.), Pipistrellus mimus (P.m.), Hipposideros speoris (H.sp.) and Hipposideros bicolor (H.b.). The data are from Neuweiler et al. (1984).

It is clear that echolocating bats have to compromise on two aspects of information retrieval. One trade is between good reflectivity and good imaging, the other between long and short distances of detection. Good imaging (= high frequency) means short range, while poor imaging (= low frequency) means longer range.

Steep F M Sweeps

Search phase echolocation calls dominated by steep F M sweeps (Table l) are short (< 5 ms) and have broad bandwidths. They appear to be well suited for precise target location in clutter and are more typical of many small vespertilionoids and some emballonurids which forage in confined situations. There appear to be two mechanisms for achieving broad bandwidth, one the addition of harmonics, the other sweeps of greater magnitude. The two approaches are typified in some Myotis (Vespertilionidae), with M. lucifugus, M. yumanensis, M. daubentoni, M. californicus and M. leibii using the harmonic approach, and Myotis myotis, M. septentrionalis, M. evotis, and M. auriculus the other approach.

Broadbandwidth calls of low intensity and short duration (< 2 ms) are typical of many species of bats (Tables l and 2) which hunt in heavy clutter. Vocalizations of this nature have been reported from nycterid, megadermatid, phyllostomine and some vespertilionid bats. The situation in these species is often complicated because the bats may refrain from echolocating in some situations (e.g., Megaderma lyra (Megadermatidae); Macrotus californicus (Phyllostomidae); Antrozous pallidus (Vespertilionidae)) or continue to produce echolocation calls when obviously relying on other cues to find their targets (e.g., Trachops cirrhosus (Phyllostomidae)- Barclay, Fenton, Tuttle and Ryan 1981, Nycteris thebaica and N. grandis (Nycteridae) - Fenton, Gaudet and Leonard 1983).

The audiograms of species using this category of echolocation calls show that the bats have good to excellent sensitivity in the lower frequency range (10 - 30 kHz). There is more and more evidence from field studies that sounds associated with prey (e.g., Fiedler 1979; Tuttle and Ryan 1981; Bell 1982, 1985; Fenton, Gaudet and Leonard 1983; Link et al. in press) or even habitats where prey may occur (Buchler and Childs 1981) may be used to locate food or feeding areas. Indeed, much of the accumulating data suggests good sensitivity to sounds below 10 kHz (e.g., Bell 1982; Ryan, Tuttle and Barclay 1983) which is strongly influenced by the pinnae (Rubsamen et al. in press).   Low frequency hearing in bats is not just associated with communication.

Harmonics

Echolocation calls in each of the categories discussed above may include harmonics whose functional significance may be unclear. For example, long, shallow FM signals often include up to four harmonics bringing into question the view that these signals are narrowband. In P. dormeri, P. mimus and R. hardwickei, the audiograms show distinct peaks of sensitivity at the frequencies of the harmonic containing the most sound energy (e.g., 35 kHz in R. hardwickei, 54 kHz in P. mimus Neuweiler et al. 1984). These data favour the interpretation that the harmonics are less important because the bat treats its signals as narrowband.

O'Neill and Suga (1982) showed that in P. parnellii, the first harmonic serves as a marker triggering range processing mechanisms.   Roverud (this volume), demonstrated that in R. rouxi presentation of the first and/or second harmonic disrupts performance in range discrimination tasks. Furthermore, foraging R. rouxi vary the harmonic components of their echolocation calls (Neuweiler et al. in press).

To date there is no plausible explanation of why flutter-detecting bats alter the energy distribution of harmonics in their echolocation calls. The bats could increase range of operation by emitting lower frequencies, but with the exception of P. parnellii, there is no evidence of increased auditory sensitivity at the frequency of the first harmonics.

Species-specific Echolocation Calls

Fenton and Bell (1979, 1981) and Ahlen (1981) showed that one could identify different species of bats by their echolocation calls, a situation also explored by others (e.g., Miller and Andersen 1984; Fenton 1982; Fenton, Merriam and Holroyd 1983). Thomas and West (1984) pointed out that studies such as faunal surveys which depend upon identifications of bats by their echolocation calls required properly calibrated equipment and several different bat detecting systems are available (reviews by Pye 1983; Fenton 1986a). There are obvious limitations to the technique, some associated with calls of different intensity (Fenton and Bell 1981), but also with variability in echolocation calls, a largely unexplored subject (but see Habersetzer 1981; Barclay 1983; Fenton 1986b; Thomas, Bell and Fenton in press). In spite of the limitations of the technique it promises to provide a great deal of data about the distribution and behaviour of some bats in the field (e.g., Baagoe in press; Miller and Andersen 1984; Fenton, Tennant and Wyszecki 1986).

FORAGING STRATEGY

Most recent data on this topic deals with animal-eating bats. The role of echolocation in the lives of frugivorous phyllostomids remains unclear, and the discovery that some vampires (Desmodus rotundus - Phyllostomidae) have infra-red detecting abilities (Kurten and Schmidt 1982) in addition to good echolocation, hearing, visual, and olfactory senses only clouds the picture.

Is it possible to identify particular echolocation call designs with specific foraging strategies? The answer is clearer now than it was in 1979, but the data base is still small. Another facet of this situation now being explored is wing structure, its influence on maneouvrability in flight and the implications of this for habitat use and echolocation call design (Aldridge and Rautenbach, submitted). The foraging strategies of animal-eating bats and the associated echolocation calls are shown in Table 2.

Bats with echolocation systems designed to exploit Doppler shifted echoes through specialized auditory systems appear to hunt fluttering targets (Table 2). They may hunt in a flycatcher mode (making short flights from perches), or from continuous flight, and detect targets in a range of settings. The important aspect of their foraging behaviour is the specialization for flutter-detection, and it is inappropriate to label them as gleaners or chasers of flying insects.

Species using long calls with shallow FM components (Tables 2 and 3) appear to be able to exploit glint from insects (Schnitzler in press) to increase their effective range. These are often species which forage in uncluttered situations and operate at long range. By comparison, species with short calls (< 5 ms) of broad bandwidth and steep FM sweeps operate at shorter range exploiting calls which are more clutter-resistant (Table 2) and perhaps working in swarms of insects (Fenton and Bell 1979).

Gleaning bats, those taking prey from surfaces, have received a great deal of speculative attention in the literature. As noted above, these species often produce short (< 2 ms) calls of low intensity and broadbandwidth. Some may use echolocation to locate targets (e.g., Myotis auriculus, M. myotis), while others produce echolocation calls even when orienting on other cues (see above and Table 2). Still others refrain from echolocating.

Associated with foraging strategy is the question of prey selection. Swift and Racey (1983) showed that Myotis daubentoni and Plecotus auritus (Vespertilionidae) used different foraging strategies, foraged in different habitats and took different prey, while Herd and Fenton (1983) found that Myotis lucifugus and Myotis yumanensis used the same foraging strategies in different habitats and took different prey. Bell and Fenton (1984) found no evidence that H. caffer selected one species of moths over others, although they fed mainly on moths, as does Cloeotis percivali (Hipposideridae – Whitaker and Black 1976). Schnitzler (in press) makes a convincing case that some flutter-detecting bats should be able to recognize preferred prey by wingbeat signatures, but whether they have the chance to do so in the wild remains to be determined. At this stage details about the association between foraging strategy and prey selection are scanty, the best being those of Barclay (1985) who found differences in the diets of the short-range Lasionycteris noctivagans and the long-range L. cinereus.

COMMUNICATION

Habersetzer (1981) demonstrated how R. hardwickei altered the frequencies of its echolocation calls depending upon whether it was hunting by itself or in a group of conspecifics, and Barclay (1983) documented call variation in emballonurids in Panama. Together these studies re-enforced earlier suggestions (summarized in Fenton 1980) that the vocalizations one individual used for orientation, could simultaneously serve a communication function.

Barclay (1982) used playback experiments with free-flying Myotis lucifugus in the field to demonstrate that these bats used the echolocation calls of others as aggregation cues, whether in a feeding, roosting, or mating situation. In contrast, Leonard and Fenton (1984) found that free-flying Euderma maculatum (Vespertilionidae) appeared to exploit echolocation calls of conspecifics as a means for maintaining spacing in foraging habitats.

Matsumura (1981) reported antiphonal and then synchronous calling by mother and infant R. ferrumequinum and she noted that mother-young reunions in this species involved no other specialized vocalizations common in other bats (review in Fenton 1985). Recent work by Rubsamen (this volume) casts doubt on the generality of this situation for he found strong sexual dimorphism in calls in R.

rouxi which would preclude females successfully finding their sons in the scenario described by Matsumura (1981).

Thomson, Fenton and Barclay (1985) found that echolocation calls served a communication role in mother-young reunions in Myotis lucifugus, a species in which females and young produce additional specialized calls (double notes, and isolation calls, respectively) before successful reunions. There is increasing evidence that mother bats selectively nurse their own young (even the molossid Tadarida brasiliensis - McCracken 1984) and that vocalizations are important distal cues in mediating mother-young recognition.

Echolocation calls can serve communication functions. Differences in call design may sometimes reflect species recognition patterns as much as resolution of problems involving orientation.

SENSORY ECOLOGY

Since echolocation is an active mode of orientation direct costs associated with its use will include cost of producing vocalizations and processing echoes, as well as information leakage (Fenton 1984). Fiedler (1979) demonstrated that M. lyra facultatively switched from echolocation to orientation using acoustic cues originating from their targets, according to the situation in which they were operating. Since then, Bell (1982, 1985) has shown that A. pallidus and M. californicus also avoid using echolocation in some settings, relying instead on acoustic (both species) or visual (M. californicus) cues (Bell and Fenton 1986). As noted earlier, some other echolocating species (e.g., T. cirrhosus, N. thebaica and N. grandis) may continue to produce echolocation calls even when clearly using acoustic cues emanating from their targets (Barclay et al. 1981; Fenton, Gaudet and Leonard 1983). There is unpublished evidence (Gaudet 1982) that M. lucifugus and E. fuscus continue to produce echolocation calls even when making visual discriminations.

Schmidt and his colleagues have begun to explore some aspects of the sensory ecology of bats, considering how different cues are used in different settings (this volume). Bats have the potential for providing us with excellent examples of sensory flexibility given their access to cues in visual, olfactory, echolocation and acoustic modes.

VARIATION

The problem of variability in behaviour haunts studies of echolocation. Marler (1985) identified variability as one of the building blocks of evolution, and yet we are still a long way from measuring, let alone understanding how bats can vary their echolocation behaviour. On one level, the ability of an individual to vary the design of its calls (or to rely on other cues) depending upon the setting in which it is hunting influences our efforts to categorize the foraging behaviour of bats. Neuweiler (1983) remarked on the important case of Myotis myotis where individuals use calls of the same design when challenged with different discrimination tasks (Habersetzer and Vogler 1983). In some bats, there may be little variation in call design. On the next level, the flexibility individuals show in foraging strategies over time can affect their selection of prey and their success. When questions (and we now have few answers) about variability are not addressed, we are failing to place our work in a broader evolutionary perspective.

Some of the studies in this volume show that workers are beginning to measure variability. When we remember that bats live a long time and have the capacity to learn by observation (Gaudet and Fenton 1984), the need for measures of variability is more obvious.

LITERATURE

Ahlen, I. 1981. Identification of Scandinavian bats by their sounds. Swedish Univ. Agric. Sci. Dept. Wildl. Ecol. Report 6, Uppsala.

Aldridge, H.D.J.N. and L.L. Rautenbach. Morphology, echolocation and resource partitioning in insectivorous bats. submitted to Journal of Animal Ecology.

Baagoe, H.J. in press. The Scandinavian bat fauna: adaptive wing morphology and free flight in the field. pp.   IN Recent advances in the study of bats (M.B. Fenton, P.A. Racey and J.M.V. Rayner, eds.). Cambridge Univ. Press, Cambridge, U.K.

Barclay, R.M.R. 1982. Interindividual use of echolocation calls: eavesdropping by bats. Behav. Ecol. Sociobiol., 10:271-275.

-           1983. Echolocation calls of emballonurid bats from Panama. J. Comp. Physiol., 151:515-520.

-           1985. Long- versus short-range strategies of hoary (Lasiurus cinereus and silver-haired (Lasionycteris noctivagans) bats and the consequences for prey selection. Can. J. Zool., 63:2507-2515.

Barclay, R.M.R., M.B. Fenton, M.D. Tuttle and M.J. Ryan. 1981. Echolocation calls produced by Trachops cirrhosus (Chiroptera : Phyllostomatidae) while hunting for frogs. Can. J. Zool., 59:750-753.

Bell, G.P. 1982. Behavioral and ecological aspects of gleaning by the desert insectivorous bat Antrozous pallidus (Chiroptera: Vespertilionidae). Behav. Ecol. Sociobiol., 10:217-223.

-           1985. The sensory basis of prey location by the California leaf-nosed bat Macrotus californicus (Chiroptera: Phyllostomidae). Behav. Ecol. Sociobiol., 16:343-347.

Bell, G.P. and M.B. Fenton 1984. The use of Doppler-shifted echoes as a flutter detection and clutter rejection system: the echolocation and feeding behavior of Hipposideros ruber (Chiroptera : Hipposideridae). Behav. Ecol. Sociobiol., 15:109-114.

-           1986. Visual acuity, sensitivity and binocularity in a gleaning insectivorous bat, Macrotus californicus (Phyllostomidae). Anim. Behav., 34:409-414.

Belwood, J.J. and J.H. Fullard. 1984. Echolocation and foraging behaviour in the Hawaiian hoary bat, Lasiurus cinereus semotus. Can. J. Zool., 62:2113-2120.

Bradbury, J.W. 1970. Target discrimination by the echolocati bat, Vampyrum spectrum. J. Exp. Zool., 173:23-46.

Brown, P.E. and R.D. Berry. 1983. Echolocaion behavior in a 'flycatcher' bat, Hipposideros diadema. J. Acoust. Soc. Am. Suppl., 1:74:32.

Buchler, E.R. 1980. The development of flight, foraging and echolocation in the little brown bat (Myotis lucifugus). Behav. Ecol. Sociobiol., 6:211-218.

Buchler, E.R. and S.B. Childs. 1981. Orientation to distant sounds by foraging big brown bats (Eptesicus fuscus). Anim. Behav., 29:428-432.

Busnel, R-G. and J.F. Fish (eds). 1980. Animal sonar systems. NATO Advanced Study Institutes Vol. A28, Plenum Press, NY.

Fenton, M.B. 1980. Adaptiveness and ecology of echolocation in terrestrial (aerial) systems. pp. 427-446 IN Animal Sonar Systems (R-G. Busnel and J.F. Fish, eds.), NATO Advanced Study Institutes Series, Vol A28, Plenum Press, N.Y.

-           1982. Echolocation calls and patterns of hunting and habitat use of bats (Microchiroptera) from Chillagoe, north Queensland. Aust. J. Zool., 30:417-425.

-           1984. Echolocation: implications for ecology and evolution of bats. Quart. Rev. Biol., 59:33-53.

-           1985. Communication in the Chiroptera. Indiana Univ. Press, Bloomington.

-           1986a. Detecting, recording and analyzing the vocalizations of bats. pp. IN Ecological and behavioral methods for the study of bats (T.H. Kunz, ed.). Smithsonian Institution Press, Washington.

-           1986b. Hipposideros caffer (Chiroptera : Hipposideridae) in Zimbabwe: morphology and echolocation calls. J. Zool. (London), in press.

Fenton, M.B. and G.P. Bell. 1979. Echolocation and feeding behaviour in four species of Myotis (Chiroptera). Can. J. Zool., 57:1271-1277.
- 1981. Recognition of species of insectivorous bats by their echolocation calls. J. Mamm., 62:233-243.
Fenton, M.B., G.P. Bell and D.W. Thomas. 1980. Echolocation and feeding behaviour of Taphozous mauritianus (Chiroptera : Emballonuridae). Can. J. Zool., 58:1774-1777.
Fenton, M.B., C.L. Gaudet and M.L. Leonard. 1983. Feeding behaviour of Nycteris grandis and Nycteris thebaica (Nycteridae) in captivity. J. Zool. (London), 200:347-354.
Fenton, M.B., H.G. Merriam and G.L. Holroyd. 1983. Bats of Kootenay, Glacier, and Mount Revelstoke National Parks in Canada: identification by echolocation calls, distribution and biology. Can. J. Zool., 61:2503-2508.
Fenton, M.B., D.C. Tennant, and J. Wyszecki. 1986. Using echolocation calls to measure the distribution of bats: the case of Euderma maculatum. J. Mamm., in press.
Fiedler, J. 1979. Prey catching with and without echolocation in the Indian false vampire (Megaderma lyra). Behav. Ecol. Sociobiol., 6:155-160.
Gaudet, C.L. 1982. The behavioural basis of foraging flexibility in three species of insectivorous bats: an experimental study using captive Antrozous pallidus, Eptesicus fuscus and Myotis lucifugus. M.Sc. Thesis, Department of Biology, Carleton University, Ottawa, Canada.
Gaudet, C.L. and M.B. Fenton. 1984. Observational learning in three species of insectivorous bats (Chiroptera). Anim. Behav., 32:385-388.
Goldman, L.J. and O.W. Henson Jr. 1977. Prey recognition and selection by the constant frequency bat, Pteronotus p. parnellii. Behav. Ecol. Sociobiol., 2:411-419.
Grinnell, A.D. and S. Hagiwara. 1972a. Adaptations of the auditory nervous system for echolocation. Z. Vergl. Physiol., 76:41-81.
- 1972b. Studies of the auditory neurophysiology in non-echolocating bats, and adaptations for echolocation in one genus Rousettus. Z. Vergl. Physiol., 76:82-92.
Gustafson, Y., and H-U. Schnitzler. 1979. Echolocation and obstacle avoidance in the hipposiderid bat Asellia tridens. J. Comp. Physiol., 131:161-167.
Habersetzer, J. 1981. Adaptive echolocation sounds in the bat Rhinopoma hardwickei, a field study. J. Comp. Physiol., 144:559-566.
- 1983. Ethookologische Untersuchungen an echoortenden Fledermausen Sudiniens. Doctoral Thesis, Fachbereich Biologie Univ. Frankfurt.
Habersetzer, J., G. Schuller and G. Neuweiler. 1984. Foraging behavior and Doppler shift compensation in echolocating hipposiderid bats, Hipposideros bicolor and H. speoris. J. Comp. Physiol., 155:559-568.
Habersetzer, J. and B. Vogler. 1983. Discrimination of surface-structured targets by the echolocating bat, Myotis myotis, during flight. J. Comp. Physiol., 152:275-282.
Herd, R.M. and M.B. Fenton. 1983. An electrophoretic, mophological and ecological investigation of a putative hybrid zone between Myotis lucifugus and Myotis yumanensis (Chiroptera: Vespertilionidae). Can. J. Zool., 61:2029-2050.
Jen, P-H. S. and R.A. Suthers. 1982. Responses of inferior colliculus neurons to acoustic stimuli in certain FM and CF-FM paleotropical bats. J. Comp. Physiol., 146:423-434.
Kunz, T.H. (ed) 1982. Ecology of bats. Plenum Press, New York.
Kurten, L. and U. Schmidt. 1982. Thermoperception in the common vampire bat (Desmodus rotundus). J. Comp. Physiol., 146:223-228.
Lawrence, B.D. and J.A. Simmons. 1982. Measurements of atmospheric attenuation of ultrasonic frequencies and the significance for echolocation by bats. J. Acoust. Soc. Am., 71:585-590.
Leonard, M.L. and M.B. Fenton. 1984. Echolocation calls of Euderma maculatum (Chiroptera: Vespertilionidae): use in orientation and communication. J. Mamm., 65:122-126.

Link, A., G. Marimuthu, and G. Neuweiler. 1986. Movement as a specific stimulus for prey catching behavior in rhinolophid and hipposiderid bats. J. Comp. Physiol., 159:403-414.

Marler, P. 1985. Forward, pp. ix to xi IN The tungara frog, a study in sexual selection and communication by M.J. Ryan. Univ. of Chicago Press, Chicago.

Matsumura, S. 1981. Mother-infant communication in a horseshoe bat (Rhinolophus ferrumequinum nippon): development of vocalizations. J. Mamm., 60:76-84.

McCracken, G.F. 1984. Communal nursing in Mexican free-tailed bat nursery colonies. Science, 223:1090-1091.

Miller, L.H. and B.B. Andersen. 1984. Studying bat echolocation signals using ultrasonic detectors. Z. Saugetierk., 49:6-13.

Neuweiler, G. 1983. Echolocation and adaptivity to ecological constraints, pp. 280 - 302 IN Neuroethology and behavioral physiology, roots and growing points (F. Huber and H. Markl, eds). Springer-Verlag, Berlin.

    -    1984. Foraging, echolocation and audition in bats. Naturwissenschaften, 71:446-455.

Neuweiler, G., W. Metzner, U. Heilmann, R. Rubsamen, M. Eckrich, and H.H. Costa. in press. Foraging behavior and echolocation in the rufus horseshoe bat, Rhinolophus rouxi, of Sri Lanka. Behav. Ecol. Sociobiol.

Neuweiler, G., S. Singh, and K. Sripathi. 1984. Audiograms of a south Indian bat community. J. Comp. Physiol., 154:133-142.

O'Neill, W.E. and N. Suga. 1982. Encoding of target range and its representation in the auditory cortex of the mustached bat. J. Neurosci., 2:17-31.

O'Neill, W.E., G. Schuller and S. Radtke-Schuller. 1985. Functional and anatomical similarities in the auditory cortices of the old world horseshoe bat and neotropical mustached bat for processing of similar biosonar signals. Ass. Res. Otolaryngol. Midwinter Meeting.

Pettigrew, J. 1986. Flying primates? Megabats have the advanced pathway from eye to midbrain. Science, 231:1304-1306.

Pye, J.D. 1980a. Adaptiveness of echolocation signals in bats, flexibility in behaviour and in evolution. Trends Neurosci., October 1980:1-4.

    -    1980b. Echolocation signals and echoes in air, pp. 309-353. IN Animal Sonar Systems (R-G. Busnell and J.F. Fish, eds.). Plenum Press, New York.

    -    1983. Techniques for studying ultrasonic sound, pp. 39-65 IN Bioacoustics (B. Lewis, ed.). Academic Press, New York.

Racey, P.A. and S.M. Swift. 1985. Feeding ecology of Pipistrellus pipistrellus (Chiroptera: Vespertilionidae) during pregnancy and lactation. I. Foraging behaviour. J. Anim. Ecol., 54:205-215.

Roverud, R.C. and A.D. Grinnell. 1985. Discrimination performance and echolocation signal integration requirements for target detection and distance determination in the CF/FM bat, Noctilio albiventris. J. Comp. Physiol., 156:447-456.

Rubsamen, R., G. Neuweiler, and K. Sripathi. in press. Pinnae produce high sensitivity in the gleaning bat, Megaderma lyra. J. Comp. Physiol.

Ryan, M.J., M.D. Tuttle, and R.M.R. Barclay. 1983. Behavioral responses of the frog-eating bat, Trachops cirrhosus, to sonic frequencies. J. Comp. Physiol., 150:413-418.

Schmidt, S., B. Turke, and B. Vogler. 1984. Behavioral audiogram from the bat, Megaderma lyra. Myotis 22:62-66.

Schnitzler, H-U. in press. Echoes of fluttering insects - information for echolocating bats, pp. —— IN Recent advances in the study of bats (M.B. Fenton, P.A. Racey and J.M.V. Rayner, eds.). Cambridge Univ. Press, Cambridge, U.K.

Schnitzler, H-U. and J. Ostwald. 1983. Adaptations for the detection of fluttering insects by echolocation in horseshoe bats, pp. 801-827 IN Advances in vertebrate neuroethology (J-P. Ewart, R.R. Capranica, and D.J. Ingle, eds.). Plenum Press, New York.

Schnitzler, H-U., H. Hackbarth, U. Heilmann and H. Herbert. 1985. Echolocation behavior of rufous horseshoe bats hunting for insects in the flycatcher-style. J. Comp. Physiol., 157:39-46.

Simmons, J.A. and M.J. O'Farrell. 1977. Echolocation by the long-eared bat, Plecotus phyllotis. J. Comp. Physiol., 122:201-214.

Simmons, J.A. and R.A. Stein. 1980. Acoustic imaging in bat sonar: echolocation signals and the evolution of echolocation. J. Comp. Physiol., 135:61-84.

Simmons, J.A., S.A. Kick and B. D. Lawrence. 1984. Echolocation and hearing in the mouse-tailed bat, Rhinopoma hardwickei: acoustic evolution of echolocation in bats. J. Comp. Physiol., 154:347-356.

Simmons, J.A., W.A. Lavender, B.A. Lavender, J.E. Childs, K. Hulebak, M.R. Rigden, J. Sherman, B. Woolman, and M.J. O'Farrell. 1978. Echolocation by free-tailed bats (Tadarida). J. Comp. Physiol., 125:291-299.

Swift, S.M. and P.A. Racey. 1983. Resource partitioning in two species of vespertilionid bats (Chiroptera) occupying the same roost. J. Zool. (London), 200:249-259.

Taniguchi, I. 1985. Echolocation sounds and hearing in the greater Japanese horseshoe bat (Rhinolophus ferrumequinum nippon). J. Comp. Physiol., 156:185-188.

Thomas, D.W., G.P. Bell and M.B. Fenton. in press. Variation in echolocation call frequency recorded from North American vespertilionid bats: a cautionary note. J. Mamm. in press.

Thomas, D.W. and S.D. West. 1984. On the use of ultrasonic detectors for bat species identification and calibration of QMC Mini bat detectors. Can. J. Zool., 62:2677-2679.

Thomson, C.E., M.B. Fenton and R.M.R. Barclay. 1985. The role of infant isolation calls in mother-infant reunions in the little brown bat (Myotis lucifugus; Chiroptera: Vespertilionidae). Can. J. Zool., 63:1982-1988.

Thompson, D. and M.B. Fenton. 1982. Echolocation and feeding behaviour of Myotis adversus (Chiroptera: Vespertilionidae). Aust. J. Zool., 30:543-546.

Tuttle, M.D. and M.J. Ryan. 1981. Bat predation and the evolution of frog vocalizations in the neotropics. Science, 214:677-678.

Vater, M. in press. Narrow band frequency analysis in bats, pp. —— IN Recent advances in the study of bats (M.B. Fenton, P.A. Racey, and J.M.V. Rayner, eds.). Cambridge Univ. Press, Cambridge.

Vaughan, T.A. 1976. Noctural behavior of the Africn false vampire bat (Cardioderma cor). J. Mamm., 57:227-248.

———— 1977. Foraging behaviour of the giant leaf-nosed bat (Hipposideros commersoni). East Afr. Wildl. J., 15:237-249.

Vaughan, T.A. and R.P. Vaughan. 1986. Seasonality and the behavior of the African yellow-winged bat. J. Mamm., 67:91-102.

Vogler, B. and G. Neuweiler. 1983. Echolocation in the noctule (Nyctalus noctula) and horseshoe bat (Rhinolophus ferrumequinum). J. Comp. Physiol., 152:421-432.

Wenstrup, J.J. and R.A. Suthers. 1984. Echolocation of moving targets by the fish-catching bat, Noctilio leporinus. J. Comp. Physiol., 155:75-90.

Whitaker, J.O. Jr. and H.L. Black. 1976. Food habits of cave bats from Zambia, Africa. J. Mamm., 57:199-204.

INTERACTION BETWEEN ECHOLOCATING BATS AND THEIR PREY

Annemarie Surlykke

Institute of Biology
Odense University
DK-5230 Odense M. Danmark

I INTRODUCTION

The echolocation system of microchiropteran bats is more sophisticated than that of any other terrestrial vertebrate (Schnitzler & Henson, 1980). Most bats are obligate echolocators but interaction between bats and their prey is not necessarily mediated solely by echolocation. Bats may rely heavily on passive hearing, i.e. detection of sounds produced by the prey (Barclay, 1982; Barclay et al., 1981; Buchler & Childs, 1981; Tuttle et al., 1985). Other sensory modalities (sight, olfaction, thermoreception) may also be involved in prey detection and localization (Fenton, 1984; Goldman & Henson, 1977; Kurten & Schmidt, 1982; Joerman, 1984; Bell, 1982). Bats exploiting such cues may cease echolocation in the final phase (Fiedler, 1979), hence, minimizing the prey's potential possibility of detecting the echolocation signals of the predator.

Around 540 of the more than 600 microchiropteran bat species eat mainly animal prey. The prey may include frogs, fish, reptiles, small mammals, and even other bats, but most microchiropteran bats are insectivorous. This may be the reason why ears sensitive to bat cries have evolved independently more than once among the nocturnal insects (Roeder, 1974; Michelsen, 1979; Yager & Hoy, 1986).

II HUNTING BEHAVIOR OF BATS

II.1 Bats hunting insects in free air

Bats may use a variety of echolocaton strategies. Signal design is determined by species-specific differences as well as by ecological constraints (Fenton & Bell, 1981; Neuweiler, 1984). The differences in echolocation behavior may reveal important theoretical aspects of bat biosonar. No prey, however, can detect for example the minor variations in emitted frequency which disclose the ability to compensate for Dopplershifts in certain CF-bats (Schnitzler, 1968). Thus with respect to predatorprey interactions the differences in echolocation strategy seem less interesting than the fact that all bats catching insects on the wing go through strikingly similar changes in their acoustic behavior during a hunting sequence irrespective of species and habitat.

The cues theoretically available to tympanate prey are 1) cry intensity, 2) time patterns (cry duration and repetition rate), and 3) frequency content. Among the different bats the time cues seem most consistent. From search to terminal phase the repetition rate increases from 2-5 Hz to 150-200 Hz and the duration of the cries decreases from 10-100 ms to 0.5-10 ms (Neuweiler, 1980; Miller, 1982; See also Neuweiler & Fenton, this volume). The frequency content also changes systematically through the pursuit: The bandwidth increases from search to terminal phase but bandwidth may be an ambiguous cue since the variation among species is large compared to the changes one bat evinces when hunting. The frequency of the main energy output may be as low as 8.8 kHz (Euderma maculatum, Woodsworth et al., 1981) or as high as 212 kHz (Cloeotis percivali, Fenton & Bell, 1981). Most bats have main energies between 30 and 60 kHz (Fenton, 1980). The source level (the cry intensity at 10 cm measured as dB (rms) re. 20 µPa. See Møhl, this volume) varies from below 75 to over 110 dB SPL (Fenton & Bell, 1981; Fenton, 1982), and cry intensity decreases in the terminal phase.

It would also be important to know whether bats hunt selectively or opportunistically. A number of laboratory investigations indicate that bats can identify their insect prey based on either the frequency or amplitude modulations of the echo induced by the fluttering of the wings (Schnitzler et al., 1983; Schnitzler & Flieger, 1983; Schuller, 1984) or on spectral cues caused by the filter properties of the target (Habersetzer & Vogler, 1983; Neuweiler, 1984). Field observations have indicated that some bats select certain prey (Goldman & Henson, 1977; Bell & Fenton, 1984; Fenton & Bell, 1979; Woodsworth et al., 1981). Whether there really are specialists among bats is difficult to say partly because of biases in methods for estimating insect abundance. Moreover, it seems that even bats clearly selective in some situations may be opportunistic in other situations like when insects are less abundant (Black, 1974; Belwood & Fenton, 1976; Barclay, 1985). Hence, if bats do not specialize the safest strategy for the prey must be to assume that all echolocators are potential predators.

II.2 Gleaning bats

About 30% of the bat species preying on animals take their prey from the ground or surfaces (Bell, 1982). Echolocation behavior of gleaning bats differs from that of bats catching insects in open air: their sounds are generally of low intensity, broad bandwidth and very short duration and the stereotyped changes in time patterns, described above, are not found (Neuweiler, 1984; Bell, 1982; Bell & Fenton, 1984; Fenton, 1982;1984). For example the facultative gleaner Myotis auriculus does not increase repetition rate when gleaning but it does while pursuing flying insects (Fenton & Bell, 1979).

III INSECT PREY

The diet of insectivorous bats consists mainly of nocturnal insects. Moths and beetles normally constitute a large proportion but the diet reflects local changes in insect abundance such as dense swarms of emerging termites or of chironomid flies (Belwood & Fenton, 1976). The insect prey may be divided in two groups on basis of the ability to detect the cries of bats:
1) Insects with ears sensitive to ultrasound
2) Non-hearing insects.

Bats hunt non-hearing insects in large numbers. For example Diptera and Coleoptera are very common prey (Barclay, 1985; Whitaker & Tamich, 1983; Woodsworth et al., 1981). Some nocturnal insects, such as large sphingids, are protected by their size and high flight speeds and others (Manduca) may be protected by their big spurs. The majority of these insects appear, however, to be without acoustical or structural defenses. Their population dynamics, for instance synchronized emerging, may protect the species. It would be interesting to know how large a proportion of those insect populations are in fact eaten by bats compared to for example the proportion of moths eaten by bats.

Ears sensitive to the echolocation calls of bats have so far been found in species from 4 orders of nocturnal insects: 1) Lepidoptera (Moths), 2) Neuroptera (Lacewings), 3) Orthoptera (Crickets), and 4) Dictyoptera (Praying mantises).

III.1 Moths

III.1.1 Moth hearing and defenses. Moths of the superfamilies Noctuoidea, Geometroidea, Pyraloidea, and Sphingoidea have ears sensitive to ultrasound (Roeder, 1974; Spangler & Takessian, 1983). The majority of these moths are silent so their ears probably serve only to detect bats. Hence, the ears of moths may constitute an example of a predator-specific defense.

The ear in the Choerocampinae (Sphingidae) is formed from modified mouthparts (the palp and the pilifer) (Roeder et al., 1968), while in other moths the ear is located on the thorax or abdomen and has from 1 to 4 sensory cells (Roeder, 1974). The metathoracic ear of noctuoid moths has been the subject of most physiological investigations. Normally it has 2 sensory cells and the typical noctuid audiograms (Fig. 1) illustrate the variations in sensitivity. The less sensitive of the 2 sensory cells, A_2, has a threshold approximately 15 to 20 dB higher than the A_1 cell. Since both cells have a dynamic range of about 20 dB the total dynamic range of

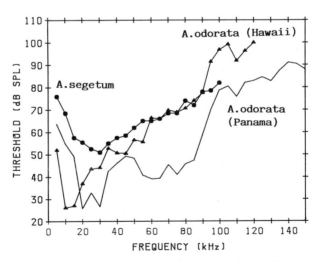

Fig.1. Audiograms from the moths Agrotis segetum (-●-) from Danmark and Ascalapha odorata from Panama (——) and from Hawaii (-▲-). (From Surlykke & Miller, 1982; and Fullard 1984b). A.odorata is one of the most sensitive moths investigated whereas A.segetum is a typical example of the generally less sensitive temperate moths.

this ear is 40 dB. Based on the ability to detect bats moths exhibit a number of different defensive strategies. Evasive maneuvers in response to ultrasound have been observed in sphingid, pyralid and geometrid moths but most behavioral knowledge stem from noctuoids (Roeder et al., 1968; Belton, 1962; Agee, 1969; Roeder, 1974).

i) Negative Phonotaxis. Faint ultrasound (distant bats) elicits a steered turning-away from the sound source accompanied by flight accelera-tion (Roeder, 1962). The directionality of the ear appears to be caused by diffraction which gives differences in the sound level at the two ears (Payne, Roeder & Wallman, 1966; Surlykke, 1984). Roeder (1967a) suggested that a rough measure of the distance to the bat was based on intensity cues. I.e. if a "typical bat" cries with a source level of 110 dB SPL at 30 kHz, the more sensitive cell of A. segetum (Fig.1) will be excited at a distance of 15 m (See Fig.5) whereas the A_2 cell (threshold 67 dB SPL) is not excited untill the distance is shorter than about 6 m. However, intensity cues alone do not give an unequivocal measure of distance. The source level of bat cries varies. Besides, a moth close to a bat but "off-axis" may experience as low a SPL as a moth far away and "on-axis" because of the sound radiation pattern of bat cries (Mogensen & Møhl, 1979). Moths might also use time cues of bat cries to help determine the distance (danger).

ii) Loops and dives. Moths flying close to a sound source respond with a variety of unpredictable, non-directional maneuvers like passive and power dives, loops, and sudden turns, often ending on the ground (Roeder, 1962). Tight turns may outmaneuver the faster and heavier bats and passive falling may be succesful against flutter detecting bats (Goldman & Henson, 1977; Schnitzler & Henson, 1980; Bell & Fenton, 1984; Vogler & Neuweiler, 1983; Schuller, 1984). Roeder (1967a; 1974) suggested that the shift in behavior from steered turning-away to non-directional evasive flight coincided with the sound threshold for the least sensitive cell in the ear, the A_2 cell.

However, notodontid moths (Noctuoidea) seem to show both negative phonotaxis and non-directional evasive flight even though their ears have only a single sensory cell (homologous to the A_1 cell) (Eggers, 1919; Surlykke, 1984). The total dynamic range of the notodontid ear equals the 20 dB of the A-cell. Above 25 kHz the directionality is 10-15 dB. Consequently, sound intensities exceeding the threshold by more than 35 dB will saturate both ears independent of direction to the bat and a non-directional evasive flight is expected. Steered turning-away should be the response to lower SPL´s where the moth can judge the direction to the sound source by comparing the intensity coding in the two ears (Surlykke, 1984).

But the role of A-cell physiology in evasive responses of both noctuids and notodontids is speculative since the actual SPL has never been measured at a reacting moth´s position in the field. Neither has the importance of time cues been investigated systematically although Roeder (1967a) showed that moths react more readily to high repetition rates and short pulses.

iii) Defense against gleaning bats. Non-volant tympanate moths freeze when exposed to ultrasound (Werner, 1981). Bats can discriminate even minor differences in surface texture (on the basis of spectral cues) (Habersetzer & Vogler, 1983), but in spite of this, freezing may be an effective strategy since many gleaners listen for rustling sounds from the prey (Bell, 1982; Fiedler, 1979; Vaughan, 1976). Furthermore, the CF-bat Hipposideros caffer only detects prey on surfaces if the prey moves since it exploits the flutter induced modulations of echo amplitude and frequency in order to reject background clutter (Bell & Fenton, 1984).

iv) Clicks of arctiid moths. Many arctiid moths produce clicks in response to ultrasound (and touch). The clicks are produced by metathoracic

554

tymbal organs (Fenton & Roeder, 1974; Fullard, 1979; Fullard & Barclay, 1980). Since many arctiids are toxic Blest (1964) and later Dunning (1968) suggested that the clicks are warning signals i.e. acoustic aposematic signals.

Fullard, Fenton & Simmons (1979), however, suggested that the clicks are percieved by bats as "extra" echoes giving the impression of multiple targets in the vicinity of the moth. This conclusion was based on the resemblance between frequency/time spectra of <u>Cycnia</u> <u>tenera</u> clicks and of terminal cries of <u>Eptesicus</u> <u>fuscus</u>. The click threshold of <u>C.tenera</u> is high (Fullard 1979) so the bat is in the terminal phase when the clicks are produced and the distance to the moth so short that there is only time for veering away.

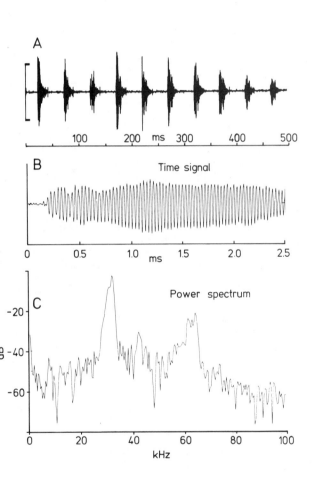

Fig.2.
The stridulalory sounds of <u>Thecophora</u> <u>fovea</u> (Noctuidae). The pulse train (2A) lasts several minutes. The vertical scale in 2A is 84 dB SPL at 1 m. Each pulse has a duration of 10-12 ms and 2B shows the first part of a pulse. The frequency is constant throughout the pulse and the power spectrum (2C) shows that the carrier frequency is around 30 kHz. (From Surlykke & Gogala, 1986).

A warning function of the clicks, on the other hand, was supported by experiments with <u>Pipistrellus</u> <u>pipistrellus</u> (Surlykke & Miller, 1985). Play-back of clicks recorded from two species of arctiids <u>Phragmatobia</u> <u>fuliginosa</u> and <u>Arctia</u> <u>caja</u> did not change the ranging acuity of the bats (1.5 cm) in a Two-Alternative-Forced-Choice experiment. Hence, the bats could discriminate between clicks and echoes. The bats also easily learned to associate clicks with toxins and to avoid clicking (toxic) rewards. Both arctiid moths showed click thresholds between 60 and 75 dB SPL corresponding to a "typical" bat being 6-7 m away. Estimates of detection ranges for bats varies from 1 to 10 m (Fenton, 1982; See also Fig.5). Hence, the clicks are produced at distances where they may actually reveal the moth to the bat which has ample time to react in. This would seem without value for the moth unless the clicks are warning signal. Some non-toxic arctiid moths produce clicks (Dunning, 1968; Fenton & Roeder, 1974; Fullard & Fenton, 1977) so acoustical mimics (Müllerian as well as Batesian) may be found.

Alternatively, clicks might function by startling the bats. The first presentations of clicks of nymphalid butterflies elicited strong startling responses but after a short while the bats reacted to the sounds as to a "dinner bell" (Møhl & Miller, 1976). Effective startling probably depends on a low likelihood of encountering clicks. Therefore an estimate of the relative abundance of clicking toxic, clicking non-toxic, and non-clicking moths would be interesting.

v) Bat free niche. Some moths inhabit niches temporally or spatially isolated from bats. Thereby bat predation is avoided but usually at the cost of living where it is cold and where the vegetation is much scarcer (Surlykke, 1986).

Thecophora fovea (Noctuidae) from Central Europe emerges in late October. The moth is exceptional among noctuids because the males produce sounds by rubbing the 1. tarsal segment of the metathoracic leg against a stridulatory swelling on the hindwing. The sounds have main energy around 30 kHz and an intensity of ca. 85 dB SPL at 1 m (Fig. 2). T.fovea has a relatively low auditory threshold at 30 kHz (35 dB SPL) permitting long-distance acoustic communication. But the frequency and intensity of the sounds imply that they are also audible to bats within a large radius and T.fovea is readily eaten by bats. However, the emergence of T.fovea may be synchronized to the hibernation of bats, isolating the moths temporally from bats and thus permitting the moths to communicate acoustically (Surlykke & Gogala, 1986). The only arctiid moth known to use the clicks for intra-specific communication is a day-flying species, Endrosa aurita (Peter, 1912). This may be another example of temporal isolation.

The resemblance of the audiograms of T.fovea to those of other noctuids (Fenton & Fullard, 1979; Roeder, 1974) indicates that also the hearing of T.fovea originally evolved to detect bats. Presumably the hearing ability secondarily "shaped" the power spectrum of the communication sounds.

III.1.3 Tuning of moth ears. The defense responses of moths depend on the ability to detect the sonar signals of bats and some investigations indicate that moth ears are tuned specifically to the frequencies emitted by sympatric bats (Fullard, 1982; Fenton & Fullard, 1981; Fullard & Thomas, 1981). But the acoustic characteristics of both moths and bats vary a lot among the species in a given area, which weakens the conclusions. However, a co-adaptation is clearly demonstrated by the hearing of the black witch moth (Ascalapha odorata). The black witch lives in Panama where the bat fauna is diverse and on Hawaii where only one bat, the Hawaiian hoary bat, Lasiurus cinereus semotus, is found. Audiograms of Panamanian A.odorata are broadly tuned with high sensitivity especially at high frequencies and the Hawaiian specimens are more sensitive to lower frequencies in the range of the social calls of the Hawaiian hoary bat (Fig. 1) (Fullard 1984b).

Generally tropical moths are rather sensitive (Fullard & Thomas 1981, Fenton & Fullard 1981, Fullard 1982, Fullard 1984b) reflecting a general higher predation pressure and a higher bat (i.e. echolocation) diversity.

Roeder (1967b) estimated that ears provide moths with 40% better chance of survival, but this may vary from habitat to habitat.

III.1.4 Bat countermeasures. The auditory characteristics of moths may form a selection pressure influencing the echolocation strategies of bats. Low cry intensities or use of frequencies outside the range of maximum sensitivity of most moth ears may for instance reduce the conspicuousness of bats (Fenton & Fullard, 1979; 1981; Fullard & Thomas, 1981). Some suggested countermeasure strategies of bats are discussed below considering physiological and physical constraints.

i) High frequency and low intensity. Most moths are less sensitive to frequencies above ca. 65 kHz. However, the echolocation range is reduced at high frequencies since the atmospheric attenuation increases with increasing frequency (Fig. 3). The influence of humidity, in contrast, is only minor at frequencies up to about 100 kHz. The curves are calculated for 20°C but changes with temperature are small within physiological ranges (Evans & Bass, 1972).

The echolocation ranges for a 20 mm sphere were calculated for bats emitting signals with source levels of 110 dB SPL (high intensity bats) and 70 dB SPL (whispering bats) (Fig.4). Behavioral echo threshold determinations vary between about 0 and 20 dB SPL (Kick, 1982; Wenstrup, 1985; See also Møhl this volume) and when calculating the curves the threshold was assumed to be 10 dB SPL at any emitted frequency.

Frequency content heavily affects the range when the source level is 110 dB SPL. The echolocation range at 50% relative humidity is 6.4 m at 10

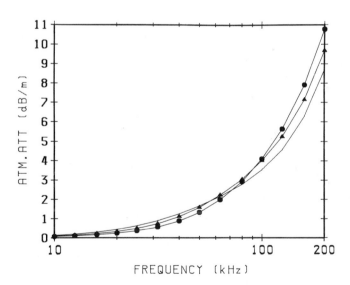

Fig. 3. Atmospheric attenuation at 20°C as a function of frequency at 50% (——), 70% (-▲-), and 100% (-●-) relative humidity. The figure is redrawn from Evans & Bass (1972) concentrating on the frequency range of importance for bat echolocation. The values are calculated but the 50% humidity curve corresponds well with measured values (Lawrence & Simmons 1982).

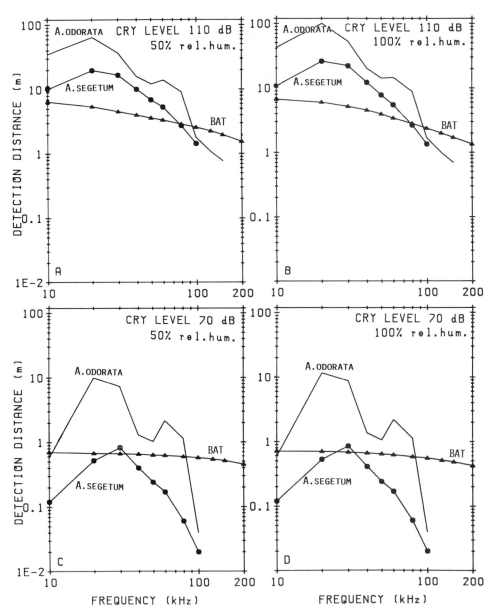

Fig. 4. The figure shows the maximum distances at which an insensitive moth
Agrotis segetum (-●-) and a sensitive moth _Ascalapha odorata_ (——) can
detect a bat at different frequencies. The third curve in each figure
(-▲-) is the echolocation range for a bat detecting a moth-sized target
(a 20 mm sphere). The curves were derived using source levels of bat
cries as given in the figures and atmospheric attenuations at the
relevant frequency and humidity from Fig. 3. The moth detection
distances were calculated using the thresholds in Fig. 1 and assuming a
geometric spreading loss of 6 dB per distance doubled. The bat
echolocation ranges were calculated using sonar equations (See Møhl
this volume) and assuming an echo threshold at the emitted frequency of
10 dB SPL.

kHz, 2.5 m at 100 kHz, and 1.5 m at 200 kHz (Fig. 4A) and the corresponding ranges at 100% relative humidity are 6.7 m, 2.3 m, and 1.3 m at 10, 100, and 200 kHz respectively, (Fig. 4B). Humidity, on the other hand, seems only of minor importance, since 4B and 4D are almost identical with 4A and 4C respectively. Therefore there is no reason to expect higher proportions of bats emitting low frequency cries in the humid tropical air than in the dryer temperate areas (Griffin, 1971).

Comparing the curves of echolocation ranges for the bats with the curves of detection distances for the moths shows that the bats gain over the moths by increasing the frequencies both when emitting high intensity and low intensity cries. At high intensities the cost is shorter echoloca- tion ranges but at low intensities the range is almost unaffected by fre- quency as indicated by the nearly horizontal curves of echolocation range in 4C and 4D. This is due to the fact that the atmospheric attenuation is linear with range while geometric spreading loss is logarithmic with range which means that at short ranges (low intensities) the relative influence on echo intensity of atmospheric attenuation is small even when shifting to high frequencies.

Lowering the intensity of the cries without changing the frequency is another strategy for the bats to reduce the conspicuousness but again at the cost of shorter echolocation ranges. But echo intensity falls off with the 4th power of distance and sound intensity only with the distance squared so lowering the cry intensity gives the bat a net gain over the moth (Compare Fig. 4A and 4B with 4C and 4D). 4C and 4D indicate that a whispering bat may detect a moth before the moth detects the bat.

Based on the curves in Fig. 4 one should expect to find bats using cries with low intensity and cries with high frequencies (both giving short detection distances) in habitats where prey density is so high that the bats can "afford" having short detection ranges, i.e. in the tropics. Moreover, shifting to high frequencies should be most advantageous for whispering bats since they can use frequencies above the most sensitive region of the tympa- nate prey without "paying" with reduced echolocation range.

The above predictions are confirmed when analysing the data of Fenton & Bell (1981). They studied 39 species of bats from North America (16) and Zimbabwe (23). All low intensity bats (source level < 75 dB SPL) were found in the tropics. The highest frequencies were found among the bats with lowest intensities. By rearranging their data into two groups, one with the 11 species using low and intermediate intensity cries (source level < 90 dB SPL) and one with the bats using high intensity cries (28 species) one can show that the frequencies of the low intensity group ranged from 43 to 212 kHz with a median of 78 kHz, whereas main frequencies of the high intensity bats ranged from 11 to 70 kHz with a median of 33 kHz. (The difference is statistically significant, P<0.001, Mann-Whitney U test).

ii) Low frequencies. Echolocation at low frequencies is another strategy to avoid the best frequency (BF) of moths. Euderma maculatum (Woodsworth et al., 1981) from Canada, and the European free tailed bat Tadarida teniotus (pers.obs.) are examples of bats echolocating at frequencies so low that it is possible to hear them with the unaided ear. Low frequencies propagate farther through air than high frequencies but as Fig. 4A and 4B show a sensitive moth still has much advance warning. Even though for example 10 kHz is outside the most sensitive range of A.odorata (Fig.1), the decreased frequency dependent atmospheric attenuation is even more advantageous to the moth than to the bat since overall detection distance for the moth is long. Besides, the target strength is reduced at frequencies with wavelengths longer than approximately the diameter of the target (ca. by a factor 2 at 10 kHz for a 20 mm sphere) (Morse, 1948; Urick,

1975) so the potential prey is probably restricted to very large insects. Therefore low frequencies are supposedly used by large fast bats in areas where moths are generally insensitive and where insect abundance is so low that detection is the major problem. This applies mainly to temperate habitats but all high intensity bats hunting large prey should profit by emitting low frequencies.

The advantage for the bat of changing the frequency may be counteracted by the adaptation of moth hearing to frequencies outside the "normal" range, e.g. by pinnae-like cuticular structures around the ear (Fullard, 1984a; 1984b). Even in cases where the bat detects the prey before the prey hears the bat the prey is not without chances since there still is time for outmaneuvering the bat The "last chance responses" of lacewings were elicited at distances down to 10 cm from the bat (Miller & Olesen, 1979; Miller, 1983).

iii) Long duration. One strategy apparently without drawbacks for the bats is to use very long cries. The integration time of moths is around 20 ms (Fig. 5), whereas integration time of most mammals is in the order of 200 ms (Dooling, 1980). Integration time has only been measured in one bat Rousettus aegyptiacus (Megachiroptera) (Suthers & Summers, 1980) and these results indicated values as short as 50 ms. However, firstly, Megachiroptera and Microchiroptera may not be very closely related .(See Pettigrew, this volume), and secondly the click sounds of Rousettus are invariably brief (from less than 1 ms to about 2 ms) and the results may not be valid for CF-bats emitting signals of up to 100 ms duration. All sonar models show that long narrow banded signals are well suited for detection (but not for

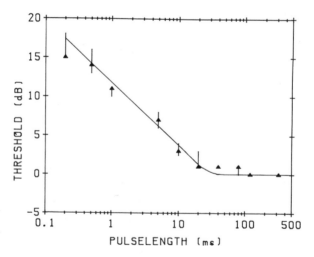

Fig. 5. Temporal integration time for a noctuid moth (A.segetum). The threshold was measured as a function of pulselength at a carrier frequency of 30 kHz. The curve shows median values with the interquartile distances as vertical bars (n=9). At pulselengths shorter than ca. 10 ms the curve has a slope of -2.5 dB per doubled duration which is close to the -3 dB expected for an energy detector. The curves cross at ca. 25 ms (Surlykke unpublished observations).

ranging). If the integration time of insectivorous microchiropterans resembles that of other mammals or is at least comparable to the duration of their own search cries the bat should be able to integrate the total energy emitted in the long CF cries. Thus by decreasing the amplitude and increasing the duration of its signal the bat emits the same amount of energy thereby keeping the echolocation range constant. Such a signal would give the bat an advantage since the distance at which the moth can detect the bat would be reduced owing to the short integration time of moths. The sensory response of the moth depends only on the amplitude of signals at durations over 20 ms. This difference between hearing physiology of predator and prey may enhance the benefit of using long constant frequency sounds in the search phase and thus be a contributing reason for the widespread use not only by CF-bats but also by many FM-bats.

iv) Passive hearing. Some bats hunting insects in free air listen for feeding buzzes from other bats (Barclay, 1982; Bell, 1980) and for insect sounds (Buchler & Childs, 1981; Tuttle et al., 1985). In this way they increase their detection range without advertising their existance to the prey. The use of passive hearing in gleaning bats may also minimize detection by the prey, but the behavior probably also reflects the difficulties with clutter when echolocating against a dense background.

III.2 Lacewings

Green lacewings (Chrysopa (Chrysoperla) carnea, Neuroptera) also have tympanal organs The ears are located on the wings and are morphologically very different from ears of noctuid moths though sensitive to the same frequencies (Miller & McLeod, 1966; Miller, 1971). Normally ultrasound elicit passive nose dives in green lacewings but occasional last chance wing flips apparently make the evasive behavior less predictable to bats (Miller & Olesen, 1979). Lacewings do not produce sounds and the resemblance between hearing physiology and behavior in moths and lacewings supports the hypothesis that ears are primarily bat detectors in these insect orders. Lacewings are less sensitive (thresholds around 60 dB SPL at BF) than moths but they may also give smaller echoes because of their smaller size. However, echo intensity predictions are complicated since their wings are rather large (wing span 25 mm) and may give large acoustical glints (Schnitzler et al., 1983).

III.3 Crickets

The hearing of crickets, in contrast to moths and lacewings, functions in rather complex intraspecific communication (Michelsen & Nocke, 1974). Communication is mediated by sonic frequencies in Gryllidae, but Popov & Shuvalov (1977) and Moisseff & Hoy (1978) showed that flying crickets are sensitive also to ultrasound and that they react bimodally to sound stimulation: at calling song frequencies (below ca. 12 kHz) they are attracted (positive phonotaxis) whereas they show negative phonotaxis to ultrasound. In the field Popov & Markovich (1982) often saw crickets drop to the ground when bats were around. Apparently crickets exhibit evasive behavioral responses resembling those of moths and lacewings. However, the development of evasive behavior in crickets probably came after acoustic communication, in contrast to the noctuid, T.fovea, where hearing and evasive behavior supposedly preceeded acoustic communication. The calling songs of orthopterans may attract bats (Buchler & Childs, 1982; Tuttle et al., 1985) and some katydids apparently counteract this by using less conspicious (less frequent) songs. This interaction is not directly echolocation mediated and whether the katydids hear and react directly to bat cries is not yet known (Belwood, 1985).

III.4 Praying mantises

Recently an additional insect order has been included in the group of insects with bat detectors (Yager & Hoy, 1986). Asian hymenopid mantises respond to bat-like ultrasound with a sudden full extension of the forelegs and a strong dorsiflexion of the abdomen. The response is only elicited during flight and it results in a abrupt and dramatic change in flight path bringing the insect to the ground. The single (2 tympana) ear mediating this response is located ventrally on the metathorax between the coxae. It is tuned to frequencies between 20 and 50 kHz (threshold 60 dB SPL). No acoustic communication has been reported in this insect, so also this ear may be an example of a specific bat detector.

IV VERTEBRATE PREY

IV.1 Frogs

Trachops cirrhosus (Phyllostomidae) from Central America hunts frogs. The bats are attracted to the mating calls of sympatric frogs and to tape recordings of frog calls and they are maximally sensitive to sonic frequencies at the carrier frequency of the calls of sympatric frogs (Ryan et al., 1983). During hunting T.cirrhosus produces echolocation calls of low intensity (below 70 dB SPL), with short durations (less than 1 ms) and a frequency sweep from 100 to 50 kHz (Barclay et at., 1981).

Sympatric frogs probably cannot hear those echolocation calls (Megala-Simmons et al. 1985) (which may explain why T.cirrhosus continue to echolocate). Behaviorally the frogs defend themselves by minimizing the acoustic output. The full complex advertisement call, which elicits the maximal female attraction but which also attracts the bats maximally is only produced when another frog is approaching. In this way some predation may be avoided (Ryan, 1980; Tuttle & Ryan, 1981). Additionally, frogs may mask their sounds by calling in noisy surroundings around waterfalls (Tuttle & Ryan, 1982). Some frogs may also mimic the calls of some poisonous toads since the more the calls sound like those of the toads the less is T.cirrhosus attracted (Ryan & Tuttle, 1983).

IV.2 Small mammals

In contrast to frogs the small rodents which are prey for Megaderma lyra for example have very good ultrasonic hearing (Kelly & Masterton, 1977) and this may have "forced" M.lyra to turn off the echolocation when hunting (Fiedler, 1979). M.lyra can catch dead prey and relies then on echolocation but when catching live prey it locates the prey by passive hearing and M.lyra is extremely sensitive from 17 to 30 kHz (thresholds around -10 dB SPL, Schmidt et al., 1983). No evasive reactions of the prey have been described.

IV.3 Other vertebrates as prey

Bats eat fish, reptiles, and blood of mammals but no reactions to bats or bat-like sounds are known from any of these prey-groups. Fenton (1980) forwarded the obvious prediction that theoretically carnivorous bats should have an easy time detecting other bats by listening to their sonar sounds. However, there are only few reports of bats catching bats and apparently they do not listen for the echolocation calls of their victims (Vaughan, 1976; Fenton et·al., 1983). Neither do other predators on bats (bat-hawks) listen for the sonar signals of the bats.

V CONCLUSION

Although the prey of bats include many different animals it is among the nocturnal insects that the specific defenses against bats are found. This emphasizes how strong a predation pressure bats excert on nocturnal insects. Vice versa, the countermeasures to insect defenses shown by bats reflect how heavily many bats rely on insect prey.

No tympanate prey has been shown to exploit all the cues theoretically present in the sonar signals of bats. Only sound vs. no-sound has unequivocally been shown to be important. Some effects of time cues have been demonstrated but no systematic investigation has yet been carried out.

ACKNOWLEDGEMENTS

I thank M. Brock Fenton for valuable suggestions. Lee Miller, Uli Schnitzler, Axel Michelsen, and Asher Treat also gave constructive criticism.

VI REFERENCES

Agee HR (1969) Acoustic sensitivity of the European corn borer moth, Ostrinia nubilalis. Ann Entomol Soc Amer 62: 1364-1367
Barclay RMR (1982) Interindividual use of echolocation calls: Eavesdropping by bats. Behav Ecol Sociobiol 10: 271-275
-"- (1985) Long- versus short-range foraging strategies of hoary (Lasiurus cinereus) and silver-haired (Lasionycteris noctivagans) bats and the consequences for prey selection. Can J Zool 63: 2507-2515
Barcley RMR, Fenton MB, Tuttle MD & Ryan MJ (1981) Echolocation calls produced by Trachops cirrhosus (Chiroptera: Phyllostomatidae) while hunting for frogs. Can J Zool 59: 750-753
Bell GP (1980) Habitat use and response to patches of prey by desert insectivorous bats. Can J Zool 58: 1876-1883
-"- (1982) Behavioral and ecological aspects of gleaning by a desert insectivorous bat Antrozous pallidus (Chiroptera: Vespertilionidae). Behav Ecol Sociobiol 10: 217-223
Bell GP & Fenton MB (1984) The use of Doppler-shifted echoes as a flutter detection and clutter rejection system: the echolocation and feeding behavior of Hippossideros ruber (Chiroptera: Hipposideridae). Behav Ecol Sociobiol 15: 109-114
Belton P (1962) Responses to sound in pyralid moths. Nature 196: 1188-1189
Belwood JJ (1985) Effects of bat predation on the calling behaviour of neotropical katydids. Paper presented at the Seventh International Bat Research Conference in Aberdeen.
Belwood JJ & Fenton MB (1976) Variation in the diet of Myotis lucifugus (Chiroptera: Vespertilionidae). Can J Zool 54: 1674-1678
Black HL (1974) A north temperate bat community: Structure and prey populations. J Mamm 55: 138-157
Blest AD (1964) Protective display and sound production in some New World arctiid and ctenuchid moths. Zoologica: 161-181
Buchler ER & Childs SB (1981) Orientation to distant sounds by foraging big brown bats (Eptesicus fuscus) Anim Behav 29: 428-432
Dooling R (1980) Behavior and psychophysics of hearing in birds. IN: Comparative studies of hearing in vertebrates (Eds. Popper & Fay) pp. 261-288
Dunning DC (1968) Warning sounds of moths. Z Tierpsychol 25: 129-138
Eggers F (1919) Das thoracale bitympanale Organ einer Gruppe der Lepidoptera Heterocera. Zool Jahrb 41: 273-376
Evans LB, Bass HE (1972) Tables of absorption and velocity of sound in still air at 68°F (20°C). Wyle Labs Huntsville Ala. Report WR 72-2
Fenton MB (1980) Adaptiveness and ecology of echolocation in terestrial (aerial) systems. IN: Animal Sonar Systems (Eds. Busnel & Fish) Plenum Press pp. 427-446

-"- (1982) Echolocation, insect hearing, and feeding ecology of
    insectivorous bats. IN: Ecology of Bats. (Ed. T.H. Kunz) Plenum Press
    pp. 261-285
-"- (1984) Echolocation: Implications for ecology and evolution of bats.
    The quarterly Review of Biology 59: 33-53
Fenton MB & Roeder KD (1974) The microtymbals of some Arctiidae. J
    Lepidopterists' Soc 28: 205-211
Fenton MB & Bell GP (1979) Echolocation and feeding behaviour in four species
    of Myotis (Chiroptera). Can J Zool 57: 1271-1277
Fenton MB & Fullard JH (1979) The influence of moth hearing on bat
    echolocation strategies. J Comp Physiol 132: 77-86
-"- (1981) Moth hearing and the feeding strategies of bats. Amer Sci 69:
    266-275
Fenton MB & Bell GP (1981) Recognition of species of insectivorous bats by
    their echolocation calls. J Mamm 62: 233-243
Fenton MB, Gaudet CL & Leonard ML (1983) Feeding behavior of the bats
    Nycteris grandis and Nycteris thebaica (Nycteridae) in captivity. J
    Zool 200: 347-354
Fiedler J (1979) Prey catching with and without echolocation in the Indian
    false vampire (Megaderma lyra). Behav Ecol Sociobiol 6: 155-160
Fullard JH (1979) Behavioral analyses of auditory sensitivity in Cycnia
    tenera Hübner (Lepidoptera: Arctiidae). J Comp Physiol 129: 79-83
-"- (1982) Echolocation assemblages and their effects on moth auditory
    systems. Can J Zool 60: 2572-2576
-"-(1984a) External auditory structures in two species of neotropical moths.
    J Comp Physiol A 155: 625-632
-"-(1984b) Acoustic relationships between tympanate moths and the Hawaiian
    hoary bat (Lasiurus cinereus semotus). J Comp Physiol A 155: 795-801
-"-(1984c) Listening for bats: pulse repetition rate as a cue for a defensive
    behavior in Cycnia tenera (Lepidoptera: Arctiidae). J Comp Physiol 154:
    249-252
Fullard JH, Fenton MB & Simmons JA (1979) Jamming bat echolocation: the
    clicks of arctiid moths. Can J Zool 57: 647-649
Fullard JH & Barclay RMR (1980) Audition in spring species of arctiid moths
    as a possible response to differential levels of insectivorous bat
    predation. Can J Zool 58: 1745-1750
Fullard JH & Thomas DW (1981) Detection of certain African insectivorous
    bats by sympatric tympanate moths. J Comp Physiol 143: 363-368
Goldman LJ & Henson OW (1977) Prey recognition and selection by the constant
    frequency bat, Pteronotus p. parnellii. Behav Ecol Sociobiol 2: 411-420
Griffin DR (1971) The importance of atmospheric attenuation for the
    echolocation of bats (Chiroptera). Anim Behav 19: 5561
Habersetzer J & Vogler B (1983) Discrimination of surface-structured targets
    by the echolocating bat Myotis myotis during flight. J Comp Physiol 152:
    275-282
Joermann G (1984) Recognition of spatial parameters by echolocation in the
    vampire bat, Desmodus rotundus. J Comp Physiol 155: 67-74
Kelly JB & Masterton B (1977) Auditory sensitivity in the albino rat. J Comp
    Physiol Psychol 91: 930-936
Kick SA (1982) Target-detection by the echolocating bat, Eptesicus fuscus.
    J Comp Physiol 145: 431-435
Kürten L & Schmidt U (1982) Thermoperception in the common vampire bat
    (Desmodus rotundus). J Comp Physiol 146: 223-228
Lawrence BD & Simmons JA (1982) Measurements of atmospheric attenuation at
    ultrasonic frequencies and the significance for echolocation by bats. J
    Acoust Soc Am 71: 585-590
Megala-Simmons A, Moss CF & Kimberley M (1985) Behavioral audiograms of the
    bullfrog (Rana catesbeiana) and the green tree frog (Hyla cinerea). J
    Acoust Soc Am 78: 1236-1244

Michelsen A (1979) Insect ears as mechanical systems. Am Sci 67: 696-706
Michelsen A & Nocke H (1974) Biophysical aspects of sound communication in insects. Advances in Insect Physiol 10: 257-296
Miller LA (1971) Physiological responses of green lacewings (Chrysopa, Neuroptera) to ultrasound. J Insect Physiol 17: 491-506
-"- (1982) The orientation and evasive behavior of insects to bat cries. IN: Exogeneous and endogeneous influences on metabolic and neural control, vol. 1 (Eds. Addink ADF & Spronk N) Pergamon Press pp. 393-405
-"- (1983) How insects detect and avoid bats. IN: Neuroethology and Behavioral Physiology (Eds. Huber & Markl) Springer pp. 251-266
Miller LA & McLeod EG (1966) Ultrasonic sensitivity: A tympanal receptor in the green lacewing Chrysopa carnea. Science 154: 891-893
Miller LA & Olesen J (1979) Avoidance behavior in green lacewings I. Behavior of free flying green lacewings to hunting bats and ultrasound. J Comp Physiol 131: 113-120
Mogensen F & Møhl B (1979) Sound radiation patterns in the frequency domain of cries from a Vespertilionid bat. J Comp Physiol 134: 165-171
Moiseff A & Hoy RR (1983) Sensitivity to ultrasound in an identified auditory interneuron in the cricket: a possible neural link to phonotactic behavior. J Comp Physiol 152: 155-167
Morse PM (1948) Vibration and Sound. McGraw-Hill. 2. Ed.
Møhl B & Miller LA (1976) Ultrasonic clicks produced by the peacock butterfly: A possible bat-repellent mechanism. J exp Biol 64: 639-644
Neuweiler G (1980) Auditory processing of echoes: peripheral processing. IN: Animal Sonar Systems (Eds. Busnel & Fish) Plenum Press pp. 519-548
-"- (1984) Foraging, echolocation and audition in bats. Naturwiss 71: 446-455
Payne RS, Roeder KD & Wallman J (1966) Directional sensitivity of the ears of noctuid moths. J exp Biol 44: 17-31
Peter K (1912) Versuche über das Hörvermögen eines Schmetterlings (Endrosa v.ramosa). Biol Zentralb 32: 724-731
Popov AV & Shuvalov VF (1977) Phonotactic behavior of crickets. J Comp Physiol 119: 111-126
Popov AV & Markovich AM (1982) Auditory interneurones in the prothoracic ganglion of the cricket, Gryllus bimaculatus. II. A high-frequency ascending neurone (HF$_1$AN). J Comp Physiol 146: 351-359
Roeder K (1962) The behavior of free flying moths in the presence of artificial ultrasonic pulses. Anim Behav 10: 300-304
-"- (1967a) Turning tendency of moths exposed to ultrasound while in stationary flight. J Insect Physiol 13: 873-888
-"- (1967b) Nerve Cells and Insect Behavior. Harvard University Press. Rev. ed.
-"- (1974) Acoustic sensory responses and possible bat-evasion tactics of certain moths. Proc Can Soc Zool Ann Meet: 71-78
Roeder KD, Treat AE & Vande Berg JS (1968) Auditory sense in certain sphingid moths. Science 159: 331-333
Ryan MJ (1980) Female mate choice in a neotropical frog. Science 209: 523-525
Ryan MJ & Tuttle MD (1983) The ability of the frog-eating bat to discriminate among novel and potentially poisonous frog species using acoustic cues. Anim Behav 31: 827-833
Ryan MJ, Tuttle MD & Barrley RMR (1983) Behavioral responses of the frog-eating bat, Trachops cirrhosus, to sonic frequencies. J Comp Physiol 150: 413-418
Schmidt S, Turke B & Vogler B (1983) Behavioral audiograms from the bat, Megaderma lyra. Myotis 21/22: 62-66
Schnitzler H-U (1968) Die Ultraschall-Ortungslaute der Hufeisen-Fledermäuse (Chiroptera-Rhinolophidae) in verschiedenen Orientierungssituationen. Z vergl Physiol 57: 376-408

Schnitzler H-U & Henson OW (1980) Performance of airborne animal sonar systems: I. Microchiroptera. IN: Animal sonar systems. (Eds. Busnel & Fish). Plenum Press pp. 109-181

Schnitzler H-U & Flieger E (1983) Detection of oscillating target movements by echolocation in the greater horseshoe bat. J Comp Physiol 153: 385-391

Schnitzler H-U, Menne D, Kober R & Heblich K (1983) The acoustical image of fluttering insects in echolocating bats. IN: Neuroethology and Behavioral Physiology (Eds. Huber & Markl). Springer pp. 235-250

Schuller G (1980) Hearing characteristics and Doppler shift compensation in South Indian CF/FM bats. J Comp Physiol 139: 349-356

"-" (1984) Natural ultrasonic echoes from wing beating insects are encoded by collicular neurons in the CF/FM bat, Rhinolophus ferrumequinum. J Comp Physiol A 155: 121-128

Spangler HG & Takessian A (1983) Sound perception by two species of wax moths (Lepidoptera: Pyralidae). Ann Entomol Soc Amer 76: 94-97

Surlykke A (1984) Hearing in notodontid moths: A tympanic organ with a single auditory neurone. J exp Biol 113: 323-335

-"- (1986) Moth hearing on the Faroes; An area without bats. Physiol Entomol 11: 221-225

Surlykke A & Miller LA (1982) Central branchings of three sensory axons from a moth ear (Agrotis segetum, Noctuidae) J Insect Physiol 28: 357-364

-"- (1985) The influence of arctiid moth clicks on bat echolocation; jamming or warning? J Comp Physiol A 156: 831-843

Surlykke A & Gogala M (1986) Stridulation and hearing in the noctuid moth Thecophora fovea (Tr.). J Comp Physiol 159: 267-273

Suthers RA & Summers CA (1980) Behavioral audiogram and masked thresholds of the Megachiropteran bat, Rousettus. J Comp Physiol 136: 227-233

Tuttle MD & Ryan MJ (1981) Bat predation and the evolution of frog vocalization in the Neotropics. Science 214: 677-678

-"- (1982) The role of synchronized calling, ambient light, and ambient noise, in anti-bat-predator behavior of a treefrog. Behav Ecol Sociobiol 11: 125-131

Tuttle MD, Ryan MJ & Belwood JJ (1985) Acoustical resource partitioning by two species of Phyllostomid bats (Trachops cirrhosus and Tonatia sylvicola. Anim Behav 33: 1369-1370

Urick RJ (1975) Principles of underwater sound. McGraw-Hill, New York.

Vaughan TA (1976) Nocturnal behavior of the African false vampire bat (Cardioderma cor). J Mamm 57: 227-248

Vogler M & Neuweiler G (1983) Echolocation in the noctule (Nyctalus noctula) and horseshoe bat (Rhinolophus ferrumequinum). J Comp Physiol 152: 421-432

Wenstrup JJ (1984) Auditory sensitivity in the fish-catching bat, Noctilio leporinus. J Comp Physiol A 155: 91-101

Werner TK (1981) Responses of nonflying moths to ultrasound: the threat of gleaning bats. Can J Zool 59: 525-529

Whitaker JO & Tamich PQ (1983) Food habits of the hoary bat, Lasiurus cinereus, from Hawaii. J Mamm 64: 151-152

Woodsworth GC, Bell GP & Fenton MB (1981) Observations of the echolocation, feeding behaviour and habitat of Euderma maculatum (Chiroptera: Vespertilionidae) in Southcentral British Columbia. Can J Zool 59: 1099-1102

Yager DD & Hoy RR (1986) The cyclopean ear: A new sense for the praying mantis. Science 231: 727-729

LOUD IMPULSE SOUNDS IN ODONTOCETE PREDATION AND SOCIAL BEHAVIOR

Kenneth Marten, Kenneth S. Norris,
Patrick W. B. Moore*, and Kirsten A. Englund

Long Marine Laboratory
University of California
Santa Cruz, California 95064 U.S.A.

*Naval Ocean Systems Center
Kailua, Hawaii 96734 U.S.A.

Norris and Møhl (1983) presented the hypothesis that odontocete cetaceans might be able to debilitate prey with intense sound and presented
evidence to stimulate observation and experiment on this topic. This paper
outlines our research on the hypothesis since its original proposal and
extends the hypothesis by suggesting that dolphins can produce loud impulse
sounds of low median frequency and long duration which might disorient
prey. Sounds of 500 Hz to 5 kHz fall into the range of hearing or labyrinth sensitivity of prey species and the long duration (to 200 ms) could
cause overloading of the unprotected fish ear.

Predatory Sounds of Wild Odontocetes

Here we describe loud impulse sounds emitted by wild bottlenose dolphins
and killer whales feeding on fish, and by sperm whales in an unknown
context. By "loud" we mean that these sounds are louder to the human ear
than sounds before and after them. Their sound pressure level is usually
at least 2-3 times the pressure level of the "closure" click trains which
often precedes them, but we still do not know absolute intensities. The
examples we have accumulated are in Table 1. The recordings contain loud,
low-frequency impulse sounds. We have now seen and recorded Indian Ocean
bottlenose dolphin predation sequences at the surface in which the dolphin
chases fish, a loud impulse sound is heard, and then the dolphin eats a
stunned fish. But the interpretation of these sequences is difficult due
to the fact that at least some of the time the dolphins are tail-swatting
(see Smolker and Richards, this volume).

Output from our Fourier Analyzer (Fig. 1) is presented in two forms:
(1) time-amplitude train, and (2) power spectrum. "Bang" peak frequency,
duration, and frequency range can be read from the analyzer output, and
averaged to characterize the bangs of various wild odontocetes (Table 1).

In Fig. 1, the echolocation click train on the left is followed very
shortly by a bang on the right, by the same animal. The bang is much
louder than the clicks which preceded it and while clicks last 1 ms, bangs
last about 100 ms. The power spectra of bangs resemble those of clicks,

567

Table 1. Predatory impulse sounds of odontocetes. All data are from wild populations, as predatory impulse sounds are rare in captivity. S.D.=sample standard deviation; L=lower; U=upper.

Odontocete Population	PREDATORY SOUNDS		
	Average Peak Frequency	Average Duration	Average Frequency Range
Indian Ocean Bottlenose Dolphins (Tursiops aduncus)	Unimodal; n=29		
	4.0 kHz SD=2.0 kHz	148 msec SD=28msec	2.0 - 6.3 kHz $SD_{L,U}$=1.7,1.5 kHz
Shark Bay, W. Australia	Bimodal; n=61		
	pk1=972 Hz pk2=5.1 kHz	128 msec SD=40 msec	0.2 - 6.6 kHz $SD_{L,U}$=.6, .7 kHz
Pacific Bottlenose Dolphins (Tursiops truncatus) La Jolla, California	580 Hz n=15 SD=132 Hz	46 msec n=17 SD=24 msec	50 Hz - 1.0 kHz n=14 $SD_{L,U}$=116Hz,.3kHz
Killer Whales (Orcinus orca) Prince William Sound, Alaska	2.3 kHz n=11 SD=.2 kHz	37 msec n=12 SD=34 msec	1.3 - 3.0 kHz n=10 $SD_{L,U}$=.6, .3 kHz
Killer Whales (Orcinus orca) Lofoten, Norway	400 Hz n=5 SD=140 Hz	209 msec n=5 SD=53 msec	0 - 1.9 kHz n=5 SD_U=1.2 kHz
Killer Whales (Orcinus orca) Iceland	120 Hz n=8 SD=29 Hz	157 msec n=8 SD=35 msec	0 - 281 Hz n=8 SD_U=156 Hz
Killer Whales (Orcinus orca) Johnstone Strait, British Columbia	150 Hz n=9 SD=53 Hz	144 msec n=10 SD=39 msec	39 - 350 Hz n=9 $SD_{L,U}$=22, 75 Hz
Killer Whales (Orcinus orca) Monterey Bay, California	175 Hz n=2 SD=106 Hz	297 msec n=2 SD=180 msec	0 - 2.2 kHz n=2 SD_U=.3 kHz
Sperm Whales (Physeter macrocephalus) Indian Ocean	2.9 kHz n=4 SD=1.2 kHz	110 msec n=6 SD=54 msec	1.4 - 7.9 kHz n=4 $SD_{L,U}$=.6, .7 kHz

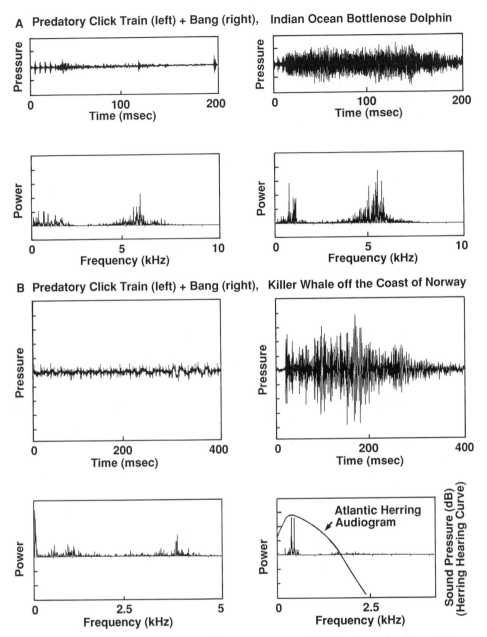

Fig. 1 Oscillogram and power spectrum of an echolocation click train
followed by a "bang" from (A) a wild Indian Ocean bottlenose
dolphin feeding on Perth herring, and (B) a wild killer whale off
the coast of Norway feeding on Atlantic herring. The low frequency
component seen in clicks is made loud in bangs. This manipulation
of the power spectrum of clicks to make bangs shows that clicks and
bangs are produced from the same source. Note: For each species,
the oscilligrams of the click train and bang have the same vertical
scale, so the intensity of the bang relative to clicks can be read.
The bang power spectrum vertical scale is 20X the click vertical
scale in A and 40X it in B. Absolute intensities are unknown. The
killer whale bang power spectrum matches the audiogram of its prey
(Enger, 1967).

suggesting it is produced by the same anatomical structures, but lower frequencies are emphasized in the bangs (Fig. 1). The hearing curve of an Atlantic herring is covered by the power spectrum of a killer whale bang in Fig. 1B. A pause between the end of the echolocation click train and the bang averages 118 ms in <u>Tursiops</u> <u>aduncus</u> (n=8, S.D.=50 msec); and values of 1.2 and 1.9 sec were measured in two killer whale samples. Time-amplitude and power spectra of the bangs (without clicks) of three other odontocetes (Fig. 2) include Alaskan killer whales taking salmon. In 12% of recorded attacks,bangs were observed.

Loud Impulse Social Sounds

To better understand the impulse sounds of odontocetes feeding in the wild, we expanded our investigation to include social impulse sounds, such as "jaw claps," and fluke-induced cavitation (Table 2). These sounds resemble predatory bangs (C.F. Figs 2A & 3A). The social sound of an Indian Ocean bottlenose dolphin is almost identical to the predatory sound of the Pacific bottlenose dolphin.

Non-Vocal Impulse Sounds

Some workers have postulated that some loud impulsive sounds of dolphins are not vocalizations. Non-vocal hydrodynamic sounds, body slaps, and fluke slaps were recorded from the captive killer whales at Marine World. One suspected non-vocal hydrodynamic sound occurred when two killer whales had been pitchpoling, their heads vertically out of the water, and we think cavitation may have occurred from the fluke movements required to elevate their bodies. The fluke-body hits are from wild Atlantic spotted dolphins (<u>Stenella</u> <u>plagiodon</u>) where excellent visibility permitted observations of the animals occasionally hitting each other with their flukes. The non-vocal hydrodynamic sounds from the captive killer whales can be distinguished from vocal bangs by slow onset and power spectrum (Fig. 2, Table 2). Killer whale surface slaps and underwater cavitations strongly resemble each other in waveforms and power spectra. The fluke-body hits of Atlantic spotted dolphins differ from these killer whale water sounds in the same way.

Training Bottlenose Dolphins to Emit High Intensity Echolocation Clicks

We trained two dolphins (Fig. 4) to emit signals that might debilitate prey, because sonic predation may have evolved from strong, long-distance echolocation. Heptuna's maximum click train averaged 219.55 dB re 1 uPa/1 m (14 clicks in train); Circe's maximum was 217.1 dB (52 clicks). The energy of their clicks was concentrated in the ultrasonic range, peaking unimodally and bimodally between 60 kHz and 120 kHz (see Ceruti et al. 1983 for a detailed spectral analysis of similar clicks from other NOSC dolphins). Comparable data from a third bottlenose dolphin showed click train averages of 208 dB and single clicks of 212 dB.

Testing the Effect of Bottlenose Dolphin Clicks on Fish

Once the bottlenose dolphins levelled out at maximum ultrasonic output, fish were put in their sound fields to assess possible detrimental effects. Each fish was restrained in a clear plastic bag that did not significantly impede the passage of sound (attenuation 0-1 dB). The fish was subjected to 25-50 click trains, where each click train consisted of on the order of 20-80 clicks, and we observed the behavior of the fish during this bombardment through an underwater video system.

We used 13 species of fish: Anchovy (<u>Stolephorus</u> <u>purpueus</u>), Whitespot goatfish (<u>Parupeneus</u> <u>porphyreus</u>), Yellow-barred jack (<u>Gnathodon</u> <u>speciosus</u>),

Table 2. Social impulse sounds and non-vocal sounds of
odontocetes. S.D.=sample standard deviation; L=lower;
U=upper.

SOCIAL SOUNDS			
Odontocete Population	Average Peak Frequency	Average Duration	Average Frequency Range
(1) Wild Indian Ocean Bottlenose Dolphins (Tursiops aduncus) Shark Bay, W. Australia	555 Hz n=11 SD=167 Hz	61 msec n=13 SD=37 msec	0 - 6.6 kHz n=11 SD_U=1.9 kHz
64% were bimodal; average pk2 @ 5.2 kHz; SD=.5 kHz			
(2) Atlantic Bottlenose Dolphins (Tursiops truncatus) N.O.S.C., Hawaii	1.8 kHz n=8 SD=.6 kHz	9 msec n=7 SD=8 msec	.5 - 3.7 kHz n=5 $SD_{L,U}$=.4, 1.2 kHz
(3) Atlantic Bottlenose Dolphins (Tursiops truncatus) Marineland, Florida	530 Hz n=5 SD=444 Hz	44 msec n=5 SD=21 msec	0 - 2.5 kHz n=5 SD_U=1.4 kHz
(4) Atlantic Bottlenose Dolphins and Pacific White-sided Dolphins (Lagenorynchus obliquidens) Marineworld, Vallejo, California	943 Hz n=6 SD=264 Hz	87 msec n=6 SD=10 msec	0 - 2.2 kHz n=6 SD_U= .5 kHz
(5) Atlantic Bottlenose Dolphins and Killer Whales (Orcinus orca) Marineworld, Vallejo, California	732 Hz n=5 SD=184 Hz	106 msec n=5 SD=17 msec	0 - 1.5 kHz n=5 SD_U= .6 kHz
(6) Atlantic Bottlenose Dolphins and Killer Whales Marineworld, Vallejo, California	814 Hz n=4 SD=517 Hz	87 msec n=4 SD=22 msec	0 - 2.0 kHz n=4 SD_U=1.5 kHz
NON-VOCAL SOUNDS			
Captive Killer Whales (Orcinus orca) Underwater Cavitation Marineworld Vallejo, California	460 Hz n=3 SD=36 Hz	127 msec n=3 SD=59 msec	0 - 1.4 kHz n=3 SD_U=.4 kHz
Captive Killer Whales (Orcinus orca) Surface Slaps Marineworld Vallejo, California	308 Hz n=4 SD=197 Hz	60 msec n=4 SD=27 msec	0 - 800 Hz n=4 SD_U=216 Hz
Wild Atlantic Spotted Dolphins (Stenella plagiodon) Fluke-Body Hits Bahama Islands	1.4 kHz n=2	130 msec n=1	1.2-3.2 kHz n=2

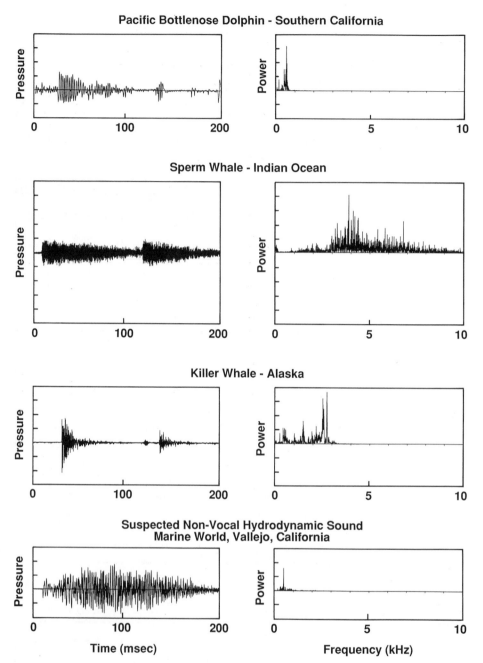

Fig. 2 Typical oscillograms (left) and power spectra (right) of bangs from
three wild odontocete populations compared with a suspected non-
vocal hydrodynamic sound from a captive killer whale with its head
vertically out of the water. The bottlenose dolphin was feeding on
Perth herring and the Alaskan killer whale was feeding on pink
salmon. The activity of the sperm whale is unknown. The vertical
scale is arbitrary, and absolute intensities are unknown.

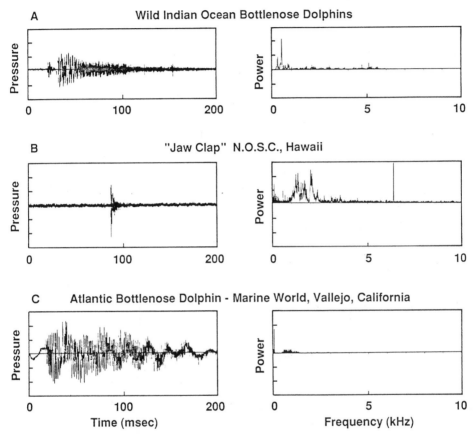

Fig. 3 Typical oscillograms (left) and power spectra (right) of impulse
sounds used in a social context. A. Wild bottlenose dolphins
engaged in social activity. B. Captive bottlenose dolphins when
first put together in the same enclosure. C. Captive bottlenose
dolphins at the time of their introduction to their new quarters at
Marine World in Vallejo, California. The vertical scale is
arbitrary, and absolute intensities are unknown.

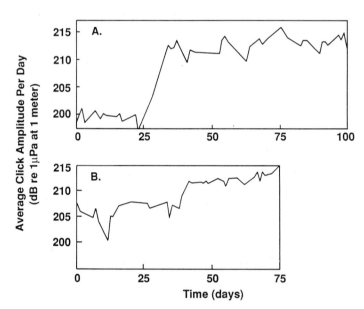

Fig. 4   Curve showing progress through time of (A) 17-year old male
(Heptuna) and (B) 8-year old female (Circe) Atlantic bottlenose
dolphins trained to make their ultrasonic clicks as loud as
possible.

Squirrelfish (sp. unknown), Parrotfish (sp. unknown), Spotted green puffer-fish (Arothron hispidus), Convict tang (Acanthurus sandvicensis), Sergeant major (Abudefduf abdominalis), White-spotted damselfish (Dascyllus albisella), Yellow-finned surgeonfish (Acanthurus xanthopterus), Millet seed butterflyfish (Chaetodon miliaris), Threadfin butterflyfish (Chaetodon auriga), and Hawaiian flag-tail fish, or aholehole (Kuhlia sandvicensis). Each dolphin bombarded a different individual of all 13 fish species. In early trials, a control fish was put in a plastic bag and given the same treatment as the test fish except for the click train bombardment, but in later trials it became apparent that this procedure was unnecessary, so it was dropped. The dolphin pulses did not influence the fish's behavior.

Testing the Effect of High Intensity High FrequencY Artificial Clicks on Fish

In the NOSC Hawaii test pool, we subjected fish to dolphin-like pulses of 225-231 dB re luPa at the fish, with the following system, where we could control pulse repetition rate, pulse frequency, and pulse duration. Two Wavetek generators (model 116 for rep. rate, model 144 for pulse frequency) played sine waves through an Electronic Navigation Industries 1140L Kilowatt Amplifier and Krohn Hite MT-56 Transformer into a NOSC custom-made Strasse transducer, which resonated at 77 kHz. Intensity was monitored with an Edo Western 6166 hydrophone (calibrated against an h-52 hydrophone), Krohn-Hite filter, and Tektronix 464 storage oscilloscope. Fish behavior was monitored through an underwater video system.

The fish were restrained in a wire mesh cage which kept them within five inches of the transducer. We subjected the test fish to various pulse frequencies between 40 and 100 kHz (transduction was weak outside this range) and repetition rates of 0-120/sec. The same 13 species listed above were tested (different individuals). In addition, Black-tailed snapper (Lutjanus fulvus) and Moorish idol (Zanclus cornutus) were subjected to the artificial clicks. As with the dolphin trials, control fish were caged in early trials, but the procedure was dropped in later trials.

The dolphin-produced and artificially-produced high intensity ultrasonic click trains had no effect on the behavior or health of the fish tested. Fish do not even appear to take much notice of ultrasonic clicks up to 231 dB re 1 uPa.

Presenting Captive Bottlenose Dolphins with Live Fish

NOSC, Hawaii. We presented live fish (free and restrained) to eleven sonar-research dolphins in the hopes of eliciting predatory impulse sounds. Because the dolphins were in floating pens, not tanks, so they did not have to suffer their own echoes. The free fish escaped from the pens before the dolphins could respond. Some of the dolphins mouthed the plastic bag around the restrained fish, but only emitted click trains. The results probably reflect the fact that these dolphins are exposed to live fish all the time and only receive prepared rations each day.

At Sea Life Park, Hawaii (3-9-84). We used three performing bottlenose dolphins "on vacation" (Kako, Mekeo, and Kaeve) and seven whitespot goat-fish, two blacktailed snapper, and one parrotfish one at a time. The dolphins seemed interested in the live fish for the first few seconds, but then paid little attention to them and sometimes even appeared to be afraid of them. They emitted only occasional click trains when near the fish and no bangs.

Long Marine Lab (8-6-86). We put 17 white croakers (Genyonemus lineatus) in with our two ex-Navy bottlenose dolphins, Arrow and Josephine,

to observe predatory behavior. Both dolphins pursued the croakers
intermittently, and one (Arrow) captured and ate nine of the 17 fish. The
dolphins emitted click trains and whistles, but no bangs.

DISCUSSION

There are several puzzles connected with the proposal of odontocete prey
debilitation. If it occurs, what kind of sound produces the effect? How
is it made? How does it affect the prey? How frequently is it used? Is
it the same sound socially?

Acoustics of Vocal "Bangs" vs. Non-vocal Hydrodynamic Sounds

"Bangs" are the loud impulse sounds associated with feeding, and they
seem to fall into two categories: vocal and suspected non-vocal hydrodynamic sounds. Vocalized bangs and suspected non-vocal hydrodynamic sounds
are broadband, but the latter are characterized by relatively slow onset
and low frequency (Fig. 2). The bangs from the killer whales in Iceland,
Johnstone Strait, and Monterey Bay are of the suspected non-vocal variety.

Fish Hearing and Prey Capture

We do not presume that the use of loud impulse sounds by odontocetes
during predation is a simple story. How sounds are used certainly must
vary from species to species, as well as within species depending on
habitat, prey species, prey size, group size, time of day, culture, and
other variables. The evidence suggests to us the hypothesis that long
duration bangs overload the fish hearing apparatus, causing nervous adaptation and temporary reduction in hearing and orientation abilities, making
the fish easier to capture. But there are many alternative hypotheses,
such as symmetrically triggering the Maulthner escape response, vibrating
the swimbladder and causing disorientation or tissue damage, the Tullio
effect, lateral lines, and others. Furthermore, sounds are obviously used
to facilitate prey capture in other ways such as co-operatively herding the
prey, or confusing them. We are impressed by how infrequently bangs can be
recorded during odontocete predation. For example, in the Gulf of
California, common dolphins (Delphinus delphis) were listened to for three
hours feeding on sardines, and false killer whales (Pseudorca crassidens)
were listened to for four hours feeding on larger fish, but no loud "bangs"
were heard (Bernie Tershy, pers. comm.). We have found that the same
population of killer whales feeding on the same species of salmon in the
same general area, sometimes use bangs and sometimes don't. Clearly
further research is needed on the incidence of loud impulse sounds during
predation, on how these sounds vary taxonomically and with prey species, on
fish hearing, and how fish respond to these sounds. Additional research is
also needed on the social use of loud impulse sounds. We have presented
here information which provides additional evidence for the sonic predation
theory of Norris and Møhl (1983), but direct observational confirmation is
still missing.

The fish ear is unprotected by an ossicular chain (except for the
Weberian ossicles of ostariophysans). Loud sounds are transduced with full
force into the "inner ear." Further, if the sound is long in duration,
temporal summation may occur, as the latency period of the nerve is completed and the sound is still in progress. Each time this occurs, which
may be a hundred or more times during a single bang, the psychoacoustic
effect is heightened, perhaps becoming unbearably high. Finally, nervous
adaptation may occur, resulting in temporary reduction in sensitivity (for
approx. 10 msec to a few seconds -- R. R. Fay, pers. comm.). This disruption of hearing sensitivity may also be accompanied by a reduction of

directional hearing and orientation ability. The extent that hearing and good orientation is important to fish for evading noisy, team-hunting odontocetes, especially where vision may be poor, "banging" fish and temporarily impairing their hearing would confer a selective advantage to the predators.

How Loud Impulse Sounds Might Be Produced

The predatory weapons of social carnivores, such as canine teeth, are often used in a restricted, sometimes ritualized, manner in social aggression. This appears to be the case with the predatory and social bangs of odontocetes. Most of the social impulse sounds described here are very loud, long-duration, and low frequency-rich, closely resembling the sounds we found associated with feeding. In social settings, these sounds are usually given by dominant dolphins during aggressive situations. High frequency (over 100 kHz), short duration (40 microseconds) echolocation clicks (Au et al., 1974), and the long, low-frequency sounds described here may be the poles of a broad continuum of emissions by a single generation mechanism. If so, some events in the production of loud impulse sounds need to be explained, especially the jaw movements that accompany the so-called "jaw clap" (Lilly, 1962). The animal opens its jaws and claps them together during the production of the sound, but these could be adventitious events associated with pressurization of the basicranial space and bony nares prior to the passage of an air stream past the generators in the nasal plug area of the forehead during phonation. Support for this possibility comes from the anatomy of throat and laryngeal musculature in the dolphin. However, there are powerful muscles for moving the larynx forward and up into the basicranial space and they are responsible for the pressurization observed by Ridgway et al. (1980), and the laryngeal movement observed by Norris et al. (1971).

We hypothesize (Fig. 5) that the opening movement of the jaw clap allows the larynx to move posteriorly on its muscular suspension to a maximum degree, opening the basicranial space to its full extent, and that closure of the jaws occurs with contraction of the muscles pulling on and swinging the larynx forward and upward into the basicranial space. These actions together should produce the largest volume of high pressure air of which the dolphin is capable, and produce the elongate high intensity signal. We also suggest that the elongate cuneiform-epiglottic spout of the larynx, with its tight seal in the laryngeal sphincter facilitates such a pressurization event. The pause of 0.1 sec in the bottlenose dolphin and 1.2 - 1.9 sec in the killer whale which we recorded between echolocation and bangs probably represents this pressurization event.

Conclusions

1. Ultrasonic sounds appear to have little effect on fish, at least for the sounds and fish species we tested.

2. Loud impulse sounds are sometimes present during odontocete predation on fish. (This is a fact, not a conclusion.)

3. Similar loud impulse sounds are present sometimes also during social behavior.

4. These sounds are characterized by relative loudness, long duration, and relatively low frequency.

5. The function, source, and intensity of these sounds remains to be demonstrated.

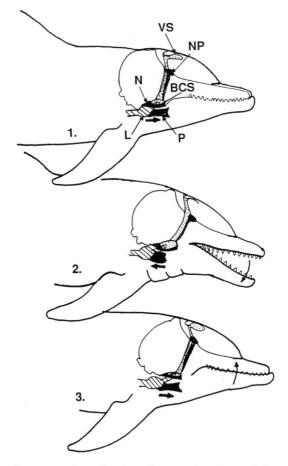

Fig. 5 Diagram of proposed mechanism for production of long duration "jaw claps" or prey debilitation noises.
(1) Normal click train production. Larynx (L) is pressed into basicranial space (BCS); nasalplug (NP) and narial muscles are metering air past sound source; and vestibular sac (VS) is filling.
(2) Jaw Clap begins. Jaw is rapidly opened, pressing muscles (P) attached to larynx posteriad, opening basicranial space by drawing nasopalatine sphincter (N) posteriad. Nasal passage is open and vestibular air is sucked inward.
(3) Sound begins as the large volume of air becomes pressurized in the basicranial space by closing the jaw and contracting the muscles attached to the larynx.

What about the Norris and Møhl sonic predation theory? Because of the theory, we examined the acoustics of wild odontocete predation. Interestingly enough, we found candidate debilitating sounds. The function of these loud impulse sounds, however, remains unknown. The next step is to try to discover their function.

ACKNOWLEDGEMENTS

We wish to express first and foremost our deep gratitude to Joseph M. Long whose inciteful interest in cetaceans and generous contributions to the University of California made this research possible. We thank Dr. Paul Nachtigall and the Naval Ocean Systems Center, Hawaii Laboratory for making dolphins available for the training work, facilities available for fish testing, and staff available for advice. The Lerner-Gray Fund of the American Museum of Natural History provided seed money. Tape recordings of odontocetes were generously shared by Rachel Smolker, Andrew Richards, Chris Johnson, Dave Bain and Gregory Silber of Long Marine Laboratory, and Virginia Cass, Thomas Lyrholm, Jonathan Gordon, Melba Caldwell, John Ford, Bertel Møhl, Craig Matkin, and Denise Herzing from other laboratories. This kind of kindred effort is what makes scientific progress on difficult subjects possible. Whitlow Au and Robert Floyd at NOSC provided technical expertise for the artificial click work, and Annette Marten assisted in those experiments. Earl Murchison, Ingrid Kang, Jeff Grovhoug, NOSC, and Sea Life Park, Hawaii facilitated presenting live fish to captive dolphins. Sue Patterson, Richard Hall, and Marion Ceruti of NOSC provided training expertise and support. And finally, Christopher Platt, Dick Fay, Art Popper, Robert Eaton, Peter Rogers, and Paul Webb provided us with knowledgeable discussion about fish hearing. We also thank Maria Choy-Vasquez, Maureen Leimbach, and Alice Russell, who prepared the manuscript.

REFERENCES

Au, W. W. L., Floyd, R. W., Penner, R. H., and Murchison, A. E., 1974, Measurement of echolocation signals of the Atlantic bottlenose dolphin, Tursiops truncatus Montagu, in open waters, J. Acoust. Soc. Am., 56(4): 1280.

Ceruti, M. G., Moore, P. W. B., and Patterson, S. A., 1983, Peak sound pressure levels and spectral frequency distribution in echolocation pulses of the Atlantic bottlenose dolphin (Tursiops truncatus), J. Acoust. Soc. Am., 74 (51) A, S73.

Enger, P. S., 1967, Hearing in herring, J. Comp. Biochem. Physiol., 22: 527.

Lilly, J. C., 1962, Vocal behavior of the bottlenose dolphin, Proc. Amer. Phil. Soc., 106 (6):520.

Norris, K. S., Dormer, K. J., Pegg, J., and Liese, G. J., 1971, The mechanism of sound production and air recycling in porpoises: a preliminary report, in: "Proceedings of the Eighth Annual Conference on Biological Sonar and Diving Mammals," Stanford Research Institute Press, Menlo Park, California.

Norris, K. S., and Møhl, B., 1983, Can odontocetes debilitate prey with sound?, Am. Nat., 122: 85.

Ridgway, S. H., Carder, D. A., Green, R. F., Gaunt, A. S., Gaunt, S. L. L., and Evans, W. E., 1980, Electromyographic and pressure events in the nasolaryngeal system of dolphins during sound production, in: "Animal Sonar Systems," R. Busnel and J. Fish, eds., Plenum Press, New York.

Smolker, R. and Richards, A., This volume, Loud impulse sounds during feeding in wild Indian Ocean bottlenose dolphins.

HARMONIC STRUCTURE OF BAT ECHOLOCATION SIGNALS

Karl Zbinden

Institute of Zoology, University of Berne
Baltzerstrasse 3, CH-3012 Berne, Switzerland

INTRODUCTION

A wide range of different harmonic patterns is found in echolocation signals of bats. Often a gradual change from single harmonic pulses to multiple-harmonic ones can be observed when FM-bats detect and approach their prey in free flight. Multiple-harmonic pulses may also be used for orientation in space under reverberant conditions or for target discrimination. The appearance of the harmonic structure of a bat pulse is influenced by the Q of vocal tract resonances. At low Q multiple harmonics increase the signal bandwidth and thus improve the ranging and discriminative properties of a pulse. At high Q dumbell pulses are created.

This study aims at demonstrating some basic concepts of harmonic structure in FM echolocation signals.

METHODS

Typical signals of several bat species were analysed with a correlation computer (Halls, 1980 and Pye, 1985) in order to determine their ambiguity function. Power spectra were calculated on a digital FFT-analyser (B&K type 2033) and sonagrams were made on a Kay 6061B sonagraph. Three signal samples with widely differing harmonic structure, each of a different bat species (Peropteryx macrotis, Tadarida teniotis and Carollia perspicillata), were selected for presentation in this paper. Tadarida teniotis is a European species, the other two are tropical.

A computer program was devised to create bat like signals on a BBC-microcomputer in order to investigate the impact of various pulse parameters on the ranging properties of the signals. The program allows the independent manipulation of most pulse parameters and plots the oscillogram, the power spectrum and the ambiguity diagram for each model signal. Crosscorrelation slices were estimated for negative velocities (approaching target) up to -13m/s (α =1.08). The pulse centre was taken as the reference point for time

---

SYMBOLS: ACR=autocorrelation, AD=Ambiguity diagram, Doppler factor α =c-v/c+v, where c=velocity of sound in air and v=target velocity, BWeff=effective bandwidth, BWs=spectral bandwidth, CF=constant frequency, EFM=exponential frequency modulation, fc(eff)=(effective)centre frequency, LFM=linear frequency modulation, LPM=linear period modulation, Q=F.reson./BW (quality factor of a resonator), T-6dB=signal duration between amplitude points of -6dB.

compression. A gauss-shaped envelope was chosen in order to minimise envelope effects on the power spectrum. The FM-sweep mode was either LFM, EFM or LPM.

RESULTS

(1) Model signals

Three different harmonic configurations having significantly different ambiguity functions can be distinguished:

(1.1) Harmonics overlapping in the time domain but not in the frequency domain (Fig.1). The AD is split into narrow ridges parallel to the α-axis indicating partial decoupling of range and velocity information. The spacing between ridges is equal to 1/fc of the fundamental. The amount of energy shift between ridges with increasing velocity is determined by the average modulation rate of the four harmonics. The width of the central ridge is given by the overall signal BW and the width of the whole surface is a function of the average BW of the single harmonics. Timing (ranging) is ambiguous within a delay given by the mean value of 1/BWeff of single harmonics H1-Hn. Thus the resolution of target range is not better than for an isolated harmonic, except when range is a priori known and the targets are of similar strength. In this case the range resolution is determined by the overall BW. The surface as a whole shows excellent Doppler tolerance regardless of the sweep mode (-1dB at 13 m/s in LPM as shown in Fig.1). The velocity accuracy is therefore very bad.

(1.2) Harmonics overlapping in the time and frequency domains (Fig.2). The AD shows a single, sharp ridge and an extended plateau at low target velocities. At high velocity the energy is shifted to a second ridge growing from the plateau. Range-velocity coupling is virtually eliminated up to moderate velocities. The potential range resolution is determined by the overall BW and is therefore very good. Doppler tolerance depends on the sweep mode. It is moderate in the case of LPM- (-3.5 dB at 13 m/s in Fig.2) and worse for EFM- and LFM-pulses. The velocity accuracy is in the order of the bat's own flight speed. By changing the contribution of single harmonics to the overall energy (e.g. H1 at half amplitude), the ratio of peak and plateau can be increased for small velocities at the expense of Doppler tolerance and of time (range) resolution. Shaping may also reduce the plateau level in the close vicinity of the main peak. This harmonic configuration allows a very accurate determination of target range when the relative target velocity is fairly low.

(1.3) Two harmonics overlapping each other in the frequency domain but not or only partially in the time domain (dumbell pulses, Figs.3 and 4). The AD of both the EFM- (Fig.3) and the LPM-pulse (Fig.4) are composed of a main ridge and two prominent side ridges. The delay between the ridges equals the spacing of the waveform peaks. Range-velocity coupling is determined by the average modulation rate only and is thus no better than for a single harmonic. In the LPM-pulse decoupling could be acheived with an ideal receiver (Johnson and Titlebaum, 1972). The range resolution is worse than for a single harmonic of the same BW, i.e. ambiguous if a bad signal-to-clutter ratio or moderate velocities are involved. The Doppler tolerance of the surface depends on the sweep mode and is best for LPM. It reflects the ripple in the power spectrum and is thus periodic. The side ridges are more Doppler tolerant than the central ridge. In EFM the central ridge has a sharp minimum (-10dB) at a velocity below 10 m/s while the side ridges stay at -3dB. Fc determines the velocity where the minimum occurs. The configuration appears suitable to discriminate moving objects from stationary ones or to estimate the relative velocity within certain a priori limits. EFM-pulses offer the best possible fit of the signal and its Doppler shifted echo and could therefore give an acceptable discrimination of velocity while maintaining optimal

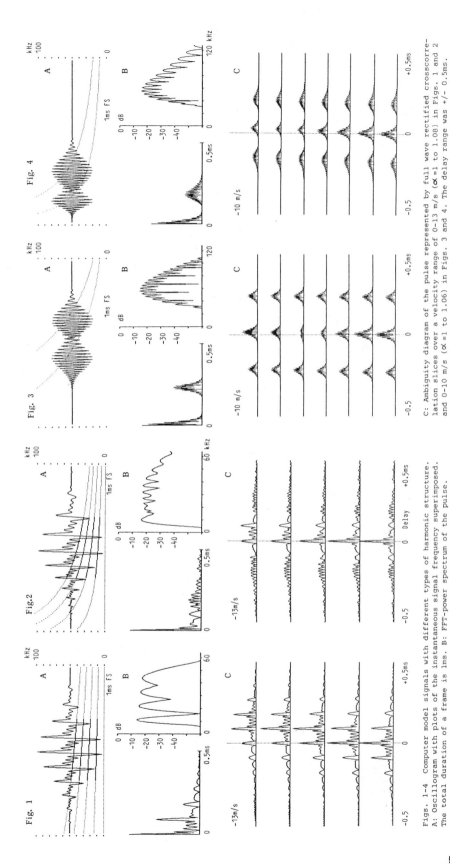

Figs. 1-4 Computer model signals with different types of harmonic structure.
A: Oscillogram with plots of the instantaneous signal frequency superimposed.
The total duration of a frame is 1ms. B: FFT-power spectrum of the pulse.

C: Ambiguity diagram of the pulse represented by full wave rectified crosscorrelation slices over a velocity range of 0-13 m/s (α=1 to 1.08) in Figs. 1 and 2 and 0-10 m/s (α=1 to 1.06) in Figs. 3 and 4. The delay range was +/- 0.5ms.

sensitivity over a large velocity range. By adjusting the sweep rate and/or
the formant frequency the range of best velocity discrimination (the position
of the minimum) is shifted. Increasing Fc in a EFM-pulse shifts the minimum
closer to zero velocity.

(2) Bat signals

(2.1) Peropteryx macrotis (Emballonuridae, insectivorous), Fig.5: The
bat was flying in a reverberant situation (laboratory room) when it emitted
the pulse which has six harmonics overlapping in time but not in frequency
(c.f. model 1.1). The overall BWs of the signal is very high (65 kHz), which
is reflected in the narrow central spike of the ACR indicating a BWeff of 53
kHz. The frequency of the centre wave of the ACR indicates that fc(eff) of
the pulse is at about 40 kHz, e.g. the fc of the third harmonic. The capability to resolve two closely spaced targets is ambiguous within a time delay
given by the mean value of 1/BWeff of the single harmonics. The AD is made up
by a number of sharp ridges parallel to the α-axis. The spacing of the ridges along the delay axis relates to fc of the fundamental. Range-velocity
coupling is negligible at very low velocities (<2m/s) only, since the ridges
are not Doppler tolerant. But as the surface as a whole is fairly Doppler tolerant (-2.5dB at v = -10m/s), the signal does only provide a velocity accuracy of less than 10m/s when the target range is not known. Depending on
the signal duration and the sweep mode this basic signal structure may be
more or less doppler tolerant. A two harmonic LPM-pulse of Lavia frons
(T-6dB = 700 µs) was perfectly Doppler tolerant (-1.8dB at -11m/s) whereas a
three harmonic pulse of Glossophaga soricina with a sigmoidal modulation
curve and a duration of 5.4ms had a velocity accuracy of about 3m/s.

(2.2) Tadarida teniotis (Molossidae, insectivorous), Fig.6: The bat
was again flying in a cluttered situation (laboratory room) when the pulse
with five harmonics overlapping in time and in frequency was emitted (c.f.
model 1.2). When hunting in the open this species usually emits long, single
harmonic, shallow-FM pulses of a very low frequency (Zbinden and Zingg,1985).
The overall BWs of the pulse is again high (approx. 40 kHz). The half power
width of the ACR is 45 µs, indicating a BWeff of 22 kHz. The frequency of the
centre period of the ACR (53 kHz) is again close to the overall spectral fc
of 48 kHz. Apart from the very narrow central ridge there are multiple side
ridges of low power (-6dB and -10dB from the ACR-peak value), lying quite far
from the main ridge. These side ridges arise when the overlapping sections of
the single harmonics match up with each other in time. The pulse does not
fully conform to model case (1.2) in that it shows a distinct range-velocity
coupling and no side ridge energy close to the main ridge. This may partly be
due to the sigmoid frequency modulation of the real pulse and partly to the
unequal energy distribution among the harmonics. As in the model the signal
energy is transferred to the right side wings of the main ridge when high velocities are applied. The main ridge splits into narrow ridges parallel to
the α-axis and the range resolution becomes ambiguous. It depends on BWeff
at which relative velocity this is going to happen. The Doppler tolerance of
the main ridge is mediocre. The ridge drops by -3dB at a velocity of approx.
-5m/s. With increasing velocity the side ridges to the right of the main
ridge degrade and those to the left (not shown in this pulse) build up. The
pulse offers a potentially excellent range resolution and ranging accuracy
with almost no ambiguities present at small flight velocities and is thus
well suited for orientation and manoeuvring within a restricted space.

(2.3) Carollia perspicillata (Phyllostomidae, frugivorous), Fig.7: The
bat was flying in a large bat house when it emitted the pulse containing two
rapidly sweeping harmonics overlapping in frequency but only partially in
time. The signal envelope has two distinct peaks (dumbell, c.f. model 1.3).
The total BW is limited to about 20 kHz (BWeff) probably by a sharp vocal

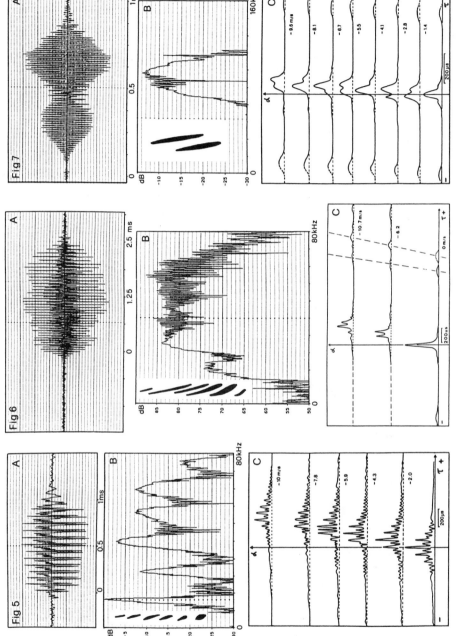

Figs. 5-7 A: Oscillogram. B: Power spectrum and sonagram. C: Ambiguity diagram. Assembly of the envelopes of crosscorrelation slices over a velocity range of 0-10 m/s.

585

tract resonance. The sweep mode is LPM. The BWeff as estimated from the ACR and the most important frequency, being 90 kHz as estimated by the fine structure frequency at the ACR peak, are well represented by the -3dB BWs and the fc in the spectrum. Due to the remaining partial overlap in time the ACR has 'shoulders' which grow into peaks when a moderate Doppler shift is applied. In the AD the energy is distributed among a skewed main ridge and two side ridges (at -8.5dB in the ACR). The main ridge is very Doppler tolerant (-1.3dB at v = -9.6m/s) and would not allow the bat to discriminate velocities even of its own flight speed on a single pulse basis. The side ridges however show a distinct relationship with velocity. The left ridge increases by up to 4dB and the right ridge decreases by up to 9dB at a velocity of -9.6m/s. If a difference between the ridges of 1dB is taken as a detectable limit, a velocity accuracy of 3m/s could be reached in a situation with rather good signal/clutter-ratio. At bad s/c the pulse would still be suited to detect obstacles in fast flight since the main ridge is Doppler tolerant. The real pulse is not identical to the model pulse and has some properties which could not be interpreted by means of the simple single formant model. The main ridge instead of the side ridges is Doppler tolerant in the real pulse. A good fit with the real signal is obtained by increasing the centre frequency of H2 by about 10 kHz and reducing its BW in the model.

DISCUSSION

Why do bats not simply use a steeper single harmonic sweep instead of multiple-harmonics to achieve a large signal bandwidth? This would result in less ambiguities and still provide the same range resolution. There may be at least four reasons why bats cannot or prefer not to adopt this solution: Bats are often interested not only in ranging, but say in target discrimination or they may have to solve a range of different tasks with the same signal. The optimisation process may in fact be multi-dimensional in most real cases. It may also be impossible to exceed a certain sweep rate and still remove the harmonics with simple formant filtering. When a bat is close to a target the pulse duration has to be restricted in order to avoid overlap of the signal and the echo. The FM-sweep rate may already be at its physiological maximum in this situation and the bandwidth necessary for accurate ranging can only be provided by adding more harmonics. Overlapping harmonics effectively decouple the range information from the velocity information which has already been pointed out by Simmons and Stein (1980). Due to their complex directivity pattern multiple-harmonics also improve the information about the direction of a target. Thus multiple-harmonics may be advantageous in some situations, in other situations they may simply be unavoidable.

Peropteryx and Tadarida emitted their signals while operating in a restricted, reverberant environment. A multiple-harmonic signal may be the only signal type emitted by bats that live and hunt exclusively in such an environment. Or it may be one of several signal types of an acoustically highly flexible bat species and is then only emitted when the bat has to manoeuvre close to obstacles or approach its hanging site.

Bats using multiple-harmonic pulses with a shallow sweep should rather more rely on the analysis in the frequency domain when they discriminate targets or try to assess target velocity. Bats using multiple-harmonic pulses with a deep sweep are more susceptible of using correlation type receivers, e.g. to perform the analysis in the time domain when they need accurate ranging and target discrimination. It is possible that some bats could change from one type of receiver to another and adjust the structure of their pulses accordingly, when the situation demands it.

Under good signal-to-clutter conditions multiple harmonics can also be favourable to retain some degree of velocity perception even with rather short FM-pulses. Bats using dumbell pulses at times may be able to switch between accurate ranging and some degree of velocity sensitivity by changing the tuning and/or the Q of the vocal tract resonator system.

Further field studies are definitely needed to investigate the striking flexibility in pulse design found in a number of bat species. For future work it will be of first importance to establish a correlation between the basic structure of the echolocation signal as emitted by the bat, the echoes received from the target and the exact behavioural situation the bat was in when emitting the signals.

ACKNOWLEDGEMENTS

The author is most grateful to Prof. J.D. Pye, London, who provided the bulk of the recordings forming the basis for this study and who gave me access to his specialised equipment for the analysis of the ambiguity functions. Thanks are also due to Dr. O. Bernath, Berne for the opportunity to use his spectrum analyser. The research was supported by the Swiss National Fund for the Promotion of Science and the Royal Society, London.

REFERENCES

Halls, J.A.T. (1980): An analogue device for the generation of sonar ambiguity diagrams. Animal Sonar Systems (eds. Busnel, R.G. and J.F. Fish), Plenum Press, N.Y. pp.909-911.
Johnson, R.A. and E.L. Titlebaum (1972): Range-Doppler uncoupling in the Doppler tolerant bat signal. Proc. 1972 I.E.E.E. Ultrasonics symposium.
Pye, J.D. (1985): Signals as clues to system performance. Systemes Sonars Aeriens Animaux: Colloque International C.N.R.S., Lyon France.
Simmons ,J.A. and R.A. Stein (1980): Acoustic Imaging in Bat Sonar: Echolocation Signals and the Evolution of Echolocation. J.Comp.Physiol. 35,61-84.
Zbinden, K. and P.E. Zingg (1985): Search and Hunting Signals of Echolocating Free-tailed Bats, Tadarida teniotis in Southern Switzerland. Mammalia 50(1), 9-25.

ACOUSTICAL VS. VISUAL ORIENTATION IN NEOTROPICAL BATS

U. Schmidt, G. Joermann and G. Rother

Zoological Institute
University of Bonn
West Germany

INTRODUCTION

In some phyllostomids visual acuity and sensitivity are remarkably well developed (Manske and Schmidt, 1976; Bell and Fenton, 1986); these bats are even capable of distinguishing simple visual patterns (Manske and Schmidt, 1979; Suthers et al., 1969). Thus, it is not surprising that in many orientation situations vision is involved. E.g., bats searching for an exit attend to light and disregard echolocation (Chase, 1983); obstacle avoidance as well is aided by vision (Chase and Suthers, 1969; Rother and Schmidt, 1982). Macrotus californicus uses vision to find prey on the ground (Bell, 1985).

These experiments were designed to compare the importance of visual and acoustical senses for orientation in the omnivorous Phyllostomus discolor and the sanguivorous Desmodus rotundus.

VISUAL VS. PASSIVE ACOUSTICAL ORIENTATION

Three Phyllostomus and three Desmodus were trained in a two-choice apparatus for food reward. During training the rewarded side was marked by acoustical (continuous broad band noise 2 - 38 kHz, 70 dB SPL RMS), visual (circular illuminated spot, ∅ 6 cm, white light, 8 cd / m²) and olfactory (grass odour) stimuli. After pretraining (P. discolor: 900 runs / animal; D. rotundus: 300 runs) this combination was used as standard; the critical trials were interspersed at the rate of 1 : 4. During the first set of experiments the 3 sensory parameters were tested separately, in the second set always 2 modalities had to be compared by the bats.

In both species the standard task was learned to nearly 100%, and when the visual or the acoustical stimulus was presented alone, the performance decreased only slightly (Fig. 1). The two species differed considerably in the bimodal tests; while Desmodus prefered the acoustical cue, Phyllostomus was highly attracted by the optical one. To evaluate the importance of these two sensory modalities, the intensity of the prefered stimulus was diminished in steps. Desmodus choose about 50% the visual stimulus when the sound level was reduced to 45 dB (contra visual standard), in Phyllostomus both modalities were equally attractive, when the

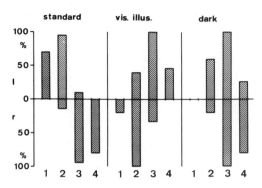

Fig. 1. Percentage of responses to the different sensory stimuli in
Phyllostomus discolor and Desmodus rotundus. (S) standard = training condition (combination of acoustical, visual and olfactory
stimuli); (A) noise alone; (O) optical sign; (A-O) choice between
acoustical and optical stimulus ($\bar{x}$ and extremes; n = three bats).

light intensity was reduced by the factor 20 (0.4 cd / m² contra 70 dB
noise).

These results clearly indicate that in Phyllostomus vision dominates
over passive acoustic cues, while in Desmodus acoustic parameters are more
important in localization.

THE USE OF VISION DURING APPROACH AND LANDING

Three Desmodus were trained to fly across a dark flight tunnel
(3 x 0.8 x 0.8 m) and land at a backlit grid (9.5 x 15.5 cm) protruding
1 cm into the tunnel. The grid was randomly changed between two positions.
We could also produce a visual illusion of the grid at the empty position;
humans could not distinguish between the real grid and the illusion. The
flight was monitored by four rows of light barriers, and in selected
flights we recorded orientation sounds (up to 40 cm in front of the landing), observed the animals with night vision scopes, or photographed the
landing phase with infrared stroboscopic light.

After two weeks of training (landing grid illuminated; about
600 flights / animal), the daily schedule included 36 standard flights
(visible target), 2 trials with invisible landing site and visual illusion
at the negative position, and 2 trials in complete darkness. In all trials
the lateral position of the bat during the approach was determined by an
array of four light barriers arranged at right angle to the flight path
40 cm in front of the target.

In the standard flights all the bats were clearly oriented towards
the target at a range of 40 cm (χ^2 - test, p < 0.01). Eliminating the visual
information diminished the early selection of the landing position
p > 0.1).

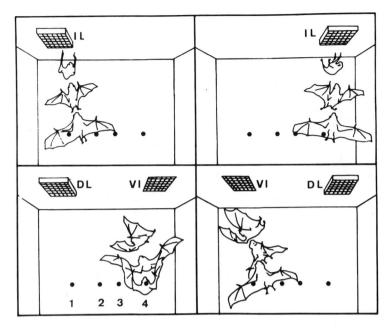

Fig. 2. Approach to the landing wall of a Desmodus. Upper row: standard
condition (landing grid illuminated from behind, IL); lower row:
landing grid dark (DL), visual illusion of the grid (VI) at the
negative position; 1 to 4: light barriers. Drawn after infrared
photographs (flash intervals: 70 ms).

To separate acoustical and visual orientation more clearly, both
parameters were presented at different locations, and in this situation the
animals aimed at the visual illusion. Only at 30 - 40 cm before the landing
they recognized the smooth wall (Fig. 2 and 3), and two of the three vampires usually manoeuvred to the grid (standard flights: 2 resp. 3% misses;
visual dummy flights: 7 resp. 27% misses). Although only few critical trials
were introduced between standard flights, there was a clear tendency for
these two bats to learn the situation, as nearly all the misses were observed during the first days. The third animal was strongly confused by the
visual dummy and usually dropped to the ground (1 vs. 93% misses).

Echolocation sounds and pattern of sound emission were not altered in
any of the situations. During dark flights speed was reduced by about 10%,
but the pulse repetition rate remained stable.

Three Phyllostomus discolor, trained under the same conditions, were
equally influenced by vision.

CONCLUSIONS

Echolocation in microchiroptera substitutes the sense of vision,
dominant in many mammalian species. As echolocation is limited in range, it
is not surprising that vision is necessary for long range orientation
(Davis, 1966; Williams et al., 1966). At medium and short ranges visual
cues may give additional information that facilitates orientation, as shown
in improved obstacle avoidance. If information of the acoustical and visual
system is available, at least the phyllostomids studied rely in many cases

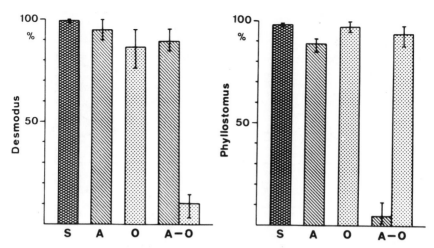

Fig. 3. Percentage of light barriers (1 - 4) activated during the different flight conditions in one vampire bat. Target grid left (1) resp. right (r).

on vision, as can be seen in experimental situations where both are contradictory. But the bats rely on their eyes only in the detection phase, the complicated motor performance of landing is fully under control of echolocation for the bats never attempted to hang up at the visual dummy.

The prevalence of vision or passive hearing depends very much on the ecological specialization of a given bat species. It can be expected that passive acoustical orientation is of utmost importance in those species that detect their prey acoustically, for instance Myotis myotis (Kolb, 1961), Megaderma lyra (Fiedler, 1979), Trachops cirrhosus (Ryan et al., 1983) and Desmodus rotundus. Our data indicate that Desmodus finds its prey primarily by passive acoustical orientation. Other species, for instance Macrotus californicus (Bell, 1985) seem to be more flexible in locating food and may switch between sensory modes depending on the situation.

REFERENCES

Bell, G.P., 1985, The sensory basis of prey location by the California leaf-nosed bat Macrotus californicus (Chiroptera: Phyllostomatidae), Behav. Ecol. Sociobiol., 16:343-347.

Bell, G.P. and Fenton, M.B., 1986, Visual acuity, sensitivity and binocularity in a gleaning insectivorous bat, Macrotus californicus (Chiroptera: Phyllostomidae), Anim. Behav., 34:409-414.

Chase, J., 1983, Differential responses to visual and acoustic cues during escape in the bat Anoura geoffroyi: cue preferences and behaviour, Anim. Behav., 31:526-531.

Chase, J. and Suthers, R.A., 1969, Visual obstacle avoidance by echolocating bats, Anim. Behav., 17:201-207.

Davis, R., 1966, Homing performance and homing ability in bats, Ecol. Monographs, 36:201-237.

Fiedler, J., 1979, Prey catching with and without echolocation in the Indian false vampire (Megaderma lyra), Behav. Ecol. Sociobiol., 6:155-160.

Kolb, A., 1961, Sinnesleistungen einheimischer Fledermäuse bei der Nahrungssuche und Nahrungsauswahl auf dem Boden und in der Luft, Z. Vergl. Physiol., 44:550-564.

Manske, U. and Schmidt, U., 1976, Visual acuity of the vampire bat, Desmodus rotundus, and its dependence upon light intensity, Z. Tierpsychol., 42:215-221.

Manske, U. and Schmidt, U., 1979, Untersuchungen zur optischen Musterunterscheidung bei der Vampirfledermaus, Desmodus rotundus, Z. Tierpsychol., 49:120-131.

Rother, G. and Schmidt, U., 1982, Der Einfluß visueller Information auf die Echoortung bei Phyllostomus discolor (Chiroptera), Z. Säugetierk., 47:324-334.

Ryan, M.J., Tuttle, M.D., and Barclay, R.M.R., 1983, Behavioral responses of the frog-eating bat, Trachops cirrhosus, to sonic frequencies, J. Comp. Physiol., 150:413-418.

Suthers, R.A., Chase, J., and Braford, B., 1969, Visual form discrmination by echolocating bats, Biol. Bull., 137:535-546.

Williams, T.C., Williams, J.M., and Griffin, D.R., 1966, The homing ability of the neotropical bat Phyllostomus hastatus, with evidence for visual orientation, Anim. Behav. 14:468-473.

ECHOLOCATION STRATEGIES OF AERIAL INSECTIVOROUS BATS AND THEIR INFLUENCE

ON PREY SELECTION

Robert M. R. Barclay

Biology Department
University of Calgary
Calgary, Alberta, Canada   T2N 1N4

There are few studies of prey selection by aerial insectivorous bats but Swift et al. (1985) noted that the available data suggests that large species feed selectively while small species are unselective. They suggested that large bats select prey, primarily on the basis of size, for reasons consistent with optimal foraging theory (i.e., to maximize the rate of net energy intake).

My research on the silver-haired (Lasionycteris noctivagans) and hoary (Lasiurus cinereus) bats (Barclay, 1985) appears to support the idea that large aerial insectivorous bats prey selectively. In my study area, (Delta, Manitoba, Canada) small insects (body length < 10 mm), primarily chironomids (Diptera), dominate the nocturnal aerial insect fauna and small bats (L. noctivagans, mass 11 g and Myotis lucifugus, 7 g) feed extensively on them (Barclay, 1985; unpublished data). The largest species, L. cinereus (27 g), on the other hand, feeds almost exclusively on moths, dragonflies and beetles and includes virtually no chironomids or other small insects in its diet (Barclay, 1985). It could be argued that individual L. cinereus are actively selecting large prey from amongst what is available.

There is another possible explanation for size-selection which has potentially important implications not only for prey selection and optimal foraging by aerial insectivorous bats, but also for the diversity of large species. I believe the important feature is not so much the size of the bat but its flight speed and maneuverability. Flight speed is generally correlated to body size in bats (Hayward and Davis, 1964) since wing loading increases with body size and increased wing loading requires greater speed to keep the animal aloft (e.g. Norberg, 1985). Wing aspect ratio also tends to increase with body size in bats (Norberg, 1985) and a high aspect ratio (i.e., long narrow wings) is associated with reduced maneuverability (Norberg, 1985). Thus big bats tend to be faster and less maneuverable than smaller species. L. cinereus for example, has a high wing loading and aspect ratio (Farney and Fleharty, 1969) and flies rapidly in straight line paths (Barclay, 1985). L. noctivagans, with a lower wing loading and aspect ratio (Farney and Fleharty, 1969), flies more slowly and maneuverably (Barclay, 1985).

Fast, unmaneuverable bats must detect prey at greater range than slower, more maneuverable species to allow time to capture prey. I hypothesize that the echolocation call features used to enhance prey detection range, reduce the detectability and thus availability of small insects. Size selection of prey thus results from differential availability of prey rather than from active selection by individual bats.

To maximize prey detection range, fast unmaneuverable bats such as <u>L. cinereus</u> could simply concentrate on large prey since these produce stronger echoes and are thus detectable at greater distances. There are, however, several echolocation call features that bats such as <u>L. cinereus</u> could alter in the search phase to increase prey detection range.

a) <u>Use high intensity calls.</u> Accurate intensity measurements of echolocation calls used in natural situations are difficult to obtain, but <u>L. cinereus</u> appears to use intense search phase calls. Foraging individuals were commonly detectable at least 40 m away using ultrasonic microphones, a much greater distance than that used by Fenton and Bell (1981) to define "high intensity" echolocators.

b) <u>Use low frequency calls.</u> Since atmospheric attenuation increases with frequency (Griffin, 1971), fast flying bats should use low frequency calls. <u>L. cinereus</u> does this (Fig. 1a and b). It's search phase calls extend down to 17 kHz, significantly lower than the calls of <u>L. noctivagans</u> (Fig. 1c and d; mean lowest frequency 25 kHz).

c) <u>Use narrow band calls.</u> Rapidly flying, open air insectivorous bats use narrowband constant frequency or shallow frequency modulated search phase calls (Simmons et al. 1979; Neuweiler, 1984). By restricting frequency sweep and the number of harmonics, energy is concentrated and bats may be able to use auditory specializations (e.g., acoustic fovea, Schuller and Pollak, 1979) allowing greater sensitivity for echo frequency. The search phase calls of <u>L. cinereus</u> fit this prediction in that they are shallow, frequency modulated calls with no upper harmonics (Fig. 1a and b). <u>L. noctivagans,</u> on the other hand, uses broader band calls with a second harmonic (Fig. 1c and d). Whether <u>L. cinereus</u> has the auditory specializations found in some other narrow band bats (e.g., Neuweiler, 1984) is not known.

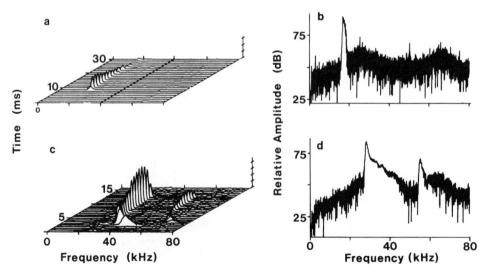

Fig. 1. Spectrograms and frequency spectra of the search calls of <u>L. cinereus</u> (a and b) and <u>L. noctivagans</u> (c and d). Note the difference in time scales between a and c.

d) <u>Increase the duty cycle</u>. By increasing repetition rate and/or call duration and thereby increasing the duty cycle, bats could enhance the chances of detecting the echo-amplitude "glint" caused by the beating wings of prey (Schnitzler et al., 1983). Although such a strategy should be beneficial to all bats, only short range (i.e., slow) bats should be able to use high repetition rates. Bats searching for prey at greater distances must wait longer between outgoing pulses to avoid eliminating the possibility of using pulse-echo interval to determine target range. Indeed, L. cinereus uses a very slow repetition rate (mean 3.3 pulses/s) compared to L. noctivagans (mean 6.1) and many other vespertilionids.

By increasing call duration, duty cycle can also be increased and this should be an option available to fast flying bats. L. cinereus uses relatively long search calls (mean 10.3 ms) compared to many other vespertilionids (e.g., Fenton and Bell, 1981), although the calls of L. noctivagans are almost as long (mean 9.4 ms).

Lasiurus cinereus search calls thus appear to be adapted for long range prey detection, something the bat requires given its rapid flight. These calls, however, have the disadvantage of reducing the detectability of small prey. In particular, low frequencies reflect poorly off small targets (i.e., targets with dimensions less than the wavelength of the calls; e.g., Pye, 1980). The calls of L. cinereus have wavelengths (16.7–19.6 mm) ten times or more greater than wing and body widths of chironomids at my study site. Preliminary experiments by Rudi Kober (pers. comm.) indicate that chironomid-sized insects reflect 20 kHz pulses so poorly that L. cinereus would likely not detect them even at a range of one meter. Insects detected at closer range will be extremely difficult to capture for a fast flying L. cinereus, particularly since the narrowband calls provide poor range information (Simmons and Stein, 1980). Further-more, the long duration of the search calls produces pulse-echo overlap for targets closer than 1.8 m and this may cause signal processing problems (H.U. Schnitzler, pers. comm.).

Thus, even if it were energetically beneficial for L. cinereus to prey on small insects, they cannot because their calls prohibit the detection of such prey or at least reduce detection range to a point where capture is not feasible. Prey selection in L. cinereus may thus be passive and not due to active choices made for optimal foraging reasons.

The echolocation call features used by L. cinereus should be advantageous to any fast unmaneuverable species and correlations should exist between flight speed/maneuverability and echolocation call design. Since measurements of bat flight speeds under natural conditions are rare, however, we must rely on wing loading and aspect ratio as indicators of speed and maneuverability. Given this, several hypotheses can be tested. Wing loading and aspect ratio should be:

a) negatively correlated to the lowest frequency of search calls;
b) negatively correlated to search call band width;
c) positively correlated to search call duration.

Data for Vespertilionids and Molossids (Ahlen, 1979; Fenton and Bell, 1981; Norberg 1981; this study) supports all three hypotheses (Table 1). Whether Vespertilionids are analyzed alone or with the few Molossids for which data is available, wing loading and aspect ratio are both negatively correlated to the lowest search call frequency and positively correlated to call duration. Weaker negative correlations exist with the degree of frequency modulation, probably partly due to problems in determining maximum frequency and to the fact that band width also depends on the number of harmonics present.

Table 1. Correlation coefficients (r) for measures of wing shape and body size versus echolocation call characteristics. Sample sizes in brackets.

	Vespertilionids	Vespertilionids and Molossids
Wing loading vs		
Lowest frequency	-0.561 (19)**	-0.700 (27)***
Frequency modulation	-0.454 (12)ns	-0.580 (20)**
Duration	0.897 (19)***	0.806 (27)***
Aspect ratio vs		
Lowest frequency	-0.469 (19)*	-0.662 (27)***
Frequency modulation	-0.535 (12)*	-0.630 (20)**
Duration	0.536 (19)**	0.750 (27)***

ns p>.05, * p<0.05, ** p<0.01, *** p<0.001

Large, fast unmaneuverable bats thus use long, narrow band, low frequency search calls which are adapted to enhance prey detection range. Such calls, however, reduce the detectability of small prey and we should find prey selection, on the basis of size, amongst large aerial insectivorous bats, whether or not such selection is beneficial from an optimal foraging standpoint. There is relatively little data available to test this prediction although the diet of L. cinereus supports it (Ross, 1967; Barclay, 1985) and Euderma maculatum, the North American Vespertilionid with the lowest frequency search calls (Fenton and Bell, 1981), also feeds on large prey (Ross, 1967).

The arguments above lead to some interesting ecological consequences. In many groups of animals (e.g., Rosenzweig, 1966) including aerial insectivorous birds (Hespenheide, 1971), niche width, in terms of prey size, increases with increasing body size. In other words, larger predators can take a greater range of prey than can smaller ones. In large aerial insectivorous bats, on the other hand, the lower limit of prey size is increased and the potential for expanded niche width is more restricted. Whether this contributes to the low diversity of large aerial insectivorous bats (e.g., Fenton and Fleming, 1976) remains to be determined.

Acknowledgements

I thank M.B. Fenton, G.P. Bell, J.H. Fullard, D.W. Thomas, H.D. Aldridge, H.U. Schnitzler and G. Neuweiler for suggestions regarding this manuscript and R. Kober for unpublished data. My research has been supported by the Natural Sciences and Engineering Research Council of Canada and the Universities of Calgary and Manitoba.

Literature

Ahlen, I., 1981, Identification of Scandinavian bats by their sounds, Swed. Univ. Agr. Sci. Dept. Wildlife Ecol., Rep. 6.

Barclay, R.M.R., 1985, Long- versus short-range foraging strategies of hoary (Lasiurus cinereus) and silver-haired (Lasionycteris noctivagans) bats and consequences for prey selection, Can. J. Zool., 63: 2507-2515.

Farney, J. and Fleharty, E.D., 1969, Aspect ratio, loading, wing span and membrane areas of bats, J. Mamm., 50: 362-367.

Fenton, M.B. and Bell, G.P., 1981, Recognition of species of insectivorous bats by their echolocation calls, J. Mamm., 62: 233-243.

Fenton, M.B. and Fleming, T.H., 1976, Ecological interactions between bats and nocturnal birds, Biotropica, 8: 104-110.

Griffin, D.R., 1971, The importance of atmospheric attenuation for the echolocation of bats (Chiroptera), Anim. Behav., 19: 55-61.

Hayward, B.J. and Davis, R., Flight speed in western bats, J. Mamm., 45: 236-242.

Hespenheide, H.A., 1971, Food preference and the extent of overlap in some insectivorous birds, with special reference to the Tyranidae, Ibis., 113: 59-72.

Neuweiler, G., 1984, Foraging, echolocation and audition in bats, Naturwissenschaften, 71: 446-455.

Norberg, V.M., 1981, Allometry of bat wings and legs and comparison with birds, Phil. Trans. Roy. Soc. Lond. B., 292: 359-398.

Norberg, V.M., 1985, Flying, gliding and soaring, in: "Functional Vertebrate Morphology", M. Hildebrand, D.M. Bramble, K.F. Liem and D.B. Wake, eds., Belknap Press, Cambridge.

Pye, J.D., 1980, Echolocation signals and echoes in air, in: "Animal Sonar Systems", R.-G. Busnel and J.F. Fish, eds. Plenum Press, New York.

Rosenzweig, M.L., 1968, The strategy of body size in mammalian carnivores. Amer. Midl. Nat., 80: 299-315.

Ross, A., 1967, Ecological aspects of the food habits of bats. Proc. Western Found. Vert. Zool., 1: 204-263.

Schnitzler, H.U., Menne, D., Kober, R., and Heblich, K., 1983, The accoustical image of fluttering insects in echolocating bats. in: "Neuroethology and Behavioural Physiology", F. Huber and H. Markl, eds., Springer, New York.

Schuller, G. and Pollak, G., 1979, Disproportionate frequency representation in the inferior colliculus of Doppler-compensating greater horseshoe bats: evidence for an acoustic fovea, J. Comp. Physiol., 132: 47-54.

Simmons, J.A. and Stein, R.A., 1980, Acoustic imaging in bat sonar: Echolocation signals and the evolution of echolocation, J. Comp. Physiol., 135: 61-84.

Simmons, J.A., Fenton, M.B., and O'Farrell, M.J., 1979, Echolocation and pursuit of prey by bats, Science, 203: 16-21.

Swift, S.M., Racey, P.A., and Avery, M.I., 1985, Feeding ecology of Pipistrellus pipistrellus (Chiroptera:Vespertilionidae) during pregnancy and lactation, II. Diet. J. Anim. Ecol., 54: 217-225.

FORAGING BEHAVIOR, PREY SELECTION, AND ECHOLOCATION

IN PHYLLOSTOMINE BATS (PHYLLOSTOMIDAE)

Jacqueline J. Belwood

Department of Entomology and Nematology
University of Florida
Gainesville, Florida 32611 U.S.A.

A thorough knowledge of food habits and feeding behavior is essential to appreciate the adaptive significance of different bat echolocation call types (Simmons et al. 1979, Neuweiler 1984). This paper describes prey selection, foraging behavior and echolocation in 12 species of phyllostomine bats (Phyllostomidae) from Barro Colorado Island, Panama (BCI). Little is known about the natural history of these bats, but they are onmivorous, feeding on insects, fruit, pollen, nectar and small vertebrates (Gardner 1977). Most species have long ears and hover, and are thought to glean prey from foliage and other substrates (Hill & Smith 1984). In addition, Trachops cirrhosus (Tuttle & Ryan 1981) and Micronycteris megalotis, M. hirsuta, and Tonatia silvicola (Tuttle et al. 1985, Belwood & Morris, in press) use prey produced sounds (frog and insect calls, respectively) to locate food.

All bats were caught in dense forest, not in clearings or in areas without trees. Fecal analysis (Table 1) and captive feedings show that M. megalotis, M. hirsuta, T. silvicola, T. bidens, Mimon crenulatum, Macrophyllum macrophyllum, and T. cirrhosus feed heavily on insects (and the latter also on vertebrates). M. nicefori and Phylloderma stenops eat fruit, and Chrotopterus auritus only vertebrates. Phyllostomus discolor and P. hastatus feed on insects and fruit, and the latter also on small vertebrates.

Orthoptera (katydids, roaches, crickets) that usually do not fly, but move along the ground or foliage (pers. obs.), dominated in the diets of wild caught T. bidens, T. silvicola, T. cirrhosus, M. megalotis, and M. hirsuta, indicating that they were probably gleaned from various substrates. This was confirmed in the lab, where T. silvicola, T. cirrhosus, and M. hirsuta gleaned and consumed prey from feeding perches, rather than in the air. These bats responded to and captured only calling (or otherwise sound-producing) prey. Live, uncaged, moving but silent insects were never approached, even when they were only a few cm from hungry bats. Large beetles (>10 mm body length (BL)) followed in importance in the diets. Only M. crenulatum and M. macrophyllum fed on non-orthopteroid insects (beetles (<5 mm BL)), and moths (size unknown), respectively, which may or may not have been gleaned.

Eleven M. hirsuta and 27 M. megalotis roosts, containing the remains of 10,944 and 12,055 insects, respectively, were found. Usually, day roosts were also used as feeding roosts, for months or years at a time, indicating that the bats foraged regularly in small familiar areas. Over two years, the major prey eaten by the 6 g M. megalotis, by number, were: roaches (10-15 mm

601

TABLE 1. Prey of wild-caught phyllostomine bats from Barro Colorado Island, Panama, based on fecal analyses. Values are percent volume.

BAT SPECIES (FOREARM (MM))	# OF BATS/FECES	ORTHOP-TERA	COLEOP-TERA	LEPIDOP-TERA	? INSECT	OTHER INSECT	BONE	SOIL	MUCOUS	FRUIT SEEDS
MACROPHYLLUM MACROPHYLLUM (34–37)	5/59	–	–	100.0	–	–	–	–	–	–
MICRONYCTERIS HIRSUTA[a] (42–45)	18/130	52.7	22.2	–	18.6	–	–	6.5	–	–
MICRONYCTERIS MEGALOTIS[b] (30–34)	18/90	66.0	18.0	0.6	–	5.0	–	–	6.4	4.0
MICRONYCTERIS NICEFORI (37–40)	4/4	–	–	–	0.5	–	–	–	–	99.5
MIMON CRENULATUM (48–52)	12/103	14.8	76.6	1.9	4.3	–	–	2.4	–	–
PHYLLOSTOMUS HASTATUS (81–85)	3/9	–	–	–	20.0	–	–	–	–	80.0
TONATIA BIDENS (55–60)	35/194	31.7	7.7	3.6	54.5	–	–	–	2.5	–
TONATIA SILVICOLA (51–53)	47/367	83.8	1.8	0.4	10.6	1.3	–	1.1	–	1.0
TRACHOPS CIRRHOSUS (57–61)	14/44	39.2	27.7	–	5.8	3.7	8.1	13.4	–	2.1

[a]call characteristics: 2–3 harmonics; peak freq. 100 ± 10 kHz; freq. range 51 ± 8 kHz; duration 0.31 ± 0.05 msec
[b]call characteristics: 1–3 harmonics; peak freq. 107 ± 0 kHz; freq. range 69 ± 15 kHz; duration 0.58 ± 0.22 msec

BL; 28%), scarab beetles (10 mm BL; 27%), and dragonflies (35-40 mm BL; 15%). Katydids (35-40 mm BL; 41%), scarabs (15-20 mm BL; 14%), and roaches (20-30 mm BL; 10%) were the most common prey of the 15 g M. hirsuta.

Regardless of diet, the calls of these bats are remarkably similar; they are 'short' (<2 msec; Simmons & Stein 1980), low amplitude, high frequency, broadband signals with multiple harmonics (Figs. 1 and 2). They are characteristic of the 'high resolution, clutter-rejecting, pursuit strategy' calls described by Simmons et al. (1979). They also resemble the calls of sympatric frugivores and nectarivores in other phyllostomid subfamilies that forage in BCI forests (Howell 1974, Novick 1977).

All the bats in this study forage in a humid, cluttered, tropical forest. Clutter-rejection calls appear well-suited to this environment as they limit the range of outgoing calls, limit echo information to nearby objects, and decrease pulse-echo overlap (Novick 1977). Compared to narrower band calls, they also provide greater acuity for the perception of obstacle (and target) position and fine structure in the bats' immediate surroundings (Vogler & Neuweiler 1983), which is also adaptive in cluttered habitats. Old World bats that glean prey in similar environments also use this type of call (Fenton et al. 1983, Neuweiler 1984).

Other factors that may affect call design as described here include:

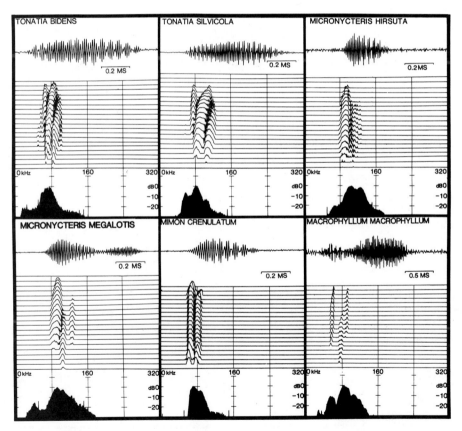

Fig. 1. Oscillograms, frequency-time patterns, and frequency spectra of echolocation calls of 6 species of insectivorous phyllostomine bats from Barro Colorado Island, Panama (BCI), for which some aspects of foraging and feeding ecology are known.

1) the presence of large carnivorous bats (<u>Vampyrum</u> <u>spectrum</u>, <u>C</u>. <u>auritus</u>) that could use other bats' echolocation calls to locate the latter as prey; 2) the need to avoid acoustical interference with loud (90-110 dB SPL at 10 cm) katydids (Orthoptera:Tettigoniidae) that are plentiful on BCI and sing in the 15 to 45 kHz range; 3) the hearing sensitivity of tympanate prey--it is probably no coincidence that <u>M</u>. <u>hirsuta</u>, who feed the most heavily on katydids, have the shortest calls (0.31 ± 0.05 msec) of the phyllostomines studied; and 4) the distance travelled from day roosts to feeding areas. Small bats such as <u>M</u>. <u>megalotis</u> and <u>M</u>. <u>hirsuta</u> (peak call frequencies 107 ± 10 and 100 ± 10 kHz, respectively), forage from their day roosts, as indicated by the presence of insect remains, in familiar areas, on insects that are abundant and evenly dispersed. In contrast, larger bats such as <u>Trachops</u> (peak frequency 79 to 53 kHz; Barclay et al. 1981), who feed on prey whose concentrations may be ephemeral and localized (e.g., frogs), probably travel greater distances in search of prey, and thus may benefit from lower frequency calls that attenuate less than higher frequncy calls.

The similarity in calls observed here indicates that foraging habitat, and not diet, may be the main factor determining echolocation call type in phyllostomines. Important questions however, remain to be answered. Can echoes from clutter rejection calls be used to distinguish potential food items (stationary insects, fruit, flowers, etc.) from the dense foliage with which the latter are associated? The inability of <u>T</u>. <u>silvicola</u> to locate

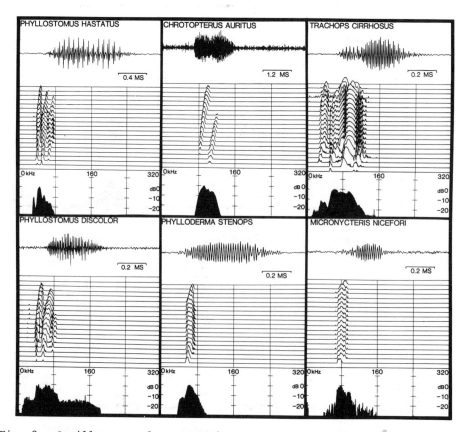

Fig. 2.   Oscillograms, frequency-time patterns, and averaged frequency spectra (n=2 to 25 pulses) of carnivorous (top), and frugivorous/ nectarivorous (bottom) phyllostomine bats from Barro Colorado Island, Panama. These bats forage in dense forest.

silent prey, and the response of bats to calling insects, which were in screen cages (Belwood & Morris, in press), strongly indicate that they do not. As pointed out by Simmons et al.(1979), the precise localization of food by these bats may result not from echolocation, but from the exploitation of other cues such as odor (in fruits and flowers) and sound (in animal prey) (see Fiedler 1979, Bell 1982). Also, how do M. megalotis and M. hirsuta, who feed on similar but different-sized prey, gauge insect size if echolocation is not used to locate prey? Last, why are moths not preyed on more heavily by the gleaning bats of BCI? In contrast, other phyllostomines (e.g., Macrotus waterhousii in the West Indies) take moths in large numbers (Belwood, unpublished data).

Acknowledgments. The recordings were made with J.H. Fullard. C.O. Handley, Jr., R. Abbey, S. Sakaluk, E. Stockwell, M. Tuttle, and R. Kunz helped with the study. Supported by NSF DEB-8211975, the Smithsonian Institution, American Association of University Women, National Academy of Sciences, Sigma Xi, American Society of Mammalogists, Florida Entomological Society (JJB) and the Natural Sciences and Engineering Research Council of Canada (JHF). I thank M.B. Fenton, J.H. Fullard, G. Neuweiler, and members of the Third Animal Biosonar Conference for comments.

REFERENCES

Barclay, R.M.R., M.B. Fenton, M.D. Tuttle and M.J. Ryan. 1981. Echolocation calls produced by Trachops cirrhosus (Chiroptera:Phyllostomidae) while hunting frogs. Can. J. Zool. 59:750-753.
Bell, G.P. 1982. Behavioral and ecological aspects of gleaning by a desert insectivorous bat, Antrozous pallidus (Chiroptera:Vespertilionidae). Behav. Ecol. Sociobiol. 10:217-223.
Belwood, J.J. and G.K. Morris. Bat predation and its influence on calling behavior in Neotropical katydids. Science: in press.
Fenton, M.B., C.L. Gaudet and M.L. Leonard. 1983. Feeding behaviour of the bats Nycteris grandis and Nycteris thebaica (Nycteridae) in captivity. J. Zool. Lond. 200:347-354.
Fiedler, J. 1979. Prey catching with and without echolocation in the Indian false vampire (Megaderma lyra). Behav. Ecol. Sociobiol. 6:155-160.
Gardner, A.L. 1977. Feeding habits, in Biology of Bats of the New World Family Phyllostomatidae. R.J. Baker, J.K. Jones, Jr., and D.C. Carter, eds. Pt. II. Texas Tech. Press, Lubbock, Texas.
Hill, J.E. and J.D. Smith. 1984. Bats, a Natural History. British Museum (Natural History), London.
Howell, D.J. 1974. Acoustic behavior and feeding in glossophagine bats. J. Mammal. 55:293-308.
Neuweiler, G. 1984. Foraging ecology and audition in bats. Naturwissenschaften. 71:446-455.
Novick, A. 1977. Acoustic orientation, in Biology of Bats, W.A. Wimsatt, ed., vol. III, Academic Press, New York.
Simmons, J.A., and R.A. Stein. 1980. Acoustic imaging in bat sonar: echolocation signals and the evolution of echolocation. J. Comp. Physiol. 135:61-84.
_____, M.B. Fenton, and M.J. O'Farrell. 1979. Echolocation and pursuit of prey by bats. Science 203:16-21.
Tuttle, M.D. and M.J. Ryan. 1981. Bat predation and the evolution of frog vocalizations in the Neotropics. Science 214:677-678.
_____, _____, and J.J. Belwood. 1985. Acoustic resource partitioning by two species of phyllostomid bats (Trachops cirrhosus and Tonatia silvicola). Anim. Behav. 33:1369-1371.
Vogler, B. and G. Neuweiler. 1983. Echolocation in the noctule (Nyctalus noctula) and horseshoe bat (Rhinolophus ferrumequinum). J. Comp. Physiol. 152:421-432.

VARIATION IN FORAGING STRATEGIES IN FIVE SPECIES

OF INSECTIVOROUS BATS - IMPLICATIONS FOR ECHOLOCATION CALL DESIGN

M. Brock Fenton

Department of Biology, York University
Downsview, Ontario, Canada M3J 1P3

INTRODUCTION

Different approaches to echolocation in the Microchiroptera, as
reflected by differences in call design (frequency, patterns of change of
frequency over time), raise an important question, namely can we equate
specific call designs with particular foraging strategies? It is difficult
to answer the question now because we do not have data on individual
variation either in call designs or in foraging strategies. In some
species, echolocation calls vary in design (e.g., Rhinopoma hardwickei,
Lasiurus cinereus, and some neotropical emballonurids - Habersetzer 1981;
Belwood and Fullard 1984; Barclay 1983, respectively), while in others
there is no evidence of variation (e.g., Myotis myotis - Neuweiler 1983).

The possible association of certain echolocation calls with specific
foraging strategies is interesting because it involves research from
different disciplines. Neuroethologists who want to put their studies in
broader behavioural and ecological settings use field data about dietary
or foraging specializations and ecologists and ethologists exploit
neurobiological data in the other direction. This situation can produce
models that are well founded in one discipline, but lack credibility in
others. Measuring variability is important for establishing the
credibility of the models.

The purpose of this paper is to examine variation in the foraging
behaviour of individuals of five species of insectivorous bats studied by
radio tracking and to consider these data in the context of flight
morphology and design of echolocation calls. Compared here are Nycteris
grandis (Fenton, Cumming, Hutton and Swanepoel, in press), Rhinolophus
hildebrandti, Scotophilus borbonicus, Tadarida midas (Fenton and
Rautenbach, 1986), and Eptesicus fuscus (Brigham and Fenton 1986). The
details of the studies are provided in the cited papers; those of sample
sizes in Table 1.

RESULTS AND DISCUSSION

Some basic morphological features and the design of echolocation
calls used by the five species are compared in Fig. 1. In aerodynamic
terms, the nycterid and the rhinolophid should be the most maneouvrable
given their wing morphologies (aspect ratios and wing tip designs) and
wing loadings. High maneouvrability in flight should enable the foraging
bat to deal with clutter. From a morphological point of view, the molossid

607

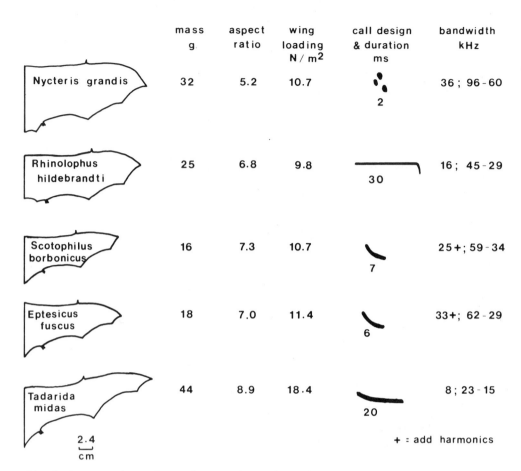

	mass g	aspect ratio	wing loading N / m²	call design & duration ms	bandwidth kHz
Nycteris grandis	32	5.2	10.7	2	36 ; 96 - 60
Rhinolophus hildebrandti	25	6.8	9.8	30	16 ; 45 - 29
Scotophilus borbonicus	16	7.3	10.7	7	25 + ; 59 - 34
Eptesicus fuscus	18	7.0	11.4	6	33 + ; 62 - 29
Tadarida midas	44	8.9	18.4	20	8 ; 23 - 15

2.4 cm

+ = add harmonics

Fig. 1. A comparison of the five species of insectivorous bats showing (to scale) the wing outline, various wing parameters, body mass, and design of echolocation calls including the pattern of frequency change over time, the duration, and the bandwidth (echolocation call data from Fenton and Bell 1981 and Schnitzler, pers. comm.; wing data from H.D.J.N. Aldridge, pers. comm.)

should be least able to deal with clutter for its high wing loading necessitates faster flight speeds and reduced maneouvrability (Pennycuick 1975). In wing morphology, the two vespertilionids are intermediate between the other two species.

The features of the echolocation calls reflect the wing morphology in all of the species except the nycterid. The vespertilionids' echolocation calls are clutter resistant because of their broad bandwidths, while the narrow band calls of the molossid are not clutter resistant (Simmons and Stein 1980). By virtue of Doppler shift compensation and flutter detection, the narrowband echolocation calls of the rhinolophid are clutter resistant.

In captivity, N. grandis produced echolocation calls during attacks on prey but obviously were not using echolocation to locate and identify their targets (Fenton, Gaudet and Leonard 1983). Their attacks could be misdirected by providing sounds from the targets in the wrong locations, leading the bats to attack the source of the sounds, even a stick or a speaker. The broadband calls of the nycterid should be clutter resistant, but their role in foraging remains unclear.

Table 1. Basis for comparison of variation in foraging behaviour of five species of insectivorous bats.

species	sample size # of bats - # of bat nights[1]	study area	source
Nycteris grandis (Nycteridae)	4 - 20	Zimbabwe	Fenton, Cumming, Hutton & Swanepoel
Rhinolophus hildebrandti (Rhinolophidae)	10 - 25	South Africa	Fenton & Rautenbach
Scotophilus borbonicus[2] (Vespertilionidae)	9 - 25	South Africa	Fenton & Rautenbach
Eptesicus fuscus (Vespertilionidae)	58 - 620	Ontario	Brigham & Fenton
Tadarida midas (Molossidae)	10 - 24	South Africa	Fenton & Rautenbach

[1] a bat with an active transmitter for one night = 1 bat night.
[2] = Scotophilus leucogaster in Barclay 1985
     = S. viridis in Fenton, Brigham, Mills and Rautenbach 1985

Foraging E. fuscus, S. borbonicus and T. midas appeared to use but one strategy, namely continuous flight in pursuit of airborne targets. There was no evidence that the marked individuals took prey from surfaces (foliage, ground, etc.) or spent any time waiting on perches for passing prey. The two vespertilionids often foraged within 0.5 m of tree canopies and sometimes foraged in the open below tree canopies. They did not venture within the canopies, but foraged above them or over fields or open water. The molossid only foraged in open areas with minimal clutter.

In contrast, the nycterid and the rhinolophid used two foraging strategies and in both species individuals varied in the proportions they used the two strategies. Both species alternated between continuous flight while hunting, and short flights from perches. In the nycterid, hunting from perches seemed to dominate, while in the rhinolophid, the reverse was true. During continuous flight the rhinolophid was usually active between the canopy and the ground cover, while the nycterid tended to fly within 1.5 m of the ground. Flights from perches by the rhinolophid were usually within $30°$ on either side of the horizontal, while those by the nycterid were usually towards the ground. The diet of the rhinolophid included beetles and moths, while the nycterid fed mainly on frogs, cicadas, crickets and moths.

The results indicate that while some species vary their foraging strategies, others do not. In R. hildebrandti, the same design of echolocation calls were used whether the animals were hunting in continuous flight or in a flycatcher mode. These data support other studies (usually not based on marked individuals) that demonstrated variation in foraging strategies (e.g., Vaughan 1976 - Cardioderma cor, or Bell and Fenton 1984 - Hipposideros caffer) and that did not (e.g., Vaughan 1977 - Hipposideros commersoni; and Vaughan and Vaughan 1986 - Lavia frons).

CONCLUSIONS

1. It is too early to definitely associate echolocation calls of particular design with specific foraging strategies in the Microchiroptera.
2. The flutter-detecting echolocation systems of rhinolophids and hipposiderids are not always associated with hunting from a perch (flycatcher strategy).
3. The flycatcher hunting strategy in Microchiroptera need not involve echolocation to locate prey.

ACKNOWLEDGEMENTS

I thank Drs. H.D.J.N. Aldridge and H-U. Schnitzler for permitting me access to their unpublished data,and H.D.J.N. Aldridge, R.M.R. Barclay, J.H. Fullard and U.M. Norberg for making comments on the manuscript. My work on bats has been supported by the Natural Sciences and Engineering Research Council of Canada.

REFERENCES

Barclay, R.M.R. 1983. Echolocation calls of emballonurid bats from Panama. J. Comp. Physiol., 151:515-520.
_____ 1985. Foraging behavior of the African insectivorous bat, Scotophilus leucogaster. Biotropica, 17:65-70.
Bell, G.P. and M.B. Fenton. 1984. The use of Doppler-shifted echoes as a flutter detection and clutter rejection system: the echolocation and feeding behavior of Hipposideros ruber (Chiroptera: Hipposideridae). Behav. Ecol. Sociobiol., 16:343:347.
Belwood, J.J. and J.H. Fullard. 1984. Echolocation and foraging behaviour in the Hawaiian hoary bat, Lasiurus cinereus semotus. Can. J. Zool., 62:2113-2120.
Brigham, R.M. and M.B. Fenton. 1986. The influence of roost closure on the roosting and foraging behaviour of Eptesicus fuscus (Chiroptera: Vespertilionidae). Can. J. Zool., 64:1128-1133.
Fenton, M.B. and G.P. Bell. 1981. Recognition of species of insectivorous bats by their echolocation calls. J. Mamm., 62:233-243.
Fenton, M.B. and I.L. Rautenbach. 1986. A comparison of the roosting and foraging behaviour of three species of African insectivorous bats (Rhinolophidae, Vespertilionidae and Molossidae). Can. J. Zool., 64: in press.
Fenton, M.B., C.L. Gaudet, and M.L. Leonard. 1983. Feeding behaviour of Nycteris grandis and Nycteris thebaica (Nycteridae) in captivity. J. Zool. London, 200:347-354.
Fenton, M.B., R.M. Brigham, A.M. Mills and I.L. Rautenbach. 1985. The roosting and foraging areas of Epomophorus wahlbergi (Pteropodidae) and Scotophilus viridis (Vespertilionidae) in Kruger National Park, South Africa. J. Mamm., 66:461-468.
Fenton, M.B., D.H.M. Cumming, J.M. Hutton and C.M. Swanepoel. 1986. Foraging and habitat use by Nycteris grandis (Chiroptera : Nycteridae) in Zimbabwe. J. Zool., London, in press.
Haberstezer, J. 1981. Adaptive echolocation sounds in the bat Rhinopoma hardwickei, a field study. J. Comp. Physiol., 144:549-556.
Neuweiler, G. 1983. Echolocation and adaptivity to echological constraints, pp. 280-302 IN Neurethology and behavioral physiology, roots and growing points (F. Huber and H. Markl, eds.). Springer-Verlag, Berlin.
Pennycuick, C.J. 1975. Mechanics of flight, pp. 1-75 IN Avian Biology vol. 5 (D.S. Farner, J.R. King and J.C. Parkes, eds.). Academic Press, London.
Simmons, J.A. and R.A. Stein. 1980. Acoustic imaging in bat sonar:

echolocation signals and the evolution of echolocation. J. Comp. Physiol., 135:61-84.

Vaughan T.A. 1976. Nocturnal behavior of the African false vampire bat, Cardioderma cor. J. Mamm., 67:227-248.

_____ 1977. Foraging behaviour of the giant leaf-nosed bat (Hipposideros commersoni). East Afr. Wildl. J., 15:237-259.

Vaughan, T.A. and R.P. Vaughan. 1986. Seasonality and the behavior of the African yellow-winged bat. J. Mamm., 67:91-102.

DETECTION OF PREY IN ECHOCLUTTERING ENVIRONMENTS

G. Neuweiler, A. Link, G. Marimuthu, and R. Rübsamen

Zoologisches Institut der Universität München
Luisenstraße 14, D-8000 München 2, FRG

Department of Animal Behaviour, School of Biological
Sciences, Madurai Kamaraj University, Madurai 625 021
India

INTRODUCTION

Echolocating bat species foraging above canopies will face no severe
difficulties in detecting flying insects. Bats foraging close to vegeta-
tions, over the ground or water surfaces, have to detect their prey within
a clutter of time-smeared echoes reflected from foliages, grass, etc.
By field observations, behavioural and neurophysiological investigations
we have studied how two different groups of bats detect prey in echo cluttering environments: the gleaning bat, Megaderma lyra and three rhinolophoid species, Rhinolophus rouxi, Hipposideros speoris and Hipposideros
bicolor which forage insects close to and within vegetation.

PREY DETECTION IN THE GLEANING BAT, MEGADERMA LYRA

The Indian False Vampire is a large bat (bw. ca. 35g) which skims
either low over the ground and water surfaces or spends its night time
hanging on rock faces or low branches waiting for any prey of suitable
size. Megaderma intercepts in air larger insects flying by, picks up
arthropods, frogs, lizards, geckoes, birds and rodents from the ground
and even catches fishes from ponds and rivers. We performed detection
experiments with frogs as prey in an outdoor cage (Marimuthu and Neuweiler,
in press). First we offered a dead and a live frog placed on the earthen
floor of the cage. In 31 presentations the bats never reacted to the
dead frogs, and to the live ones only when they moved. All jumping frogs
were detected and caught within 5 s even in darkness. Freshly killed
frogs were also detected and captured (100% in 54 experiments) only when
we briskly pulled them over the floor with a fine string tied to a hindleg
of the frog. Apparently Megaderma lyra only detected moving targets.
Since the bats continuously emitted their brief, multiharmonic echolocation pulses (Möhres and Neuweiler, 1966) they could have detected the
moving frogs either by echolocation or by listening to the faint noises
produced by moving frogs. We again offered dead frogs but now briefly
pulled over a watered glass plate five times and subsequently pulled
over the earthen floor. In 62 experiments the bats never reacted to
the five pulls over the glass plate which generated no noise, but were
immediately alarmed and caught the frogs when pulled over the floor (except in 4 cases of 62). From these results we conclude that Megaderma
lyra only detects moving targets, however, not by echolocation but only
by listening to the noises produced by the moving prey. Noiseless prey
is not detected.

The auditory system of Megaderma lyra is adapted to the detection of faintest noises. Neuronal (Neuweiler et al., 1984) and behavioural (Schmidt et al., 1984) audiograms disclose extraordinary low thresholds of -15 to -20 dB SPL in the frequency range of 15-25 kHz which is a strong component of rustling noises (Fig. 1). This specific high sensitivity is lost when the large pinnae of Megaderma are folded back to the head and reappears when they are repositioned (Fig.1, Rübsamen et al.in press). Therefore the specific sensitivity to faint noises is largely due to the huge and medially fused pinnae. The tonotopy of the inferior colliculus reflects the broad band character of the audiogram in Megaderma lyra. All frequencies up to about 110 kHz are represented in fairly equal isofrequency slices in a dorsolateral to ventromedial order. Many units with BFs in the range of 10 to 30 kHz featured upper thresholds and responded more vigorously to noise than to pure tones. This specific responsiveness to noise and the low upper thresholds (40-50 dB SPL) underscore that not only the peripheral but also the neuronal system of Megaderma lyra is adapted to the detection of faint noises. Thus, Megaderma lyra circumvents the problem of echoclutter by discarding echolocation for prey detection and resorts to passive auditory prey detection.

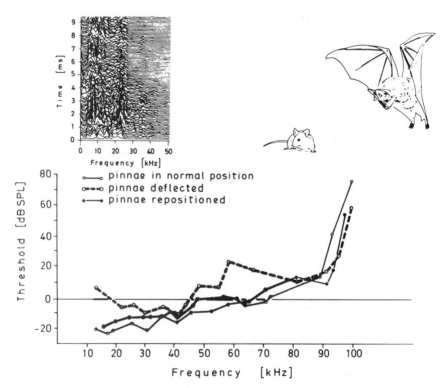

Fig. 1. Audiogram and sensitivity gain by the pinnae in Megaderma lyra. The audiograms are derived from multiunit recordings in the inf.colliculus. When the contralateral pinna was folded back severe sensitivity losses occurred (pinna deflected). When the pinna was repositioned the sensitivity peaks recurred. Inlet: Spectrogram of noise produced by briskly pulling a dead frog over the floor.

FLUTTERING TARGET DETECTION BY RHINOLOPHOIDS

It is well known that FM/CF/FM-bats which emit long (10-60 ms) CF- components are adapted to fluttering target detection (Neuweiler et al., 1980). This was interpreted as an adaptation to prey detection in echo cluttering environments. Hipposiderids which only emit brief CF-components (3-7 ms) face a similar echocluttering problem. We therefore did a comparative study on the foraging behaviour and prey detection in the FM/CF/FM bat Rhinolophus rouxi (CF 74-79 kHz) and the two CF/FM bats Hipposideros speoris (CF: 130-140 kHz) and Hipposideros bicolor (CF: 150-157 kHz).

In Madurai H. speoris circles around the foliages of trees and bushes in stereotyped flight paths hunting insects on the wing. The smaller H. bicolor preferably forages flying insects inside the foliages and even manoeuvres through thick hedges with unbelievable skill. Unlike H. speoris, H. bicolor also picks up insects from the ground and from the walls (Habersetzer et al., 1984).

Rhinolophus rouxi is a forest bat. In Sri Lanka we found them consistently foraging only in dense forests and never in clearings, open forests or farmlands (Neuweiler et al., in press). The horseshoe bats caught insects either flying along or around bushes or trees or they used a 'sit and wait' strategy. They perched on leafless twigs protruding below canopy at places where there was some space between the foliages and their vantage points. They continuously turned their body around and probed the space for fluttering targets, mainly with pure CF-signals continuously emitted at an average rate of close to 10/s without a pause.

Individual horseshoe bats have restricted foraging areas to which they return many nights. Within this area of about 200-400 qm the bats randomly moved between a few distinct vantage points (e.g. 48 times/h between 7 vantage points). Catching flights occurred about once every 4 min. The bat announced a take off by 2 - 3 longer echolocation tones. Catching flights were brief and rarely took the bat farer away than 5 m which fits into the time window described by Roverud (this volume). In all instances the bats emitted echolocation sounds which were mainly CF-signals. Only during flight prominent initial and final FM-components were added. During searching for fluttering insects the duty cycle of the sound sequences were close to 50% and rose up to 74% in catching flights by lengthening sound duration from 45 to 50 ms or by 10% and shortening intervals from 57 to 29 ms or by 50%. Thus the duty cycles were increased mainly by reducing intervals.

These field studies show that horseshoe bats and hipposiderids catch flying insects. In behavioural experiments with all three species we tested if fluttering wings are the only cue for prey detection. When we simultaneously offered stationary insects and insects running on the floor of the cages Rh. rouxi and H. speoris never reacted to this potential prey. This is in sharp contrast to H. bicolor which also did not respond to stationary targets but immediately pursued and caught insects running over the floor or along the sides of the cages. It also flew to any noise source and inspected it. Interestingly, among the three species H. bicolor has the largest pinnae. This and the behaviour described suggest to us that H. bicolor also detects prey by prey generated noises whereas H. speoris and Rh. rouxi seem to detect fluttering wings only (Link et al., 1986).

However, when we presented wingless, tethered cockroach nymphs vibrating vertically up and down by less than 1 cm and at a rate of about

10/s all three species were immediately attracted and caught the prey. Therefore oscillating movements and not specifically fluttering wings are detected and induce catching responses in all three species. In a simultaneous choice experiment between vertically oscillating and wing beating cockroaches invariably the wingbeating target was preferred by all three species. This indicates that wing beats are the strongest, but not exclusive cue for prey detection.

With a wing beat dummy we tested the catching responses to different wing beat rates. All three species no more reacted to wing beats lower than 2 to 1/s. From experiments with various combinations of wing beat amplitudes and rates we realized that speed of the wing and not the wing beat rate was the decisive detection cue for the bats. We found a detection threshold for wing speeds of 2 - 1 cm/s. Surprisingly the

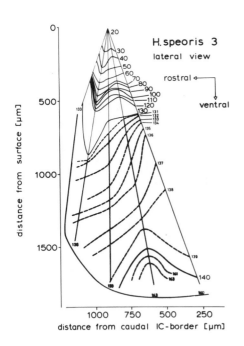

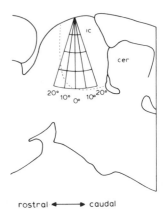

Fig. 2. Acoustical fovea in Hipposideros speoris (shaded area). Left graph shows the tonotopy (5 electrode tracks, BFs of multiunit recordings in steps of 50 µm) in the inferior colliculus of one specimen which emitted CF-components of 140 kHz. Upper graph shows position of electrode tracks in a mid-sagittal plane. IC Inferior colliculus, CER Cerebellum.

hipposiderid species which only emit brief CF components performed as well as horseshoe bats wich emit a CF component 10 times longer. This raises some doubts if a pure tone signal is really such an imperative prerequisite for fluttering target detection. However, it might be a mandatory prerequisite for prey detection in echo-cluttering environments.

Clutter resistent echolocation with narrow band echolocation signals requires a narrowly tuned receiving filter as it has been described as "acoustic fovea" in horseshoe bats and Pteronotus parnellii (see Vater, this volume). Since the hipposiderid species performed as well as horse-shoe bats in prey detection we checked for an acoustical fovea in H.speoris. In the neuronal auditory system a fovea shows up as a vast overrepresentation of the narrow frequency range of the species-specific CF echolocation signal and by a narrowly tuned filter in the audiogram.

Audiograms from the colliculus inferior in H. speoris showed a narrow filter tuned to the species specific frequency range of 130 - 140 kHz. Q-values of single units with BFs in this frequency band reached values of 140. A detailed multi-unit scanning of the inferior colliculus disclosed that the ventromedial part of the inferior colliculus consists of a huge overrepresentation of the echofrequency range (Fig.2). However, there is a distinct difference to the fovea of horseshoe bats: the fovea is not centered around the frequency of the CF-component emitted by handheld bats but skewed towards lower frequencies. The overrepresented frequency range comprises about 2/3 of the frequency range of the final FM-sweep. Thus the acoustical fovea in H. speoris is broader than in horseshoe bats and the CF-echofrequency is represented in its ventralmost layers.

These studies show that hipposiderids also have an acoustical fovea and that they detect fluttering prey as well as horseshoe bats. Why then horseshoe bats emit CF-signals which are ten times longer than those of hipposiderids? Duration of the pure tone signal is not correlated with the capacity to detect fluttering insects. Perhaps minimal sound duration is correlated with the cochlear filter: as narrower the cochlear filter as longer the signal analyzed should be. Therefore the obligatory implementation of constant frequency components into the echosignal in rhinolophoid bats might originate from the need for a narrow receiving filter in order to overcome echo clutter. The minimal duration of the CF-signal might then correlate with the sharpness of the filter. Fluttering target detection does not necessarily require pure tone echolocation since echo glints from moving wings are also effectively carried by FM-sweeps.

REFERENCES

Habersetzer J, Schuller G, Neuweiler G (1984) Foraging behavior and Doppler shift compensation in echolocating hipposiderid bats, Hipposideros bicolor and Hipposideros speoris. J Comp Physiol 155:559-567.
Link A, Marimuthu G, Neuweiler G (1986) Movement as a specific stimulus for prey catching behaviour in rhinolophid and hipposiderid bats. J Comp Physiol A 159:403-413.
Marimuthu G, Neuweiler G (in press) Acoustical cues used by the Indian False Vampire, Megaderma lyra, to detect prey. J Comp Physiol A.
Möhres FP, Neuweiler G (1966) Die Ultraschallorientierung der Großblatt-Fledermaus (Megadermatidae). Z Vergl Physiol 53:195-227.
Neuweiler G, Bruns V, Schuller G (1980) Ears adapted for the detection of motion or how echolocating bats have exploited the capacities of the mammalian auditory system. J Acoust Soc Am 68:741-753.
Neuweiler G, Metzner W, Heilmann U, Rübsamen R, Eckrich M, Costa HH (in press) Foraging behaviour and echolocation in the rufous horseshoe bat Rhinolophus rouxi, of Sri Lanka. Behav Ecol Sociobiol.
Neuweiler G, Satpal Singh, Sripathi K (1984) Audiograms of a South Indian bat community. J Comp Physiol A 154:133-142.
Schmidt S, Türke B, Vogler B (1984) Behavioral audiogram from the bat Megaderma lyra. Myotis 22:62-66.
Rübsamen R, Neuweiler G, Sripathi K (in press) Pinnae produce high sensitivity gain in the gleaning bat, Megaderma lyra. J Comp Physiol A.

HOW THE BAT, PIPISTRELLUS KUHLI, HUNTS FOR INSECTS

Hans-Ulrich Schnitzler, Elisabeth Kalko, Lee Miller and
Annemarie Surlykke

Lehrbereich Zoophysiologie
University of Tübingen
D-7400 Tubingen, Federal Republic of Germany

Institute of Biology
Odense University
DK-5230 Odense, Denmark

INTRODUCTION

Most studies on the insect catching behavior in bats describe either
echolocation behavior or capture techniques and flight behavior. Only in a
few laboratory studies have both behaviors been studied simultaneously. We
were able to do this in the field following the development of a battery
operated 6 flash strobe system, which could be synchronized with the sound
recordings. This equipment is still very bulky and cannot be moved during
a recording session. Therefore we concentrated our efforts on a bat
species, which reliably hunted in one area and could therefore be easily
photographed. In the village Gornije Okrug near Trogir in Yugoslavia we
found a site where several Pipistrellus kuhli regularly hunted for insects
near a street light. The data from 10 nights of recording are presented
here.

METHODS

The echolocation sounds of hunting bats and the synchronization pulses,
which indicated the triggering of the flashes of our strobe system, were
recorded on a portable magnetic tape recorder (Lennartz electronic
6000/607) and analyzed at reduced speed (1/64) with a Nicolet UA 500 real
time spectrum analyzer (range 10 kHz, terminated at 2 kHz). The resulting
spectra were characterized by a frequency range of 128 kHz, a frequency
resolution of 1280 Hz, a time difference of 0.16 ms between consecutive
spectra. Each spectrum contains 20% new information. The spectra were
displayed on an oscilloscope screen. The frequency was shown on the y-
axis, the amplitude on the z-axis and the screen was photographed (Tönnies
oscilloscope camera) on 60 mm wide paper film using streak photography.
This resulted in sonagram-like real time spectrograms with a dynamic range
of about 40 dB. A Nikon F2 35 mm reflex camera was aimed at the area where

619

the bats were likely to hunt. When a bat flew into this area the shutter was opened and the sequence of 6 flashes (Metz Mecablitz 45CT3 with long lens adapter 45-33) was triggered by a custom made flash sequencer. For most photographs the following settings were used: Focus 8-10 m, F-stop 8 or 5.6, Zoom lens 80 mm, flash duration 1 or 0.4 ms. Sometimes we tethered a moth on a string attached to a long pole and thus attracted bats into the study site.

RESULTS

The three behavioral categories: search, approach and terminal phase used by Griffin et al. (1960) to describe the pulse pattern of <u>Myotis lucifugus</u> can also be used to describe the echolocation behavior of hunting <u>Pipistrellus kuhli</u>. Fig. 1 and 2 give examples of sound sequences from hunting bats.

In <u>search flight</u> high flying bats produce shallow sweeps with

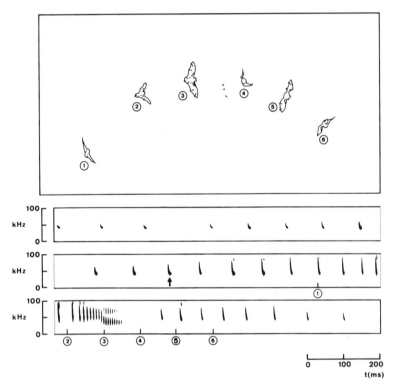

Fig. 1. <u>Pipistrellus kuhli</u> pursuing and catching an unidentified flying insect in the field. The corresponding sound sequence is given in the lower three panels. The arrow marks the beginning of the approach phase. The numbers indicate synchronizations between the sound sequence and stroboscopic flashes.

a duration of 8-12 ms and a sweep width of 3-6 kHz. When approaching the recording site the bats reduce the sound duration to 8-6 ms and add an initial downward sweep of up to 35 kHz in bandwidth to the shallow sweep. The terminal frequency of the search pulses is kept rather constant and seems to be characteristic for each individual bat. The terminal frequencies of different sequences were mostly between 35-40 kHz. The time interval between pulses is mostly around 100 ms but sometimes near 200 ms. This suggests that the bats produce one pulse per wingbeat and sometimes skip one cycle. At an average of 8 pulses/s the bats will have a duty cycle of 6-10%.

The approach phase is characterized by a continuous reduction of pulse interval from about 70 ms to about 35 ms and pulse duration from about 7 ms to about 3 ms. The bandwidth of the pulse is increased and a second harmonic is often introduced. First and second harmonic start at about 95 kHz and sweep down to the individual terminal frequency of the bat or its second harmonic. The duty cycle is kept at about 8% throughout the whole approach phase.

The terminal phase or buzz can be separated into two sections

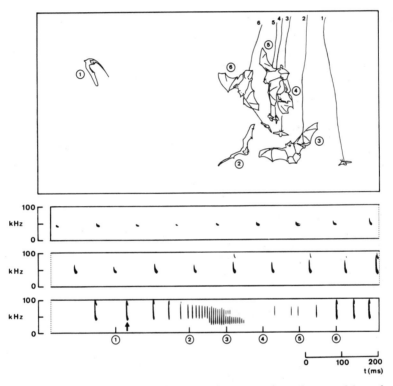

Fig. 2. Pipistrellus kuhli pursuing a tethered noctuid moth.

(Fig. 1 and 2). In the first section sweep width and terminal frequency is similar to that of sounds in the approach phase. In the second section sounds have terminal frequencies of between 19 and 24 kHz and sweep widths of between 20 and 40 kHz. A second harmonic may be present in both sections. In the first section the pulse duration decreases from about 2 ms to about 1 ms and the pulse interval is reduced from about 18 ms to 6 ms. In the second section the pulse durations and intervals are kept constant at about 0.3 ms and about 5 to about 6 ms respectively. Approach and terminal phases are rather variable. Total durations for approach and terminal phases were between about 250 ms and 1000 ms, thus indicating large differences in time spent during insect pursuits. The pause after the terminal phase had a duration of between 50 ms and 212 ms. In successful catches the pause was more than 100 ms long. Shorter pauses suggest unsuccessful catches.

The photographs of hunting bats allowed a comparison of echolocation and hunting behavior (Fig. 1 and 2). From the photographs we estimated a flight speed of about 3.5 to 5.3 ms in search flight. They also indicate a correlation between sound emission and wingbeat cycle. In the search phase the bats normally make about 10 wingbeats/s and produce one single sound near the top of the stroke. The reaction to a target is sometimes indicated by a strong turn, which suggests that the bats must have a rather wide search cone. This point of first reaction corresponds well with the beginning of the approach phase, as we define it.

From 3 suitable photos we estimated a reaction distance of about 70-120 cm. If we assume that the bats detect an insect in the pulse just prior to the first reaction, the detection distance can be calculated if we add the distance flown in the last interval (ca. 40 cm). This would result in detection distances of about 110-160 cm in the three cases discussed above. During the approach the bat's head is pointed at the target. At the beginning of the terminal phase at an estimated distance of about 30-50 cm the bat tilts upward, extends its wings and forms a pouch with its interfemoral membrane. The insect is most probably caught in the pouch since the capture is finished by bending the head forward into the tail membrane probably to grab the insect.

Afterwards the bats return to search flight. The buzz ends when the bat is very close to the insect or maybe even when it has made contact with the insect. The pause after the buzz indicates how fast a bat gets an insect out of the tail pouch. The insects we captured near the study site were mainly moths of the families Lymantridae (Lymantria dispar) and Noctuidae. Our photographs indicate that the bats captured mostly small to medium sized insects.

DISCUSSION

The nearly constant frequency pulses in the search phase improve prey detection as the sound energy is concentrated onto the bandwidth of individual neurons (Grinnell and Hagiwara, 1972). The rather high duty cycle of 6-10% also improves the chance to receive an echo with a glint from the prey. Such glints, which have high amplitude, are produced when an echolocation pulse hits the insect when the wings are nearly perpendicular to the impinging sound waves (Schnitzler et al., 1983). For instance, if the wingbeat rate of the insect is 100 Hz the bats would perceive one glint in nearly every echo, which would result in a marked increase in the detection distance.

If long pulses are so advantageous for bats why are the searching cries of P. kuhli restricted to durations below 12 ms? We assume that pulses longer than 12 ms would have the disadvantage in overlapping with the returning echo from the insect, thus reducing the chance of detection. The bats would therefore have to adapt the duration of the search cry to the detection range and the average duration of the cries would be an indicator for the average lower limit of the detection distance where pulse echo overlap is avoided. In P. kuhli this lower limit would be at 136-170 cm at pulse durations of 8-10 ms. This corresponds well with the estimated detection distances of 110-160 cm.

The principle that the prevention of overlap determines the pulse duration in search flight can probably be generalized for other species that also produce shallow sweeps like noctules, tadarids, other pipistrelle bats and Eptesicus.

When nearing the study site and in the approach and terminal phase of an insect pursuit the bats produce broadband signals thus increasing the accuracy of range and angle determination and perhaps also to get more spectral information from the target. Our estimations of the distance between bat and target during a pursuit suggest that the bats reduce the sound duration parallel to the decreasing distance so that no overlap between outgoing pulse and returning echo occurs. This compares well with results in <u>Myotis lucifugus</u> (Cahlander et al., 1964).

ACKNOWLEDGEMENT

The work was supported by the Deutsche Forschungsgemeinschaft (SFB 307) and by Grants from the Danish Research Council.

REFERENCES

Cahlander, D.A., McCue, J.J.G., Webster, F.A., 1964, The determination of distance by echolocating bats. <u>Nature</u> 201: 544-546.

Griffin, D.R., Webster, F.A., Michael, C.R., 1960, The echolocation of flying insects by bats. <u>Anim.Behav.</u> 8: 141-154.

Grinnel, A.D., Hagiwara, S., 1972, Adaptations of the auditory nervous system for echolocation studies of New Guinea bats. <u>Z.vergl. Physiol.</u> 76: 41-81.

Schnitzler, H.-U., Menne, D., Kober, R., and Heblich, K., 1983, The acoustical image of fluttering insects in echolocating bats, <u>in:</u> Neurophysiology and behavioral physiology, F. Huber, and H. Markl, eds., Springer, Berlin Heidelberg New York.

THE COMMUNICATION ROLE OF ECHOLOCATION CALLS IN

VESPERTILIONID BATS

Jonathan P. Balcombe and M.B. Fenton

Department of Biology, Carleton University, Ottawa, Canada
(present address: Department of Biology, York University
North York, Ontario, Canada M3J 1P3)

INTRODUCTION

It has been demonstrated by playback presentations that some bats exploit the echolocation calls of conspecifics in a communication mode (Barclay 1982; Leonard and Fenton 1984). Barclay's (1982) study showed that Myotis lucifugus responded positively to the calls of conspecifics, artificial representations of same, and the calls of the sympatric Eptesicus fuscus.

The purpose of this study was to assess the influence of call design on the communicative aspect of echolocation calls by using playback presentations. Bell (1980) had reported positive associations among some foraging bats in an Arizona community. If design of echolocation call reflects foraging strategy, we predict that foraging bats will respond more to calls similar in design to their own.

MATERIALS AND METHODS

Field work was conducted in the Okanagan Valley, in British Columbia from May to the end of July 1985, and at Pinery Provincial Park, Ontario in August 1985 and from May to the end of July 1986. In the Okanagan Valley, playback presentations were performed along the Okanagan River and at the edge of a 5 ha lake where swarms of bats, mainly Myotis lucifugus and M. yumanensis foraged. Other presentations were performed outside an abandoned warehouse which housed a colony of over 600 M. yumanensis. In Pinery Provincial Park the presentations were conducted at five spotlight locations where Lasiurus borealis foraged, and outside a building colony of 150 M. lucifugus.

Tape recorded calls were played from a Racal Store 4D tape recorder operated at 76 cm/s, amplified and broadcast through an 8.5 cm diameter electrostatic speaker (Simmons, Fenton, Ferguson et al. 1979) erected 3 m above the ground. Outgoing stimuli were monitored on a Telequipment D32 oscilloscope, and surrounding bat activity was continuously monitored using a QMC mini bat detector (QMC Instruments, Mile End Road, London).

Stimuli presented in the Okanagan included: 1) echolocation calls recorded from a feeding group of about 50 M. lucifugus and M. yumanensis; 2) foraging calls of Eptesicus fuscus; 3) of Lasiurus cinereus and 4) of Rhinolophus megaphyllus. These same stimuli were presented to M. lucifugus in Ontario, along with control stimuli of reverse congeneric calls, "white noise", and artificially produced conspecific calls. At Pinery Provincial Park, foraging calls of <u>Lasiurus borealis</u> were also presented, along with control stimuli of "white noise" and reverse <u>L. borealis</u> calls. Another set of playback presentations to <u>Lasiurus borealis</u> included 1) foraging calls of conspecifics; 2) conspecific feeding "buzzes"; 3) and 4) each of these stimuli played backwards.

Each playback trial consisted of a 2 min silent period and a 2 min stimulus period, with the order of these periods assigned randomly in each trial. We monitored the responses of free-flying bats by observing them and the speaker against the night sky. We scored bats flying within 2 m of the speaker, assigning a value of <u>1</u> to individuals that did not change their flight paths with respect to the speaker, <u>2</u> to individuals that swerved as they passed the speaker, and <u>3</u> to bats circling the speaker.

The results of the 1985 data are based on the computation of difference scores. For each trial we derived a single response score by calculating the difference between the square root of the stimulus period count and the square root of the silent period count. This transformation allowed a more meaningful application of a 4-way MANOVA test by normalizing the data, minimizing a positively skewed distribution, and reducing large standard deviations. The 1986 data have not yet been analysed statistically.

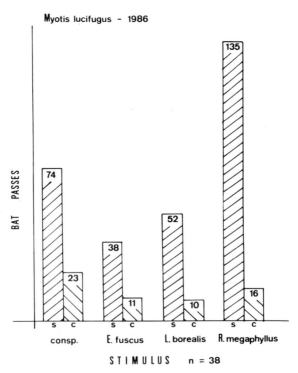

Fig. 1 The total number of bat passes for 38 x 4 presentations over 7 nights to <u>Myotis lucifugus</u> at the Ontario roost in 1986.
s = stimulus half of presentations   c = silent half of presentations

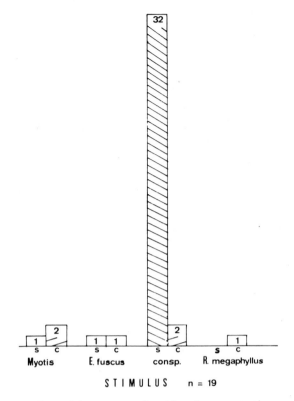

Lasiurus borealis - 1986

Fig. 2 The total number of bat passes for 19 x 4 presentations over 4 nights to
Lasiurus borealis at foraging sites in Pinery Provincial Park in 1986.
s = stimulus half of presentations   c = silent half of presentations

RESULTS

    At both the feeding and roost sites, Myotis bats in the Okanagan Valley were
significantly more responsive to the calls of conspecifics than to any
other stimulus we presented (4-way MANOVA, $p < 0.01$; $p < 0.001$). Outside the
roost, responses typically involved one or more bats flying in tight circles around
the speaker and some individuals hovering in front of it. In the feeding area, the
responsiveness of Myotis was less pronounced and responding individuals were those
that appeared to be arriving on site or bats feeding in low densities. None of the
bats feeding in the swarm was observed to investigate the stimuli. At the Ontario
roost M. lucifugus in 1986, calls of R. megaphyllus elicited the greatest response
followed by the calls of conspecifics (Fig. 1). The responsiveness of Myotis bats in
both British Columbia and Ontario increased dramatically (>50%) when young bats
became volant in early July, and by late July, all stimuli except "white noise" had
produced responses from these bats on at least one occasion.

In 1986 L. borealis responded only to the conspecific stimulus (Fig. 2) except for two occasions when individuals approached a speaker from which L. cinereus feeding buzzes were presented. This generally reflects the 1985 data when statistically significant responses (4-way MANOVA, p < 0.01) were recorded only to presentations of conspecific feeding buzzes. Responding L. borealis swooped down towards the speaker and occasionally circled it one or more times. Observation of individually marked bats showed a high level of site fidelity to feeding areas around lights, and we observed some chases at these sites, usually when two bats pursued the same insect.

Lasiurus cinereus often foraged with L. borealis but never responded to our presentations. Once we observed an L. cinereus attack the same insect an L. borealis was pursuing, but otherwise the two species did not interact during our study.

DISCUSSION

The results generally support the suggestions that echolocation calls can serve a communication function (Barclay 1982; Leonard and Fenton 1984). The lower responsiveness of Myotis in feeding situations, particularly where feeding bat densities were high, corroborates Barclay's (1982) suggestion that Myotis eavesdrop to locate good feeding sites. When we presented a playback near a feeding swarm we did not significantly alter a bat's acoustic surroundings as we did when presenting playbacks near the roost.

The strong response of M. lucifugus to R. megaphyllus calls does not fit the prediction that foraging bats will respond more to calls of similar design. Other than curiosity, we cannot explain this responsiveness. The levels of responses increased in July when the young became volant, probably reflecting the naivete of juveniles.

Because individual L. borealis return to the same foraging sites each night, there is no reason to expect them to eavesdrop to find feeding areas. The significant responsiveness to feeding buzzes of conspecifics and observations of aerial chases between bats suggests that L. borealis attend to other individuals pursuing prey. Other workers have suggested that L. borealis are territorial (e.g., Barbour and Davis 1969) but in spite of site loyalty and aerial chases we observed, our data provide no evidence of territoriality.

ACKNOWLEDGEMENTS

This research has been supported by Natural Sciences and Engineering Research Council Operating and Equipment Grants to MBF and a Sigma Xi Grant-in-aid-of-Research to JPB. We are grateful to H.D.J.N. Aldridge for reading the manuscript and making helpful suggestions.

REFERENCES

Barbour, R.W. and W.H. Davis. 1969. Bats of America. Univ. of Kentucky Press, Lexington.
Barclay, R.M.R. 1982. Interindividual use of echolocation calls: eavesdropping by bats. Behav. Ecol. Sociobiol., 10:271-275.
Bell, G.P. 1980. Habitat use and response to patches of prey by desert insectivorous bats. Can. J. Zool., 61:528-530.
Leonard, M.L. and M.B. Fenton. 1984. Echolocation calls of Euderma maculatum (Chiroptera: Vespertilionidae): use in orientation and communication. J. Mamm., 65:122-126.
Simmons, J.A., M.B. Fenton, W.R. Ferguson, M. Jutting and G. Palin. 1979. Apparatus for research on animal ultrasonic signals. R. Ont. Mus. Misc. Pub., pp. 1-32.

AUDTIORY INPUT TO THE DORSAL LONGITUDINAL

FLIGHT MOTOR NEURONS OF A NOCTUID MOTH

Lee A. Miller and Bent M. Madsen

Institute of Biology
Odense University
DK-5230 Odense, Denmark

SUMMARY

The motor neurons (mn) innervating the dorsal longitudinal muscles (dlm) receive synaptic inputs activated by auditory pathways. Quiescent mn's in a "non-flying" preparation receive either subthreshold depolarizing or hyperpolarizing synaptic inputs in response to ultrasound. On the other hand, sound stimuli presented to a "flying" preparation can modulate motor neuronal responses in different ways depending of some unknown state of the nervous system. Sound intensity is the stimulus parameter that triggers responses. The modulation of neuronal responses has significance for the insect's avoidance behavior to bats.

INTRODUCTION

All known insectivorous bats that capture their prey on-the-wing emit a rather stereotyped pattern of cries. Cry rates range from a few per s to over 200 per s during prey capture. Cry durations can be as short as 0.5 ms.

Certain tympanate insects detect and evade bats (Miller, 1983). The most extensively studied of these are moths (Roeder, 1975; Fenton and Fullard, 1981). Freely flying noctuid moths show a variety of behavioral responses when confronted by hunting insectivorous bats or when stimulated with artifical bat cries. Moths will turn and fly away from distant bats, or show power dives, looping, zig-zag flight, and passive falls to bats at close range.

Two auditory cells (A1 and A2) in each ear trigger avoidance behaviors in noctuid moths. Both cells have identical tuning curves, but the A1 cell is about 20 dB more sensitive than the A2 cell. Acoustical shadowing by the body and wings provide directional information.

Roeder (1975) made extracellular studies of the responses of auditory interneurons to ultrasonic stimuli. To date there are limited electrophysiological studies of moth mn's. Thus, one of our objectives was to record intrasomal responses from flight mn's to auditory stimuli, especially from preparations in feigned flight.

MATERIALS AND METHODS

One to seven day old adult moths (Mamestra brassicae) were obtained from culture. The anatomy of mn's was studied using standard intensified colbalt chloride techniques (Bacon and Altman, 1977). Adequate stainings from 45 preparations were obtained. Individual mn's were stained with a fluorescent dye, Lucifer Yellow, following intrasomal recordings. Intra-somal recordings were made using glass microelectrodes from dlm mn's in minimally dissected preparations. Successful stainings and intrasomal recordings were obtained from 28 preperations. To determine the auditory threshold of the A1 and A2 sensory cells extracellular recordings were made from the tympanic nerve in about 20 preparations. Ultrasonic stimuli were electronically generated in a standard manner and presented through a small (diameter = 15 mm) electrostatic loudspeaker of our own design. Stimuli consisted of sound pulses having rise-fall times of 0.5 ms, durations of from 1.5 to 10 ms, pulse repetition rates of from 10 per s to 125 per s, carrier frequencies of from 10 kHz to 100 kHz. Echoes were at least -30 dB re. the incident sound.

RESULTS

The tuning characteristics of the auditory afferents resemble those for other nocutid moths. The A2 cell threshold is consistantly -20 to -25 dB re. the A1 cell. Maximum sensitivity is 35 dB SPL at 25 kHz. The aud-itory threshold for M. brassicae is from 10 to 15 dB lower than that reported for most other noctuids.

Five pairs of large mn's and three pairs of small mn's found in the pterothoracic ganglia innervate the dl muscles of the mesothorax. The morphologies of these mn's are similar to those described in other moths (Kondoh and Obara, 1982).

Before starting the dissection for intrasomal recordings, we checked the insect's behavior while in stationary and restrained flight. Our methods could not reveal subtle responses like tendencies to turn during ultrasonic stimulation. 90% of the insects tested showed some form of response and lack of response in the remaining 10% was not due to deafness (n = 48). In all cases the threshold for behavioral responses occurred at a sound level of at least +20 dB relative to the A1 threshold. Consequen-tly, the input from two or three auditory cells triggered responses.

We recorded intrasomally from 28 dlm mn's in 28 preparations and found no quantitative or qualitative differences in responses from large or small cells to pulsed ultrasound. Motor neurons from 23 preparations were quies-cent and responded with either depolarizing or hyperpolarizing synaptic poten-tials, but never with both. Motor neurons in 5 other preparations showed spontaneous rhythmic activity, which could be modulated by ultrasound.

Depolarizing input to quiescent cells in response to acoustic stimula-tion had peak amplitudes of from 1 to 12 mV, durations of from 20 to 60 ms and latencies of from 18 to 30 ms (Fig. 1A). The threshold for depolariz-ing input ranged from +18 dB to +36 dB (re. the ipsilateral A1 threshold), and only rarely caused such cells to spike. Interestingly, the latency did not decrease with increasing sound intensity.

Hyperpolarizing responses of quiescent cells had maximum peak ampli-tudes of 3 mV, durations of about 100 ms, and latencies of from 9 to 15 ms (n=6) (Fig 1B). The threshold for hyperpolarizing responses was the same as for depolarizing responses.

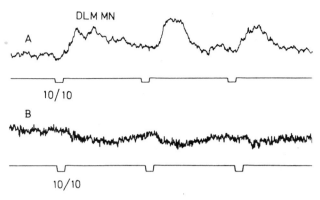

Fig.1A,B. Synaptic input recorded intrasomally from large dorsal longitudinal motor neurons (DLM MN) in response to pulsed ultrasound. The upper
panel (A) shows depolarizing responses and the lower panel (B) hyperpolarizing respones. Sound pressure ca. 87 dB SPL, sound frequency = 25 kHz,
pulse duration = 10 ms, pulse rate = 10 per s.

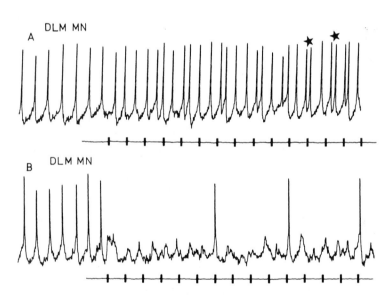

Fig.2A,B. Modulation of the activity of a flight motor neuron (DLM MN) in
response to an identical stimulus presented at different times. In A the
neuron began to fire paired spikes (the stars) several hundred ms after the
onset of an 8 s pulse train (lower trace). In B the spiking ceases
but rhythmic, synaptic input continues. Other stimulus parameters were:
pulse duration = 10 ms, pulse rate about 7 per s., sound frequency = 25
kHz, and sound pressure = +37 dB re. to the A1 auditory cell threshold.

Synaptic drive to rhythmically active dlm mn's most likely originates
from the flight pattern generator. Spiking activity occurred at frequencies of from 14 to 19 Hz, or rather close to the normal wing beat frequency

(Fig. 2). Acoustic stimulation modulated rhythmic activity and the same neuron could respond differently to the same stimulus presented at different times, in contrast to the responses of quiescent neurons. A rhythmically spiking cell firing one action potential per cycle could be excited to fire two spikes per cycle on some cycles following ultrasonic stimulation (the stars in Fig. 2A). On other occasions the same stimulus to the same preparation could produce a cessation of spiking, but rhythmic synaptic input continued (Fig. 2B). Pulsed ultrasound could re-start rhythmic "flight" activity in a neuron that had spontaneously ceased spiking shortly prior to stimulation. Response latencies were from 10 to 200 ms and seemed to depend on the cycle of synaptic drive. In all cases the stimulus intensity needed to modulate rhythmic activity was at least +30 dB re. to the ipsilateral A1 cell threshold. The responses were independent of stimulus repetition rate. There were likewise no differences between responses from rhythmically active large and small motor neurons. In sum, a spiking motor neuron could respond differently to the same acoustic input without affecting the rhythmic synaptic drive.

DISCUSSION

How auditory inputs modulate responses from motor neurons depends on the state of the nervous system. Auditory stimulation produces only one type of synaptic response in quiescent mn's for the duration of an experiment. However, if the neural flight pattern generator is active the same acoustic stimulus can elicit two types of responses from dlm mn's: an augmented response giving two spikes per cycle (Fig. 2A) or a diminished response causing the cessation of spiking but not the cessation of cyclic synaptic drive (Fig. 2B). Whether an augmented or a diminished response will occur cannot be predicted. This property of the central nervous system would have selective value since the predator (a bat) would not be able to predict exactly what the moth will do thus increasing the prey's chances for survival. Roeder (1975) called this the "evitability" of behavior.

The behavior of green lacewings (Neuroptera) in the presence of hunting bats is also unpredictable (Miller and Olesen, 1979; Olesen and Miller, 1979). Green lacewings show many of the same behaviors as moths (Lepidoptera), with passive nose-dives being the prime avoidance behavior. Olesen and Miller (1979) suggest a physiological mechanism that involves the "uncoupling" of motor neurons from the neural flight pattern generator. Some unknown property of the nervous system determines when, which, and how many muscles are uncoupled. The behavior is unpredictable and variable even when the same stimulus is presented at different times. They propose that the pattern generator itself is not affected by the stimulus. Our recordings from moth mn's during feigned flight support their proposal. Upon stimulation the synaptic drive to the neuron can fall below the threshold level for spike generation, but the rhythmic input from the flight pattern generator persists (Fig. 2B). The neural mechanism underlying variability remains a mystery.

Behaviorally relevant informantion on how sensory input modulates neural responses can only be obtained from "flying" preparations. Such preparations have not only provided valuable information on the neural mechanisms of flight in locusts (Robertson and Pearson, 1984), but also on sensory modulation of these mechanisms (Reichert and Rowell, 1986). Hopefully these techniques can be more extensively applied to moths in the future allowing for a fuller understanding of the neural mechanisms of avoidance behavior.

ACKNOWLEDGEMENT

We thank the Danish Natural Science Research Council for support.

REFERENCES

Bacon, J.P., Altman, J.S., 1977, A silver intensification method for cobalt-filled neurons in wholemount preparation, Brain Research, 138:359-363.

Fenton, M.B., Fullard, J.H., 1981, Moth hearing and the feeding strategies of bats, Amer. Sci., 69:266-275.

Kondoh, Y., Obara, Y., 1982, Anatomy of motoneurones innervating mesothoracic indirect flight muscles in the silkmoth, Bombyx mori, J. Exp. Biol., 98:23-37.

Miller, L.A., 1983, How insects detect and avoid bats, in: "Neuroethology and behavioral physiology," F. Huber and H. Markl, eds., Springer-Verlag, Berlin, pp. 251-266.

Miller, L.A., Olesen, J., 1979, Avoidance behavior in green lacewings I. Behavior of free-flying green lacewings to hunting bats and ultrasound, J. Comp. Physiol., 131:113-120.

Olesen, J., Miller, L.A., 1979, Avoidance behaivor in green lacewings II. Flight muscle activity, J. Comp. Physiol. 131:121-128.

Reichert, H., Rowell, C.H.F., 1986, Neuronal circuits controlling flight in the locuts: how sensory information is processed for motor control, Trends Neurosci., 9:281-283.

Robertson, R.M., Pearson, K.G., 1984, Interneuronal organization in the flight system of the locust, J. Insect. Physiol. 30:95-101.

Roeder, K.D., 1975, Neural factors and evitability in insect behavior, J. Exp. Zool. 194:75-88.

DISRUPTING FORAGING BATS : THE CLICKS OF ARCTIID MOTHS

M.G. Stoneman and M.B. Fenton

Department of Biology, Carleton University, Ottawa, Canada*

INTRODUCTION

Dunning and Roeder (1965) demonstrated how the clicks of some arctiid moths interfered with the ability of flying Myotis lucifugus to catch mealworms thrown into the air. Dunning (1968) subsequently demonstrated that some M. lucifugus learned to use the clicks of bad tasting arctiids to distinguish them from those of arctiids that lacked chemical protection. Surlykke and Miller (1985) found that Pipistrellus pipistrellus also learned to associate arctiid clicks with bad taste. Three hypotheses have been proposed to explain the responses of the bats to the moth clicks: 1. the moth clicks are aposematic signals (Dunning 1968; Surlykke and Miller 1985); 2. the moth clicks are startle displays (Humphries and Driver 1970); and 3. the moth clicks jam the bats' echolocation (Fullard, Fenton and Simmons 1979).

The three hypotheses depend upon different responses by the bats. The aposematic hypothesis presumes that if the bats avoided prey presented with clicks, the animals have prior experience with clicks and bad tasting moths. The startle (= deimatic behaviour of Edmunds 1974) hypothesis is based on the animals' responses to unexpected sounds, while the jamming hypothesis uses interference of the clicks with the bats' echolocation behaviour. In our experimental design, the jamming hypothesis predicts that only bats orienting by echolocation should be influenced by the clicks.

The purpose of this study was to test the jamming hypothesis by comparing the effects of arctiid clicks on foraging captive Megaderma lyra and Macrotus californicus, two species which orient by vision when hunting in light (D. Audet, pers. comm.; Bell and Fenton 1986) or by echolocation in the dark (Fiedler 1979; Bell 1985). An advantage of these species is that we could use the same individuals hunting in different conditions to ascertain the influence of the moths' clicks. Furthermore, our study involves flying bats hunting in their normal mode (i.e., searching for prey on the ground).

MATERIALS AND METHODS

Three Megaderma lyra and seven Macrotus californicus which were adult and whose prior experience with the clicks and tastes of arctiids was unknown, were trained to take mealworms from a screen-bottomed dish on the floor of a flight cage 3 m long by 1 m wide by 2 m high. The bats learned to roost at one end of the cage and the mealworms were presented at the other end. When the bats flew within 0.5 m of the dish, they broke a set of light beams which activated a softswitch on an

---

*   present address: Department of Biology, York University, Downsview
     Ontario, Canada M3J 1P3

ultrasonic amplifier (Simmons, Fenton, Ferguson et al. 1979), feeding recorded arctiid clicks or tape noise to a Panasonic EAS-10TH400B speaker. the recordings were presented from tape loops operated on an Ampex PR500 tape recorder operated at 76 cm/s. The speaker and a QMC S200 microphone were located under the dish containing the mealworms. The clicks were from one of <u>Cycnia tenera</u>, Euchaetias <u>egle</u>, <u>Phyrrharctia isabella</u>, or <u>Hypoprepia fucosa</u>.

Each trial consisted of a bat leaving the roosting end of the cage, flying to within 0.5 m of the dish (breaking the light beams) and returning to the roosting end of the cage. A <u>complete</u> trial was scored when the bat attempted to take a mealworm from the dish; a <u>balk</u> when the bat terminated its approach to the dish and returned to the roost. We used three treatments: 1. <u>baseline</u> with the amplifier turned off; 2. <u>control</u> with the amplifier receiving tape noise; and 3. <u>experimental</u> with the amplifier receiving recorded moth clicks. The clicks were the same intensities as those from live moths (c. 70 dB SPL at 2 cm). We ran half of the trials in total darkness, the rest under dim room lighting. In the former treatments we observed the bats through a Javelin Night Vision scope, in the latter, directly. The proportions of the treatments were assigned as follows: 60% baseline; 20% each, control and experimental, and the actual sequences of presentations were determined using a random digit table (Rohlf and Sokal 1981).

RESULTS

We ran 8414 trials, 5071 baseline, 1674 control and 1669 experimental. The results of the trials were summed over individuals and days and maintained separately for each bat. In baseline trials the bats typically flew directly to the dish, hovered briefly just above it (20 cm), dropped into the dish, took a mealworm, and flew back to the roost. In control trials the bats slowed their flight speeds and sometimes hovered higher above the dish, and occasionally balked. In almost all of the experimental trials (98.4% of 1669) bats responded to click presentations by snapping back their heads, executing a quick pirouette and flying back to the roost. On 27 experimental trials (1.6%) bats took mealworms from the dish, but these occurrences were not predictable from individual bats or trials. The behaviour of the two species was similar, although <u>M. californicus</u> appeared to be more easily disturbed than <u>M. lyra</u>. In most dark trials (>80%) the bats produced echolocation calls detectable by the QMC S200 microphone, and in about 20% of the light trials, echolocation calls also were produced. The experiment results are shown in Fig. 1.

When the proportions of balks under each set of conditions are compiled, there are no significant differences between light and dark trials (x^2 = 0.08 to 3.5; p = 0.9 to 0.1). There were significant differences (x^2 = 170 to 1000; p << 0.001) between treatments. In other words, the hypothesis that bats are more affected in dark than in light is rejected by the data which also show that moth clicks interrupted the attacks of bats significantly more often than control or baseline presentations. The data permit us to reject the jamming hypothesis.

DISCUSSION

Fullard, Fenton and Simmons (1979) based their jamming hypothesis on the matching of the intensity and spectral characteristics of the clicks of some moths to the feeding buzz calls of <u>Eptesicus fuscus</u>. Surlykke and Miller (1985) rejected the jamming hypothesis on the responses of trained <u>P. pipistrellus</u> doing target discriminations. The descriptions of the initial reactions of the <u>P. pipistrellus</u> to moth clicks (Surlykke and Miller 1985) appears to be consistent with a startle effect. We reject the jamming hypothesis because of the responses of the bats foraging in light and dark conditions.

It is clear that bats learn to associate moth clicks with bad taste (Dunning 1968; Surlykke and Miller 1985), but this does not prove the aposematic hypothesis. To

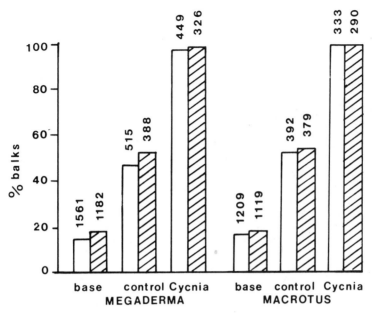

Fig. 1 The responses of attacking Megaderma lyra And Macrotus californicus in base (no sound), control (tape and instrument noise), and experimental (Cycnia) trials in light (open bars) and dark (diagonal lines) conditions. The clicks were from Cycnia tenera but trials with the clicks of three other arctiids (Euchaetias egle – 21 trials; Phyrrarctia isabella – 25 trials; and Hypoprepia fucosa – 24 trials) were not significantly different from the responses to C. tenera clicks. The numbers of trials are shown on the tops of the bars.

challenge the aposematic explanation requires trials involving random presentations of moth clicks to naive bats (not previously experienced with arctiid clicks or tastes). On this basis, neither our experiments nor those of Dunning (1968) nor Surlykke and Miller (1985) bear on the aposematic hypothesis. The aposematic explanation is more complicated than the startle hypothesis, and the moths' displays, which occur just before the attack, are more compatible with a startle display as defined by Edmunds (1974).

Startle displays appear to work by interfering with information processing and they may be acoustically or visually based. the startle response in many mammals is reflexive, causing the animal to clench its jaw muscles and retract and dorsally flex the neck (Davis 1984). The responses of the hunting bats we studied to moth clicks agrees with at least part of the classical startle response in mammals.

ACKNOWLEDGEMENTS

We thank Prof. Dr. G. Neuweiler and Dr. G.P. Bell for their assistance in obtaining the bats, and Dr. J.H. Fullard who provided the recordings of Cycnia tenera clicks. We thank D. Audet for permitting us access to her unpublished observations on M. lyra and H.D.J.N. Aldridge and D.D. Yager for commenting on the manuscript. This study was supported by operating and equipment grants from the Natural Sciences and Engineering Research Council of Canada to MBF.

REFERENCES

Bell, G.P. 1985. The sensory basis for prey location by the California leaf-nosed bat, Macrotus californicus (Chiroptera : Phyllostomidae). Behav. Ecol. Sociobiol., 16:343-347.

Bell, G.P. and M.B. Fenton. 1986. Visual acuity, sensitivity and binocularity in a gleaning insectivorous bat, Macrotus californicus (Chiroptera : Phyllostomidae). Anim. Behav., 34:409-414.

Davis, M. 1984. The mammalian startle response, pp. 123-147 IN Neural mechanisms of startle behavior. (R.C. Eaton, ed.). Plenum Press, NY.

Dunning, D.C. 1968. Warning sounds of moths. Z. Tierpsychol., 25:129-138.

Dunning, D.C. and K.D. Roeder. 1965. Moth sounds and the insect-catching behavior of bats. Science, 147:173-174.

Edmunds, M. 1974. Defence in animals. Longmans Press, Essex.

Fiedler, J. 1979. Prey catching with and without echolocation in the Indian false vampire bat, Megaderma lyra. Behav. Ecol. Sociobiol., 6:155-150.

Fullard, J.H., M.B. Fenton and J.A. Simmons. 1979. Jamming bat echolocation : the clicks of arctiid moths. Can. J. Zool., 57:647-649.

Humphries, D.A. and P.M. Driver. 1970. Protean defense by prey animals. Oecologia, 5:285-302.

Rohlf, F.J. and R.R. Sokal. 1981. Statistical tables, second edition. W.H. Freeman and Co., San Francisco.

Simmons, J.A., M.B. Fenton, W.R. Ferguson, M. Jutting and G. Palin. 1979. Apparatus for research on animal ultrasonic signals. Misc. Pub. R. Ont. Mus., Toronto, 32 pp.

Surlykke, A. and L.A. Miller. 1985. The role of arctiid clicks on bat echolocation: jamming or warning? J. Comp. Physiol., 156:831-843.

THE ECHOLOCATION ASSEMBLAGE: ACOUSTIC ENSEMBLES IN A NEOTROPICAL HABITAT

James H. Fullard and Jacqueline J. Belwood

Department of Zoology, University of Toronto

Department of Entomology, University of Florida

Simmons et al. (1979), Fenton and Fullard (1979), Fenton (1984) and Neuweiler (1984) suggest that the echolocation patterns of bats may be affected by certain ecological factors (e.g. foraging strategies, prey defences). The echolocation calls of 37 species of bats from the island of Barro Colorado (BCI), Republica de Panamá were recorded and the acoustic characteristics of these calls are being co-related with the natural histories of the bats in collaboration with the studies of C.O. Handley, Jr. of the U.S. Museum of Natural History.

To introduce the potential usefulness of these data we propose, as a model, the acoustic relationship between echolocation frequency design and the auditory characteristics of sympatric, tympanate moths. Moths have ears sensitive to the echolocation frequencies generated by the bats that hunt them (Roeder, 1970; Fullard, 1982; 1984; Surlykke, 1984). Assuming that bat-detection is the main selective force acting on the physiological design of moth ears (Fullard, 1986) their auditory threshold curves (audiograms) should have two regions of sensitivity. A moth's primary sensitivity, where its ear is best tuned, will match the range of echolocation frequencies most strongly affecting it, that is, those bats which form the heaviest predation threat. Bats presenting less of a threat, either because they are less common or feed less on moths, will emit echolocation frequencies that affect the moths to a lesser extent. This reduced pressure results in secondary sensitivities at frequencies lower and higher than its primary tuning. Bats whose frequencies are within a moth's primary sensitivity are termed syntonic echolocators and those whose calls lie outside of this range are termed allotonic (Fullard, 1986). Moths are unable to discriminate frequencies to identify particular bat species to determine their likely predatory threat. Therefore, what matters to a moth is the combined echolocation spectrum that represents the total predation potential of its habitat. This spectrum will consist not only of the echolocation frequencies of the bats within that habitat but, ideally, some measure of the dietary prefences and population levels of those bats. This acoustic predatory ensemble is the echolocation assemblage (Fullard, 1982) and should be a direct measure of the selection pressure exerted on the moths sympatric with it.

Figure 1 illustrates averaged echolocation spectra from the recorded calls of the BCI bats. Complexities in harmonic design, durations and intensities are evident within the community but, for the present, we shall only examine the frequency characteristics of this assemblage (for further infor-

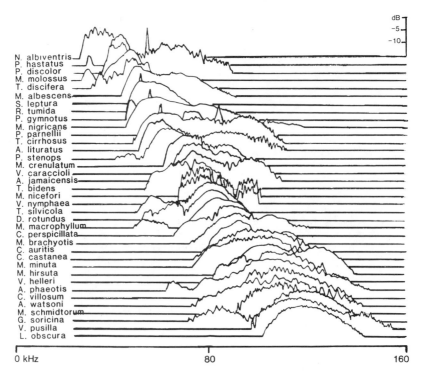

Fig. 1. The individual frequency spectra (Fast Fourier Transform) of 37
species of neotropical bats discussed in this paper. Spectra
were each averaged from echolocation calls (n = 2 to 31) of bats
recorded as they flew within screened enclosures on BCI, normalized (peak dB = 0) and ordered from lowest to highest.

mation see Belwood, 1986). Figure 2A is a computer-assembled composite spectrum of all 37 normalized species spectra and represents the total habitat
echolocation assemblage, that is the total frequency output of the 37 species
recorded. Compared to this assemblage is the median audiogram (neurally derived) of 25 sympatric notodontid moths from 6 species. The audiogram predicts that these moths face primary selection pressure from 30 to 60 kHz syntonic bats and secondary selection from 10 to 30 kHz and 60 to 100+ kHz allotonic bats. The total habitat assemblage portrays the moths' selection pressure assuming that all of the bats were equally distributed and all fed
equally heavily on moths with similarly intense echolocation calls. The poor
match between the notodontid audiogram and the total habitat assemblage reflects the fact that many of the island's bats are not primarily insectivorous let alone lepidoptivorous. If we remove the exclusively (Gardner, 1977)
frugivorous species the assemblage narrows somewhat (Figure 2B). Using only
insectivorous species the assemblage (Figure 2C) shifts to lower frequencies
more syntonic with the moths' primary sensitivity. Figure 2D is the assemblage of 6 BCI species suspected of preying heavily on moths. Here the match
is closer and one begins to observe the actual predatory assemblage to which
these insects are exposed (i.e. the pertinent predation pressure). Within
this 6 species assemblage are both primary and secondary moth predators, with
vespertilionids (e.g. Myotis nigricans) and mormoopids (e.g. Pteronotus parnellii) forming the primary pressure owing to their greater population numbers and higher frequency, less common phyllostomids (e.g. Macrophyllum macrophyllum) forming the allotonic secondary selection. Adjusting the lepidoptivorous assemblage to reflect differences in population levels and precise feeding habits should improve the matching and could be used to predict
the actual predation potential of each species.

At first glance, using computer manipulations of echolocation assemblages to best fit sympatric moths' audiograms appears to prove the obvious: moths listen for bats. This conclusion, however, is prejudicially based on the uniquely well understood sensory relationships between moths and bats. In addition, BCI probably represents the most intensively studied community of tropical bats and our knowledge of dietary habits in these bats allows for a quantitative examination of the sensory ecology of this habitat's moths and bats. Less well understood are bat/prey interactions in other geographic areas as well as the acoustic relationships arising from other ecological factors such as a habitat's climate and vegetation. Echolocation assemblage examinations can provide us with testable hypotheses and predictions about the foraging ecologies of bats. For example, Figure 3 compares the assemblages of clearing bats (e.g. unforested areas, open water) to those that forage in the BCI forest. This figure suggests that forest bats use higher frequencies, perhaps as clutter-resistant acoustic mechanisms (Simmons and Stein, 1980) while clearing bats emit lower, more distance-efficient frequencies.

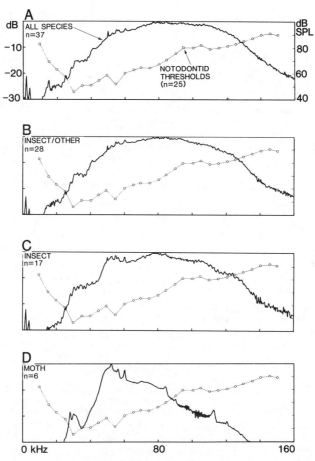

Fig. 2. A comparison of the echolocation assemblages of BCI bats and the median audiogram of sympatric notodontid moths. 2A: total habitat assemblage from spectra in Fig. 1., 2B: assemblage of species reported to feed on insects and fruit, 2C: bats reported to feed primarily on insects, 2D: assemblage of bats thought to feed heavily on moths.

Since the echolocation structures of bats are complex, the use of simple variables such as peak frequency may not be sufficient to explain subtler relationships between the echolocation design of bats and their ecological constraints. The use of assemblage examinations should broaden our understanding of these relationships.

Acknowledgements

We thank C.O. Handley, Jr., R. Abbey, E. Stockwell, S. Sakaluk and R. Kunz for their assistance and the Smithsonian Tropical Research Institute for permission to use their facilities on BCI and for funding (JJB). Drs M.B. Fenton and R.M.R. Barclay provided helpful comments. This study was funded by the Natural Sciences and Engineering Research Council of Canada (JHF), the National Science Foundation, Sigma Xi, National Academy of Sciences, American Society of Mammalogists and the Florida Entomological Society (JJB).

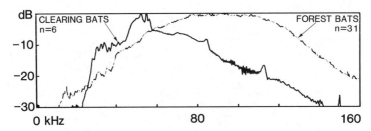

Fig. 3. Comparison between the echolocation assemblages of bats that commonly fly in clear sites and those that have been trapped in the BCI forest (Handley, unpublished data).

References

Belwood, J.J., 1986, Foraging behaviour, prey selection and echolocation in phyllostomine bats (Phyllostomadae), (this issue).

Fenton, M.B., 1984, Echolocation: implications for ecology and evolution of bats, Quart. Rev. Biol. 59:33–53.

Fenton, M.B., and Fullard, J.H., 1979, The influence of moth hearing on bat echolocation strategies, J. Comp. Physiol., 132:77–86.

Fullard, J.H., 1982, Echolocation assemblages and their effects on moth auditory systems, Can. J. Zool., 11:2572–2576.

Fullard, J.H., 1984, Acoustic relationships between tympanate moths and the Hawaiian hoary bat (Lasiurus cinereus semotus), J. Comp. Physiol., 155:795–801.

Fullard, J.H., 1986, Sensory ecology and neuroethology of moths and bats: interactions in a global perspective, in: "Recent advances in the study of bats", M.B. Fenton, P.A. Racey and J.M.V. Rayner, eds, Cambridge University Press, Cambridge.

Gardner, A.L., 1977, Feeding habits, in, "Biology of bats of the new world family Phyllostomatidae. Part II", R.J. Baker, J. Knox Jones Jr., and D.C. Carter, eds., Texas Tech Press, Lubbock.

Neuweiler, G., 1984, Foraging, echolocation and audition in bats, Naturwissenschaften, 71:446–455.

Roeder, K. D., 1970, Episodes in insect brains, Amer. Sci., 58:378–389.

Simmons, J.A., and Stein, R.A., 1980, Acoustic imaging in bat sonar: echolocation signals and the evolution of echolocation, J. Comp. Physiol. 135:61–84.

Simmons, J.A., Fenton, M.B., and O'Farrell, M.J., 1979, Echolocation and
    pursuit of prey by bats, <u>Science</u>, 203:16-21
Surlykke, A., 1984, Hearing in notodontid moths: a tympanic organ with a
    single auditory neurone, <u>J. Exp. Biol.</u>, 113:323-335.

MICROBAT VISION AND ECHOLOCATION

IN AN EVOLUTIONARY CONTEXT

John D. Pettigrew

Neuroscience Laboratory
Dept. of Physiology and Pharmacology
University of Queensland, St. Lucia 4067

INTRODUCTION

The aims of this chapter are three-fold:

i. To provide a survey of spatial visual abilities in a range of
microbats,
ii. To relate visual and sonar spatial abilities in the task of aerial
insect capture
iii. To present a speculative scenario of the origins of these sonar and
visual abilities in the first microbats.

MICROBAT VISION

The vision of microbats gets a poor press, despite the efforts of some
investigators (e.g. Chase, 1972; Suthers and Braford, 1980) to redress the
imbalance. There are at least two problems here.

The first problem is the general lack of appreciation of the physical
constraints which limit visual systems in dim illumination. What may
appear to be very poor spatial resolution, by the standards of diurnal
organisms like ourselves, can often prove to be optimal design for
specialized tasks like detecting the silhouette of a rapidly-moving target
against a dim background. These issues have been examined in detail for a
number of invertebrate visual systems (e.g. Snyder et al., 1977; Land,
1981) but there is little data yet available to decide just how well
designed are microbat eyes for particular visual tasks, such as detection
of high velocity targets at low quantum flux. Some of the results below
suggest that they are better than most investigators have suspected.

The second problem concerns the technical difficulties of working on
microbat vision. I have certainly been deterred from working on the small
eyes of rhinolophids and hipposiderids, two families of microbats
representing the extremes of specialization for sonar, and therefore
perhaps a correspondingly decreased reliance on vision. But I am no longer
happy to discount the possibility of spatial visual capabilities in such
microbat families. Some vespertilionids and emballouroids, with eyes
hardly larger than rhinolophids, have quite passable visual resolution, so
one must be cautious until the data are in. Let us now look at some of the
data on spatial vision in microbats.

Retinal Topography in Microbats

The topographic arrangement of retinal ganglion cells is an unrivalled source of quantitative information about the visual capabilities of a species (Hughes 1985). Since ganglion cells provide the only link between eye and behavioural output, via the "informational bottleneck" of the optic nerve, the spatial resolving power of retinal ganglion cells determines the spatial resolving power of the whole animal in a direct way. The predictability of the relationship between behavioural visual acuity and retinal ganglion cell acuity has held up in a variety of different vertebrates and there is no reason to suppose it will fall down when behavioural acuities of microbats become available for the visual tasks they perform in their natural environment. In the meantime, I have been examining the retinal topography of ganglion cells in a variety of microbats. The results from four representative species in three families are shown in Table I (from Pettigrew et al., 1986 in which can be found the maps of retinal ganglion cell density).

TABLE I

Visual abilities in 4 microbats, 2 megabats & rat. Column 1 shows maximal density of retinal ganglion cells in thousands of cells per square mm; second column shows axial length of the eyeball in mm; third column shows the minification factor calculated assuming that the posterior nodal distance of a nocturnally adapted eyeball is 0.5X axial length; fourth column is the linear density of ganglion cells in the high density region, calculated from columns 1 & 3; Column 5 is the visual acuity limit, in cycles/deg, given by sampling theory and column 4; column 6 is wavelength of the dominant high frequency in the sonar call of each of the 4 microbats.

SPECIES	RGC/sq mm (x1000)	A.L. (mm)	Minifn. (deg/mm)	RGC/mm	Acuity (c/deg)	Lambda
Nyctophilus gouldi	6	1.9	60	1.3	0.6	3
Megaderma lyra	7	4.2	27	3.1	1.6	3
Macroderma gigas	3.5	7.0	16	3.7	1.9	5
Taphozous georgianus	8	3.7	31	2.9	1.5	17
Pteropus scapulatus	6	11	10	7.7	3.9	
Pteropus poliocephalus	8	13.3	8.6	10	5.0	
Rattus norvegicus	3	6.1	17	3.5	1.7	

All three microbat families were found to have rather similar patterns of retinal organization with a specialization looking forwards (and slightly upwards in some species) which has the appropriate spatial orientation for aiding prey capture. The spatial resolution provided by this specialization is in the order of 0.5-2 cycles/deg. For those readers unfamiliar with this way of referring to spatial visual acuity, here are some facts to help grasp these figures. One's thumb nail is about 1 degree across if held at arm's length. Normal human visual acuity is 60-70 cycles/deg, which means that a black-white grating would still be visible as composed of separate bars when 60 such equally-spaced bars are fitted side by side across the thumb nail. If the visual acuity of microbats seems poor compared with our own, note that the cat hunts successfully with a visual acuity of about 10 cycles/deg.

COMPARISON OF VISION AND SONAR IN MICROBATS

In making this comparison, I shall confine my attention to the spatial resolving power of the two systems, ignoring aspects like ranging. [Information about target range is provided by echo-delay sensitivity in sonar systems (e.g. Suga and O'Neill, 1979) and by a variety of complicated mechanisms in visual systems (Collett and Harkness, 1982).]

The wavelengths used for sonar by microbats range from about 2 cm to about 2 mm. At the long end of the spectrum are some of the larger emballonuroids such as Tadarida australis whose loud cry around 14 kHz is audible, even to my failing auditory system, around waterholes at night in the Australian outback. At the short end of the spectrum are vespertilionids like Kerivoula whose 160 kHz calls explain the ability of this genus to pluck spiders from their webs without becoming entangled themselves, not to mention its genius at avoiding at avoiding mist nets! (Woodside and Taylor, 1985).

This one order of magnitude range of acoustic ability is somewhat greater than the range of visual acuities in Table 1, although there are some as yet unstudied phyllostomoids which may extend the upper limit of visual acuity and the tiny eyes of rhinolophids will very likely prove to have visual acuity lower than the 0.5 cycles/deg found in Nyctophilus unless the density of ganglion cells is enormously greater to make up for the difference in eye size.

The second difference between the spatial resolution of the two systems relates to the site where diffraction of the wave-front is limiting. This limit occurs within the eye in the case of vision but occurs at the target in the case of sonar (making the assumption here that the receiver is optimally designed to detect wavelengths in the range emitted, as are all sonar systems we know). For this reason, the wavelength of the sonar signal is the limiting factor in the determination of the absolute size of target not its distance (discounting for the moment the effects on the magnitude of the signal which will limit detection at long range). Visual systems on the other hand resolve angles subtended at the eye and so resolution for absolute size will improve as the target approaches. To illustrate this difference, let us take some actual examples of prey-catching by microbats.

Capture by an emballouroid. With most of the power of its call at 20 kHz, Taphozous georgianus can use sonar to achieve good discriminability of spherical targets around 2 cm circumference or greater (the wavelength at this frequency in air), so long as it is close enough to the target to hear an echo. On the other hand, the visual system can resolve 1.4 cycles/degree which would correspond to 7 mm diameter (corresponding to a circumference about the same as the sonar wavelength) at a distance of around 1 m, a range which would be sufficient for a 10 cm bat to start executing an aerial prey capture program. Provided that there are a few photons about, the visual system of this species can clearly perform as well as the sonar system over the range of insect sizes known to be taken (2-12 mm diameter). This result took me by surprise, but is entirely consistent with the behaviour of this species which is to hunt above the canopy under open starlight which would help silhouette prey.

Emballonuroid visual abilities were dramatically confirmed recently when I had the good fortune to observe the behaviour of Craseonycteris thonglongyae in the field at the study area of Surapon Dangkhae in Thailand. Craseonycteris feeds in two short bouts, one just before dark and another just before sunrise in a manner not too dissimilar from the Indian nightjars, Caprimulgus indicus, which can be seen hawking for insects in the same area at the same time. Both bird and bat appear to swoop upon their aerial prey from below, apparently taking advantage of the twilight sky to provide a brighter backdrop for the prey silhouette. While we could see the winged insects on which the nightjars were feeding if we also were able to view them against the skylight, the airborne invertebrates on which Craseonycteris was presumably feeding were too small to be seen from our observation distance of 3-10 m. By monitoring Craseonycteris continuously with a bat detector tuned to its sonar, we could verify that most (but not all) of the sudden upward swoops were unaccompanied by any increase in the pulse repetition rate, even though a terminal buzz is a feature of some swoop events as well as obstacle avoidance by this bat. It was clear to me that this living representative of an ancient branch of the emballonuroid lineage feeds visually.

Capture in a megadermatid: As a second example, let us look at Macroderma gigas, about which much new information has been accumulated recently. Most of the energy in the call of M. gigas is in the second harmonic which sweeps from 60-40 kHz (Guppy et al., 1985). Under some circumstances the third harmonic is more pronounced so that this bat could have access to information in the 100 kHz band although the most sensitive region of the audiogram is also around 60 kHz. Spatial resolution of the sonar system will thus be around 3-4 mm circumference or 1 mm across. In comparison the visual system, operating at 2 cycles/deg (Table 1), will have this kind of resolution at a distance of 0.3 m. Given the large prey items taken by M. gigas (from 5 cm grasshoppers to 20 cm birds), it is apparent that vision and sonar are both well matched. One might therefore predict that either could be used alone for prey capture. This has been confirmed in the field (Pettigrew et al., 1986).

There is one respect in which sonar and vision do not match up in M. gigas, and this concerns directionality. An acoustic axis can be defined, both for its pinna which is a highly directional amplifier, and for the nose leaf which tends to concentrate sonar pulses emanating from the nostrils into a beam. Both of these acoustic axes point upwards at about 50 deg. to the bat's horizontal if we consider the second harmonic of the call. The acoustic axis of the pinna is horizontal for frequencies lower than 20 kHz where it is aligned with the visual axis through the specialized region of increased ganglion cell density. The functional significance of this finding is difficult to establish, but given that passive auditory localization is also used for prey capture by this bat (Fiedler 1979) it seems possible that alignment of the passive auditory localizing machinery in visual coordinates is more important than aligning vision and sonar. This leads to the further suggestion that sonar may play a more important role in obstacle avoidance. For carrying out this task on the wing inside a totally-dark cave, an upwards and forwards axis for the sonar system is entirely appropriate, just as the other two passive systems are irrelevant.

In Megaderma lyra, vision and sonar are also well-matched as one might expect from another member of the same family. In this respect it is interesting to note that <u>Megaderma</u>, with a smaller eye than <u>Macroderma</u> has a much higher ganglion cell density which compensates. This is what one might expect if both closely-related species have the same total number of ganglion cells.

In the vespertilinoid, <u>Nyctophilus gouldi</u>, sonar abilities clearly outstrip visual abilities, since the sonar wave limit is the same as for <u>Megaderma</u>, yet the visual acuity is worse by a factor of three.

VISION AND SONAR IN EARLY MICROBATS

Discussions about bat origins have often been muddled by attempts to squeeze both megabats and microbats into the same evolutionary scenario. There is not space here to deal with the probability of a diphyletic origin for bats, except to say that the bulk of evidence points to a comparatively recent origin for megabats from dermopterans as opposed to an ancient origin for microbats from a separate evolutionary line (Pettigrew, 1986; Pettigrew and Cooper, 1986). While I consider the gliding-flying transition a reasonable scenario for a phytophagous lineage, I have difficulties imagining such a sequence for an insectivore because gliders do not have sufficient maneuverability to take flying insects as the first microbats seem likely to have done. I am attracted to the idea (Caple et al., 1985) that leaps after aerial insects were gradually perfected under the constraints of natural physical laws, such as conservation of angular momentum, until mid-leap maneuverability crossed the border from acrobatic to aerobatic. I discount the possibility of gleaning in the origin of the first microbats since I do not see the pressure that would drive an insect gleaner into true powered flight. I likewise think it unlikely that sonar would have been sufficiently developed in the first microbat to enable much useful guidance toward an airborne insect. The short range of ultrasound pulses makes it unlikely that it was an echo from an insect that first enticed our early microbat off its branch, as does the required neural processing which seems unlikely to have been sufficiently sophisticated on the first try. Vision seems a more likely candidate to provide the appropriate resolution and range, just as it does in many living microbats today. The primary role of sonar would then have been in the detection of obstacles, particularly in the cave roost where there is no alternative sensory channel and where three other flying vertebrates have sought shelter, independently inventing sonar on each occasion (swiftlets, oilbirds and rousette megabats). If this is correct, the use of sonar for insect capture occurred as a modification of the avoidance system after flight was achieved. Some further clues to the mechanism of sonar-based insect capture may therefore be provided by an examination of the way in which the sonar system interacts with the visual system in the guidance of aerial prey capture.

REFERENCES

Caple, G., Balda, R.P. and Willis, W.R. (1983) The physics of leaping animals and the evolution of preflight. Amer. Nat. <u>121</u>: 455-467.

Chase, J. (1972) The role of vision in echolocating bats. Unpubl. Ph.D. Thesis, Indiana University, 214 pp.

Collett, T.S. and Harkness, L.I.K. (1982) Depth vision in animals. In: Analysis of visual behaviour, Ingle, D.J., Goodale M.A., Mansfield, R.J.W. (eds), M.I.T. Press, Cambridge, Mass., pp 111-176.

Fiedler, J. (1979) Prey catching with and without echolocation in the Indian False Vampire (<u>Megaderma lyra</u>). Behav. Ecol. Sociobiol.,6: 155-160.

Guppy, A., Coles, R.B. and Pettigrew, J.D. (1985) Echolocation and acoustic communication signals in the Australian Ghost Bat, Macroderma gigas (Microchiroptera: Megadermatidae). Australian Mammalogy, 8:299-308.

Hughes, A., (1985) New perspectives in retinal organization. in: Progress in retinal research, Osbourne, N.N. and Chader, G.J., (eds), Pergamon Press, Oxford, vol., 4 pp 243-313

Land, M.F., (1981) Optics and vision in invertebrates. in: Handbook of Sensory Physiology. Vol. VII/6B. Autrum H. (ed), Springer, Berlin, pp 471-592

Pettigrew, J.D., (1986) Megabats have the advanced pathway from eye to mid-brain. Science 231: 1304-1306.

Pettigrew, J.D., and Cooper, H.M., (1986) Aerial Primates: Advanced visual pathways in megabats and flying lemurs. Soc. for Neuroscience Abstr. (in press).

Pettigrew, J.D., Hopkins, C. and Dreher, B. (1986) Retinal topography in microbats. MS submitted.

Pettigrew, J.D., and sixteen other authors (1986) The Australian Ghost Bat, Macroderma gigas, at Pine Creek, Northern Territory. Macroderma, 2: 10-19.

Snyder, A.W., Laughlin, S.B. and Stavenga, D.G., (1977) Information capacity of eyes, Vision Research, 17: 1163-1175.

Suga, N. and O'Neill, W.E., (1979) Neural axis representing target range in the auditory cortex of the mustached bat. Science, 206: 351-353.

Suthers, R.A. and Braford, M.R., Jr., (1980) Visual systems and the evolutionary relationships of the Chiroptera. in: Proceedings of the Fifth International Bat Research Conference, Wilson D.E. and Gardner A.L. (eds) Texas Tech, Lubbock, pp. 331-346.

Woodside, D.P. and Taylor, K.J., (1985) Echolocation calls of fourteen bats from eastern N.S.W. Australian Mammalogy, 8: 279-298.

650

FUTURE DIRECTIONS

M. Brock Fenton

Department of Biology
York University
North York, Ontario, Canada M3J lP3

On the last afternoon of the meeting, a group including F. Awbry, R.M.R. Barclay, J.J. Belwood, P. Brown, R. Coles, W. Evans, M.B. Fenton, J.H. Fullard, A. Guppy, K. Marten, L. Miller, J. Pettigrew, and K. Zbinden met to discuss some of the work which needed to be done on the ecology of echolocating animals. Here are the important areas they identified:

1. What do echolocating animals eat? Is there any evidence that echolocating animals terminate attacks on some prey which means rejecting some items without making physical contact? Evidence of this kind would provide a clear indication of the discrimination powers of the echolocating animal. How much variability is there in the diets of echolocating animals and how does this mirror prey availability. These matters bear on the discrimination ability animals can achieve through echolocation.

2. We need evidence obtained from moving animals responding to moving targets, the normal situation for hunting bats or dolphins under many circumstances. For the same reason, we need to know how echolocating animals perform in clutter and how (if?) they adjust their sounds according to the situation. Central to this matter is the question of variability in call design.

3. The question of call design and variation must be considered in the context of the role of echolocation calls in communication. How much of the variation in call design can be attributed to communicative effects?

4. What is the role of echolocation in the lives of facultative echolocators?

5. Do fish and other marine organisms show parallels to the hearing-based defense common in many nocturnal insects?

6. There are excellent opportunities for collaborative field work involving researchers studying echolocators and their prey. Such collaboration often can produce more effective results than studies approaching the question from one point or another.

On a more practical note, the group identified the need for a
tape library, a resource from which interested parties could obtain
recordings of different species of echolocating animals under different
circumstances. No specific move was made to achieve this worthy goal.

SECTION V

ECHOLOCATION AND COGNITION

Section Organizer: Kenneth S. Norris

ON THE EVOLUTION OF ACOUSTIC COMMUNICATION SYSTEMS
IN VERTEBRATES
Part I: Historical Aspects

Kenneth S. Norris and Evan C. Evans III

Joseph M. Long Marine Laboratory, University of California
Santa Cruz, California

Naval Ocean Systems Center, Kaneohe, Hawaii

INTRODUCTION

This paper seeks to trace the evolution of the processes of vertebrate communication and echolocation. To do this we have drawn information from the evolution of receptors, from animal communication and we refer to the origins of the supporting cognitive systems. The second paper reviews the cognitive systems underlying communication. The two papers are meant to be considered together.

There appear to be two simultaneous and quite different trends in the evolution of information systems in animals. One is the perceptual or cognitive history, in which organisms have been able to deal with more and more of their world by drawing increasingly fine-tuned generalizations from the essentially infinite data pool surrounding them. This involves restricting attention to things of importance and thereby rejecting massive amounts of available data. For data that are admitted, ways of structuring and processing information that save time and space further augment the development. This process of rejecting much and generalizing the remainder is in fact the basic task that all sensing animals must perform.

The second is the evolution of increasingly capable sensory systems. Vertebrate sense organs have come to process ever wider ranges of physical data with high sensitivity. While even simple hearing organs such as the sacculus of frogs are very sensitive (Lewis and Narins 1985), the frequency range is modest compared to the hearing of some higher vertebrates. The ears of mammals can detect displacements of a third the diameter of a hydrogen atom in the cochlear fluid (Bekesy 1962), while at the same time receiving signals ranging across wide bandwidths that may range from low sonic sounds to those of more than 100 kHz.

Yet this has produced no super-organism perceiving all. Instead, each organism is fitted to see and hear what concerns it most. The refinements for none of us are complete.

On the cognitive side no sharp distinction can be made between sensory organs and the brain. Indeed, there exists a continuum between these two ends of the nervous system. Generalization is performed at various locations along the way.

When faced with assessments in any sensory modality we, and other animals, tend to "satisfice" (Newell and Simon 1972). Approximate solutions sufficient for the moment are used. If more refinement is needed animals tend to resort to learning or take longer to deal with the message. The basic business of all animals is to deal effectively with one-time assessments of significant patterns in nature rather than to make a logical analysis of possible options. Time is too short. This is far less costly in terms of information processing than more refined analysis.

Another aspect is the speed in message processing. We submit that the central nervous system approached a neurophysiological ceiling in the speed of information processing long ago. More recent increases in processing speed have been achieved by "tricks" <u>sensu</u> Minsky (1985).

Acoustic <u>signals</u>, often concerned with the emotional state of the sender, are typically brief. Acoustic <u>messages</u> carrying more complex information tend to be slower. It is not surprising, therefore, to find that higher animals use signals to denote the context of more complex messages. We cry in pain when we put a hand on a hot stove and talk about it later. The rapid but primitive system still serves its ancient function but it remains unspecific except in context.

Thus the evolution of messages seems to have been largely the addition of new layers of abstraction to an animal's repertoire, rather than a replacement of old methods with new. In higher social mammals what appear to be all steps in the evolution of communication may be used together in a single redundant system.

When we write of communication we concern ourselves with the process of behavioral information transfer. Thus we speak of data as "the heterogeneity of the environment," and information as "a difference that makes a difference" (Bateson 1966). Information is thus internal to the organism. Our definition of communication includes both intentional and unintentional levels. This breadth of definition allows us to look broadly at all contributory aspects of communication.

It is impossible to order the communication signals of animals into a single series representing stages of evolution, except that birds and especially mammals show more complexity and flexibilty than other animals and ultimately mammals develop language. As we will show, language is not a unique development but precursors for all aspects of it exist in the communication of other animals.

Insect, fish, and amphibian calls tend to be structurally similar to one another, though insects often use much higher frequencies than the lower vertebrates. This appears to result to a significant extent from ancient commonalities of the nervous system, and similar environmental constraints on the use of sound.

We begin our discussion by describing the features of sound as a medium for information exchange, then outline acoustic sensory evolution. We then trace acoustic communication in the various phylogenetic groups and briefly trace the origins of echolocation. Finally we describe the remarkable evolutionary parallelism in these events across the breadth of the higher vertebrates.

SOUND AS A MEDIUM FOR INFORMATION EXCHANGE

Information exchange in the living world occurs at all levels, from genetic instruction to an elaborate statement uttered by man. The five major senses may be divided into the contact senses (touch, taste, and smell) and what we call the teleportive senses (vision and hearing). All of the former require direct stimulus contact and thus have little potential for communication at a distance.

The other two senses are called teleportive because information can be exchanged over large distances by means of wave trains, without such direct contact. These modes also facilitate exchange between organisms simultaneously engaged in other activities.

Of the three contact senses, olfaction does indeed have distance capability. It is widely used by insects for finding mates, and by other animals for identification, for signaling temporary states (e.g., oestrus, injury of school-mates), and for marking territory (Mueller-Schwarze and Silverstein 1980). In social insects it is centrally involved in many aspects of social function. Even so such use is of a signaling or unitary nature that employs a lock-and-key mechanism involving only a relatively few molecules.

Olfaction does not lend itself to the transmission of long, complex messages. While odors may be persistent, they also tend to be wafted by winds. Some odors can remain in "patches" for long periods. But the lock-and-key nature of odor reception also allows animals to follow specific chemicals through a complex odor environment, as for example when homing salmon can find their spawning stream through pulp mill waste by tracking specific molecules (Hasler 1966), and considerable complexity in the use of odor for regulating social behavior is found in social insects.

Because of its distance capability, olfaction can be regarded as a bridge between the contact and the teleportive senses. Olfaction indicates a large external world "out there." The refined paired sense organs of the waveform-based teleportive senses allow much more refined sensory interpretation of external events. Both teleportive senses encourage internal comparisons of data from a larger external world, one ultimately large enough to contain the concept of self. When sounds are made they are heard and felt as if emanating from inside, and the internal generation of acoustic signals involves knowledge on the part of the sender that a specific message has been initiated, and hence involving a "self." A well developed sense of self is, we think, a prerequisite for the elaboration of language.

Both vision and hearing have huge potential for inter-organism communication. Both tend to deal with more complex, faster moving external events than do the contact senses. This fact is reflected in the intricate innervation serving these senses. Exquisitely elaborate front-end processors (retina and cochlea) send information to complex in-line processors such as the lateral geniculate or the cochlear nuclei.

Vision seems to involve the greatest instantaneous information processing load due to highly complex spatial analysis involved with such activities as the production of cognitive maps and search images. It probably taxed the electro-chemical possibilities of the CNS long ago. Much such processing is in the retina. In the human retina about 134 million rods and cones converge on one million nerves in the optic tract. In hearing where the bandwidth of processed signals may be much greater than in vision, there is a 1 to 2 expansion: 25 thousand hair cells diverge into 50 thousand auditory fibers in the auditory-vestibular tract (Uttal 1973).

Vision is definitely an input sense. One can interpret what one sees, but, except in special cases such as firefly courtship or the activities of biolum-inescent fish, one cannot visually modify external events with it. Although vision tends to reinforce focal consciousness strongly, this effect is minimal in comparison to the acoustic mode. By focal consciousness we mean that anything happening to an organism is appreciated as happening to it and not to something else. The acoustic channel is unique in possessing a pronounced two-way aspect and a potential for manipulation by the individual, hence it is the modality of choice for complex intentional communication. Even though visual signal systems abound (signs, gestures, body language, for example), they tend to be restricted to stereotyped signaling of modest complexity.

Desmond (1979) and Menzel (1986) have suggested that as the demand for complex, high speed information exchange increased, the "language of the eyes" became inadequate and the selection pressure toward development of the acoustic mode increased. We believe that the main forcing function for vertebrates was the trend toward complex learning-based vertebrate society.

The special advantages of acoustic information exchange are impressive. The earth's principal media (air, soil, rock, water) conduct or transmit sound rather well, night or day, through fog and through vegetation. There are problems, such as noise, interference, masking, and specular returns, but these are not unique to the acoustic modality. Animals can select frequency bands that are least absorbed by forest trees or that are relatively free of ambient noise. Sound is ephemeral (self-erasing) thus largely eliminating autocontamination (except for acoustic clutter) while facilitating high speed manipulation. The bandwidth is wide, permitting both a huge number of distinctive combinations as well as acoustic multiplexing (one can hear birds, wind rustling in trees and carry on a conversation simultaneously). All these properties nicely fit the temporal compression requirements set by short-term memory (STM). Sound is energy efficient. It can be used for communication without interfering with other bodily activities. Last, acoustic transmission provides a strong reinforcement of self because in higher vertebrates it typically involves the conscious production of signals based significantly or wholly on memory. Thus, sound is the prime candidate for abstract, complex, flexible and intentional information exchange systems.

THE REFINEMENT OF VERTEBRATE HEARING

The evolution of the vertebrate ear seems to have been a rather typical case of refining evolution. Old structures often used originally for other things became coopted into the building of complex middle and inner ears of high sensitivity and increasingly broad frequency range.

We recognize five structural grades in the evolution of vertebrate ears. By structural grades we mean general levels of structural advancement that may have been reached more than once, in different lineages and at different times.

We list these grades as follows. First came a mechanoreceptor sensitive only to near-field sound in the aquatic environment, as is found in early aquatic vertebrates. Second, it became provided with a transducer (probably a gas bubble) that allowed pick up and transduction of far-field pressure waves, as in fish with swim bladders or other gas spaces near the labyrinth (Popper 1974). Third, as the amphibian-reptile transition began an air ear developed. Fourth, as endothermy was achieved a sensitive air ear developed in both the mammalian and avian lineages, but built on quite different plans. And finally, the fifth grade occurred as some mammalian lineages reentered the water and the air ear was fitted again for aquatic hearing, sometimes with an additional ossicle.

This development of hearing was correlated with the development of extensive acoustic communication. But in the midst of this evolution lay the nearly silent reptilian grade in which acoustic communication seems to have been widely confounded by body temperature fluctuations in an air environment.

The First Grade

The anlage of the acoustico-lateralis system, whose structures underlie all vertebrate hearing, probably arose in invertebrate taxa. In today's animals, according to van Bergeijk (1966), structures that may be like a possible precursor include the echinoderm water canal receptors, and invertebrate statocysts. He calls the acoustico-lateralis system a "general-purpose hydrodynamic displacement detector." At any rate modern vertebrates contain sensory elements, the cupulae and hair cells with their stereo- and kinocilia like those of this ancient system.

The oldest and most unvarying portion of the higher vertebrate hearing system is the labyrinth, the fluid-filled organ of inertial navigation in vertebrates. In simplified form it is found in jawless vertebrates, as we see today in the hagfish Myxine as single canal and two cristae (van Bergeijk 1966), and then in lampreys as an organ of two canals, and three otoliths (de Burlet 1934). Still within the Paleozoic, the modern labyrinth of three canals appeared in fishes. It is capable of assessing the acceleration of the contained fluids in three orthogonal planes and with minor modification is found today throughout the higher vertebrates (Lowenstein 1957).

The Second Grade

The early labyrinth is usually buried deep in the tissue of the head where environmental sound could have been attenuated. The first true hearing was probably achieved by adventitious transduction of far-field sound (particle displacement pressure waves) in the region of the labyrinth. Van Bergeijk (1966) believes that this took place in a swim bladder whose primary adaptive role was either to allow aerial respiration or hydrostatic balance. Whichever was the case, it would have acted as a transducer for pressure waves, converting them into near-field displacement waves close enough to the labyrinth to produce sensation.

Lowenstein (1957) lists 53 teleost fish species that have been found to hear. In these contemporary forms the labyrinth and the ear are usually functionally separate as a small portion of labyrinth tissue has been coopted from the inertial guidance organ to specialize on environmental displacement reception.

That van Bergeijk's hypothesis is probably correct is shown in endocasts of the fossil rhipidistean crossopterygian, Eusthenopteron (Jarvik 1954), and by experimental work such as that of Popper (1974). Remarkably preserved specimens of Eusthenopteron, thought to be close to the lineage between fishes and the earliest amphibia, have allowed many soft anatomy structures to be outlined, including a probably physostomous air sac, probably used in aerial respiration. Thus the transducer of pressure to displacement waves was probably present as fish made their original transition onto land.

Two major groups have developed special hearing adaptation. The first is an assemblage of fish families, the Ostariophysi, that possess "Weberian Ossicles," or modified vertebral spines that connect the swim bladder with the sacculus of the labyrinth.

The second is the herring family Clupeidae, which has developed an independent adaptive complex in which long tube-like anterior branches of the swim bladder extend to the labyrinth and embrace it in bony bullae utilizing uniquely, the utriculus as a sensor (O'Connell 1955). Thus any system of cupulae, ossicles and fluid-filled canals can probably receive near-field sound providing that the displacement waves enter from the proper direction relative to the possible motion of the ossicle.

The Weberian Ossicles clearly have given the ostariophysans an extra wide hearing range among fishes, ranging from a low of about 16-25 Hz to 7 kHz (von Frisch and Stetter 1932) in the European minnow, Phoxinus, and more usually conferring an upper range of 2-3 kHz (Tavolga and Wodinsky 1963). More usual upper ranges for non-ostariophysan fishes are from about 600 Hz to 3 kHz (Lowenstein 1957).

The Third Grade

The transition to land involved compensating for the radically changed acoustic impedance relations of hearing in air as compared to water, and the development of the middle ear. Some insight into this transition can be gained by looking at the tadpoles and adults of modern frogs. The tadpoles of Rana, for example, are aquatic, while the adult is also capable of air hearing (Geisler et al, 1964). The labyrinth is present in the tadpoles and, during the massive reorganization of metamorphosis, is maintained intact while the middle ear is entirely revised. Witschi (1949) has shown that tadpoles hear using a bronchial columella, specialized bronchial membranes and lung vibrations, which changes during metamorphosis to the more familiar frog ear sensitive to airborne sound.

To be sure, the evolutions of the middle ear in both the ancient amphibians and reptiles were complicated adaptive radiations, in which the ear anatomy more closely fits the ecotypes of the animals involved than it does the taxonomy (Olson 1966). It involved emergence of the cochlea and the development of a "pressure difference receiver" in which the two ears are connected by an air passage. This condition occurs in frogs, reptiles and birds, and confers directional capability using relatively low frequency sound (Lewis 1983). Hearing of modern lizards is good over modest frequency ranges, and snakes may be exquisitely sensitive to earth-borne vibration. Certainly some frogs are remarkably sensitive to seismic signals as Narins and Lewis (1984) have shown. They found that certain frogs press their throat pouches against wet substrate, call, and others can pick up the seismic vibrations thus created as far away as 6 m. They calculate that the frog saccule and the mammalian cochlea have roughly equal sensitivities.

A major problem for these newly emergent vertebrates was probably that their body temperatures rose and fell with that of the thermally heterogeneous diurnal environment, affecting the reliability of both hearing and acoustic social signaling. As a reptile's body temperature drops the frequency range of hearing narrows, sensitivity reduces and is finally extinguished as the animal descends into torpidity (Werner 1983). Time-based frequency discrimination may be confounded. Other systems such as muscles, enzymes and vision are also involved. The functional animal is literally assembled and disassembled each day. Since the diurnal terrestrial environment of reptiles is usually highly thermally heterogeneous, social communication may also be confounded. At night, in shaded places such as forests, or in water, thermal uniformity is much greater.

It is not surprising, therefore, that social message transmission by acoustic signals is notably impoverished in most modern reptiles and probably in their ectothermic ancestors. Only where facultative or physiologic endothermy intervened as may have occurred in some ancient reptiles (Bakker 1972) or where the environment was relatively thermally uniform may hearing have become more reliable.

The lizard ear seems uncommonly prone to damage from intense sound. Bondello (1976) and Bondello et al (1979) have produced apparently Permanent deafness by subjecting lizards to vehicle noise from motor cycles and dune buggies. It may be that the fish ear is similarly unprotected since circumstantial evidence suggests that dolphins may use long-duration low frequency sound in the hearing range of fishes to debilitate prey (Marten et al, this volume).

The Fourth Grade

The middle ear of mammal-like reptiles showed a clear trend toward the mammalian condition as jaw musculature moved forward on the jaw. The old jaw joint was left without suspensory function and its reduced bones partly became the middle ear ossicular chain (Hopson 1966).

The ossicular chain. The transition from the reptilian to mammalian middle ear has been suggested as a process in which the new ossicles provided mechanical advantage not found in the reptilian grade, allowing increased sensitivity. But Manley (1972), and Werner and Wever (1972) concur that reptile and bird columellas are very efficient up to about 6-8 kHz. The hinging of the columella and extracolumella seems to provide considerable mechanical advantage in translating motions of the tympanum to that of the columella.

Such an effect is absent or modest in terrestrial mammals; a multiplier of between 1 and 2.6, depending upon frequency, was found by Wever and Lawrence (1954). They suggest that the majority of amplification appears to be produced hydraulically by the ratio of areas of the tympanum and the footplate of the stapes (ratios vary from about 18-26 in humans to over 40 in some other mammals).

The new mammalian ear included development of the external ear including a recessed tympanum and perhaps a pinna. The three bone ossicular chain and a coiled cochlea appeared (Stebbins 1980) both probably related to broadened frequency response and sensitivity of the ear.

Ecological factors. Cynodont reptiles, the lineage leading to the earliest mammals, included mouse-sized creatures, some remarkably rodentlike in jaw structure and dentition. An example is Bienotherium, (see Hopson 1966) an herbivorous Late Triassic tritylodont with a diastema and incisorlike anterior teeth.

Jerison (1973) has suggested that these mammalian ancestors might have been primarily nocturnal. If so, because of their small size they were probably also physiological endotherms or heterotherms. They could not have depended upon high thermal lag, as the large ruling reptiles may have done, to maintain operating temperature. If they were homeothermic these little animals probably must have had access to energy-rich food to compensate for the inevitable high heat loss at night. They probably also had fur or some other thermal protection. And acute hearing into higher frequencies would have been at a premium at night.

The late Triassic-early Jurassic may have been propitious for their evolutionary emergence. New energy-rich food sources may have been prevalent. The gymnosperm evolution was in full flood at that time, including the development of most modern genera of conifers (see Smith et al 1942). Gymnosperm seeds provide energy-rich food in small bite-sized packets usually broadcast in huge numbers.

The pinnae and ear canal. These features enhance directional hearing in an ear not equipped for detecting pressure differentials. The pinna funnels and amplifies sound like an ear trumpet and its complex folding produces patterns of reverberation.

Since directionality depends upon comparison between two acoustically separate ears, a number of acoustic isolation mechanisms develop in later mammals. In the aquatic environment the ears of cetaceans and pinnipeds become isolated from each other and from the skull by air spaces (Norris 1980).

The coiled cochlea. The coiling of the cochlea allows its lengthening in a confined space, and the extra length provides an extension of the frequency range of the animal. As the basal turn grows hearing may extend to higher and higher frequencies. Such extensions must have underlain the development of mammalian echolocation. A "cochlea" has developed in birds by convergence with the mammalian condition, from a branch of the labyrinth (Welty 1982) but the pressure differential ear was poorly adapted to high frequency hearing hence frequency range remains modest.

Protection of the Inner Ear. The insertion of the outer two ossicles into the mammalian middle ear conduction system probably ultimately allowed mammals to refine their hearing, in terms of frequency range. A key development seems to have been the evolution of protection for an increasingly broad band cochlea (Fleischer 1978), through the middle ear reflex and neural protection. The broad-band cochlea is subject to damaging sound over its entire range, which may face it with a much greater chance of damage than for a narrow-band cochlea.

The two suspensory muscles for the ossicular chain are involved in the mammalian "middle ear reflex" in which various kinds of disturbances, including loud sounds, cause the two muscles to contract, markedly reducing transmitted sound (see Wever and Lawrence 1954; 179). The second is inhibition of sound transmission at the nervous level (Arthur, Pfeiffer and Suga 1971).

Birds, which retain the columella and extracolumella have nevertheless become highly vocal animals. In them there is a single middle ear muscle (Young 1981), and a middle ear reflex (Oeckinghaus 1985).

The Fifth Grade

Reentry of mammals into the sea reimposed all the impedance matching problems that their ancestors had faced upon emergence onto land. Directional hearing was nearly lost. In fact little water borne sound penetrated the impedance barrier, and coupling between underwater sound production and propagation was also poor (Norris 1968). Recovery, for cetaceans and pinnipeds has been complete. These animals hear and propagate sound as well or better than their terrestrial counterparts.

Fleischer (1978) points out that cetaceans and sirenians have included the tympanic bone in the middle ear chain. In cetaceans high frequency sound reaches it via fatty channels of the lower jaw and throat (Norris 1968). In the bulla a topography of thickness and contour seems to act as a complex resonator in transmission of sound to the ossicular chain. The entire very heavy tympanic bone may move as a unit under the influence of very low frequency sound. This, according to Fleischer (1978), gives cetaceans "by far the greatest frequency range of hearing in nature for which their elaborate middle ear construction is prerequisite." Concurrent changes in the cochlea and brain have also occurred (McCormick et al 1970, Bullock and Ridgway 1972) allowing very high order acoustic performance (Herman 1980).

Because low frequency sounds have such long wavelengths relative to the receiving apparatus of an aquatic animal the emphasis is generally on higher frequency signals, sometimes wholly in the ultrasonic range. But many sonic and even subsonic sounds are used by marine mammals. There has been controversy about how the lowest of these could be received, given wavelengths considerably in excess of the animal's total length (see for example, Barham 1973). Recently Lipatov (1980) has performed experiments that suggest that the everted tympanum which lies buried in the blubber of mysticetes, and the ancient buried pinna of odontocetes may pick up such sound, as Fraser and Purves (1960) earlier suggested. The blubber coat of the animal may be the primary receiver while the two structures just mentioned, which penetrate it more or less at right angles to the axis of the animal, may cause vibrations of the entire bulla. This may account for the exceptionally low frequencies that some cetaceans use (to about 12.5 Hz for the blue whale; Cummings and Thompson 1971).

Thus in cetaceans reception points may have been split in two parts: one for low and one for higher sounds, and a new ossicle added to the chain. Many aspects of neural processing have been changed in the solution of problems presented by reentry into the sea, and by the evolution of echolocation.

LEVELS OF ANIMAL COMMUNICATION

Few fossils beyond endocasts of the nervous system relate to the evolution of communication systems (see Edinger 1964, Jerison 1976). Only general features, such as the remarkable general enlargement and restructuring of mammalian brains through the Cenozoic can be cited.

One exception is the work of Fleischer (1976) who was able to pinpoint the approximate time at which early odontocete cetaceans achieved high frequency hearing by preparing sections of fossil cochleas that allowed measurement of the width of the basilar membrane along its length.

But mostly one must infer history from the communication of the various living animal groups. This record shows only the broadest outlines of an evolutionary series. Fairly complex communication is found in insects which represent a complex radiation of acoustic communication in typically very small animals. Even though they rely upon sense organs that are basically different in plan from vertebrate ears their acoustic signals are reminiscent of those of lower vertebrates. It is only in birds and mammals that more advanced communication appears. It seems to differ mostly in terms of flexibility of message structure and in some birds and mammals, the complexity of messages and generality of their use. In the interest of space our review will be very brief.

Invertebrates

In the invertebrates acoustic communication is best developed in several groups of insects; especially the moths, lacewings, locusts, cicadas, and crickets. A number of structures serve as ears, and hearing ranges from acute vibration sensitivity to ultra-sound (Hutchings and Lewis 1983). Most sounds are made by males during sexual behavior. Most orthopteran signals are stereotyped rhythmic click series produced by stridulation that are often amplitude modulated. Pure tone signals are produced by some crickets using impacts of a stridulatory organ timed to match the resonance of an elytral resonator (Dumortier 1963, Alexander 1963, Elsner 1983).

Click trains may be simple series or they may be patterned signals in which the message is carried as a pulse pattern, or by the interspaces between clicks. In some insects such as some grasshoppers, two click series may be produced simultaneously by two hind legs, and the emission alternated to produce a single constant tone whose length is related to mate selection (Elsner 1983). Some insect sounds are very loud, such as the snapping sounds of the cicada made by popping the tymbal organs in and out (Elsner 1983), and some, such as the clicks of noctuid and arcteid moths may reach very high frequencies (see Stone and Fenton, this volume).

The complexity of bee communication, which in the hive is partly tactile, partly odor-mediated, and partly acoustic (Wenner 1968), has been shown to reach remarkable levels of complexity. This communication, which concerns such matters as food sources, directions and amounts, and details of nonfood-related behavior such as swarming, satisfies most criteria established to typify vertebrate communication. For instance the waggle dance includes such features as symbolism and displacement (see Griffin 1981 for a recent review). The main difference between bee communication and that of higher vertebrates seems to be greater flexibility in the communication of the latter.

The subtlety of bee communication is probably aided by communicants being at similar body temperatures in hives where the bees themselves warm the air. In the diurnal terrestrial environment outside the hive microclimatic effects will frequently bring such small ectotherms to different thermal states at any given time, even though they are spatially close to one another.

Fish

Fish signals are generally reminiscent of those of insects. Hawkins and Myrberg (1983) mention that about 50 families make some sound but those to which communicative functions can be attributed are much fewer (see Lowenstein 1957).

Most social signals are pulsed sounds produced by male stridulation or muscular excitation of the swim bladder (Fish 1972). Repetition rate may increase as courtship proceeds. In some the interpulse interval carries the message (Myrberg et al 1978). Communal calling by muscular drumming on swim bladders, especially in the families Sciaenidae and Haemulidae, seems to be related to social facilitation during courtship and spawning. These collective calls may be very loud and their very intensity may be stimulatory. Fish and Cummings (1972) have recorded such calls from the yellow-mouth corvina, Cynoscion xanthulus, in the Salton Sea, California, and said, "This increase in noise level is the greatest sustained effect ever reported from underwater natural causes." As is true in many insects, most signals seem quite stereotyped, and restricted in the context of their use.

Amphibia

Frogs and toads are, of course, often highly vocal. Their signals are mostly single or pulsed calls given by males, sometimes with modest levels of frequency modulation. They are usually functional in mating choruses (Blair 1968) or in territorial defense between males (Martof 1953), but defense sometimes involves females as well, for example in certain members of the Leptodactlyidae (Barrio 1963). There is a small number of stereotyped context-specific calls known, such as those used between males to discourage amplexus (Bogert 1960) and some low level female calls are also known. Typically the frequency range is modest, and matches the hearing range of the emitting animal. Once again, social synchrony is probably aided by frogs calling from the same pond, where uniform body temperature can be achieved.

Reptiles

Few reptiles are soniferous. Most lizards and essentially all snakes seem not to use airborn sound in communication, but rather to depend heavily upon visual cues and odor. Rattlesnakes, which vibrate rattles apparently as a warning to animals that might harm them, apparently do not signal each other with these sounds. Important exceptions are the tropical and subtropical lizard family Gekkonidae in which territorial barks are a feature of many species, and the crocodilians, a lineage near that of birds. Both sexes of these large reptiles, especially the alligator, Alligator mississippiensis, use a variety of vocal signals denoting aggression, defense of territory, and signals emitted by mothers guarding incubating sites and young (Campbell 1973, Garrick and Lang 1977). Note that these reptiles are not only large enough to damp out most body temperature fluctuation but live in an aquatic environment where uniform temperatures are the rule, and the geckos live primarily in the "homeostatic" tropical night. Whether snakes and other reptiles may communicate by substrate vibrations seems unassessed.

Birds

Thermal homeostasis, at least during activity periods, is the rule among birds, and in them vocalization has achieved structural complexity equivalent to or exceeding that of many mammals. Frequency modulation and complex harmonic structure have become prominent. Calls range from single signals, to group social facilitation calls, to patterned signals of various kinds. Lewis (1983) suggests that the frequency dependence of the pressure difference ear may restrict the frequencies that occur in bird song, in order that directional hearing may be maintained.

One complex call series is that of the yellow-throated vireo with a basic repertoire of eight different units (Smith et al 1978). These units can be modulated, produced in couplets or triplets or higher combinations giving a huge potential message carrying capacity, probably far larger than used by the bird. Even more complicated repertoires are mentioned in Kroodsma and Miller (1982).

Simple signals may be modulated with additional emotional information, just as is the case in many mammalian calls. Further, calling between birds is frequent and often complex and precisely timed, as in antiphonal singing (Thorpe 1973). This allows "negotiation" between animals to occur. Learning is frequently involved in modifications of bird calls (Baker et al 1981, Marler and Mundinger 1971, Nottebaum 1972). The calls of one bird may result in the modification of the answering calls of another bird (Kroodsma 1978).

Echolocation appears in birds (see Suthers and Hector, this volume) and because it involves precise assessments of the timing and frequency composition of pulses of sound (Griffin 1958), is probably dependent upon their thermal homeostasis. It is of modest refinement, however, because of the relatively low frequencies involved, probably restricted because of the avian pressure difference method of assessment of acoustic directionality.

The rise of social signaling in birds is probably also a correlate of the synchrony allowed by thermal homeostasis, and of their usually three dimensional habitat where communicants are frequently out of sight of one another, but where sound is a reliable medium. Although male vocalization predominates in territorial and aggressive calling by birds, assembly calls, social facilitation calls, and nurturant signals are emitted by both sexes in social species such as jungle fowl, starlings and pinyon jays.

That much cognitive complexity underlies these simple descriptions is shown by the work of Pepperberg (1986) who has been able to teach allospecific vocal patterns (human speech) to an African gray parrot. She says, "The bird now employs English vocalizations in order to request, refuse, identify, categorize, or quantify more than fifty items, including objects which vary somewhat from the training exemplars." Her success in training this sort of behavior in a parrot while others have failed is apparently dependent upon using objects and circumstances familiar to the bird in other contexts, and to trainers who talked about things they wanted the parrot to learn. Nonetheless the capabilities thus revealed are impressive in their complexity.

Mammals

The acoustic communication of mammals varies widely in complexity, which seems directly correlated with the emergence of complex learning-dependent social behavior in some groups, which, we think underlies human language. The parallels with birds are striking. Mammals (other than man) and bird signals deal mostly with context-related sounds, often modifiable by learning, based upon the modulation of a limited series of basic, rather invariant signals. Mimicry is a feature in both groups, and in dolphins it becomes extremely rapid (Richards, Wolz and Herman 1984). A capacity like this clearly must underlly human speech, since we order words in our minds from memory and immediately vocalize the result.

A few mammals chorus or "sing songs" like birds. The ways of life of these mammals are also like those of birds. Two mammal groups are involved; arboreal-aerial and aquatic species. Most other mammals are primarily terrestrial, and much subtle communication can be visually or odor mediated. Birds, and these few mammals living much of the time out of sight of one another in air, water, or in trees, rely heavily on the acoustic mode. The complicated many-minutes-long songs of humpback whales come to mind (Payne and McVay 1971), and the "songs" of the fruit bat, Sypsignathus (Fenton 1985). The dawn choruses of birds (Henwood and Fabrick 1979) are paralleled by mammals living in a similar habitat. Howler monkeys (Alouatta) living in dense treetop dwelling groups may chorus at dawn (Eisenberg 1981), and dolphin whistling rises to a crescendo at dawn and dusk (Powell 1966).

Many mammals use what we call metaphorical communication as their most advanced means. That is, signaling occurs as a metaphor of the communicated message. The animal acts out that which it communicates, or uses a symbol of an event at the time it occurs. Such messages when in the social context are frequently couched in terms of the contingencies of relationship. A subordinate wolf may bare its neck to the dominant animal as a symbol of subordinance. The mammal may act out that which it communicates or use symbols for events at the time they occur. The howling of wolves pursuing prey carries precise information about the exact state of the chase at any moment to a listener. The message is an analogue representation; that is, a proportionality regulates meaning.

The "pep rallies" of wild dogs, occurring before the hunt (Estes and Goddard 1967) may also be metaphors of the chase, but without the prey present. That is, the animals may "act out the chase" with increasing excitement as they twitter and sing to each other presumably to assure social cohesion for events ahead.

The primordea of culture may exist in these patterns; that is, the animals may be discussing events in the future and preparing for them. Such cultural primordea involving the increasing dominance of learned patterns within animal societies, is a feature of the communication of various social mammals such as primates, social carnivores and dolphins. Such signal+context calling is doubtless very old, since communications as diverse as frog and fish choruses and cricket calls also function because they are in context.

The discourse of some higher social mammals reaches a level that has been called "animal politics" (de Waal 1982). The import is that communication about relationships, between individuals in complex social groups such as those of chimpanzees, involves subtle levels of manipulation, especially between sexes (see also Trivers 1985).

A few quite precise symbolic calls have been identified in mammalian communication. The calls of vervets that are specific to certain kinds of predators seem clearly to be abstractions in no basic sense different from many human words (Seyfarth, Cheney and Marler 1980a,b), except that they seem to be only emitted in the immediate context of events. Another case is that of the sperm whale, which makes only clicks. The patterns of these clicks are frequently arranged in codas, or patterns that are exchanged between animals in "social situations" (Watkins and Schevill 1977). Different sperm whales appear to have their own "identity coda". Another such pattern, a five-click call of two separated clicks and three grouped clicks, is perhaps given in agonistic contexts. It has been heard from many whales and over several months' time (Watkins, Moore and Tyack 1985).

The evolutionary path taken in the development of words, then, might be envisioned first as the stripping away of the context of a metaphoric signal, bit by bit, until a symbol is left for a state, action or thing that can be used in an open-ended fashion. Then in an animal society with learned cultural elements these symbols could be used in the abstract discussion of future or past events.

Mammals that have been given language training, such as chimpanzees, orangutans, gorillas, sea lions and dolphins, certainly can respond to symbols representing various parts of speech (Savage-Rumbaugh et al 1985, Herman et al, 1984). The substrate of understanding "words" is certainly there for these animals, even if they seem not yet usually freed from the contexts of their own symbols.

One thing that strikes us from this review is that the convergence between mammals and birds with regard to the evolution of communication is profound. Rising from different points of origin in the almost silent reptiles both have achieved homeothermy and a protected middle ear. Both have developed extensive vocal communication, both have developed at least protosyntax, both use a relatively invariant collection of signals and both have come to use learned modulations of these signals. Finally, both have developed cognitive capabilities that allow language learning.

THE EMERGENCE OF ECHOLOCATION

At what stage in the evolution of communication did the capacities needed for well-developed echolocation come together? We list the following required capabilities:

* near-field hearing for feedback control of emission
* far-field hearing for analysis of incoming signals
* sharp onset signals and high speed processing ability
* good directionality of sound emission, and hence at least reasonably high frequency
* high acoustic sensitivity with overload protection
* stable time base for uniform analysis
* outgoing sound must not neurally suppress echo

Where in our evolutionary scheme do these requisites occur together? Click trains, as we have seen, are ubiquitous, and so must auditory feedback be. Hearing of far-field sound is present in all animals that use social phonation. Sound high enough to produce marked directionality appears in insects, birds and mammals, while vertebrate hearing of both high frequency range and sensitivity appears only with the evolution of birds and mammals.

The simplest level of echolocation, namely the use of echoes to assist in obstacle avoidance, seems accessible to any soniferous vertebrate or invertebrate with a stable body temperature, either from endothermy or from living in a thermally stable environment. Indeed, Tavolga (1976) reports acoustic obstacle avoidance in a sea catfish.

In the homiothermic birds the cave swiftlets and oil birds echolocate (Novick, 1959, Konishi and Knudsen 1979, and Suthers, this volume). Their pressure difference ears provide directionality using relatively low frequency sound and because this directionality is essential to an echolocating animal they are precluded from using high frequencies that might allow refined environmental assessment, such as is performed by bats and dolphins. Therefore it is not surprising that these bird systems appear to be modestly developed navigational aids.

Mammalian systems are most highly evolved in the two groups where need for refined systems is greatest; animals feeding in open space, and in darkness; the microchiropteran bats and the odontocete cetaceans. Other mammals, such as rats, tenrecs and shrews perform obstacle avoidance (Riley and Rosenzweig 1957, Gould et al 1964, Sewell 1970).

REFERENCES

Alexander, R. D., 1968, Arthropods, in Animal Communication: Techniques of Study and Results of Research, T. A. Sebeok (ed), Indiana Univ. Press, Bloomington, Indiana.

Baker, M. C., Spitler-Nabors, K. J., and Bradley, D.C., 1981, Early experience determines song dialect responsiveness of female sparrows, Science 214, 819.

Bakker, R., 1972, Anatomical and ecological evidence of endothermy in dinosaurs, Nature 238, 81.

Barham, E. G., 1973, Whales' repiratory volume as a possible resonator for 20 Hz signals, Nature 245, 220.

Barrio, A., 1963, Consideraciones sobre comportamiento y grito agresivo propio de algunas especies de Ceratophrynidae (Anura), Physis 24, 143.

Bateson, G., 1966, Problems in cetacean and other mammalian communication, in Whales, Dolphins and Porpoises, K. S. Norris (ed), Univ. Calif. Press, Berkeley, California.

Bekesy, G. von, 1962, The gap between the hearing of external and internal sounds, Symp. Society Exper. Biol. 16 Biological Receptor Systems, 267.

Bergeijk, W. A. van, 1966, The evolution of vertebrate hearing, Present Contributions to Sensory Physiology, vol. 2, 105.

Blair, W. F., 1968, Amphibians and reptiles, in Animal Communication, T. A. Sebeok, ed., Indiana Univ. Press, Bloomington, Indiana.

Bogert, C. M., 1960, The influence of sound on the behavior of amphibians and reptiles, in Animal Sounds and Communication, W. E. Lanyon and W. N. Tavolga (eds), American Inst. Biol. Sci., publ 7.

Bondello, M. C., 1976, The effects of high-intensity motorcycle sounds on the acoustical sensitivity of the desert iguana, Dipsosaurus dorsalis, Master's Thesis, California State Univ., Fullerton, California.

Bondello, M. C., Huntley, A. C., Cohen, H. B., and Brattstrom, B., 1979, Pt. II, The effects of dune buggy sounds on the telencephalic auditory evoked response in the Mojave fringe-toed lizard, Uma scoparia. Final Rept., Bureau of Land Management, Contract CA-060-CT7-2737.

Bullock, T. H., and Ridgway, S. B., 1972, Evoked potentials in the central auditory system of alert porpoises to their own and artificial sounds, Jour. Neurobio. 3:11, 79.

de Burlet, H. M. 1934, Vergleichende anatomie des statoacustisohen organs, in Handbuch der Vergleichenden Anatomie der Wirbeltiere, BolkGoppert-Kallius-Lubosch, eds., vol 2, Urban and Schwarzenberg, Berlin.

Campbell, H. W., 1973, Observations on the acoustic behavior of crocodilians, Zoologica, Spring, 1-11.

Cowles, R. B., and Bogert, C. M., 1944, A preliminary study of the thermal requirements of desert reptiles, Bull. Amer. Mus. Nat. Hist. 83, 265.

Cummings, W. C., and Thompson, P. O., 1971, Underwater sounds from the blue whale, Balaenoptera musculus, Jour. Acoust. Soc. America 50:4, 1193.

Desmond, A., 1979, The Ape's Reflexion, Blond and Briggs, Ltd., London.

Dumortier, B., 1963, The physical characteristics of sound emissions in Arthropoda, in The Acoustic Behavior of Animals, R-G Busnel, (ed), Elsevier, Amsterdam.

Edinger, T., 1964, Recent advances in paleoneurology. Progress in Brain Research, W. Bargmann and J. P. Schade (eds), 147.

Eisenberg, J. F., 1981, The Mammalian Radiations: An Analysis of Trends in Evolution, Adaptation, and Behavior. Univ. Chicago Press, Chicago.

Elsner, N., 1983, Insect stridulation and its neurophysiological basis, in Bioacoustics: A Comparative Approach, B. Lewis (ed), 69-92.

Estes, R. D., and Goddard, J., 1967, Prey selection and hunting behavior of the African wild dog, Jour. Wildlf. Mgt. 31, 52.

Fenton, M. B., 1985. Communication in the Chiroptera. Indiana Univ. Press, Bloomington, Indiana.

Fish, J. F., and Cummings, W. C., 1972, A 50 dB increase in sustained ambient noise from fish (Cynoscion xanthulus), Jour. Acoust. Soc. America 52:4, 1266.

Fleischer, G., 1976, Hearing in extinct cetaceans as determined by cochlear structure, Jour. Paleont. 50:1, 133.

Fleischer, G., 1978, Evolutionary principles of the mammalian middle ear, Adv. Anat. Embryol. Cell Bio. 55:5, 1.

Fraser, F. C., and Purves, P. E., 1960, Hearing in cetaceans: Evolution of accessory air sacs and the structure and function of the outer and middle ear in recent cetaceans, Bull. British Mus. (Natural History) 7:1, 1.

von Frisch, K., and Stetter, H., 1932, Untersuchungen uber den Sitz des Gehorensinnes bei der Elritze. Z. vergl. Physiol. 17, 686.

Garrick, L. D., and Lang, J. W., 1977, Vocal signals and behaviors of adult alligators and crocodiles, Amer. Zool. 17, 225.

Geisler, C. D., van Bergeijk, W. D., and Frishkopf, L. S., 1964, The inner ear of the bullfrog, Jour. Morph. 114, 43.

Gould, E., Negus N. C., and Novick, A., 1964, Evidence of echolocation in shrews, Jour. Exp. Zool. 156:1, 19.

Griffin, D. R., 1958, Listening in the Dark, Yale Univ. Press, New Haven, Conn.

Griffin, D. R., 1981, The Question of Animal Awareness, in Evolutionary Continuity of Mental Experience, Rockefeller Univ. Press, New York.

Hasler, A. D., 1966, Underwater guideposts; homing of salmon. Univ. Wisconsin Press, Madison.

Hawkins, A. W., and Myrberg, A. A. Jr., 1983, Hearing and sound communication under water, in Bioacoustics: A Comparative Approach, B. Lewis, ed., Acad. Press, New York.

Henwood, K., and Fabrick, A., 1979, A quantitative analysis of the dawn chorus: temporal selection for communicatory optimalization, Amer. Nat. 114:2, 260.

Herman, L. M., 1980, Cognitive characteristics of dolphins, in: Cetacean Behavior: Mechanisms and Functions, L. Herman (ed), John Wiley and Sons, New York.

Herman, L. M., Richards, D. G., and Wolz, J. P., 1984, Comprehension of sentences by bottlenosed dolphins, Cognition 16, 129.

Hopson, J. A., 1966, The origin of the mammalian middle ear, Amer. Zool. 6:3, 437-450.

Jarvik, E., 1954, On the visceral skeleton in Eusthenopteron, with a discussion of the parasphenoid and palatoquadrate in fishes. Kungl. Svensk. Vetensk, Handl. 5, 1.

Jerison, H. J., 1976, Paleoneurology and the evolution of mind, Scient. Amer. 234:1, 90.

Konishi, M., and Knudsen, E. I., 1979, The oilbird: hearing and echolocation, Science 204, 425.

Kroodsma, D. E., 1978, Aspects of learning in the ontogeny of bird song: where, from whom, when, how many, which, and how accurately?, in The Development of Behavior: Comparative and Evolutionary Aspects, Garland STPM Press, New York.

Kroodsma, D. E., and E. H. Miller, 1982, Acoustic Communication in Birds, vol. 2, Acad. Press., New York.

Lewis, B. 1983. Directional cues for auditory localization. in: Bioacoustics; a comparative approach, Acad. Press., New York.

Lipatov, N. V., 1980, Underwater sound: peculiarities and bioacoustical modelling, in Animal Cultures and a General Theory of Cultural Evolution, P. C. Mundinger (ed), Ethol. Sociobiol. 1, 183.

Lowenstein O., 1957, The sense organs: acoustico-lateralis system, in The Physiology of Fishes, M. Brown (ed), vol. 2, Acad. Press, New York.

Manley, G. A., 1972, A review of some current concepts of the functional evolution of the ear in terrestrial vertebrates, Evolution 26:4, 608.

Marler, P., and Mundinger, P., 1971, Vocal learning in birds, in Ontogeny of Vertebrate Behavior, Academic Press, New York.

Marten, K., Marler, P., and Quine, D., 1977, Sound transmission and its significance for animal vocalization: II. tropical forest habitats, Behav. Ecol. Sociobiol. 2, 291.

Martof, B. S., 1953, Territoriality in the green frog, Rana clamitans, Ecology 24, 165.

Menzel, E., 1986, How can you tell if an animal is intelligent?, in Dolphin Cognition and Behavior: A Comparative Approach, R. J. Schusterman, J. A. Thomas and F. G. Wood (eds), L. Erlbaum Assoc., Hillsdale, New Jersey.

Minsky, M., 1985, Communication with an alien intelligence, Byte 10:4, 127-138.

Muller-Schwarze, D, and Silverstein, R. M. (eds), 1980, Chemical Signals: Vertebrates and Aquatic Invertebrates, Plenum Press, New York.

Myrberg, A. A., Jr., Spanier, E., and Ha, S.J., 1978, Temporal patterning in acoustic communication, in Contrasts in Behavior, E. S. Reese and F. J. Lighter (eds), John Wiley and Sons, New York.

McCormick, J. O., Wever,E.G., Palin, J., and Ridgway, S.H., 1970, Sound conduction in the dolphin ear, Jour. Acoust. Soc. America 48:6, 1418.

Narins, P. M., and Lewis, E. R., 1984, The vertebrate ear as an exquisite seismic sensor, J. Acoust. Soc. Amer. 76:5, 1384.

Norris, K. S., 1968, The evolution of acoustic mechanisms in odontocete cetaceans, in Evolution and Environment, E. Drake (ed), Yale Univ. Press, New Haven, Conn.

Norris, K. S., 1980, Peripheral sound processing in odontocetes, in Animal Sonar Systems, R-G Busnel and J. F. Fish (eds), Plenum Press, New York.

Nottebaum, F., 1972, Neural lateralization of vocal control in a passerine bird, II. Subsong, calls and a theory of vocal learning, Jour. Exper. Zool. 179:1, 35.

Novick, A. 1959, Acoustic orientation in the cave swiftlet. Biol. Bull. 117, 497.

O'Connell, C. P., 1955, The gas bladder and its relation to the inner ear in Sardinops caerulea and Engraulis mordax. Fish. Bull. 104:56, 505.

Oeckinghaus, H. 1985, Modulation of activity in starling cochlear ganglion units by middle ear muscle contractions, perilymph movements and lagena stimuli, Jour. Comp. Physiol., A, 157, 643.

Olson, E. C., 1966, The middle ear-morphological types in amphibians and reptiles, Amer. Zool. 6:3, 399.

Payne, R. S., and McVay, S., 1971, Songs of humpback whales, Science 173:3997, 585.

Pepperberg, I. M., 1986, Acquisition of anomalous communicatory systems: implications for studies of interspecies communication, in Dolphin Cognition and Behavior: A Comparative Approach, R. J. Schusterman, J. A. Thomas and F. G. Wood (eds), L. Erlbaum Assoc., Hillsdale, New Jersey.

Popper, A. N., 1974, The response of the swim bladder of the goldfish (Carassius auratus) to acoustic stimuli, Jour. Exper. Biol. 60, 295.

Powell, B. A., 1966, Periodicity of vocal activity of captive Atlantic bottlenose dolphins: Tursiops truncatus, Bull. So. California Acad. Sci., G5 4, 237.

Richards, D. G., Wolz, J. P., and Herman, L. M., 1984, Vocal mimicry of computer-generated sounds and vocal labeling of objects by a bottlenosed dolphin, Tursiops truncatus, Jour. Comp. Psych. 94:1, 10.

Riley, D. A. and M. Rozenzweig, 1957, Echolocation in rats, Jour. Comp. Physiol. Psych. 50, 323.

Savage-Rumbaugh, E. S., Sevick, R. A., Rumbaugh, D. M., and Rubert, E., 1985, The capacity of animals to acquire language: do species differences have anything to say to us?, Phil. Trans. Roy. Soc. London B308, 177.

Seyfarth, R. M., Cheney, D. L., and Marler, P., 1980a, Monkey responses to three different alarm calls: evidence of predator classification and semantic communication, Science 210, 801.

Seyfarth, R. M., Cheney, D. L., and Marler, P., 1980b, Vervet monkey alarm calls: semantic communication in a free-ranging primate Animal Behav. 28, 1070.

Smith, G. M., Gilbert, E. M., Evans, R. I., Duggar, B. M., Bryan, G. S., and Allen, C. E., 1942, A textbook of general botany, Macmillan Co., New York.

Smith, W. J., Pawlukiewicz, J., and Smith, S. T., 1978, Kinds of activities correlated with singing patterns of the yellow-throated vireo, Animal Behav. 26, 862.

Stebbins, W. C., 1980, The evolution of hearing in mammals, in Comparative Studies of Hearing in Vertebrates, A. N. Popper and R. R. Fay (eds), Springer-Verlag, New York.

Tavolga, W. N., and J. Wodinsky, 1963, Auditory capacities in fishes: pure tone thresholds in nine species of marine teleosts, Bull. Amer. Mus. Nat. Hist. 126, 179.

Tavolga, W. N., 1976, Acoustic obstacle avoidance in the sea catfish (Arius felis), in Sound reception in fish, 185-203, A. Shuif and A. D. Hawkins (eds), Elsevier, Amsterdam.

Thorpe, W. H., 1973, Duet-singing birds, Sci. Amer. 229, 70.

Trivers, R., 1985, Social Evolution, Benjamin/Cummings Publ., Menlo Park, California.

Uttal, W. R., 1978, The Psychology of Mind, L. Erlbaum Assoc, Hillsdale, New Jersey.

de Waal, F. B. M., 1982, Chimpanzee Politics: Power and Sex among the Apes, Harper and Row, New York.

Watkins, W. A., and Schevill, W. E., 1977, Sperm whale codas, Jour. Acoust. Soc. America 62:6, 1485.

Watkins, W. A., Moore, K. E., and Tyack, P., 1985, Sperm whale acoustic behaviors in the southeast Caribbean, Cetology 49, 1.

Welty, J. C., 1982, The Life of Birds, 3rd ed., Saunders College Publ, Philadelphia.

Wenner, A. M., 1968, Honey bees, in Animal Communication: Techniques of Study and Results of Research, T. A. Sebeok (ed), Indiana Univ. Press, Bloomington, Indiana.

Werner, Y. L., 1983, Temperature effects on cochlear function in reptiles: a personal review incorporating new data, in Hearing and Other Senses: Presentations in Honor of E. G. Wever, Fay, R. H., and Gourevitch, G. (eds), Amphora Press, Conn.

Werner, Y. L., and Wever, E. G., 1972, The function of the middle ear in lizards: Gekko gecko and Eublepharis macularius (Gekkonoidea), Jour. Exp. Zool. 179, 1.

Wever, E. G., and Lawrence, M., 1954. Physiological Acoustics, Princeton Univ. Press, Princeton, New Jersey.

Witschi, E., 1949, The larval ear of the frog and its transformation during metamorphosis, Z. Naturforsch. 4b, 230.

Young, J. Z., 1981, The Life of Vertebrates, 3rd Ed., Clarendon Press, Oxford.

ON THE EVOLUTION OF ACOUSTIC COMMUNICATION SYSTEMS IN VERTEBRATES

Part II: Cognitive Aspects

Evan C. Evans III and Kenneth S. Norris

Naval Ocean Systems Center, Kaneohe, Hawaii
Joseph M. Long Marine Laboratory, University of California
Santa Cruz, California

INTRODUCTION

This second paper concentrates on the cognitive aspects of acoustic communication systems. It contrasts sharply with the history of acoustic sensory and communication systems just outlined in that cognitive processing seems generally designed to avoid information overload. Also, because most cognitive analysis depends on behavioral observations of living animals, a history of development is difficult to outline.

This paper amplifies our point made briefly in the first paper that the central nervous system (CNS) seems to have approached a ceiling in information processing speed early in its history. This limitation forced the development of other means of providing better access to environmental information. It also amplifies Miller's (1956) comment: "In order to survive in a constantly fluctuating world it is better to have a little information about a lot of things than to have a lot of information about a small segment of the environment"

DATA REJECTION AND LIMITS OF THE SYSTEM

The first and simplest response to the vast excess of data contained in the environment is filtering. Animals attend to only small fractions of the data available. This major reduction is accomplished in two ways, first through limiting peripheral sensors to relatively narrow bands of stimuli (simple filtering), and second by focusing attention on data of immediate interest (controlled filtering). Simple filtering is under genetic control as opposed to cognitive control. The process is of little further interest except for one underappreciated fact. Although it is widely understood that the global mosaic of mutually interdependent ecosystems could not exist without extensive commonality in biochemical composition and processing, the fact that a similar commonality must exist for peripheral sensors in general is less often noted. To achieve the remarkable global coordination so often taken for granted, all must see, hear, feel, taste, and smell in roughly the same narrow bands. Were this not true, communication of any sort between or within species would be difficult. Indeed, such difficulty is employed in many predator-prey situations where the prey communicates in frequencies (or other modalities) to which the predator is minimally sensitive.

What is least appreciated is that biochemical commonality interacts strongly with data-band commonality. Most organisms must perform whatever information processing they do within the well known electrochemical constraints of living nervous systems. That is for fast action, they are limited by the go/no-go dichotomy of nervous action potentials, by transmission speeds between 0.1 and 100 m/sec, with dead times set by 1 msec spike durations followed by up to 10 msec recovery times, and by a firing frequency ceiling of about 1 Hz (Uttal

1973; Kandel and Schwartz 1985). Learning, thought, memory do involve slower graded potentials and network plasticity, but these too have definite biochemical and bioelectrical limits. These simple facts underly our earlier statement that the nervous system approached neurophysiological ceiling long ago in evolutionary time.

Of greater cognitive interest are the other forms of filtering that are under at least the partial control of the animal. These include attention (mentioned above), forgetting, and various forms of generalization such as feature detection, chunking, and metachunking. We realize that feature detection is considered preattentive (Neisser 1967; Treisman 1986). It is here included among things of cognitive interest because it has been shown (Treisman 1986) to determine neural coding dispatched to higher levels in the CNS for further processing. The activity, therefore, differs significantly from the simple filtering mentioned above, which prevents data from entering peripheral sensors in the first place. Instances of feature detection include nervous transforms designed to enhance contours, detect moving spots, integrate general luminous intensity (Lettvin et al 1959) or that sharpen modal frequencies (Suga et al 1976). Most such generalizations are achieved through lateral inhibition, often influenced by efferent signals from higher nervous centers. Chunking and metachunking are discussed later.

There is, however, another important information-limiting mechanism that acts like a filter but is, in reality, a limitation of the organism itself. This is short-term memory (STM) or short-term store (Atkinson and Shiffrin 1968; Broadbent 1958). Short-term memory has variously been referred to as the brain's working buffer or scratch pad. The literature and opinions regarding STM are great. Fortunately, only two relatively uncontroversial aspects of STM are relevant to our discussion. First, the time constant (time for the content of STM to decay to 1/e of its initial value) is short, 100 to 150 milliseconds (Uttal 1978; Norman 1982). Second, the capacity of STM is small, generally estimated at 7 plus or minus 2 chunks (Miller 1956; Simon 1974, 1985). Chunking is, as said above, a form of generalization, but it is best understood as the procedural trick of collecting a number of items under a single supercategory (e.g. carrots, potatoes, beets, etc. under vegetables) or a number of entities (roots, leaves, trunk, branches or redwoods, oaks, sycamores, firs) under a single concept (trees). The exact number of items in a chunk and the exact capacity of STM are unimportant. What is important is the relative smallness of that capacity in the light of what it "should" be, given the information overload any animal regularly encounters.

We suggest that this curious smallness represents a compromise (optimized by the evolutionary process) between admitting larger amounts of data for processing and protecting a complex nervous system from overload. The smallness of STM limits the size of both externally presented and internally proposed problems for solution. Although the properties of STM undoubtedly varies between species, its capacity is uniformly small and its time constant is uniformly short. All sensory input is thought to be relayed to STM, where it must be recognized as important (attention) and passed on to other centers for further processing, or ignored and within a short time forgotten or overwritten. Short-term memory, therefore, operates as a necessary protective bottleneck in the nervous system. Both STM and the higher brain's primary function of prediction (discussed later) require rapid initial processing (acquisition, partial analysis, and reallocation). This fact has important implications for communication.

AVAILABLE "TRICKS" AND AIDS

Although slower information processing speeds are unquestionably important, frequently encoutered survival situations, then as now, continued to "reward" rapid and accurate response. The most assured rewards came from prediction of events 200 to 500 msecs into the future (Craik 1943; Dushkin 1970; Gevins et al 1983) and the issuance of motor commands producing action reasonably compatible with those predictions, all from a "snap shot" of the immediate situation. This predictive activity may be thought of as automated motor satisficing (as defined in our first paper), allowing the animal to avoid crashing into things or biting itself without engaging high-level information processing. A principal evolutionary strategy was, and is, fine-tuning the accuracy and extending the temporal range of this basic prediction ability.

672

We suggest, therefore, that the brain is primarily a predicting machine and only secondarily a thinking machine. The larger implications of our point are beyond the scope of this discussion. Its importance here is that both the limitations necessarily set by short-term memory (STM) and the brain's primary activity of prediction place severe time boundaries on perception in general.

Household Economy

The only routes for further improvement lay in exquisite household economy, in innovative neural "wiring" and in the perfection of as large a repertoire of procedural "tricks" as possible. In the matter of household economy, the multiple use of existing structures evolved for other purposes is obvious (lungs, nasal cavities as resonating or amplifying devices, for example). Innovative "wiring" includes such things as positive and negative feedback, or parallel processing in three-dimensional, self-modifying networks. These techniques, possibly not in their full neural magnificence, are well known to electrical engineers. Procedural "tricks," such as feature detection, mentioned earlier, are largely achieved by lateral inhibition, an ingenious spatial arrangement of negative feedback (Bullock, Orkand and Grinnell 1977). These are examples of "wiring" sophistication. We wish, however, to emphasize other procedural "tricks" that usually receive less attention.

Basic to the household economy just mentioned is the fact that there is a vast amount of internal communication going on continuously within any organism, from exchanges between cells and the discourse of muscles to the myriad electrochemical activities of the nervous system. Much of this internal communication is kept from chaos by a number of CNS rhythms (alpha, beta, delta, theta), by other "household metronomes" such as reflex arcs (Sherrington 1906) or oscillators (von Holst 1939), and by external circadian or annual rhythms.

Rhythmicity

The basic patterns or rhythms of neural control are often reflected in behavior, especially locomotion and communication. The same is true of music, a rhythmic and often emotion-laden form of communication. In this regard, it is worth noting that humans who stutter can sing words clearly whereas they cannot speak them (Van Riper 1971). The pronounced external meter of the song apparently can substitute for an internal meter that has been lost. Simmons and Baltaxe (1975) have proposed a regulatory mechanism involving rhythmic and prosodic perception as a basis for encoding and decoding language.

The same rhythmic neural processing involved in locomotion will serve for most animal communication. The footfalls of an quadruped repeat with metronomic regularity, and between their beat other more subtle events of coordination are carried out. The zebra veers to avoid the lion. In the same way many animal communications are rhythmic and are modulated with secondary messages between the beat. Both rhythm and locomotion are controlled to a significant degree by reflex arcs at the level of the spinal column. The zebra's heartbeats, its breathing too, are keyed into the rhythm of its hoofbeats, as is its emotional state. Less obvious, however, is the fact that the information processing the zebra is using to predict the lion's next act is also keyed to that same rhythm. The animal, its predictions, its actions, its emotions are a single unit under rhythmic control, an example of what we would call "animal jazz." The zebra-lion event in no fundamental way differs from musicians "jamming" together. Both are units under the automated control of rhythm that not only coordinates syncopated interaction but also carries heavy emotional freight. Both draw from a very ancient and common source. The great economy of rhythmicity is seen in movement, in information processing, in communication.

At this juncture, rhythmicity is simply listed as a load-splitting "trick" that redirects routine information processing loads to automated centers thus freeing more sophisticated areas for complex processing. It is worth noting here that strikingly rhythmic acoustic exchanges between insects, amphibians, birds and animals are widely observed. Whether these exchanges are related to language or music is an important question that is entertained later when we consider echolocation and the development of language. It is also noteworthy that locomotion, breathing and other bodily actions have the important effect of breaking up acoustic utterances into rhythmic subunits, which later may be related to the origins of syllables, words, or phrases.

Supramodal Identity

Other economies serving the fundamental need for reducing the information processing load should be mentioned. Of particular importance is the matter of supramodal identity (Geschwind 1965; Davenport 1977). A single object, say an animal, possesses a characteristic set of aspects, sounds, and smells. If approached closely, such a set may include touch and taste as well. A great economy is achieved if all these sensory reports can be lumped as attributes of a single object, as opposed to logging each separately and then determining singularity through crossmodal comparisons. The intersection of modes does not, of course, occur at the peripheral sensors. At some point in the central nervous system a cross-modal "name," or what we prefer to call a supramodal identity, may occur. This mental entity seems to us to be an excellent candidate for the proto-word, and, since the object's attributes are lumped, for proto-chunking.

Internalization

One final economy, or "trick," needs special mention, and that is internalization (Shepard 1984; Gibson 1979). Internalization represents the excellent economy of establishing in an animal's nervous system more or less automatic analysis-action routines matched to frequently recurring external events. When such events happen, they need not be processed as novel occurrences using the full complement of nervous processing capability, but rather assigned to an automated subroutine producing reasonably compatible reactions. For example, long ago all animals internalized the global day-night cycle. They do not waste nervous system commitment determining whether or not it is time to go to sleep or wake up. Internalization, then, is another load-distribution trick.

Since internalization most often involves the repeated observance of event sequences in the external world, the most usual result is an ingrained concept of causality, that is, that a reaction follows an action, not the reverse. Thus, the concept of linear time sequences tends to be internalized. This circumstance, we submit, is a step in the development of proto-syntax. When and if more complex sequences of external events are internalized, the basic linear notion of causality is enhanced to include the sense that different arrangements of the same units can produce significantly different results. This is an important second step in the development of proto-syntax. The linear nature of acoustic communication reinforces both stages of proto-syntax development.

COMMUNICATION HIERARCHIES

Hierarchical organization is one of the simplest effective economies. Such organization may be seen in nervous information processing and in animal communication. Communication hierarchies probably proceeded from simple signals through various modifications or combinations of the basic set to enhance message detail. The most complicated exchanges seem to reside in birds and mammals.

Smith (1986), working with birds and mammals, describes several such hierarchies: "The basis of formalized communication in any species is a repertoire of signaling acts: vocalizations, postures and movements, nudges, ways of depositing scents, and so on." He notes that additional information may be carried by a range of variation or modulation of these basic signal acts and by interactive signaling rituals between two or more animals (see also Hailman 1977).

The basis of all Smith's hierarchies is the "signal act" repertoire. Signal options for a given species may be quite limited and are often correlated with the urgency of their use. Calls that are used in attack, escape, copulation, or some fundamental social act tend to be stereotyped; others with less trenchant function tend to vary more widely.

Smith notes that no species seems to have a very large signal act repertoire, by conservative estimate no more than a total of 40 to 45 signal acts (Moynihan 1970; Smith 1969, 1986). We note that the International Phonetic Alphabet contains only 90 phonemes and that most human languages use less than half of these (Berlitz 1982). Others have noted that the context of such signals can frequently be read across species boundaries (Morton 1977). One of us has found this to be true for the spinner dolphin (Norris et al 1985) where the general context of signals related to aggression, intimate social contact, or fear can be understood easily by a human. The value of this basic invariance is two-fold. It allows the animal to define the domain of the message (court-ship, aggression, warning) and also makes possible the recognition of modula-tions within that domain as more refined information. Invariance allows basic messages and to some extent their modulations to be transmitted across species boundaries.

Smith's second repertoire is that of "variations of signal form." These variations may be characteristic to an entire species, or they may be restricted to subpopulations in which case they are called "dialects." We would say that the basic signal act is modulated to carry additional information. Smith notes that these second level variations or modulations are again widely recognized across species boundaries.

To give an example, Smith notes that the "zit call" of the kingbird, a basic unit in several kinds of aerial maneuvering, including attack, is modu-lated by a quaver as the bird suddenly changes its flight path when swooping on an intruder. This same quavering modulation is applied to its "T-zee call", another basic unit uttered in "loose" flying about or as it flies to a perch, but again the quaver is associated with a change in flight path. Smith suggests that this modulation signals "imminent likelihood of quitting some current activity." Quavering variants occur and apparently are similarly understood in many species of both passerine and non-passerine birds.

Smith's third repertoire consists of "patterned combinations of signal units," which may be either basic or modulated and, in addition, may obey set combinatorial rules. For instance, prairie dogs convey information by regula-ting the intervals between their barks. Intervals of equal duration signal sustained attentive behavior. Shortened intervals appear to presage cessation of vigilance and withdrawal into the burrow, while irregularly lengthened inter-vals indicate the probablility of discontinuance of vigilance in favor of some behavior other than withdrawal.

Smith's fourth repertoire is that of formalized interaction between two or more animals. These can be short ritual greetings, or long complexly structured events. Such extended transactions contain a regular sequence of units, imply-ing rules or frameworks, which Smith calls "nonhuman grammar." Such formalisms permit shared control of an event, as in human dialogue. They facilitate coop-eration by "defining the contract;" i.e. they make interaction more predictable and thus undergird sociality. We note the interesting parallel with Bruner's approach to cognition in which long sequences of events called "strategies" are examined (Bruner et al 1956).

Bain (in press) has recently pursued such an approach in the analysis of killer whale vocalizations. Bain's work draws heavily from the hierarchical analysis of visual information in the retina by Marr (1982). The basis of Bain's hierarchical analysis is sound intensity, a parameter from which informa-tion can be extracted. He speculates that higher levels of information may be incorporated through regular variation of intensity to produce a frequency, then through varying frequencies, and lastly through patterns of frequency varia-tions. Bain uses a series of recursive Gabor transforms and hierarchical clus-ter analysis to study killer whale calls. For his initial runs, he feels that he can successfully classify at least 80% of their calls. We consider the parallism between Smith's behavior observations and Bain's recursive analytical techniques significant. Not only is the hierarchical sequence of transforms seen in these two approaches, but also it may be seen in the transformation of rhythmic motor sequences into particular locomotion and in Marr's transformation of a 2-D feature map into 3-D visual perception.

COGNITIVE ASPECTS OF ANIMAL COMMUNICATION

Language and echolocation are two of the most sophisticated animal uses of audition known. Some comparisons are of value here. In both echolocation and language the outgoing signal is in part controlled by instantaneous feedback which is largely near-field sound (displacement and radial propagation). The returned message is far-field sound (pressure and straight-line propagation). Thus in echolocation the analysis must involve the comparison of a near-field acoustic "image" with its reflected far-field aspect. This is not the case for language. Humans monitor their utterances in the near-field while interpreting far-field returns from an external source. The speaker listens to output in the control mode but not as a comparison with the resultant returned message. Thus, while echolocation and language both involve patterned acoustic strings and their analysis, they are subtly different.

There are other differences. Directionality of echolocation is much greater. The directionality index (DI) for <u>Tursiops</u> is about 18 db (Au and Floyd 1984), versus an estimated less than 3 db for human speech. In echolocation, narrow-beam, high-intensity ensonification is required for effective, energy-efficient detection at a distance. High repetition rates and associated high-speed nervous processing are also required. Repetition rates of 736 clicks/sec have been observed for <u>Phocoena phocoena</u> (Busnel and Dziedzic 1966) compared to an estimated maximum of 18 phonemes/sec for champion fast talker John Moschitta speaking at 552 words/minute (Honolulu Star Bulletin 1983; Miller 1966). Thus, directionality, repetition rate, and processing speeds are at least an order of magnitude greater in echolocation than in linguistic communication.

Echolocation is typically closely tied to locomotion whereas in linguistic communication the rhythmic roots derived from locomotion are more subtle and it has evolved into a more or less static activity. An echolocating animal emits clicks with great precision, placing them carefully with regard to returned echoes. Such control is much less evident in speech.

Both echolocation and language may require very rapid modification or formation of signals during the life of a message string. The dolphin may change the character of its clicks within a click train in response to the targets it inspects (Penner, this volume). In a somewhat similar manner, humans shape words from memory while uttering a sentence. Dolphins, too, can mimic with great rapidity, beginning the mimic before the pattern is completely emitted (Richards, Wolz, and Herman 1984).

In both birds and mammals cooperative societies in which learning is prominent have emerged among animals living in open habitats such as the ocean, the savannas, and the forest canopy. The stability of such groups depends on information exchange not only on matters of survival (food, shelter, protection from predators) but also on relationships between members of the group. The latter exchange becomes increasingly crucial to maintaining both group and individual fitness as the society grows more complex (Axelrod and Hamilton 1981). Symbolic exchanges (both acoustic and by other means) are known to exist among social insects as well as higher animals (Griffin 1984). As social complexity increases, older communication systems become too limited in the sophistication of information that can be exchanged. Social balance in the more complex societies is maintained by a constant testing of the relationship between distantly related group members (Trivers 1985). Such testing places inordinate demands upon every aspect of communication.

We have argued that proto-words, proto-syntax, and message hierarchies are widespread within the animal kingdom. Yet animals tend to continue unitary information exchange through olfaction and through visual or acoustic signaling. Metaphoric communication most often takes the form of animals acting out the context of a message. It is, however, too slow and too emotionally transparent to answer all the demands of complex societies. We believe this provided the selective pressure toward linguistic communication.

Such stable social groups function as a school for self, for intentionality, and for language. Groups provide such necessities as a nursery for younger animals and a library of learning thus far attained. The interaction of self with similar selves represents a forcing function for full conscious awareness from which the open-ended word strings of language are constructed. Similarly, the interaction of individual intent as a transactional exchange in group activities must foster both a collective intent and a collective knowledge, in short a common body of codified meanings. Furthermore, societies tend toward greater complexity which, in turn, places increasing demands on what must be known and understood (membership recognition and accepted behavior, misunderstandings and mistakes, individual versus collective goals). This incipient information overload places a premium on increasingly refined communication both for individual safety and for continued group stability.

What forced the overloaded metaphoric and paralinguistic communication of body language (Reusch and Bateson 1951) to flip (using Desmond's terminology (1979)) to the ordered and open-ended strings of acoustic symbols that is language? The subject has, of course, been much speculated upon. Two factors, both the result of the formation of social groups, seem most plausible. As suggested by some (Humphrey 1976; Desmond 1979), information overload was one. The second is more interesting in the light of the sixteen proposed "design features" characteristic of human language (Hockett and Altmann 1968; Hockett 1960); that is the need for prevarication, also for its detection and control.

These factors would certainly be operative in any complex social group. The limited capacity of short-term memory and the need to temporally compress perception favors a flexible high-speed system that lends itself to chunking and meta-chunking (higher order abstraction made possible through the chunking of chunks). Arbitrary strings of acoustic symbols controlled by a set of combinatorial rules (grammar) and sequential order (syntax) would appear to be ideal. Dolphins and other higher mammals that have been tested seem to possess the motor abilities and the cognitive processes to produce and deal with such sequences, yet (except in humans) only the primordea seem to have been achieved.

The need for prevarication is particularly interesting. Trivers (1985) puts it well when he says: "One of the most important things to realize about systems of animal communication is that they are not systems for the dissemination of the truth. An animal selected to signal another animal may be selected to convey correct information, misinformation, or both." Prevarication has been listed as a design feature of human communication systems by Hockett and Altman, and yet its social value is only now becoming appreciated. As social complexity, crowding, and cross-purposes increase, the individuals most rewarded are those best versed in bluff, politics, politeness, and imagination. These, especially the last, are the positive aspects of prevarication (thinking or communicating about things that don't actually exist). As Humphrey (1976) put it, any heritable trait that increases the ability of an individual to outwit his fellows will spread through the gene pool. As we have said, metaphoric exchange is heavily influenced by emotional state and hence intrinsically transparent. For example, facial expressions and body language generally tend to be influenced by the limbic system (Hockett 1978). A good example of this is the fact that it is nearly impossible to lie to aphasics (Sacks 1985). The intentionally misleading semantics of word strings has no meaning for them and thus no effect. Most aphasics have trained themselves to read body language and intonation (the music of language), thus the true intent is apparent. Spoken language conveys information through the controlled arrangment of learned, abstract acoustic symbols and is, therefore, more under cortical rather than limbic control. It is a medium having good potential for information exchange combined with emotional opacity, the very thing needed for the "animal politics" of social groups.

All but two of Hockett and Altmann's design features have been found in animals (Savage-Rumbaugh and Hopkins 1986): DF1 vocal-auditory channel, DF2 broadcast transmission and directional reception, DF3 rapid fading, DF4 interchangeability, DF5 complete feedback, DF6 specialization, DF7 semanticity, DF8 arbitrariness, DF9 discreteness, DF11 openness, DF12 tradition, DF13 duality of patterning, DF15 reflexiveness, and DF16 learnability. Better demonstration of DF15 reflexiveness is, in our opinion, needed in animals since it is a most compelling argument for language in human form. Recently evidence has improved for the remaining two; DF14 prevarication (Seyfarth et al 1980a and 1980b; Dennett 1983; Griffin 1981) and DF10 displacement (Herman 1986). Thus, again we would expect incipient language among mammals forming social groups, and to a somewhat lesser extent among birds.

Linguistic communication requires a significant overlap of culture among the participants. Within any stable group such overlap is a given. It is represented in group goals and behavior, in collective knowledge, in similar needs and exposures. This common body of codified meaning operates as the default knowledge base underpinning all human language. The specialization for language represents an adaptation that ameliorates the information-load problem directly. Language operates not as a filter but as a tool with an innate potential to handle problems of any complexity. As a repository of collective knowledge and memory, the group is an excellent medium for traditional transmission (Hockett and Altmann's DF12). Language reinforces group culture which in turn reinforces language. We do not insist that language must arise in any social group. We do emphasize, however, that all necessary factors are simultaneously present in a situation where information overload creates a strong selection gradient for enhanced communication skills.

ORIGIN OF HUMAN LANGUAGE

We can go no further on the matter of animal language. We can, however, add our hypothesis to the others regarding the origin of language in man. Some (Hamilton 1973) have suggested man's shift to hunting/gathering life-style as the forcing function. Others (Parker and Gibson 1979; Hewes 1973, 1976) have favored expanded tool-using technology. No doubt both were important. Both speculations imply the existence of stable groups larger than the unitary family. We emphasize that the very existence of such stable groups is itself an important forcing function. Furthermore, hunting (except herding into traps or stampeding over cliffs) is largely silent. Silence prevails again with tool use, with the notable exceptions of rhythmic or repeated use, which interestingly is often accompanied by song. Instruction in tool use is largely demonstrational. The thumping inadequacy of language for this purpose can be exquisitely experienced by anyone accosting a new computer armed only with the manual.

We cannot, of course, state explicitly how language evolved. However, certain facets of human and animal behavior provide hints. Rhythm seems particularly important. Rhythmic patterns divide the acoustic string into subunits more easily handled by short-term memory, thus aiding message comprehension. Rhythm also aids by keeping the minds of the sender and receiver in synchrony. Furthermore, the reflex loops of human motor control centers for speech would seem to be set at iambic pentameter (Jaynes 1976). Babies begin the learning of language by attending to the rhythm and intonation (music) of their mother's voice (Roederer 1982; Bruner 1983). At this early period, rhythm has been proposed as an exercise for speech motor reflexes in preparation for learning language (Roederer 1982).

Thus, in our view both the social group and rhythmicity represent the cradle of language. How might development have proceeded from there? The beginnings might have been in ritual dance and song (Richmond 1980; Roederer 1982). Such events not only amuse are also rich in gesture, body language, rhythm, and intonation. All these accompaniments facilitate the elaboration of acoustic symbolism that can be generally understood, i.e. codified to have an agreed meaning. Man's intellect is primarily suited to thinking about people (Humphrey 1976). Like any other animal, man's approach to life is transactional. This stance applies to his own species, to hunted animals, even to plants and nature in general. His transactions were acted out, the metaphoric communication format common to all animals. The rhythm and music, the repetition and gesture, the mystery and magic of such group enactments would tend to be strongly imprinted in individual and corporate memory. Furthermore, the social group automatically provides an audience, the self-reinforcing justification for such elaborate amusements.

Language development probably proceeded from enactive expression to iconic expression to symbolic expression, much as it does in young children today (Bruner 1966). The constraints of short-term memory (STM) may have forced duality of acoustic patterning (Hockett and Altmann's DF13), i.e. a limited set of phonemes that are combined in different ways to form many words. Recursion and linear syntax, in early form at least, may be seen in the hierarchical patterning of bird and mammal call sequences. Short-term memory limitations and rapid recall requirements are seen in the small number of grammatic rules characteristic of all present-day human languages (Dennett 1983; Chomsky 1975; Minsky 1985). The next stage in the elaboration of that abstract, flexible, open-ended acoustic information exchange system known as human language was the invention of writing. This is another fascinating development, but one beyond the scope of the present discussion.

CONCLUSIONS

From a cognitive standpoint, the evolution of acoustic communication systems among vertebrates required the bypassing of certain innate limitations of the central nervous system. These were the necessarily small capacity of short-term memory and the biochemical ceiling on information-processing speeds. The first was initially circumvented by various forms of filtering (attention, feature extraction, chunking, forgetting) and the second, by a number of "tricks" that reduced processing loads (rhythmicity, supramodal indentity, internalization, and chunking again). The primary function of the brain is seen as prediction, and the devices just named are coordinated so that perception itself is temporally compressed. The coordinate whole is nicely demonstrated in the acoustic systems that have evolved.

Most, if not all, animal communication systems show rhythmic patterning which is thought to break the exchange into subunits suitable for efficient cognitive processing and to keep the parties of an exchange in synchrony. More complex messages are achieved by modifying and combining a small set of basic signals in a hierarchical manner. An animal can act out an event individually and often does. A powerful forcing function for the further development of sophisticated communication is found in the formation of complex social groups. Social groups operate as both a nursery and a school for the further development of advanced acoustic exchange leading ultimately to human language. We argue that all the necessary basics for human language must be widespread among social animals, while noting that there is a strong tendency for social animals to stabilize just short of such development.

Dolphin echolocation and human language represent the two of the most highly developed acoustic information systems thus far known.

LITERATURE

Atkinson, R. L. and Shiffrin, R. M., 1968. Human Memory: A Proposed System and Its Control Processes, in The Psychology of Learning and Motivation, Spence and Spence (eds), Academic.

Au, W. and Floyd, R. W., 1984. Receiving Patterns and Directivity Indices of the Atlantic Bottlenose Dolphin, JASA 75:1, 255-262.

Axelrod, R. and Hamilton, W. D., 1981. The Evolution of Cooperation, Science 211, 1390-1396.

Bain, D., in press. Acoustical Behavior of Orcinus: Sequences, Periodicity, Behavioral Correlates and an Automated Technique for Call Classification, in Behavioral Biology of Whales, Kirkevold and Lockhard (eds), Alan Liss Press.

Berlitz, C., 1982. Native Tongues, Grosset & Dunlap.

Broadbent, D. E., 1958. Perception and Communication, Pergamon Press.

Bruner, J. S., Goodnow, J. and Austin, G., 1956. A Study of Thinking, John Wiley.

Bruner, J. S., Oliver, R. R., Greenfield, P. M. et al, 1966. Studies in Cognitive Growth, John Wiley.

Bruner, J. S., 1983. Child's Talk, W. W. Norton & Co.

Bullock, T. H., Orkand, R. and Grinnell, A., 1977. Introduction to Nervous Systems, W. H. Freeman & Co.

Busnel, R.-G. and Dziedzic, A., 1966. Résultats Métrologiques Expérimentaux de L'Écholocation Chez le Phocaena Phocaena, et leur Comparison avec ceux de Certaines Chauves-Souris, pp 307-335 in Animal Sonar Systems, Busnel (ed), INRA-CNRZ, Jouy-en-Josas.

Chomsky, A. N., 1975. Reflections of Language, Pantheon.

Craik, K., 1943. The Nature of Explanation, Cambridge Univ. Press.

Davenport, R. K., 1977. Cross-Modal Perception: A Basis for Language, p 81 in Language Learning by a Chimpanzee: The Lana Project, Rumbaugh (ed), Academic.

Dennett, D. C., 1983. Intentional Systems in Cognitive Ethology: the "Panglossian Paradigm" Defended, Behav. and Brain Sci. 6:3, 343-390.

Desmond, A., 1979. The Ape's Reflexion, Blond & Briggs, Ltd.

Dushkin, D. A., 1970. Psychology Today, Chap 17, CRM Books.

Geschwind, N., 1965. Disconnexion Syndromes in Animals and Man (Parts 1 and 2), Brain 88, 585-644.

Gevins, A. S., Scheaffer, R. E., Doyle, J. C., Cutillo, B. A., Tannehill, R. S. and Bressler, S. L., 1983. Shadows of Thought: Shifting Lateralization of Human Brain Electrical Patterns During Brief Visuomotor Task, Science 220:4592, 97-99.

Gibson, J. J., 1979. The Ecological Approach to Visual Perception, Houghton & Mifflin.

Griffin, D. R., 1981. The Question of Animal Awareness: Evolutionary Continuity of Mental Experience, Rockefeller Univ. Press.

Griffin, D. R., 1984. Animal Thinking, Harvard Univ. Press.

Hailman, J. P., 1977. Optical Signals: Animal Communication and Light, Indiana Univ. Press.

Hamilton, W. J., 1973. Life's Color Code, McGraw-Hill.

Herman, L. M., 1986. Cognition and Language Competencies of Bottlenosed Dolphins, pp 221-252 in Dolphin Cognition and Behavior: A Comparative Approach, Schusterman, Thomas and Wood (eds), L. Erlbaum Assoc.

Hewes, G. W., 1973. Primate Communication and the Gestural Origin of Language, Current Anthropology 14, 11.

Hewes, G. W., 1976. The Current Status of the Gestural Theory of Language Origin, Annals of New York Acad. of Sciences 280, 488-499.

Hockett, C. F., 1960. The Origin of Speech, Sci. Amer. 203, 89-96.

Hockett, C. F., 1978. In Search of Jove's Brow, Amer. Speech 53:4, 243-313.

Hockett, C. F. and Altmann, S. A., 1968. A Note on Design Features, pp 61-72 in Animal Communication: Techniques of Study and Results of Research, Seboek (ed), Indiana Univ. Press.

von Holst, E., 1939. Die Koordination als Phänomen und als Methode Zentralnervosen Funktionsanalyse, Ergebnisse der Physiologie 42, 228-306.

Honolulu Star Bulletin, 1983. March 9.

Humphrey, N. K., 1976. The Social Function of Intellect, pp 303-317 in Growing Points in Ethology, Bateson and Hinde (eds), Cambridge Univ. Press.

Jaynes, J., 1976. The Origin of Consciousness and the Breakdown of the Bicameral Mind, Houghton Mifflin.

Kandel, E. R. and Schwartz, J. H., 1985. Principles of Neural Science, 2nd ed, Elsevier.

Lettvin, J. Y., Maturana, H. R., McCullough, W. S. and Pitts, W. H., 1959. What the Frog's Eye Tells the Frog's Brain, Inst. of Radio Engineers Proceedings 47, 1940-1951.

Marr, D., 1982. Vision: A Computational Investigation into the Human Representation and Processing of Visual Information, W. H. Freeman & Co.

Miller, G. A., 1956. The Magic Number Seven, Plus or Minus Two: Some Limits on our Capacity for Processing Information, Psych. Rev. 63:2, 81.

Miller, G. A., 1966. Speech and Language, pp 789-810 in Handbook of Experimental Psychology, Stevens (ed), John Wiley.

Minsky, M., 1985. Communication with an Alien Intelligence, BYTE 10:4, 127-138.

Morton, E. S., 1977. On the Occurrence and Significance of Motivation-Structural Rules in Some Bird and Mammal Sounds, Amer. Nat. 3:981, 855.

Moynihan, M., 1970. Control, Suppression, Decay, Disappearance and Replacement of Displays, J. Theor. Biol. 29, 85.

Neisser, U., 1967. Cognitive Psychology, Appleton-Century-Crofts.

Norman, D. A., 1982. Learning and Memory, W. H. Freeman & Co.

Norris, K. S., Wursig, B., Wells, R. S., Wursig, M., Brownlee, S. M., Johnson, C., and Solow, J., 1985. The Behavior of the Hawaiian Spinner Dolphin, Stenella longirostris, Admin. Rept., National Marine Fisheries Service, Southwest Fisheries Center, LJ-85-06C.

Parker, S. T. and Gibson, K. R., 1979. Developmental Model for the Evolution of Language and Intelligence in Early Hominids, Behavioral and Sciences 2:3, 367-408.

Penner, R. L., this volume. Biosonar Attending Behavior in Tursiops truncatus.

Reusch, J. and Bateson, G., 1951. Communication: The Social Matrix of Psychiatry, W. W. Norton.

Richards, D. G., Wolz, J. P. and Herman, L. M., 1984. Vocal Mimicry of Computer-Generated Sounds and Vocal Labeling of Objects by a Bottlenosed Dolphin, Tursiops truncatus, J. Comp. Psych. 94:1, 10.

Richmond, B., 1980. Did Human Speech Originate in Coordinated Vocal Music?, Semiotica 32:3-4, 233.

Roederer, J. G., 1982. The Search for a Survival Value in Music, Proc. 104th Meeting of Acoustical Soc. of Amer., Phychological Acoustics Section, Orlando, FL.

Sacks, O., 1985. The Man Who Mistook His Wife for a Hat and Other Clinical Tales, Summit Books.

Savage-Rumbaugh, E. S. and Hopkins, W. D., 1986. Awareness, Intentionality and Acquired Communicative Behaviors: Dimensions of Intelligence, pp 303-313 in Dolphin Cognition and Behavior: A Comparative Approach, Schusterman, Thomas, and Wood (eds), L. Erlbaum Assoc.

Seyfarth, R. M., Cheney, D. L. and Marler, P., 1980a. Monkey Responses to Three Different Alarm Calls: Evidence of Predator Classification and Semantic Communication, Science 210, 801.

Seyfarth, R. M., Cheney, D. L. and Marler, P., 1980b. Vervet Monkey Alarm Calls: Semantic Communication in a Free-Ranging Primate, Animal Behav. 28, 1070.

Shepard, R. N., 1984. Ecological Constraints on Internal Representation: Resonant Kinematics of Perceiving, Imagining, Thinking and Dreaming, Psych. Rev. 91:4, 417.

Sherrington, C. S., 1906. The Integrative Action of the Nervous System, Yale Univ. Press.

Simmons, J. Q. and Baltaxe, C., 1975. Language Patterns of Adolescent Autistics, J. Autism and Childhood Schizo. 5:4, 333.

Simon, H. A., 1974. How Big Is a Chunk?, Science 183, 482-488.

Simon, H. A., 1985. The Parameters of Human Memory, Proc. of Ebbinghaus Symposium, Berlin, FRG.

Smith, W. J., 1969. Messages of Vertebrate Communication, Science 165, 145-150.

Smith, W. J., 1986. Signalling Behavior: Contributions of Different Repertoires, pp 315-330 in Dolphin Cognition and Behavior: A Comparative Approach, Schusterman, Thomas and Wood (eds.), L. Erlbaum Assoc.

Suga, N., Neuweiller, G. and Moller, J., 1976. Peripheral Auditory Tuning for Fine Frequency Analysis by the CF-FM Bat, Rhinolophus ferrumequinum. IV. Properties of Peripheral Auditory Neurons, J. Comp. Physiol. 106, 111-125.

Treisman, A. M., 1986. Features and Objects in Visual Processing, Sci. Amer. 255:5, 114B-125.

Trivers, R., 1985. Social Evolution, Benjamin/Cummings Publ. Co.

Uttal, W. R., 1973. The Psychobiology of Sensory Coding, Harper & Row.

Uttal, W. R., 1978. The Psychobiology of Mind, L. Erlbaum Assoc.

Van Riper, C., 1971. The Nature of Stuttering, Prentice Hall.

COGNITIVE ASPECTS OF ECHOLOCATION

Donald R. Griffin

The Rockefeller University
1230 York Avenue
New York, N. Y.   10021-6399

Echolocation is an active mode of perception. Rather than passively waiting for information to arrive at its sense organs, an echolocating animal emits probing signals and listens for important information supplied by the resulting echoes. A comparably active process is the electric orientation that has independently evolved in several groups of fish. They also broadcast informative energy and gather important information by sensing alterations in the electric fields in which they find themselves. What we might call "active perception" of this sort requires that the brain must both generate the appropriate type of probing signal, which often differs according to the animal's situation, and then selectively process the raw afferent input from its receptors to obtain the information needed at the moment. This amounts to a sort of dialog between the animal and its environment. Furthermore such active probing for important information suggests spontaneous control by the animal of its own behavior, rather than a set of fixed and automatic responses to external stimulation.

Correlating stimuli with responses, or inputs with outputs, used to seem sufficient to permit understanding, prediction and control of animal, and perhaps even human behavior. This comfortably familiar behavioristic approach had channelled our conceptions of animals into a tidy "black box" format. But it has become increasingly obvious that restricting attention to stimuli and responses fails to do justice to the rich versatility of animal behavior. Therefore ethologists and comparative psychologists are paying more and more attention to what goes on inside the black box, as discussed by Walker (1983) and Griffin (1984). This swing of the pendulum away from strict behaviorism is well described by Roitblat, Bever and Terrace (1983):

"Animal cognition is concerned with explaining animal behavior on the basis of cognitive states and processes, as well as on the basis of observable variables such as stimuli and responses. For a time it appeared, at least to some, that discussion of cognitive states was not necessary, either because they were exhaustively determined by environmental events, or because they were epiphenomenal and without causal force. In any case, it was assumed that a sufficiently detailed description of overt events would suffice for explanation. A great deal of research into animal behavior has made it clear, however, that such cognitive states are real and necessary components of any adequate theory that seeks to explain animal behavior."

Does the complexity and versatility of echolocation tell us anything significant about cognitive states and processes in the brains (or minds) of bats or dolphins? Many of the simpler questions about the behavior of echolocating animals have now been answered, at least to a first approximation, as amply documented in the proceedings of the three international symposia on animal sonar, especially this third one. It is therefore time to move on to inquire about cognitive processes that must occur in the brains of animals specialized for echolocation. Perhaps we can eventually learn a little about what it is really like to be an echolocating bat or dolphin, despite the technical philosophical arguments that this is impossible advanced by Nagel (1974). These arguments apply almost as forcibly to inferring what it is like to be another person, a sort of inference without which our social relationships would be virtually impossible. Without claiming absolute, logically watertight proofs or totally complete and error-free inferences, we can nevertheless attempt to appreciate what it is probably like to be an echolocating animal. But to do so will require that our scientific perceptions attain a versatility comparable to the attainments of the animals we are studying.

Questions such as these lead us into tangled thickets of deeply significant problems that cannot be resolved in any satisfactory fashion on the basis of available data, as discussed in detail by Midgley (1978, 1983), Terrace (1983), Walker (1983), Wasserman (1983), and Griffin (1984). But inadequacy of available data has been generally been seen by scientists as a reason to inquire more deeply, rather than to give up in despair. This has been especially true of research on echolocation, a subject that was not even imagined 50 years ago.

Information-seeking behavior by a wide variety of animals also includes actions that increase the likelihood that significant information will reach the animal's sense organs. Obvious examples include scanning movements of the head and eyes or ears to obtain visual or acoustic information from different directions, accommodation to bring objects at different distances into focus, sniffing to bring air to the olfactory epithelium, and "feeling around" with the paws after reaching into a place where something important such as food may be hidden from view. In all these cases animals act to enhance the effectiveness of their various sense organs. But echolocation is different in that the informative energy is generated by the animal itself. Acoustic echolocation also differs from electrolocation as practiced by weakly electric fishes in the brief but significant time delays ordinarily required for the return of echoes of orientation sounds. This distinction may not be an absolute one inasmuch as bats using relatively long constant frequency orientation sounds often receive and process echoes that arrive during sound emission.

It is an interesting fact that while bats employ a wide variety of signal patterns in their orientation sounds, the toothed whales seem to rely exclusively on brief and almost invariant clicks. These of course have broad frequency spectra, and differences between echoes from various targets can theoretically be recognized by spectral analysis of any broadband signal. I have often wondered whether spectral analysis of broadband clicks requires more voluminous neural machinery than detection of spectral differences among the echoes of the orientation sounds used by many bats that contain approximately an octave of frequency sweep. Perhaps bat brains are not large enough to accomplish this effectively, although I know of no theoretical reason why this should be the case.

The dual, "first-active-then-passive", nature of echolocation by dolphins and those bats that use short duration orientation sounds requires that the brain must operate in two distinct modes, not quite simultaneously as in long CF bats and electric fishes, but with the receptive processing

slightly delayed relative to signal generation. Furthermore, the two
processes must be closely linked and coordinated. For example, Suga and
Schlegel (1972) and Suga and Shimozawa (1974) have analyzed how auditory
sensitivity is temporarily reduced during and a few milliseconds after
emission of the orientation sounds of certain bats. This neural
inhibition, together with similarly coordinated contractions of the large
intraaural muscles, protects the auditory system from overloading during
the emission of intense orientation sounds - serving the same function as
the transmit-receive switches of artificial radar and sonar systems. As
described at this conference by Simmons, the whole system appears to
provide a sort of gain control that keeps echo loudness nearly constant as
a bat approaches a small target despite great changes in actual echo
intensity. Interacting regulatory functions are of course not unique to
echolocation. Virtually all coordinated motor activities involve
comparable linkages between separate systems of motor control and
modulation by afferent feedback. What is special about the brain
mechanisms of echolocation is the precise coordination between motor and
sensory activities, and, in many bats and virtually all echolocating
cetaceans, the controlled time delays between the two.

The refinement of appropriately interacting neural systems involved in
insect pursuit and capture by bats of the family Vespertilionidae
illustrates the special demands on an active brain posed by specialized
behavior that is guided by echolocation. Clear examples are available from
high speed photographs of Myotis lucifugus catching fruit flies and tossed
mealworms (Webster and Griffin, 1962; Webster, 1967), together with more
recent detailed analyses of insect catching behavior by other species
(Miller and Degn 1981). After detecting one or more echoes from the insect
at a moderate distance, the motor system regulating sound emission
increases pulse repetition rate, shortens pulse duration, and varies the
degree of frequency modulation. Auditory information from the two ears
leads to turning of the head and slightly later to differential action of
the two wings to turn the bat towards the insect. The latter meanwhile has
ordinarily moved, and perhaps turned sharply in both horizontal and
vertical directions. Some insects may even emit interfering sounds that
complicate still further the work of the auditory brain either by
interfering with echolocation, by warning of a distasteful insect, or both
both (Surlykke and Miller 1985). In the final stages of the pursuit, the
bat may extend one wing to make contact with the insect, then fold it into
the wingtip and convey it to the mouth. These maneuvers convert the
forearm temporarily from a wing into a prehensile hand. This has drastic
aerodynamic effects, and in most of Webster's original high speed
photographs the bat ended up heading in the opposite direction, but with
the insect firmly in its teeth.

One aspect of these early experiments that is pertinent in connection
with possibly cognitive aspects of echolocation is the degree of learning
involved. Insectivorous bats, at least those of the family
Vespertilionidae with which I am most familiar, will ordinarily pursue
almost any small moving object thrown up in their vicinity when they are in
search of insect prey. This is scarcely surprising since under natural
conditions almost any small moving target encountered at night is very
likely to be an insect. I wonder whether bats that specialize in
"flycatcher style" insect hunting show comparable ability to discriminate
among insects and insect-like targets. Constant frequency searching
signals may be less suited for this than FM pulses. This is simply one
more significant ramification of echolocation behavior that deserves
careful study.

Webster and I set out to study the degree to which bats could
discriminate between small targets, and we found that captive Myotis

lucifugus which had first learned to catch mealworms tossed into their flight paths would initially pursue and seize disks of roughly the same size as the edible mealworms. But within a few days they learned that only some of the moving and tumbling targets they encountered were edible, and they rejected almost all of the disks while catching most of the mealworms (Griffin, Friend and Webster 1965). Our principal interest at the time was the nature of the differences between the echoes of the two categories of target, a subject that has since been investigated much more thoroughly in the elegant experiments of Simmons and others. But from a cognitive viewpoint, it is important to recognize that in all of these experiments bats have demonstrated a highly developed ability to learn that certain targets are edible (and worth catching in our early experiments), or (in Simmons' discrimination experiments) that approaching them causes the experimenter to provide food. No one familiar with the results of these experiments would be tempted to think of discriminative echolocation as a rigidly stereotyped form of behavior.

Recent investigations of many closely related species have now shown that although Myotis lucifugus rarely if ever abandons the octave of frequency sweep in its orientation sounds, many other insect hunting bats use a type of searching signal which has little or no frequency modulation (Miller and Degn 1981). In these species the sound generating mechanisms of the midbrain must change from emitting nearly constant frequency signals during the search phase to the progressively shorter and more prominently frequency modulated pulses characterizing approach and capture of insect prey. These are only a few of the coordinated neural mechanisms that are necessary for the efficient insect capture by echolocation. Further investigations are likely to reveal additional refinements and complexities, but these are enough to illustrate the versatility of coordinated motor and sensory mechanisms involved in echolocation.

But does any of this involve cognition, unconscious or conscious? It certainly requires fairly complex information processing within the central nervous system; but is that enough to warrant using the term cognition? Those reluctant to take this semantic step can argue that many biological functions, both in the central nervous system are other organs, involve complex regulation of information flow and resulting physiological activities. For instance we do not apply the term cognition to the regulation of blood chemistry by our kidneys. Where and how can a meaningful line be drawn? This is one of the thorny, philosophical questions that arise when one pursues questions of animal cognition. Yet we know from our own subjective experiences that in at least one species, conscious states and thoughts are significant and that they sometimes influence our future behavior. Unless we assume, rather unparsimoniously, that our species is unique in experiencing conscious thoughts and subjective feelings, the question of animal consciousness inevitably arises once we begin to inquire about animal cognition.

While the narrowest forms of strict behaviorism are no longer defended, the subtle residues of behavioristic inhibitions and constraints on scientific thinking still remain potent, though not fully recognized, influences on the study of animal behavior. For example, the recently rekindled interest in animal cognition, as exemplified in the above quotation from Roitblat, Bever and Terrace (1983) has included few attempts to clarify just what is meant by "cognitive states and processes" - except that they are something other than overtly observable stimuli and responses. They are evidently internal processes that occur in animal brains and that selectively organize the otherwise almost infinite flux of neural excitation and inhibition that could in theory lead to almost any sort of behavior or lack of it. Yet we are strongly admonished by Terrace (1983), Wasserman (1983) and many others that a renewed interest in animal cognition should not be mistaken for a relapse into pre-Watsonian mentalism or the suggestion that animals might experience conscious, subjective thoughts or feelings. I do not share these inhibitions, because they seem to be justified by dogmatic assertions rather than persuasive evidence, as discussed elsewhere (Griffin, 1984, 1985a, 1985b). But this philosophical issue need not inhibit the consideration of cognitive processes that may occur in the course of behavior guided by echolocation. Whether unconscious or conscious, a great deal of cognition is required for many types of behavior in which echolocation plays a significant role.

At this point it is appropriate to consider human echolocation where cognition and consciousness are sometimes decoupled in a significant fashion. As thoroughily established long ago by Supa, Cotzin and Dallenbach (1944), and later analyzed parametrically by Rice (1967a, 1967b), many blind people employ a form of echolocation to detect obstacles before colliding with them. Unfortunately the echolocation of the human blind is far less sensitive and efficient than that of bats and dolphins, a scientific and humanitarian challenge that has attracted regrettably little investigation. There is no doubt that many experienced blind people detect, classify and react appropriately to many types of objects in their immediate vicinity on the basis of echoes from sounds of various sorts that they themselves generate. After extensive practice sighted subjects wearing blindfolds can learn to do almost as well as the most expert blind people. Yet a large proportion of these reasonably successful human echolocators do not even realize that the information telling them of the proximity of obstacles comes through their ears. The usual term for the ability of the blind to detect obstacles before contact with them is "facial vision;" and when asked how they know that something is there, many blind people say they sort of feel that there is a solid object just ahead.

This subjective impression of tactile perception is perhaps understandable because many objects are sensed by touch, and those first detected by echolocation are often felt shortly afterwards. Human echolocation certainly requires effective information processing when significant echoes from obstacles are distinguished from background noise and when sounds are generated at times when echo information is needed. If one is looking for reasons to denigrate consciousness, as has long been fashionable in many scientific circles, one could argue that echolocation is inherently an unconscious process comparable to many of our postural reflexes based on input from the receptors of the inner ear. Since these reflexes operate smoothly and efficiently without any conscious awareness of their sensory basis, they show that efficient physiological regulation is not enough to demonstrate conscious awareness of what our bodies are doing.

The lack of conscious awareness of the auditory basis of obstacle detection by some, but by no means all, the human blind may well be a special case from which excessive generalization would be misleading. Our brains are far from being specialized for echolocation, and when deprived of one of the principal sensory systems, it may be a natural tendency for information arriving by channels not normally exploited to be linked with mechanisms of perception ordinarily employed in vision or tactile feeling of objects. If so, bats or dolphins deprived of the ability to echolocate might imagine that they were detecting food by this, to them, predominant sensory channel when they were actually seeing or even tasting it.

One striking aspect of bat behavior involves what superficially seems a failure of echolocation, but nevertheless it demonstrates the presence of cognitive maps in their brains or minds. This is what Moehres and Oettingen-Spielberg (1949) called Wiederorientierung, which they contrasted with Erstorientierung when bats orient themselves for the first time in a novel situation. This sort of spatial memory is strikingly demonstrated when bats collide with newly erected obstacles in a place with which they have become thoroughly familiar during repeated flights. They also tend to turn back from familiar obstacles that have suddenly been removed, leaving an open space through which they could easily fly. When obstacle avoidance tests are used to measure the acuity or echolocation, bats often learn the location of test obstacles and avoid them by memory rather than by echolocation, as demonstrated by less successful avoidance of the same obstacles after they have been shifted in position (Griffin 1958).

While the spatial memory of bats has not been studied as thoroughly as would be desirable, they obviously learn the details of familiar areas and carry out complex flight maneuvers on this basis, even when information from their echolocation conflicts radically with the remembered situation. An extreme and dramatic case of this kind is described by Moehres and Oettingen-Spielberg. One of their bats had been roosting for a long time in a small cage hung from the ceiling near the center of a laboratory room. The cage had a door on one side through which the bat was accustomed to fly in and out. When the cage was rotated 90 or 180 degrees while the bat was elsewhere it would try to fly back through the side that had been open. Furthermore when the cage was removed altogether, the bat flew through the customary position of the door and then executed the complete and elaborate landing maneuver at the point where it had been accustomed to cling to the top of the cage!

Managing all this remembered spatial information certainly requires something that can reasonably be called cognition. We have learned about it from the mistakes bats make while ignoring information available from echolocation, rather than from correct orientation based on spatial memory that has led to the concept of cognitive maps in laboratory rats (Tolman, 1948; O'Keefe and Nadel, 1978) or honeybees (Gould 1986). But the amusing mistakes of bats should not mislead us into overlooking fact that this reliance on spatial memory demonstrates the presence of quite detailed cognitive maps of familiar territory. While the experiments discussed above were carried out in laboratory rooms, observations of wild bats strongly suggest that they patrol the same hunting areas repeatedly; and it is reasonable to infer that they must also have cognitive maps of outdoor areas through which they fly night after night. If suitable circumstances could be found or established, it might be interesting to test this possibility by experiments analogous to those of Moehres and Oettingen-Spielberg. Would sudden changes in familiar night roosts or customary landing spots lead to similarly revealing mistakes, or would echolocation and other senses operate more effectlvely under natural outdoor conditions?

While it does not directly involve echolocation, no consideration of cognition in echolocating bats would be complete without discussion of the observations reported by Porter (1979) of social recognition in Carollia perspicillata. She maintained a breeding colony of these small neotropical fruit bats where reasonably normal social relationships developed. The social unit is the harem with a dominant adult male, several adult females and their subadult offspring of all ages. As in most bats nursing infants sometimes dropped to the floor of the cage and emitted distress calls. Often the mother would fly to the infant and retrieve it; but when she did not do so, the harem male or other adult females would often approach, nudge and vocalize at the mother of the fallen baby. This sort of behavior was always directed at the mother of the calling infant, indicating that the other adult members of the colony knew which infant belonged with which mother or vice versa.

In short, many of the complex and versatile kinds of behavior carried out by bats and toothed whales with the aid of echolocation require a degree of neural regulation and adaptation to changing circumstances that corresponds well to what is considered evidence of cognition in other animals. A more challenging question is whether any of these instances of cognition linked to echolocation also entail conscious awareness on the part of the echolocating animal of what it is doing. While we cannot yet answer this level of scientific question conclusively, it is well to recall how many surprising and significant discoveries have resulted from the open minded investigation of animal behavior in general and of echolocation in particular. The talent and ingenuity exemplified by the papers presented at in this and earlier animal sonar symposia could well lead, in the future, to a significant cognitive ethology of echolocation, and to a meaningful if not perfect understanding of what it is really like to be an echolocating bat or dolphin.

REFERENCES

Gould, J. L., 1986, The locale map of honey bees: Do insects have cognitive maps? Science, 232:861-863.
Griffin, D. R., 1958, "Listening in the dark, the Acoustic Orientation of Bats and Men," Yale University Press, New Haven, Conn. (Reprinted by Cornell Univertsity Press, 1986, with preface by J. A. Simmons).
Griffin, D. R., 1984, "Animal Thinking," Harvard University Press, Cambridge, Mass.
Griffin, D. R., 1985a, The cognitive dimensions of animal communication, Pp. 471-482 in: "Experimental Behavioral Ecology and Sociobiology, in Memoriam Karl von Frisch 1886-1982," B. Hölldöbler and M. Lindauer eds., Gustav Fischer (Fortschritte der Zoologie 32), New York.
Griffin, D. R., 1985b, Animal consciousness, Neuroscience & Biobehavioral Reviews, 9:615-622.
Griffin, D. R., Friend, J. H., and Webster, F. A., 1965, Target discrimination by the echolocation of bats, J. Exptl. Zool., 158:155-168.
Midgley, M., 1978, "Beast and Man, the Roots of Human Nature" Cornell University Press, Ithaca, N. Y.
Midgley, M., 1983, "Animals and Why they Matter," University of Georgia Press, Athens, Ga.
Miller, L. A. and Degn, J., 1981, The acoustic behavior of four species of vespertilionid bats studied in the field, J. Comp. Physiol., 142:67-74.
Moehres, F. P., and Oettingen-Spielberg, T., 1949, Versuche über die Nahorientierung und das Heimfindevermogen der Fledermäuse, Verhandlungen der deutschen Zoologen in Mainz, 1949 pp. 248-252.
Nagel, T., 1974, What is it like to be a bat? Philos. Rev., 83:435-450. (Reprinted in Nagel, T., 1979, "Mortal Questions," Cambridge University Press, New York).
O'Keefe, J., and Nadel, L., 1978, "The Hippocampus as a Cognitive Map," Oxford University Press, New York.
Porter, F. L., 1979, Social behavior in the leaf-nosed bat Carollia perspicillata II. Social communication, Z. Tierpsychol., 50:1-8.
Rice, C. E., 1967a, Human echo perception, Science, 155:656-664.
Rice, C. E., 1967b, The human sonar system, in: "Animal Sonar Systems," R-G. Busnel, ed., Laboratoire de Phsyiologie Acoustique, Jouy-en-Josas, France:
Roitblat, H. L., Bever, T. G., and Terrace, H. S., (eds.), 1983, "Animal Cognition," Erlbaum, Hillsdale, N. J.

Suga, N., and Schlegel, P., 1972, Neural attentuation of responses to emitted sounds in echolocating bats, _Science_, 177:82-84.

Suga, N., and Shimozawa, T., 1974, Site of neural attenuation of responses to self-vocalized sounds of echolocating bats, _Science_, 183:1211-1213.

Supa, M., Cotzin, M., and Dallenbach, K. M., 1944, "Facial vision." The perception of obstacles by the blind, _Am. J. Psychol._, 57:133-183.

Surlykke, A., and Miller, L. A., 1985, The influence of arctiid moth clicks on bat echolocation; jamming or warning? _J. Comp. Physiol._, 156:831-843.

Terrace, H. S., 1983, Animal Cognition, Ch. 1 _in_: "Animal Cognition" H. L. Roitblat, T. G. Bever, and H. S. Terrace, eds., Erlbaum, Hillsdale, N. J.

Tolman, E. C., 1948, Cognitive maps in rats and men, _Psychol. Rev._, 55:189-208.

Walker, S., 1983, "Animal Thought," Routledge and Kegan Paul, London.

Wasserman, E. A., 1983, Animal intelligence: Understanding the minds of animals through their behavioral "ambassadors," Ch. 3 _in_: "Animal Cognition," H. L. Roitblat, T. G. Bever, and H. S. Terrace, eds., Erlbaum, Hillsdale, N. J.

Webster, F. A., and Griffin, D. R., 1962, The role of the flight membranes in insect capture by bats, _Anim. Behav._, 10:332-340.

Webster, F. A., 1967, Some acoustical differences between bats and men, Pp. 63-88 _in_: "Proc. Intl. Conference on Sensory Devices for the Blind," R. Dufton, ed., St. Dunstan's, London.

COGNITION AND ECHOLOCATION OF DOLPHINS

Ronald J. Schusterman

Institute of Marine Sciences, Univ. of California
Santa Cruz, CA 95064 and Department of Psychology
California State Univ., Hayward, CA 94542

INTRODUCTION

In this conference, we have already been given several examples of
sonar hunting bats relying on an internal representation of some past
experience as a basis for action. These include experiments demonstrating the predictive tracking of horizontally moving targets by a fishing
bat (Campbell & Suthers, this volume) and the extracting of a species-
specific concept from acoustical "glint" cues by greater horseshoe bats
(von der Emde, this volume). Internal representation involves the
encoding of information about specific features of stimuli as well as
about relations among stimuli (Schusterman and Krieger, in press).

The purpose of this paper is to explore some aspects of the interaction between the cognitive skills of dolphins and their echolocation
capabilities. I will first discuss the concept of cognition by briefly
reviewing the differences between behavioristic psychology and cognitive
psychology. I will then review some aspects of dolphin cognition,
initially outside the context of echolocation and then within the context of echolocation. Before getting into the text of this paper, the
reader should be aware that every time the word "dolphin" appears, I am
usually referring to captive studies with the genus of delphinid known
as Tursiops, the one Wood (1986) calls "the adaptive bottlenose" and the
one that we know best from seeing its performances in oceanariums, aquar-
iums and zoos and hearing about experiments dealing with its cognitive
skills and its echolocation abilities.

WHAT IS COGNITION?

The history of cognitive psychology as an interdisciplinary effort,
which included not only philosophy, linguistics and artificial intelligence, but the neurosciences and ethology as well, is brilliantly described in Howard Gardner's new book, "The Mind's New Science" (1985).
The extreme reductionism, functionalism and operationalism of a stimulus-
response (S-R) behavioristic psychology eschewed such topics as mind and
thinking, and such concepts as plans, desires or intentions. Moreover,
S-R behaviorists never talked about any form of construct related to
mental representations. In general, they insisted on the Lockian point
of view, that animals are fundamentally blank slates to be written upon

by experience (Roitblat, 1986), and that "laws of learning" could be applied to virtually all organisms in most, if not all, tasks without regard to specific learning skills or cognitive structures tailored by natural selection to a particular task. At any rate, the goal of much of S-R behavioristic psychology was to develop an elaborate machinery detailing how much the triadic contingency of stimulus/response/reinforcement could describe and explain the learning and shaping of any behavior. In general, questions about an organism's <u>umwelt</u> or perceptual world were largely ignored. Only on rare occasions did S-R behaviorists consider questions about language acquisition, planning, reasoning, and intentionality, rule learning and learning sets, concept formation, memory, serial pattern learning, problem solving, etc. in human or nonhuman animals. The relative sterility of many forms of behavioral psychology became most apparent in the late 1950s when several behaviorists like Karl Lashley, Edward Tolman and Harry Harlow spoke of men and animals as behaving not as passive reactors to environmental pushes and pulls, but behaving as they do because of expectancies, plans, hypotheses and strategies about the world around them. [For theory and research in the late 1950s and early 1960s, which has structured much current thinking about strategies and rule learning in animals; see Levine (1959), Restle (1958) and Schusterman (1962)].

During the 1960s, several books were published which called for the abandonment of the constricted conceptual view of stimulus-response in favor of more mentalistic conceptions. The impact of this approach became evident about 20 years ago with the publication of Ulric Neisser's influential textbook "Cognitive Psychology" (1967), in which he refers to cognition as "all processes by which sensory input is transformed, reduced, elaborated, stored, recovered, and used." Although the term "cognition" as originally used by Neisser and other experimental psychologists referred to the processing of information by relatively linguistically sophisticated humans, comparative psychologists also conceptualize nonverbal animals as active processors of environmental information.

As Figure 1 shows, Neisser's definition can be readily viewed within the context of an information processing approach in which the acquisition, storage, retrieval and utilization of information involves a number of different stages. Cognitive theories attempt to identify what happens in each of the stages and to identify interactions between the stages. After a brief exposure to a visual or auditory stimulus (e.g., 5 msec), sensory input may be maintained in a sensory store for a considerably longer time period (e.g. 250 msec). Transformation of the sensory input means that animals or, to be more precise, their CNS's, are not passive recipients of environmental information, but reduce and enhance or elaborate on incoming information. Thus, an animal's representation of the world depends not only on the incoming information, limited by its attention and by its short-term memory (STM), but also on

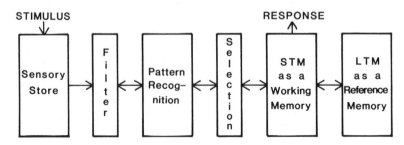

Fig. 1. Stages of an information-processing model (after Reed, 1982).

the elaboration of this new information by previously stored knowledge
[which has been momentarily retrieved from long-term memory (LTM)].
According to Jerison (1973; 1986), the construction of perceptual worlds
is information processing performed by large and complex neural networks
and that such construction accounts for the gross enlargement of the
brain in "higher" vertebrates (birds and mammals), enlargement beyond the
requirements for controlling the body--the so-called "structural encephalization." An encephalization quotient (EQ), as defined by Jerison
(1973), is the ratio between the observed brain weight and the expected
brain weight for a defined body weight. Wood and Evans (1980) discussed
the relationship between echolocation in odontocete cetaceans, adaptability in "niche" hunting or feeding strategies and EQ at the last Animal
Sonar Conference in 1979, and most recently Jerison (1986) has speculated
that echolocation in the high EQ dolphin may have resulted in dolphin
schools having a shared reality.

In applying an information processing model to dolphin cognition and
echolocation, imagine if you will an echolocating dolphin comparing the
complex patterns of echoes (and possibly reflected light patterns as
well) emanating from objects in its current environment with its stored
memory of complex echo patterns (and also possibly reflected light patterns) from previously captured palatable prey. Indeed, the term "hunting by searching image" has been widely adopted in the ethological and
behavioral ecology literature (Krebs, 1973) and fits very nicely into an
information processing model of foraging behavior. Furthermore, the
study of predation behavior in general and specifically hunting by means
of echolocation in bats and dolphins may be profitably studied by using a
comparative cognitive approach. Hypotheses deriving from ecologically
oriented research on hunting behavior are well suited to laboratory
manipulation and analysis (i.e., hunting by searching image, hunting by
expectation, area-restricted search and even "niche" hunting and optimal
foraging).

Returning to Figure 1, note that environmental input appears to be
processed in separate sequential stages from sensory input to response
output. This is, of course, not the case. Information from the environment enters a sensory buffer, it's filtered and selected, classified,
processed in STM, and eventually stored in LTM. Although the stages are
arranged discreetly and sequentially, there is a great deal of reciprocal
interaction between the processing stages, as indicated by the two-headed
arrows, so that an earlier stage can be influenced by information in a
later stage and vice versa. Flow from input through sensory store, pattern recognition and memory stages is often called bottom-up or sensory-
driven processing. The reverse flow from memory through the pattern
recognition stage is often called top-down or concept-driven processing.
In top-down processing, sensory information is analyzed in a manner
controlled by the animal's knowledge and expectancies.

Getting back to Neisser's definition of cognition, perhaps the
final part is the most significant (Reed, 1982). Following the various
stages of information processing, the information must be used in an
adaptive manner--for example, to make complex decisions about sensory
events and to solve problems regarding feeding, reproduction and predator
avoidance. (Perhaps in the case of dolphin biosonar, the integration of
sensory events from different modalities also includes cross modal transfer between sonar and vision.) At the last Animal Sonar Conference in
1979, Schusterman (1980) reported that on a variety of signal detection
tasks (primarily nonecholocating), the decision criteria of dolphins and
pinnipeds changed predictably and rapidly as a function of the animal's
expectancy of signal occurrence and as a function of knowledge about
reinforcement contingencies or the payoff matrix. Recently, some of

these results have been replicated in an experiment in which a dolphin used biosonar (Au and Turl, 1984).

In summarizing about what cognition refers to, or more specifically, what is meant by animal cognition, I will quote from Herb Roitblat's remarkably lucid, rather thorough and brand new textbook, "Introduction to Comparative Cognition" (1986).

"Comparative cognition is the study of the minds of organisms. Mind is the set of cognitive structures, proces- ses, skills, and representations that intervene between experience and behavior. Comparative cognition views animals as intelligent processors of information capable of adapting to their environments through expression of varied cognitive skills. These skills include learning, remembering, problem solving, rule learning, concept formation, perception, pat- tern recognition, and others. Comparative cognition seeks to explain behavior in terms of the skills, representations, and processes that organisms use as they interact with their environment" (Roitblat, 1986, pp. 1-2).

DOLPHIN COGNITION SANS ECHOLOCATION

Although the cognitive skills of dolphins based on anecdotal and naturalistic evidence may be found in Greek and Roman literature, McBride and Hebb (1948) gave the first objective modern account of the cognitive abilities of dolphins albeit in the context of play and motivation and they placed them as high or higher than apes in any comparative ranking (see also Kellogg, 1961). John Lilly (1961) supported this notion, but may have gone overboard in suggesting that dolphins communicate in a sonic language as sophisticated as human language. Nevertheless, Lilly's insights into such potentially significant cognitive/communicative skills as mimicry and intentionality and his special efforts in pointing out the importance of social bonding between these animals and their trainers in the application and formal demonstration of such skills is now recognized by many scientists who have worked closely with dolphins (Jerison, 1986). However, from the standpoint of rigorous experimental methodology, the best demonstration of such cognitive skills as memory and rule learning in dolphins comes from the recent work of Louis Herman and his colleagues (for a summary of these studies, see Herman, 1980; Herman, 1986; Richards, 1986).

Working Memory as STM

Given two discriminative stimuli, D1 (a triangle) and D2 (a square), an animal selects D1 if conditional stimulus C1 (a triangle or a white disc) is provided, and selects D2 if the conditional stimulus is C2 (a square or a black disc). The conditional relations are: If C1, then D1; if C2, then D2. When the conditional or "sample" stimuli are physically the same as the discriminative stimuli, the task or performance is called "identity" matching, and when sample and discriminative stimuli are phys- ically different, the performance is called "nonidentity," "arbitrary" or "symbolic" matching (see Sidman and Tailby, 1982 for a more detailed account of the relationship between conditional discriminations and matching to sample). A delay between presentation of the conditional or sample stimulus (e.g., C1) and the availability of the discriminative or choice cues (e.g., D1 and D2) defines a "delayed matching to sample" (DMTS) procedure and is considered a versatile technique for studying STM, or "working" memory, in animals. The DMTS task can be used to illustrate the distinction between working memory and LTM or "reference" memory. When the animal is given cues (D1 and D2), it has to remember or retain a mental representation of the sample (C1) in order to choose the appropriate discriminative stimulus, and such information need be retained only long enough to complete a particular trial, after which the information is best discarded since it is not needed and it may interfere with the memory requirements of the next trial (e.g., presentation of C2). This is a type of working memory and illustrates retention for a limited time of recently acquired information within the context of enduring knowledge. Memory for C1 or C2 will not work if there is no enduring knowledge of "if C1, then D1; if C2, then D2." In contrast to information in working memory that may be disposed of following each trial, information about the relationship between the conditional stimulus, or sample, and the discriminative stimuli, or choice cues, had to be retained in memory on all trials. Such memory is called "reference" memory or LTM. (For a more detailed description of methodology, comparative results and theory about animal memory, see Honig, 1978; Roitblat, 1986; Domjan and Burkhard, 1986).

A variety of DMTS procedures have been modified for testing the working memory of a single dolphin (Kea) for particular acoustic signals generated electronically and presented under water (Herman, 1980). The most important feature of the acoustical modification of visual DMTS is that following the conditional, or sample, sound, the two choice sonic cues are presented successively with a 0.5 sec interval between the termination of the first cue and the onset of the second cue. The results of these tests showed that the auditory STM of a dolphin compares favorably with the visual STM of monkeys (D'Amato, 1973; Wright, Santiago, Sands & Uruicoli, 1984) and is superior to the visual STM of pigeons (Grant, 1976; Santiago and Wright, 1984). For example, when "novel" sample sounds were played for 2.5 sec, the dolphin had virtually a perfect memory for the sounds, even with delays as long 120 seconds, whereas pigeons have much poorer retention after delays of only 20 seconds-- particularly with visual samples having a duration < 4 sec. In other experiments using just two familiar sounds differing in frequency and amplitude modulation, Kea's memory in either identity or symbolic DMTS was stable and accurate out to 50 sec delays. Moreover, identity "matching" remained accurate even when the sample duration was reduced to only 3 msec! However, evidence from several experiments showed that Kea's retention of sounds was degraded by proactive interference--probably because she "confused" the two sample sounds--as well as by retroactive interference--probably because Kea's "rehearsal" of the sample sound was disrupted when a different sound was inserted between the sample and the choice cues for almost the entire 15-sec delay interval. Finally, the dolphin Kea was given a serial probe recognition (SPR) task, i.e., a multiple memory task, in which a list of different sounds (drawn from a pool of 800) was presented as a "multiple" sample with the last or most recent item followed by a single test sound. The dolphin was trained to press either a "Yes" or a "No" paddle to indicate whether or not the test sound was in the list of items. Kea was correct on about 85% to 95% of the trials when list length consisted of three or fewer items, and her performance deteriorated to about 70% correct when a maximum list length of six items was given. The dolphin's serial position curve showed a pronounced recency effect, i.e., there was superior performance on terminal items, suggesting that new items displace or degrade the quality of older items held in a dolphin's working memory.

In summary then, a single dolphin has, thus far, been shown to have memory skills for acoustical signals which converge with the those of monkeys for visual signals and suggest a specialized sonic memory related to social communication and echolocation. The memory of a dolphin is degraded by both proactive and retroactive interference, as has been shown with all other vertebrates thus far tested. The ability of the dolphin Kea to maintain in memory a 3-msec duration sound for more than 50 seconds also indicates a specialized sonic memory.

Learning Set and Rule Learning

The initial idea of "learning to learn" or learning set (LS) formation by monkeys was developed by Harry Harlow (1959) in order to explain all kinds of transfer of learning between problems of a single type. These included Pavlovian learning as well as Kohler's "insight" learning (1927), where chimpanzees, after being previously stymied in their attempts to reach food, suddenly hit upon the solution of using a variety of tools to extend their reach. In a typical LS paradigm, monkeys are given a series of two-choice visual discrimination problems and typically they solve the initial ones slowly in "trial and error" fashion before solving novel problems of the same type following a single information trial. It was first shown empirically and unequivocally by Schusterman (1962; 1964) that the solution to such problems by chimpanzees depends on the development of an abstract rule (the same rule sometimes used in solving discrimination-reversal problems); a rule which goes beyond the concrete characteristics of the visual stimuli. Such an abstract rule or strategy has been termed "win-stay, lose-shift" with respect to the discriminative stimuli.

Conceptual abstractions, like win-stay, lose-shift, which are unrelated to specific perceptual relations of immediate experience, are also involved in generalized matching-to-sample and sameness-difference problems. For example, although pigeons can treat the relationship between A and A as the same, or B and B as the same (a first-order relationship), they have difficulty recognizing that the relationship existing between A and A is the same as the relationship existing between B and B, i.e., pigeons may be incapable of conceptualizing a relationship about a relationship (a second-order relationship). However, chimpanzees and some other nonhuman primates appear quite capable of rapidly learning to classify almost any stimulus pair as the same, regardless of whether or not they lie along the same stimulus generalization gradient (Premack, 1978; 1983). [For a summary of different viewpoints on the ability of pigeons and chimpanzees to acquire concepts, see Roitblat (1986)].

The dolphin Kea, much like Schusterman's chimpanzees on visual problem-solving, was also shown to be capable of developing a win-stay, lose-shift strategy, and thus eventually could use a single information trial to solve any one of a number of two-choice auditory discrimination or discrimination reversal problems (Herman, 1980). Furthermore, the same dolphin later learned auditory matching and demonstrated a generalized auditory matching-to-sample capability analogous to that shown by nonhuman primates with visual stimuli (Herman, 1980). A different type of rule learning which emphasized reinforcement training was demonstrated with a rough-toothed dolphin. Only those motor patterns which were novel were selected for reinforcement by the trainers. Following a period of frustration and confusion, the dolphin started performing novel behaviors, such as spitting and swimming in corkscrew patterns, demonstrating rule learning and some degree of creativity (see Pryor, 1986 for a review of reinforcement training and cognition in dolphins). Yet still another example of a dolphin learning set comes from studies of vocal mimicry with the dolphin Ake. She was trained to mimic specific whistle sounds and then tried to mimic any novel sound she heard (Richards, 1986). Such accomplishments by dolphins suggest that, at least within the acoustical modality, they may be capable of coding environmental information or "thinking about the world" in relatively abstract ways unencumbered by perceptions arising from immediate experience.

696

From a cognitive standpoint, one of the features of animal sonar systems that is particularly intriguing is that, unlike the more passive ways of perceiving the environment, doing it actively with echolocation allows a researcher with the appropriate equipment to monitor changes or adjustments in an animal's echolocating signals and thus look for signs of active information processing prior to overt behavioral decision-making. By designing experiments in animal sonar which seek to determine the nature and causal sequence of the processes operating on representations, we will most certainly advance our current theories of cognition as well as identify the kinds of information conveyed to a given species by the echoes of its sonar emissions (Simmons, 1980). The consequences of such a research program should give us a much better understanding of the types of interactions occurring between concept-driven and sensory-driven processes.

Concept-Driven Processes

There are several recent examples demonstrating that a dolphin's knowledge and expectancies affected its performance on an echolocation task. In one of the studies, the goal was to induce a cognitive state in the dolphin that can best be described as "failure expectancy" (as reflected in response bias) and then eliminate or "extinguish" the failure expectancy. Details of the study can be found in the original paper (Schusterman, Kersting and Au, 1980). Suffice it to say that in a biosonar study of material discrimination, the dolphin, Sven, could distinguish between a small hollow aluminum cylinder and a small glass cylinder with virtually perfect accuracy, but could not distinguish between the same types of large cylinders (twice the size of the small ones). By striking one of two response paddles, the dolphin indicated which of the cylinders was present, aluminum or glass. Throughout the insolvable problem, the animal biased its response, nearly always choosing the paddle associated with glass, suggesting it was echolocating in a perfunctory fashion. We therefore reasoned that, if Sven was given test sessions (consisting of 64 trials) with the solvable problem which were intermittently and unpredictably preceded by sessions with the unsolvable problem, then the dolphin might keep its sonar turned down or off during the initial stages of the solvable problem and thus err in favor of the paddle associated with glass. Figure 2 is a summary of the results which confirmed the hypothesis. The dolphin did indeed make significantly more errors by biasing its choice response during the first half of the sessions with the solvable problem. Interestingly, the failure expectancy was so persistent that it took 20 consecutive sessions of the solvable problem to extinguish it the first time and 10 consecutive sessions of the solvable problem to extinguish failure expectancy the second time. Unfortunately, the dolphin's click trains were only sporadically monitored during a single session of the solvable problem, and with this sparse data we could discern no clear relationship between click trains and performance. We therefore interpreted our results as meaning that the dolphin's failure expectancies diminished its attention for listening to previously established distinctive echoes. This interpretation is similar to one suggesting that bats having learned to rely on spatial memory pay less attention to echoes in an obstacle avoidance task (see Griffin, this volume). However, there is a high likelihood that the dolphin Sven did turn his sonar down when he expected the insolvable problem. Evidence for this effect comes from some recent work on detection tasks.

In experimental studies of echo ranging by dolphins (Au, Penner and Kadane, 1982), click emissions were monitored more assiduously than

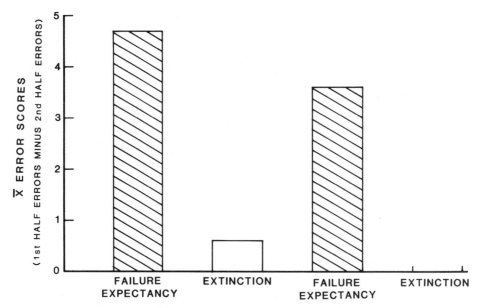

Fig. 2. Summary of results showing a dolphin's errors on a solvable
discriminative echolocation task as a function of induced
"failure expectancy," extinction of the "failure expectancy,"
re-induction of "failure expectancy," followed again by
extinction.

described above, and click intervals (which were greater than the two-way
transit time) were found to be similar during correct detections (target
present) and correct rejections (target absent), demonstrating that dolphins clearly expected to hear an echo at a specified time following
pulse emission. In fact, like the failure expectancy study, a dolphin
may fail to report target presence for a few trials after the range has
been changed, even though the dolphin had previously perfected detection
at that range. In a psychophysical study on the noise limitations of
echo-ranging in two dolphins, Au, Penner and Kadane (1982) continuously
monitored click emissions. Among other things, they found that during an
insolvable target detection task (masked by noise) in which the animals
showed response bias, one dolphin did not emit any detectable clicks on
about 30% of the trials while the other dolphin failed to emit clicks on
14% of the trials despite behaving as if they were echolocating. These
results do indeed show that when food-motivated, echolocating dolphins
have knowledge, they use that knowledge in a global way, they adjust
motor output as well as sensory input and they will place themselves on a
50% partial reinforcement schedule rather than refuse to work at a task.

Perception and Conception

Recently, Nachtigall and Patterson (1980) in a brief report described training a dolphin on what they called a sameness-difference
echolocation problem. The animal was required to report whether or not
two simultaneously presented targets were the same or different. The
targets were constructed to be aspect independent, and differences between targets were made as obvious as possible by varying the size, shape
and material. After the dolphin perfected the original discrimination
with two pairs of targets, two novel pairs were introduced and perfect
transfer ensued. This work suggests that, perhaps like chimpanzees in
the visual modality, dolphins by means of echolocation may be able to
conceptualize abstract relationships. However, the evidence in this

report of dolphins forming second-order relationships about targets ensonified by echolocation clicks is not without some ambiguity. For
example, it is not clear what stimulus relationships were perceived by
the dolphin, i.e., whether the sonar targets belonged to similar or
different perceptual classes (see Premack 1978 for a full discussion of
the problem of distinguishing between perceptual relations or first-order
relations and conceptual relations or second-order relations). There is
always a difficulty in understanding the nature of stimulus equivalence.
Pattern recognition, stimulus generalization and concept formation in
echolocating dolphins have been discussed by Schusterman (1980), who
showed that even in a well controlled study of dolphin echo-recognition
by Hammer (1978), the animal probably used a simple detection of a single
set of features to classify a whole range of targets differing in materials, such as aluminum, bronze, glass and steel. Thus, except for
detection tasks like target size, the more complex psychophysics of
recognition and classification by dolphin sonar has, as yet, not been
clearly elucidated, and the dolphin's world of echo information is still
relatively poorly understood. Moreover, the intermodal equivalence of
sonar cues and reflected light cues from common objects remains a rich,
untapped source for cognitive research with the high EQ dolphins.

ACKNOWLEDGMENTS

The writing of this paper was fully supported by Office of Naval
Research Contract N00014-85-K-0244. I thank Forrest G. Wood for his
cogent and "fussy" comments on the original manuscript. I also thank all
the people in Helsingor, Denmark, who almost literally blew my mind at
the Animal Sonar Systems Conference, 1986.

REFERENCES

Au, W. W. L. and Turl, C. W., 1984, Dolphin biosonar detection in
    clutter: Variation in the payoff matrix, J. Acoust. Soc. Am.,
    76:955.
Au, W. W. L., Penner, R. H. and Kadane, J., 1982, Acoustic behavior of
    echolocating Atlantic bottlenose dolphins, J. Acoust. Soc. Am.,
    71:1269.
D'Amato, M. R., 1973, Delayed matching and short-term memory in mon-
    keys, in: "The Psychology of Learning and Motivation: Advances
    in Research and Theory," G.H. Bower, ed., Academic, N.Y.
Domjan, M. and Burkhard, B., 1986, "The Principles of Learning and
    Behavior," Brooks/Cole, Monterey.
Gardner, H., 1985, "The Mind's New Science," Basic Books, N.Y.
Grant, D. S., 1976, Effect of sample presentation time on long-delay
    matching in the pigeon, Learning and Motivation, 7:580.
Hammer, C. E., 1978, Echo-recognition in the porpoise (Tursiops
    truncatus): An experimental analysis of salient target
    characteristics, Naval Ocean Systems Center, San Diego, Tech. Rep.
    192.
Harlow, H. F., 1959, Learning set and error factor theory, in:
    "Psychology: A Study of a Science," S. Koch, ed., McGraw-Hill,
    N.Y.
Herman, L. M., 1980, Cognitive characteristics of dolphins, in:
    "Cetacean Behavior: Mechanisms and Functions," L.M. Herman, ed.,
    Wiley-Interscience, N.Y.
Herman, L. M., 1986, Cognition and language competencies of bottlenosed
    dolphins in: "Dolphin Cognition and Behavior: A Comparative
    Approach," R.J. Schusterman, J.A. Thomas, and F.G. Wood, eds.,
    Lawrence Erlbaum Associates, Hillsdale, N.J.

Honig, W. K., 1978, Studies of working memory in the pigeon, in: "Cognitive Processes in Animal Behavior," S. H. Hulse, H. Fowler, and W. K. Honig, eds., Erlbaum, Hillsdale, N.Y.

Jerison, H. J., 1973, "Evolution of the Brain and Intelligence," Academic Press, N.Y.

Jerison, H. J., 1986, The perceptual worlds of dolphins, in: "Dolphin Cognition and Behavior: A Comparative Approach," R. J. Schusterman J. A. Thomas and F. G. Wood, eds., Lawrence Earlbaum Associates, Hillsdale, N.J.

Kellogg, W. N.1, 1961, "Porpoises and Sonar," University of Chicago Press, Chicago.

Kohler, W., 1927, "The Mentality of Apes," Reprinted by Vintage Books, N.Y.

Krebs, J. R., 1973, Behavioral aspects of predation, in: "Perspectives in Ethology," P. P. G. Bateson and P. H. Klopfer, eds., Plenum Press, N.Y.

Levine, M., 1959, A model of hypothesis behavior in discrimination learning set, Psychol. Rev., 66:353.

Lilly, J. C., 1961, "Man and Dolphin," Doubleday, N.Y.

McBride, A. F. and Hebb, D. O., 1949, Behavior of the captive bottlenose dolphin, Tursiops truncatus, J. Comp. Physiol. Psychol., 41:111.

Nachtigall, P. E. and Patterson, S., 1980, Training of a sameness/ difference task in the investigation of concept formation in echolocating Tursiops truncatus, in: "Proceedings of the International Marine Animal Trainer Association Conference, October 28-31, 1980," J. Pearson and J. Barry, eds., Boston: New England Aquarium.

Neisser, J., 1967, "Cognitive Psychology," Appleton-Century-Crofts, N.Y.

Premack, D., 1978, On the abstractness of human concepts: Why it would be difficult to talk to a pigeon, in: "Cognitive Processes in Animal Behavior," S. H. Hulse, H. Fowler, and W. K. Honig, eds., Erlbaum, Hillsdale, N.J.

Premack, D., 1983, The codes of man and beasts, The Behavioral and Brain Sciences, 6:125.

Pryor, K., 1986, Reinforcement training as interspecies communication, in: "Dolphin Cognition and Behavior: A Comparative Approach," R. J. Schusterman, J. A. Thomas and F. G. Wood, eds., Lawrence Earlbaum Associates, Hillsdale, N.J.

Restle, F., 1958, Toward a quantitative description of learning set data, Psychol. Rev., 65:77.

Richards, D. G., 1986, Dolphin vocal mimicry and vocal object labelling, in: "Dolphin Cognition and Behavior: A Comparative Approach," R. J. Schusterman, J. A. Thomas and F. G. Wood, eds., Lawrence Earlbaum Associates, Hillsdale, N.J.

Roitblat, H. L., 1986, "Introduction to Comparative Cognition," W. H. Freeman and Co., San Francisco.

Schusterman, R. J., 1962, Transfer effects of successive discrimination reversal training in chimpanzees, Science, 137:422.

Schusterman, R. J., 1964, Successive discrimination-reversal training and multiple discrimination training in one-trial learning by chimpanzee, J. Comp. Physiol. Psychol., 58:153.

Schusterman, R. J., 1980, Behavioral methodology in echolocation by marine mammals, in: "Animal Sonar Systems," R. G. Busnel and J. F. Fish, eds., Plenum, N.Y.

Schusterman, R. J., Kersting, D. and Au, W. W. L., 1980, Response bias and attention in discriminative echolocation by Tursiops truncatus, in: "Animal Sonar Systems," R. G. Busnel and J. F. Fish, eds., Plenum Press, N.Y.

Schusterman, R. J. and Krieger, K., in press, Artificial language comprehension and size transposition by a California sea lion (Zalophus californianus), J. Comparative Psychol.

Sidman, M. and Tailby, W., 1982, Conditional discrimination vs. matching to sample: An expansion of the testing paradigm, J. Exp. Anal. Behav., 37:5.

Simmons, J. A., 1980, The processing of sonar echoes by bats, in: "Animal Sonar Systems." R. G. Busnel and J. F. Fish, eds., Plenum Press, N.Y.

Wood, F. G., 1986, Social behavior and foraging strategies of dolphins, in: "Dolphin Cognition and Behavior: A Comparative Approach," R. J. Schusterman, J. A. Thomas and F. G. Wood, eds., Lawrence Earlbaum Associates, Hillsdale, N.J.

Wood, F. G. and Evans, W. E., 1980, Adaptiveness and ecology of echolocation in toothed whales, in: "Animals Sonar Systems," R. G. Busnel and J. F. Fish, eds., Plenum Press, N.Y., 381.

Wright, A. A., Santiago, H. C., Sands, S. F. and Urcuioli, P.J., 1984, Pigeon and monkey serial probe recognition: Acquisition, strategies, and serial position effects, in: "Animal Cognition," H. L. Roitblat, T. G. Bever and H. S. Terrace, eds., Lawrence Earlbaum Associates, Hillsdale, N.J.

LOUD SOUNDS DURING FEEDING IN INDIAN OCEAN BOTTLENOSE DOLPHINS

Rachel Smolker and Andrew Richards

Long Marine Laboratory
University of California
Santa Cruz, California 95064 U.S.A.

INTRODUCTION

The hypothesis that odontocetes use intense sounds to debilitate their prey, was put forth in detail by Norris and Mohl in 1983. While they were able to draw together a strong body of indirect evidence in support of this hypothesis, definitive observations of prey debilitation in action have been lacking from studies of wild or captive dolphins. In the course of fieldwork on the behavior and vocalizations of wild Indian Ocean bottlenose dolphins (Tursiops sp.), we observed dolphins producing loud sounds while feeding. We present these observations and discuss them as evidence for acoustic prey debilitation. We also present other hypotheses as to their origin.

METHODS

These observations were conducted in Shark Bay, Western Australia during four months in the austral winter of 1985 and three months in the austral winter of 1986. At a small campground called Monkey Mia, eight dolphins have become habituated to contact with humans and will accept fish fed by hand and physical contact from visitors to the camp. These eight are part of a much larger population in the area, many of which we have learned to recognize by their dorsal fins. Approximately 60 dolphins can be observed regularly at close range and for many hours at a time. Observations were made from a small boat and while standing in shallow water near shore, at distances ranging from 2 to 60 meters.

All recordings of dolphin sounds were made using audio-range equipment. A Sony TC-D5M cassette recorder with TDK SA-X or TDK Metal tapes was used during both 1985 and 1986. In 1985 a Y^2 limited, yak-yak hydrophone with a Barcus Berry preamplifier was used for all recordings, and in 1986 a Magnavox Q41B sonobuoy hydrophone was used for all but one recording session. Both systems were responsive from 20 Hz to 18 kHz, with the Magnavox system known to be flat ± 3 dB in that range. In 1986 video recordings with concurrent audio were made with a Sony CCD-V8-AFu camcorder, the Y^2 hydrophone, the Barcus Berry preamplifier, and Sony metal particle videotapes. This system was responsive from 20 Hz to 15 kHz, but unfortunately it had automatic gain control which minimized amplitude differences. A Hewlett-Packard 5451C fourier analyzer was used for sound analysis.

RESULTS

A total of 41.5 hours of recordings of dolphin vocalizations were made
from groups of dolphins engaged in various activities, 28 hours in 1985 and
13.5 hours in 1986. These activities were defined broadly as feeding, resting
and social interaction. These recodings contained a total of 1,041 relatively
loud impulsive sounds, reffered to here as "bangs". These bangs were generally much louder than any other sound at a given instrumentation setting
and usually startled the person monitoring the recording. Qualitatively, the
bangs ranged from a sharp crack to a loud "tch", with most of the sounds at
the "tch" end of the spectrum. Eight percent of the bangs were recorded
either during social interactions or in ambiguous contexts. This category
encompassed almost all the crack type bangs. Bangs which occurred during
social interactions often occurred in volleys of several bangs in rapid
succession which we have counted as one event. The remaining 92% of the bangs
occurred during behavioral contexts which clearly involved dolphins feeding
on fish. The following discussion will be limited to these 954 bangs, 582
from 1985 and 372 from 1986.

Ninety-eight percent of the bangs associated with feeding were recorded
during 12 sessions in 1985 and 14 sessions in 1986, each of which contained
five or more bangs. During 19 of the 26 sessions, fish were visible below the
water surface or were observed jumping or tumbling out of the water from
amongst the dolphins. During the remaining sessions feeding was inferred from
the behavior of the dolphins and the presence of seabirds catching fish among
them. Often the fish could be identified as Perth herring (Nematolosa
vlaminghi), a clupeid of 10 to 20 cm in length, common in these waters. The
dolphins, in groups of 2 to 20 individuals, fed on the fish in waters ranging
in depth from less than 1 m to 10 m, and sometimes within 1 m of shore.

The dolphins typically milled around and through a school of fish,
emitting click trains and some whistles. Occasionally a dolphin broke into
a very fast burst of speed, apparently in pursuit of a fish isolated from
the school. These chases often terminated with a loud bang and sometimes a
surface disturbance, although bangs were also heard without any clearly
associated events. There were some differences between the observations of
1985 and 1986. In 1985 this behavior was usually observed in water 1 to 2 m
deep, where surface activity was more frequent. There were 131 instances of
surface disturbances, all sprays of water, which appeared to be produced
simultaneously with the bang. On 57 occasions we observed a fish emrging
from these splashes and tumbling through the air for a short distance. On
three occasions, two fish tumbled out together. On four occasions these fish
were observed on landing to lie motionless or quivering, too stunned to swim
away before being grabbed by a dolphin. In 1986, the dolphins tended to feed
in deeper water, where it was more difficult to observe their behavior. While
we saw an occasional event similar to those of 1985, surface disturbances
consisted most often of uwellings or "flukeprints" appearing immediately
after the bang. Only twice did we see fish tumbling out of the water.

The precise details of behavior at the instant in which the bang was
produced were extremely difficult to observe, since it occurred very quickly
and was often either obscured by splashing, or took place out of sight, below
the water surface. In 1985 we observed 17 cases of chases terminating in a
fast pinwheeling motion by the dolphin, without being able to tell if this
occurred before, during or after the production of the bang. On 10 occasions
in 1986 we clearly observed fast tail swiping motions associated with the
production of bangs. Typically, a dolphin swam very fast for a short distance
and swung the tail flukes around suddenly in a horizontal pinwheeling motion
through roughly 60 to]80 degrees. The bang was produced sometime during the
swipe, after the dolphin had already begun to turn. Nine of these events were
videorecorded. Observations which were incomplete, but strongly suggestive

of tail swiping were made on an additional 18 occasions.

Random samples of 110 bangs from 1985 and 62 bangs from 1986 were chosen and oscillograms and power spectra were examined. The oscillograms showed a prolonged oscillatory waveform of roughly constant amplitude. Full amplitude was reached from within 1 ms to more than 20 ms. Decay was more gradual than the onset. The mean duration of the 1985 sample was 133 ms, (SD=32 ms), while the mean duration of the 1986 sample was 103 ms, (SD=25 ms). The significant difference between the means remains unexplained. It may indicate generally different causes for the sounds from the two samples, or it may be the result of some environmental phenomenon such as the generally shallower water of the 1985 events. The power spectra of these sounds varied to the extent that it was difficult to typify them. In general, the recorded sound energy was concentrated between 200 and 6000 Hz. The common patterns found in the spectra included spectra with energy concentrated below 1 kHz, spectra with energy concentrated around 5 kHz, and a bimodal distribution combining these two. However there were many that did not fit this description and more importantly there was little correlation with known events. For example, the spectra of sounds associated with water sprays encompassed all the above types and more. The nonbiological possibilities for the sources of this variation include multiple path effects varying with water depth, hydrophone placement, and bottom type, large temperature and salinity gradients in the bay, and equipment effects such as hydrophone ringing, tape saturation, amplifier clipping, and restricted (audio) bandwidth. Calculations of the intensity of these sounds were not made due to lack of the appropriate data.

DISCUSSION

Three major hypotheses have been proposed for the origin to these sounds. One is that they are internally produced by a dolphin and are an example of acoustic prey debilitation as discussed by Norris and Mohl. The second is that the sounds are produced by cavitation during a tail swipe. The third is that they are the sound of an impact between a dolphin's flukes and a fish. Because we were not able to see the details of most of the behavior involved in the production of these bangs, circumstantial evidence becomes the key in the attempt to elucidate the origin of these sounds. However these hypotheses are not necessarily mutually exclusive and we may eventually find that they are all operative under different circumstances.

Captive observations incicate that dolphins can produce loud impulse sounds via some internal mechanism (Caldwell et al. 1962, Lilly 1962, Tavolga & Essapian 1957). Au et al. (1974) reported that two Tursiops were capable of emitting sounds with sound pressure levels as high as 228.6 dB re 1 uPa at 1 yd. Such sound pressure levels are equal to the lethal threshold of fish (Norris & Mohl 1983). Thus it is reasonable to hypothesize that the sounds we have recorded were internally produced and were stunning fish. However, additional evidence in some cases contradicts this hypothesis. The numerous instances of water sprays and fish ejected from the water associated with sound production, are much more plausibly explained with tail swipes than as displacement effects from a sound field. In ten instances sound production was clearly during a tail swipe when the dolphin was faced away from the presumed direction of its prey.

The displacement problem and some of the tail swipes can be dealt with in a modification of this hypothesis (Norris pers. comm.). In this the sound is produced by the dolphin internally and it debilitates but does not kill the prey. The prey's remaining ability to avoid capture lies in its integration with its school, and this is easily disrupted by the rapid swinging of flukes in the vicinity of the prey. At times the prey may even be thrown from the water with such a maneuver. This is consistent with all the observations except the ten instances of sound production during a tail swipe.

The second hypothesis proposes that bangs are produced during tail swipes by cavitation. Here debilitation may be produced by the sound itself, by water displacements producing pressure gradients and shear forces either acting on tissue or causing severe disturbance to mechanoreceptors, or by physically hitting the prey. This is of course cosistent with all the observations. The cavitation mechanism is supported by two instances recorded on video, of tail swipes in a social situation, that apparently only contact water yet produce some sound, although not clearly a bang. However there are other cases of tail swipes with no exceptional sound.

The third hypothesis, that debilitation and sound production are both results of physical contact between flukes and fish, prehaps best explicates all the observations, although we have never been able to see this directly. We have, however, observed a few cracks and bangs produced in social situations which were clearly caused by tail hits against cospecifics. A volly of such sounds is produced by rapid mutual hitting among several dolphins. Wild Atlantic bottelnose dolphins have been observed to hit fish with their tails by Wells (1986), and have been observed pinwheeling by Leatherwood (1975). Hult (1982), described a captive Atlantic bottlenose dolphin hitting a fish with her flukes. Hult mentioned that the "sound of the blow was immediately discernible", through the walls of the tank. He also reported 15 swipes that missed, without mention of any sound. This dolphin was also observed to "jaw clap" at the fish, though Hult does not mention any sound associated with this either, and did not report any effect on the fish.

To our minds the contact hypothesis offers the most parsimonious explanation for some of the bangs, and provides a concrete mechanism for debilitation. However, given the variation in sounds subsumed under the term bang, and the incomplete nature of the observations, it is not now possible nor desirable to ascribe one cause to all of them. Acoustic prey stunning may not involve any overt motion whatsoever and is likely to be very difficult to document, even in captivity. There may well be more than one way to debilitate a fish, and if there are, they are all likely to be used by the opportunistic Tursiops.

REFERENCES

Au, W. W. L., Floyd, R. W., Penner, R. H., and Murchison, A. E., 1974, Measurement of echolocation signals of the Atlantic bottlenose dolphin, Tursiops truncatus Montague, in open waters, J. Acoust. Soc. Am., 56(4):1280.
Caldwell, M. C., Haugen, R. M., and Caldwell, D. K., 1962, High-energy sound associated with fright in the dolphin, Science, 138:907.
Hult, R., 1982, Another function of echolocation for bottlenose dolphins (Tursiops truncatus), Cetology, 47:1.
Leatherwood, S., 1975, Some observations of feeding behavior of bottlenose dolphins (Tursiops truncatus) in the northern Gulf of Mexico and (Tursiops cf. T. gilli) off southern California, Baja California and Nayarit, Mexico, Marine Fisheries Review, 37(9):10.
Lilly, J. C., 1962, Vocal behavior of the bottlenose dolphin, Proc. Am. Philos. Soc., 106(6):520.
Norris, K. S., and Mohl, B., 1983, Can odontocetes debilitate prey with sound?, Am. Nat., 122:85.
Tavolga, M. C., and Essapian, F. S., 1957, The behavior of the bottlenose dolphin (Tursiops truncatus); mating, pregnancy, parturition, and mother-infant behavior, Zoologica, 42:11.
Wells, R. S., 1986, Structural aspects of dolphin societies, Ph.D. thesis, University of California at Santa Cruz.

ATTENTION AND DETECTION IN DOLPHIN ECHOLOCATION

Ralph H. Penner

Naval Ocean Systems Center

A systematic and predictable relationship exists between the distance to detected targets and the time between dolphin echolocation pulses. Two-way acoustic travel time accounts for the major part of the interpulse interval at detection distances beyond 10 meters. An additional time over the necessary acoustic two-way travel time completes the interval and remains relatively constant as interpulse interval lengths increase or decrease in adjustment to increasing or decreasing target distance (Morozov et al., 1972; Penner, 1981). Real-time monitoring of dolphin echolocation pulses, allows an experimenter to know the distance at which an animal is focusing its attention. Previous studies on human attention (e.g. Posner et al., 1979) have demonstrated that detection of signals can be directly effected by similar attention mechanisms.

Echolocation detection was studied using a two alternative forced choice procedure with successive presentation of targets. The dolphin (Tursiops truncatus) was required to station within a circular hoop, echolocate, and indicate the presence or absence of a target by hitting one of two manipulanda located immediately to its right or left (figure 1). Dolphin attention in this study was examined by changing target distance conditions. Three target conditions were presented: (1) targets presented at only one distance per session, (2) random blocks of ten trials with targets presented at one of five distances per block, and (3) trial-to-trial random sessions with five distances presented with no sequential trials at the same distance. Each condition was made up of multiple 100 trial sessions. The five target distances were 40, 60, 80, 100, and 120 m and the targets were acoustically bright hollow aluminum cylinders in diameter and high with target strengths equal to -11dB. Target presence or absence was randomized with a .5 probability of occurrence.

Dolphin pulses were monitored by a hydrophone located 2 m in front of the animal's stationing position. Distances to the targets were measured from that station. Information from the hydrophone was fed into an Apple II microcomputer that provided both stored and real time information on the number of pulses produced, the intervals between pulses, and the amplitude of the pulses (Kadane et al, 1980).

RESULTS AND DISCUSSION

Attention studies have generally shown that concentration on a position where signals are likely to occur results in improved detection performance. Figure 2 shows the animal's overall percent correct performance on both target-present and target-absent trials for the five distances across the three target conditions. A comparison of correct rejection performances reveals that the animal was nearly always correct at reporting that targets were in fact not present. Comparisons of correct-detection performances however show that if only one target distance was presented per session (figure 2a) the animal easily detected targets out to 100 m but if many distances were presented, and the animal could not predict the focusing distance, performance began to drop off at 60 meters (figures 2b and 2c). Clearly, being able to attend to only one distance allowed the animal the opportunity to more readily detect targets. Psychophysical detection thresholds are influenced by the manner in which the data are collected (Nachtigall, 1980; Schusterman, 1980). In this case, if an animal only needs to attend to one distance it is much more likely to detect the target than if it is required to search from 40 to 120 m.

Dolphin echolocation is unique in allowing an objective look at the attention process contributing to the performance differences shown in figure 2. The distance at which an animal is attending may be observed by examining the intervals between echolocation pulses. Or more specifically, by examining the mode (the most frequently occurring) interpulse interval in the three target conditions. Figure 3 displays the modes of the interpulse interval distributions under the three target distance conditions. Data on figures 3a and 3b show that under condition 1, where targets are presented at only one distance per session, and under condition 2 where targets are presented at one distance for at least 10 trials in a row, the animal quite logically attends to only one distance. Figure 3a particularly shows that the animal gates his echolocation pulses to the distance whether targets are present or absent. Contrast that to the data presented in figure 3c. The data from figure 3c were taken under the trial-to-trial random procedure. The animal had no way of knowing in advance whether the target would be presented at 40, 60, 80, 100, or 120 m. On those trials where a target was actually detected (correct detection) the modes of the click trains appear similar to the fixed distance data presented in figure 3a. When, however, no target was presented and the animal reported no target present (correct rejection), the interpulse intervals demonstrate that the animal normally scanned out to the farthest distance regardless of where a previously presented target appeared. This finding is further demonstrated in figure 4. Every interpulse interval in which the animal correctly detected a target at 40 m or correctly rejected on a trial following a 40 m target presentation is displayed in cumulative distributions of interpulse intervals. Figure 4a shows that when all targets in a session were presented at 40 m the dolphin searched at the 40 m distance whether targets were present or absent. The most frequently occurring interpulse intervals peaked sharply around 80 msec whether targets were presented or not indicating that the animal was sharply focusing on one distance. These data may be most vividly contrasted with those shown in Figure 4c. When targets occurred at any of the five distances, and the animal was required to detect targets between 40 and 120 m, the interpulse interval distribution peaked at 160 msec during correct rejection trials but peaked near 80 msec. when targets were in fact present at 40 m. The attention focus is most dramatically seen on the correct rejection trials and is directly influenced by the probable distance of previously presented echolocation targets.

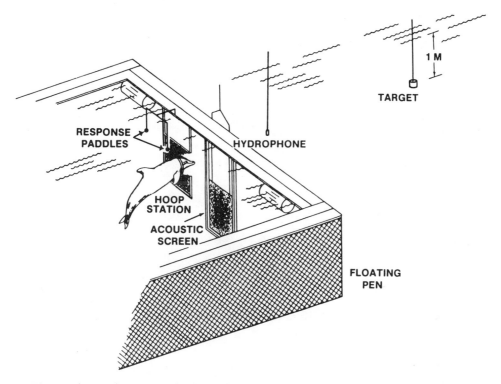

Figure 1.   Schematic of the dolphin's pen and experimental apparatus.

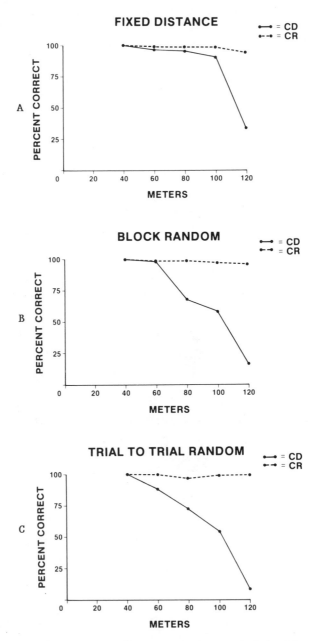

Figure 2. Percent correct performance on correct detection (CD) and a correct rejection (CR) trials with targets at five distances under three different conditions.

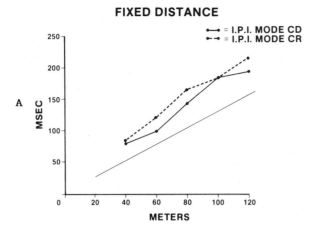

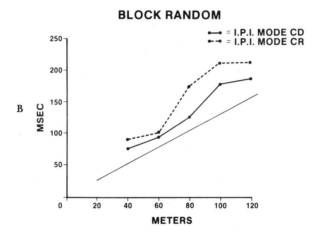

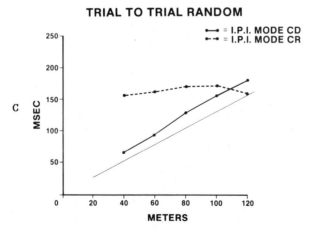

Figure 3. The modes of distributions of all intervals between echolocation clicks. The inter pulse interval (IPI) modes for correct detection (CD) and correct rejection (CR) trials.

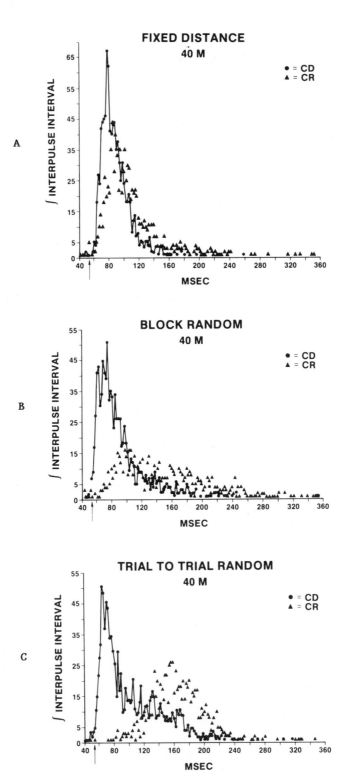

Figure 4. Number of inter pulse intervals (IPI) for both correct detection (CD) and correct rejection (CR) trials ranging from 40 to 360 milliseconds. Two-way travel time for sound presented with an arrow on the abscissa.

REFERENCES

Kadane, J. Penner, R. H., Au, W. W. L. and Floyd, R., 1980, Microprocessors in the collection and analysis of _Tursiops truncatus_ echolocation data, _J. Acoust. Soc. Am._, Suppl. 1, 68:58.

Morozov, V. P., Akopian, A. I., Burdin, V.L., Zaytseva, K. A., and Sokovykh, Y. A., 1972, Repetition rate of ranging signals of dolphins as a function of distance to target, Biofisika, 17(1):49-55.

Nachtigall, P. E., 1980, Odoncete echolocation performance on object size, shape and material, _in_: "Animal Sonar Systems," R. G. Busnel and J. F. Fish, eds., Plenum Press, New York.

Penner, R. H., 1981, Biosonar attending behavior in _Tursiops truncatus_, Abstracts, Fourth Biennial Conference on the Biology of Marine Mammals, San Francisco, CA.

Posner, M. I., Snyder, C. R. R., and Davidson, B. J., 1979, Attention and Detection of Signals, _J. Exp. Analysis of Behavior_, 32:363-372.

Schusterman, R. J., 1980, Behavioral Methodology in Echolocation by Marine Mammals, _in_: "Animal Sonar Systems," R. G. Busnel and J. F. Fish, eds, Plenum Press, New York.

ODONTOCETE SONAR SYSTEMS RESEARCH - FUTURE DIRECTIONS FROM A ETHOLOGIST'S

PERSONAL POINT OF VIEW

R. P. Terry

Information Theory Group
Delft University of Technology 2600 GA Delft, Netherlands

INTRODUCTION

The last decade of research on dolphin sonar capabilities expanded and clarified basic information previously in existence. Significant progress was achieved in understanding aspects of biosonar such as radiation patterns of sonar emission, physical components of odontocete sonar signals and the limitations of their discriminatory capabilities. Fundamental aspects of cetacean sonar capabilities are, however, still only partially understood. Elusive topics yet to be adequately addressed include the role of sonar in communicative behavior, the morphological basis for sound production and the information value of cetacean vocalizations.

The decade between 1965 and 1975 produced a tremendous surge of dolphin acoustical research. A review of the literature published since 1978 and a look at current work dealing with cetacean biology suggests that progress in the field of cetology involving medicine, husbandry, and dolphin social organization has outpaced the investigative work done on odontocete sonar systems. The trend to organize and carry through field investigations as compared to captive studies increased faster in areas of cetacean biology other than sensory and acoustical studies. Certainly financing, availability of time, equipment and facilities, all have restrictive influences on the variety and quantity of work being done. Other factors, however, may play a more subtle role in the pace of research.

New directions in applied research are needed to address some of the continuing gaps in information on dolphin sonar systems. Basic experimental design and behavioral methodology in sonar research continues to improve with experience. Equipment and techniques also continue to improve to meet expanded understanding concerning the range and limitations of cetacean vocalizations. Traditional avenues of research, some already in place two decades ago, may, however, limit advances by imposing a form of innovative "tunnel vision" on investigative patterns. In doing so they also limit a broader perspective on the problems under investigation. Experience since 1978 at several research facilities around the world suggest several positive directions for future research on dolphin sonar systems.

COOPERATIVE RESEARCH INTERACTION: CROSS-DISCIPLINE

Cooperation between researchers is fundamental to research. Cooperative, cross-discipline interaction between specialists in divergent fields increases the likelihood of success in attacking complex problems concerning cetacean biosonar. These problems are numerous. One that exemplifies the need for such cooperative interaction, however, involves the controversy over the source of sound production in odontocetes. Systematic attempts to trace the origin of sound produced by dolphins began in the 1960's and have since continued sporadically. The problem produced a number of inventive approaches to a solution, but there are currently still no absolute answers. All of the data gathered to date tends to lump the theories of sound production into two main divergent groups.

The first group points to the odontocete nasal system as the source of sound production. Ridgway (1983) gave the most recent review of this explanation which predominates in the United States. According to this theory, the nasal plugs, under muscular control, produce the sounds as air passes between the plugs and the nasal walls (similar to stretching the opening of an inflated balloon). The two nasal plug regions could produce sounds independently and simultaneously (including both sonar clicks and pure tone whistles). Although the experimental evidence reviewed by Ridgway (1983) points to the nasal passages and associated air sacs as the general source of most sound emissions, it still does not pinpoint the primary vibrating source or whether that source is the same for whistles and clicks.

The theory adopted by the second group traces the origin of sound production to the larynx of the dolphin. Purves and Pilleri (1983) outlined the evidence for a laryngeal source of dolphin vocalizations, a view still prevalent in some parts of Europe. This theory suggests that a series of membranes associated with the larynx (the rima glottidis between arytenoid cartilage) are the source of both whistles and clicks. Air from the lungs passes across the membranes which are under muscular control. The sound is created as the membranes stretch and relax. The air re-cycles through air sacs associated directly with the laryngeal membranes. Most of the physical evidence for this current explanation stems from much older investigative work by Purves (1967).

Promising work is currently in progress on the problem in California using X-ray computed tomography to computer image the process of sound production in a dolphin's head (Cranford, 1987), and in Sweden using 3-D computer graphics (Amundin, 1987). The problem, however, still awaits a definitive answer which will have wide implications in considering other aspects of sonar use by dolphins. The precise source of sound production also reflects directly on the method of sound propagation through the dolphin head (primarily via bone or soft tissue), the possible acoustic significance of fatty tissues in the melon, and the nature of sonar radiation patterns. Further, locating a precise mechanism for sound production will help with interpretation of sonar emissions from other odontocete species that vary greatly in functional anatomy - including the largest challenge, the Sperm Whale (Physeter macrocephalus).

It was evident at an early stage that anatomical dissection would not be sufficient to solve the problem. The experimental evidence for the source of dolphin vocalizations presented to date comes from a number of fields: anatomy, physiology, morphology, neurology, physics, electronics, engineering and even biochemistry. Unfortunately, the evidence has often come from separate disciplines working independently over long periods. A team approach to the same problem would ideally include specialists from each of the major fields involved to attack the problem simultaneously.

716

Several centers of cetacean research around the world have already found an interdisciplinary team approach valuable and productive.

The integration of specialists from different fields of study to concentrate on sonar systems allows optimal results with a minimum waste of time and maximum efficiency. A team working on problems involving the nature and use of echolocation by odontocetes might typically include specialists from the fields of acoustical physics, bioengineering, information theory, comparative psychology, behavioral biology, sensory physiology and taxonomy. Each member of the team contributes the ability to generate the characteristic insights derived from prolonged experience with one specialty area. All interact in compiling the final results. As biosonar research moves into some of its most complex realms, including cognitive research and the possible multi-varied uses of sonar, interdisciplinary cooperative research may become more than desirable; it may become absolutely necessary.

COOPERATIVE RESEARCH INTERACTION: INTERNATIONAL

Cooperative interaction in scientific research involves geography as well as specialty expertise. The present and past conferences on Animal Sonar Systems demonstrate this eloquently. Emphasis on increased international cooperation is still needed, however, as well as a continuing recognition of its value.

A striking example for this need comes from a recent, extensive review of cetacean sound recordings. Watkins and Wartzok (1985) presented a list of published sound recordings from 43 odontocete species which included ". . . the most comprehensive references for studies on the vocalizations". An omission is apparent in an otherwise competent and complete overview of sensory biophysics in marine mammals. Several significant and recent (post 1980) references to work completed outside the United States are not given. These include more recent sound analyses of recordings from the Boutu, Inia geoffrensis (Kamminga and Wiersma, 1981), the Tucuxi, Sotalia fluviatilis (Wiersma, 1982) and Commerson's Dolphin, Cephalorhynchus commersonii (Kamminga and Wiersma, 1982). In their place older recordings are listed. The most recent reference given for Sotalia is 1972. The predominance of recordings listed come from North American researchers, a poignant reminder of the early growth of research in this field and its domination geographically by researchers from America. The publication illustrates the role which subtle regional bias may still play in obtaining information even in an age of computer data banks. It is becoming increasingly easier to overlook data references from global research, and it is already past time to establish a central data bank specifically for cetacean research. The need certainly still exists for a continued awareness of the extent to which sonar research has spread across the globe, an expansive change, even in a decade, from the days when major investigations centered in America and isolated regions of Europe.

However the future value of international cooperation lies not only in keeping informed and communicating, it extends to simultaneously sharing experimentation. To overcome natural tendencies toward regional (even national) bias, shared experimental work is of significant value. Researchers often talk to one another within the international community but do not always "hear" each other due to differences in cultural backgrounds and experiences in approaches to the acquisition and dissemination of knowledge. A future emphasis on international joint projects could take advantage of exotic odontocete species available worldwide for research purposes and at the same time combine resources and close cultural gaps. This form of cooperative investigation can not only

verify (or undermine) tentative conclusions to a plethora of questions
concerning dolphin sonar capabilities - it is also simply good scientific
technique. Further, experiments done in conjunction on a global basis,
with similar results, may finally put to rest some of the oldest unsolved
controversies while strengthening the validity of previous results.

SONAR USE AND ADAPTIVE CAPABILITIES

The traditional preoccupation of most dolphin sonar research has focused
on the physical nature of the sonar signal, its capabilities and
limitations. With information now available on the physical parameters of
sonar emissions a shift in research intensity from the physical to the
biological realm is appropriate. The focus of investigative attention
needs to move out of the so-called "black box" that connects the production
of a sonar signal with a subsequent behavioral response. A new emphasis
needs to be placed on the use of sonar signals in ecological and behavioral
contexts.

There is still virtually no data concerning the environmental influences
concerning adaptability of cetacean sonar. Species with apparently
separate riverain and coastal marine populations, providing very different
ecological and behavioral challenges to sonar use, need to be singled out
for studies on sonar system capabilities. Such species include the Tucuxi
(Sotalia fluviatilis), the Irrawaddy Dolphin (Orcaella brevirostris) and
the Finless Porpoise (Neophocena phocaenoides). Members of each of these
species already exist in captive locations around the world. Even the
simplest discrimination experiments performed on Tursiops two decades ago,
may provide valuable missing information on sonar use and adaptability.

Areas devoid of or deficient in systematic research on the use and
application of sonar emissions in cognitive and behavioral domains are
numerous. They include the following:

(1) The origin of sonar production.

Few attempts have been made to trace the development of sonar in the
first year following birth. Early attempts to record sonar use
following birth suggest a latent period of several weeks before the
appearance of sonar signals and the possible role of learning in their
development (Caldwell and Caldwell, 1972). As success continues with
captive dolphin births, this area of investigation is becoming more
readily accessible. Two recent captive studies have shed some new light
on this old problem (Reiss, 1987; Schultz, 1987). Further work needs to
clarify any differences in the physical parameters of sonar signals from
immature to mature individuals, and even between females and males,
especially in species that exhibit any degree of sexual dimorphism.

(2) Adaptability and/or cognitive control in the use of sonar.

Only one paper to date (Au et al, 1985) suggests that there is some
degree of adaptability in the use of a dominant, or peak, frequency
utilized by a cetacean species (interestingly, the species used was the
Beluga, Delphinapterus, and not the ubiquitous Tursiops). With the
quantity of work done on Tursiops, however, no studies have attempted to
determine the extent of cognitive control over sonar emissions. The
problem could be approached during routine discrimination work by
masking the preferred dominant frequency of the species with interceding
noise of the same frequency.

718

(3) The use of sound cues to coordinate group behaviors.

It has been traditionally assumed that captive dolphins use sound
signals to coordinate trained behaviors. Simultaneous jumps, flips and
other coordinated group tasks point to the need for vocal cues. To
date, however, even with the long experience of trained show
performances, there is no confirmation of such signals. Conclusive
answers to the question would greatly aid in understanding social
organization and coordinated group behavior in the wild, including
feeding and protective strategies.

Sonar discrimination work on a variety of odontocete species, captive
research on sonar development and investigations into its use in behavioral
domains are possible now at numerous locations. Recent evidence for the
production of loud impulse sounds by dolphins during predation and social
interactions has lead to several possible new lines of research (Martin et
al, 1987). Work begun in the early 1970's (Caldwell and Caldwell, 1972) on
the intraspecific (and possibly interspecific) transfer of information via
sound also needs to be continued in light of present understanding of sonar
capabilities.

MULTI-SPECIES APPROACH TO SONAR SYSTEMS RESEARCH

Progress in understanding the sonar of cetaceans demands the expansion
of research to dolphin species other than the Bottlenose Dolphin. Tursiops
sp. has become both the experimental workhorse and the working model for
dolphin sonar capabilities. With 65 known odontocete species (31 of which
are delphinids), a relatively small number of Tursiops individuals have
provided the vast bulk of information on odontocete sonar systems. A
multi-species approach is now necessary for continued progress in the
understanding of dolphin sound production and use.

The sensory biophysics review of odontocete vocalizations by Watkins and
Wartzok (1985), as well as an earlier review by Evans (1973), demonstrate a
continuing problem in the systematic study and analysis of dolphin sonar.
The data accumulated come from a broad time span, a variety of equipment
and probably a number of recording techniques under both captive and wild
conditions. In the Watkins and Wartzok review, the latest references given
for sound recordings from individual odontocete species range from 1962
(Grampus griseus) to 1983 (Tursiops truncatus). Under these conditions,
the validity of the data for comparison purposes becomes a function of the
time elapsed between recordings versus the advances in recording technology
over the same time period. The use of high quality, standard equipment and
techniques for cataloging vocalizations from a wide range of dolphin
species allows less variation and more certainty in the interpretation and
comparison of results. Interspecies comparisons of different sonar signal
components correspondingly gain increased validity. This systematic
approach is already being applied (Kamminga and Wiersma, 1981). It now
needs to be expanded to sample variations and similarities between sonar
signals from a large number of riverain, littoral and pelagic species.
Only with the background of such data can valid, broad conclusions be drawn
concerning odontocete sonar use and capabilities.

Other odontocete genera and species must yet undergo the same research
scrutiny previously applied only to Tursiops. The Tucuxi, Sotalia
fluviatilis, provides a good example of the need for such a systematic and
multispecies approach to sonar investigations. Sotalia is a small
delphinid species with separate and apparently isolated populations living
along the coast and in the rivers of northern South America. Sonar signals
of the marine population from the coast of Columbia (Sotalia fluviatilis

guianensis) was recorded under optimal conditions in captivity (Wiersma, 1982). The sonar clicks showed a two component nature with simultaneous high and low dominant frequencies at 95 kHz and 30 kHz respectively. The sonar signal from the riverain population (Sotalia fluviatilis fluviatilis), was recorded under less than optimal conditions in the Rio Negro of the Amazon (Norris et al, 1972). These signals showed a "paired click" structure in some sonar trains, but the recordings were unable to give accurate frequencies for the emissions. Sotalia thus remains one of those species of special interest to investigators due to its two-component sonar system and the speculative ecological implications of its structure. Work to repeat a hollow-sphere discrimination experiment already performed with Tursiops (Kamminga and van der Ree, 1976), however, quickly demonstrated the need for experience with more exotic species of dolphins. A tentative behavioral profile for the coastal Sotalia in captivity shows wide differences in temperament and trainability between this species and Tursiops truncatus (Terry, 1986).

Differences between species demand the development of new techniques for continuing with sonar discrimination work. At the same time, however, the experience with Sotalia shows that work with smaller delphinid species is possible. The need for testing species other than Tursiops is becoming more apparent. The Beluga, Delphinapterus, is one of the few species besides Tursiops that has undergone systematic testing for its detection and target recognition capabilities. Initial results indicate a variability in sonar use for object detection not observed in Tursiops (Au, 1987). At present, species readily available for such work include relatively rare captive species such as the Boutu (Inia geoffrensis) and Commerson's Dolphin (Cephalohynchus commersonii) and more common captive species such as the False Killer Whale (Pseudorca crassidens) and the Pacific White-sided Dolphin (Lagenorhynchus obliquidens). Recent studies on the sonar emissions of Pseudorca suggest an acoustic repertoire during echolocation that is varied and quite different from the patterns known for Tursiops (Thomas et al, 1987). Preliminary echolocation discrimination studies have also been performed on other odontocete species (including Delphinus, Phocaena, and even Inia) but not with the experimental thoroughness achieved with Tursiops truncatus (Nachtigall, 1980). An intensive look at sonar capability and use by dolphin species from different genera and a wide variety of divergent habitats is essential to a more complete understanding of what has already proven to be a complex informational system.

DISCUSSION AND CONCLUSIONS

The directions outlined for future research on odontocete sonar systems should serve as a reminder of the need for dynamic interaction in research to maintain the impetus for insight and innovation. The various lines of research directly interface with and feed off of one another. Success in approaching the difficult questions involved concerning the role of sonar in cognition and its adaptability will depend upon the degree to which cross-disciplinary and international cooperation occur. These in turn will be better realized through a multi-species approach to the problems under consideration. Insufficient funding has been a pervasive and compelling argument for the lack of progress in dolphin sonar research since the last Animal Sonar Systems Conference in 1979. Like most intricate issues, however, it is not the only answer to the problem. Implementation of different approaches to research requires first and foremost a commitment to their value. The rest, as difficult, time-consuming and expensive as it may be, is mechanical.

Even with the implementation of new approaches to sonar research, changes must also occur in the processing of accumulated data. After the systematic measuring of animal performance in discrimination tasks, signal frequencies and the estimation of acoustic properties, questions still must be addressed to the real significance of the data. What does a critical interval in the integration time of signal processing have to do with the physiology of sound production or the elicited behavioral response of an echolocating dolphin? To shed light on what is clearly a wide and complex subject, to stimulate research and move investigative activity in the field to a higher level, more time and energy must be spent on the interpretation of the masses of data collected to date. In the end, it will be the insights generated by human interpretation of the data compiled - the job that computers cannot do - that will provide final answers.

Finally, it is becoming apparent that future scientific cooperative interaction, locally and globally, must at some stage also include the layman. Modifications to traditional methodology in captive research while opening up new view points, can, at the same time, capture the public imagination. Increased contact with the public, open lines of communication, a free flow of information on what is being done, why, and the results are becoming increasingly important. Researchers need to keep in mind the information needs of an increasingly aware, but often misinformed, public toward the nature and goals of research. The field of cetacean research, the infrequent use of invasive methods of research, and especially captive studies exemplified by most sonar system investigations are particularly vulnerable to misinformation and mistrust by the public. Gaining public confidence is not only practical in terms of financing and support; making knowledge widely available is at the heart of science itself.

Recent and future developments in creative approaches will hopefully give impetus to the solution of long-term problems and rejuvenate research into the next decade. Understanding in the field of odontocete sonar systems research, even at a slowed investigative pace, has come a long way in the last decade. Investigative results have provided tentative answers to questions about which pioneers in the field of biosonar, like Donald Griffin and Rene-Guy Busnel, could only dream. Directions for future research, realized with renewed motivation, may provide answers to the visionary dreams of current researchers.

REFERENCES

Amundin, M., 1987, The sound production apparatus in the harbour porpoise, _Phocoena phocoena_, the jacobita, _Cephalorhynchus commersoni_, and the bottlenose dolphin, _Tursiops truncatus_: 3-D computer graphics studies and the effects of helox atmospheres, this volume.

Au, W. W. L., Carder, D. A., Penner, R. H., and Scronce, B. L., 1985, Demonstration of adaptation in beluga whale echolocation signals, J. Acous. Soc. Am., 77 (2): 726-730.

Caldwell, D. K., and Caldwell,M. C., 1972, Senses and communication, _in_: "Mammals of the Sea, Biology and Medicine", S. H. Ridgway, ed., C. C. Thomas, Springfield, Illinois.

Cranford, T. W., 1987, Anatomic basis for high frequency sound transduction in the dolphin head using x-ray computed tomography and computer graphics, this volume.

Evans, W. E., 1973, Ecolocation by marine delphinids and one species of fresh-water dolphin, J. Acoust. Soc Am., 54(1): 191-198.

Kamminga, C., and Wiersma, H., 1982, Investigations on cetacean sonar V: the true nature of the sonar sound of Cephalorhynchus commersonii, Aq. Mammals, 9(3): 95-104.

Kamminga, C., and Wiersma, H., 1981, Investigations on cetacean sonar II: acoustical similarities and differences in odontocete sonar signals, Aq. Mammals, 8(2): 41-60.

Kamminga, C., and van der Ree, A. F., 1976, Discrimination of solid and hollow spheres by Tursiops truncatus (Montagu), Aq. Mammals, 4:1-10.

Martin, K., Norris, K., Moore, P. W. B., and Englund, K., 1987, Loud impulse sounds in odontocete predation and social behavior, this volume.

Nachtigall, P. E., 1980, Odotocete echolocation performance on object size, shape and material, in: "Animal Sonar Systems", R.-G. Busnel and J. F. Fish, eds., Plenum Press, New York.

Norris, K. S., Harvey, G. W., Burzell, L. A., and Kartha, T. D. K., 1972, Sound production in the freshwater porpoises Sotalia cf. fluviatilis Gervais and Deville and Inia geoffrensis Blainville in the Rio Negro, Brazil, Invest. on Cetacea IV: 251-249.

Purves, P. E., and Pilleri, G. E., 1983, "Echolocation in Whales and Dolphins', Academic Press, New York.

Purves, P. E., 1967, Anatomical and experimental observations on the cetacean sonar system, in: "Animal Sonar Systems: Biology and Bionics,: R.-G. Busnel, ed., Lab de Physiol. Acoust., Jouy-en-Josas, France.

Reiss, D., 1987, Observations on the development of echolocation in young bottlenose dolphins, this volume.

Ridgway, S. H., 1983, Dolphin hearing and sound production in health and illness, in: "Hearing and Other Senses: Presentations in Honor of E.G. Wever", R.R. Fay and G.Gourevitch, eds., Amphora Press, Groton, Conneticut.

Schultz, M.L., 1987, Sonar clicks from infant, juvenile and adult bottlenosed dolphins, Tursiops truncatus, in captivity, this volume.

Terry, R. P., 1986, The behaviour and trainability of Sotalia fluviatilis guianensis in captivity: a survey, Aq. Mammals, 12 (3): 71-80.

Thomas, J., Stoermer, M., Bowers, C., and Garver, A., 1987, Detection abilities and signal characteristics of echolocating false killer whales (Pseudorca crassidens), this volume.

Watkins, W. A., and Wartzok, D., 1985, Sensory biophysics of marine mammals, Mar. Mamm. Sci., 1(3): 219-260.

Wiersma, H., 1982, Investigations on cetacean sonar IV: a comparison of wave shapes of odontocete sonar signals, Aq. Mammals, 9(2): 57-67.

SECTION VI

ECHOLOCATION THEORY AND APPLICATIONS

Section Organizers: Whitlow W. L. Au and Jeffrey Haun

SOME THEORETICAL CONCEPTS FOR ECHOLOCATION

Richard A. Altes

Chirp Corporation
8248 Sugarman Dr.
La Jolla, CA 92037

INTRODUCTION

Some fundamental questions follow from recent developments
in animal sonar research and signal processing: (1) What is the
meaning of ordered neuronal maps of various echolocation
parameters in the context of (a) sampling and interpolation
theory, and (b) parameter estimation? (2) What is the impact of
recent behavioral experiments upon viable detection models for
echolocation? (3) What is the relevance to animal sonar of new
time-frequency signal representations?

INTERPOLATION, PARAMETER ESTIMATION, AND NEURAL MAPS

Neurophysiological investigations on bats have revealed
ordered cortical maps of neurons that are tuned to different
values of an echo feature [1]. The physical location of a
neuron within such a map is correlated with its responsivity to
a particular value of the relevant feature. There are
amplitopic maps to represent echo amplitude, Doppler maps, and
delay maps composed of range-tuned neurons. The more familiar
tonotopic maps, some of which are highly distorted [1], exist at
the periphery of the nervous system due to mechanical filtering
in the inner ear, as well as at other sites [2].

The accuracy of an animal's internal representation of
important echolocation features can perhaps be inferred from the
widths of neuronal response curves and the differences between
the "best" stimulus for neighboring neurons in a map. This idea
can be analyzed by studying the relation between cortical maps,
interpolation concepts, and parameter estimation theory.

From the viewpoint of interpolation, an ordered set of
neurons could represent sample values of a function that is
dependent upon the mapped parameter, e.g. a function of range,
range-rate, etc. Neuronal tuning curves with nonzero width
could be a basis for interpolating between the samples, i.e.,
for evaluating the function between the sample points.
Interpolation has not been actively discussed with respect to
hearing and echolocation, but interpolation based on

zero-crossing representations is central to some recent theories of visual processing [3].

A simple conceptualization of neuronal response curves in an ordered map is shown in Figure 1. The figure is actually an illustration of "hat (chapeau) functions" from interpolation theory [4]. Hat functions are basis functions for linear interpolation (piecewise linear fits). Smoother functions with continuous first (and higher order) derivatives are of course more true to life, and are also associated with more accurate interpolation. Hat functions are chosen for their mathematical convenience and easy visualization. Note that each hat function in Figure 1 is zero where the others are maximized. A broken line approximation to any function $g(x)$ is obtained by sampling the function at $x = \tau_1, \tau_2, \ldots, \tau_n$ and forming a sum of weighted hat functions, where the hat function $H_k(x)$ is weighted by $g(\tau_k)$, the sample at $x = \tau_k$. Alternatively, weights can be determined so as to yield a best (minimum mean-square error) approximation to $g(x)$ by the sum of weighted hat functions.

An important theorem of de Boor [4,5] for functions $g(x)$ with bounded second derivative is that there is an optimum placement of the sample points $\{\tau_k\}$ for representing $g(x)$ with hat functions. The sample points should be positioned such that

$$\int_{\tau_1}^{\tau_k} |g''(x)|^{1/2} \, dx = \frac{k-1}{n-1} \int_{\tau_1}^{\tau_n} |g''(x)|^{1/2} \, dx , \qquad (1)$$

for $k = 2,3,\ldots,n-1$, where $g''(x)$ is the second derivative of $g(x)$. From Figure 1, it is apparent that Eq.(1) not only defines the points of best excitation for the neuronal response curve models, but also the widths of the simplified response curves.

If $g(x)$ equals $\log(x)$, for example, then (1) yields an exponentially increasing sampling interval with proportional width (constant Q) tuning curves; $g(x)=\log(x)$ yields

$$\tau_k = \tau_1 \exp[(k-1)\log(\tau_n/\tau_1)/(n-1)]. \qquad (2)$$

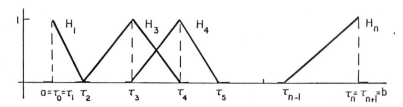

Figure 1. Basis functions for piecewise-linear interpolation from de Boor [4]

In (2), $\tau_{k+1} - \tau_k$ is proportional to τ_k. The maximum difference between g(x) and the piecewise linear interpolation between any two sampling points defined by (2) is independent of k, i.e., the non-uniform sampling in (2) results in uniform maximum error between sampling points.

Another example of biological interest is $\hat{g}(x)$ equal to log(1+x), which is linear for small x and logarithmic for large x. In this case, (1) results in sampling points

$$\hat{\tau}_k = \tau_k - 1 \tag{3}$$

where τ_k is the same as in (2). For small values of k, sample spacing and response curve widths are approximately uniform, while for large values of k, sample spacing is proportional to x, as for interpolation of g(x) = log(x).

Assuming optimal interpolation of a parameter in a cortical map, the map can be used to infer nonlinear transformations that have been introduced by the auditory or nervous systems. If the sample spacing on the ordered map is small enough such that

$$\int_{\tau_k}^{\tau_{k+1}} |g''(x)|^{1/2} dx \approx \left| g'' \left[\frac{\tau_{k+1} + \tau_k}{2} \right] \right|^{1/2} (\tau_{k+1} - \tau_k) \quad , \tag{4}$$

then the difference in best stimulus levels for neighboring neurons can be used to infer the function g(x), since (1) and (4) imply that

$$\tau_{k+1} - \tau_k \approx \text{(constant)} \times \left| g'' \left[\frac{\tau_{k+1} + \tau_k}{2} \right] \right|^{-1/2} \quad . \tag{5}$$

Eq. (5) implies that the distance between samples should vary inversely with the square root of $|g''(x)|$ for x in the neighborhood of the sampling points. A rapid change in the slope of g(x) requires a high sampling rate.

"Best" stimulus levels in a topographic arrangement of neurons can be used to infer the function that is best represented by the ordered map, if the neurons are used for linear interpolation. For example, if the difference between best levels, $\tau_{k+1} - \tau_k$ is proportional to τ_k, then (5) implies that $|g''(x)|$ is proportional to $1/x^2$, and g(x) is proportional to log(x). Nonlinear transformations of various input features can thus be inferred from topographic arrangements of neurons. Nonlinear transformations of amplitude are useful for increasing the dynamic range of a measurement system, and are also associated with optimum detection and estimation in nongaussian noise [6-8]. Nonlinear transformations of frequency, time, and position parameters can be used to obtain scale invariant representations [9].

Another interpretation of cortical maps is based on parameter estimation concepts. Neuronal maps can be related to parameter estimation by assuming that the response of each cortical neuron is the output of a hypothesis test for a specific parameter value. A map of parameter-tuned neurons is then composed of an ordered set of hypothesis tests or feature detectors for various values of the mapped parameter.

For the sake of argument, assume that the hypothesis tests are implemented with a log-likelihood function

$$\ell(\underline{r}|x) = \log p(\underline{r}|x) \tag{6}$$

where $\underline{r}$ is a set of observed data samples (perhaps internal to the animal) and x is again the mapped parameter (range, range rate, amplitude, etc.). The function $p(\underline{r}|x)$ is the probability of observing a given data set $\underline{r}$ for a specific value of the parameter x. Finding the value of x that maximizes $\ell(\underline{r}|x)$ for a given data set $\underline{r}$ is the same as finding the most likely parameter commensurate with the observed data, i.e., the maximum likelihood estimate.

The analysis is taken further by assuming a specific form for the function g(x) that is presumably represented by a cortical map. If an ordered map consists of a sequence of hypothesis tests, then a parameter can be estimated by observing the point of maximum excitation on the neuronal map, i.e., the largest value of $\ell(\underline{r}|x)$. Therefore, let

$$g(x) \equiv \ell(\underline{r}|x) \ . \tag{7}$$

Interpolation between hypothesis tests for specific values of x is helpful for accurate determination of the x-value that maximizes g(x). In fact, interpolation results such as (1) indicate the specific parameter values that should be tested. These hypothesized parameter values are the same as the values of x at the optimum sampling points, $\{\tau_k\}$.

At high signal-to-noise ratio (where spurious peaks in g(x) due to ambiguities can be neglected), the expected (root-mean-square) error between the estimated parameter value $\hat{x}$ and the correct x-value is given by the Cramer-Rao lower bound [10]:

$$\text{RMS error} = \{ \ E \ [(x-\hat{x})^2] \ \}^{1/2}$$

$$= | \ E \ [(\partial^2/\partial x^2)\ell(\underline{r}|x)] \ |^{-1/2}$$

$$= | \ E \ [g''(x)] \ |^{-1/2} \ . \tag{8}$$

The maximum likelihood estimator is efficient, i.e., its RMS error is given by (8). Inefficient estimators have larger RMS errors.

Comparison of (5) and (8) indicates that the optimum sampling interval for interpolating g(x) is proportional to the expected error in estimating the parameter x when g(x) equals $\ell(\underline{r}|x)$. If the data permit a highly accurate estimate of x, then the sampling points should be close together. Many hypotheses should be made in an interval where x can be accurately estimated. In biological terms, a cortical map should be composed of neurons with sharp excitation curves and with closely spaced best excitations, if the corresponding parameter values are accurately estimated at the sensor level. Conversely, inaccurate peripheral estimation should be correlated with relatively broad, widely spaced cortical excitation curves. This requirement may be related to "over-representation" in cortical maps. Over-representation occurs in the Doppler-shifted-CF cortical processing area of Pteronotus, as well as in cortical representations of the foveal region of the eye in mammalian vision [11].

Suga and Tsuzuki [12] have reported sharpening in frequency-tuned neurons, such that Doppler-sensitive cortical neurons are more finely tuned than peripheral ones because of lateral inhibition. How is such sharpening to be explained in terms of interpolation/estimation theory? If no relevant information is lost in peripheral coding, then peripheral sampling and interpolation must be sufficient for functional representation of Doppler information. Why should a different interpolation method with more narrow basis functions and higher sampling rate be needed in the cortex if the same function is represented in the cortex as in the periphery?

A reasonable explanation of cortical sharpening uses the observation that asymmetric lateral inhibition (on one side of a tuning curve) is a kind of differentiation operation, while symmetric lateral inhibition (on both sides) is associated with a second derivative [3]. If neuronal tuning curves are interpolating basis functions for g(x), then g(x) is represented as a weighted sum of such functions. The derivative g'(x) is then represented by a set of differentiated basis functions, and g''(x) is represented by twice-differentiated (sharpened) basis functions. Interpolation theory suggests that sharpened CF tuning curves do not represent Doppler hypotheses per se, but rather the changes in Doppler that are introduced by acceleration and rate of acceleration, e.g., from fluttering insect wings.

The correspondence between optimum sampling intervals and RMS error is based on the assumption that neurons in an ordered map implement an approximation to a log-likelihood function for estimating the mapped parameter. Even if a log-likelihood function is not implemented, it seems to follow by definition that a given unit within a neural map forms some kind of hypothesis test for a specific value of the mapped parameter. If this interpretation is correct, then signal processing models can be tested with neurophysiological methods, as well as with behavioral techniques. For example, the effect of added external noise and variations of stimulus level on the response of a neuron can be used to assess the level of internal noise in the signal processing system. Knowledge of this internal noise level is critical to distinguishing between different receiver models with behavioral tests.

If a neuron in an ordered map has a log-likelihood function response or implements some other test for a specific value of the mapped parameter, then observation of this response as a function of various signal and noise levels may allow different candidate signal processing models to be distinguished. Three candidates for the log-likelihood ratio are

$$\ell_1(\underline{r}|x) = \int r(t) \, u^*(t;x) \, dt \qquad (9)$$

$$\ell_2(\underline{r}|x) = \left| \int r(t) \, u^*(t;x) \, dt \right|^2 \qquad (10)$$

$$\ell_3(\underline{r}|x) = \iint S_{rv}(t,f) \, \frac{\overline{S_{uv}(t,f;x)}}{(N_o/2) + \overline{S_{uv}(t,f;x)}} \, dt \, df \, . \qquad (11)$$

The models in (9 - 11) are associated with log-likelihood ratio tests for detecting a signal with a hypothesized parameter x. The test involves calculating the log-likelihood function, subtracting $\log p(\underline{r}|H_o)$, the log-probability of observing the same data in the presence of noise alone, and comparing the result with a threshold. Since $\log p(\underline{r}|H_o)$ is independent of the signal parameter x, the log-likelihood ratio can be used instead of the log-likelihood function for estimating x.

In (9) and (10), r(t) is the input time series composed of either noise alone or noise plus the parameter-dependent signal u(t;x). In (11), $S_{rv}(t,f)$ is the data spectrogram obtained with a unit-energy window function v(t), $\overline{S_{uv}(t,f;x)}$ is the ensemble-average spectrogram of a random, parameter-dependent signal u(t;x), and $N_o/2$ is the power spectral density of the noise at the input to the hypothesis tester.

The model in (9) is a fully coherent correlation receiver for a signal that is known except for the parameter x, which is hypothesized [13,14], and for additive, white Gaussian noise (WGN). The model in (10) is an envelope detected version of $\ell_1(\underline{r}|x)$, and has been called a "semi-coherent" receiver [15]. The semi-coherent version is relevant when the signal is known except for a random phase shift in addition to the hypothesized parameter x. The third model is a spectrogram correlator, which pertains to an input signal that is random but is still dependent on x [16, 17], e.g. a delayed or Doppler shifted version of a Gaussian stochastic signal in WGN. A random echo is obtained when a sonar scatterer is a randomly time-varying, range-extended reflector. Such targets can sometimes be described by a scattering function [18]. In this case, the spectrogram of the echo is the spectrogram of the transmitted signal convolved with the target scattering function [17].

In summary, interpolation and parameter estimation concepts lead to mutually consistent results for interpretation of neuronal maps that are ordered with respect to an echo parameter. This consistency is based on an assumption that the response of each neuron in the map is determined by a hypothesis test for a particular value of the mapped parameter. If this assumption is correct, then neuronal responses in ordered maps can be used to test various signal processing models for animal echolocation.

As one would intuitively expect, the granularity of a

cortical map and the accuracy of a parameter estimate are related, as shown by (5) and (8). The accuracy of a parameter estimate may nevertheless be better than it appears from the granularity of a cortical map, since interpolation between hypothesized values of x may yield an accurate representation of the log-likelihood function $\ell(\underline{r}|x)$.

DETECTION MODELS AND THE EFFECT OF UNCERTAINTY

Detection by bats of a time-reversed FM signal, as well as the signal itself, has recently been reported by Mohl [19]. The results are not in agreement with any of the models in (9 - 11), since all three models predict degraded detection performance when a signal is time-reversed, and such degradation was not observed. Even if the models are allowed to adapt to a time reversal, the surprisingly high signal to noise ratio that was required for reliable detection seems to eliminate (9) and (10) from further consideration.

If either of the matched filter models is valid, then detectability may be conditioned on stimuli that were not present in the time-reversal experiment. One missing stimulus in [19] is motion; the simulated echoes correspond to a motionless planar reflector. Since the bats were also motionless (or at least not in flight), such a reflector is typical of clutter, i.e., of a class of echoes that are not of interest. M. Zakharia has discussed the importance of pulse-to-pulse echo differences in practical sonar systems [20]. One of the most basic differences between two echoes in a realistic situation is the delay change associated with time-dependent range. In man-made radar systems, moving target indication (MTI) exploits this observation by filtering out clutter echoes with zero delay difference [21]. The presence of an MTI process in bats would explain the high SNR required for detection in the time-reversal experiments, since neither the bat nor the simulated target was moving.

Another possible explanation of Mohl's results is the performance degradation of an optimum receiver that must entertain multiple hypotheses. The results of Nolte and Jaarsma, as summarized in [23], demonstrate that multiple hypotheses degrade detection performance, especially near threshold, at low SNR. Suga's cortical maps indicate an enormous number of hypotheses for joint estimation of amplitude (or target cross section), direction of arrival, Doppler, range, etc. The number of different hypothetical echoes is obtained by computing the product of the number of hypothesized values for each parameter, e.g., the number of amplitude values x the number of resolvable directions x the number of resolvable range intervals x .... H.-U. Schnitzler has remarked that performance degradation caused by multiple hypotheses could explain the relatively poor performance of bats in field observations, compared with controlled laboratory conditions.

A well controlled experiment can be defined as one that eliminates all hypotheses but those associated with the variable that is being tested. The basic question is how well an animal succeeds in ignoring hypotheses that are relevant to survival in nature but irrelevant to the experiment.

The need to reduce uncertainty in order to optimize performance implies that Doppler tolerance [24] and Doppler compensation [36] can be interpreted as different means to the same end, i.e., reduction of the number of different filters or hypotheses that must be used in the receiver [61]. Other, similar strategies are scale invariant representations [9], the use of time-varying gain [60] and level tolerant frequency tuning [12] in order to reduce the effect of amplitude variations, and the use of range-tracking neurons to focus on a given target and ignore others [1]. The reduction of uncertainty seems to be a primary consideration in animal sonar systems, and it could be a useful principle for design and interpretation of future experiments.

If the time-reversal experiment is really indicative of the detection performance of FM bats, then it supports the conclusion of Menne and Hackbarth [15], who did a careful computer simulation of a jittering target discrimination experiment performed by Simmons [22]. The simulated jitter experiment seems to indicate that neither of the models in (9) and (10) are valid for echolocation. The spectrogram correlator model (11) still seems to be viable. Coherent MTI capability (if it exists in bats) and multiple hypotheses or uncertainty also affect the Menne/Hackbarth results, as discussed below.

MTI can be performed coherently or non-coherently (with envelope detected receiver outputs); coherent MTI is more sensitive than the non-coherent version. If coherent MTI is an allowable (albeit unlikely) hypothesis for echolocation, then coherent pulse-to-pulse processing must be considered in candidate receiver models. This type of processing was not considered in the Menne/Hackbarth jitter simulation [15]. A pulse-to-pulse coherent processor can construct hypotheses from pairs of echoes which are viewed as part of the same signal. An echo pair from a target that does not jitter is

$$u_0(t) = u(t-D) + u(t-D-T), \tag{12}$$

where $u(t)$ is the transmitted signal (assumed identical for the two pulses), D is the echo propagation delay, and T is the interval between transmissions. An echo pair from a jittering target is

$$u_1(t) = u(t-D+\tau/2) + u(t-D-T-\tau/2) \tag{13}$$

where τ is the delay extent of the jitter.

The log-likelihood ratio test to discriminate between jitter (H_1) and no jitter (H_0) is

$$\ell(\underline{r}|\tau) = \int_{-\infty}^{\infty} r(t) \ [u_1(t) - u_0(t)] \ dt \tag{14}$$

where the data time series $r(t)$ equals $u_1(t)$ plus white, Gaussian noise (WGN) under hypothesis H_1, and $u_0(t)$ plus WGN under H_0. The output signal-to-noise ratio (SNR) for this test is

$$d^2_{\ell|\tau} = \frac{\{E_r[\ell(r|\tau,H_1)] - E_r[\ell(r|\tau,H_0)]\}^2}{Var_r[\ell(r|\text{noise alone})]} \qquad (15)$$

where $E_r[\ell(\underline{r}|\tau,H_k)]$ is the expected value (with respect to $\underline{r}$) of the log-likelihood statistic in (14) when τ is given and H_k is true, and $Var_r[\ell(\underline{r}|\text{noise alone})]$ is the variance (again with respect to $\underline{r}$) of the linear discriminant (14) when $r(t)$ consists only of WGN.

Letting

$$R(\tau) = \int_{-\infty}^{\infty} u(t) \, u(t + \tau) \, dt = R(-\tau) \qquad (16)$$

denote the autocorrelation function of each transmitted pulse, and observing that $R(\tau)$ is zero for τ equal to T, T$\pm\tau$/2, and T$\pm\tau$, where T is the interval between transmitted pulses, (13 - 15) yield

$$d^2_{\ell|\tau} = (2/N_o) \int_{-\infty}^{\infty} [u_1(t) - u_0(t)]^2 \, dt$$

$$= (8/N_o) \, [R(0) - R(\tau/2)] \, . \qquad (17)$$

For equally likely hypotheses, the probability of a correct decision for the above value of d^2 is [10]

$$Pr[\text{correct}|\tau] = (2\pi)^{-1/2} \int_{-\infty}^{d_{\ell|\tau}/2} \exp(-x^2/2) \, dx$$

$$\equiv erf_*(d_{\ell|\tau}/2) \qquad (18)$$

which is a monotone function of $R(0) - R(\tau/2)$. This result predicts a dip in the bat's performance curve when the jitter excursion coincides with a signal ambiguity peak, as observed by Simmons [22]. No dip is observed for the hypothesis test in Menne and Hackbarth [18], which assumes independent range estimates for each transmitted pulse (no pulse-to-pulse coherent processing).

Menne and Hackbarth have pointed out that $erf_*(d/2)$ becomes insensitive to variations in d when SNR is large. The bats in the jitter experiment were apparently operating at a sufficiently high signal-to-noise ratio that the decrease in Pr[correct] caused by autocorrelation sidelobes should be negligible. Although this argument may apply to the result in (17), the effect of an unknown jitter parameter reduces the effective SNR and increases the effect of sidelobes on d^2 and Pr[correct], as outlined below.

The performance prediction in (18) assumes that the

magnitude and sign of the jitter parameter τ are known. For an unknown τ value, the mth moment (with respect to τ) of a log-likelihood ratio that has been ensemble averaged with respect to $\underline{r}$ can be defined as

$$E_\tau\{[E_r[\ell(\underline{r}|H_k)]]^m\} \equiv \int_{-\infty}^{\infty} [E_r[\ell(\underline{r}|\tau,H_k)]]^m p(\tau) \, d\tau \qquad (19)$$

where $p(\tau)$ is the probability distribution of the unknown parameter τ. The above moments for m = 1 and m = 2 can be used to calculate the mean and variance of the log-likelihood statistic in order to determine the output signal-to-noise ratio as in (15) and hence the performance measure in (18). When $p(\tau)$ is uniform over the width of the signal autocorrelation function $R(\tau)$, it turns out that

$$d_\ell^2 = \frac{2 \, [R(0) - R(\tau/2)]^2}{N_o \, R(0)} \qquad (20)$$

where τ is the actual delay extent of the jitter. The right hand side of (20) has a stronger dependence on the sidelobes of the signal autocorrelation function than in (17) and is smaller by a factor of four when $R(\tau/2)$ equals zero. A larger dip in the percentage of correct responses is thus predicted in (18) when jitter excursion coincides with a sidelobe of $R(\tau)$. Again, such a degradation occurs in the performance of a real Eptesicus [22], but not in the performance of the computer simulation in [15]. The degradation predicted by (20) illustrates the effect of uncertainty or multiple hypotheses upon optimum receiver performance.

In summary, both the time-reversal experiment of Mohl [19] and the jitter simulation experiment of Menne and Hackbarth [15] are implicitly dependent on the assumptions (1) that pulse-to-pulse coherent processing does not condition or otherwise affect hypothesis testing in animal echolocation, and (2) that nearly all the hypothesis tests indicated by Suga's cortical maps can be ignored by an animal in a controlled experiment. Despite these assumptions, it is evident that the validity of the matched filter hypothesis has been seriously challenged by the experiments in [15] and [19].

SIGNAL DESIGN AND AMBIGUITY ANALYSIS FOR SPECTROGRAM CORRELATORS

If the matched filter hypothesis is rejected, then what is the relevance of sophisticated bionic signal design based on a matched filter assumption, and the occurrence of such signals in echolocation [24-26]? The signal design results are based on analysis of a wideband ambiguity function. The ambiguity function is a plot of the matched filter response for various hypothesized ranges and Doppler scale factors, when the input is a single echo with specific range and scale factor (usually a replica of the transmitted signal with range zero and scale factor one).

The ambiguity function depends on range and Doppler mismatch between hypothesized and actual echo parameters. It exhibits a main peak for zero range-Doppler mismatch, and sidelobes at erroneous range and Doppler hypotheses that happen to cause a large matched filter response despite the fact that they differ from the correct values. A signal with good range and Doppler resolution has a matched filter that responds strongly only when the hypothesized range and Doppler parameters are correct or nearly correct. The associated ambiguity function exhibits a peak at the point where the range and Doppler hypotheses are correct, and relatively small response for all other range and Doppler hypotheses that are not close to the correct values.

A range-resolvent, Doppler tolerant signal with no range-Doppler coupling has an ambiguity function that is large for many different Doppler hypotheses and at the correct range hypothesis, and is small for range hypotheses that are not close to the correct value. Some FM signals used by bats seem to be optimally designed to achieve Doppler tolerance with no range-Doppler coupling, as well as good range resolution. Such signals result in a knife-edge ambiguity function with the blade along the Doppler axis of the ambiguity plane [24,27,58].

Equations (9) and (10) indicate that, if matched filtering is used, the log-likelihood function $\ell(\underline{r}|\tau,s)$ depends upon the ambiguity function when the transmitted signal is the input to the receiver. The curvature of the ambiguity function in the neighborhood of $\tau = 0$ and $s = 1$ determines the required sampling rate in this neighborhood, in accordance with (5). The interpolation results thus suggest that relatively few Doppler hypotheses are needed to represent the response to a Doppler tolerant signal, i.e., relatively few matched filters are required.

The ambiguity concept can be applied to alternative receiver models by using the generalized ambiguity analysis introduced in [28]. Instead of examining the response of a matched filter for many range and Doppler hypotheses, the log-likelihood function, $\ell(\underline{r}|\tau,s)$, is examined for various values of the hypothesized range and Doppler parameters τ and s. This procedure generates the conventional magnitude-squared ambiguity function when $\ell(\underline{r}|\tau,s)$ is the same as (10), and the unsquared ambiguity function (sometimes called the uncertainty function or time-frequency autocorrelation function) when $\ell(\underline{r}|\tau,s)$ is the same as (9). When the conditional log-likelihood function is given by (11), however, a new, generalized ambiguity function is obtained from the spectrogram correlator responses to correct and incorrect parameter hypotheses.

The generalized ambiguity function for a spectrogram correlator will often yield the same results as conventional ambiguity analysis. Suppose, for example, that the signal has instantaneous frequency f(t). A point (planar) target coming toward the sonar will compress echoes in time and dilate them in the frequency domain. The effect is to transform f(t) into sf(st), where s is the Doppler scale factor: s equals $(1+v/c)/(1-v/c)$; v is radial velocity toward the sonar, and c is the speed of sound. From the viewpoint of spectrogram correlation, a Doppler tolerant frequency modulation with no range-Doppler coupling is such that the echo and signal spectrograms are highly correlated regardless of s for zero range mismatch, i.e.,

$$f(t+\tau) = s \; f(st) \qquad\qquad (21)$$

for any value of s and for $\tau = 0$. Equation (21) is satisfied if

$$f(t) = \frac{constant}{t} \qquad\qquad (22)$$

which corresponds to hyperbolic frequency modulation (HFM) and linear period modulation. The instantaneous frequency in (22) is the Doppler tolerant FM that has been observed in bat echolocation.

Some published HFM waveforms seem to exhibit range-Doppler coupling [59], but this effect is caused by the choice of time origin. Time t = 0 in (22) is at the point where f(t) becomes infinite and the instantaneous period becomes zero. The actual transmission has no power at this point, but the signal displays a linear increase in instantaneous period $f^{-1}(t)$ over the length of the pulse where power is nonzero. Extrapolating the linear period vs time plot back to zero period yields the time origin for no range-Doppler coupling.

Sidelobes of the ambiguity function of a matched filter receiver have been used by Simmons to predict performance degradations in bat sonar [22]. An important question is whether ambiguity sidelobes of a spectrogram correlator are comparable to those of a matched filter. Equation (A19) in [17] implies that the ambiguity function of a spectrogram correlator is the same as that of a matched filter, except that it is smeared in time and frequency by convolution with the ambiguity function of the filter or time window that is used to compute the spectrogram. In animal hearing, this window has comparatively large time-bandwidth product due to the shape of the tuning curves of primary auditory neurons [18, 39]. For spectrogram correlation, the window ambiguity function in animals may create much less distortion of the signal ambiguity function than would occur with conventional filters in man-made systems.

ECHOLOCATION, HUMAN AUDITION, AND SPECTROGRAMS

Mohl's time-reversal experiment was also tried with human listeners, who were found to perform slightly better than the bats [19]. This result is similar to a Dolphin target discrimination experiment, which also found that human listeners could perform as well or better than the animals [29]. These results suggest that a realistic receiver model should be consistent with human audition as well as with echolocation experiments. Some fundamental results in human audition, such as Green's energy summation experiment [30], are consistent with a spectrogram correlator model. In addition to auditory channels that are tuned to CF tones, the human auditory system apparently contains separate perceptual channels that are tuned to different amplitude or frequency modulated versions of each CF tone [31,32]. The following argument shows that such channels are commensurate with a generalized spectrogram correlator which uses eigenvector data analysis.

The spectrogram correlator is a special case of a detector that uses eigenvector (Karhunen-Loeve, or principal component) analysis to detect random signals in noise [33]. If the eigenvectors of the signal covariance matrix are sinusoids (i.e., if the signal is quasi-stationary over the spectrogram window duration) then the eigenvector-based detector becomes a spectrogram correlator as in (11). For a randomly modulated signal m(t)u(t), let C_m be the covariance matrix of the random modulation m(t) alone. The eigenvectors of C_m , multiplied by the nonrandom "carrier" signal u(t), are the eigenvectors of the randomly modulated signal m(t)u(t) [34]. An eigenvector-based detector for sinusoids or chirps that are randomly amplitude or frequency modulated should thus contain filters (eigenvectors) which are sensitive to different AM/FM versions of the sinusoids or chirps. Such modulation occurs when bat signals are reflected from flying insects [35,36]. Separate perceptual channels that are sensitive to specific rates of amplitude or frequency modulation [31,32] and collicular neurons tuned to different types of AM and FM [37] can be interpreted as components of a generalized spectrogram correlator.

A growing set of evidence can thus be assembled in support of a spectrogram correlator model (or a generalized eigenvector version of the model) for both human and animal audition. Of course, such a model could exist in parallel with other detectors like matched filters. Such a parallel configuration is obtained when data consist of both deterministic and random signal components [10]. The matched filter configuration in (10) can be approximated by appropriate processing of a spectrogram [17,18,38], so a spectrogram could be used as a "front end" for a system that uses both kinds of detectors.

The spectrogram correlator in (11) and the semi-coherent detector in (10) are identical for an important special case. A simple random amplitude modulation is to multiply each observation of a nonrandom signal by a constant with unit amplitude and random phase shift. The resulting signal covariance matrix has a single eigenvector, which is the nonrandom signal itself. The eigenvector-based (Karhunen-Loeve) detector in this case forms the statistic in (10), i.e., a semi-coherent detector.

The spectrogram correlator is simply a new realization or canonical form of an optimum receiver for random signals. The concept of an "optimum receiver" in echolocation is thus still applicable. Different optimum receiver structures are obtained for different assumptions about signal and noise.

The utilization of spectrograms or "neurograms" [39] for detection and estimation implies the use of separate filters that may possess overlapping bandwidths as in Figure 1. Interpolation concepts may thus prove useful for the analysis of spectrogram processors. Indeed, Hackbarth [40] has experimentally determined a "best overlap" or optimum density of fixed-bandwidth filters when spectrogram cross-correlation is used for binaural analysis. Optimum filter bandwidth for time delay estimation with overlapping filters has been investigated by D. Menne [this volume].

SPECTROGRAMS AND OTHER TIME-FREQUENCY REPRESENTATIONS

Spectrograms are relevant to echolocation at ultrasonic frequencies because of mechanical filtering in the cochlea [41-43], and the apparent loss of phase sensitivity for frequencies above 8 KHz [44]. In recent engineering literature, however, the spectrogram is being augmented by other time-frequency representations. The two main categories of such representations are ARMA-based short-time spectra and Wigner distributions.

ARMA representations assume that the observed random time series is generated by passing white noise through a linear filter with poles (resonances) and zeroes. For a nonstationary process, the poles and zeroes are time-varying. Since the white noise input has constant power spectral density, one can generate a chirp with a filter that has a time-varying resonance near the instantaneous frequency of the chirp. The high amplitude, specular parts of sonar echoes can be modelled as a weighted sum of the transmitted signal and integrated or differentiated versions of the transmitted signal [25]. Differentiation is associated with zeroes and integration is associated with poles of the target transfer function at zero frequency. "Natural modes" or "ringing" in the echo are associated with poles at nonzero frequencies.

An all-zero filter can be synthesized by forming a weighted sum of delayed input samples. Such a weighted sum forms the moving average (MA) part of the ARMA process. An all-pole filter can be synthesized by putting another all-zero filter in a feedback loop, such that the filter output is subtracted from the input process. Feedback introduces the autoregressive (AR) part of the process by causing future samples of the time series to be correlated with past samples, with regression coefficients determined by the weights of the filter in the feedback loop. If both filters are used, the resulting auto-regressive, moving average (ARMA) random process has a power spectrum that is easily expressed in terms of the filter coefficients, i.e., the weights that are applied to delayed input samples in the two filters.

By estimating the ARMA filter coefficients from the time series data in a sequence of intervals, one obtains a sequence of local power spectral estimates that can be used to construct a time-frequency representation. The most general form of such a representation is obtained when the ARMA coefficients are explicitly assumed to be time dependent. The associated coefficient estimation problem has been investigated by Grenier [44,45]. This model is well-suited to the description of echoes from targets with moving parts (insect wings, fish tails) and is more general than the scattering function [17]. The scattering function description assumes that sonar reflectors behave as all-zero filters (delay, weight, and sum) with statistically uncorrelated weights that have random, wide-sense stationary time variation [18].

The Wigner distribution was originally introduced to the engineering literature as a representation with optimum instantaneous frequency estimation and minimum extraneous smearing of energy on the time-frequency plane, at least with respect to a single, FM component [46,47]. Further applications and limitations of the Wigner distribution have been investigated by Martin, Flandrin, and Escudie [48-50]. The Wigner distribution provides a good illustration of the difference between accuracy and resolution, since the distribution yields a highly accurate representation of a single component, and yet is not well-suited (without modification) to resolving two or more components that are closely spaced in time or frequency.

Figure 2a, 2b, and 2c show pseudo-Wigner distributions (on a logarithmic amplitude scale) of a linear FM chirp, a constant frequency (CF) tone, and an impulse. Figure 2d shows the Wigner distribution of a superposition of the chirp, tone, and impulse. Although each component alone is well-represented, the sum of the three components yields a confusion of peaks, some of which are large-negative. Figure 2 implies that, if Wigner distributions are to be used for signals with harmonics or for multiple echoes encountered in echolocation, then the signals should somehow be separated into separate components, and the Wigner distribution should be calculated for each component separately. A superposition of different Wigner distributions, one for each signal component, would be highly resolvent, but one Wigner distribution for all of the superposed components is often undesirable.

The best utilization of Wigner distributions thus depends on a method for separating different components of a complicated signal. There is much to learn from animal audition with respect to this separation problem. Animal auditory systems exhibit admirable signal separation capability, as evidenced by the cocktail party effect, the specificity of neural feature detectors, and the nearly total separation of transients and longer signals in the dolphin brain, where transient sensitivity appears in the inferior colliculus, and sensitivity to longer, communication-like sounds occurs in the cortex [51].

A classification and subtraction method for separating signal components has been used for sonar target characterization with bionic signals [52], and the same method can be applied to separating signal components for a time-frequency representation [53]. Another classical method for signal separation is eigenvector (principal component) analysis, which has already been mentioned in connection with generalized spectrograms. A third method (currently being investigated by the author) is based on association and tracking of spectral peaks in a sequence of short-time spectral estimates.

A method for classifying different signal components, which is a first step to separating them, is to construct a number of different spectrograms with different filter widths. The signal component that is best matched to a particular filter width will have maximum response with minimum smearing on the corresponding spectrogram. An example of this technique is shown in Figures 3a, 3b, and 3c. The three figures show three different spectrogram representations of the same composite signal that confounded the Wigner representation in Figure 2d. The three spectrograms (again with logarithmic amplitude scales) are calculated with three different filter bandwidths. A narrow bandwidth (Figure 3a) suppresses the chirp and impulse relative to the CF component. An intermediate bandwidth (Figure 3b)

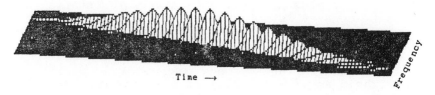

Figure 2a. Pseudo-Wigner distribution of a linear FM (quadratic
phase) chirp. Amplitude scale is logarithmic, i.e.,
proportional to $\log[1+W_u(t,|)]$.

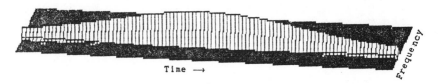

Figure 2b. Pseudo-Wigner distribution of a CF (constant
frequency) tone.

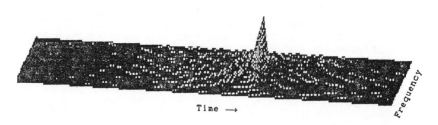

Figure 2c. Pseudo-Wigner distribution of an impulse.

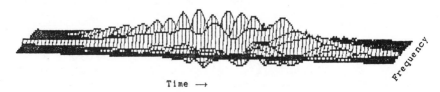

Figure 2d. Pseudo-Wigner distribution of a superposition of the
chirp, tone, and impulse used to construct Figures
2a - 2c.

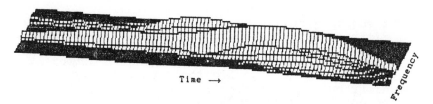

Figure 3a. Spectrogram of the superposition of chirp, tone, and
impulse used in Figure 2d. Spectrogram time window is wide and
filter bandwidths are narrow, causing the tone to be
accentuated. Amplitude scale is logarithmic as in Figure 2.

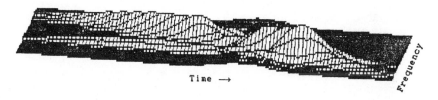

Figure 3b. Spectrogram of the superposition of chirp, tone, and
impulse used in Figure 2d. Spectrogram time window width and
filter bandwidths are at an intermediate setting, causing the FM
chirp to be accentuated. Amplitude scale is logarithmic as in
Figure 2.

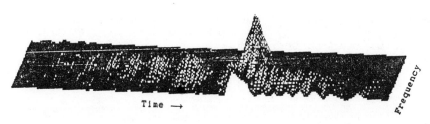

Figure 3c. Spectrogram of the superposition of chirp, tone, and
impulse used in Figure 2d. Spectrogram time window is short and
filter bandwidths are wide, causing the impulse to be
accentuated. Amplitude scale is logarithmic as in Figure 2.

accentuates the chirp. A wide bandwidth (Figure 3c) accentuates the impulse. This method seems to be relevant to audition, since at least four different, interconnected tonotopic maps are found in the cat cortex. Constituent neurons exhibit different frequency selectivity in different maps [54].

The spectrogram can be written as a convolution of the Wigner distribution of the signal with the Wigner distribution of the window function that is used to construct the spectrogram [46]. The use of multiple spectrograms with different windows to identify different signal components thus suggests that the Wigner distribution in Figure 2d can also be used if appropriate smoothing operations are applied.

BINAURAL DETECTION AND ESTIMATION USING THE WIGNER REPRESENTATION

Equation (11) describes a detector for a random signal in Gaussian noise, using a spectrogram representation of the sampled input time series, $\underline{r}$. Since the Wigner distribution is an alternative form of time-frequency representation, detectors that use a Wigner representation of $\underline{r}$ may be of interest. The semicoherent operation in (10) can be written as a weighted Wigner distribution [55] by using Moyal's formula [56]:

$$\ell_2(\underline{r}|x) = (2\pi)^{-1} \int_{-\infty}^{\infty} \int_{-\infty}^{\infty} W_r(t,\omega) \, W_u(t,\omega;x) \, dt \, d\omega \tag{22}$$

where $W_r(t,\omega)$ is the Wigner distribution of the observed data $r(t)$, and $W_u(t,\omega;x)$ is the Wigner distribution of the signal with hypothesized parameter x.

An eigenvector-based detector for Gaussian stochastic signals forms a weighted sum of magnitude-squared inner products as in (10). Each term in the weighted sum is the magnitude-squared integral of the data r(t) multiplied by a different eigenvector of the signal covariance matrix, i.e., the semi-coherent (envelope detected) output of a different matched filter. By using Moyal's formula as in (22), such a detector can be written in the form:

$$\ell_3(\underline{r}|x) = (2\pi)^{-1} \int_{-\infty}^{\infty} \int_{-\infty}^{\infty} W_r(t,\omega) \, G_u(t,\omega;x) \, dt \, d\omega \tag{23}$$

where

$$G_u(t,\omega;x) = \sum_n [\lambda_n/(\lambda_n + N_o/2)] \, W_{\phi_n}(t,\omega;x) \, . \tag{24}$$

In (24), λ_n is the eigenvalue corresponding to the nth eigenvector ϕ_n of the signal covariance matrix, or the expected power at the output of a filter matched to ϕ_n when only the signal is present. Similarly, $\lambda_n + N_o/2$ is the expected power at the filter output when both signal and noise are present. The function $W_{\phi_n}(t,\omega;x)$ is the Wigner distribution of the nth eigenvector when the eigenvector is written as a continuous function of time on the interval $(t-T/2,t+T/2)$:

$$W_{\phi_n}(t,\omega;x) = \int_{-T}^{T} \phi_n^*(t-\tau/2;x)\ \phi_n(t+\tau/2;x)\ e^{-j\omega\tau}\ d\tau \quad . \tag{25}$$

In (23-25), the covariance function of the stochastic signal u(t) is assumed to depend on a parameter x, and this parameter therefore conditions the eigenvectors of the signal covariance matrix.

Another version of a Wigner-based detector follows naturally from the Jeffress/Licklider model of binaural interaction, which is based on a "neural coincidence network." This version is not only of interest as a model for sound localization in animals, but also as a first step toward application of Wigner distributions to array processing in man-made systems.

A linear version of the Jeffress/Licklider model is shown in Figure 4a. The time series r(t) consists of a random signal with added external noise, which is assumed to be white and Gaussian. Two versions of this time series, $r_1(t)$ and $r_2(t)$ are obtained from the two ears, after passage through a critical bandwidth filter at each ear. Beam forming by the pinnae and critical band filtering both tend to reduce the number of different signal components that confounded the Wigner representation in Figure 2.

Sampled versions of $r_1(t)$ and $r_2(t)$ are passed through tapped delay lines or shift registers, which partly model the Jeffress/Licklider neural coincidence network [57]. For a linear model (Fig. 4a), signal samples at corresponding positions in the network are summed, and independent samples of internal noise are added to each sum. The sums form a vector of samples $\underline{z}$, where

$$z_1 = r_1(t_1) + r_2(t_N) + n_1$$
$$z_2 = r_1(t_2) + r_2(t_{N-1}) + n_2$$
$$\cdot$$
$$\cdot \tag{26}$$
$$\cdot$$
$$z_N = r_1(t_N) + r_2(t_1) + n_N \quad .$$

In (26), $n_1, n_2, \ldots, n_N$ are independent samples of internal noise, and $r_k(t_m)$ is the mth time sample from ear number k. The summing operations in (26) are analogous to the E-E binaural interaction neurons reported by Suga [11], although Suga has remarked that the interaction between E-E neurons may be multiplicative as well as additive.

A detector can easily be designed for the new data vector

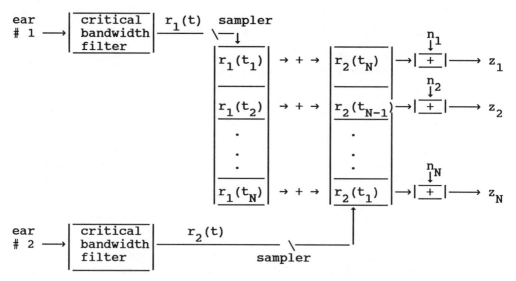

Figure 4a. A linear version of the Jeffress/Licklider model for binaural interaction. Vertical columns of time samples are tapped delay lines, shift registers, or a sequence of interconnected neurons. Neural coincidence is represented by horizontal summation of samples. Samples of internal noise are added after summation. Resulting data vector is $\underline{z}$. For detection of a random signal, a nonlinear operation (quadratic transformation) is applied to $\underline{z}$.

```
ear          critical     r₁(t)    sampler
# 1  ──→     bandwidth ──────────   \─↓                    n₁
             filter                                        ↓
                        ┌─────────┐              ┌─────────┐  ┌───┐
                        │ r₁(t₁)  │ → x →         │ r₂(tN)  │→│ + │──→ y₁
                        ├─────────┤              ├─────────┤  └───┘
                        │ r₁(t₂)  │ → x →         │r₂(tN-1) │→│ + │──→ y₂
                        │    .    │              │    .    │
                        │    .    │              │    .    │       nN
                        │    .    │              │    .    │       ↓
                        │ r₁(tN)  │ → x →         │ r₂(t₁)  │→│ + │──→ yN
                        └─────────┘              └─────────┘
ear          critical     r₂(t)
# 2  ──→     bandwidth ──────────────────  \───────┘
             filter                         sampler
```

Figure 4b. A nonlinear version of the Jeffress/Licklider model for binaural interaction. Neural coincidence is represented by horizontal multiplication of samples. Resulting data vector is $\underline{y}$. A pseudo-Wigner or weighted pseudo-Wigner representation is obtained by a linear transformation (weighted sum) of the elements of $\underline{y}$.

744

z, assuming that the time samples at the outputs of the critical band filters are uncorrelated under both hypothesis H_1: $r(t)$ = a random signal plus external noise, and hypothesis H_0: $r(t)$ = external noise alone. In matrix notation, the detector is

$$\ell(\underline{z}) = \underline{z}^* \; [C_N^{-1} - C_{S+N}^{-1}] \; \underline{z} \tag{27}$$

where $\underline{z}^*$ is the conjugate-transpose of $\underline{z}$, C_N is the covariance matrix of the data vector $\underline{z}$ when H_0 is true (only noise is present), and C_{S+N} is the data covariance matrix when H_1 is true (signal and noise are present). Because of the transformation in (26) and the assumption that the time samples are uncorrelated under both hypotheses, both C_N and C_{S+N} have nonzero elements only along the two diagonals connecting opposite corners of the matrices. This pattern will be called an "X-diagonal form".

An interesting property of the X-diagonal form is that the matrix inverse (when it exists) also has X-diagonal form. The existence of an inverse depends on the addition of the independent internal noise samples in (26), since these make the elements on the main diagonal larger than those on the cross-diagonal, assuring a nonzero matrix determinant.

The sum of inverse covariance matrices $[C_N^{-1} + C_{S+N}^{-1}]$ in (27) thus has X-diagnonal form. The resulting detection statistic has the form

$$\ell(\underline{z}) = \sum_{k=-N_-}^{N_-} q_{N_++k,N_++k} \; |z_{N_++k}|^2$$

$$+ \sum_{k=-N_-}^{N_-} q_{N_+-k,N_++k} \; z_{N_+-k}^* \; z_{N_++k} \tag{28}$$

where N_- equals $(N-1)/2$, N_+ equals $(N+1)/2$, and N, the number of samples in (26), is defined to be an odd number. The weights $q_{i,j}$ in (28) are determined from the matrix $[C_N^{-1} + C_{S+N}^{-1}]$, and the particular form of the sum results from the X-diagonal form of this matrix.

The first sum in the detection statistic (28) is a weighted energy measure. The second term is a sum of the products $z_1 z_N$, $z_2 z_{N-1}$, $z_3 z_{N-2}, \ldots, z_N z_1$. If the sums are written as continuous integrals, (28) is equivalent to

$$\ell(\underline{z}|t) = \int_{-T/2}^{T/2} f(t+\tau) \; |z(t+\tau)|^2 \; d\tau$$

$$+ \int_{-T/2}^{T/2} z^*(t-\tau) \ h(t-\tau,t+\tau) \ z(t+\tau) \ d\tau \qquad (29)$$

where the notation $\ell(\underline{z}|t)$ refers to the fact that the statistic is computed at a specific time. At the next time instant, the arrays in the two shift registers in Figure 4a are translated by one increment (downward in the first shift register and upward in the second), yielding a new z-vector and a new value for $\ell(\underline{z}|t)$. Each shift register stores T seconds of data. If the statistic in (28) and (29) is modified such that the center point of the sum is moved up or down in Figure 4a, then interaural time delay can also be hypothesized.

Let the Fourier transform of $h(t-\tau,t+\tau)$ with respect to the τ variable be a function $H(t,-\omega\ /2)$, and

$$h(t-\tau,t+\tau) \equiv \int_{-\infty}^{\infty} H(t,-\omega\ /2) \ e^{j\omega\tau} d\omega$$

$$= 2 \int_{-\infty}^{\infty} H(t,\omega) \ e^{-j2\omega\tau} d\omega \quad . \qquad (30)$$

The second sum in (29) is then

$$2 \int_{-\infty}^{\infty} H(t,\omega) \ [\int_{-T/2}^{T/2} z^*(t-\tau) \ z(t+\tau) \ e^{-j2\omega\tau} \ d\tau] \ d\omega$$

$$= \int_{-\infty}^{\infty} H(t,\omega) \ [\int_{-T}^{T} z^*(t-\tau/2) \ z(t+\tau/2) \ e^{-j\omega\tau} \ d\tau] \ d\omega \quad . \qquad (31)$$

The function of t and ω in brackets on the right hand side of (31) is a pseudo-Wigner distribution of z(t) [46-50]. The whole right hand side of (31) is a pseudo-Wigner distribution that is weighted in time and frequency by the function $H(t,\omega)$, which is obtained from the matrix inverses in (27). The covariance matrices that determine $H(t,\omega)$ are second order statistical descriptions of the random signal and noise.

Substituting (30) and (31) into (29) yields

$$\ell(\underline{z}|t) = \begin{cases} \text{a weighted energy detector in the interval} \\ (t-T/2,t+T/2) \\ \\ + \\ \\ \text{a weighted pseudo-Wigner distribution .} \end{cases} \qquad (32)$$

In (32) $\underline{z}$ is a symmetrized version of the original vector $\underline{r}$ of input time samples. The data samples are symmetrized by adding the input to a time reversed version of itself. For uncorrelated samples, only a weighted energy detector would be obtained without symmetrizing the data. Symmetrizing converts the data covariance matrix from diagonal to X-diagonal form, and leads to addition of the weighted Wigner distribution to the detection statistic. In the case of a chirp with instantaneous frequency that changes linearly with time, symmetrization literally converts the time-frequency representation from a diagonal stripe (\ or /) to an X.

Figure 2d indicates that a superposition of arbitrary signals yields a highly oscillatory, confusing Wigner representation. The detection result, however, implies that a symmetrized version of a single component may yield a distribution that is useful for extracting the component from noise. The CF and impulsive components in Figures 2b and 2c are invariant under the symmetrizing operation, and thus yield Wigner distributions that are similar to Figures 2b and 2c after the waveforms are symmetrized. The chirp, however, is converted into a superposition of upward and downward chirps that cross (the X pattern). Figure 5 shows the Wigner distribution (again on a logarithmic amplitude scale) of a symmetrized version of the chirp depicted in Figure 2a. Unlike the superposition of arbitrary signals, the superposition of a signal and a time-reversed version of the same signal apparently produces a well-behaved Wigner distribution. The only anomaly is the large peak in the center of the distribution, an attribute that facilitates detection.

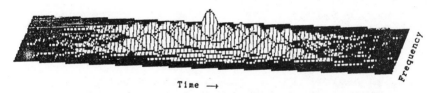

Time →

Figure 5. Pseudo-Wigner distribution of a symmetrized linear FM chirp, created by superposition of the signal and the time-reversed signal as in Fig. 4a. Amplitude scale is logarithnmic as in Figs.2 and 3.

The spectrogram correlator, $\ell_3(\underline{r}|x)$ in (11), has a form similar to the Wigner detector, i.e., it is a weighted version of the data spectrogram. The weighting function in (11) is obtained from the ensemble average spectrogram of the random signal. Similarly, the statistics in (22) and (23) are weighted versions of the Wigner distribution. It is thus conjectured that the weighting function $H(t,\omega)$ in the Wigner detector for symmetrized data depends on the ensemble-average Wigner distribution of the symmetrized random signal.

If neural coincidence is measured by E-E neurons, then the interaction between corresponding shift register samples in Fig. 4a should perhaps be represented by multiplication rather than summation [11], as shown in Fig. 4b. The network in Fig.4b computes the product

$$y(t) = r_1^*(t-\tau/2) \; r_2(t+\tau/2) \; . \tag{33}$$

The product in (33) can be linearly transformed to obtain the pseudo-cross-Wigner distribution and either of the receiver models in (22) or (23) if $r_1(t)$ equals $r_2(t)$, i.e., if interaural delay is correctly hypothesized. <u>Linear</u> binaural interaction as in Fig. 4a must be followed by <u>nonlinear</u> (quadratic or pairwise product) operations for detection of random signals. Conversely, <u>nonlinear</u> binaural interaction as in Fig. 4b and (33) is followed by strictly <u>linear</u> operations for Gaussian random signals.

A similar observation about linearity and nonlinerity can be made for spectrogram correlation. Formation of the spectrogram is a nonlinear process involving envelope detection of filter outputs. Subsequent processing of the data spectrogram $S_{rv}(t,f)$ is linear, as in (11).

SUMMARY AND CONCLUSION

Ordered maps of parameter-tuned neurons can be considered as functional representations. From this viewpoint, neuronal tuning curves represent basis functions for interpolation between sample values. The sampling points are modelled as the parameter values that maximally excite each neuron in the map. The system can estimate a mapped parameter to much higher accuracy than one might expect from the sampling interval, i.e., the difference in parameter values that maximally excite neighboring neurons in the map. Accuracy depends on the order of the interpolating functions. The simple example in Figure 1 pertains to very low order interpolation (a piecewise linear fit). Higher orders and more accurate representations are obtained with smoother basis functions that more closely resemble actual tuning curves.

Each parameter-tuned neuron in an ordered map acts as a feature extractor (hypothesis test) for a specific value of the mapped parameter. An ordered sequence of such tests can be used to estimate the parameter by finding the hypothesis test with the largest response. Under high signal-to-noise ratio conditions, the best accuracy of a parameter estimator (the lowest RMS error) is given by the Cramer-Rao bound. Succeeding hypotheses for slightly different parameter values should intuitively be separated by the a parameter increment that is proportional to the RMS error of the estimator. From a parameter estimation viewpoint, the sampling interval in a neuronal map should be proportional to the expected error of the estimator for the values of the hypothesized parameter in the neighborhood of the samples. If this error is proportional to the Cramer-Rao bound, then the sampling interval should vary inversely with $|E\{\partial^2 \ell(\underline{r}|x)/\partial x^2\}|^{1/2}$, the square root of the magnitude of the ensemble-average second derivative of the log-likelihood function for the specified parameter value, x. This result is also obtained (more rigorously) from an interpolation theorem of de Boor, if the function that is represented by the map is the log-likelihood function $\ell(\underline{r}|x)$.

The idea that an ordered map of parameter-tuned neurons contains hypothesis testers which may (or may not) approximate log-likelihood statistics leads to a proposal that such neurons be used to test various receiver models and concepts. In particular, it may be possible to obtain an estimate of the internal noise in the auditory system by examining the response of (say) range-tuned neurons as a function of signal amplitude for various levels of added external noise.

Recent behavioral experiments seem to indicate that detectors designed for random signals are the most likely candidates for echolocation. Random echoes are obtained when nonrandom (or random) signals are passed through filters with randomly time-varying coefficients. Such filters can be used to model scattering of sound from a fluttering insect.

For FM signals, the experiments seem to imply that the fully coherent matched filter model is still viable for bats only if (i) MTI and pulse-to-pulse coherent processing are implemented, or (ii) laboratory experiments have not adequately restricted the number of echo parameters hypothesized by the bats under test, thus allowing uncertainty-induced performance degradation. These possibilities have not been ruled out for echolocation, although preliminary experiments by B. Mohl imply that the uncertainty argument is more relevant than MTI.

For CF signals, the fine tuning of the highly specialized inner ear of some bats appears to allow for matched filtering of narrowband echoes, but the matched filter is probably followed by an envelope detector. A matched filter followed by an envelope detector is a special case of a random signal detector for signals that are modulated with a time invariant random phase shift.

A convenient implementation of a random signal detector is a spectrogram correlator. Random signal detectors can also exploit other time-frequency representations, such as the Wigner distribution. Detectors that use the Wigner distribution implement signal transformations which are very similar to the Jeffress/Licklider model for binaural interaction.

REFERENCES

1. N. Suga, "The extent to which biosonar information is represented in the auditory cortex," in Dynamic Aspects of Neocortical Function, G.M. Edelman, W.E. Gall, and W.M. Cowan, eds., Wiley, New York (1984) 315-373.
2. G. Neuweiler, "Auditory processing of echoes: Peripheral processing," in Animal Sonar Systems, R.-G. Busnel and J.F. Fish, eds., Plenum, New York (1980) 519-548.
3. D. Marr, Vision, W.H. Freeman, San Francisco (1982) 41-98.
4. C. de Boor, A Practical Guide to Splines, Springer-Verlag, New York (1978) 39-48.
5. C. de Boor, "Good approximation by splines with variable

knots," in <u>Spline Functions and Approximation Theory</u>, A. Meir and A. Sharma, eds., Birhauser Verlag, Basel (1973) 57-72.

6. A.D. Spaulding and D. Middleton, "Optimum reception in an impulsive interference environment (parts I and II)," IEEE Trans. on Commun. COM-25 (1977) 910-934.

7. H.V. Poor and J.B. Thomas, "Locally optimum detection of discrete-time stochastic signals in non-Gaussian noise," J. Acous. Soc. Am. 63 (1978) 75-80.

8. A.S. Kassam and H.V.Poor, "Robust techniques for signal processing: A survey," Proc. IEEE 73 (1985) 433-481.

9. R.A. Altes, "The Fourier-Mellin transform and mammalian hearing," J. Acous. Soc. Am. 63 (1978) 174-183.

10. H.L. Van Trees, <u>Detection, Estimation, and Modulation Theory, Part I</u>, Wiley, New York (1968) 46-52.

11. N. Suga and W.E. O'Neill, "Auditory processing of echoes: Representation of acoustic information from the environment in the bat cerebral cortex," in <u>Animal Sonar Systems</u>, op. cit., 589-611.

12. N. Suga and K. Tsuzuki, "Inhibition and level-tolerant frequency tuning in the auditory cortex of the Mustached bat," J. Neurophys. 53 (1985) 1109-1145.

13. P.M. Woodward, <u>Probability and Information Theory with Applications to Radar</u>, Pergamon, Oxford (1964).

14. C.E. Cook and M. Bernfeld, <u>Radar Signals</u>, Academic, New York (1967).

15. D. Menne and H. Hackbarth, "Accuracy of distance measurement in the bat <u>Eptesicus Fuscus</u>: Theoretical aspects and computer simulations," J. Acous. Soc. Am. 79 (1986) 386-397.

16. R.A. Altes, "Texture analysis with spectrograms," IEEE Trans. on Sonics and Ultrasonics SU-31 (1984) 407-417.

17. R.A. Altes, "Detection, estimation, and classification with spectrograms," J. Acous. Soc. Am. 67 (1980) 1232-1246.

18. R.A. Altes, "Echolocation as seen from the viewpoint of radar/sonar theory," in <u>Localization and Orientation in Biology and Engineering</u>, D. Varju and H.-U. Schnitzler, eds., Springer-Verlag, Berlin (1984) 234-244.

19. B. Mohl, "Detection by a pipistrelle bat of normal and reversed replica of its sonar pulses," Acustica 61 (1986) 75-82.

20. M. Zakharia, "Sonar evaluation in natural environment," in <u>Air-borne Animal Sonar Systems</u>, B. Escudie and Y. Biraud, eds., CNRS, Lyon (1985), 14.1 - 14.12.

21. W.W. Shrader, "MTI Radar," in <u>Radar Handbook</u>, M.I.Skolnik, ed., McGraw-Hill, New York (1970) 17-1 - 17-60.

22. J.A. Simmons, "Perception of echo phase information in bat sonar," Science 204 (1979) 1336-1338.

23. R.J. Urick, <u>Principles of Underwater Sound</u>, 2nd ed, McGraw-Hill, New York (1975) 348.

24. R.A. Altes and E.L. Titlebaum, "Bat signals as optimally Doppler tolerant waveforms," J. Acous. Soc. Am.48 (1970) 1014-1020.

25. R.A. Altes, "Sonar for generalized target description and its similarity to animal echolocation systems," J. Acous. Soc. Am. 59 (1976) 97-105.

26. R.A. Altes, "Methods of wideband signal design for radar and sonar systems," Univ. of Rochester (1970), Nat'l Tech. Info. Service No. AD 732-494.

27. R.A. Altes, "Some invariance properties of the wideband

ambiguity function," J. Acous. Soc. Am. 53 (1973) 1154-1160.

28. R.A. Altes, "Target position estimation in radar and sonar, and generalized ambiguity analysis for maximum likelihood parameter estimation," Proc. IEEE 67 (1979) 920-930.

29. J.F. Fish, C.S. Johnson, and K.K. Ljungblad, "Sonar target discrimination by instrumented human divers," J. Acous. Soc. Am. 59 (1976) 602-606.

30. D.M. Green, "Detection of multiple component signals in noise," J. Acous. Soc. Am. 30 (1958) 904-911.

31. B.W. Tansley and J.B. Suffield, "Time course of adaptation and recovery of channels selectively sensitive to frequency and amplitude modulation," J. Acous. Soc. Am. 74 (1983) 765-775.

32. R.H. Kay and D.R. Mathews, "On the existence in human auditory pathways of channels selectively tuned to the modulation present in frequency modulated tones," J. Physiol. (London) 225 (1972) 657-667.

33. R.A. Altes, "Spectrograms and generalized spectrograms for classification of random processes," ICASSP 84 Proceedings, IEEE No. 84CH1945-5, vol 3 (1984) 41 B.8.1 - 41 B.8.4.

34. R.A. Altes, "Detection and classification phenomena of biological systems," in Adaptive Methods in Underwater Acoustics, H.G. Urban, ed., D. Reidel, Dordrecht (1985) 537-554.

35. H.-U. Schnitzler, D. Menne, R. Kober, and K. Heblich, "The acoustical image of fluttering insects in echolocating bats," in Neuroethology and Behavioral Physiology, F.Huber and H.Markl, eds., Springer-Verlag, Berlin (1983) 235-250.

36. H.-U. Schnitzler and O.W. Henson, Jr.,"Performance of airborne animal sonar systems: I. Microchiroptera," in Animal Sonar Systems, op. cit., 154-160.

37. G.D. Pollack, "Organizational and encoding features of single neurons in the inferior colliculus of bats," in Animal Sonar Systems, op. cit., 549-587.

38. R.A. Altes, "Possible reconstruction of auditory signals by the central nervous system," J. Acous. Soc. Am. 64 (1978) S137

39. N.Y.S. Kiang, D.K. Eddington, and B.Delgutte, "Fundamental considerations in designing auditory implants," Acta Otolaryngol. 87 (1979) 204-218.

40. H. Hackbarth, in preparation.

41. G. von Bekesy, Experiments in Hearing, McGraw-Hill, New York, 1960.

42. R.R. Pfeifer and D.O.Kim, "Cochlear nerve fiber responses: Distribution along the cochlear partition," J. Acous. Soc. Am. 58 (1975) 867-869.

43. E.F. Evans, "Peripheral processing of complex sounds," in Recognition of Complex Acoustic Signals, T.H. Bullock, ed., Abakon Verlagsgesellschaft (Dahlem Konferenzen), Berlin (1977)

44. Y. Grenier, "Time-dependent ARMA modeling of nonstationary signals," IEEE Trans. Acous., Speech, Sig. Proc. ASSP-31 (1983) 899-911. Also: ICASSP 84, op. cit. ref. 33, 41 B.5.1 - 41 B.5.4.

45. Y. Grenier, "Nonstationary signal modeling with application to bat echolocation calls," in Air-borne Animal Sonar Systems, op. cit., 19.1 - 19.35.

46. T.A.C.M. Claasen and W.F.G. Mecklenbrauker, "The Wigner

distribution - a tool for time-frequency signal analysis, Parts I - III," Philips J. Res. 35 (1980) 217-250, 276-300, and 372-389.

47. P. Flandrin and B. Escudie, "Sur la localisation des representations conjointes dans le plan-temps-frequence," C.R. Acad. Sc. Paris 295 (1982) 475-478.

48. P. Flandrin and B. Escudie, " An interpretation of the pseudo-Wigner-Ville distribution," Signal Processing 6 (1984) 27-36.

49. P. Flandrin, "Some features of time-frequency representations of multicomponent signals," ICASSP 84, op. cit., 41B.4.1 - 41B.4.4.

50. W. Martin and P. Flandrin, "Analysis of nonstationary processes: short time periodograms versus a pseudo Wigner estimator," in Signal Processing II: Theories and Applications, H.W. Schussler, ed., North-Holland, 1983, pp. 455-458.

51. T.H. Bullock and S.H. Ridgway, "Evoked potentials in the central auditory system of alert porpoises to their own and artificial sounds," J. Neurobiology 3 (1972) 79-99.

52. D.P. Skinner, R.A. Altes, and J.D.Jones, "Broadband target classification using a bionic sonar," J. Acous. Soc. Am. 62 (1977) 1239-1246.

53. R.A. Altes, "Overlapping windows and signal representations on the time-frequency plane," in Air-borne Animal Sonar Systems, op. cit., 17.1 - 17.47. Also to appear in Acustica.

54. T.J. Imig and A. Morel, "Organization of the thalamocortical auditory system in the cat," Ann. Rev. Neurosci. 6 (1983) 95-120.

55. S. Kay and G.F. Boudreaux-Bartels, "On the optimality of the Wigner distribution for detection," ICASSP 85 Proceedings, IEEE No. CH2118-8/85 (1985) 1017-1020.

56. J.E. Moyal, "Quantum Mechanics as a Statistical Theory," Proc. Cambridge Phil. Soc. 45 (1949) 99-132.

57. D.M. Green and W. A. Yost, " Binaural Analysis," in Handbook of Sensory Physiology, Vol. 5, Pt. 2. New York, Springer-Verlag (1975) 462-480.

58. R.A. Johnson, "Energy spectrum analysis as a processing mechanism for echolocation," doctoral dissertation, Univ. of Rochester, Rochester, N.Y. (1972).

59. R.E. Williams and H.F. Battestin, "Time coherence of acoustic signals transmitted over resolved paths in the deep ocean," J. Acous. Soc. Am. 59 (1976) 312-328.

60. S.A. Kick and J.A. Simmons, "Automatic gain control in the bat's sonar receiver and the neuroethology of echolocation," J. Neurosci. 4 (1984) 2725-2737.

61. R.A. Altes, "Bioacoustic systems: Insights for acoustical imaging and pattern recognition," Internat'l Symp. on Pattern Rec. and Acous. Imaging, SPIE 768 (1987) 61-68.

DETECTION AND RECOGNITION

MODELS OF DOLPHIN SONAR SYSTEMS

Whitlow W.L. Au

Naval Ocean Systems Center

Kailua, Hawaii 96734

INTRODUCTION

Research has shown that dolphins possess highly sophisticated sonar capabilities. Several reviews and papers presented at the second international meeting on Animal Sonar Systems (Busnel and Fish, 1980) elucidated the sonar capabilities of dolphins. Nactigall (1980) discussed object size, shape and material composition discrimination by echolocating odontocetes. Murchison (1980) presented data on target detection and range resolution capabilities of Tursiops truncatus. Au (1980) discussed signals used by Tursiops in open waters. Data on target recognition of cylinders with varying wall thicknesses and material composition (Au and Hammer, 1980) and on sphere-cylinder discrimination (Au et al., 1980) were also presented. These and other studies indicate that dolphin sonars are superior to any man-made sonar systems for short ranges (two to three hundred meters), shallow water (typical of bays, inlets, and coastal waters). The dolphin sonar may be considered the premier sonar for the detection and recognition of slow moving or stationary targets in shallow waters where the reverberation and noise background levels are high.

The objective of this paper is to examine the dolphin sonar system from theoretical and empirical prospectives. Appropriate ideas and models which may improve our understanding of the echolocation process will also be considered. Human listerning results using simulated dolphin echolocation signals to insonify targets will also be used to gain insights into available target cues on which target recognition and discrimination may be made by echolocating dolphins.

I. APPLICATION OF THE SONAR EQUATION

The noise-limited form of the sonar equation can be used to describe the target detection performance of a sonar. It equates the detection threshold (DT) to the echo-to-noise ratio when the target is just being detected, and can be expressed in dBs as (Urick, 1983)

$$DT = \underbrace{SL - 2TL + TS}_{\text{Echo level}} - \underbrace{(NL - DI)}_{\text{Noise level}} \qquad (1)$$

where:
SL = source level
TL = transmission loss
TS = target strength
NL = background noise level
DI = receiving directivity index

The equation is written in terms of intensity or the average acoustic power per unit area. However, when dealing with transient-like signals such as dolphin signals, echoes may be many times longer than the transmitted signal. An example of a simulated dolphin echolocation signal and an echo from a 7.62-cm diam. water-filled sphere are shown in Fig. 1. Therefore, eq. 1 should be transformed to a more generalized form involving energy flux density (Urick, 1983). This can be achieved by using the relation between intensity (I) and energy flux density (E) and the definition of TS,

$$I = \frac{1}{T} \int_0^T \frac{p^2(t)\ dt}{\rho c} = \frac{E}{T} \tag{2}$$

$$TS = 10\ \text{Log}\ \frac{\text{echo intensity 1 m from target}}{\text{incident intensity}} \tag{3}$$

Using eq. 2, the source level term of eq. 1 can now be written as

$$SL = 10\ \text{Log}\ I = SE - 10\ \text{Log}\ \tau_i \tag{4}$$

where the source energy flux density $SE = 10\ \text{Log}\ E$ and τ_i is the duration of the projected signal. Similarly, TS of eq. 4 can be rewritten as

$$TS = TS_E - 10\ \text{Log}\ (\tau_e/\tau_i) \tag{5}$$

where TS_E is the target strength based on the ratio of the energy in the echo over the incident energy, and $_e$ is the duration of the echo. However, if we assume that dolphin detect signals in noise like an energy detector having a specific integration time of τ_{int}, then τ_{int} should be used in place of τ_e in eq. 5. Substituting eqs. 4 and 5 into eq. 1, we get the transient form of the sonar equation applicable to a dolphin

$$DT_E = DT - 10\ \text{Log}\ \tau_{int} = SE - 2TL + TS - (NL - DI) \tag{6}$$

The detection threshold, DT_E corresponds to the energy-to-noise ratio used in human psychophysics and is equal to $10\ \text{Log}(E_e/N_o)$, where E_e is the echo energy flux density and N_o is the noise spectral density level. DT is the signal-to-noise ratio used in sonar engineering.

Peak-to-peak sound pressure level (SPL_{pp}) rather than energy flux density is normally measured. A simple relationship between E and SPL_{pp}

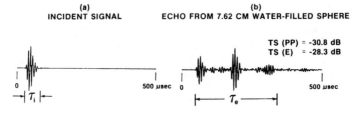

(a)
INCIDENT SIGNAL

(b)
ECHO FROM 7.62 CM WATER-FILLED SPHERE

TS (PP) = -30.8 dB
TS (E) = -28.3 dB

0   500 µsec   0   500 µsec

τ_i τ_e

Fig. 1. Example of a simulated dolphin echolocation signal and the echo from a 7.62-cm diameter water-filled sphere.

can be derived by letting the acoustic signal equal A·s(t), where A is the peak amplitude and s(t) is the waveform function ($|s(t)| \leq 1$).   Then,

$$SPL_{pp}(dB) = 20 \, Log \, (2A) = 20 \, Log \, A + 6 \qquad (7)$$

$$E_{pp}(dB) = SPL_{pp}(dB) + 10 \, Log \, (\int_0^T s^2(t) \, dt) - 6. \qquad (8)$$

The Log integral term does not vary much for dolphin signals in Kaneohe Bay, and is approximately -52 $\pm$ 1 dB.   Equation 8 can now be expressed as

$$E(dB) = SPL_{pp}(dB) - 58 \qquad (9)$$

Equation 8 along with the source level data of Au (1980) shown in Fig. 2 can be used to obtain estimates of SE in the sonar equation.

Values of the receiving directivity index, DI, were calculated by Au and Moore (1984) from measured receiving beam patterns of Tursiops.   Their results are shown in Fig. 3 for frequencies of 30, 60 and 120 kHz.   The linear curve fitted to the results can be expressed as

$$DI(dB) = 16.9 \, Log \, f(kHz) - 14.5 \qquad (10)$$

Au and Penner (1981) determined the detection threshold for two Tursiops using the maximum SE per trial.   Large fluctuations in SE occurred in most click trains and it was not clear which clicks were used by the dolphins for target detection.   Therefore, conservative estimates of

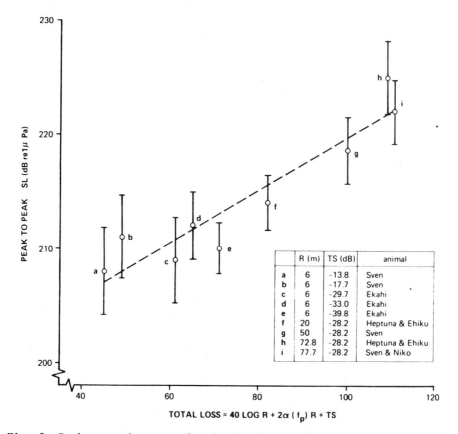

Fig. 2. Peak-to-peak source level of dolphin echolocation signals versus total loss for five animals performing different tasks.

755

DT$_E$ representing the best S/N available to the dolphin were made using the largest SE. The average SE per trail was typically 4 to 5 dB below the maximum SE. Au (this volume) included the results of three other experiments with detection thresholds (DT$_E$) between 7.2 and 12.7 dB.

In comparing detection sensitivities between humans and dolphins we need to be aware that detectability of a signal is a function of the signal duration, frequency, and bandwidth, and the psychophysical testing procedure. Two-alternative forced-choice (2AFC) experiments with humans subjects indicate DT$_E$ of 7.9 dB (Green et al., 1957) to 5 dB (Jeffress, 1968). McFadden (1968) using a Yes/No paradigm measured a DT$_E$ of 10 dB for a 400 Hz, 125-msec signal in continuous and burst noise. Martin and Au (1986), using a Yes/No paradigm and simulated dolphin signals time streched into the human auditory range obtained DT$_E$ of 4.5 dB. Dolphin and human signal detection thresholds in noise seem to be similar despite differences in the methods and signals used.

II. DOLPHIN SONAR MODELED AS AN ENERGY DETECTOR

The threshold versus signal duration and the critical ratio experiments of Johnson (1968a,b) with <u>Tursiops truncatus</u> indicate that the animal's inner ear functions like the human inner ear and that the animal integrates acoustic energy in the same way as humans. Green and Swets (1966) have shown that an energy detector is a good analogue of the human auditory detection process. Therefore, it seems reasonable to approach the dolphin auditory detection process as an energy detector. Au and Moore (This volume) used an electronic "phantom target" to study signal

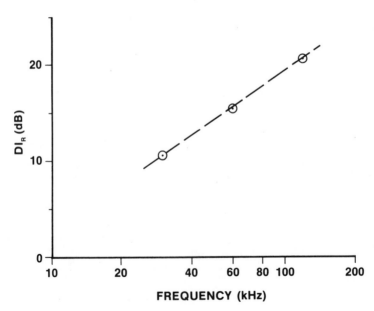

Fig. 3. Receiving directivity index for <u>Tursiops truncatus</u> as a function of frequency (From Au and Moore, 1984).

detection processes of a dolphin. One, two or three replicas of each click emitted were projected back to the animal. All pulses were within the integration time of the dolphin (Au and Moore, this volume). Their results are shown in Fig. 4, along with an energy detector response curve. The dolphin's performance followed the response of an energy detector.

Urkowitz (1967) examined the detection of a deterministic signal in white Gaussian noise using an energy detector and derived expressions for the correct detection and false alarm probabilities. The probability of a false alarm for a given threshold V_T is given by

$$P(FA) = 1 - Pr(V_T \leq \chi^2{}_{2TW]})$$ (11)

where Pr is the area under the chi-square distribution curve with 2TW (time-bandwidth) degrees of freedom. For the same threshold level V_T, the probability of a correct detection is given by

$$P(D) = 1 - Pr(V_T/G \leq \chi^2{}_D)$$ (12)

where: $$D = (2TW + E/N_o)^2/(2TW + 2E/N_o)$$ (13)

$$G = (2TW + 2E/N_o)/(2TW + E/N_o).$$ (14)

Pr is now the area under the noncentral chi-square distribution with a modified number of degrees of freedom D and a threshold divisor G.

These expressions derived by Urkowitz (1967) were applied to dolphin detection in noise data by assuming an unbiased detector in determining the probability of a correct response P(C) from P(FA) and P(D). The calculation was done by first choosing desired values of P(FA) and 2TW and then determining V_T by an iterative procedure. Then with the iterated value of V_T,

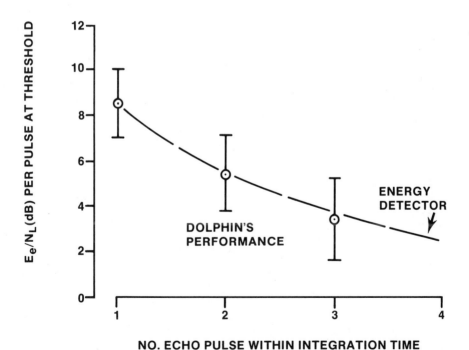

NO. ECHO PULSE WITHIN INTEGRATION TIME

Fig.4. Dolphin detection threshold in noise versus number of pulses within an integration time (From Au andMoore, this volume).

P(D) was calculated for different values of E/N_0. The procedure was continued for different 2TW degrees of freedom, until the values of P(C) were
obtained as a function of E/N_0 which best fitted the dolphin data. The
performance data for <u>Tursiops</u> detecting targets in masking noise in three
different studies are shown in Fig. 5 along with the results of Urkowitz
energy detection model for 2TW = 22.

Urkowitz energy detection model agrees well with the dolphins' performance results, further supporting the notion of the dolphin being an
energy detector. The unbiased detector assumption used to derive P(C) is
a good one for signal-to-noise conditions at or above the 75% correct
threshold. <u>Tursiops</u> tends to be unbias for high signal-to-noise conditions (Au and Synder, 1980; Au and Penner, 1981; Au and Moore, this
volume).

McGill (1968) considered the detection of a signal known except for
phase using an energy detector in a 2AFC procedure. Au and Penner (1981)
applied McGill's expression to a Yes/No paradigm by determining d'(2AFC)
from the tables of Elliot (1964), and by using the relation d'(2AFC) = 2
d'(Y/N) (Tanner and Sorkin, 1972). The values of P(C) was then calculated
as a function of d'(Y/N) assuming an unbiased detector. McGill's formula
also fitted the dolphins' performance.

III TARGET RECOGNITION AND DISCRIMINATION

Hammer and Au (1980) examined targets used in a recognition experiment
with simulated dolphin signals. Echoes were passed through a matched
filter having the transmitter signal as the reference. The envelopes of
the matched filter provided information to explain the dolphin's performance. Times of arrival and shapes of the matched filter envelope for the
different echo highlights, especially the second component, were found to
be important. They suggested that dolphins may use time-separation pitch
(TSP) cues caused by the interaction of highly correlated highlights.

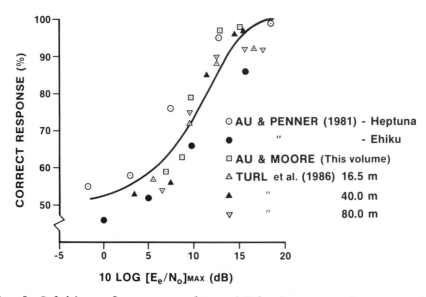

Fig. 5. Dolphin performance results and Urkowitz energy detector model.

Altes (1980) proposed a viable dolphin echolocation model which involved the time-frequency energy density or spectrogram of echoes. The time-frequency energy density function contains information on both highlight distribution and energy spectrum. Target detection, recognition, and discrimination would be achieved by correlating the spectrogram of echoes with previously measured spectrograms of known targets. Altes spectrogram model seems consistent with the TSP model of Hammer and Au (1980), and with the use of energy detection.

Martin and Au (1982, 1986) took a different approach in studying target discrimination by dolphins. They performed experiments using the excellent discrimination and pattern recognition capabilities of the human auditory system to analyze target echoes from the same targets used in dolphin experiments. Fish et al. (1976) previously demonstrated that instrumented human divers, listening to echoes produced by projecting dolphin-like signals, could discriminate the thickness and material composition of metallic plates as well as or better than dolphins. The echoes were time-stretched into the human auditory range. Martin and Au (1982, 1986) used a similar technique but performed human listening experiments in a sound booth. Their specific objective was to determine what acoustic cues were available for discrimination.

Targets were examined acoustically using a monostatic sonar system (see Au and Synder, 1980) that projected a broadband simulated dolphin signal. The echoes were digitized at a sample rate of 1 MHz and stored on magnetic tape. The digitized echoes were played back to humans at a sample rate of 20 Khz, which effectively stretched the echoes by a factor of 50 and lowered the frequency by a factor of 50. For each trial, one of ten echoes was presented to the subjects at a rate of 4 echoes per second. The specific echo in a set of ten echoes per target was randomly selected for each trial.

A. Discrimination of cylindrical targets

Human subjects could easily discriminate standard aluminum cylinders from probe targets used in the general discrimination and material composition experiment of Hammer and Au (1980). For the general discrimination task, the two standard aluminum cylinders had outer diameters of 3.81 cm and 7.62 cm with wall thicknesses of 0.64 and 0.95 cm, respectively. The probe targets are described in Table 1. The discrimination task was found to be trivial even though some of the probes were aluminum cylinders, solid and hollow.

In the material composition discrimination task, subjects had to discriminate two aluminum standards from targets made of steel, bronze and glass. The targets had one of two wall thicknesses, 0.32 and 0.40 cm and outer diameters, 3.81 and 7.62 cm, respectively. The dolphin could discriminate the aluminum standards from the bronze and steel probes but classified the glass probes with the aluminum standards. The average

Table 1. Probe targets used in the general discrimination experiment.

Reference	IP1	IP2	IP3	IP4	IP5	IP6	IP7	IP8
Composition	Al	CPN	Al	CPN	Al	CRK	Al	PVC
Wall (cm)	0.48	solid	solid	solid	0.64	solid	solid	0.79
O.D. (cm)	6.35	6.35	3.81	4.06	11.43	11.43	7.62	7.62

performance of three human subjects in the aluminum versus steel and aluminum versus bronze was 98% and 95% correct, respectively. Subjects first determined whether an echo originated from a large or small cylinder based on duration and TSP cues. Echoes from the large cylinders had lower TSP and longer durations. Subjects reported that the small aluminum had lower TSP than the small bronze cylinder and the large aluminum had a TSP cue whereas the large bronze did not. Examples of the target echoes are shown in Fig. 6 for the small cylinders. One can see that the aluminum should have a higher TSP than the small bronze target. The aluminum versus steel discrimination was based on hearing clearly perceptible TSP with the aluminum targets and less perceptible TSP with the steel targets.

Schusterman, et al. (1980) trained the dolphin used by Hammer and Au (1980) to discriminate between the small aluminum and glass targets using a two-alternative forced-choice paradigm. They were not able to train the animal to discriminate between the large aluminum and glass cylinders. In the human listerning experiment, four subjects discriminated between the aluminum and glass targets with performance accuracy varying from 74% to 94% correct. The main discrimination cue was difference in echo durations between the aluminum and glass echoes. Typical examples of the echoes from the targets are shown in Fig. 7. The glass echoes damped out approximately 14 and 5 ms before the aluminum echoes for the small and large targets, respectively. The duration difference may not have been perceptible to the dolphin, but could be perceived by humans as a result of the time expansion of 50. The ambient noise was not a problem for the dolphin since it typically operated with E/N_0 in excess of 40 to 50 dB.

The duration cues for the large aluminum versus glass discrimination were examined further by truncating the echoes between groups of high-lights, indicated by the tick marks in Fig. 7. The truncation caused the signals to be of equal duration, eliminating any duration cues. The

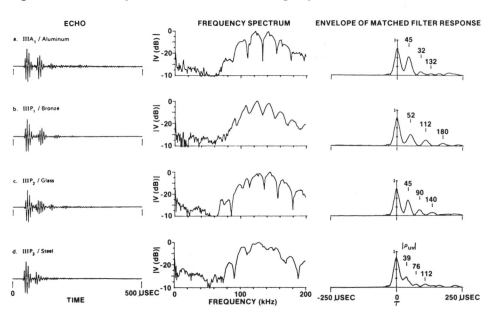

Fig. 6. Backscatter measurements of the 3.81-cm cylinders. Highlight arrival times are indicated on the matched filter results. (From Hammer and Au, 1980).

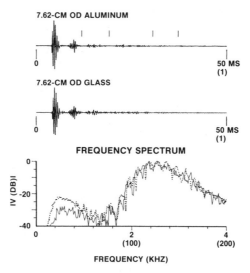

ECHO WAVEFORM

7.62-CM OD ALUMINUM

0   50 MS
                                       (1)

7.62-CM OD GLASS

0   50 MS
                                     (1)

FREQUENCY SPECTRUM

Fig. 7. Echoes from 7.62-cm aluminum (solid) and glass (dashed) cylinders (From Martin and Au, 1986).

performance of two subjects as a function of the signal length is shown in
Fig. 8. Discrimination accuracy decreased as the signals became shorter.
The final truncation eliminated all but the first two echo components, yet
the subjects were able to discriminate the signals above 70% correct. The
time between the first and second echo components was virtually the same
for both targets. Thus, the discrimination was based on cues other than
TSP differences. The subjects indicated that the glass target had a
slightly higher "click pitch" than the aluminum. This cue was difficult to
extract and was not always reliable.

B. Discrimination of Cylindrical Target in Noise

Martin and Au (1986) next performed material composition discrimination in noise. The results are shown in Fig. 9 for the aluminum versus
glass discrimination. For the 3.81-cm cylinders, E/N_o had to be at least
10 dB greater than the detection threshold in noise before a subject could
discriminate above 75% correct. For the 7.62-cm cylinders, E/N_o had to be
at least 30 dB greater than the detection threshold. Subject NN gained a
sudden insight into discriminating the 7.62-cm cylinders at the 25 and 30
dB signal-to-noise ratio. In general, learning the discrimination was
insightful, as subjects began to utilize different cues.

C. Sphere - Cylinder Discrimination

Human discrimination between spheres and cylinders was measured using
foam, solid aluminum and water-filled steel targets. Au et al. (1980)

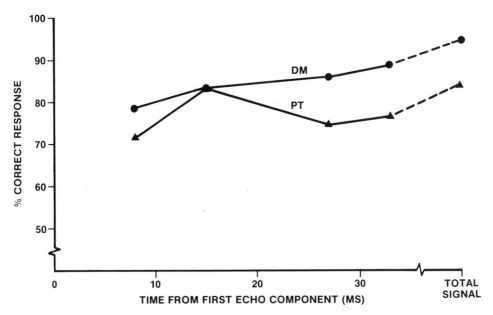

Fig. 8. Human discrimination of 7.62-cm aluminum and glass cylinders versus signal duration (From Martin and Au, 1982).

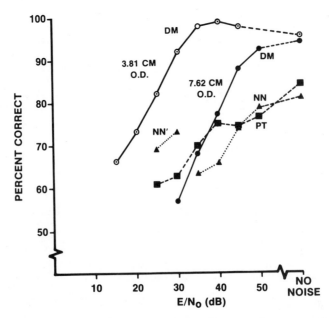

Fig. 9. Aluminum versus glass discrimination results as a function of the signal energy-to-noise ratio (From Martin and Au, 1986).

762

demonstrated that an echolocating dolphin could discriminate between foam spheres and cylinders. They speculated that echoes returning via a surface reflected path were larger in amplitude for spheres than for cylinders. However, when the surface reflected echo component was blocked with an absorbent "horsehair" mat, the dolphin still performed the discrimination. The same foam targets were also used with humans. Discrimination results pooled across subjects are given in Table 2. The average probability of correct discrimination varied between 84% and 96%. The windowed results indicated that surface echo components did provide cues but both the dolphin and humans could perform the discrimination task without these cues.

Subjects reported that two cues used for discrimination were a higher pitch associated with cylinder echoes and the presence of low-frequency reverberation in the sphere echoes. The target strength of a cylinder increases with frequency and is constant with frequency for a sphere (Urick, 1983). An example of echoes from a foam sphere and cylinder is shown in Fig. 10. The sphere echo has a low amplitude portion following the main reflection; the cylinder echo does not. The spectrum for the cylinder has more energy than the sphere's at higher frequencies.

Performance of two subjects in the solid aluminum sphere-cylinder and the water-filled steel sphere-cylinder was between 94% and 100% correct. Subjects reported that the metal target echoes did not sound like foam echoes. However, the same discrimination cues were used: a higher click pitch for cylinders, and more reverberation for spheres.

D. Target Detection in Reverberation

The clutter screen experiment of Au and Turl (1983) was examined with the targets in the plane of the clutter screen. The human and dolphin performance as a function of the peak-to-peak echo-to-reverberation ratio are shown in Fig. 11. The cue used by the humans was the presence of a "click" sound from the targets. The echoes from the clutter screen sounded diffused, whereas the echoes from the aluminum cylinders sounded like compact clicks with a definite TSP. Subjects learned to integrate only over the duration of the target echo.

Table 2. Foam sphere versus cylinder dolphin discrimination results. The windowed results refer to echoes which had the air-water surface reflected components truncated.

Spheres (Dia.)	Cylinders (Dia. x Length)	Presentation Schedule	Total Signal	Windowed Signal
S1: 10.2 cm	C1: 1.9 x 4.9 cm	S2 vs C4	96%	88%
S2: 12.7 cm	C2: 2.5 x 3.8 cm	S2/S3 vs C3/C4	93%	85%
S3: 15.2 cm	C3: 2.5 x 5.1 cm	S2/S3 vs C1/C4	88%	81%
	C4: 3.8 x 5.4 cm	S1/S2 vs C4/C5	84%	--
	C5: 3.8 x 5.1 cm	S1/S2 vs C2/C4	91%	83%

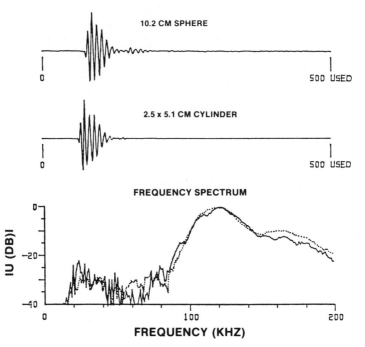

ECHO WAVEFORM

10.2 CM SPHERE

0   500 USED

2.5 x 5.1 CM CYLINDER

0   500 USED

FREQUENCY SPECTRUM

FREQUENCY (KHZ)

Fig. 10. Echoes from foam sphere (solid) and cylinder (dotted). (From Martin and Au, 1986).

E. Aspect Independent Discrimination of Cylinders

The final experiment involved a material composition discrimination of cylinders at various aspects. The procedure involved training the dolphin to discriminate between two targets at the baseline aspects of 0, 45, and 90°, where 0° was the broadside aspect. After the dolphin could perform the discrimination at 90% correct for the baseline aspects, probe trials at different aspects was used. In the first discrimination task, aluminum versus steel, the dolphin performance did not reach the 90% criterion for 45 and 90° aspect. An easier discrimination of aluminum versus coral rock was then chosen. The dolphin's results are shown in Table 3, which includes both the baseline (shaded) and probe results.

A similar procedure was used with the humans but with a wider variety of targets. Subjects were trained to discriminate between pairs of targets for the baseline angles. After achieving near 100% correct performance on the baseline angles, sessions were made more complex by presenting echoes from seven aspects, 0, 15, 30, 45, 60, 75 and 90°. The human results pooled over the four subjects are shown in Table 4. As with the dolphin, the humans could perform the aluminum versus rock discrimination readily but had problems with the aluminum versus steel discrimination (4th column of Table 4). When the subjects were given the possibility of having any one of seven aspects presented for a given trial, performance at the baseline angles deteriorated. No specific cues could be focused on in this experiment since the echoes were so complex. However, the human

764

Table 3. Dolphin discrimination results as a function of aspect angle
of the targets (From Martin and Au, 1986).

BASELINE PERFORMANCE

0°		45°		90°	
AL	ROCK	AL	ROCK	AL	ROCK
100%	94%	91%	89%	100%	96%

PROBE SESSIONS

0°		15°		30°		45°		60°		75°		90°	
AL	ROCK	AL	ROCK	AL	ROCK	AL	ROCK	AL	ROCK	AL	ROCK	AL	ROCK
100%	100%	98%	97%	96%	100%	100%	100%	100%	100%	100%	100%	100%	96%

subjects indicated that echoes at the 45° aspect angle contained the most
information pertinent to the other angles.

IV. DISCUSSION AND CONCLUSIONS

An echolocating dolphin's ability to detect target echoes in noise and
to make fine discrimination of target features seems similar to human
auditory detection and pattern recognition capabilities. Although human

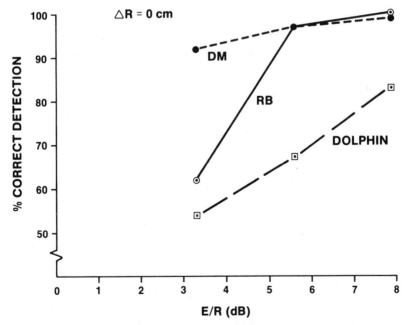

Fig. 11. Target detection in clutter performance results for humans and
dolphin (From Martin and Au, 1986).

Table 4. Human discrimination results as a function of target aspect angle for different target pairs. The values in each column are the percent correct results in identifying a particular target in a target pair (From Martin and Au, 1986).

	HOLLOW ALUM	CORAL ROCK	HOLLOW ALUM	SOLID ALUM	SOLID ALUM	CORAL ROCK	HOLLOW ALUM	HOLLOW STEEL	ALUM WATER	ALUM AIR
0°	89	92	56	93	93	94	93	97	90	100
15°	98	93	99	76	88	91	97	57	100	80
30°	89	93	77	73	94	92	53	91	69	91
45°	88	98	83	93	81	93	84	87	95	91
60°	81	85	81	90	80	87	53	69	71	92
75°	98	80	91	58	95	70	68	71	83	98
90°	97	96	97	98	97	92	100	95	99	98

hearing experiments were performed differently and with different stimuli, detection thresholds of dolphin and man are comparable. The question that comes immediately to mind is "why does the dolphin sonar perform better than man-made sonars in shallow water environments?" There are several factors inherent in dolphins that may provide advantages over man-made sonars. A dolphin is a highly mobile aquatic mammal that is capable of using its sonar while in motion, going to different depths and locations, and looking at objects from various aspects. Reverberation and noise observed from one aspect may be different when observed at other aspects, and locations. The dolphin may also have good long term auditory and spatial memory which would be effective in recognizing desired targets in specific positions and in spatial pattern recognition of echoes from various locations.

Most man-made active sonars have been designed to eliminate the human auditory capabilities from the system. Therefore, the excellent analysis and pattern recognition capabilities of the human auditory system are ignored. When a listening capability is included in an active sonar, less than ideal kinds of signals are often used: narrowband pulse tones or continous transmission frequency modulated signals (CTFM). Neither of these signals possess the time resolution capabilities of broadband transient-like pulses that dolphins use.

Human listening experiments using dolphin-like sonar signals can be useful in understanding target cues and processing methods needed to extract them. The human listening experiments of Martin and Au (1982, 1986) indicated that differences in time-separation pitch associated with correlated echo highlights and differences in echo duration were the predominant discrimination cues in almost all of the tasks. Duration cues as much as 30 dB below the peak level of the primary echo component were found to be important and useful. Only in the sphere-cylinder and truncated aluminum-glass discriminations was spectral information in the form of click pitch an important cue. These cues used by the human listeners may be the same cues used by echolocating dolphins. It seems that time domain processing of highlight separation and highlight amplitudes within echoes may provide most of the target information to the dolphin. Therefore, broadband transient-like echolocation signals with good time resolution would be most useful to dolphins and humans.

REFERENCES

Altes, R. A. 1980, Model for Echolocation, in "Animal Sonar Systems", R.G. Busnel and J.F. Fish, ed., Plenum, New, pp. 625-671.

Au, W. W. L. and Snyder, K. J., 1980, Long-range Target Detection in Open Waters by an Echolocaing Atlantic Bottlenose Dolphin (<u>Tursiops truncatus</u>), <u>J. Acoust. Soc. Am.</u>, 68, 1077-1084.

Au, W. W. L., Echolocation Signals of the Atlantic Bottlenose Dolphin (<u>Tursiops truncatus</u>) in Open Waters, in "Animal Sonar Systems", R. G. Busnel and J.F. Fish, ed., Plenum, New York, pp. 251-282.

Au, W. W. L., Schusterman, R.J. and Kersting, D.A., 1980, Sphere-Cylinder Discrimination via Echolocation by <u>Tursiops truncatus</u>, in "Animal Sonar Systems," R.G. Busnel and J.F. Fish, Plenum, New York, pp. 859-862.

Au, W. W. L. and Penner, R. H., 1981, Target Detection in Noise by Echolocating Atlantic Bottlenose Dolphins, <u>J. Acoust. Soc. Am.</u>, 70, 687-693.

Au, W. W. L. and Moore, P. W. B., 1986, Echo Perception and Detection by Dolphins, this volume.

Busnel, R. G. and Fish, J. F., 1980, "Animal Sonar Systems", Plenum, New York).

Elliot, P. B., 1964, Appendix 1-Tables of d', "Signal Detection and Recognition by Human Observers", J.A. Swets, ed. Wiley, New York, pp. 651-684.

Fish, J. F., Johnson, C. S. and Ljungblad, D. K., 1976, Sonar Target Discrimination by Instrumented Human Divers, <u>J. Acoust. Soc. Am.</u>, 59, 602-606.

Green, D. M., Birdsall, T. G. and Tanner, W. P., Jr., 1957, Signal Detection as a function of Signal Intensity and Duration, <u>J. Acoust. Soc. Am.</u>, 29, 523-531.

Green, D. M., and Swets, J. A., 1966, "Signal Detection and Psychophysics" (Krieger, Huntington, N.Y.).

Hammer, C. E. and Au, W. W. L., 1980, Porpoise Echo-recognition: an Analysis of Controlling Target Characteristics, <u>J. Acoust. Soc. Am.</u> 68, 1285-1293.

Jeffress, L. A., 1969, Mathematical and Electrical Models of Auditory Detection, <u>J. Acoust. Soc. Am.</u>, 44, 187-203.

Johnson, C. S., 1968a, Relation between Absolute Threshold and Duration-of- Tone Pulses in the Bottlenosed Porpoise, <u>J. Acoust. Soc. Am.</u>, 43, 757-763.

Johnson, C. S., 1968b, Masked Tonal Thresholds in the Bottlenosed Porpoise, <u>J. Acoust. Soc. Am.</u>, 44, 965-967.

McFadden, D., 1966, Masking-Level Differences with Continuous and with Burst Masking Noise, <u>J. Acoust. Soc. Am.</u>, 40, 1414-1419.

McGill, W. J., 1967, Variations on Marill's Detection Formula, J. Acoust. Soc. Am., 43, 70-73.

Martin, D. W. and Au, W. W. L., 1982, Aural Discrimination of Targets by Human Subjects Using Broadband Sonar Pulses, Naval Ocean Systems Center, San Diego, Ca., NOSC TR 847.

Martin, D. W. and Au, W. W. L., 1986, Investigation of Broadband Sonar Classification Cues, Naval Ocean Systems Center, San Diego, Ca., NOSC TR 1123.

Murchison, A.E., 1980, Detection Range and Range Resolution of Echolocating Bottlenose Porpoise (Tursiops Truncatus), in "Animal Sonar Systems", R. G. Busnel and J. F. Fish, ed., Plenum, New York, pp. 43-70.

Nachtigall, P. E., 1980, Odontocete Echolocation Performance on Object Size, Shape and Material, in "Animal Sonar Systems", R. G. Busnel and J. F. Fish, ed., Plenum, New York, pp. 71-95.

Schusterman, R. J., Kersting, D. and Au, W. W. L., 1980, Response Bias and Attention in Discriminative Echolocation by Tursiops truncatus, in "Animal Sonar Systems," R. G. Busnel and J. F. Fish, ed., Plenum, New York, pp. 983-986.

Tanner, W. P., Jr. and Sorkin, R. D., 1972, The Theory of Signal Detectability, in "Foundation of Modern Auditory Function", Vol II, J. V. Tobias, ed., Academic, New York, pp. 63-97.

Turl, C. W., Penner, R. H. and Au, W. W. L., 1986, Comparison of the Target Detection Capability of the Beluga Whale and Bottlenose Dolphin, J. Acoust. Soc. Am., 82, 1487-1491.

Urick, R. J., 1983, "Principles of Underwater Sound", 3rd ed, McGraw Hill, New York).

Urkowitz, H., 1967, Energy Detection of Unknown Deterministic Signals, Proc. I.E.E.E., 55, 523-531.

A BRIEF HISTORY OF BIONIC SONARS

C. Scott Johnson

Naval Ocean Systems Center
San Diego, CA 92152-5000

The word "bionics" is not a contraction of the words biology and
electronics. The word was coined in 1959-60 by COL. Jack Steele to
identify some of the programs at the Wright-Patterson Air Force Base
(von Gierke 1986, private communication). The word comes from the Greek;
bion is the unit of life in Greek and the ending ics indicates life-like.
The U. S. Air Force stopped using the term when the TV shows "The Six
Million Dollar Man" and "The Bionic Woman" were shown on ABC television
(Hogan 1986). While the term "bionic sonar" means different things to
different people, for the purposes of this discussion we will interpret
it broadly to mean any attempt to build an electro-mechanical sonar
using our knowledge of animal sonar systems.

While not inspired by thoughts of animal sonar systems the Continuous
Transmission Frequency Modulated sonar (CTFM) should be mentioned. CTFM
sonars were being made before we knew very much about biological sonars,
and it was suggested (Kay 1961, and Pye 1960) that biological sonars might
operate on the CTFM principle. We now know this is not the case, but
several sonars built on the CTFM principle and utilizing human listeners
to interpret the signals have been built for use by blind people (Kay 1979).
The U. S. Navy's AN/PWS-1 and AN/PQS-2 diver-operated sonars work in
essentially the same way as the devices for the blind.

Bionic sonars historically have been one of a kind experimental
prototypes. One of these was designed and constructed by the Electromagnetic Systems Laboratories (ESL) under contract to the Naval Ocean
Systems Center (NOSC). This sonar is designated the AN/SQQ-27 XN-2 and
was based on a theory of R. A. Altes (Jones et al. 1975, Floyd 1980).
This system used a broadband signal, and a computer to store the
parameters associated with targets. While the AN/SQQ-27 worked
reasonably well it suffered from neglecting certain target features, and
in addition, there doesn't appear to be any biological system that uses
such a scheme (Floyd 1980). No further work has been done on this
design.

Another sonar intended for divers and requiring the operator to
interpret the signals aurally, is called simply the "Broadband Diver Sonar
(BDS)." The BDS was designed, built, and tested at NOSC. This system uses
a broadband pulse. After a pulse is sent out, echoes returning in a
certain time or range interval are digitized, stored in memory, and
played to the operator slowed down in time, as in the Fish et al. (1976)

experiment. This sonar performed somewhat better than the Navy's new AN/PQS-2 diver sonar during both pool and open water tests.

There are currently four signal processing models that show promise in bionic sonar development (Floyd 1980). They are: (1) matched filtering or replica correlation, (2) simple energy detection in a bandpass filter, (3) energy detection with a bank of constant Q filters, and (4) time separation pitch (TSP).

Models 2 and 3 do not require knowledge of the outgoing signal other than bandwidth, but 1 and 4 do. Models 3 and 4 are essentially the same (Johnson 1980) if the outgoing signal is analyzed with the returning target echo. It has been suggested that bats use a time domain analysis (Simmons 1980) however current analysis of the bat data does not confirm this (Meene and Hackbarth 1986). Dolphins may use a frequency domain analysis (Murchison 1980). This would mean that bats might be using a theory 1-type processing system while whales might be using a theory 3- or 4-like system.

We know that the auditory system of Atlantic bottlenose dolphins (Tursiops truncatus) can be modeled as a bank of constant Q filters (Johnson 1980) similar to humans but of much greater total bandwidth. If we could determine whether or not whales perceive TSP we could make a choice as to whether Model 3 or 4 would be effective.

REFERENCES

Fish, J. F, C. S. Johnson, and D. K. Ljungblad, 1976, Sonar target discrimination by human divers, J. Acoust. Soc. Am. 59, 602-606.

Floyd, R. W., 1980, Models of cetacean signal processing, In: "Animal Sonar Systems," R. G. Busnel and J. F. Fish (Eds.) Plenum Press, New York.

Hogan, J., 1986, Robotics aim to ape nature, IEEE Spectrum Feb. 1986:66-71.

Johnson, C. S., 1968, Masked tonal thresholds in the bottlenose porpoise, J. Acoust. Soc. Am., 44, 965-967.

Johnson, R. A., 1980, Energy spectrum analysis in echolocation, In: "Animal Sonar Systems," R. G. Busnel and J. F. Fish (Eds.) Plenum Press, New York.

Jones, R. N, B. J. Sudderth, R. J. Eitriem, R. J. Giancoli, P. C. Chestnut, R. A. Altes, and D. L. Wangness, 1975, AN/SQQ-27 (XN-2) Final Report, Electromagnetic Systems Laboratories, Project 1762 Final Report No. ESL-TM17, 21 July 1975.

Kay, L., 1962, A plausible explanation of bat echolocating acuity, Animal Behavior 10:34-41.

Meene, D. and H. Hackbarth, 1986, Accuracy of distance measurements in the bat Eptesicus fuscus, J. Acoust. Soc. Am. 79, 386-397.

Murchison, A. E., 1980, Detection range and range resolution of echolocating bottlenose porpoise (Tursiops truncatus), In: "Animal Sonar Systems," R. G. Busnel and J. F. Fish (Eds), Plenum Press, New York.

Pye, J. D., 1960, A theory of echolocation by bats, J. Laryng. Otal. 74:718-729.

Simmons, J. A., 1980, The processing of sonar echoes by bats, In: "Animal Sonar Systems," R. G. Busnel and J. F. Fish (Eds), Plenum Press, New York.

BIOSONAR SIGNAL PROCESSING APPLICATIONS

Robert W. Floyd

Naval Ocean Systems Center
Kailua, Hawaii 96734

   Dolphins have a sophisticated echolocation, or sonar, system that is
well adapted to their environment. Given that it serves the animal well,
one might ask how the dolphin sonar compares to man made sonars, and if
there is anything useful one might apply from knowledge of the dolphin's
echolocation system. In this paper we will compare the capabilities of
dolphin sonar to some existing manufactured sonars, identify important
differences between the two systems, and describe methods by which manmade
sonars could be improved using our current knowledge of dolphin sonars.

PERFORMANCE COMPARISON

   Although the basic parameters of the dolphins' sonar suggest that it is
better suited for classification than for detection, the actual detection
capabilities of the dolphin sonar compare favorably with those of
conventional sonars. Figure 1 shows a comparison of a dolphin and a Straza

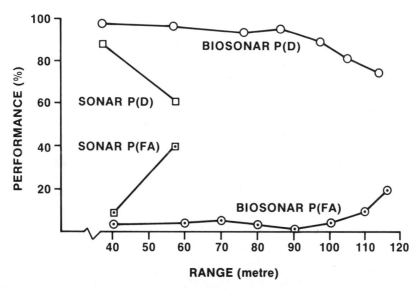

Fig. 1. A comparison of the performance a dolphin and a CTFM sonar in
        detecting a 7.62 cm water filled steel sphere in Kaneohe Bay.

500 continuous transmission, frequency modulated (CTFM) sonar in detecting a 7.6 cm water filled steel sphere in Kaneohe Bay, Hawaii. Note that the performance of the CTFM sonar dropped off rapidly after 40 metres, while the performance of the dolphins was good out to 113 meters (Au and Snyder, 1980). In this experiment, the range at which the target was presented was constant throughout the experimental session, and there was an equal probability of a target present or target absent trial. In figure 1, P(D) is the percent of correct detections, and P(FA) is the percent of responses that a target was present when it was in fact absent.

The figure of merit is a standard method of comparing conventional sonars. It is based on easily measured physical parameters. The two parameters most generally used are transmit energy, or power, and the receiver directivity index. The parameter list in table 1 shows that the dolphin is poorer by 25 dB in transmit energy, and poorer by 5 dB in directivity index. Therefore, by conventional standards the dolphin would have a 30 dB poorer figure of merit than the CTFM sonar, and yet it achieved far better performance.

Table 1. Comparison of the Basic Parameters of the Dolphin, CTFM (Straza 500) and a Typical Pulse Sonar (WQS-1)

Parameter	Dolphin	CTFM	Pulse Mode
Transmit Energy	157 dB	182 dB	176 dB
Power	204 dB	191 dB	209 dB
Time	15 μsec.	120 msec.	0.5 msec
Directivity Index (Receiving)	20 dB	25 dB	25 dB
Range Resolution	1 cm	3 metre	37.5 cm
Beamwidth			
Horizontal	17°	2.5°	2°
Vertical	17°	17°	23°

The only clear advantage that the dolphin has over the other two sonars is range resolution. This provides no processing gain in a noise limited environment, but does provide significant gain over the other two sonars in a reverberation limited environment. Reverberation is caused by reflections of the transmitted beam from random bottom, surface, for volume scatterers. The intensity of the reverberation is proportional to the scattering strength of the medium or interface, and the volume or area of the ensonified scatterers (Urick, 1967). The area that is ensonified is in turn proportional to the beamwidth and the range resolution of the signal. In sonar terminology the reverberation level is expressed as:

Volume reverberation:
$$RL_V = SL - 40 \log r + S_V + 10 \log V \qquad (1)$$
$$V = ct/2 \ \Psi \ r^2$$

Surface (or bottom) reverberation:
$$RL_S = SL - 40 \log r + S_S + 10 \log A \qquad (2)$$
$$A = ct/2 \; \emptyset \; r$$

where:

RL	=	Reverberation level
SL	=	Source level
S_v, S_s	=	Volume and surface scattering coefficients
r	=	Range in metres
V	=	Ensonified volume
A	=	Ensonified area
Ψ	=	Solid angle of beam pattern
Ø	=	Horizontal beam pattern
ct/2	=	Range resolution of signal

Thus, for all other variables being equal, the reverberation level is directly proportional to the range resolution of the signal. The geometry of the range detection experiment was such that the sonar beam for both the CTFM sonar and the dolphin intercepted both the surface and the bottom at the longer detection ranges. If both sonars were reverberation limited, then the dolphin sonar would gain 25 dB relative to the CTFM sonar in reverberation suppression. This would explain most of the difference in performance between the two systems.

The dolphin achieves excellent resolution by using a short, wide band echolocation signal (Au et al. 1974). The potential resolution of the signal can be calculated from its bandwidth or its autocorrelation function (ACF). The ACF gives more information about possible models of echolocation, because various hypotheses can be made depending on what parts of the ACF are used.

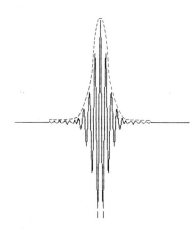

Fig. 2. The digital autocorrelation function of a dolphin echolocation click sampled at 1 MHz. The two short vertical bars represent 1 cm range resolution.

The following table lists the range, azimuth and altitude resolution capabilities of the dolphin that have been measured and those that have been estimated by using various physical paramaters of the dolphin echolocation system. The measured range resolution is from Murchison, 1980; the azimuth and altitude from Renaud and Popper, 1975; and the beam pattern from Au and Moore, 1984.

Table 2. Dolphin Sonar Resolution, Measured and
          Estimated.

	Range (at 1 Meter)	Azimuth	Altitude
Measured	1 cm	0.9°	0.7°
Estimated from range resolution	--	4.8°	--
Estimated from c/2BW	2 cm	9.6°	--
Estimated from correlation function	0.25 cm	1°	--
Estimated from critical bands	--	--	1°
Estimated from beam pattern	--	16°	16°

The measured absolute range resolution of 1 cm at 1 metre is better than
one would predict using either a bandwidth or envelope of the
autocorrelation function. However, the resolution is not as good as would
be predicted using the central peak of the coherent ACF. If we assume that
azimuth is estimated by time of arrival differences, the measured azimuth
resolution is consistent with what one would predict using the coherent
form of the ACF. In addition, target classification experiments indicate
that the dolphin easily discriminates time differences of less than 7
μsecs. This suggests that for signals occuring within a short period (<
200 μsecs.), ACF equivalent processing can occur. A possible mechanism for
this is Time Separation Pitch (TSP) suggested by Floyd (1980), and Au and
Hammer (1980) as a method of target classification.

Azimuth resolution requires a wideband signal also, but the mechanism
for determing azimuth differences probably relies entirely on spectral
cues. When Au and Moore (1984) measured a dolphin's beam pattern, they
discovered that the maximum angle of sensitivity was not at 0° referenced
to the rostrum, but was elevated at about 10° relative to the rostrum. If
we use the approximation of the dolphin's receiving aperture proposed by Au
et al. (1974), we can calculate what the spectral response of the dolphin
receiver was for the angles at which Renaud made her measurements.

For a rectangular array hypothesized by Au (1974), the beam pattern in
the sagital plane as a function of wavelength and vertical angle, or
altitude, can be expressed as:

$$A(\emptyset) = \left\{ \frac{\sin[(L/\lambda) \sin \emptyset]}{(L/\lambda) \sin \emptyset} \right\}^2 \qquad (3)$$

For a frequency of 100 kHz and an aperture of 7.3 cm., the receiving
sensitivity versus angle would have the pattern shown in figure 3. Note
that there is a notch in the sensitivity at about 11°. This pattern
will be similar for other frequencies, although the notches will occur at
slightly different angles.

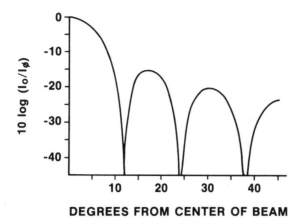

Fig 3. Predicted sensitivity of the dolphins receiving
        array versus angle at 100 kHz.

   We can also plot the results of equation 3 as a function of frequency
for several different angles. Figure 4 shows the intensity of the received
broadband signals for 10, 11, and 12°. This is the same angular range in
which Renaud and Popper made their vertical sensitivity measurements. At
these angles, and at the frequencies of the dolphin's echolocation signals,
the notches in the spectrum are spaced at approximately 10 kHz per degree
change in altitude. This spacing represents one critical band (Johnson,
1968), and the loss of energy in the critical band could be used to
determine that a sound source was at a particular altitude. This
hypothesis of vertical angle resolution has not been tested, and Au and
Moore (1984) have also suggested that it may be done by off-axis intensity
differences.

   Spatial resolution by the methods discussed thus far require a
relatively high signal to noise ratio (SNR). In particular, the azimuth
resolution by spectrum dips would appear to require at least a 30 dB SNR.
In most practical cases (for the dolphin, at least) this is not a very
stringent requirement. For instance, to improve the SNR for the dolphin in
the detection experiment by 30 dB, the target range should be reduced by a

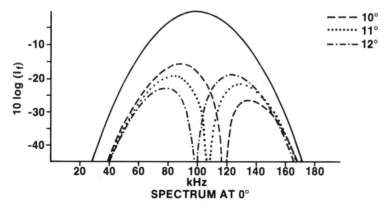

Fig 4. The spectrum of a wide band signal convolved with the
        hypothetical frequency response of the dolphins receiving array
        at several angles. The smooth curve on top is the response at 0°.

777

factor of 5 to approximately 20 metres. Since precise azimuth resolution
would only be needed as the dolphin approached a tartet or obstacle, a 20
metre range at which the altitude resolution could be used would be more
than sufficient.

Spatial resolution, particularly in range, appears to be essential for
target classifications. Figure 5 shows an example of three target echoes.
Note that the bronze cylinder has its second highlight displaced by seven
μsec from the second highlight of the aluminum target. These two echoes
were easily discriminable by the dolphin. On the other hand, the second
highlight of the glass echo occurs at the same time delay as that of the
aluminum echo. These echoes were very difficult for the animal to
discriminate (Au and Hamer, 1980). These results indicate that time
structure, or spectral changes caused by time structure, is the salient cue
in identification of these types of targets.

Martin and Au (this volume) have constructed an automatic target
recognition algorithm that is based exclusively on the time domain features
of the target echoes. This method appears to duplicate the dolphin's
performance for the echoes studied.

SUMMARY OF DIFFERENCES

The primary difference between dolphin and conventional sonars is that
the dolphin uses a wide band echolocation signal and binaural receivers.
This combination provides for superior target recognition capabilities,

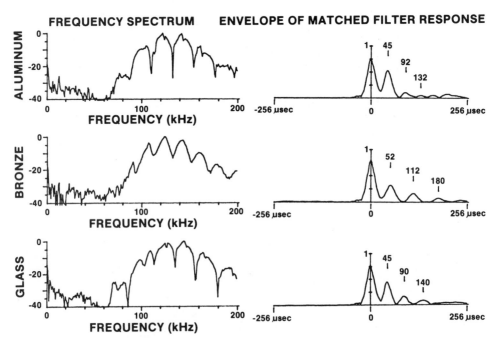

Fig 5. Spectra and replica correlator outputs for aluminum,
        bronze and glass echoes.

better target detection in reverberation, and far better spatial resolution
than conventional sonars with similar beamwidths. However, the dolphin
sonar provides no performance gain over conventional sonars in target
detection in isotropic noise.

The dolphin sonar has a conical beam 17° wide with a one degree resolution. It is possible to build a conventional sonar with similar coverage and resolution. However, it would have 200 beams and be enormously expensive. It would, however, perform much better than the dolphin in isotropic noise and somewhat better in most reverberation limited environments.

IMPLEMENTATION

An advantage of the bionic approach is the relative simplicity of the transducers. One can use a separate transmitting transducer and two receiving hydrophones, or a pair of transducers, one of which can act as both a transmitter and a hydrophone. One requirement is that the transmitter be relatively broadband, or low Q. This can be difficult, particularly for underwater transducers, because the efficiency at the center frequency is usually greater at higher Q's. The effective bandwidth of the signal may be increased by precompensation of the transmitted signal (Au and Floyd, 1979), using multiphase backing material for the transducer elements (Bar-Cohen et al., 1984), using quarter wave matching layers between the piezoelectric material and the load (Desilets et al., 1978), or by shaping the piezoelectric material to broaden the resonant peak (Heyser, 1986).

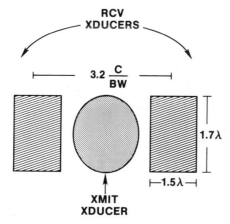

Fig 6. A binaural sonar array that mimics the dolphin's beamwidth and resolution normalized to wavelength and sound velocity in the acoustic medium.

The geometry in figure 6 is a good compromise for many possible sonar applications. This arrangement could be modified for specialized applications. For instance, if the target was confined to the horizontal plane, increasing the transducers' vertical dimensions would narrow the beam's vertical extent and provide greater directivity gain against noise and reverberation. Decreasing the transducers' horizontal dimension would increase the beam's horizontal extent and thus eliminate the need for mechanical scanning.

Processing of the echoes is a more complicated problem. One current hypothesis is that the dolphin uses TSP to resolve small differences in range. Since it is the actual range or time information, rather than the spectrum, that contains the desired information, a more direct approach is to measure the time information directly. The most efficient way to extract time domain information is to process the incoming signal with a replica correlator, which is a special case of a finite impulse response filter (FIR).

The replica correlator operates by taking in a sample of data and then multiplying and summing the latest datum and the n-1 previous samples with n samples of the stored replica. This process is computationally very intensive. The number of multiply/accumulate operations per unit time is:

$$N = L R^2 \qquad\qquad (4)$$

     N = number of operations per microsecond
     L = length of replica signal
     R = sample rate in millions of samples per second

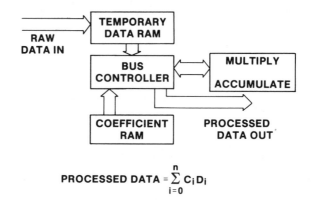

$$\text{PROCESSED DATA} = \sum_{i=0}^{n} C_i D_i$$

Fig 7. Block diagram of a replica correlator.

Typical values we have used are lengths of 64 microseconds and sample rates of 1 MHz, which require 64 operations per μsec. Until recently this computation rate has been impossible to accomplish in real time except with a supercomputer. From equation 4 it is obvious that the required computation rate is very sensitive to the sample rate. With proper filtering, the sample rate can be reduced to 400 KHz, which requires 10 operations per microsecond. This processing speed is within the capability of the Motorola DSP56200 signal processing chip. The Zoran ZR33881 signal processing chip has even more impressive capabilities. It has 8 multiply/accumulate units in parallel, each working at 20 million operations per second. Of course, the effective processing speed can be increased by adding more chips in parallel.

Although current processors can handle dolphin-like signals, a single processor is unable to handle longer signals. Since transducers are normally peak power limited, long signals have more transmit energy and can be used to detect targets at greater ranges. This is particularly important in air, a medium that has a much higher attenuation coefficient than water. We can process longer signals by either using more signal processing chips or by reducing the sample rate.

In principal we have been over-sampling the echoes. A more efficient method would be to take the signal, which has a 40 kHz bandwidth, and demodulate down to base band (0-40 kHz). Since the minimum rate at which continuous signals can be sampled is 1/2BW (Burdic, 1968), this would allow us to sample the signal at 100 kHz. This sample rate would require 1/16 of the processing speed of a 400 kHz sample rate for the same time and bandwidth signal. This would allow us to use a signal that is 16 times longer with the same speed of processors. This would increase the signal energy by 12 dB, which could nearly double our range in a noise limited environment.

There is, however, a disadvantage to the demodulation technique when trying to achieve binaural localization. Although the envelope of the replica correlator output is the same for both the demodulated signal and original signals, the central peak of the coherent output is broader. Since the results of the Renaud and Popper experiments are consistent with the hypothesis that the coherent output is used in binaural localization, using demodulation may degrade the azimuth resolution of this system. A practical compromise might be to use a longer signal with demodulation at greater ranges where the extra signal strength is needed, and a shorter signal without demodulation to achieve better spatial resolution at shorter ranges.

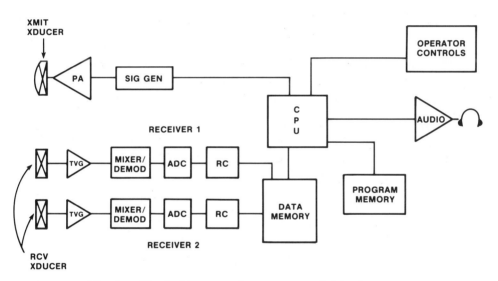

Fig 8. Block diagram of a conceptual bionic sonar.

The parts of a conceptual bionic sonar are described in the block diagram in figure 8. This sonar use is based on the signal processing methods we believe the dolphin uses, but the functional block applies to in-air sonars also. The transmit transducer is a broadband transducer with a conical beamwidth of 17°. It is driven with a signal that is loaded into the signal generator from a controlling computer (CPU). This signal can be dolphin-like to achieve maximum resolution at short ranges, or a larger time bandwidth product (TBW) signal to provide maximum signal to noise ratio at greater ranges. The binaural receiver consists of two identical receiving sections. Each receiving transducer has the dimensions described in figure 6. The received signal is amplified through the time varying gain (TVG) amplifier. The large TBW signals pass through the mixer/demodulator, while this section is bypassed for the small TBW signals. The conditioned signals are digitized by the analog to digital converter (ADC) and the digitized data stream is filtered by the replica correlator. The processed output can then be analyzed to obtain the spatial resolution listed in table 2, and to obtain target identification using the methods described by Martin and Au. If the sonar is a manned system, the filtered echoes may also be stretched in the time domain in a manner similar to that used by Fish et al. to provide additional classification information to the operator.

REFERENCES

Anonymous, 1986, The vector signal processor, Electronics, July 24, 59-66.

Au, W. W. L., Floyd, R. W., Penner, R. H., and Murchison, A. E., 1974, Measurement of echolocation signals of the Atlantic bottlenose dolphin, Tursiops truncatus, in open waters, J. Acoust. Soc. Am., 56: 1280.

Au, W. W. L., Floyd, R. W., and Haun, J. E., 1978, Propagation of Atlantic bottlenose dolphin echolocation signals, J. Acoust. Soc. Am., 64: 411.

Au, W. W. L. and Hammer, C. E., 1980, Target recognition via echolocation by Tursiops truncatus, in: "Animal Sonar Systems," R.-G. Busnel and J. F. Fish, eds., Plenum Press, New York

Au, W. W. L., Schusterman, R. J., and Kersting, 1980, Sphere-cylinder discrimination via echolocation, in: "Animal Sonar System," R.-G. Busnel and J. F. Fish, eds., Plenum Press, New York

Au, W. W. L. and Snyder, K., 1980, Long-range target detection in open waters by an echolocating Atlantic bottlenose dolphin, J. Acoust. Am., 68: 1077.

Au, W. W. L., and Penner, R. H., 1981, Target detection in noise by echolocating Atlantic bottlenose dolphins, J. Acoust. Soc. Am., 70: 687.

Au, W. W. L., Penner, R. H., and Kadane, J., 1982, Acoustic behavior of echolocating Atlantic bottlenose dolphins, J. Acoust. Soc. Am., 71: 1271.

Au, W. W. L., and Moore, P. W. B., 1984, Receiving beam patterns and directivity indices of the Atlantic bottlenose dolphin, J. Acoust. Soc. Am., 75: 255.

Bar-Cohen, Y., Stubbs, D. A., and Hoppe, W. C., 1984, Multiphase backing materials for piezoelectric broadband transducer, J. Acoust. Soc. Am., 75: 1629.

Batberger, C. L., 1965, Lecture notes on underwater acoustics, Air Warfare Res. Dept. Rep. No. NADC-WR-6509.

Burdic, W. S., 1968, "Radar Signal Analysis," Prentice-Hall, Englewood Cliffs, New Jersey.

Desilets, C. S., Fraser, J. D., and Gordon, S. K., 1978, The design of efficient broadband piezoelectric transducers, IEEE Transactions on Sonics and Ultrasonics, su-25, 115.

Fish, J. F., Johnson, C. S., and Ljungblad, D. K., 1976, Sonar target discrimination by instrumented human divers, J. Acoust. Soc. Am., 59: 62.

Heyser, R. C., 1986, Broadband ultrasonic transducers, Jet Propulsion Laboratory invention report NPO-16590/5909

Johnson, C. S., 1968, Masked tonal thresholds in the bottlenose porpoise, J. Acoust. Soc. Am., 44: 956.

Johnson, R. A., and Titlebaum, E. L., 1976, Energy spectrum analysis: A model of echolocation processing, J. Acoust. Soc. Am., 60: 484.

Martin, D. M., and Au, W. W. L., 1982, Aural discrimination of targets by human subjects using broadband sonar pulses, Naval Ocean Systems Center TR 847.

Martin, D. W., and Au, W. W. L., 1986, Investigation of broadband sonar classification cues, Naval Ocean Systems Center TR 1123.

Menne, D., and Hackbarth, H., 1896, Accuracy of distance measurement in the bat Eptesicus fuscus: Theoretical aspects and computer simulations, J. Acoust. Soc. Am., 79: 386.

Murchison, A. E., 1980, Detection range and range resolution of echolocating bottlenose porpoise (Tursiops truncatus), in: "Animal Sonar Systems," R. G. Busnel and J. F. Fish, eds., Plenum Press, New York.

Renaud, D. L., and Popper, A. N., 1975, Sound localization by the bottlenose porpoise Tursiops truncatus, J. Exp. Biol., 63: 569.

TAKE OFF SIGNALS EMITTED BY <u>Myotis</u> <u>mystacinus</u>:

THEORY OF RECEIVERS AND MODELLING

B. Escudié

Signal Processing Laboratory - U.A 346 CNRS
I.C.P.I.
Lyon, France

INTRODUCTION

Since 1976, our group carried systematical annual experiments on
Myotis mystacinus in very convenient sites: Montellier (Ain) France, 20 km
N.E of LYON. During June 1980 (16/05/80) records were made of SONAR signals
emitted by Myotis mystacinus during take off and the early part of flight.
Bats were taking off from behind a shutter of an old barn (see fig. 1). We
only register during this experiment the take off signals and the cruise
ones emitted when the bat is flying to the microphone ($\mu 6$). The sequence
of 25 Sonar signals have been analyzed by the various techniques of signal
processing that we described formerly [1] [2] :

$$S(t) \longrightarrow Z(t) = S(t) + iQ(t) \ , \ |Z(t)|e^{i\Phi(t)} = Z(t), \nu_i(t) = \frac{1}{2\pi} \frac{d\Phi}{dt}$$

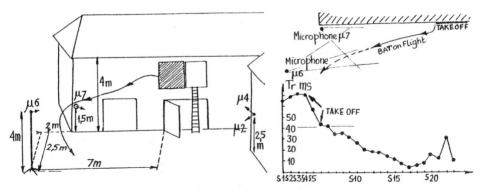

Fig. 1 . experiment on the sight in Montellier (Ain) France and repetition
rate of recorded signals.

1) Main Characteristics of the Emitted Signals

Each of the 25 signals is described by spectral analysis $\hat{\nu}_S(\nu)$, amplitude and frequency modulation $A(t), \nu_i(t)$ (see fig. 2). The bandwidth B is

defined by the values $\mathcal{T}_m/100$, $\mathcal{T}_m$ being the maximum value of the spectral density $\mathcal{T}_S(\nu)$. A similar definition is used for duration T. The signals are high bandwidth duration product BT as shown by figure 3. Duration T decreases in a monotonic way from (S1) to (S25) except for signals (S6) and (S11). Under the assumption of coherent reception (matched filter or equivalent receiver using a time frequency analysis [3] [4] [5]), 1/B may be accepted as an estimation of range accuracy. This parameter exhibits a strong increase for signals (S15) to (S25). For the same set of signals, ν_c the centre frequency decreases in connection with an increase of α_m, the modulation rate (figures 3 and 4). The last parameters are defined as below :

$$\nu_m \leqslant \nu_i(t) \leqslant \nu_M, \, 0 \leqslant t \leqslant T \, , \quad \nu_c = \frac{1}{2}(\nu_M + \nu_m)$$

$$\alpha_m = \frac{\nu_M}{\nu_m} \, , \quad \frac{B}{\nu_c} = 2\,\frac{\nu_M - \nu_m}{\nu_M + \nu_m} = 2\,\frac{\alpha_m - 1}{\alpha_m + 1}$$

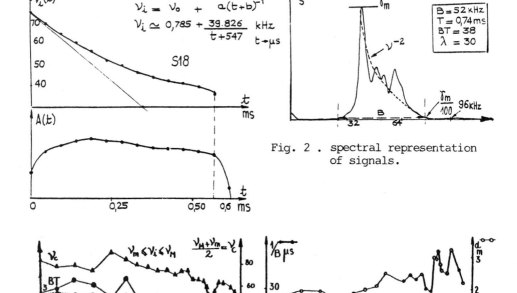

Fig. 2 . spectral representation of signals.

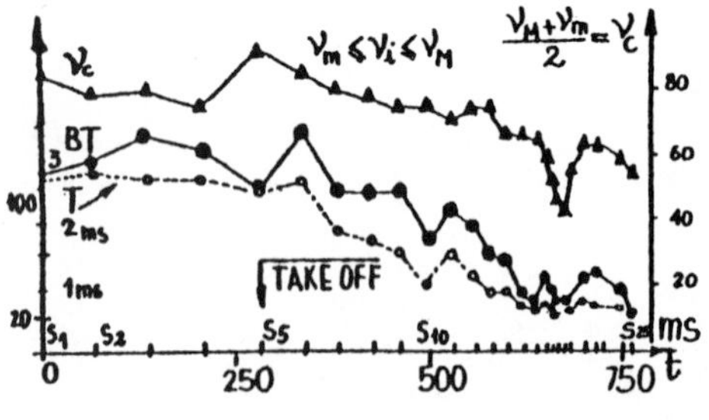

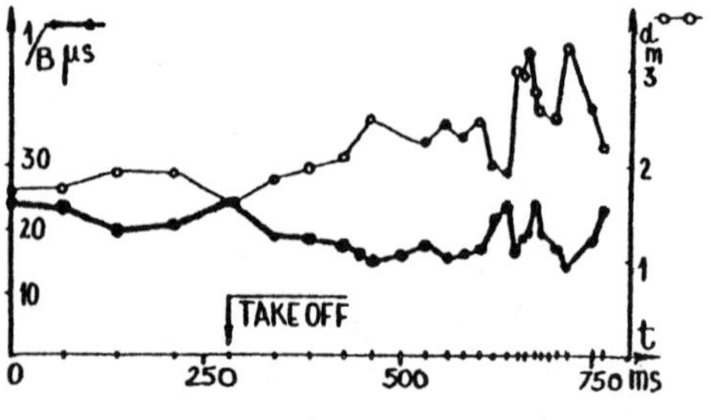

Fig. 3 . evolution of signals parameters.

Fig. 4 . evolution of range resolution and modulation rate.

Figure 5 indicates that signals (S15) to (S25) emitted during the final part of the sequence have high modulation rates and relative bandwidth B/ν_c. Such a feature may be explained by taking into account that a tentative model for angle estimation by bats may be an "interferometric receiver". The angular response R(u) being related to the spectrum density : [6] [7]

$$R(u) = \int_R \mathcal{T}_S(\nu)\,\big|P(\nu u)\big|^2 \cos 2\pi \nu \mu D \, d\nu \, , \quad \mu = \frac{\sin\alpha}{c_o}$$

where $P(\nu u)$ is the space frequency response of each receiver (see fig. 6a). Under these constraints the response may be enhanced by cancelling side lobes. This improvement is due to the increase of the modulation rate α_m and relative bandwidth $B/\nu c$ [6] [7]. A few signals, as (S9), have harmonic components which may be used in a similar way to get a better resolution for angle or bearing estimation [8].

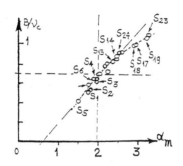

Fig. 5 . relative bandwidth and
modulation rate.

2) Frequency Modulation and SONAR Performances

Figure 7 displays various frequency modulations ν_i of the emitted signals related with time location in the sequence. The main features appear as :

α) before take off : (S1) to (S6). Signals are high BT ones very similar to cruise or detection signals. The frequency modulation $\nu_i(t)$ is a parabolic one (S1), (S2), or a cubic one ($at^3 + bt^2 + ct + d$) with a convenient definition of time origin : (S3) to (S6). Such frequency modulations have been described by different authors [9] [10].

β) During take off and early part of flight : (S7) to (S11). The instantaneous frequency is altered from a "cubic" law to "hyperbolic" one. This last feature is indicated by a linear period modulation.

γ) Steady state for flight : (S11) to (S25). Signals are very similar to those emitted by Myotis mystacinus during cruise, identification and pursuit [9] [2] [1]. Interpretation of such features is possible by the use of the coherent receiver model [2] [5] [11] [12]. At rest the bat has to receive echoes due to walls, trees, insects which are motionless or low Doppler distorded echoes. The use of the optimum Doppler tolerant modulation is not required in such a situation [7]. A parabolic modulation (symmetric or even function) is able to ensure a decoupling effect for range and range rate estimation at low Doppler effects [7]. The high value of center frequency is related to short wave-lenghts ensuring a good accuracy for angle display of targets. When going on take off and steady state flight, a Doppler tolerant SONAR signal is required with a good angle resolution related to a high modulation rate α_m.

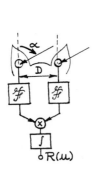 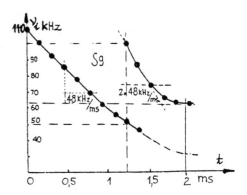

Fig. 6a . interferometric model.   Fig. 6b . frequency modulation and
harmonic components.

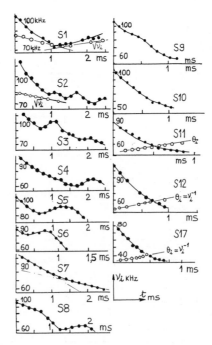

Fig. 7 . instantaneous frequency of take-off sonar signals.

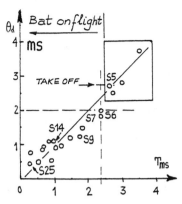

Fig. 8 . decoupling delay and duration

3) θ_d as a Doppler Tolerance Parameter

θ_d, the so called "decoupling delay", may be defined as : [11] [12]

$$E\{\hat{\tau}\} = 0 \quad , \quad \theta_d = -\delta_z . \lambda_z^{-2}$$

with the emitted signal $Z(t-\theta_d)$, which can be interpreted as an unbiased estimation of delay $\hat{\tau}$ derived from the delayed emitted signal $Z(t-\theta_d)$.
θ_d is an "intrinsic parameter" for signals related to the properties of $A(t)$ and $\nu_i(t)$. For the analyzed sequence θ_d is a positive valued parameter decreasing in a rather monotonic way related to duration T. Figure 9 exhibits the strong correlation between T, Tr, θ_d when the bat is on flight. This strong dependence may be explained as follows : Trk is the repetition period defined by $t_{k+1} - t_k$, t_k being the occuring time of signal Sk. T has the same variation as θ_d, but delayed. This suggests, as in a previous result, that the bat is using the kth range estimation to predict the $(k+1)^{th}$ one [13].

The parameters such as θ_d, or $\tau_m(\eta_o)$ and $\tau_m(\alpha_o)$ range bias due to speed and an acceleration, were computed by an optimal fitting of polynomials to $A(t)$ and $\nu_i(t)$, obtained by amplitude and frequency demodulation. Figure 10 exhibits the relative values of $\tau_m(\eta_o)$ and $\tau_m(\alpha_o)$ which are very low valued bias: [14]

$$\eta_o \simeq 1 - \frac{2V_R}{C_o} \quad , \quad \alpha_o = \frac{C_o - V_R - \delta_c \tau/2}{C_o + V_R + \delta_c \tau/2} \quad , \quad V_R \text{ radial velocity, } C_o = 340 \text{m/s}$$

Under this practical conditions it appears that bats are emitting "suboptimum signals" very near the optimum ones defined by detection and estimation theory [13] [12]. An important feature has to be pointed out : this result does not imply that there is no coupling effect between range and range rate estimates. In such a suboptimum situation the coupling coefficient is very low but the covariance (of estimation) matrix is not a diagonal one [13]. Regardless of this peculiar case, accelaration tolerant signals may be found among SONAR signals emitted by bats when they are faced with fast evolutive situations, as pursuit or prey capture [12] [14].

CONCLUSION

These results are in accordance with observations by J.A SIMMONS on an American bat [15]. From these results, one can only indicate that coherent reception or time and frequency receivers are tentative models in agreement with the processed signals. A further comment may be found in the paper due to R.A ALTES [16] : "It is a capital mistake, WATSON, to theorise before one has data. Insensibly, one begins to twist facts to suit theories, instead of theories to suit facts". (A.C DOYLE "A scandal in Bohemia").

ACKNOWLEDGMENTS

B. ESCUDIE is very indebted to his colleagues of the French group (R.C.P. 445 CNRS) and mainly Y. TUPINIER and Y. BIRAUD for their active preparation of experiments in Montellier. He wants to acknowledge J. CHAVE, J.M. COURBET, P. FLANDRIN and M. MAMODE to their suggestions and critical analysis of signals.

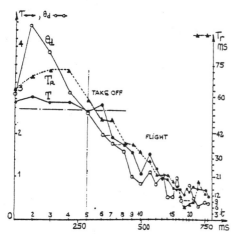

Fig. 9 . time parameters.   Fig. 10 . Doppler and acceleration bias.

REFERENCES

[1] Biraud, Y., 1985, French group activities, in : "Air borne animal Sonar systems CNRS Conf.", Lyon.
[2] Lin, Z.B., 1983, Ambiguité compression des signaux Sonar des chauves-souris, in : "9th GRETSI Meeting", Nice, France. Vol. 2 p. 691.
[3] Altes, R.A., 1985, An M.T.I receiver model for bats ?, in : "Tech. Mem. TM 353 ORINCON".
[4] Altes, R.A., 1985, Overlapping Windows and signal representation on the time frequency plane, in : [1], p. 17.1.
[5] Flandrin, P., 1986, On detection estimation procedures in time frequency plane, in : "Proc. ICASSP Tokyo, Vol. 4", p. 23.31.
[6] Bard, C., 1978, Réponse angulaire des interferomètres à correlation, in : "Acustica, Vol 40, n° 3", p. 139.
[7] Escudié, B., 1979, "2nd animal Sonar system symposium", Plenum Press, p. 715.
[8] Chiollaz, M., 1979, Thèse Doct. Ing. I.N.P. Univ. Grenoble.
[9] Tupinier, Y., 1985, Signaux d'écholocation de Myotis mystacinus in : [1], p. 26.1.
[10] Zbinden, K., Zingg, P., 1985, Cruising and hunting signals of echoloca-ting European free-tailed bat in : [1], p. 161.
[11] Mamode, M., 1985, Vers une définition adaptée de la date d'arrivée d'un écho Sonar en vue de son estimation, in : [1], p. 13.1.
[12] Escudié, B., Mamode, M., 1985, Tolérance à l'effet Doppler et signaux Sonar émis par les chauves-souris, in : [1], p. 23.1.
[13] Mamode, M., 1982, Bats' Sonar signals and acceleration tolerance, in : "Proc. ICASSP Paris, Vol. 2", p. 11.32.
[14] Mamode, M., 1981, Thèse Doct. Ing. I.N.P. Univ. Grenoble.
[15] Simmons, J.A., Private communication.
[16] Altes, R.A., 1979, in : [7], p. 630 and following.

NOSELEAVES AND BAT PULSES

J. David Pye

Queen Mary College
Mile End Road, London E1 4NS

One characteristic of noseleaves is that interference between the two
nostril-sources influences the sound radiation pattern. Some implications
of this are presented and demonstrated by a simple model; similar features
are found in some bat pulses, thus accounting for much variation in their
observed structures.

Constant frequency bats. Möhres (1953) observed the half-wavelength
spacing of nostrils in Rhinolophus. In theory this gives a half-power
(-3 dB) beamwidth of 60° (Fig. 1) and is fully symmetrical (rotationally
invariant) about the axis through the nostrils. But the evidence is confusing. Airapet'yants and Konstantinov (1970, translation 1973) gave no
units but presumably plotted intensity for they quoted "a width of 54° at
the level 0.5, close to the theoretically calculated value of 60°". But
Schnitzler and Grinnell (1977) plotted amplitude in the same species and
found a -3 dB beamwidth of only 30° (as if there were four nostrils), with
strong constraints also in the vertical plane. This suggests that diffraction is produced by parts of the leaf structure which is not a simple
baffle.

Since the two nasal channels are separated by a medial septum, there
is the possibility of independent resonances. Off-tuning of the two sides
by only -3 dB but in opposite directions will then produce a phase difference of 90°, swinging the beam by 30° to one side but without requiring
any visible physical movement of the bat's head or even of the noseleaf.
Beamwidth is then degraded and a second lobe appears on the other side
(Fig. 1) but such a spatial pattern may be useful for rapidly improving
the signal-to-noise ratio for a target to one side. If real bat beams
are actually half this width then beam steering would be even more useful.

Frequency sweeps. For a half wavelength spacing, in-phase sources
produce a null along the lateral axis (at 90° and passing through the
sources). This also occurs with spacings of any other odd number of half
wavelengths because destructive interference then occurs. Any number of
whole wave-lengths allows the two components to add and so produces a
peak. The general result is a multi-lobed filter curve as in Fig. 2.
The range of 2-4 mm is fairly representative of internarial spacings
for most bats since average values measured on cadavers were: Glossophaga
soricina 1.5 mm, Carollia perspicillata 2.15 mm and Phyllostomus hastatus

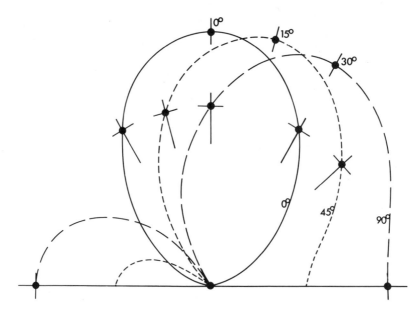

Fig. 1.   Sound pressure amplitude for two sources in a plane baffle and
half a wavelength apart. Solid line: 0° phase difference, beam
axis at 0°; dotted line: 45° phase difference, main beam at 15°;
dashed line: 90° phase difference, main beam at 30°. Crosses
mark the half intensity (-3 dB) points. All patterns rotate
around the baseline to give the distributions in 3-D.

3.58 mm (first lateral nulls at 111 kHz, 79 kHz and 47.5 kHz respectively).

      In the median plane, at 0°, all frequencies add in-phase to give a
flat response. For other angles the frequency scale is simply compressed
by the factor sin θ as shown in Fig. 2. A noseleaf should therefore constitute a filter whose frequency response curve is directionally dependent.
This is confirmed by a model noseleaf, consisting of two holes in a metal
plate and fitted over a loud-speaker emitting fm sweeps (Fig. 3). Two
microphones side-by-side at 0° give similar traces but when one is moved
to an angle laterally, sharp minima appear in the trace as predicted. The
model used here had holes of 5 mm diameter and 1 cm between centres. Close
to 90°, nulls appeared at 17 kHz, 51 kHz and 85 kHz and were associated
with abrupt changes of phase.

      When fm noseleaf bats are recorded by two microphones in different
directions, some pulses are similar (Fig. 4 left) while others are strikingly different (Fig. 4 centre and right) in the relative amplitudes and
phases of their components. It is assumed that in the former case the zero
axis of the bat's beam passes midway between the microphones, giving symmetrical responses, while in the other cases the microphones lie at
different angles and "see" a different filter response. Differences
between the characteristics of the two microphones cannot be responsible
or the results would be consistent.

      Clearly care must be taken in describing such pulses when they have
been recorded by a single microphone. Variations between pulses are often
undoubtedly real and imposed by the bat but the same pulse is different if
observed from different angles. A noseleaf must therefore be regarded as

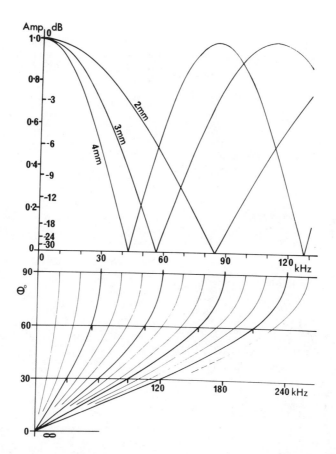

Fig. 2.   Theoretical filter characteristics of paired, in-phase sources
spaced 2, 3 and 4 mm apart. Above: amplitude plots if observed
on the internarial axis, at 90° to the median plane. Below:
abscissa scales for other directions. In the median plane (0°)
itself the characteristic is flat at all frequencies.

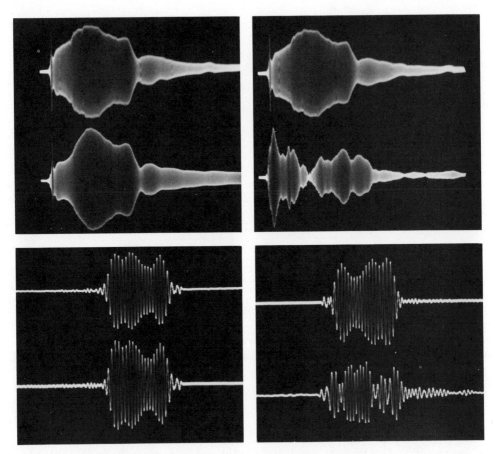

Fig. 3. Model frequency sweeps recorded from two holes, 1 cm apart in a
         metal plate. Top: slow sweeps; below: fast sweeps. Left: two
         microphones close together at 0° (normal to the plate). Right:
         one microphone moved to 85°, close to the line joining the holes;
         nulls appear at the expected frequencies. QMC Instruments micro-
         phones with S-100 detectors as signal amplifiers.

Fig. 4. Phyllostomatid pulses recorded with two microphones spaced apart. Top: <u>Carollia perspicillata</u> (23°); middle: <u>Phyllostomus discolor</u> (53°); bottom: <u>Artibeus jamaicensis</u> (53°). B & K ¼-inch microphones, type 4135, and type 2604 amplifiers, with PI-6100 recorder. Michael Hardy collected and shipped the bats from Trinidad, Larry Roberts assisted with the recordings.

the final element of the vocal tract formant filter but one whose effect varies with direction. The possible significance of such directionally dependent factors to the bat itself is now under consideration. It is also hoped to investigate the diffraction properties of a variety of noseleaves, which will complicate the simple interference theory given above.

References

Airapet'yants, E.Sh. and Konstantinov, A.I. 1970. "Echolocation in Animals", Nauka, Leningrad (English translation IPST, Jerusalem, 1973).

Möhres, F.P. 1953. Uber die Ultraschallorientierung der Hufeisennasen (Chiroptera : Rhinolophinae), Z.vergl.Physiol., 34: 547.

Schnitzler, H.U. and Grinnell, A.D. 1977. Directional sensitivity of echolocation in the horseshoe bat, Rhinolophus ferrumequinum, J.comp.Physiol. A, 116: 51.

TIME-FREQUENCY PROCESSING OF BAT SONAR SIGNALS

Patrick Flandrin

Laboratoire de Traitement du Signal
UA 346 CNRS ICPI 25 rue du Plat
69288 Lyon Cedex 02 France

ABSTRACT

It is known that time-frequency distributions can be used for the demodulation of bat sonar signals and, in some cases, for detection-estimation tasks via a 2-D correlation process. In this paper, we formalize the intuitive notion of correlation of time-frequency structures in the simplified case of AM + FM signals. This results in a general time-frequency receiver structure depending only on a time-frequency smoothing function. According to the choice of this smoothing, it is shown that the receiver can vary continuously from a semi-coherent (matched filter and envelope detector) to a non-coherent (energetic) one. This suggests that both can be viewed as limiting cases of a unique time-frequency processing. The proposed approach is illustrated on the detection of echoes (over water) of signals emitted by Myotis daubentoni and recorded in the field.

INTRODUCTION

In bat sonar studies, the use of time-frequency distributions can be aimed at two different tasks:
- the first one is related to signal analysis, in order to get a description of the time-varying structure of bat sonar signals (see e.g. [1]);
- the second one is related to signal processing, in order to hypothesize some possible bat's receiver models via a time-frequency formulation (see e.g. [2]).

In both cases, optimal time-frequency distributions can be derived, with respect to prescribed requirements. This paper is intended to shortly present such distributions, with special reference to receivers which can be based on.

TIME-FREQUENCY ANALYSIS

A common way of performing time-frequency analyses is to make use of spectrograms [3]:

$$(1) \quad S_x(t,\nu) = \left| \int_{-\infty}^{+\infty} x(u)\, h(u-t)\, e^{-i2\pi\nu u}\, du \right|^2$$

where $k(.)$ is a short-time analysis window.

In fact, though very intuitive, spectrograms are not the unique time-frequency distributions but only a special case of the general "Cohen's class" defined as [4]:

$$(2) \quad C_x(t,\nu;f) = \iint_{-\infty}^{+\infty} e^{i2\pi n(u-t)} f(n,\tau) x(u+\tfrac{\tau}{2}) x^*(u-\tfrac{\tau}{2}) e^{-i2\pi\nu\tau} dn\, du\, d\tau$$

where $f(n,\tau)$ is some arbitrary time-frequency window.

It can then be shown that (1) is obtained from (2) by choosing $f(n,\tau)$ as the ambiguity function of the analysis window:

$$(3) \quad A_k(n,\tau) = \int_{-\infty}^{+\infty} k(u+\tfrac{\tau}{2}) k^*(u-\tfrac{\tau}{2}) e^{i2\pi nu} du$$

One of the major advantages of the general formulation (2) is that suitable time-frequency distributions can be derived by choosing f's which satisfy some given requirements. If we consider for instance AM + FM waveforms, as those emitted by different species of bats:

$$(4) \quad x(t) = a_x(t) e^{i\varphi_x(t)}$$

where $a_x(t) \geq 0$ corresponds to the envelope (AM law) and:

$$(5) \quad \nu_x(t) = \frac{1}{2\pi} \frac{d}{dt} \varphi_x(t)$$

to the instantaneous frequency (FM law), a reasonable requirement is that both modulation characteristics could be evidenced by the desired distribution and exactly computed from it. Unfortunately, it is known [4] that such requirements are incompatible with (3) and, hence, that spectrograms only give access to smoothed versions of the AM and FM laws. Exact demodulation is achieved, however, by the so-called Wigner-Ville distribution [4-5]:

$$(6) \quad W_x(t,\nu) = \int_{-\infty}^{+\infty} x(t+\tfrac{\tau}{2}) x^*(t-\tfrac{\tau}{2}) e^{-i2\pi\nu\tau} d\tau$$

corresponding to $f(n,\tau) = 1$.

The Wigner-Ville distribution is retained since it possesses many other desirable properties and corresponds to the most ideal time-frequency concentration in the case of a linear FM:

$$(7) \quad \nu_x(t) = \nu_0 + \alpha t \Rightarrow W_x(t,\nu) = \delta(\nu - \nu_x(t))$$

without any trade-off on the modulation slope α.

Moreover, the Wigner-Ville distribution can be considered as the prototype of all time-frequency distributions since (2) can be equivalently written as

$$(8) \quad C_x(t,\nu;f) = \iint_{-\infty}^{+\infty} \Pi(t-t',\nu-\nu') W_x(t',\nu') dt'\, d\nu'$$

where Π is the 2-D Fourier transform of f.

The Wigner-Ville distribution plays therefore a central role in the time-frequency analysis of time-varying signals and, in most cases, a very versatile approximation of it can be implemented via (8) by choosing the time-frequency smoothing function Π as a separable function of time and frequency [5].

798

As far as time-varying signals are concerned, a very intuitive way of comparison is to deal with time-frequency structures, generally by means of a 2-D correlation of their time-frequency distributions [3,6]. Taking into account what has been said about the optimality of the Wigner-Ville distribution for signal analysis, the following approach can be proposed: given a reference signal of Wigner-Ville distribution $W_x(t,v)$, and an observed signal $y(t)$ of time-frequency distribution $C_y(t,v;f)$, the degree of resemblance between the two signals can be measured by means of the 2-D correlation:

(9)
$$\ell = \iint_{-\infty}^{+\infty} W_x(t,v) \, C_y^*(t,v;f) \, dt \, dv$$

which takes on the equivalent form [7]:

(10)
$$\ell = \int_{-\infty}^{+\infty} \Pi(\tau,n) \cdot \left| A_{xy}(n,\tau) \right|^2 dn \, d\tau$$

where A_{xy} is the cross-ambiguity function between x and y.

Therefore, if we address the binary decision problem:

(11)
$$\begin{cases} H_0 : & y(t) = b(t) \\ H_1 : & y(t) = b(t) + x(t) \end{cases}$$

where $b(.)$ is white Gaussian noise, it appears that comparing (9) to a threshold realizes exactly the optimum detection of the known deterministic signal with unknown random delay and Doppler shift of prior probability density function $\Pi(\tau,n)$ [3].

In a different perspective, (9) can be viewed as defining a general class of receivers, which are classified in terms of the arbitrary function [7]. We have, for instance, the following special cases:

1. $\Pi(\tau,n) = \delta(\tau) \, \delta(n)$

(12)
$$\ell = \left| \int_{-\infty}^{+\infty} x(t) \, y^*(t) \, dt \right|^2$$

which is nothing else but the classical matched filter followed by an envelope detector;

2. $\Pi(\tau,n) = \delta(\tau) \cdot H(n)$

(13)
$$\ell = \iint_{-\infty}^{+\infty} \left[x(t) y^*(t) \right] \left[x(t') y^*(t') \right]^* h(t'-t) \, dt \, dt'$$

and:

(14)
$$\lim_{H(n) \to 1} \ell = \int_{-\infty}^{+\infty} |x(t)|^2 \cdot |y(t)|^2 \, dt$$

which corresponds to an intensity correlator;

3. $\Pi(\tau,n) = 1$

(15)
$$\ell = \left[\int_{-\infty}^{\infty} |x(t)|^2 dt \right] \left[\int_{-\infty}^{+\infty} |y(t)|^2 dt \right]$$

which leads to an energy-based detector.

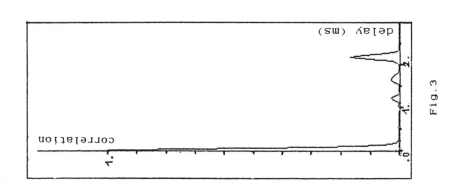

Fig.1

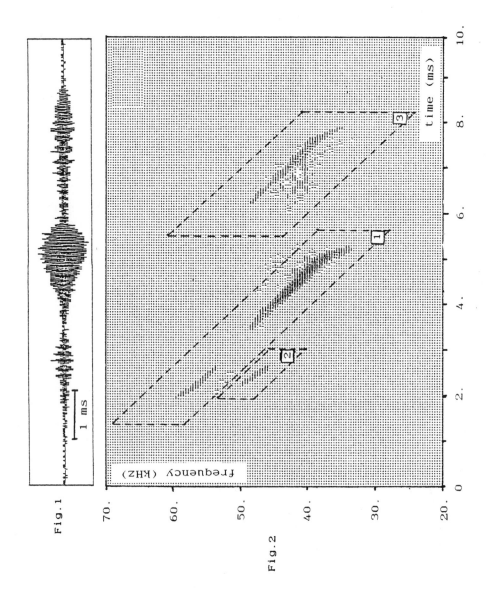

Fig.2

Fig.3

Therefore, (9) defines a general class of receivers, which are generally in between semi-coherent and non-coherent ones. Conversely, in such a time-frequency formulation, semi-coherent and non-coherent receivers are not conflicting ones but only appear as different limiting cases of a unique receiver.

AN EXAMPLE

Figs. 1 and 2 present a waveform recorded in the field when bats (Myotis daubentoni) were flying over water. In the considered example, two animals are simultaneously present and the global waveform corresponds to the superposition (in both time and frequency) of three different parts: a short catch cry (part 2) is superimposed to a long cruise call (part 1) of another bat, and echoes over water (part 3) of this long call are overlapping at the end of the waveform.

Since echolocation calls of Myotis daubentoni are AM + FM signals, this structure is evidenced in Fig. 2 by making use of a smoothed Wigner-Ville distribution.

A simplified version [7] of the correlation process (9) has been implemented between the parts 1 and 3 of the waveform. The result is plotted in Fig. 3: it clearly shows that the intricate structure of part 3 is in fact due to the superposition of three delayed replica of the main echolocation call.

CONCLUSION

It has been argued that both analysis and processing of bat sonar signals can be tackled by means of a time-frequency formulation. The most efficient and convenient approach to do so makes use of the Wigner-Ville distribution as the central time-frequency tool. A large class of receivers can be based on it, which both admits a very natural physical interpretation (in terms of correlation of time-frequency structures) and allows to handle in a unique formulation receivers which, in other approaches, could be thought of as conflicting ones (8-9).

ACKNOWLEDGEMENTS

This work was supported by the CNRS program RCP 445 and in part by Sté Métravib (Ecully, France).

G. Venet, A. de Melo and R. St Dizier are gratefully acknowledged for their interest and their help. Thanks are also given to Y. Biraud and Y. Tupinier who conducted the experiment and R. Emmerich who preprocessed the data.

REFERENCES

[1] W. Martin, K. Kruger-Alef,
    Application of the Wigner-Ville spectrum to the spectral analysis
    of a class of bio-acoustical signals blurred by noise, in: Proc.
    CNRS Conf. Airborne Animal Sonar Systems, pp. 18.1-18.25, Lyon,
    1985.

[2] R. A. Altes,
    Echolocation as seen from the viewpoint of radar/sonar theory, in:
    Localization and Orientation in Biology and Engineering,
    Varju/Schnitzler, eds., pp. 234-244, Springer, 1984.

[3] R. A. Altes,
Detection, estimation and classification with spectrograms, J.
Acoust. Soc. Am., 67 (4), pp. 1232-1246, 1980.

[4] T.A.C.M. Claasen, W.F.G. Mecklenbrauker,
The Wigner distribution - A tool for time-frequency signal
analysis, Philips J. Res., 35, pp. 217-250, 276-300, 372-389, 1980.

[5] W. Martin, P. Flandrin,
Wigner-Ville spectral analysis of non-stationary processes, IEEE
Trans. on ASSP, ASSP-33 (6), pp. 1461-1470, 1985.

[6] S. Kay, G. F. Boudreaux-Bartels,
On the optimality of the Wigner distribution for detection, in:
Proc. IEEE ICASSP'85, pp. 1017-1029, Tampa, 1985.

[7] P. Flandrin,
On detection-estimation procedures in the time-frequency plane, in:
Proc. IEEE ICASSP'86, pp. 2331-2334, Tokyo, 1986.

[8] B. Møhl,
Detection by a Pipistrelle bat of normal and reversed replica of
its sonar pulses, in: Proc. CNRS Conf. Airborne Animal Sonar
Systems, pp. 7.1-7.22, Lyon, 1985.

[9] D. Menne, H. Hackbarth,
Accuracy of distance measurement in the bat Eptesicus fuscus:
Theoretical aspects and computer simulations, J. Acoust. Soc. Am.,
79 (2), pp. 386-397, 1986.

ECHOES FROM INSECTS PROCESSED USING

TIME DELAYED SPECTROMETRY (TDS)

Lee A. Miller and Simon Boel Pedersen

Institute of Biology, Odense University, Odense
Acoustics Laboratory, Technical University, Lyngby
Denmark

INTRODUCTION

TDS uses a linearly frequency modulated (FM) signal for system response analysis. The technique was first described by Heyser (1967) and has been applied to microwave and ultrasonic imaging systems (Heyser and LeCroissette 1974). Briefly, the technique utilized an FM probe signal where the instantaneous frequency of the signal received by the transducer is a function of the FM sweep rate and path length, or time delay, during transmission. Reflected signals (echoes) arriving at the transducer are delayed relative to the direct signal. Each time delay is associated with a frequency change hence the name, Time Delayed Spectrometry. TDS has several useful analytical aspects. For example, the nature of the probe signal allows for easy conversion between time and frequency domains. Also, one can show that, under certain restrictions, the echo is a (linear) FM signal modulated with the transfer function of the system (target). Hence the envelope of the echo is the numerical value of the transfer function for the target. If the phase information is available the impulse response can be obtained and the system can be uniquely described.

One implication for bats using FM signals is the possibility to identify targets based on information in the envelope of the echo. This information is available to the bat without further processing provided the bat can "hear" the envelope. We therefore analysed echoes from various insects using FM signals to see if there were obvious and predictable changes in echo envelopes at particular angles of incidence.

METHODS

We analyzed the echoes from individual moths of five different families constituting 15 species (n = 27) and from a beetle (Tenebrio). The insects were killed and waxed to a needle (diameter 0.5 mm) at the mesothorax. The wing pointing towards the loudspeaker was waxed at an angle of 40 to the horizontal. The needle with the insect was placed in a holder (stainless steel tubing, diameter 0.9 mm) 580 mm in front and normal to the center of an electrostatic loudspeaker (diameter 60 mm). The head of the insect always pointed to the right. the insect could be pivoted around a point at the mesothorax, and echoes were recorded five angles: 0 (broad side), 10 and 20 right (head turned away from the loudspeaker) and 10 and 20 left. There were no extraneous echoes within the time window used for analysis. The holder alone gave no measurable echo. One second-ary reflection of the echo signal from a nearby object gave rise to a high frequency (approximately 10 kHz) ripple artifact on the last part of the envelope (Fig. 3). The envelopes were therefore low-pass filtered at 5 kHz, which does not remove information relevant to insect echoes. We used echoes from a table tennis ball (diameter 37.7 mm) to check the setup and to compare its echoes with those from insects (Fig. 1).

Linear FM signals (100 kHz to 66 kHz, or 100 kHz to 50 kHz; 2.5 ms in duration) were generated by a microcomputer controlled system and gated to start and stop at zero. Fig. 1a shows the instantaneous frequency and amplitude (envelope) of the FM signal recorded at the position of the target where the sound pressure level was 99 1 dB (100 kHz to 50 kHz). Echoes were received by a 1/4" Brüel & Kjær microphone (grid removed) placed 15 mm to one side and 15 mm in front of the loudspeaker, and record-ed at 30 ips on a Racal Store 7 magnetic tape recorder. Echoes were re-recorded at 1/8th or 1/16th the original speed. The re-recorded signals were first low pass filtered (3 dB = 128 kHz, effective frequency) and digitized ($\Delta T = 3\ 1/8\ \mu s$, effective A to D conversion). Envelopes were obtained by processing the echo (time signal) with a pair of digital band-pass filters. The effective center frequency of each filter was 80 kHz with a band width of 80 kHz (± 40 kHz). Since the impulse response of the second filter was designed to be the Hilbert transform of the impulse re-sponse of the first filter, the output of the second filter is phase shifted by 90° relative to the output of the first filter. The envelope, a(t), of the echo was then computed from:

$$\left|a(t)\right| = (g(t)^2 + gH(t)^2)^{1/2} \qquad\qquad 1.$$

where g(t) and gH(t) are the output signals from the filters. Each envelope in Fig. 1b-d, Fig. 2 and Fig. 3 is the average of 8 echoes. Since linear FM signals are used, the envelope can be interpreted as the numeric-al value of the transfer function of the object. The frequency scale of this transfer function is proportional to rt where r is the sweep rate (Hz/s) and t is time. High frequencies are found to the left and low frequencies to the right in the "time" scales of Figs. 1 b-d, 2 and 3.

RESULTS

Fig. 1b shows the envelopes of echoes from a table tennis ball mounted in the position otherwise occupied by the insects. There was no change in the spectra or amplitudes of echoes with changes in angle through 20. Small amplitude modulations on the envelope may be due to the elastic re-sponse of the thin shelled table tennis ball (Hickling 1967) or to reson-ance, which is a function of the wavelength and circumference.

Envelopes from echoes of a moth (Odontopera bidentata, Geometridae; body length 19 mm; forewing length, base to tip, 22 mm; live weight 110 mg) and a beetle (Tenerbrio, body length 13 mm; elytrum length, base to tip, 9 mm; live weight 85 mg) are shown in Fig. 1c and 1d. Echoes from all angles have reasonably good signal to noise ratios. Echoes from the moth are strongly frequency filtered as a function of angle (Fig. 1c, the large fluctuations in the envelope). Echoes from the beetle show frequency filtering mostly at 0° and 10° R (Fig. 1d). There is no great difference in the target strength of the two insects even though the beetle has a considerably smaller (echo reflecting) surface owing to its shorter and narrower wings. On the other hand, the beetle has a hard, exposed exo-skeleton while the moth's body is covered with fluffy scales. The envel-opes for all angles shown in Fig. 1d contain small amplitude modulations as does the envelope for 20° L in Fig. 1c.

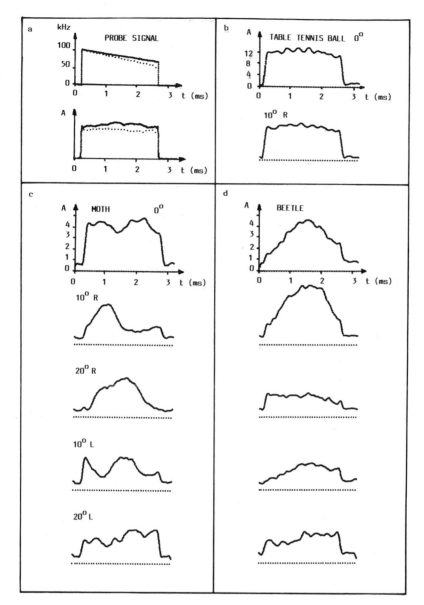

Fig. 1. The FM signal and envelopes of echoes from objects rotated $\pm$ 20° in the sound field. Echoes from a table tennis ball are not frequency filtered, but those from insects show frequency filtering as a function of angle.

Fig. 2 shows the envelopes of echoes from four individual moths of the species <u>Crocallis elingnaria</u>, Geometridae (body lengths 19-20 mm; forewing lengths (base to tip) 18-20 mm). Rather strong echoes occurred at 0 and 10° R with fluctuations of up to 6 dB in the magnitude of a given envelope. Almost no echoes appeared for individual 3 at 10 L and individuals 2, 3 and 4 at 20° L (head turned towards the loudspeaker). Variations in the envelopes as a function of angle may result from slight individual differences in weights, in body and wing dimensions, and in handling and mounting of the insects.

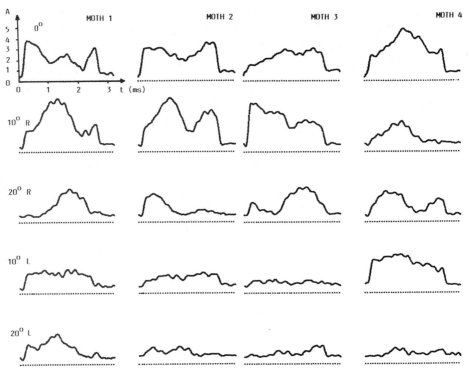

Fig. 2. Envelopes of echoes from four individual moths of the species
Crocallis elingnaria, Geometridae. Weights of the individuals were: moth 1
- 83 mg, moth 2 - 140 mg, moth 3 - 111 mg, moth 4 - 84 mg. FM sweep:
100-50 kHz.

Fig. 3 The envelope of a moth echo
and a reflection from a nearby object.
The interference frequency is about
10 kHz. Compare to Fig. 1c 0.

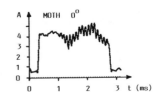

Fig. 3 shows the envelope from a moth (0, see Fig. 1d) and a secondary reflection from a nearby object. The two echoes interfered in such a
way as to produce an amplitude modulation of about 10 kHz, which starts at
about 0.7 ms from the onset of the envelope. The (constant) frequency of
the amplitude modulation is proportional to the (linear) FM sweep rate and
the distance between the two objects.

DISCUSSION

"FM" bats can discriminate targets, but the cues they use are unknown
(see Schnitzler and Henson 1980). Experienced bats, Myotis lucifugus,
could select mealworms from discs projected into the flight path.
Habersetzer and Vogler (1983) trained bats to discriminate two square
plates, one containing minute changes in hole depths relative to hole
depths of the second plate.

Echoes from targets probed with linear FM signals have envelopes whose numeric values are the transfer functions of the respective targets. Using non-linear FM sweeps changes the envelope structure predictably. The signals emitted during the search phase by many "FM" bats are not linear, but become more so during the approach phase. Some bats, like Myotis myotis, use a nearly linear FM sweep in searching signals.

The envelope of an echo is a signature of the target at a specified position (linear and angular coordinates). The echo will occupy a specific place on the basalar membrane and if an "FM" bat can "hear" the envelope the bat is provided with momentary cues regarding the orientation, size, and type of prey. Echoes from the insects we studied do show some species specific differences (Fig. 1c and d), especially when compared at different angles. The variation in echoes among individuals of the same species (Fig. 2) coupled with additional variations expected when the prey is flying may limit the usefullness of the envelope as a species specific signature. Experiments are needed to test envelope detection in "FM" bats.

The processing of amplitude and frequency modulations of echoes is known for some bats. Grinnell (1970) could record evoked responses from the brain of the "FM" bat Saccopteryx bilineata to amplitude modulations of beat notes when two signals overlapped. Also, Henson (et.al. this symposium) using the "CF-FM" bat, Pteronotus p. parnellii, recorded amplitude modulations in the cochlear microphonic potentials created by echo-echo overlap. Some central (collicular) neurons in the brain of the "CF-FM" bat, Rhinolophus ferrumequinum, respond to slight amplitude and frequency modulations (Schuller 1979). Amplitude modulations of 16% could readily evoke responses in collicular neurons. Both "CF-FM" bats mentioned above seem to use Doppler and amplitude information in the echo to identify targets (see Von der Emde, and Ostwald, Schuller and Schnitzler this symposium). If similar neurons are found in "FM" bats and if the envelope can be detected then amplitude modulations caused by the overlap of two FM echoes (Fig. 3) should be detected. The modulation index of the ripple on the envelope of the echo shown in Fig. 3 is about 13%. The ripple freqency is proportional to the time difference between the two objects assuming the sweep rate is constant. This means the bat can use either the time or frequency domain for "calculating" the distance between the two targets.

ACKNOWLEDGEMENT

We thank the Danish Natural Science Research Council for support.

REFERENCES

Grinnell, A.D., 1970, Comparative auditory neurophysiology of neotropical bats employing different echolocation signals, Z.vergl.Physiol., 68:117-153.

Habersetzer, J., Vogler, B., 1983, Discrimination of surface-structured targets by the echolocating bat Myotis myotis during flight, J.comp.Physiol. 152: 275-282.

Heyser, R.C., 1967, Acoustical measurements by time delay spectrometry, J.Audio Eng.Soc., 15:370-382.

Heyser, R.C., Le Croissette, D.H., 1974, A new ultrasonic immaging system using time delay spectrometry, Ultrasound Med.Biol., 1:119-131.

Hickling, R., 1967, Echoes from spherical shells in air, J.Acoust Soc.Am., 42:388-390.

Schnitzler, H.-U., Henson, O.W., Performance of airborne animal sonar systems: I. Microchiroptera, in: "Animal Sonar Systems," R.G. Busnell, J.F. Fish, eds., Plenum Press, New York (1980).

Schuller, G., 1979, Coding of small sinusoidal frequency and amplitude modulations in the inferior colliculus of "CF-FM" bat, Rhinolophus ferrumequinum, Exp.Brain Res., 34:117-132.

SONAR DISCRIMINATION OF METALLIC

PLATES BY DOLPHINS AND HUMANS

Whitlow W. L. Au and Douglas W. Martin

Naval Ocean Systems Center
Kailua, Hawaii 96734

In the first Animal Sonar Systems symposium held in Frascati, Italy, Evans and Powell (1967) demonstrated that a blind-folded echolocating dolphin could discriminate the composition and thickness of metallic plates. Mackay (1967) orginally suggested this experiment to test if a dolphin could detect phase differences. An example of a typical sonar run showing the animal's position at 1 s intervals is presented on the left side of Fig. 1. Weisser, et al. (1967) examined the plates by insonifying them at normal incident angle using simulated dolphin echolocation signals. No obvious differences between echo waveforms were observed. Fish et al. (1976) demonstrated that instrumented human divers wearing a helmet with a sending and two receiving transducers could perform the discrimination task as well as or better than the dolphin. Broadband simulated dolphin clicks with energy centered at 60 kHz were projected and the echoes time-streched by a factor of 128 before presentation to the divers.

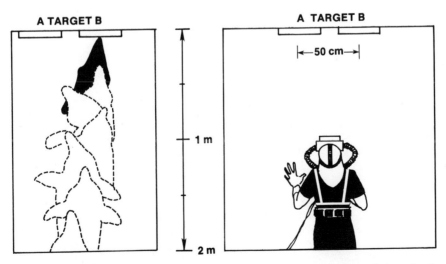

Fig. 1. An example of the dolphin's position (1 s intervals) swimming towards the plates, and geometry of the human diver experiment.

An example of a diver's orientation to the plates is shown on the right of Fig. 1. No explanations were given on the cues used by the divers to perform the discrimination task.

In this paper, the results of another acoustic examination of the same plates used by Evans and Powell (1967) are discussed. Simulated dolphin signals at various incident angles were used. The diagrams of Fig. 1 indicate that both the dolphin and human divers did not usually insonify the plates at a normal incident angle. The geometry for the human diver experiment indicated an incident angle of approximately 14°.

Procedure

Backscatter measurements of the plates were made at the Naval Ocean Systems Center's, Hawaii Laboratory test pool using a computer controlled monostatic echo ranging system discussed by Au and Synder (1980). The plates were suspended from two lines connected to a 1.9 m arm attached to a rotor. The distance between the transducer and plates was 1 m, the same as in the instrumented diver experiment. The echoes were digitized with an 8-bit A/D converter operated at 1 MHz sample rate, and the results stored on digital magnetic tape. Human listening experiments were conducted playing back echoes through a D/A system to subjects in a sound isolation booth. A sample rate of 40 kHz was used during the playback which stretched the original signal by a factor of 25 and reduced the peak frequency from 60 kHz to 2.4 kHz.

Discrimination performance was measured using 64-trial sessions. For each session, the echoes from the standard 0.22-cm thick copper plate and echoes from one of the other 9 plates were used. Subjects were required to push the A-switch whenever they were presented with echoes from the standard, and the B-switch for other echoes. The stimulus was repeated at four pulses per second for 15 seconds or until the subject responded, whichever occurred first. Echoes obtained at normal incident and at a 14° incident angle were used.

RESULTS AND DISCUSSION

Backscatter results at normal incident for the standard copper and three other plates are shown in Fig. 2. The echo waveforms are shown individually and the spectra are plotted on the same graph. Very little difference can be observed in the echoes. Two human subjects could not perform above chance in discriminating the standard 0.22-cm copper plate from the other plates.

Echoes from the same four plates at the 14° incident angle are shown in Fig. 3. The echo waveforms are shown on the left and the envelopes of the matched filter response are shown on the right. The reference signal for the matched filter was the transmitted signal. Differences in the echoes are obvious from the figure. Such obvious differences also existed between the standard copper and the other plates, including the other copper plates of different thicknesses. The numbers above the matched filter envelopes are the times of arrival of different echo highlights in μs, relative to the first echo highlight. Two human subjects had no difficulty discriminating the standard from the other plates. The task was trivial and the performance level was close to 100%. The primary cue indicated by the subjects was the difference in time-separation pitch (TSP) of the various echoes. Arrival time difference between correlated echo components can give rise to the perception of TSP. Echoes with different highlight structures will produce different TSP.

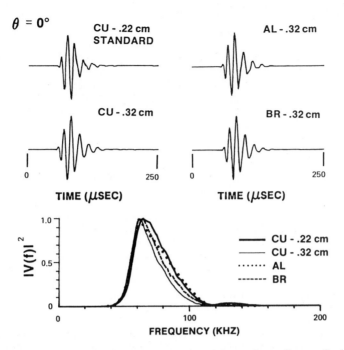

Fig. 2.   Backscatter results at normal incident from four of the plates
          used by Evans and Powell (1967).

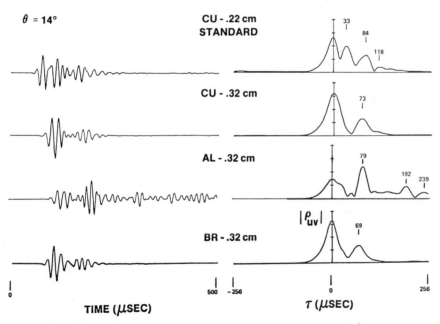

Fig. 3.   Backscatter results at 14° incident angle for the same four
          plates of Fig. 2.

At normal incidence the echoes resembled the incident signal and did not seem to contain much useful information for discrimination. However, as the incident angle increased to 10°, the echoes began to have multiple highlights which could be used for discrimination. Although the echoes at 14° were 20 to 30 dB lower in amplitude than echoes at normal incidence, they still provided useful discrimination cues. Two scattering processes were suspected of producing the multiple highlight echoes: "leaky Lamb waves" and internal trapped waves. The two scattering processes are described schematically in Fig. 4. The trapped wave situation is for the longitudinal wave. Transverse waves of lower velocity will also be excited in the plates and converted to longitudinal waves at a boundary upon exiting the plate.

In order to understand the backscattering process better, two 0.79-thick aluminum plates, 30.5 and 61.0 cm in diameter, were acoustically examined with a higher frequency (120 kHz peak frequency) transient pulse. The results of the backscatter measurement for an incident angle of 15° are shown in Fig. 5. Below each of the echoes are the envelopes of the matched filter responses showing the relative times of arrival of the highlights. The first highlight, A, is probably a leaky wave component since it is always present even as the incident angle increased to 40°. Component B is the result of the trapped longitudinal wave reflecting off the edge of the plate as depicted in Fig. 4b. For a trapped mode with an edge reflection, the time between the first and second highlights for the larger plate should be twice that of the smaller plate, since its diameter was twice that of the smaller plate. The matched filter results confirmed the presence of the edge reflected trapped mode component. Component C is the result of the trapped longitudinal wave reflecting off the same edge as component B and then off the opposite edge, and back off the original edge. For this mode, travel time should be 3 times the travel time for component B. Again, the matched filter results confirmed the presence of the triple-reflected trapped component. Component D for the 30.5-cm plate is a trapped mode component that experienced five edge reflections. It should have an arrival time that is 5 times that of component B. The modes of propagation for the other highlights have not been determined.

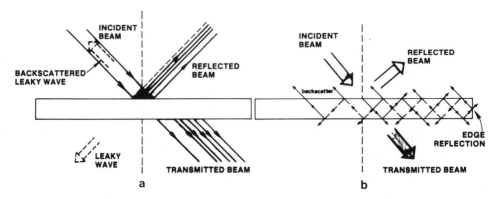

Fig. 4.   Schematic representation of (a) leaky lamb wave backscatter and (b) trapped longitudinal with an edge reflection.

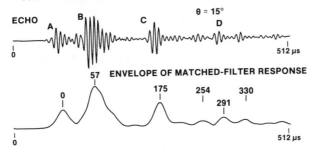

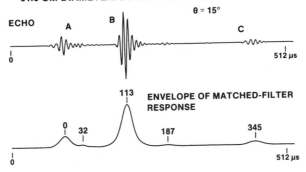

Fig. 5. Backscatter results for the 30.5 and 61.0 cm diameter plates at
an incident angle of 15°.

REFERENCES

Au, W. W. L., and Synder, K. J., Long-range Target Detection in Open
    Waters by an Echolocating Atlantic Bottlenose Dolphin (<u>Tursiops
    truncatus</u>, <u>J. Acoust. Soc. Am.</u>, 68:1077-1084.

Evans, W. E., and Powell, B. A., 1967, Discrimination of Different
    Metallic Plates by an Echolocating Delphinid, in "Animal Sonar Systems:
    Biology and Bionics", R. G. Busnel, ed., Laboratoire de Physiologie,
    Jouy-en-Josas, 78, France, pp. 363-382.

Fish, J. F., Johnson, C. S., and Ljungblad, D. K., 1976, Sonar Target
    Discrimination by Instrumented Human Divers, <u>J. Acoust. Soc. Am.</u>, 59:
    602-606.

Mackay, R. S., 1967, Experiments to Conduct in Order to Obtain Comparative
    Results, in "Animal Sonar Systems: Biology and Bionics", R.G.Busnel,
    ed., Laboratoire de Physiologie, Jouy-en-Josas, 78, France, pp. 1173-
    1196.

Weisser, F. L., Diercks, K. J., and Evans, W. E., 1967, Analysis of Short
    Pulse Echoes From Copper Plates, <u>J. Acoust. Soc. Am.</u>, 42:1211(A).

ECHOES FROM SOLID AND HOLLOW METALLIC SPHERES

Theodorus A.W.M. Lanen and Cees Kamminga

Information Theory Group
Delft University of Technology
Delft, the Netherlands

SUMMARY

A very successful pattern recognition experiment has been carried out with a dolphin identifying a solid sphere versus a hollow one. Results suggest that the dolphin utilized a time difference cue to distinguish between the two spheres. A theoretical evaluation of the physics of the targets, involving a model of the dolphin sonar, has been carried out. The ensonifying signal follows the model of an elementary signal as proposed by Gabor, i.e. a harmonic signal with a Gaussian envelope. The numerical values of the 3 needed parameters are estimated from the sonar recordings of the live experiment. The target modelling of the sphere is done with the quantities inner-to-outer diameter, density of the metal, and longitudinal and transversal wave velocity. A good match between the computational model and the actual experiment shows up.

EXPERIMENTAL SET-UP

Subject

The dolphin in the experiment[1,2] was an adult female bottlenose dolphin, named Doris, who has been in captivity since September 1969. She was one of the best subjects in previous echolocation experiments, and showed herself to be a very willing participant in the experiment.

Experiment

The dolphin was faced with the task of distinguishing between two externally identical, cast iron spheres, one of which was solid and the other hollow. The outer diameter of the spheres was fixed at 15 cm.

In the initial stage of the experiment the inner-to-outer diameter of the hollow sphere was set to 0.6. It appeared that the dolphin did not succeed on her own initiative in detecting the difference between the two targets. For this purpose, the solid sphere was marked with a piece of tape, its size being gradually reduced in successive trials. After this period of training the dolphin managed to distinguish by means of its sonar between the solid and the hollow sphere in the absence of the visual cue. The reliability with which the dolphin now carried out its discrimination task appears from a detection score of 95 out of 100 trials.

In the second stage the diameter ratio of the hollow sphere was reduced to a value of 0.4, giving it a very close resemblance of about 94 percent to the solid sphere in volume. The echo functions of these two spheres show a

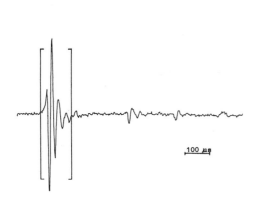 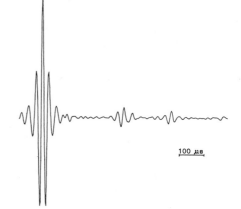

Fig. 1. Sonar pulse with reflections from solid sphere.

Fig. 2. Cross-correlation result.

very small difference-of-time interval between the first (specular) and the secondary echo. The detection score now dropped to a statistically lower value of 77 out of 100 trials.

Results

The sonar pulses the dolphin emitted while performing its discrimination task were recorded by means of a miniature hydrophone, mounted in a rubber suction cup, placed on the dolphin's rostrum. Figure 1 shows an example of a sonar pulse together with the first and secondary reflections as recorded in this manner. For a proper measurement of the time duration between the two echoes it is necessary to enlarge the signal-to-noise ratio. A common method is to apply a cross-correlation technique between the emitted signal and the received echoes. In the case of white Gaussian noise the correlation technique is a theoretically optimal means of detecting signals buried in noise. To this end a cross-correlation is performed between the emitted sonar pulse (indicated between two vertical bars) and the complete time function. The result of this filtering process for the signal in figure 1 is shown in figure 2.

As will be clear after the introduction of Gabor's model, the most promising method is to measure the time differences between the echoes making use of the maxima occurring in the envelope of the cross-correlation result. This strategy yields values for the time differences occurring in the experiment as presented in table 1.

Table 1. Experimentally and computationally determined time differences (μs) between first and secondary echoes.

sphere	solid	0.4 hollow	0.6 hollow
experiment	185.6	175.5	148.0
comuputation	186.5	182.9	160.1

MODELLING

Gabor model

A closer look at the sonar signal emitted by the dolphin, while performing its discrimination task, reveals that it follows the model of an elementary signal as proposed by Gabor[3,4], i.e. a harmonic signal with a Gaussian envelope. Figure 3 shows the striking resemblance between the model (left) and a typical sonar pulse (right).

A mathematical formulation of the Gabor model is given by following formula:

$$p(t) = A \exp\{-\alpha^2(t-t0)^2\} \exp\{j(2\pi f0(t-t0)+\varphi)\} \qquad (1)$$

This expression states that it is possible to describe the <u>form</u> of the sonar pulse by (just) <u>three</u> parameters, i.e.:

   α : 'sharpness' of the pulse,
   f0 : dominant frequency, and
   φ : phaseshift relative to the peak in the envelope.

It must be remarked that 'α' is equal to the -3 dB bandwidth (bw) apart from a multiplicative constant. The other two parameters A and t0 are less essential. They define the magnitude of the pulse and the epoch of its peak relative to the origin of a time coordinate system, respectively.

   Taking the Fourier transform of p(t) yields following expression, apart from a multiplicative constant:

$$p(f) = \exp\{-(\pi/\alpha)^2(f-f0)^2\} \exp\{-j(2\pi f0t0-\varphi)\} \qquad (2)$$

The absolute value of p(f), $|p(f)|$, is shown in figure 4. This corresponds to a symmetrical spectrum centred around the dominant frequency f0. Obviously the form of the time signal p(t) is preserved in taking the Fourier transform.

   Let us now further investigate the properties of the Gabor model. To this end we introduce 'Δt' as the 'effective time duration' and 'Δf' as the 'effective frequency width' of the model. If these two parameters are defined in the proper way it can be proved[4] that only for the Gabor model does their product equal the value 1, 'Δt' being $\sqrt{\pi}/\alpha$ and 'Δf' being $\alpha/\sqrt{\pi}$. That is to say, we have found an example in nature that makes the relation

$$\Delta t * \Delta f >= 1, \qquad (3)$$

which is at the root of the fundamental principle of communication, pass into an identity. Concluding, it may be stated that the dolphin communicated by means of a signal that is an optimum in terms of 'use of time' and 'use of bandwidth', their product being minimal.

Calculation of echoes

   The process of ensonifying the sphere by the dolphin can be represented schematically by figure 5[5,6,7]. In this figure the centre of the sphere of radius 'a' is fixed at the origin of a spherical coordinate system. The scattered pressure p(τ), τ being a time parameter, is to be calculated at

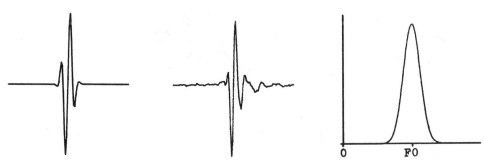

Fig. 3. Sonar pulse and its model equivalent.   Fig. 4. Gabor model in
                                                           frequency domain.

point P in the fluid surrounding the sphere. The incident wave considered here is a plane pressure pulse, that is a plane wave of finite duration, proceeding down the Z-axis of a Cartesian coordinate system that also has its origin at the centre of the sphere.

Suppose the sphere is described by its form function f(ka), where ka is a dimensionless frequency parameter expressed in the sphere's dimensions. Stated simply, this function describes the sphere's response for a range of frequencies[6,7]. To calculate this function the following parameters descri-bing the sphere are required:

ρ : density of surrounding fluid
$\rho 1$: density of sphere material
c0 : longitudinal velocity in surrounding fluid
c1 : longitudinal velocity in sphere material
c2 : transversal velocity in sphere material
b/a : inner-to-outer diameter ratio.

We will not go into further details concerning the computation of this function because this is beyond the scope of this article.

Furthermore, a description of the incident sound wave is needed. Using for this wave the signal model as introduced in the previous section, with values for the parameters extracted from the actual experiment, we are able to calculate the scattered sound wave at point P. To this end we must first rewrite the frequency spectrum of the Gabor model (equation 2) in terms of the frequency parameter ka, yielding g(ka) as a frequency description of the incident sound wave. Now the scattered wave at point P is given by the Fourier integral form[5]:

$$p(\tau) = \int_{-\infty}^{\infty} g(ka)\ f(ka)\ \exp\{jka\tau\}\ d(ka) \tag{4}$$

It may be clear that p(τ) depends only on the spherical coordinates 'r' and 'θ' and is independent of 'φ' because of the presence of a geometrical symmetry. The computations may be simplified by supposing 'θ' to be 180 degrees, inducing P to be situated on the Z-axis. This situation is common-ly known as 'monostatic backscattering'.

Calculated results

Substituting values extracted from the experiment for the sphere pa-rameters, the respective form functions of the solid, and the 0.4 and 0.6 hollow spheres were calculated. Figure 6 shows the modulus of the form function of the 0.6 hollow sphere, with ka ranging from 0 to 30, enclosing the frequency range in which the sonar signal is active. In figure 7 the modulus of the frequency spectrum of the incident sonar pulse, 'cast' in Gabor's model, is drawn. Making use of these results in equation 4 yields an echo structure as shown in figure 8, in which 5 separate echoes can be

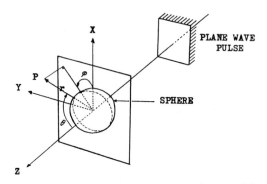

Fig. 5. Geometry of the scattering problem.

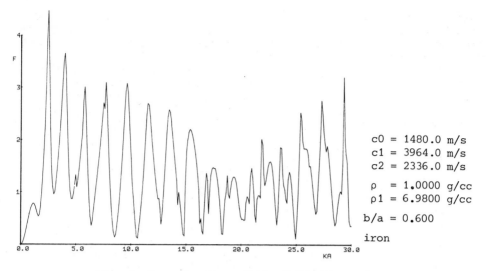

```
c0 = 1480.0 m/s
c1 = 3964.0 m/s
c2 = 2336.0 m/s

ρ  = 1.0000 g/cc
ρ1 = 6.9800 g/cc

b/a = 0.600

iron
```

Fig. 6. Form function of the 0.6 hollow sphere.

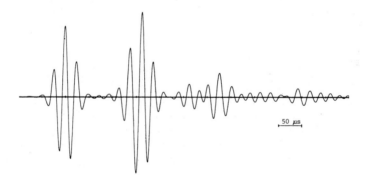

```
f0 = 40.0 kHz
bw = 20.0 kHz
φ  = 0.00

c0 = 1480.0 m/s

a  = 0.075 m

gabor
```

Fig. 7. Frequency spectrum of the incident sound wave.

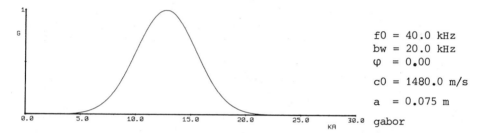

50 μs

Fig. 8. Resulting echo structure ....

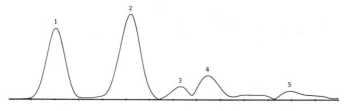

Fig. 9. .... and its envelope

distinguished. As is clear, for example, from this figure the supposition by Mackay[8] that the first and secondary echoes always are phase-inverted in the case of a hollow sphere and show no phaseshift in the case of a solid sphere is not confirmed.

Because the Gabor model, a signal model with an exactly defined envelope is exploited as the incident sonar pulse, it is also possible to calculate the envelope of the echo structure. The envelope for the echo structure in figure 8 is shown in figure 9.

Echo formation

As can be concluded from figures 8 and 9 the echo structure consists of a sum of time- and phase-shifted, amplitude-modulated replicas of the incident sonar pulse. If $p(\tau)$ represents the incident sonar pulse, the echo structure $e(\tau)$ may be written in formula as follows:

$$e(\tau) = \sum_i A_i \exp\{-j\varphi_i\}\, p(\tau - \tau_i) \qquad (5)$$

A question that arises is how successive echoes are formed. It may be clear that the first echo in figures 8 and 9 results from a specular reflection. Echoes 2 through 5 require a more complex explanation.

Because the wavelength of the incident sound wave is in the order of the shell thickness a special type of surface wave, called 'Lamb' wave, will be generated at an angle slightly greater than the critical angle α_{crit} defined for transversal waves (figure 10). This critical angle is calculated by the following formula:

$$\alpha_{crit} = \arcsin(c0/c2) \qquad (6)$$

These Lamb waves travel around the sphere at velocity $c2'$, being close to c2, while emitting energy at an angle α_{crit} with the normal. The velocity $c2'$ cannot be found easily. It depends not only on the elastic constants of the material but also on the plate thickness (i.e. shell thickness) and - in contrast to the longitudinal and transversal waves - on the frequency. Moreover it depends on the ratio of curvature to wavelength. Furthermore, $c2'$ is a constant for equal values of the product of frequency and shell thickness. This mechanism is responsible for the formation of echoes 2, 4 and 5 in figure 9.

Echo 3 is the result of a different wave path. As this echo is formed by travelling along path number 3 (figure 10) without penetrating the spherical shell, its velocity remains c0, the whole path. An explanation for

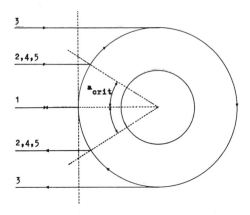

Fig. 10. Wave paths.

the presence of this echo is the fact that each elastic sphere bears in it (more or less) the properties of an acoustically rigid sphere.

COMPARISON WITH EXPERIMENT

As the resulting echo structure can be written according to equation 5 the time differences between the first and secondary echoes are based on the epochs at which maxima occur in the envelope of the echo structure (figure 9). Computations of these differences for the solid, and the 0.4 and 0.6 hollow spheres yields the results exhibited in table 1. A comparison with the experimentally obtained values shows a maximum deviation of 8 percent for the 0.6 hollow sphere.

As we have seen in the preceding section the time difference between the first and the secondary echo depends heavily on c_2'. This parameter in turn is dependent on the dominant frequency. In the computations a value of 40 kHz (averaged over the whole experiment) was substituted for the dominant frequency without making distinction to the sphere's inner-to-outer diameter. There are indications[1], however, that the dolphin might have changed its dominant frequency while 'interrogating' different spheres. This phenomenon will be one of the subjects of future research.

REFERENCES

1. Hol, W.A. and C. Kamminga, 1979. "Investigations on Cetacean sonar I. Some results on the threshold detection of hollow and solid spheres performed by the Atlantic bottlenose dolphin, Tursiops truncatus". Aq. Mammals, 7(12): 41-64.
2. Kamminga, C. and A.A.M. v.d. Krogt, 1984. "Time difference perception in bio-sonar from the Atlantic bottlenose dolphin". Delft Progress Report, 9(1984): 174-183.
3. Gabor, D., 1946. "Theory of communication". J. Inst.Elect.Eng., 93: 429-459.
4. Gabor, D., 1947. "Acoustical quanta and the theory of hearing". Nature, vol. 159: 591-594.
5. Rudgers, A.J., 1969. "Acoustic pulses scattered by a rigid sphere immersed in a fluid". J. Acoust. Soc. Am., vol. 45, no. 4: 900-910.
6. Hickling, R., 1964. "Analysis of echoes from a hollow metallic sphere in water". J. Acoust. Soc. Am., vol. 36, no. 6: 1124-1137.
7. Hickling, R., 1962. "Analysis of echoes from a solid elastic sphere in water". J. Acoust. Soc. Am., vol. 34, no. 10: 1582-1592.
8. Mackay, R.S., 1966. Discussion on: "The acoustics of small tanks", in: Marine Bio-Acoustics, vol. 2, W.N. Tavolga, ed., Pergamon Press, Oxford: 12.

TARGET IDENTIFICATION IN A NATURAL ENVIRONMENT:

A STATISTICAL VIEW OF THE INVERSE PROBLEM

Manell E. Zakharia

I.C.P.I. Lyon, Laboratoire de Traitement du Signal
UA 346b, CNRS; 25, Rue du Plat, 69288, Lyon Cedex 02
France

INTRODUCTION

In a natural environment, the sonar system of bats or dolphins has three main roles: detection, location and recognition (classification) of targets and obstacles. To get the information on their prey (identification and size estimation), the animals using a biological sonar are faced with solving an inverse sonar problem: starting with an echoe's sequence, how to extract characteristic parameters that can be used for target identification.

The echo formation mechanism, for a sonar target, can be modeled as a linear filtering of the emitted signal [1], [2], [5]; this filtering is a combination of spatial and temporal filtering. It is a random one, mainly because of the medium fluctuations and the movement of the whole sonar system (emitter and receiver movement, target movement and deformation).

The echo is also "spoiled" by the noise, the reverberation, and by the undesirable target's echoes.

A biological sonar system must take into account all these perturbating phenomena, while trying to identify a target.

GENERAL SONAR SITUATION

The space and frequency transfer function of an emitter, a receiver or a target can be defined as a combination of the beam pattern and the temporal impulse response [7], [9], [10]. The duality between the spectral and the spatial frequencies increases the problem complexity and is one of the major reasons why the inverse problem solution unicity is hard to get (without any a priori information).

A sonar system can be given by defining three main parts:
    *an emitter with a space impulse response h_e
    *a receiver with a space impulse response h_r
    *a target with a space impulse response h_t

The sonar situation can be defined by the polar position of the acoustic centers of the emitter O_e and the receiver O_r, respect to the target:

$R_e, \theta_e, \varphi_e, R_r, \theta_r, \varphi_r$

To get a complete definition of the situation, we have also to fix the angular position of both emitter and receiver with respect to the axes $O_t O_e$ and $O_t O_r$:

$\theta_{te}, \varphi_{te}, \psi_{te}, \theta_{tr}, \varphi_{tr}, \psi_r$

Figure 1 gives an example of such a situation [6], [9], [10]. We have plotted the beam pattern of the emitter, receiver and target for a given frequency f_0.

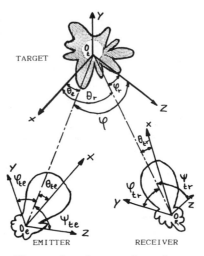

Figure 1. Sonar situation.

Commonly, in sonar, we are used to locating the target respect to the sonar, and not the opposite, as we did. We have chosen the target center as the coordinates center, only for the display clarity.

In a general case (natural environment), all the sonar system components are moving and all the directivity patterns can be time-variant, mainly the target one. The coordinates and the space and frequency responses can also be time-varying.

Let us try to decompose the various phenomena involved in the echo formation, in man-made sonar [3], [4], [5], [7], [9]. For animal sonar, the phenomena involved are very similar.

The electrical signal at the emitter can be considered as the input signal and the electrical signal at the output of the receiver can be considered as the output signal (for animal sonar, the input can be the nervous stimulus and the output can be the signal delivered by the ear, after spatial and spectral filtering):

-The electrical signal (nervous stimulus) is converted, by the emitter (mouth), to an acoustical signal with an electro-acoustical transfer function, H_{EAe}.

-The acoustical signal is emitted with a given directivity (for each frequency) corresponding to the emitter spatial impulse response, H_e.

-It reaches the target through an acoustical path, which can be roughly considered as a linear filter with a time-varying complex transfer function, H_{Pet}.

-It is reflected by the target, which possesses its own space and frequency transfer function H_t. H_t is the function we want to characterize, by the sonar, in order to identify and to classify the target.

-The echo reaches the receiver (ear) through another acoustical path with a transfer function, H_{Ptr}.

-The echo is received by a transducer (ear) with a given space and time impulse response, H_r.

-It is converted to an electrical signal (nervous stimulus) through an electro-acoustical transfer function, H_{EAr}.

The general expression of the echo can be written:

$$E(t) = Z(t) *H_{EAe}(t)$$
$$*H_e(t, r_e(t), \theta_e(t), \varphi_e(t), \theta_{et}(t), \varphi_{et}(t), \psi_{et}(t))$$
$$*H_{Pet}(t, r_e(t), \theta_e(t), \varphi_e(t))$$
$$*H_t(t, \theta_e, \varphi_e, \theta_r, \psi_r)$$
$$*H_{Ptr}(t, r_r(t), \theta_r(t), \varphi_r(t))$$
$$*H_r(t, r_r(t), \theta_r(t), \varphi_r(t), \theta_{rt}(t), \varphi_{rt}(t), \psi_{rt}(t))$$
$$*H_{EAr}(t)$$
$$+N(t) = REV(t) \qquad (1)$$

where $N(t)$ is an additive noise and $REV(t)$ is the reverberation.

$*$ is a convolution operator. This convolution should be made in both time and space domain (where the space variables are functions of the time) on all the variables!

The only observation we can get is the echo $E(t)$. The information we would like to get is H_t, the target impulse response.

PARAMETER SEPARATION

The general convolution, in equation (1), cannot be inverted easily, in order to reach the target transfer function. We need a priori information that can help in reaching an approximation of the solution. We can set a separation between the various parameters involved and get a simpler expression of the echo:

$$E(t) = \{Z(t)*H_{EA}(t)*H_S(t)*H_T(t, r_e, \theta_e, \varphi_e, r_r, \theta_r, \varphi_r)\}$$
$$+N(t) +R(t) \qquad (2)$$

where:
$*H_{EA}$ includes all the electro-acoustic transfer functions.
$*H_S$ includes the propagation filtering and all the spatial convolutions, due to the directivity; the directivity convolution can be split in two: distance and angular position influence.

*H_T is the target transfer function.

Even in such a case, we can see that the target transfer function is time-variant. It depends on the target movement and deformation, during the signal duration, and on the target attitude, respect to the sonar system. A "static" look at the target is not sufficient, for a recognition, its "cross-section" depends strongly on its "attitude" and on the moment of observation. We need a modeling of the target that can take into account its movement and its attitude variations.

TARGET MODELING

The target model presented by R. A. Altes [1] takes into account the various acoustical phenomena involved in the echo formation [2]. It can give interesting results in the recognition of simple shape still targets [5], [8].

In this model, the target can be represented as a generalized transversal filter: the echo can be modelled as a sum of weighted and delayed copies of the emitted signal and of its n-order derivatives and integrals. These weights and delays can be used for target characterization. An optimum receiver (in the case of white Gaussian noise) will be a set of filters matched to each echo component (integrals or derivatives of the emitted signal) [1], [5]. The receiver can be reduced to a "constant-Q" filtering, in the time-frequency plane [1].

The weights and the delays can be estimated by computing the amplitude and the position of the maxima of the envelope at the output of each filter and used for target classification.

In the case of a natural target, this model can be generalized to each target position or (and) deformation. The parameters can then be considered as variables depending on both the time and the echo number.

Let us first consider the case of a target without deformation, moving slowly enough, so we can consider it as still during the signal duration (the only effect of the movement will be the Doppler effect). In this case, an echo sequence can give various observations of the (still) target for different observation angles. This number of observations is usually limited (few directions) and the animals cannot always turn around their prey before catching them. In such a situation, we can get a set of characteristic parameters, for each direction of observation. We can use all these parameters to identify a target. Not only the values of these parameters can be characteristic, but also their variations, from one echo to the other.

Let us now consider the case of target that can move (and rotate) and be deformed, during the signal duration (case of flies or butterflies). The influence of the movement and of the deformation cannot be expressed simply, the interpretation of the echo fine structure can be very hard, as long as these perturbating phenomena can induce non-linear echo filtering and transformations.

In a general case, we can have a combination of both effects, in the same echo sequence.

CONCLUSION

Considering the target strength as a level attenuation can only be interesting in detection problems; it cannot be used directly for target recognition because of the target movement and deformation. Any receiver operating on natural targets should take the various movements into account. The fine structure of a time-frequency representation of the echoes (set of constant-Q matched filters) can help in solving this problem. Nevertheless, it seems clear that any information given by a single echo is not sufficient for a robust classification. We will need a set of echoes of the same target to establish a robust classification.

REFERENCES

[1] Altes, R. A., "Sonar for a generalized target description and its similarities to animal echolocation systems". J. Acous. Soc. Am. 59: 1, Jan 1976.

[2] Freedman, A. "A mechanism of acoustic echo formation". Acustica, 12:10, 1962.

[3] Ol'Shevskii V.V., "Statistical methods in sonar". Consultants Bureau, Plenum Pub., 1978.

[4] Urick, R.J., "Principles of underwater sound". McGraw Hill Ed. 1983.

[5] Zakharia, M.E., "Contribution a la caracterisation et a l'identification de formes simples par sonar actif: application a un sonar de peche". Docteur-Ingenieur Thesis Dissertation, Universite d'Aix-Marseille II, Luminy, June 1982.

[6] Zakharia, M.E., "Problemes inverses en sonar de peche: possibilites et limitations". Revue du CETHEDEC, 20e annee, 3e trimestre, No. 76, 1983.

[7] Zakharia, M.E., "Sonar evaluation in natural environment". International CNRS Conference on "Air-borne Animal Sonar Systems". Lyon Feb 1984. To appear in Acustica.

[8] Zakharia, M.E., Sessarego, J.P., "Sonar target classification using a coherent echo processing". I.C. ASSP, Paris, May, 1982

[9] Zakharia, M.E., "Statistical time varying look at sonar target identification". Advances in Signal Processing. Workshop on Acoustic, Speech and Signal Processing, Peking, China, April 1986.

[10] Zakharia, M.E., "Reflexions sur la mesure de la response spatio-frequentielle d'une cible sonar". Colloque J.E.S.P.A., Lyon, France, June 1986.

AN AUTOMATIC TARGET RECOGNITION

ALGORITHM USING TIME-DOMAIN FEATURES

Douglas W. Martin and Whitlow W.L. Au

Naval Ocean Systems Center
Kailua, Hawaii, 96734

The capabilities of dolphins and humans to discriminate target properties from target echoes was discussed by Au (this volume). Nachtigall
(1980) previously presented a similar review of dolphin discrimination
capabilities. Predominant cues seem to be time-domain features related to
differences in the highlight structure of target echoes. This paper
describes a time domain feature extraction and pattern recognition algorithm which uses cues previously identified as salient for humans (Martin
and Au, 1982, 1986). Results are also compared with those of a spectral
processing model developed by Chestnut et al. (1979, 1981).

An example of a simulated dolphin echolocation signal and the subsequent echo from a 3.81-cm-diameter water-filled aluminum cylinder are shown
in Fig. 1. Echo waveforms contain highlights from multiple internal
reflections, with the differences in highlight arrival times caused by
different acoustic paths in the cylinders. When two highly-correlated
broadband pulses are separated by time T, a time separation pitch (TSP) can
be perceived with a frequency 1/T (Small and McClellan, 1963). Echoes with
more than two highlights of unequal amplitudes produce an interacting or
complex TSP. Hammer and Au (1980) first suggested that time-separation
pitch associated with highly correlated echo highlights was a probable cue
used by dolphins in size and material composition discrimination tasks.
The saliency of TSP for human echo discrimination was further supported in
studies by Martin and Au (1982, 1986). They also identified cases where

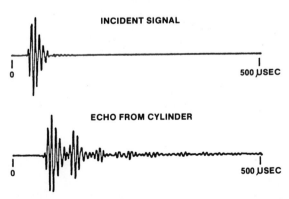

Fig. 1. Example of a simulated dolphin echolocation signal and the echo
from a 3.81-cm diameter water-filled alumunum cylinder.

differences in echo duration were salient cues. The use of TSP by echolocating animals to perceive distance was first suggested by Normark (1961).

FEATURE EXTRACTION AND PATTERN RECOGNITION

Software was developed to extract highlight separation and highlight
amplitude ratios from echo envelopes to form target feature sets. These
two features are necessary determinants of both TSP and echo duration.
Echoes were synthesized from the extracted feature sets, and usually the
synthetic echoes contained the same discrimination cues as the real echoes
(Martin and Au, 1986). Feature sets were also used to synthesize signals
for discrimination tests, to determine the relative importance of the
features. Two types of automatic pattern recognition algorithms were
tested. The first used the time-domain features and the second used the
filter bank model developed by Chestnut et al. (1979, 1981).

The software that extracted time domain features was a highlight
detector. The software measured probability of occurrence, time separation
from the largest highlight, and amplitude ratio relative to the largest
highlight, for each highlight. The first stage of the processor was a peak
detector that stored information about every point in a signal where the
slope changed from positive to negative. A series of criteria selected the
extrema which were defined as highlights. Small-amplitude extrema in the
immediate neighborhood of a larger maximum were rejected. After obtaining
a list of highlights for a given signal, the absolute maximum was assigned
a time separation of zero and an amplitude ratio of one. The other
highlights were assigned negative or positive time separations according to
position before or after the largest highlight. Amplitude ratios for each
highlight were calculated with respect to the maximum.

In the second stage, reference vectors were created from a set of 10
echoes by aligning the features across the signals. Absolute maxima were
aligned, so that every group of signals had a highlight occurring with
probability 1.0, amplitude ratio 1.0, and time separation 0. Other
highlights were aligned, and amplitude ratios and time separations became
statistical quantities represented as means and standard deviations for the
group of signals. The probability of occurrence for each highlight was
calculated; i.e., those which occurred in only one signal had probability
1/n, where n was the number of signals in the input. Means and standard
deviations for both time separation and amplitude ratio were also
calculated.

Feature lists statistically defined each highlight in a group of
signals. Each target was represented by a feature vector of twenty-five
elements. The time axis was partitioned into twenty-five time bins, each
bin representing 20 μs of the original signal. If the bin did not contain
any highlights, it was assigned a value of zero. If a highlight occurred
within the time bin, its amplitude ratio was assigned to the vector
element. If two highlights occurred in the same time bin, the largest
amplitude value was used. The next stage of software processing determined
whether the extracted features defined separable target classes. The
features were the input to an automatic pattern recognition algorithm and
were also used to create synthetic target echoes for human discrimination
test. Fig. 2 shows real and synthetic echoes for a solid aluminum
cylinder. While minor differences existed in the fine structure of the
echoes, the overall echo similarity is visually obvious.

PATTERN RECOGNITION

The accuracy of the extracted feature sets was tested using automatic
pattern recognition algorithms. The algorithms classified target echoes by

830

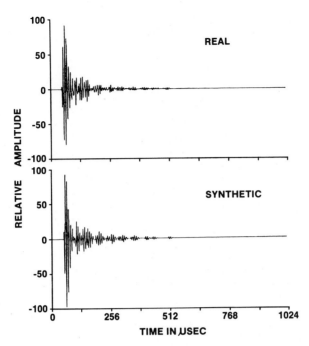

Fig. 2. An example of a real and a synthetic echo for a solid aluminum
cylinder.

calculating the Euclidian distances between sets of test and reference
feature vectors. Reference and test data were represented by vectors of 25
features; each feature represented the time-separation and relative
amplitude of an echo highlight. Reference data were means from ten echoes.
Because each feature vector contained 25 elements, time separation
as large as 500 µs (25x20) were represented. Within the constraint of a
twenty-point time resolution, the Euclidian distance between two vectors
constructed in this way is a measure of stimulus similarity in terms of the
extracted feature sets.

Performance of the feature extraction and pattern recognition
algorithms for material composition discrimination is shown in Tables 1
and 2 as confusion matrices. Rows of the matrices represent test echoes;
columns represent reference echoes (mean vectors). The algorithm
calculated the distances from each test echo to each reference vector, and
identified each with the minimum distance reference class. Elements on the
matrix diagonal represent correct responses; off-diagonal elements
represent confusions. The algorithm's performance was 90% correct for
material composition (Table 1) and 100% for internal structure (Table 2).
Chance performance was 14.3% (1 in 7) for material composition and 25% (1
in 4) for internal structure. When echoes from two different glass
cylinders were added to the data of Table 1., to make a 9x9 confusion
matrix, performance dropped to 62% correct; chance = 11%. Many glass test
echoes were incorrectly identified, and some echoes from other cylinders
were wrongly identified as glass. Identification of echoes from
the glass cylinders was also poor for both humans and dolphins. (Schusterman et al., 1980; Martin and Au, 1982).

A spectral feature extraction algorithm used by Chestnut et al. (1979,
1981) was also used to test target recognition performance. The model
consists of a bank of parallel, contiguous constant-Q filters, and the
extracted features are samples of the target's frequency response.

Table 1. Confusion matrix for cylinder material composition discrimination. Cyl-1, 3.8 cm O.D. and 0.32 cm wall thickness. Cyl-2, 7.6 cm O.D. and 0.40 wall thickness.

REFERENCE TARGET ECHOES

TEST TARGET ECHOES	ALUM CYL-1	STEEL CYL-1	BRONZE CYL-1	ALUM CYL-2	STEEL CYL-2	BRONZE CYL-2	SOLID ALUM CYL
ALUM CYL-1	97%					3%	
STEEL CYL-1		100%					
BRONZE CYL-1			100%				
ALUM CYL-2				93%		7%	
STEEL CYL-2				37%	63%		
BRONZE CYL-2						100%	
SOLID ALUM CYL							89%

Table 2. Confusion matrix for internal structure discrimination. All cylinders had outer diameters of 7.6 cm.

REFERENCE TEST TARGET

TEST TARGET ECHOES	ALUM CYL-2 AIRFILLED	CORAL CYL-2 SOLID	ALUM CYL-2 WATERFILLED	ALUM CYL-2 SOLID
ALUM CYL-2 AIRFILLED	100%			
CORAL CYL-2 SOLID		100%		
ALUM CYL-2 WATERFILLED			100%	
ALUM CYL-2 SOLID				100%

Thirty constant-Q filters over the frequency range of 50 to 200 kHz were used. As above, the algorithm calculated the Euclidian distance between test and reference feature vectors, and identified test echoes with the minimum-distance reference class. Prior to feature extraction, the echo spectra were normalized by the spectrum of the incident signal. The resulting signals represented the targets' transfer function in the frequency domain.

This filter bank model was tested with material composition and sphere-cylinder discriminations using the same targets as in the ealier test. When test and reference echoes were measured on the same day, performance of the filter bank model was 90 to 100% correct for material composition discriminations. Discrimination between metal spheres and cylinders was 90% correct. However, when test and reference echoes were measured on different days, the spectral feature algorithm incorrectly classified all the echoes. The time-domain algorithm's performance was not degraded by using stimuli collected on different days.

The filter bank model seemed to be highly sensitive to minor variations in placement and orientation of both the transducer and target, and to signal fluctuations caused by slight target motion. Since the time-domain model was not affected significantly by these minor variations, it can be considered more robust than the spectral processing model.

REFERENCES

Au, W.L., 1986, Detection and Recognition Models of Dolphin Sonar Systems, this volume.

Chestnut, P, Landsman, H and Floyd, R.W., 1979, A Sonar Target Recognition Experiment, J. Acoust. Soc. Am., 66:140-147.

Chestnut, P.C. and Floyd, R.W., 1981, Aspect-Independent Sonar Target Recognition Method, J. Acoust. Soc. Am., 70:727-734.

Hammer, C.E. Jr and Au, W.W.L., 1980, Porpoise Echo Recognition: an analysis of Controlling Target Characteristics, J. Acoust. Soc. Am., 68:1285-1293.

Martin, D.W. and Au, W.W.L., 1982, Aural Discrimination of Targets by Human Subjects Using Broadband Sonar Pulses, Naval Ocean Systems Center, San Diego, Ca., TR 847.

Martin D.W. and Au, W.W.L., 1986, Investigation of Broadband Sonar Classification Cues, Naval Ocean Systems Center, San Diego, Ca., TR 1123.

Nachtigall, P.E., 1980, Odontocete Echolocation Performance on Object Size, Shape and Material, in "Animal Sonar Systems," R.G. Busnel and J.F. Fish, eds., Plenum Press, New York, pp. 71-95.

Normark, J., 1961, Perception of Distance in Animal Echolocation, Nature, 190, 363-364.

Schusterman, R.J., Kersting D.A. and Au, W.W.L., 1980, Response Bias and Attention in Discriminative Echolocation by Tursiops truncatus in "Animal Sonar Systems," R.G. Busnel and J.F. Fish, eds., Plenum Press, New York, pp. 983-986.

Small, A.M. Jr, and McClellan, M.E., Pitch Associated with Time Delay Between Two Pulse Trains, 1963, J Acoust Soc Am, 35:1246-1255.

A MATCHED FILTER BANK FOR TIME DELAY ESTIMATION IN BATS

Dieter Menne*

Department of Zoology
National University of Singapore
10, Kent Ridge Crescent
Singapore 0511

INTRODUCTION

Most bats possess a short frequency sweep as a component of their echolocation call, and it is widely accepted in neuroethology that this component is essential for range estimation. Neurons responding with short bursts to frequency sweeps could therefore be involved in range processing (Feng et al., 1978; Suga and O'Neill 1979; Pollak, 1980; Bodenhamer et al., 1979; Vater, 1981; O'Neill and Suga 1982). Many time-marking neurons fire with very little time jitter when a sweep passes through their characteristic frequency. A set of such neurons individually tuned to all frequencies of the fm-sweep make up a filter bank, and neurophysiologists often assume that the time information from this ensemble of filters, suitably averaged, is directly used by bats for time delay estimation.

Theorists have discussed filter banks as models for time delay estimation, but they did not believe that the achievable precision is high enough to explain the behavioral data of Simmons (1979), because in contrast to crosscorrelation models, the filter bank model does not make optimal use of all information in the sonar signal. Hackbarth (1984, 1986) has conducted computer simulations of filter banks; as she varied only a limited number of parameters it is not possible to derive general conclusions from her plotted data.

In this paper I will derive the precision of time delay estimation for a bank of filters excited by a frequency sweep. The following major questions will be discussed.

1. Is there a bandwidth of a single filter in the bank that is optimally matched to the sweep such that a maximum of precision can be achieved?

2. What is the minimum standard deviation of time delay estimation with this type of receiver?

---

* on leave from: Lehrstuhl Zoophysiologie, Institut für Biologie III
D-7400 Tübingen, F.R. Germany

CONCEPTS AND ASSUMPTIONS

A frequency sweep will excite a bandpass filter while the sweep frequency matches the bandpass frequency. The envelope of the filter output will be the shorter the narrower the filter bandwidth is. When no pulse-compression techniques are used, at a given signal-to-noise ratio

$\sqrt{2E/N_o}$ the standard deviation of time estimation ('timing error')

obtained with a pulse is proportional to its duration: the shorter the pulse, the lower the timing error. It could be followed that to minimize the timing error the single filter's bandwidth should be as low as technically realizable, but this is not correct. The duration of the envelope at the output of a bandpass filter cannot become shorter than the filter time constant, which is about the reciprocal of the filter bandwidth. If the input signal is shorter than the filter time constant, the pulse is effectively broadened by the filtering. One may therefore expect that for a given sweep rate there exists an intermediate bandwidth for which the filter is optimally matched to the echolocation signal. I will derive the value of this bandwidth using a simplified model of the echolocation call and of the filter bank in the bat's ear.

The assumptions on which the derivation is based can be divided into two types. Essential assumptions make up the backbone of the model; changing them will lead to another model. Non-essential assumptions make the derivation easier; changing them will modify some numerical results, but has no influence on the major conclusions.

Essential assumptions

1. The filter-bank system is built to measure the time of arrival of a frequency sweep.

2. The system consists of a series of bandpass filters covering the whole range of frequencies in the echolocation sound.

3. The output from the filters after envelope detection is the only information that the processing system can use.

4. The system is able to average time estimates from separate channels.

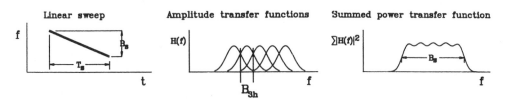

| Linear sweep | Amplitude transfer functions | Summed power transfer function |

Fig. 1. (a) A linear sweep of duration T_s and bandwidth B_s is a used as a simplified echolocation call. (b) The filter bank model of the bat's peripheral hearing system consists of a set of bandpass filters with 3-dB bandwidths of B_{3h}, that overlap at their 3-dB points. (c) The power transfer function of the system is almost rectangular and has the same bandwidth as the linear sweep.

Non-essential assumptions

of a single bandwidth and sweep time for a species is arbitrary. I assume
that both bat species produce the same sound pressure level which

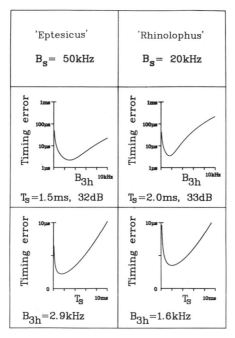

Fig. 2.   Upper row: Standard   deviation of time delay estimation σ_t
          ('timing error') for a filter bank with bandwidth B_{3h} of the single
          filter.   For Eptesicus, the linear sweep had a bandwidth of B=50kHz
          and a duration of T_s=1.5ms; SNR was 32dB.   Corresponding values for
          Rhinolophus are B_s=20kHz, T_s=2.0ms and SNR 33dB.   Values for SNR
          are different, because signal powers were assumed to be equal and
          the sweeps have different durations. Curves have minima at
          B_{3h}=2.9kHz (Eptesicus) and B_{3h}=1.6kHz (Rhinolophus).   Lower row:
          Timing error as a function of signal duration.   The bandwidths B_{3h}
          of the filters in the bank are those with minimum timing error in
          the upper graphs. Because signal power, not energy, was assumed to
          be constant, the SNR in these graphs increases with increasing
          time.   Scale of the timing error in the upper graph is logarithmic
          and in the lower graph linear.

leads to a slightly different value of the SNR for the two sounds
(32dB/33dB).   For Eptesicus fuscus a SNR of 32dB ($d_s \approx$ 40) has been chosen,
for comparison with the behavioral data of Simmons (1979) (see Schnitzler
et al., 1985; Menne and Hackbarth, 1986).   In Fig. 2, the timing error
given by (10) is plotted against the bandwidth B_{3h} of the filters in the
bank. The curves have minima at 2.9kHz (Eptesicus) and 1.6kHz
(Rhinolophus).   For Rhinolophus, this corresponds to $Q_{10db} \approx$ 24 at 70kHz,
which agrees well with electrophysiological results (Suga, Neuweiler,
Möller, 1976).   At these frequencies and the assumed SNR, timing error is
2.2μs for Eptesicus.   This is somewhat higher than the timing error in

5. The peak power of the emitting system is limited to N_s.

6. The power density of external or internal noise N_0 is constant over the frequency range of the sweep.

7. The system estimates the arrival time of the returning echo only; the time of signal emission is assumed to be known. This assumption is not limiting, because even if a bat has to estimate the time of emission with the same system, the returning sound will have a much lower SNR; its error will be the dominant part of the total timing error.

8. The transfer function of each filter is Gaussian with a constant 3-dB bandwidth B_{3h}.

9. Filters in the bank overlap at their 3-dB points (Fig. 1). With this choice of overlap, the summed power transfer function of all filters is quite flat over the whole frequency band. This assumption contradicts electrophysiological results that show a much higher degree of overlap in the mammalian auditory system. However, overrepresentation does not lead to an improved precision of signal parameter estimation in the presence of external noise (Carter et al., 1973; Altes, 1980). When a frequency band is represented by several filters, the noise in them is highly correlated and information gained by averaging is reduced. On the other hand, overrepresentation in the auditory system may reduce the influence of statistically independent noise produced in separate channels of the peripheral system. In the model presented here, the noise correlation (Harris, 1978) of two overlapping filters is 0.06, which indicates that the filter outputs are almost uncorrelated. With Assumption 9, it is therefore feasible to estimate the lower limit of timing error that is reached when the processing system does not introduce additional noise.

10. The echolocation sound used is a linear frequency sweep of duration T_s, bandwidth B_s, which is the difference between highest and lowest frequency, and constant power N_s (Fig. 1). The sweep rate, measured in Hz/s, is B_s/T_s.

MATHEMATICAL DERIVATION

My aim is to find the bandwidth B_{3h} of a single filter that minimizes the overall error of time measurement σ_τ of n filters in the bank.

The total energy in the signal is

$$E_s = N_s T_s. \tag{1}$$

The SNR of the signal is

$$d_s = \sqrt{\frac{2E_s}{N_0}} = \sqrt{\frac{2N_s}{N_0}} \sqrt{T_\varepsilon} \tag{2}$$

I assume that the filters in the bank overlap at their 3-dB points (Fig. 1), in such a way that the power transfer function of the whole filter bank closely approximates the rectangular spectrum of the input sweep. The number of filters in the bank is

$$n = \frac{B_s}{B_{3h}} . \tag{3}$$

At this frequency spacing, the noise correlation between two adjacent filters (equation 17, Harris 1978) is 0.06., which means that the noise output of pairs of filters may be considered uncorrelated. By averaging over all independent time estimates (Assumption 4), the standard deviation of time estimation ('timing error') for the filter bank σ_t is

$$\sigma_t = \frac{\sigma_{t1}}{\sqrt{n}} , \tag{4}$$

where σ_{t1} is the standard deviation of time estimation with one filter.

For the echolocation signal, a linear frequency sweep was assumed, and the n filters share equal time slices of the whole sweep or equal parts of the signal energy E_s. The efficiency η_f (Barton and Ward, 1969, p. 56) of one filter thus is

$$\eta_f = \frac{1}{n} . \tag{5}$$

The linear sweep excites the filters in the bank one after the other. The output envelope of one filter is the same as it would be if the filter was excited by a signal with the filter's center frequency, a Gaussian envelope and a 3-dB duration of

$$\tau_{3a} = \frac{T_s}{B_s} B_{3h} \tag{6}$$

The ratio r (Table 3.5, Barton and Ward, 1969) between filter bandwidth B_{3h} and signal envelope bandwidth B_{3a} is

$$r = \frac{B_{3h}}{B_{3a}} = \frac{\pi}{2\ln 2} \frac{T_s}{B_s} B_{3h}^2 . \tag{7}$$

The standard deviation of the time measurement using one filter of the bank is given by Barton and Ward (1969, equation 3.20 and Table 3.5)

$$\sigma_{t1} = \sqrt{\frac{1}{8\eta_f \ln 2} \frac{1+r^2}{r} \frac{\tau_{3a}}{d_s}} . \tag{8}$$

After inserting η_f, τ_{3a} and d_s

$$\sigma_{t1} = \sqrt{\frac{n}{8\ln 2} \frac{\sqrt{T_s} B_{3h}}{B_s \sqrt{2N_s/N_0}} \frac{1+r^2}{r}} . \tag{9}$$

839

With (4), after averaging over all filters the timing error σ_t is

$$\sigma_t = \frac{1}{\sqrt{8\ln 2}} \frac{\sqrt{T_s}}{B_s} \frac{B_{3h}}{\sqrt{2N_s/N_0}} \frac{1+r^2}{r} \; ,$$

(10)

where r is defined in equation (7). This function has a minimum at a 3-dB bandwidth B_{3min}

$$B_{3min} \approx 0.505 \sqrt{\frac{B_s}{T_s}} \; .$$

(11)

To convert from 3-dB bandwidth to 10-dB bandwidth, which is more common in neurophysiology, the following approximation holds for the Gaussian amplitude transfer function

$$B_{10min} = 1.82 B_{3min} \; .$$

(12)

The minimum standard deviation of time measurement σ_{tmin} with the filter bank

$$\sigma_{tmin} \approx \frac{0.50}{\sqrt{B_s}\sqrt{2N_s/N_0}} = \frac{0.50}{d_s}\sqrt{\frac{T_s}{B_s}} \; .$$

(13)

CONCLUSION AND DISCUSSION

There are three major conclusions. Firstly, for a given duration of the sweep T_s and B_s, there exists a bandwidth B_{3min} of a single bandpass filter that will lead to minimum timing error with a filter bank. A bank with filters of this bandwidth is matched to the linear sweep signal.

Secondly, B_{3min} is proportional to the square root of the sweep rate B_s/T_s (equation 11): the faster the frequency sweep, the higher the optimal bandwidth of a single filter in the bank.

Thirdly, if one assumes that a bat's signal is power limited, then the minimum error of time measurement is inversely proportional to the square root of the bandwidth (equation 13). Compare this with the crosscorrelation receiver, where the timing error is inversely proportional to the bandwidth itself (Menne and Hackbarth, 1986).

To illustrate these conclusions with numerical examples, I have chosen two combinations of sweep durations and bandwidths from the literature as characteristic of the lower harmonic of Eptesicus fuscus and of the fm-part of Rhinolophus ferrumequinum (Beuter, 1976). The examples will be identified by the bat's genus, even though it is clear that the selection

840

Simmons' (1979) jitter-experiment (1.2μs). But the SNR in these experiments was estimated very conservatively and may have been higher; in addition, the bat may easily have improved its precision by averaging over several sounds. The forms of the upper two curves in Fig. 2 gives an impression of the change in timing error when bandwidth and sweep rate are not optimally matched. The maximum is sharper for <u>Rhinolopus</u> than for <u>Eptesicus</u>, therefore the performance of the latter should be less affected by imperfect matching.

The bandwidth of peripheral filters are fixed by evolution for a species, but individual bats have a certain freedom to choose the parameters of their echolocation signal. To show the influence of different sweep rates on timing error, I will assume that the bandwidth B_s of the signal is fixed, but that the bat is able to change the duration T_s, of the sweep. In the lower half of Fig. 2, timing error is plotted against duration of the signal. The bat has much freedom to increase the duration of the signal over the optimal value without losing too much precision, but it must be taken into account that the power of the signal was assumed to be constant for this plot, therefore lengthening of the sound will increase the energy and improve the SNR. So the graph may be interpreted such that an increase of the SNR by lengthening the sound does not reduce the timing error.

To simplify the derivation, I have chosen filters with equal bandwidth and an echolocation call with a linear frequency sweep. A more realistic model of the peripheral hearing system is a bank of filters with a bandwidth proportional to the center frequency (constant-Q filters). According to (11), a lower bandwidth is optimal for a slower sweep rate. A signal matched to a constant-Q filter bank would therefore start with a high sweep rate and end with a lower one. This type of echolocation signal can be observed in many fm-bats, and it may be an evolutionary adaptation to the frequency representation in the mammalian hearing system.

ACKNOWLEDGEMENTS

I thank Dr. J. Counsilman, National University of Singapore, for his commenting on the manuscript. The author was supported by a 'Langzeitdozentur' of the German Academic Exchange Service (DAAD).

REFERENCES

Altes, R.A., 1980, Detection, estimation and classification with spectrograms, J. Acoust. Soc. Am., 67:1232.
Barton, D.K., Ward, H.R., 1969, Handbook of Radar measurement, Prentice Hall, Englewood Cliffs, N.J.
Beuter, K.J., 1980, Systemtheoretische Untersuchungen zur Echoortung der Fledermäuse, Thesis, Universität Tübingen.
Bodenhamer, R.D., Pollak, G.D., 1981, Time and frequency domain processing in the inferior colliculus of echolocating bats, Hearing Research, 5:317.
Carter, G.C., Knapp, C.H., Nuttall, A.H., 1973, Estimation of the magnitude-squared coherence function via overlapped fast Fourier transform processing, IEEE Trans. Audio Electroacoust. AU-21:337.
Feng, A.S., Simmons, J.A., Kick, S.A., 1978, Echo detection and target ranging neurons in the auditory system of the bat <u>Eptesicus fuscus</u>, Science 202:645.
Hackbarth, H., 1984, Systemtheoretische Interpretation neuerer Verhaltens- und neurophysiologischer Experimente zur Echoorturng der Fledermäuse, Thesis Universität Tübingen, F.R.G.

841

Hackbarth, H., 1986, Phase evaluation in hypothetical receivers simulating ranging in bats, Biol. Cybern., 54:281.

Harris, F.J., 1978, On the use of windows for harmonic analysis with the discrete Fourier transform, Proc. IEEE 66:51.

Menne, D., Hackbarth, H., 1986, Accuracy of distance measurement in the bat Eptesicus fuscus: Theoretical aspects and computer simulations, J. Acoust. Soc. Am., 79:386.

O'Neill, W.E., Suga, N., 1982, Encoding of target range and its representation in the auditory cortex of the mustached bat, J. Neuroscience, 2:17.

Pollak, G.D., 1980, Organizational and encoding features of single neurons in the inferior colliculus of bats, in: "Animal Sonar Systems", R.-G. Busnel, J. Fish eds., Plenum Press, New York.

Schnitzler, H.-U., Menne, D., Hackbarth, H., 1985, Range determination by measuring time delay in echolocating bats, in: "Time resolution in auditory systems," A. Michelsen, ed., Springer, Berlin.

Simmons, J.A., 1979, Perception of echo phase information in bat sonar, Science 204:1336.

Suga, N., O'Neill, W.E., 1979, Neural axis representing target range in the auditory cortex of the mustache bat, Science 206:351.

Suga, N., Neuweiler, G., Moller, J., 1976, Peripheral auditory tuning for fine frequency analysis by the cf-fm bat, Rhinolophus ferrumequinum 1V. Properties of peripheral auditory neurons, J. Comp. Physiol., 106:111.

Vater, M., 1981, Single unit responses to linear frequency-modulations in the inferior colliculus of the Greater Horshoe bat, Rhinolophus ferrumequinum, J. Comp. Physiol. 141:249

PARTICIPANTS

ALTES, Richard A.   Chirp Corporation
                            8248 Sugarman Drive
                            La Jolla, CA   92037   USA

AMUNDIN, Mats   Kolmardens Djurpark
                            Kolmarden   Sweden

AU, Whitlow W. L.   Naval Ocean Systems Center
                            P.O. Box 997
                            Kailua, HI   96734-0997   USA

AWBREY, Frank   Hubbs-Sea World Research Institute
                            1700 South Shores Road
                            San Diego, CA   92109   USA

BARCLAY, Robert M. R.   Department of Biology
                            University of Calgary
                            Calgary, Alberta   T2N 1N4   Canada

BELWOOD, Jacqueline   Entomology and Nematology
                            University of Florida
                            Gainesville, FL   32611   USA

BISHOP, Allen   Department of Anatomy
                            School of Medicine
                            University of North Carolina
                            Chapel Hill, NC   27514   USA

BJORNO, Leif   Industrial Acoustics Laboratory
                            Institute of Manufacturing
                            Engineering
                            Technical University of Denmark
                            Building 352
                            DK-2800 Lyngby   Denmark

BRILL, Randy   Brookfield Zoo
                            Brookfield, IL   60513   USA

BROWN, Patricia   Maturango Museum
                            P.O. Box 1776
                            Ridgecrest, CA   93555   USA

BUSNEL, Marie-Claire   Chemin de la Butte au diable
                            Bievres   France

BUSNEL, Rene-Guy   Chemin de la Butte au diable
                            Bievres   France

CAMPBELL, Karen   Indiana University
                                  Medical Sciences Program
                                  Physiology Section
                                  Myers Hall 263
                                  Bloomington, IN 47405   USA

CARDER, Donald A.   Naval Ocean Systems Center
                                  Code 514
                                  San Diego, CA 92152-5000   USA

CASSEDAY, John H.   P.O. Box 3943
                                  Duke University Medical Center
                                  Durham, NC 27710   USA

CRANFORD, Ted   Long Marine Laboratory
                                  University of California
                                  Santa Cruz, CA 95060   USA

DALSGAARD, Jacob   Biologisk Institut
                                  Odense Universitet
                                  Campusvej 55
                                  5230 Odense M   Denmark

ESCUDIE, Bernard   Laboratoire de Traitment du
                                  Signal (U.A.C.N.R.S. 346)
                                  Institute de Chimie et de
                                  Physique Industrielles
                                  31, Place Bellecour
                                  69228 - Lyon Cedex 02   France

EVANS, William   National Marine Fisheries Service
                                  Universal Building, #1011
                                  1825 Connecticut Ave., NW
                                  Washington, D.C. 20235   USA

FENG, Albert S.   Department of Physiology and
                                  Biophysics
                                  University of Illinois
                                  Urbana, IL 61801   USA

FENTON, M. Brock   Faculty of Science
                                  York University
                                  4700 Keele Street
                                  North York, Ontario M3J 1P3   Canada

FLANDRIN, Patrick   Laboratoire de Traitment
                                  du Signal (U.A.C.N.R.S. 346)
                                  Institute de Chimie et de
                                  Physique Industrielles
                                  31, Place Bellecour
                                  69228 - Lyon Cedex 02   France

FLOYD, Robert W.   Naval Ocean Systems Center
                                  Hawaii Laboratory, Code 513
                                  P.O. Box 997
                                  Kailua, HI 96734-0997   USA

FRIEDL, William A.   Naval Ocean Systems Center
                                  Hawaii Laboratory, Code 513
                                  P.O. Box 997
                                  Kailua, HI 96734-0997   USA

FULLARD, James   Department of Biology
Erindale College
University of Toronto
Mississauga, Ontario   L5L 1C6   Canada

FUZESSERY, Zolton   Department of Neurophysiology
University of Wisconsin
Medical School
1300 University Avenue
Madison, WI   53706   USA

GOOLER, David M.   Department of Physiology and
Biology
College of Liberal Arts and Sciences
524 Buxrill Hall
407 South Goodwin Avenue
University of Illinois at
Urbana-Champaign
Urbana, IL   61801   USA

GRIFFIN, Donald   The Rockefeller University
1230 York Avenue
New York, NY   10021   USA

GRINNELL, Alan D.   Jerry Lewis Center
UCLA Medical Center
University of California
Los Angeles
Los Angeles, CA   90024   USA

GUPPY, Anna   Zoologisches Institut
der Universität
Luisenstrasse 14
D-800 München 2   FRG

HARTLEY, David   Medical Science Program
Physiology Section
Indiana University
Bloomington, IN   47405   USA

HAUN, Jeffrey E.   Naval Ocean Systems Center
Hawaii Laboratory, Code 513
P.O. Box 997
Kailua, HI   96734-0997   USA

HENSON, O. W.   Department of Anatomy
University of North Carolina
Chapel Hill, NC   27514   USA

JEN, Phillip   College of Arts and Science
Division of Biological Sciences
213 Lefevre Hall
University of Missouri
Columbia, MO   65201   USA

JOHNSON, C. Scott   Naval Ocean Systems Center
Code 514
San Diego, CA   92152-5000   USA

JOHNSON, Richard

Department of Math and
Computer Science
Western New Mexico University
Silver City, NM 88062

USA

JORGENSEN, Morton

Biologisk Institut
Odense Universitet
Campusvej 55
5230 Odense M

Denmark

KAMMINGA, Cees

Technische Hogeschool Delft
Afdeling Der Elektrotechniek
2600 GA Delft

Netherlands

KOBER, Rudi

Behringstreasse 10
7410 Reutlingen

FRG

LANEN, Theodor

Information Theory Group
Delft University of Technology
Mekelweg 4 Postbus 5031
2600 Delft

Netherlands

LINDHARD, Morton

Department of Zoophysiology
University of Aarhus
DK-8000 Aarhus C

FRG

MACKAY, Stuart

2083 16th Avenue
San Francisco, CA 94116

USA

MARTEN, Ken

Long Marine Laboratory
Santa Cruz, CA 95060

USA

MASTERS, Mitch

Department of Zoology
Ohio State University
1735 Neil Avenue
Columbus, OH 43210

USA

MENNE, Dieter

Department of Zoology
National University of Singapore
10 Kent Ridge Crescent
05111

Singapore

MILLER, Lee

Biologisk Institut
Odense Universitet
Campusvej 55
5230 Odense M

Denmark

MØHL, Bertel

Department of Zoophysiology
University of Aarhus
DK-8000 Aarhus C

Denmark

MOORE, Patrick W. B.

Naval Ocean Systems Center
Hawaii Laboratory, Code 512
P.O. Box 997
Kailua, HI 96734-0997

USA

MOSS, Cindy

Department of Psychology
Brown University
Providence, RI 02912

USA

NEUBAUER, Werner	4693 Quarter Charge Drive Annandale, VA 22003	USA
NEUWEILER, Gerhard	Zoologisches Institut Der Universität Luisenstrasse 14 D-8000 München 2	FRG
NORRIS, Ken	Long Marine Laboratory University of California Santa Cruz, CA 95064	USA
O'NEILL, William	Center for Brain Research University of Rochester Medical School Rochester, NY 14642	USA
OSTWALD, Joachim	Institut fur Biologie der Universität Auf Der Morgenstelle 28 D-7400 Tübingen	FRG
PENNER, Ralph	Naval Ocean Systems Center Hawaii Laboratory, Code 512 P.O. Box 997 Kailua, HI 96734-0997	USA
PETTIGREW, Jack	Department of Physiology and Pharmacology University of Queensland St. Lucia, Queensland	Australia
POLLAK, George	Department of Zoology University of Texas Austin, TX 78712	USA
PORTER, Hop	Naval Ocean Systems Center Biosciences Division, Code 51 San Diego, CA 92152	USA
POWELL, Bill A.	National Marine Fisheries Service University Building #1011 1825 Connecticut Ave., NW Washington, D.C. 20235	USA
PYE, David	School of Biological Sciences Queen Mary College Mile End Road London, E1 4NS	England
RIESS, Diane	2338 Franklin Street, #2 San Francisco, CA 924123	USA
RIDGWAY, Sam	Naval Ocean Systems Center Code 514 San Diego, CA 92152-5000	USA
ROSS, Linda	Department of Zoology University of Texas Austin, TX 78712	USA

ROVERUD, Roald

Department of Biology
University of California
Riverside, CA 92521

USA

RUBSAMEN, Rudolf

Lehrstuhl Allegmeine Zoologie
Ruhr Universitat Bochum
Postfach 10 21 48
4630 Bochum 1

FRG

SCHMIDT, Sabine

Zoologisches Institut
Der Universität
Luisenstrasse 14
D-8000 München 2

FRG

SCHMIDT, Uwe

Zoologisches Institut
Poppelsdorfer Schloss
D-5300 Bonn 1

FRG

SCHULLER, Gerd

Zoologisches Institut der
Universität München
Luisenstrasse 14
D-8000 München

FRG

SCHUSTERMAN, Ronald J.

1629 Mariposa Street
Palo Alto, CA 94306

USA

SIMMONS, James A.

Department of Psychology
Brown University
Providence, RI 02912

USA

SMOLKER, Rachel

Long Marine Laboratory
100 Shaffer Road
Santa Cruz, CA 95060

USA

SUGA, Nobuo

Department of Biology
Washington University
St. Louis, MO 63130

USA

SURRLYKKE, Annemarie

Institute of Biology
Odense University
DK-5230
Odense M

Denmark

SUTHERS, Roderick

Medical Sciences Program
Indiana University
Bloomington, IN 47405

USA

TERRY, Paul

Technische Hogeschool Delft
Afdeling der Elektrotechniek
Mekelweg 4, Postbus 5031
2600 GA Delft

Netherlands

THOMAS, Jeanette A.

Naval Ocean Systems Center
Hawaii Laboratory, Code 512
P.O. Box 997
Kailua, HI 96734-0997

USA

VATER, Marianne

Zooligisches Institut LMU
Luisenstrasse 14
8 München 2

FRG

von der EMDE, Gerhard

Lehrbereich Zoophysiology
Universitat Tübingen
Auf der Morganstelle 28
7400 Tübingen   USA

WATKINS, William

Woods Hole Oceanographic
Institute
Woods Hole, MA 02543   USA

WENSTRUP, Jeffrey

Department of Physiology
and Anatomy
University of California
Berkeley, CA 94720   USA

WIERSMA, Henk

Technische Hogeschool Delft
Afedling der Elektrotechniek
Mekelweg 4 Postbus 5031
2600 GA Delft   Netherlands

WOOD, F. G.

Naval Ocean Systems Center
Code 514
San Diego, CA 92152-5000   USA

WOODWARD, Donald

Chief of Naval Research
(ONR-1142BI)
800 N. Quincy Street
Arlington, VA 22217   USA

ZAKHARIA, Manell

Laboratoire de Traitment du
Signal (U.A.C.N.R.S. 346)
Institut de Chimie et de Physique
Industrielles
31, Place Bellecour
69228 - Lyon Cedex 02   France

ZBINDEN, Karl

Abteilung Wirbeltiere
Hallerstrasse 6
3012 Bern   Switzerland

ZOOK, John M.

Department of Zoological
and Biomedical Science
Ohio University
Athens, OH 45701   USA

Dolphin (continued)
  species: (continued)
  Delphinus delphis, 82, 328, 576
    D. sp., 525, 720
  Globicephala scammoni, 526
    G. sp., 521
  Inia geoffrensis, 18, 132, 135,
    328, 526–528, 717, 720
  Lagenorhynchus albirostris,
    132, 137
    L. obliquidens, 526, 571, 720
  Lipotes vexillifer, 18, 526
  Neophocoena asiaeorientalis,
    528
    N. phocaenoides, 17–19, 718
  Orcaella brevirostris, 18, 19,
    137, 138, 718
  Platanista indi, 18, 521, 526,
    528
    P. minor, 526
  Phocoena phocoena, 12–15, 61–66,
    132, 139, 140, 314, 328,
    528, 531, 676, 720
  Phocoenoides dalli, 17, 18,
    528, 529
  Pontoporia blainvillei, 526
  Sotalia fluviatilis, 18–20, 132,
    138, 139, 717, 718, 719
  Stenella coeruleoalba, 312–314
    S. longirostris, 67–77,
    328, 525, 526
    S. plagiodon, 570, 571
  Tursiops aduncus, 132, 136,
    568–571
    T. scylla, 525
    T. truncatus, 10–13, 82–85,
    109–113, 121–127, 132, 135,
    136, 161–168, 281–287, 295–
    299, 312, 313, 317–321, 451–
    579, 703–713, 753–758
    T. sp., 325, 328, 521–538,
    676, 694–699, 703–706,
    718–720
  target
    detection, 453–461
      in noise, 454–458
      in reverberation, 458–461
    discrimination, 758–765
    recognition, 461–463, 758–765
    shape discrimination, 461–463
    vocalization, recorded, 703
    whistling, 83, 704
Doppler
  compensation in bat, 93, 307,
    308–310, 372–374
  coupling, 400
  and echo, *see* Echo
  sensitivity, 400–401
  shift, 265, 372–374, 480
    compensation, 307–310
  tolerance, 401

Ear of vertebrate, 659–661
  of bat, large, 289–293
Echo
  acuity, fine, for, 366–371
  backscattering of, 809–813
  calculation, 817–818
  complex, 295–299
  delay perception, 366–371
  discrimination for metal plates,
    809–813
  and Doppler shift, 150–153
  encoding, 414–415
  filter bank, matched, 835–842
  and fluctuation of target,
    417–418
  and fluttering of insect wings,
    477–481
  formation, 820, 823–827
  from insect, 477–481, 803–807
  Lamb wave, 820
  pattern recognition, 830–833
  and plate, metallic, 809–813
  from sphere, metallic, 815–821
  signal-to-noise ratio, 371
  and surface material, 415
  target
    natural, 416–418
    parameters, 414–418
    range-extended, 414–415
  time-delayed, 803–807, 835–842
  time domain feature, 829–833
Echolocation
  adaptability, 162
  application, 723–842
  assemblage, acoustical, 639–643
  attention, 162
  and auditory sysyem, 147–350
  bandwidth, 129–145
  of bat, 353–411, 435–450,
    551–560, 603, 619–628,
    645–650, 687, 688
  call
    analysis, 400
    of bat, 603, 619–628
    of dolphin, 683–701
  in cluttering noise, 613–617
  and cognition, 653–711
  and computer manipulation,
    640–641
  concepts, theoretical, 725–752
  definition, 683
  detection, 387–411, 731–734
  directionality, 374–379
  of dolphin, 9–22, 161–168,
    683–701
  Doppler shift, 372–374
  echo delay, 355–361, 366, 371–372
  emergence, 665–666
  estimation, 387, 411
  experiment design, critical,
    387–411

Echolocation (continued)
  feeding strategy, 521-534
  fluctuation, 388-393
  and fluttering target, 615-617
  frequency design, 639
  and habitat, 521-534
  history, natural, 519-652
  of human when blind, 687
  and image, acoustic, 353-361,
    371-372
  improper, 141
  and insect prey, 551-566,
    619-623, see Moth
  interpolation, 725-731
  literature survey, 10-17
  and map, neural, 725-731
  of marine mammal, 521-534
  of microbat, 645-650
  and noise, 388-393, 613-614
    -to-echo ratio, 371-372
  oscillogram, 58, 478, 479, 603
  parameter estimation, 725-731
  perception of image, acoustic,
    353-385
  performance of bat, 353-385
  phase sensitivity, 404-407
  range discrimination, 361-366
  signal, 619-623
    design, 734-736
    energy, 389-391
    power, 389-390
    production, 7-145
  sound frequency, 513, 619-623
  spectrogram, 734-742
  and system, auditory, 147-350
  target, 613-617
    detection, 435-450
    location, 374-379
    velocity, 372-374
  theory, 723-842
  time duration, 129-145
  type of, 9-22, 129-145
  uncertainty, 731-734
  and vision of bat, 645-650
  waveform, 11
Echolocator of bat
  allotonic, 639
  syntonic, 639
Emission, cochlear, 228-229
Energy detector, model of, 758
Eptesicus fuscus (bat), 23-25,
      29-34, 115-120, 178-180,
      183, 184, 189, 271-274,
      358-360, 363, 366-379,
      398, 401, 404, 407, 418,
      437, 442, 443, 445, 489,
      501, 541, 545, 607-609,
      625, 636, 837, 840-841
  cerebellum, role of, 271-274
  colliculus, inferior, 183, 184,
      189

Eptesicus fuscus (continued)
  echolocation, 271-274
  electrode punctures in the
      cerebellum, 271-274
  larynx cross-section, 24
  lemniscus, lateral, 178-180
  neuron sensitivity, 271-274
  ontogeny, 115-120
  signal, vocal, 115-120
  space, auditory, frontal, 271-274
  spectrogam, 117-118
    vermis, cerebellar, 271-274
  E. pumilus, 226, 541
  E. serotinus, 437, 444
  E. sp., 225, 232, 441, 622, 837
Euchaetias egle (moth), 636, 637
Euderma maculatum (bat), 544, 598
Eusthenopteron sp., 659

False killer whale, see *Pseudorca
      crassidens*
Fast blue dye, fluorescent, 347-349
Filter
  acoustic of bat, 35-38
  bank, 833-842
    mathematics, 838-840
Fish
  ear, 576-577, 658-659
  signal, 662-663
Fishing bat, see *Noctilio leporinus*
Flight, target-directed, 150, 151,
      620-621
Fluttering, 536-540 see Glint
Foraging behavior of bat, 535-549,
      601-605, 635-638
Formant frequency of sonar, 89-90
Frequency of sound, see Sonar

Gabor model, 816-817
Ganglion cell of retina
  of microbat, 646
Gate, glottal, of bat, 31-33, 36-38
Generator, laryngeal, of bat, 23-24,
      29-31
Geniculate body, medial, 185-186
Gleaning bat, see *Megaderma lyra*
Glint of fluttering insect, 417-418,
      477, 480, 481
Glischropus tylopus (bat), 491, 492
Globicephala scammoni (dolphin), 526
  G. sp., echolocation, 521
Glossophaga soricina (bat), 584, 791
Goldfish decompression sickness, 81
Grampus griseus (whale) echolocation,
      521, 526, 719
Greater horseshoe bat, see *Rhino-
      lophus ferrum-equinum*
Greater mustache bat, see *Pteronotus
      parnellii*

Hagfish, see *Myxine* sp.

Target (continued)
  of bat (continued)
    discrimination, 413-434
    echo parameters, 414-418
    localization, 374-379
    natural, 414-418, 442-443
    range, 253-258, 355-357,
      371-372
    recognition, 413-434
    velocity, 372-374
  cylinder, metallic, as, 829-833
  discrimination, 413-434, 758-765
  of dolphin
    detection, 451-465
    range, maximum, 451-454
    in reverberation, 458-461
    shape, 459, 461-463
  fluctuating, 419-421
  non-fluctuating, 418-419
  recognition, 758-765, 829-833
    algorithm, 829-833
    time-domain features, 829-833
Tenebrio sp. (insect), echo,
    803-807
Tetramethylbenzidine, 244
Thecophora fovea (moth), 555, 556
Time for sound
  delay, 835-842
  window, 513-517
Tipulo oleracea (insect), 495-499
  T. sp., 416
Tonatia bidens (bat), 601, 602
  T. silvicola, 601-604
Tonotopy, 170-187
Trachops cirrhosus (bat), 539,
    542, 545, 562, 592,
    601-604
Tract of bat
  acoustic, 289-293
  central, 186-187
  vocal, 149
Tursiops aduncus (dolphin), 132,
    136, 568-571
  T. scylla, 525
  T. sp., 325, 328, 521-534,
    676, 694-699, 703-706,
    718-720
  T. truncatus, 10-13, 82-85,
    109-113, 121-127, 132,
    135, 136, 161-168, 281-
    287, 295-299, 312, 313,
    317-321, 451-479, 703-713,
    753-758
  audition, 161-168
  bandwidth of sonar, 132,
    135, 136
  behavior, mother/infant,
    121-127

Tursiops aduncus (continued)
  T. truncatus (continued)
    in captivity, 10, 109-113
    click production, 109-113,
      121-127, 318, 164-165
    discrimination, temporal,
      317-321
    echolocation, 161-168, 295-299,
      707-713
      experiment, 707-713
    feeding, 703-706
      and loud sound, 703-706
    jaw-hearing, evidence of,
      281-287
    signal detection theory, 318
    sonar, 10-13, 109-113
    sound debilitating prey, 703-706
Tylonycteris pachypus (bat), 490-492
  T. robustula, 490-492
Tyto alba (bat), 291

Vampyrum spectrum (bat), 541, 604
Vespertilio murinus (bat), 541,
Vision of
  bat, 590, 645-650
  rat, 646
Vocal tract of bat, 35-38, 430
Vocal tract of bird, 40-42, 87-91

Wavelength of sound
  resonant frequency formula, 88
Weberian ossicles, 659
Whale
  behavior, social, 567-579
  click train, 567-570
    bangs, 569, 572, 576
  head, 79-86
    and beam, ultrasonic, 82
  oscillogram, 572, 573
  predation, 567-579
  sound, 567-579
    for disorientation of prey
      567-579
  see separate whales
Wheat germ agglutinin, 243, 244
  and horseradish peroxidase,
    243. 244
Whistling by
  dolphin, 83
  human, 83
White-sided dolphin, *see*
    Lagenorhynchus obliquidens
White whale, *see Delphinapterus*
    leucas
Wigner representation, 738-739
Wigner-Ville distribution of
    time frequency, 798